A DICTIONARY OF LAKE DISTRICT PLACE-NAMES

BY

DIANA WHALEY

This book is a guide to the place-names of the English Lake District
from the earliest times to the twentieth century.
It is based on previous scholarship and fresh research
and combines detailed and authoritative
commentary on the names with insights into the setting,
languages and history which gave rise to them.
It will prove absorbing reading and an important resource
for anyone interested in the past of the
region and the origins of its names.

Cover photograph: Great Gable from Innominate Tarn, Haystacks
(Ian Whaley)

ENGLISH PLACE-NAME SOCIETY
REGIONAL SERIES
VOLUME 1

GENERAL EDITOR

RICHARD COATES

A DICTIONARY OF LAKE DISTRICT PLACE-NAMES

A DICTIONARY OF LAKE DISTRICT PLACE-NAMES

BY

DIANA WHALEY

NOTTINGHAM
ENGLISH PLACE-NAME SOCIETY
2006

Published by the English Place-Name Society
School of English Studies,
University of Nottingham,
Nottingham NG7 2RD
Tel. 0115 951 5919
Fax. 0115 951 5924
Registered Charity No. 257891

© English Place-Name Society 2006
Photographs © Ian Whaley

All rights reserved.
No part of this publication may be reproduced,
stored in a retrieval system or transmitted
in any form or by any means, without prior permission of the
English Place-Name Society.

ISBN 0 904889 72 6

Distributed by
Paul Watkins / Shaun Tyas Publishing,
1 High Street, Donington,
Lincolnshire PE11 4TA

Typeset by Paul Cavill and Printed in Great Britain
by Woolnough Bookbinding, Irthlingborough, Northants.

To
Ian, Robin & Matthew

The author and publishers would like warmly to
acknowledge the support and encouragement
of
The Arts and Humanities Research Council
and
The British Academy
in funding the research and production of this book.

CONTENTS

Preface	xi
List of Plates	x
Introduction	xv
A Guide to the Dictionary	xxxv
Symbols, Terms and Abbreviations	xl
Bibliography, with Abbreviations	xlv
Map 1 The Lake District (Physical)	lix
Map 2 The Lake District (Administrative)	lx
A DICTIONARY OF LAKE DISTRICT PLACE-NAMES	1
List of Common Elements in Lake District Place-Names	385

LIST OF PLATES

Cover photograph: Great Gable from Innominate Tarn, Haystacks

1. Trusmadoor, looking north to the Solway Firth
2. Truss Gap and Swindale Beck
3. Mosedale, Loweswater
4. Scarth Gap (to R) from Ennerdale
5. Buttermere lake and village, with (L to R) Low Bank, Grasmoor, Whiteless Pike, Wandope and Crag Hill
6. Buttermere and Crummock Water
7. The Bishop, Barf
8. Beckstones Gill, Above Derwent, showing potential 'bakestones'
9. Braithwaite, 'the broad clearing', with Keswick in distance
10. Crosthwaite Church, Keswick
11a. Whitewater Dash or Dash Falls 11b. Aira Force
12. Two onetime shieling sites, Gutherscale and Skelgill, in the trees below Catbells (R); Derwentwater beyond.
13. River Derwent, Grange and Greenup (the small hollow above the village)
14. Stockley Bridge and Grains Gill
15. Derwentwater and Skiddaw
16. Saddleback or Blencathra, seen from the east
17. The Cockpit, stone circle on Barton Fell
18. Celtic, Anglian and Scandinavian naming in north-east Lakeland
19. Thirlmere from above Wythburn
20. Helvellyn: Red Tarn, with part of Swirral Edge leading to Catstycam
21. Patterdale Church
22. Grisedale, Patterdale
23. Yewbarrow, Great Gable and Lingmell, from Wastwater
24. The Screes, Wastwater
25. Mickledore (L), between Scafell Pike and Scafell
26. Langdale Pikes from The Band, with Pike O'Stickle to L
27. Viking Age hogbacks at Gosforth Church
28. Walls, Roman bath-house at Glannoventa, Ravenglass
29. Beacon Tarn, with Coniston fells in background
30a. Coppicing at Bouth Fall Stile. 30b. Charcoal-burner's hut near Satterthwaite
31. Windermere at dawn, from Low Wray
32. Whitbarrow Scar, with wooded valley of Witherslack to L

Photographs are ordered mainly by location, from north to south and west to east. All are by Ian Whaley, except for 3, 10 and 30 (Diana Whaley), 25 (Matthew Whaley).

PREFACE

The idea of the Dictionary — born very early one morning on the quiet north-western shores of Windermere — has been to fulfil the need for a comprehensive single-volume survey of Lake District place-names. My aim has been to record and explain, as far as possible, the place-names shown on the Ordnance Survey One Inch map of the Lake District (1994 printing) and falling within the boundaries of the Lake District National Park. The need for such a volume arises since anyone curious about the Lakeland names has hitherto been confronted with the choice of using books which are designedly short and selective (Brearley 1974, long out of print, or Gambles 1994) or with picking out the relevant parts from six specialist volumes: the three English Place-Name Society volumes for the former Cumberland (1950–52), the two for Westmorland (1967), and Eilert Ekwall's *Place-Names of Lancashire* (1922). It also has to be said (though not without fear of throwing stones inside a glass house) that place-names everywhere have a way of attracting unlikely explanations, and these abound in the very extensive literature about the Lake District. In producing this Dictionary I have tried to lay out the more well-based interpretations, while making clear the limits of knowledge, and I hope that it will provide useful and accessible guidance to readers with various degrees of prior knowledge.

The Dictionary is profoundly indebted to previous scholarship, and especially to the six volumes just mentioned, while at the same time offering new information and interpretations where appropriate and covering a large number of names not previously discussed in print. Traditional place-name surveys have often given priority to names considered to be of particular historical or linguistic interest — mainly those recorded before c.1550 — and the pre-modern names undoubtedly tend to be the more fascinating ones. However, both as a result of using the One Inch map as a base for the corpus, and as an article of faith, I have also tried to give serious attention to names only recorded in modern times. The result is a weightier volume than the slim 'rucksack book' I originally envisaged, but to have done otherwise would have misrepresented the continuing creativity of naming in the Lake District, for many of these names tell their own stories, reflecting recent history or man's interaction with nature, while some of them may be a good deal older than their first appearance in documents. I have not attempted systematic coverage of historic names that are no longer in use, and for these the six volumes mentioned should be consulted.

It is a pleasure to record a number of debts accumulated during the production of this book. Fieldwork and archive work in 1999 was assisted by an award from the Small Grants scheme of the British Academy, and an award from the Research Leave scheme of the Arts and Humanities Research Board (now AHRC) enabled me to work uninterruptedly on the Lake District material in the first half of 2001. A research grant from the University of Newcastle upon Tyne in 1999 provided for a laptop computer which has been an indispensable aid, and the School of English Literature, Language & Linguistics at Newcastle funded valuable clerical assistance with the transfer of material from database to text files.

I gladly acknowledge the unfailing helpfulness of the staff of the Cumbria Record Office at Carlisle and Kendal; and for permission to cite unpublished material from the archives of the Leconfield and Lowther estates and Carlisle Dean & Chapter I would like to thank, respectively, the Earl of Egremont, the Earl of Lonsdale and Canon David Weston, and well as David Bowcock of Carlisle CRO. Thanks are also due to Carlisle City Library, the British Library (Maps and Rare Books), the Public Record Office, the National Library of Scotland (including the Map Library), and my 'home' library, the Robinson Library in the University of Newcastle upon Tyne (above all Special Collections, and Arts Liaison Librarian Jessica Plane). Expert help of other kinds has been generously given by Eleanor Kingston, archaeologist with the Lake District National Park Authority, who provided access to the Lake District Historic Environment Record (formerly Sites and Monuments Record) and answered a number of specific queries, by Angus Winchester of Lancaster University, and by Keith Clark of the Matterdale Historical and Archaeological Society.

The English Place-Name Society, the Society for Name Studies in Britain and Ireland and the Scottish Place-Name Society do great work in advancing onomastic study within these islands and beyond, and in providing a forum for mutual encouragement and exchange of ideas, and I have pleasure in recording my appreciation to the members of all three for the interest they have shown in the present work and for, knowingly or unknowingly, suggesting various avenues of thought. Some names are mentioned within individual dictionary entries, and I would like particularly to thank Oliver Padel who generously gave of his time to consider some of the Brittonic names, Paul Cullen who read a final draft of the List of Common Elements, and Mary Atkin (also mentioned below). I am deeply grateful for general support and particular comments to Richard Coates, Hon. Directory of the English Place-Name Survey and General Editor of this volume, and to his predecessor, Victor Watts, whose untimely death in December 2002 was a grievous loss to scholarship and to all who knew him. Paul Cavill, Principal Research Fellow of the English Place-Name Society, set up the volume for printing with his customary care and discernment, and I have greatly appreciated his genial professionalism. The difficult task of producing the two maps was nobly executed by Darcey Gillie and Tom Burslem, and I have special pleasure in thanking Ian Whaley for his photographs, both in themselves and as an outward sign of support at all stages.

I have been delighted and touched at the helpfulness of farmers and other Cumbrian residents who have taken a good deal of trouble to answer unexpected queries (mainly written) about local conditions and about relatively recent names, and I am extremely grateful to them. Without their information the motivation for several names, including some that appear rather anomalous in their locations, would have remained obscure, and individuals are gladly acknowledged under the entries for the names concerned.

To Mary Atkin and the late Bill Atkin of Levens, Kendal, I owe thanks not only for kind hospitality but also for their infectious dedication to the study of the Cumbrian past; similarly Gillian Scarborough and Roger Haines have not only offered a haven from which to visit London libraries, but also friendship and encouragement. Warm thanks also go to Edward and Sheila Wallis for 'test-driving' a draft sample of the Dictionary.

Despite best efforts and help gratefully received, the scope of the book is such that slips and omissions are all but unavoidable, and any errors are due to nothing other than my own fallibility. It may well be that further light can be thrown on some names by additional documentary evidence or local knowledge, and it is also the case that although the Ordnance Survey maps are a wonderful resource they are not an infallible guide to local place-name usage. I would be extremely grateful to hear from anyone able to offer further information on any relevant points.

Finally, many of the best times of my life have been spent in the Lake District, and it is impossible to thank all those who sparked my enthusiasm and shared the special experiences of lakes and high fells, but I would especially like to mention, from the early days, my mother, my late brother and the Hodge family, and from more recent years my husband and sons, best of companions, to whom this book is dedicated.

School of English Literature, Language & Linguistics,
University of Newcastle upon Tyne
(D.C.Whaley@ncl.ac.uk)

INTRODUCTION

'What are the names of the ridges that run that divide St John's vale & Wanthwaite . . . Surely this must be White pike? and then what is the name of that high round hill . . .' (Coleridge, CN 798). Striding those ridges or contemplating the view from his window in Greta Hall, Keswick, Coleridge represents all those, whether residents or visitors, whose love of the Lake District includes a fascination with its names. To them, the names of becks, gills, fells, crags, dales, meres, tarns and thwaites are as much part of the experience as the rush of slate-green water over rocks, or the dawn mist shrouding a lake. But what do the names mean, and what can they reveal about the region and its history?

THE MEANING OF LAKE DISTRICT PLACE-NAMES
Many or most place-names start off life as descriptive phrases — 'the dairy farm', 'the hamlet by the stream', 'pig valley', 'the green hill', 'Ulf's lake' — but as they become accepted by communities as the normal way of referring to a particular place this function of reference takes over as the main or sole one. The name's original meaning may remain clear, but it may be lost, as the language of the community and the name itself change through the centuries. The names — KESWICK, BROUGHTON, GRIZEDALE, GREENHOW, ULLSWATER — remain, and we speak about these places or visit them without knowing what the names signified to the people who created them. However, the names are rich with insights if we can interpret them.

The literary gentlefolk who published descriptions of the Lake District from the seventeenth century onwards were often intrigued with Lakeland place-names and their histories. John and Thomas Denton, Fleming, Machell, West, Hutchinson and Coleridge are just a few of the writers whose etymologies are a delightful mix of sound historical and linguistic insights on the one hand, and misplaced erudition and sheer invention on the other. Sometimes right, sometimes wrong, their suggestions are always resourceful. Fleming thought the River BRATHAY bore the Greek name for a kind of juniper (1671, 33). His contemporary Machell thought that GRASMERE might refer to the lake's grassy banks, but alternatively, bearing in mind that it is slightly larger than its neighbour Rydal Water, that it might be a corruption of 'Grosmere (i.e. the great mere)' — and hence presumably partly French. 'Take which derivation liketh you best', he shrugs, 'for both do agree [with] the Nature of the place' (Machell, c. 1692, ed. Ewbank, 144).

Over the last century or so, generations of scholars of great learning, insight, and patience have advanced the study of English place-names, including those of the pre-1974 counties of Cumberland, Westmorland and Lancashire, parts of which comprise the Lake District. My constant debt to these scholars will be evident throughout this book. However, place-name interpretations that are very wide of the mark continue to abound in print. Two extremely useful and well-illustrated guidebooks published in 1994, for instance, contain some quite nonsensical place-name etymologies. We are told, for example, that 'the name *Dunmail* is derived from the Norse and means a cairn, or pile of stones', while in fact it is the 'Raise' not the 'Dunmail' of DUNMAIL RAISE that is Norse and has the meaning 'cairn'. SKIDDAW is explained as 'C[eltic] "ska-da": sun god; O[ld] N[orse] "skyti": crag, "haugr": height', which fails to make clear that these are two totally different interpretations and that the first, however picturesque, has no authority whatever. The frequency with which misinformation of this kind appears in print was part of the motivation for the making of the present Dictionary, and although I dare not hope that it is definitive or error-free, I hope that it will be a useful resource.

The business of interpreting place-names is indeed fraught with pitfalls, but there are clear principles and processes, the first and most important of which is this: Because place-names develop through time so that their origins are often obscured, *their interpretation has to be based on the fullest and earliest evidence possible* — ideally, a whole run of spellings beginning soon after the supposed date of coining. Modern forms are often unintelligible or deceptive. Without the form *Styalein* from 1230 we could not recognise STARLING (DODD) as a re-formation of a name meaning 'Alein's path', and similarly without their early spellings we could not know that ULLOCK meant 'wolves'-play' or BRACELET 'broad plain'. In cases where a name appears interpretable but no early spellings are known, therefore, we cannot know whether it is as straightforward as it seems, or whether some corruption or re-formation has taken place. Unless there is reason to suspect corruption, I have normally taken names of this sort at face value, in the belief that a modern name has a validity of its own even if it arose as a re-formation of an earlier one — it meant something to the speakers who reformed it. The lack of early spellings, however, should be noted and taken seriously. It must also be said that even where there are medieval spellings, they may not be early enough, plentiful enough or unanimous enough to point clearly to a single etymology, and that uncertainties can still exist even where there is a good run of early spellings. As a result, the 'p' words 'perhaps', 'possibly', 'probably' and 'presumably', not to mention 'conceivably', 'uncertain', or 'obscure', appear with rather tiresome frequency in the Dictionary. However, given the scarcity of the early documentation for Cumbria (reviewed below), it would be misleading to appear more definite than the evidence allows. It should also be noted that the identification of names in documents with known places can be problematic (see, e.g., ULLOCK and WESTRAY). On a more positive note, however, identification is often quite secure (e.g. FOLDGATE), and even relatively late spellings can save us from misconceptions: see GRAINSGILL BECK or WREN GILL for sixteenth and seventeenth century examples, while in some cases topographical and comparative evidence can aid interpretation of a name for which early spellings are lacking altogether (see TRUSMADOOR). Modern local pronunciations can also be helpful pointers.

Equipped with early spellings, we can begin to establish the linguistic history of names. This involves considering (simultaneously rather than in sequence):

a. What do the early spellings suggest about the development of the name? We can trace, for instance, the rather unlikely transformation of 'Tofi's Tarn' into Outrun (see OUTRUN NOOK). The spellings at each stage have to be interpreted with care, in the light of the known history of the language(s) concerned, and of knowledge about the spelling conventions used in the documents.
b. How is the name structured? The most common structure is a compound of two elements, e.g. KES-WICK, CROS-THWAITE, in which the second element, the 'generic', has a rather broad and basic sense while the first, the 'specific', 'qualifying element' or 'qualifier', identifies some more specialised feature. Most often, the specific will be a significant word such as an adjective or noun used descriptively, as in GREENUP or RAVEN CRAG, or a personal name as in ROGER GROUND (and occasionally it is uncertain whether a significant word or personal name is present). Exceptions to this, with generic first, are Brittonic names of a certain period such as PENRUDDOCK and probably GLENRIDDING, or Gaelic-influenced ones, the so-called 'inversion compounds' such as SEATOLLER and DALEMAIN. It is generally not difficult to see where the boundary between the two elements falls, though BRANDELHOW illustrates the uncertainty that results when it is not clear, and several names have been reformed through time because a medial -s- has been assumed to belong to a syllable other than the original one (e.g. GATESGARTH). The most frequently used elements are reviewed in the List of Common Elements.
c. What language(s) is/are likely to be involved? The range of languages, and some aspects of the relations between them, are discussed below, but it can be noted here that Greek is not one of them, and that the French adjective *gros* is unlikely to have been compounded with English *mere* 'lake', so that we can safely dismiss the suggestion from Fleming, mentioned above, that BRATHAY derives from the Greek for 'juniper' or the one from Machell that GRASMERE might mean 'the great mere'. In many problematic names the difficulty is essentially one of identifying the language(s) involved: see, for instance, the river-names LOWTHER and WINSTER.

In posing these questions we are asking from what particular word(s) the name derives, i.e. what is its etymology? To etymologise the place-name or identify its component parts — say the Old Norse words for 'white' and 'beck' — does not quite tell us all we need to know about its meaning, however. What exactly do these terms mean in the name? In this case, the answer is fairly obvious, for **white** in Lakeland stream-names normally refers to foaming, 'white water' (whereas an Icelandic or Norwegian 'White River' would often refer to the 'milky' waters from a glacier), and **beck** has a particular place in a system of terms for watercourses. (**Bold** type here and throughout indicates an entry in the List of Common Elements.)

Interpreting a place-name demands that we take it in its whole context, including the local landscape and history. Also vital to the whole process is the use of comparative material, ideally from places with comparable landscapes and/or linguistic histories such as parts of Wales, Yorkshire and Norway. More broadly, our thinking must be guided (though not utterly straitjacketed) by generally received knowledge about the range of ways in which place-names are formed and developed.

The interpretations in the Dictionary will, it is hoped, show these processes in action, but two aspects in particular need further discussion: sources for early spellings of Lakeland place-names, and the range of languages likely to enter into them, and the following two sections introduce those.

THE EVIDENCE FOR LAKE DISTRICT PLACE-NAMES

As discussed above, the interpretation of place-names must take its starting-point from the evidence of early documents, which establish both the existence of a name at a particular date, and its particular form, as known to the writer of the document. Anyone interested in the early history of Cumbria has to face the fact that early documents, being essentially pragmatic and relating especially to land ownership, tenancy or taxation, are very sparse indeed in this region, compared with more central, and more agriculturally and economically prized, parts of England. There are no surviving pre-Conquest charters apart from Gospatric's Writ (*PNCumb* III, xxvii–xxx), and next to no relevant coverage in Domesday Book (1086), and the documentary dearth is worse in the predominantly high ground of the Lake District than in the rest of Cumbria. Spellings before the twelfth century are therefore extremely rare. The evidence for the place-names tends to come in several strata. The more important settlements, especially those of the lower-lying areas, are often mentioned in central government documents of the twelfth and thirteenth centuries such as Pipe Rolls, Lay Subsidy Rolls, Feet of Fines etc., which are mainly held in the Public Record Office at Kew, London. Names of natural features will normally be preserved at this early date only if they happen to lie on or near important boundaries, as when LODORE, DOCK TARN and GLARAMARA appear in charter bounds of c. 1210 for an estate belonging to the Cistercian Abbey of Furness (*Furness Coucher* II, 2, 570). Detailed land surveys of the sixteenth century often name smaller hamlets and individual farms, and from that century onwards Parish Registers name hundreds of habitations not previously mentioned in documents. The names of fells, crags, tarns, and minor streams may well not appear until the late seventeenth century onwards, when travellers and antiquarians beginning with Fleming and Machell begin to describe the Westmorland countryside in some detail, or the late eighteenth, when the maps of Jefferys and others, and guidebooks such as those of West and Clarke give valuable evidence. Even so, a very great number of these names, especially those away from the more accessible valleys, do not (to my knowledge) appear in written form until the first Ordnance Survey One Inch and Six Inch maps of the mid-nineteenth century. A category of source material that has been seriously under-used (with significant exceptions such as Winchester 1987, Corrigan forthcoming and numerous articles in *CW*) is the wealth of locally-produced documents of the late medieval and early modern periods, especially estate records of major landowning Cumbrian families, many of which were brought into the care of the Cumbria Record Office after the English Place-Name Surveys of Cumberland and Westmorland were completed. Although a few of these are cited in the present Dictionary, to exploit them fully would be a formidable task and the way forward undoubtedly lies in detailed studies of defined localities.

LANGUAGES AND PEOPLES OF THE LAKE DISTRICT
English place-names have aptly been called 'signposts to the past', as in the title of a classic book on the subject (Gelling 1997), and, despite important cautions and complications (sketched below), the place-names of upland Cumbria have a particular value as historical evidence given the scarcity of documentary and archaeological evidence for the centuries before c.1200 AD. That same scarcity, however, makes it extremely difficult to interpret the place-names as a body, and important questions about population, politics, territorial organisation and settlement patterns remain unanswered even when all the evidence is considered. Nevertheless, the languages out of which the names are forged present clues to layers of human settlement and influence, and these are briefly reviewed here. The presence of very significant components of both Celtic and Scandinavian naming, and the fact that within Celtic both Brittonic and Gaelic influences are represented, is especially distinctive of the region.

Before the Norman Conquest
The earliest inhabitants of the Lake District, including the Neolithic people who made axe-heads on the scree-slopes above LANGDALE, or the Bronze Age ones who erected CASTLERIGG STONE CIRCLE or SAMPSON'S BRATFUL, have left no known linguistic trace, and the earliest language we can speak about is:

'Old European', a pre-Celtic but Indo-European language believed to have been current in the first two millennia BC. Little is known about this, and some scholars dispute its existence, but the river-name DERWENT is among the names for which the case has been made.

Brittonic or Brythonic was and is one of the two branches of Celtic found in Britain and Ireland, the other being Gaelic or Goidelic. It was spoken throughout what are now England, Wales and southern Scotland during the first millennium BC, through the Roman period, and until the mid-first millennium AD in England. It survived later in Wales (to the present day), Cornwall, and southern Scotland, as well as being exported to Brittany, and it seems to have been spoken in Cumbria until about the 11th century. The early phase of Brittonic is distinguished as 'British'; this was spoken up to c. 500 AD and contained forms with case-endings, which are often indicated by hyphens in reconstructed forms. In the present volume names attributed to this branch of Celtic are referred to as 'British', abbreviated 'Brit.' in the Dictionary, or 'Brittonic', according to period (though the distinction can be difficult to make in individual cases), while the people are referred to as Britons or British. Welsh, Cornish and Breton emerged as separate branches of Brittonic in or around the later sixth century (Jackson 1953, 5). Meanwhile, the names Cumberland and Cumbria contain the name of the Celtic-speaking peoples whose name is Latinised as *Cumbri* (cf. modern Welsh *Cymry* 'fellow countrymen, the Welsh'), and the Brittonic language formerly spoken in Cumbria and southern Scotland is known as 'Cumbric', though it is not clear precisely when it diverged from Welsh and Cornish, nor exactly how it differed (Jackson 1953, 9–10). This term is also used in the volume, as is 'Cumbria', for the Cumbric-speaking territories (especially those south of the Solway), or 'historic Cumbria' where it is necessary to distinguish this from the post-1974 county of Cumbria.

The dominant British tribe in this region was that of the Carvetii, a branch of the Brigantes (see Higham & Jones 1985), partly in eclipse during the Roman occupation of the first to early fifth centuries AD. Their territory may have approximated to the kingdom of Rheged which emerged after the Roman withdrawal. The extent of this, the historic Cumbria, is disputed, but it probably centred on the Solway basin and Eden valley and spanned a large tranche in the west of what is now northern England and southern Scotland (e.g. Higham 1986a, 253, Phythian-Adams 1996, 4; the Scottish border only came into being after the Norman Conquest). It seems not to have extended beyond the Duddon–Stainmore line, and hence excluded what is now South Lakeland. The fortunes of the Britons of Cumbria throughout the seventh to eleventh centuries were complex, variable (through both time and space) and interdependent with those of Anglian Northumbria, British Strathclyde, Gaelic-speaking Scotia, and the southern English kingdom of Wessex, not to mention relations with the Scandinavian or Norse-Gaelic incomers of the tenth century (e.g. see BIRKBY). It is not surprising that careful consideration of the evidence can produce rather different historical scenarios (e.g. Kirkby 1962, Higham 1986a, chs 6 and 7, Phythian-Adams 1996, and the introductions to *PNCumb*, *PNLancs* and *PNWestm*). As a single but important instance, many scholars believe in a British resurgence southwards from Strathclyde early in the tenth century (e.g. Jackson 1963, 72, 81–3, Higham 1985, 40–2 and 1986a, 318), while Phythian-Adams argues for survival rather than revival, 1996, 87 etc.). Most of the place-names associated with this theory lie north of the Lake District, but CARHULLAN is one of them.

Turning to the place-names, river-names constitute a major part of the British heritage throughout England, and these are found in all parts of the Lake District: COCKER, CRAKE, ESK, IRT, KENT, LEVEN etc., many of these now of unknown meaning, while LOWTHER and WINSTER both have alternative Scandinavian explanations. DACRE and CRUMMOCK may also be river-names by origin. Very considerable scatters of archaeological remains from the Brittonic-speaking period are found, sometimes in relatively high situations, and they coincide with Brittonic names at (MOOR) DIVOCK and DEVOKE (WATER); CARROCK (FELL) has both a spectacular British fort and a Brittonic name. Elsewhere, the visible remains of the Britons in Iron Age hill forts now carry later names which refer to the forts using Scandinavian **borg** as in BORROWDALE near Derwentwater or Middle/Modern English **castle** as in MAIDEN CASTLE and in CASTLE CRAG (Borrowdale). Old English **burgæsn*, modern **borran**, may also often refer to cairns of this period. Habitation names of Brittonic origin are rather sparse, mainly concentrated in northern Lakeland, and of a relatively late type. They are mainly named from natural features using characteristic elements with clear Welsh parallels such as **blain, blaen* 'summit, top of' as in BLINDCRAKE (cf. the mountain-name BLENCATHRA), or *pen(n)* 'head, end' as in PENRITH and PENRUDDOCK. These names belong to a structural type thought to have arisen in the fifth or sixth century (Jackson 1953, 225–7), in which the generic element comes first, with adjective or dependent noun second and strongly stressed, but beyond that the dating is difficult. (*Glen* 'valley' as in GLENAMARA, GLENCOYNE and GLENRIDDING, all near the head of Ullswater, may be another example, though Gaelic origins are also possible for these.) The distribution is mainly north of the Derwent — Eamont line, and indeed examples are more numerous beyond the northern boundaries of the National Park (e.g. Cumrew and Cumdivock).

Latin as spoken by the Roman occupying forces of the first centuries AD had very little impact on English place-names generally, although signs of the legionaries' presence remain at GLANNOVENTA (Ravenglass), MEDIOBOGDUM (Hardknott) and GALAVA (Ambleside). It is important to realise that these are Celtic, specifically Brittonic, names, adopted by the Romans and Latinised through the addition of endings. A few other names are not from the Romano-British period but recall the Roman presence by referring to their structures (CAERMOTE, WALLS at Ravenglass and BORROWDALE in Fawcett Forest) or to routeways (HIGH STREET and possibly BROADGATE FARM, and LANE HEAD in Mungrisdale).

Old English (OE) or *Anglo-Saxon,* in its Northern and Midlands variety of Anglian, was the language introduced in or around the seventh century through expansion across the Pennines from the mighty kingdom of Northumbria, via the Aire or Tyne gaps or across Stainmore. Anglian names are particularly common around the low-lying and relatively fertile fringes of the Lake District, and tend to designate places that flourished and attained parish or township status (see O'Sullivan 1984, 1985). Of the elements *hām* and *-ingahām* associated elsewhere in Cumbria and throughout England with the earliest phases of Anglo-Saxon settlement there is only one sure example in the Lake District, WHICHAM, although names such as GOSFORTH or KESWICK could also be relatively early. The most prolific OE element is **tūn** 'farmstead, village, settlement', in the Lake District as in the rest of England, and it occurs at (in clockwise order) BARTON, BAMPTON, BROUGHTON (EAST), COLTON, BROUGHTON (WEST or IN FURNESS), IRTON, LORTON and EMBLETON. References to topographical features in this stratum are rarer, but include *hop* 'enclosed valley' as in GREENUP and MEATHOP and *denu* 'valley' as originally in LANGDALE. Some names that have the appearance of Anglian-type habitation names occur in unlikely places, however, for example DUDDERWICK at the head of Haweswater, BLOWICK near ULLSWATER, STOCKLEY (BRIDGE) in upper Borrowdale or HUTTON in Dunnerdale. Some such names may be post-Anglian, but in other cases we need to remember that some apparently remote spots were on major routeways of the past, which did not chance to be turned into modern roads.

Speakers of *Old Norse* (ON) or *Old Scandinavian* are believed to have reached Cumbria via western Scotland and the Hebrides, Ireland and the Isle of Man in the early tenth century, their advent doubtless partly encouraged by the collapse of the Northumbrian kingdom east of the Pennines in the late ninth century. The Scandinavian language as represented in the Lake District appears to be mainly of the West Norse variety, i.e. brought by settlers of Norwegian rather than Danish or Swedish origin. Some evidence, meanwhile, especially the Cumbrian names in **-by** (setting aside those that are post-Conquest), has been interpreted as pointing to influence from the Danelaw (Fellows-Jensen 1985a, 67, *SSNNW*, 288–320; cf. Grant 2004, ch. 5). Whatever the putative origin of the Nordic incomers, I shall be referring to them as Scandinavian(s) and their language as Old Norse (ON).

Whether 'Viking', with its associations of piracy, violence and paganism, is also appropriate depends on the nature of the Scandinavian immigration, and conditions may well have varied. We can assume the appropriation of some existing settlements, even wholesale takeover of administrative units and their central power bases (Winchester 1985, 99), perhaps achieved by military force or the threat of it; but there is also evidence of purchase (see COPELAND), and of significant occupation of agriculturally poor land which may formerly have been populated sparsely, if at all (see further below). What is certain, from evidence including Viking Age stone sculpture (Bailey 1980), dialect (Ellis 1985) and modern genetic studies (Roberts 1981a, 1981b), is that there was profound Scandinavian influence in the region. The ON language was spoken from the ninth century until well into the eleventh and perhaps early twelfth century, and its impact on the place-names was immense. To take examples from the Derwentwater area alone, we have many names containing ON topographical terms including **á** 'river' as in GRETA, **dalr** 'valley' as in BORROWDALE and COLEDALE, **haugr** 'hill, mound' as in HOW and probably SKIDDAW, **hryggr** 'backbone, ridge', as in BIRKRIGG, **holmr** 'islet' or 'land flanked by water', as in BRIGHAM (Keswick), and **nes** 'headland', as in ASHNESS. How, Birkrigg, Brigham and Ashness are habitations named from nearby natural features, and names of this sort are exceptionally common, as in the North-West at large (Fellows-Jensen, *SSNNW* 297).

In Yorkshire and the East Midlands, major hallmarks of Scandinavian settlement are the habitative elements **by**, *thorp(e)* and *toft*, together with names in which OE **tūn** is preceded by a Scandinavian or Scandinavianised element: either a personal name (forming the so-called Grimston or Toton hybrids) or an appellative or significant word (producing the 'Carleton hybrids'). In the Lake District, by contrast, the favoured methods of naming settlements are by reference to natural features (as mentioned above), or by using the 'quasi-habitative' **þveit** and the shieling terms discussed below. The element *by* occurs, but *thorp(e)* and *toft* do not, and there are a few Carleton hybrids (CARLETON, CONISTON, KELTON) but seemingly no Grimston/Toton hybrids. The *-by* names occur on favourable sites around the low-lying perimeter, and in that respect resemble the *-tūn* names: ALDBY, SOULBY, BOMBY, KIRKBY, GUTTERBY and IREBY are all within two miles of a *-tūn* place-name. The *-by* and hybrid *-tūn* names must have arisen in contexts of close Anglo-Scandinavian contact, and may possibly point to takeover of existing estates; the tenth-century cross at GOSFORTH with its Scandinavian mythological scenes, described in Bailey & Cramp 1988, 100–4, gives material evidence for such takeovers, this time without renaming. Meanwhile, Scandinavian incomers evidently also penetrated places where there had been little or no Anglian settlement, and where any previous British settlement is all but undetectable. Thus the spread of Scandinavian settlement into upper Borrowdale (the more northerly one) may be suggested by a trail of names in **þveit** 'clearing', including ROSTHWAITE, SEATHWAITE and STONETHWAITE, while temporary shieling sites which have subsequently become permanent farms are signalled by **skáli** 'hut, shieling' as in GUTHERSCALE, SKELGHYLL and KESKADALE in the Newlands valley. Outside this area, there are other strings of **þveit** names, for example in the Furness Fells and the Kent valley, while AMBLESIDE, BOUTH, HAWKSHEAD and TORVER all seem to have originated as humble shieling sites. (On the use and distribution of the various 'shieling' terms, see **ærgi, erg** in the List of Elements, also *SSNNW* 291–7 and Whyte 1985.)

In some place-names the grammatical case endings of ON are evident in early or even modern forms, as when both HARTER FELLs preserve the genitive or possessive ending *-ar*, or BROTHERILKELD and SAWREY contain relics of a plural *-ir/-ar*. Such endings give reassurance that these names were given during the Norse-speaking period, and are not later continuations of Norse naming habits (see further below). Norse and Gaelic influences intermingle in Cumbrian place-names, so we turn now to Gaelic.

The form of *Gaelic* or *Goidelic* spoken in Ireland, the Isle of Man, mainland Scotland and the Hebrides was all but identical up to the tenth or even twelfth century (Jackson 1951, 79), and the general term 'Gaelic' is used in this volume. Only a few Lakeland names are composed entirely out of Gaelic elements: KINMONT, RAVENGLASS, BETHECAR, and possibly LOGAN (BECK), MELLBREAK, and WHINLATTER, but traces of Gaelic influence are found in all parts of Lakeland, usually in conjunction with Norse and English elements, and it seems likely that even the wholly Gaelic names arose about the tenth century in a Norse-Gaelic context, rather than showing direct influence from the Gaelic-speaking parts of what is now Scotland. The originally Gaelic word **ærgi, erg** 'shieling' was adopted as a productive element and combines with ON elements in names including MOSERGH, TORVER and probably GLARAMARA. The still more prolific element **crag** 'rocky height' may also have come in at this period and Gaelic influence may have played some part in the limited adoption of the elements **glen** and **mell**. Gaelic personal names are found in PATTERDALE, SETMURTHY and seemingly HOBCARTON, while the morphological type often known as 'inversion compound' is a celebrated feature of Cumbrian place-nomenclature which shows the extension of a common Gaelic word order to Germanic materials. It usually consists of an ON generic, often referring to a minor site connected with pasturage (*sætr* being particularly common) followed by a personal name, of various origin. Examples are BEWALDETH, BROTHERILKELD and SETMURTHY (again). Topographical generics are also common, e.g. *Bechwythop* under WYTHOP. Inversion compounds continued to be generated after the Norman Conquest, perhaps encouraged by similar formations in French, in names such as STARLING (DODD), SEATALLAN and SEAT ROBERT, and may have influenced names such as HOLEHIRD or STONE ARTHUR. (See further *Scand & Celt* 13–66, *PNCumb* III, xxiii–xxiv, Grant 2002 and Grant 2004, ch. 2 on inversion compounds.)

Exactly where and how the Gaelic-Norse confluence came about has been much discussed. The older view looked to Ireland and the Isle of Man (*PNCumb* III, xxii–xxiii), and names such as IREBY seem to bear this out, but the weight of the historical, art-historical and onomastic evidence suggests that many Scandinavian incomers came via the Hebrides and Galloway (e.g. Smyth 1975–9, I, 79–86, Bailey 1985, 58–9, *SSNNW* 306, 412, Grant 2002, especially 65–75).

After the Norman Conquest

The *French* impact on Lake District place-names is relatively modest and mainly indirect, arising from the adoption of French words into English (partly through Anglo-Norman, the distinctive variety of Old French that developed in England after the Conquest), and the use of these in names from then on, as noted below. The Norman Conquest did not bring a rapid change of culture to Cumbria, nor, anywhere in England, an abandonment of the vernacular language. The Normans' famous castle-building leaves its mark, but outside the Lake District National Park, and the 'plantation' of Norman settlers was mainly in the Carlisle area. However, it was William II (Rufus) and Henry I who parcelled out most of what is now Cumbria into great feudal baronies named from ALLERDALE, COCKERMOUTH, COPELAND and FURNESS, which may have reflected earlier territorial units and which formed the basis of later administrative divisions that persisted to the present day (Winchester 1985 and 1987, 16–17). The medieval baronies stamped their mark on the landscape and its place-names with the vast hunting **forests** which constituted the upland part of most of them (Winchester 1987, 17). The Conquest also opened the way for the establishment of great monasteries founded from mother houses in France, such as the Benedictine priory of St Bees or the Cistercian Abbey of FURNESS, both founded in the 1120s. CALDER ABBEY, founded as a daughter house of Furness in 1134, and the Premonstratensian SHAP, founded at the end of the twelfth century, are the only ones whose ruins lie within the National Park. The charters produced by such institutions provide much of the earliest evidence for Lake District place-names, and at least two linguistically French names were given under this direct French influence: GRANGE in Borrowdale, an outlying farm of the monks of Furness, and FARMERY, site of the infirmary of CALDER ABBEY. The monks are also commemorated in names such as ABBEY FLATTS and MONKS BRIDGE, and indirectly in a number of farms in High Furness with names using **park** which were part of the Furness lands before the Dissolution in the 1530s; farm-names in **ground** (q.v.) may also have Furness connections.

Middle and *Modern English* form a continuum, but are conventionally applied, respectively, to English spoken between the eleventh and fifteenth centuries (the period in which NEWLANDS and probably nearby LITTLE TOWN and CAT BELLS arose), and from c.1500 onwards (e.g. AMETHYST GREEN, JOHN BELL'S BANNER and probably ABBOT'S BAY). Middle and Modern English have been progressively enriched by words adopted from a number of languages, especially French (e.g. **bay, cause(wa)y, coppice, farm** or **reservoir**) and Latin (*district, station, plantation*), but also by more exotic items which occasionally appear in the place-names (e.g. THE BUNGALOW). Names given in the English of recent centuries reflect both Standard English and the local dialect, which is rich in words of Norse origin, though sometimes these have been replaced or remodelled through the influence of Standard English, for instance names containing *Birch*, formerly *Birk*.

Complications
Place-names throw a good deal of light on settlement history, not only revealing the peoples and languages that have been active in an area but often hinting at the complex and fluctuating relations — political, social and linguistic — between them. However, the Lake District material is ideally understood in a wider interdisciplinary and regional context than is possible here (for which see, e.g., Baldwin & Whyte 1985, *SSNNW* and Grant 2004), while at the same time, any investigations need to allow for diverse historical conditions even at quite local levels. As to the place-name evidence, even assuming that individual names are

correctly interpreted, it must be treated with great caution, and among the general points to be borne in mind the following are especially important.

A place-name may be considerably older than its first appearance in a surviving document. Hundreds of Lakeland names, for instance, have every appearance of being pre Norman Conquest, given by or under the influence of Anglian or Scandinavian incomers, yet have not been found in documentary sources predating the thirteenth to sixteenth centuries. A still more spectacular example is the Brittonic TRUSMADOOR, of which the first record known to me is from 1867.

On the other hand, place-names containing elements favoured by, say, Anglian or Scandinavian settlers are by no means certain to date back to the settlement periods concerned, since some vocabulary and naming habits continued in use for many centuries. Thus *beck* 'stream' descends from ON **bekkr**, but it became 'naturalised' in Middle and Modern English dialect, both as a common noun and as a place-name element, so that individual names in *beck* could date from the period when Norse was spoken (roughly tenth to late eleventh or early twelfth century), or from any time between then and the first documentary appearance of the name. TOM RUDD BECK, with its combination of forename and surname, cannot have been given in the Norse-speaking period. Hence we can say that the element *bekkr* is linguistically of Norse origin, but that a place-name containing it could be considerably later. There is a similar continuity of usage for much OE vocabulary. The need for fresh naming continued after the Conquest, since there was evidently widespread expansion up the lower slopes of valleys, especially in the twelfth and thirteenth centuries (see **thwaite** and Winchester 1987, *passim*). For these reasons, several entries in the Dictionary refer to the reflex, i.e. descendant, of ON or OE terms, or use phrasing such as 'from ON *xxxx* or its reflex'.

Within the names presumed to be pre-Conquest, we can to some extent distinguish between OE/Anglian place-names, such as those ending in *-ton* (OE **tūn**), and ON/Scandinavian place-names, including many of the **-thwaite** names. However, the close kinship of the two languages within the Germanic language family can make this difficult. There are many pairs of OE and ON words that are cognate, i.e. of the same origin and similar or identical meaning and form (allowing for regular differences between the languages), so that it can be impossible to know which is the main source of the place-name. For instance UZZICAR (*Husaker* c.1160) is 'cultivated field with a house', but it could equally well be of OE or ON origin; and many names apparently originating in one language have been Anglicised or Scandinavianised, e.g. ASHNESS, MEATHOP. It should also be mentioned that a Norse place-name does not necessarily imply the presence of a Scandinavian person or community, since words and naming fashions can be spread beyond the group who originated them, as well as persisting through time as mentioned in the previous paragraph.

Further, it is clear that in many cases the names that have survived to the present day are not in any sense the only or 'original' ones, and the fact that the first known place-name belongs to the Anglian, Scandinavian or later medieval period does not mean that the place was uninhabited before that time. In some cases this has been demonstrated by archaeology, for instance the well-known prehistoric settlement near THRELKELD, or the recently discovered hut circles and field systems of the 'later prehistoric period' at BALDHOW END, Matterdale (Loney & Hoaen 2000). In these cases earlier communities may have had names for the sites which were replaced by incomers. Further, even without the disruption caused by invasions, there is a great deal of wastage and replacement of place-names over the centuries, as any attempt to trace the points from a set of medieval charter bounds on a modern map demonstrates. And several medieval names do not appear on modern maps or consequently in this volume because they are 'lost', i.e. recorded in documents but no longer used. There is strong continuity, too, however. In a great many cases existing names are simply taken over, so that new land-holders, however wealthy or powerful, do not as it were announce their presence through naming. We know, for instance, of Anglian patrons at Dacre and Scandinavian ones at Gosforth and Isel from the exceptional stone carvings there, yet DACRE is a Brittonic name and GOSFORTH and ISEL Anglian ones (see Bailey 1980, Bailey & Cramp 1988, 90–3, 100–4 and 118–9). The 'impressive fusion of Anglian and Scandinavian traditions' in the sculpture at the Scandinavian-named WABERTHWAITE (Bailey & Cramp 1988, 151–2) is also suggestive of contact and continuity of some kind.

In general, the two elements of a compound name are more likely to be drawn from the same language than from disparate ones. However, there are exceptions, and here it is important to distinguish between different situations. There are:

a. Genuinely hybrid or bilingual names, which may have arisen within speech communities familiar with more than one language. Possible examples include HEWTHWAITE and LODORE, but in general these are rather rare.
b. Names originally given in one language to which explanatory or 'epexegetic' elements have been added at a later stage and in another language, often duplicating the original meaning. In SCARTH GAP ON *skarð* already means 'gap, notch in the landscape', and *gap* (ME or ModE, albeit of ON origin) has been added because the Norse term was no longer understood, because it was regarded as a proper name (hence 'the gap called Scarth'), or else simply through the pressure of the 'norm' of two elements per place-name.
c. Secondary or derivative names. A common feature is the cluster of related names, in which the name of a valley, hamlet or other major focus in the locality has given rise to further names. These are not strictly hybrids, since any awareness of the language and meaning of the basic name will have given way to its use purely as a name. Among the many examples are TROUTBECK PARK and BRIDGE or the six names in this volume derived from GRISEDALE in Patterdale.
d. Added to this, we constantly have to allow for the situation mentioned above, where elements originating in distinct languages have entered the linguistic melting pot of Middle or Modern English.

MAN AND THE LANDSCAPE
It would be difficult to find such an intense concentration of varied natural beauty anywhere in Europe. The fells, dales and lakes of the Lake District

National Park are not large in absolute terms, but the scenery, from massive rock ridges to lake-shores bowered with woodland, from peat mosses to sand-dunes, delights and inspires, and the place-names reflect the creative virtuosity of nature, from STRIDING EDGE to LOWESWATER, from MOSEDALE to ESKMEALS. Coleridge's claim, 'In the North every Brook, every Crag, almost every Field has a name — a proof of greater Independence & a Society truer to Nature' (*CN* 1578, 1803, cf. *CN* 579, 1799), is rather unfair on the South, but nevertheless has a kind of truth to it. A truth that Coleridge may or may not have meant to convey is that most Lakeland names refer to the landscape, either directly in that they denote fells, valleys, streams, lakes, woods or mosses, or indirectly in that most farmsteads, hamlets or villages are named from nearby landscape features (e.g. BUTTERMERE, CALDBECK). The predominantly pastoral economy and dispersed settlement patterns of the Lake District mean that settlement names such as those ending in *-ton* or *-wick* so prevalent in most parts of England are rather sparse except in the lowland fringes, though names in ON þveit penetrate the valleys of the high fells. Place-names referring to attributes of the landscape may well be not merely descriptive, but quite practical: BUTTERMERE and GRASMERE are not only aesthetically pleasant places, but relatively fertile ones, whereas a MOSEDALE (q.v.) is a place to make the farmer's heart sink.

The vocabulary used to describe the varied landscape is extremely diverse, not only because of the multiple strata of languages and peoples that have contributed to it, but also because of the precision and subtlety of the topographical terminology of English place-names. This has been demonstrated above all by Margaret Gelling and Ann Cole in a range of publications including *The Landscape of Place-Names* (2000), though even their energetic researches leave something to be said about the usage of typical Lakeland vocabulary such as the hill-terms **crag**, **dodd**, **fell**, **how(e)**, **knott**, **pike**, **rigg**, or **scar**.

As has been emphasised by historical geographers since the generation of W. G. Hoskins in the 1960s, the Lake District, like other regions of Britain, is far from being a primeval wilderness, but rather is the product of millennia of interaction between man and nature. Although the basic landforms were laid down in the 'deep space' of the geological eras, the scenery bears the signs of the hand of man and is constantly evolving, and not only because of dramatic interventions such as the afforestation of ENNERDALE or the creation of reservoirs out of THIRLMERE and HAWESWATER. Indeed, the dominant theme of human history in the Lake District is that of the ever-changing proportions between land that is waste and land that is cultivated by clearance (of trees, stones, heather, bracken etc.), enclosure, drainage and other methods of improvement.

To glance through the individual entries in the Dictionary will give hundreds of examples of ways in which the place-names reflect the unique landscape of the Lake District, its evolution through time, and the use of it by man and his livestock, but a few important threads will be drawn out here.

Geology
The foundation for the varied scenery of the Lake District is its underlying geology, which, setting some complications aside, is in three broad zones or bands running diagonally from south-west to north-east. Place-names, including

most of those mentioned below, often reflect the geology in some way.

i. The central zone is bounded on the north by an imaginary line which skirts the south-east corner of Derwentwater, while its southern boundary just tips the north end of Windermere. This is the zone of the Borrowdale volcanic series (with the exception of largish patches of granite towards the west), and here vulcanicity and subsequent weathering of rocks of varying toughness has produced knotted, craggy summits commanding steep precipices and deep gullies. Here are many of the most famous fells including the angular profiles of GREAT GABLE, PILLAR, STEEPLE and CONISTON OLD MAN or the craggy bumps of HAYSTACKS or CRINKLE CRAGS, and it is here that most features named with **crag**, **knott** or **pike** are found. Soils are generally thin, and relatively fertile spots may be registered in place-names, e.g. BUTTERMERE, GRASMERE and RYDAL.
ii. The northern zone stretches from here to a diagonal line just north of Cockermouth. This is the land of Skiddaw slates, of valleys and of hills like SKIDDAW itself, BLENCATHRA, GRASMOOR or CATBELLS, all characterised by sweeping and relatively smooth lines with occasional craggy outcrops.
iii. The southern zone, of Silurian flags, shales and grits, gives a gentler landscape, mainly lower and often wooded. Here are the FURNESS FELLS including the long slopes of THE HEALD and CLAIFE and COLTHOUSE HEIGHTS.

Among the complications to this three-zone map is the great limestone escarpment of WHITBARROW SCAR to the south-east.

Through dramatic episodes of uplift or mountain building, rocks across the three zones were raised to form a single upland dome, a point close to SCAFELL PIKE being the hub from which ridges and valleys radiate 'like spokes from the nave of a wheel' (Wordsworth, *Prose* II, 280).

Apart from the underlying geology, the main shaping of the geological raw materials has come about by several episodes of glaciation, the last of these scouring out valleys such as (the northern) Borrowdale and Langdale as recently as 10,000 years ago. Place-name elements such as **cove** and **comb** often announce the presence of glaciated corries, while **edge** designates, among other things, arêtes or ridges produced when two corries are gouged out back to back. The BOWDER STONE is among the most spectacular of the great erratic boulders transported by ice. Meanwhile, the continuous effects of faulting, weathering, and cutting by streams have produced features including MICKLEDORE and PIERS GILL.

Lakes and tarns

The lakes and tarns which give the Lake District its unique character and its name are, on a geological time-scale, quite recent and still evolving. The more spectacular changes, such as the separation of great lakes into two as at CRUMMOCK WATER and BUTTERMERE or BASSENTHWAITE LAKE and DERWENTWATER, are in the too distant past to register directly in place-names (except that deposits from GRETA, the 'stony river', are partly responsible for the latter separation), but the ongoing build-up of vegetation that turns pools into bogs has left some pool-names in the landscape where standing water no longer exists. Thus the WASDALE in Shap is named from a lake that has disappeared, while in KENTMERE human intervention has reduced the lake from which the valley is named, and created a larger, higher RESERVOIR. Drainage of a low, flat area also brought the NEWLANDS valley into cultivation in the late medieval period.

Woodland
Despite the general principle that place-names tend to refer to features that are distinctive or even rare in their locality, the characteristic native trees of the region are well represented in the place-names: in approximate order of frequency, **birk/birch, eik/oak, ash, yew, thorn, hollin** and **eller** (alder). Cartmel Fell alone, for instance, boasts THE ASHES, BIRKET HOUSES and BIRKS BROW, HOLLINS, OAK HEAD, ROWAN TREE HILL and RANKTHORNS, while BEECH HILL alludes to a species which, like larch, was introduced from the eighteenth century onwards (Pearsall & Pennington 1973, 134). Most references in place-names are to stands of trees, but there are also outstanding individuals such as the BORROWDALE YEWS, still standing, or the now gone eponymous FALLEN YEW or CROOKED BIRCH.

It is well known that the mainly treeless appearance of the Lakeland fells is mainly a product of human intervention, and that the clearance of the virgin forest which must have clothed many slopes up to about 2000ft/610m. was a very long and gradual process, undertaken both to open up pasture and arable land, and to exploit woodland resources. Although the details of this are obscure, pollen evidence points to the replacement of trees by crops as early as the fourth millennium BC (Millward & Robinson 1970, 74–6).

Within historic times, some place-names referring to woodland are in locations no longer wooded, e.g. GREAVES, HESKETT FARM and some names containing ON **skógr**, OE **sceaga** 'wood, copse'. Some place-name elements, moreover, explicitly refer to woodland clearance, including OE **lēah** and the ubiquitous ON **þveit**, though these may have developed broader usages. Exploitation of existing woodland and the planting of new can be traced in most parts of Lakeland (e.g. STOCKLEY and STAVELEY), but is especially concentrated in High Furness, where we find GREAT HAGG, ROGER RIDDING, BOUTH FALL STILE, and numerous names containing elements such as **coppice, hagg, plantation, ridding** and **spring**, as well as **forest** in its modern rather than medieval sense (contrast GRIZEDALE FOREST with SKIDDAW FOREST).

Waste and cultivated land
Everywhere in the Lake District the contrast can be seen between the unenclosed waste of fell, moor or marsh and enclosed land that has been cleared of stones and unwelcome vegetation and turned into pasture or arable land; a long ribbon of drystone wall usually provides the physical boundary between the two. Many place-name elements register the importance of enclosing small parcels of land in order to improve it and control livestock, among them **close, croft, garth,**

ground, **intake** and **pasture**. Widespread enclosure of common fields in the valleys and of upland waste by walls, fences or hedges did not take place until the mid eighteenth century onwards, and then not on a comparable scale to the notorious Highland Enclosures. The place-names, examined individually and in local clusters, give glimpses of the taming of the land, often progressively higher up the fell over the centuries, but as with other themes, their limitations need to be recognised, for place-names can complement, but not replace, detailed local studies based on landscape archaeology or meticulous documentary work such as found in Winchester 1987.

The land of the Lake District is graded on the Agricultural Land Classification scale as 5, the poorest grade, or at best 4 (forests and lakes are excluded). Nevertheless, arable farming was conducted even in smaller valleys as part of self-sufficient mixed agriculture, and the dominant crops of barley, oats and the rarer rye are commemorated in names such as BARTON, BIGLAND, HAVERTHWAITE (probably) and RYDAL. Meanwhile, even the open fell or common provided resources not necessarily recognised as such by modern eyes, including peat, **bracken**, and **ling** (see further Gambles 1989 and Winchester 2000). Names such as THE HUNDREDS and GREEN QUARTER are reminders of the local organisation of rights and duties in the management of sparse resources.

Animals, wild and domestic

Most of the animals and birds mentioned in Lake District place-names would have been primarily viewed either as a threat and perhaps classed as vermin in Statutes of Henry VIII and Elizabeth I, or as a resource. In the 'threat' or 'vermin' category are badger, always designated **brock**; fox or tod; eagle or erne; and above all the raven, named in no fewer than eleven RAVEN CRAG(S). Mammals now extinct in the region include wolf, wild cat and grey fell fox. Especially in the case of birds, the exact species designated by place-name elements can be uncertain (e.g. **cock**, **dove/dow**).

The ancient occupations of hunting, fishing and shooting are registered in names such as WABERTHWAITE, ANGLER'S CRAG and THE BUNGALOW — the latter illustrating the transition from subsistence to sport. Several names containing **cock** 'cock, wild bird' may relate to snaring for the pot, though COCKLAW and probably COCKLEY focus on the spectacular mating display of the black grouse. On a grander scale, the medieval **forests** and some of the early modern **parks** were preserves of deer hunting, and **hart** and *deer* appear in several names, as well as *roe* (OE *rā*) in RAYRIGG and RAYSIDE. **Buck** is ambiguous — a male deer or goat, while the **hare**, another edible quarry, is difficult to trace since the word (from OE *hara*) is difficult to separate from the reflex of OE *hār* 'hoary, grey' in names such as HARE CRAG.

It has been said that 'the chief theme of landscape history in the Lake District between the twelfth and the sixteenth centuries is the decline of the deer forest at the expense of sheep-farming' (Millward & Robinson 1970, 168), and although Lakeland farming was more diverse in former times than at present, animal husbandry has dominated in most areas. Control of stock was essential — the occasion of many a bitter local feud in early modern times — hence the designation of areas of rough pasture as the **fell** and **common** grazed from particular manors, townships or other units. A number of place-names refer to sheds or enclosures, such as FUSEDALE, BAYSBROWN and names in **fold**. **Gate** in some instances, as in two FELLGATEs, specifically refers to control of livestock movement.

Sheep, especially Herdwick sheep, may be thought of nowadays as the archetypal Lakeland animal, and the name of the breed may recall the great monastic sheep-farms of the later Middle Ages (see HERDWICK CROFT). The word *sheep* does not appear in the place-names (except for SHEEPBONE BUTTRESS), perhaps because it is too general, and the ON words *á* and *sauðr* are absent unless perhaps in SOUTHWAITE, but *wether* 'castrated ram', *lamb* and *ewe* all occur (though *ewe* is not always distinguishable from **yew**), and the numerous words for 'shieling' (see **ærgi, erg**) relate above all to sheep grazing. Cattle were also extremely important in former times, for instance on the great estates of the Cistercian monks (see Winchester 1987, 43) or among the stock of seventeenth-century statesmen or yeomen (Marshall 1973, 192, 218). Hence we have numerous names containing *calf* and two in *cow*, bulls in BOWNESS, bullocks at STOTT PARK, and oxen in OXEN- names as well as OWSEN FELL and probably AUSIN FELL. Pigs too were important, thriving on native oak woodland and giving rise to names including four GRISEDALES (two spelt thus, a GRIZEDALE and MUNGRISDALE) and seven SWIN(E)SIDES. Goats may have been domesticated or wild, and are linguistically rather elusive, illustrating some of the difficulties of deducing environmental history from place-names, since OE *gāt* and ON **geit** 'goat' are easily confused both with each other and with *gate* from ON **gata** 'road, track' or OE *geat* 'gate', while ON *hafr* 'he-goat' is difficult to separate from ON *hafri* 'oats'; the ambiguity of **buck** has already been noted. *Kid* appears unproblematically in three names in this volume. Horses are commemorated in ESKHOLME, YOADCASTLE, ROSGILL and two of the ROSTHWAITEs as well as more obviously in HORSE CRAGS, and see 'Communications' below.

Industrial activities

From the names on the map one would not guess quite how populous and busy many Lakeland valleys were in later medieval and modern times, and not merely with stock-rearing and agriculture. On the whole, specialised exploitation of mineral and other resources was carried on at sites which simply retained their established names: mining of plumbago (black lead) at SEATHWAITE, Borrowdale or of copper at CONISTON, production of iron at BACKBARROW or of gunpowder at ELTERWATER, slate quarrying at HONISTER. Nevertheless, the names themselves give glimpses of the tanning industry at BARKBOOTH, and of fulling in WALLTHWAITE; charcoal-burning is represented at EALINGHEARTH and probably COLWITH, while **kilns** and (water-driven) **mills** providing fuel or power for a variety of industries also leave their mark on place-names. Mining is commemorated at ORE GAP, quarrying at SHAP (indirectly) and the harvesting of stones to be shaped and used for particular purposes at BECKSTONES.

Communications

Despite the isolation of many Lakeland valleys, connections between them and out towards the coast or inland markets were vital for trade and communications.

Routeways favoured the main valleys, but where necessary passes threaded the gaps in mountain ramparts, and were named using elements including *skarð* and *gap* (of ON origin), English *door*, and *pass*, ultimately of French origin. A wealth of words for 'road' or 'track' is represented in the place-names, including **gate**, **lane** or **lon(n)ing**, **rake**, **steel/stile** and **sty(e)**. The OLD CORPSE ROAD commemorates toiling journeys to burials at a distant parish church, as, conceivably, does WIDOW HAUSE, while three names in *way* designate modern long-distance leisure routes (see 'Landscape and leisure' below). Horses were vital for transport, and a handful of names containing **capel** 'horse, nag' and a hill term may mark routeways.

River crossings are signalled in about a dozen names containing either **ford** or ON **vað**, including GOSFORTH and CROOKWATH, and in some forty containing **bridge**. The lakes were to some extent an obstacle to travel but also themselves constituted major routeways; thus THE FERRY crossed the 'waist' of Windermere, while goods were shipped from one end of the lake to the other, and LANDING HOW marks one of several historic landing places on the lake.

Buildings
Human impact on the landscape is at its most obvious in the form of agricultural and domestic buildings, designated by a range of terms including OE *bōðl*, OE **cot**/ON **kot**, ON **búð**, **hall**, **house**, **place**. Before the seventeenth century, when stone and slate came into general use, the less grand of these would have been simple structures of clay and wood.

The great gentry estates of early modern times, often dominated by a major country house named as a **hall,** have also generated further place-names, as when Witherslack Hall gave rise to LAWNS HOUSE and NORTH LODGE as well as PARK WOOD. Religious buildings as focal points in the landscape are alluded to in names such as CHAPEL STILE and the two KIRK FELLs, and the influence of monastic estates has been mentioned above. The modern map also offers a rather random sprinkle of inn- and hotel-names.

Landscape and leisure
In the latter half of the eighteenth century a new view of the Lake District emerged: that of the aesthete who relished the picturesque, unlike the farmer or quarryman who eked a living from slender natural resources and an unfriendly climate. The painters, poets and diarists of the Romantic era and the mass tourists of the railway era and beyond made little impact on the place-names of the Lake District, yet there are some onomastic consequences of modern tourism. The rock-climbers, starting c.1881 with Haskett-Smith and other Oxford undergraduates, bestowed a large number of minor names, a few of which appear on smaller scale maps and hence in this volume, such as SHEEPBONE BUTTRESS and NAPES NEEDLE. The word *leisure* now appears on the map (at LOWTHER), as do the long-distance recreational routes DALES WAY, CUMBRIA WAY and CUMBRIA CYCLE WAY, while the word **farm** which appears so frequently on the modern map in many cases helps to distinguish farmhouses from holiday cottages. The Lake District National Park, which gives rise to BROCKHOLE NATIONAL PARK CENTRE,

and whose boundaries define the scope of this volume, draws together all the threads by attempting to balance the needs of residents and visitors, heritage, environment and economics in this extraordinary corner of England.[1]

---

[1] Further information on the topics covered in this Introduction can be found in works including: Bailey 1980, Baldwin & Whyte 1985, Bragg 1990, Gambles 1989, Higham 1986a, Lindop 1993, Marshall & Davies-Shiel 1977, Millward & Robinson 1970, Monkhouse 1960, Pearsall 1961, Pearsall & Pennington 1973, Phythian-Adams 1996, Ratcliffe 2002, Rollinson 1967, Rollinson 1996, Shackleton 1967, Winchester 1987 and Winchester 2000 (full references in Bibliography). Fuller information on many of the place-names in this volume, their elements and their context, is to be found in *PNCumb*, *PNLancs* and *PNWestm* and Fellows-Jensen 1985a (= *SSNNW*), while for English place-names generally see Cameron 1996, Ekwall 1960 (= *DEPN*), Gelling 1997, Gelling & Cole 2000, Mills 1998 and Watts 2004. For further references on the place-names of the United Kingdom, see Spittal and Field 1990 and its forthcoming Supplement.

A GUIDE TO THE DICTIONARY

COVERAGE
The Dictionary covers all the place-names of the Lake District National Park as shown on the One Inch map of the Lake District, 1994 printing (scale 1:63,360; full details in Bibliography). For the small areas of the National Park not appearing on this map, sheets 85, 89, 96, and 97 of the Landranger series (scale 1:50,000) have been used, and in fact now that the One Inch map is no longer published, these Landranger sheets, together with 90, provide the closest approximations to it; the place-names shown differ very little. The National Park, founded in 1951, covers an area of 885 square miles (2,292 square kilometres), about one third of the modern county of Cumbria. The names of COCKERMOUTH, PENRITH and KENDAL, which lie just outside the National Park boundaries, are also included in the volume because of their historic and continuing importance in the region.

ORDER OF ENTRIES
a. The entries are in alphabetical order. Word divisions (e.g. WHIT BECK or WHITBECK) are ignored in the alphabetisation, especially since the same name can often appear as one or more words.
b. Where place-names complete in themselves have additional distinguishing prefixes, the prefix is ignored in the alphabetisation, e.g. HIGH COLEDALE is treated under COLEDALE, and LITTLE LANGDALE under LANGDALE. On the other hand, where a qualifying term such as *high* or *great* is integral to the name it is included in the alphabeticisation, e.g. HIGH STILE is under HIGH.
c. Some names are treated together in groups or sets:
   i. Groups of related names in the same locality. The basic name is given first, followed by its derivatives in alphabetical order, with the basic name abbreviated, e.g:
   COLEDALE: C~ BECK NY2122, C~ HAUSE NY1821, HIGH C~ NY2223
   Where the basic name does not itself appear on the One Inch map, it is listed without grid reference, as is the case with COLEDALE.
   ii. Sets of identical or closely comparable names in different locations, grouped in order of grid reference, West to East (see, e.g., the layout of BLEA TARN or RAVEN CRAG).

LAYOUT OF INDIVIDUAL ENTRIES
Let us take the example of ROSTHWAITE and ROSTHWAITE FELL:

> ROSTHWAITE NY2514 (hab.) Borrowdale, R~ FELL NY2511. *Rasethuate* 1503, *Rastwhat* 1563 (*PNCumb* 353), *Rosthwait* 1786 (Gilpin I, 199); *Rosthwaite Fell* 1867 (OS).
> ♦ 'The clearing with/by a cairn'; **hreysi**, **þveit**, cf. RAISTHWAITE. *Hreysi* and its reflex *raise* normally signal a man-made cairn or heap of stones, but here it could refer to The How, a rocky knoll which rises steeply from the valley floor. R~ **Fell** is a bleak height some two miles S of the hamlet of R~.

This breaks down into:

> **Place-name(s)**: ROSTHWAITE, R~ (= ROSTHWAITE) FELL
> **Grid reference(s)**: NY2514, NY2511
> **Status**: (hab.) = habitation
> **Civil Parish**: Borrowdale
> **Early spelling(s), with date(s):** *Rasethuate* 1503, *Rastwhat* 1563, *Rosthwait* 1786; *Rosthwaite Fell* 1867
> **Source(s) of early spellings:** (*PNCumb* 353), (Gilpin I, 199), (OS). Abbreviations are explained in the Bibliography.
> **Discussion:** This includes:
> - the meaning: 'The clearing with/by a cairn';
> - the constituent elements: **hreysi** 'cairn', **þveit** 'clearing' and **fell** 'high ground'; words in **bold** are in the List of Common Elements;
> - cross-reference: to RAISTHWAITE
> - further remarks, as appropriate.

To explain in more detail, each dictionary entry covers a place-name or group of related names, and comprises:

Place-name
This is given in capitals and in the spelling found on the One Inch map, followed by any common variant(s).

Grid reference
A short grid reference is given, consisting of two letters and four digits. The letters are NY for approximately the northern two-thirds of the Lake District and SD for the southern third.

Status
a. The feature denoted by the place-name is indicated where necessary, e.g. '(stream)'. The most common case where confusion is possible is that of the 'topographical habitative' names, where villages, hamlets, farms or other dwellings are named from landscape features, e.g. BUTTERMERE, CALDBECK, CANNY HILL, and these are indicated as '(hab.)'. In many cases the landscape feature still bears the name, but this is normally only noted if both landscape feature and habitation are named on the One Inch map. These distinctions are also made among the early spellings, where possible and necessary.
b. Names of parishes are labelled as such, e.g. CONISTON (hab. & parish).

Height
The height of hills, mountains or crags, where shown on the One Inch map, is given in feet and metres, e.g. BRANDRETH NY2111 (2344ft/715m.).

Civil parish
a. The history of English parish boundaries is exceedingly complex, with boundaries and the concept of a 'parish' changing a great deal through time, though alongside that there is considerable continuity of territorial or administrative boundaries across the centuries (see Winchester 1990). The parishes used in the dictionary are modern civil parishes, many of which correspond with the townships or chapelries into which the large ecclesiastical parishes were subdivided. Civil parish boundaries are currently shown on 1:25,000 OS maps, and were shown on earlier printings of the One Inch map. For the most part I use the boundaries shown on the Ordnance Survey Administrative Areas Diagram for Cumbria (South Part) 'as notified to 1-5-92'. I have, however, ignored the amalgamations of parishes represented on the 1992 map since the previous, smaller units are more helpful in identifying the location of places, and they are readily 'translatable' into the newer, larger units if necessary. This applies to (i) Langdales, Grasmere, Rydal, Ambleside and Troutbeck, now a unitary parish of 'Lakes'; (ii) Windermere and Bowness, now united as Windermere; (iii) Blawith and Subberthwaite, now a unitary Blawith and Subberthwaite.
b. Parishes with dual or treble names which resulted from these or earlier amalgamations are, for brevity, referred to by their first constituent only, e.g. Bewaldeth and Snittlegarth appears as 'Bewaldeth'. The other parishes concerned are: Crosthwaite and Lyth, Drigg and Carleton, Dunnerdale with Seathwaite, Ennerdale and Kinniside, Irton with Santon, Meathop and Ulpha, St John's Castlerigg and Wythburn, Underbarrow and Bradleyfield, Whitwell and Selside. The double name is retained in the case of Ireby and Uldale, since the part of the parish within the National Park falls mainly in the former Uldale rather than Ireby.
c. In groups of related place-names, the parish is the same for all, unless otherwise specified.
d. Where features such as rivers or mountains belong in more than one parish, either forming the boundary or straddling it, each of the parishes involved is indicated, e.g. Eskdale/Nether Wasdale.

Early spellings
a. The earliest known spelling is given, and any others (often a selection from the many available) that are of particular value for determining the origin and development of the place-name. Generally, more spellings are given for problematic names, or when spellings are previously unpublished. Abbreviations in documents are silently expanded, unless their meaning is unclear.
b. Brackets around a letter or letters mean that the name occurs with and without the bracketed letter(s).
c. Brackets are placed around whole words when these are not part of the place-name in question, e.g. *greatrig (head)* c.1692, under GREAT RIGG.
d. Where a place-name is recorded as part of a personal name, e.g. *Adam de Hulueleyc* (Adam of Ullock), this is indicated by '(p)'. Here *Hulueleyc* is part of the evidence for the place-name ULLOCK.
e. Where there is uncertainty over identification, i.e. as to whether the name in the document refers to the same place as the one under discussion, this is indicated, e.g. '*Brantwood* 1851 (OS), possibly = *Brentwode* 1356 (*PNLancs* 219)'.
f. In the case of a place that has been renamed, this is indicated, e.g. 'From 1900 (OS), previously *Slack House* 1867 (OS)'. Where the identification is less certain this is indicated. There is no attempt to provide complete coverage of superseded names.
g. Where a cluster of related names is covered in one entry, the early spellings for each name are separated by semi-colons.
h. Early spellings of related places in the neighbourhood which are not on the One Inch map but which can throw light on the name under discussion are introduced 'cf.' ('compare').

Date
a. Each early spelling is followed by its date and source, e.g. *Fieldhead* 1753 (PR) means that the first written record for FIELD HEAD in Eskdale known to me is in the Eskdale Parish Register for 1753.

The format '1170, 1408' is used when a particular spelling occurs at two dates; '1278 to 1487' means that the spelling occurs more or less frequently between those dates; '1205-10' means that the document in which the spelling appears can be dated within a range of years but not more precisely. Note that some medieval documents are not originals but copies. Where information is available, the date of both the original and the copy is given.
b. 'From (a given date)' means that the name as spelt on the One Inch map is recorded from at least as early as the date given. Where necessary, non-significant differences such as capitalisation or hyphens, as in *Helm-crag* for HELM CRAG, have been ignored.

Source
In the case of spellings from medieval or later documents which are printed in published place-name surveys of the former Cumberland, Westmorland and Lancashire, the surveys themselves are cited as the sources (in the format *PNCumb* 347, *PNWestm* II, 149 or *PNLancs* 124), and the original documents, and often a discussion of the name, can be traced from there. For the spellings that I have extracted directly from documents, published and unpublished, these

sources are identified (and abbreviations explained in the Bibliography). Where no source is specified, it is the same as the one last mentioned.

Discussion
The etymology and meaning of the place-name are discussed, including any linguistic, historical, topographical or other information that is relevant. A recurrent format is meaning, then constituent elements, e.g. for SEATHWAITE (Borrowdale):

'the clearing where sedge grows', from ON *sef* 'sedge', preserved in local dial. as *seave*, plus ON þveit 'clearing'.

As seen here, the meaning or translation of the name is in quotation marks and will, for clarity, often reverse the ordering of the elements. The standard format is varied according to circumstances such as degree of certainty in the interpretation; the principles of interpretation are outlined on pp. xiv–xvi above. Alternative interpretations are presented as (a), (b) etc. where this seems helpful. Recurrent place-name forming elements mentioned in the discussions are shown in bold, e.g. **bekkr, hall**, which indicates that further information about them can be found in the List of Common Elements. Cross-references to other place-names included in the Dictionary are in small capitals, e.g. BRAITHWAITE (but references to parishes and other areas are not shown in capitals, since they are so frequent). Where a source (e.g. *PNCumb* 357) that is cited in relation to early spellings also contains an interpretation of the name in question, the source reference is not usually repeated in the Discussion section, and if it is repeated, the page numbers are usually omitted and can be assumed to be the same. Sources are explained in the Bibliography and symbols, terms, and abbreviations in the list of that name, which follows.

SYMBOLS, TERMS & ABBREVIATIONS
used in the Dictionary

> *indicates cross-references within this list.*
For abbreviated titles of published and unpublished sources, see Bibliography.
Counties are as before the boundary changes of 1974.

SYMBOLS

General

SMALL CAPITALS — place-name appearing in the Dictionary
bold type — word is explained in List of Common Elements
~ following a letter — abbreviation of a name already mentioned, e.g. ROSSETT, R~ GILL, where R~ = ROSSETT
* — a linguistic form which is deduced (often from place-name evidence) rather than directly recorded
( ) around letter(s) — the forms occur both with and without the bracketed letters, e.g. **holm(e)**
( ) around word(s) — the bracketed word is not part of the place-name under discussion, e.g. *greatrig (head)*, containing GREAT RIGG.

Old English & Old Norse: essential normalised spellings

j — in ON, sounded like the initial sound in *young*
ð (eth) — in OE and ON, sounded like the voiced sound in Modern English *this*
Þ, þ (thorn) — in OE and ON, sounded like the voiceless sound in Modern English *thin*
ǫ — in ON, a vowel similar to that of Modern German *schön*
v — in ON, sounded closer to Modern English *w* at the time in question
¯ — the short dash or macron above Old English vowels, as in *ā* or *ō*, indicating long vowels
´ — the accent sign above Old Norse vowels, as in *á* or *ó*, indicating long vowels

Phonetic symbols: essentials, with approximate equivalents in Modern Standard English

[æ]	pat (Southern English)	[ɔi]	coin
[ai]	pile, buy	[j]	young
[ə]	about, Linda	[ŋ]	ring
[ei]	pate, lay	[ʃ]	shy, leash
[iː]	peat, lea	[θ]	thigh

TERMS & ABBREVIATIONS

adj.	adjective
affix	in general usage, a prefix or suffix, but in relation to p.ns, usually an additional word or phrase which distinguishes one place from other(s) of the same name, or parts of a place one from the other, e.g. (NEAR) & (FAR) SAWREY
Angl	Anglian, the Old English dialects of the Midlands, N. England and parts of S. Scotland, or the speakers of those dialects
appellative	a common noun, referring to a member of a class e.g. ON *ulfr* 'wolf' as distinct from the man's name *Ulfr* (see ULDALE)
back-formation	a name extrapolated from another; especially common in the case of river-names extrapolated from settlement names, e.g. GOWAN
Beds	Bedfordshire
Brit.	British (language), referring to the early phase of > Brittonic
Brittonic	the *p*-Celtic group of languages comprising > Cumbric, Welsh, Cornish, Breton, and their ancient common ancestor, British (see Introduction, p. xviii)
c.	Lat *circa* 'about'
Celtic	the language group comprising the *p*-Celtic > Brittonic and (on the Continent) Gaulish, and the *q*-Celtic > Gaelic
cent.	century
cf.	Lat *confer* 'compare'
cognate	corresponding word(s) or sound(s) in different but historically related languages, e.g. ON *hryggr* and OE *hrycg* are cognate, both meaning 'back, ridge'
compound	a name formed from two or more elements, e.g. GRASMERE, KESWICK or EAGLE CRAG, as opposed to a > simplex name. Normally, the second element is a common noun referring to some kind of natural feature or habitation; this is often termed a generic. The first element usually defines or characterises the second element and is often known as the specific. It may be a personal name, a descriptive adjective, or a noun referring to some kind of vegetation, animal, human activity, etc. In an > inversion compound generic precedes specific.
corrie	a steep-sided rock basin formed by glaciation
Cumbria	(i) 'historic Cumbria', the territory of the > Brittonic-speaking peoples of NW England and SW Scotland (see Introduction, pp. xviii–xix); (ii) the post-1974 county created from the former Cumberland, Westmorland, Lancashire-North-of-the-Sands, and the Sedbergh area of WYorks.
Cumbric	the > Brittonic language spoken in Strathclyde and Cumbria in the first millennium BC and first AD
Dan	Danish
dat.	dative, a grammatical case, often used after prepositions, in OE and ON
dial.	dialect
E	East (and NE North-East etc.)
E.	Eastern
early ModE	early Modern English (i.e. English of the 16th–17th centuries)

el.	element, a constituent part of a > compound or the sole constituent of a > simplex name. Those shown in **bold** are explained in the List of Common Elements
EPNS	English Place-Name Society
etymon	root or original word from which a name or element is derived, e.g. *bere-tūn* 'barley-farm' is the OE etymon of the p.n. BARTON
EYorks	East Yorkshire/East Riding of Yorkshire
Fr	French
ft	feet
folk etymology	the reforming of a name by speakers, often in order to make it comprehensible
Gaelic	Gaelic or Goidelic, the *p*-Celtic group of languages comprising Irish, Scottish and Manx Gaelic, and their ancient common ancestor, Common Gaelic (see Introduction, p. xxii)
gen.	genitive, a grammatical case, normally indicating possession, in OE and ON
generic	see 'compound'
Germanic	the language-group to which OE and ON (as well as languages including Dutch and German) belong
hab.	habitation: village, hamlet, farmstead or house
hybrid	a name composed of elements which originated in more than one language
hypercorrect(ion)	relating to changes motivated by a (mistaken) attempt at correctness
hypocoristic	a pet or familiar version of a forename, e.g. *Harry* is a hypocoristic version of *Henry*, and *Dob* or *Dobbin* of *Robert*
Icel	Icelandic
inversion compound	a place-name in which a personal name or other specific follows the generic or main element, e.g. DALEMAIN 'Máni's valley'; see Introduction, p. xxii
Ir	Irish
Lake District, Lakeland	defined in this volume as the area covered by the Lake District National Park
Lancs	Lancashire
larger scale maps	here used to refer to 1:25,000 or 1:10,000 maps (older 2½ Inch or Six Inch to the mile)
Lat	Latin
LDHER	Lake District Historic Environment Record (formerly Sites & Monuments Record), Lake District National Park Authority, Kendal
lit.	literally
m.	metre(s)
ME	Middle English, the language c. 1100– c.1500
ModE	Modern English, the language from 16th cent. to present
N	North
N.	Northern
n.	(in dictionary references) noun
n.	(in bibliographical references) note
NCy	North Country
nom.	nominative, a grammatical case expressing subject, in ON or OE
Norw	Norwegian

NYorks	North Yorkshire/North Riding of Yorkshire
ODan	Old Danish
OE	Old English, the language to c. 1100 (normally cited in its Anglian form)
OFr	Old French
OIr	Old Irish
ON	Old Norse, usually as represented in early Icelandic or Norwegian texts
ONorw	Old Norwegian
One Inch map	OS Touring Map and Guide 3, Lake District (refers to 1994 printing unless otherwise specified; previously One Inch to One Mile Tourist Map)
OS	Ordnance Survey (refers to Six Inch maps unless otherwise specified)
OW	Old Welsh
OWN	Old West Norse: Old Icelandic and/or Norwegian as distinct from Old Danish and/or Swedish
Oxon	Oxfordshire
(p)	place-name recorded as part of a personal name, e.g. *Hulueleyc* in *Adam de Hulueleyc* (Adam of Ullock)
pers.comm.	personal communication, either written or oral
pers.n.	personal name (a male forename, unless otherwise specified)
pl.	plural
p.n.	place-name
PrW	Primitive Welsh
PR	Parish Register
q.v.	Lat *quod vide* 'which see', used in cross-referencing
R.	River
reflex	the descendant or later stage of a word or sound, e.g. ModE *oak* is the reflex of OE *āc*
S	South
S.	Southern
SAM	Scheduled Ancient Monument
sb.	(in dictionary references) substantive, i.e. noun
Scand	Scandinavian
secondary name	one formed from an existing name, e.g. RYDAL BECK from RYDAL
shieling	outlying pasture with a hut or huts, to which livestock are brought to graze in the summer months; see List of Common Elements: **ærgi, erg**
[*sic*]	Lat 'thus', indicating that a form, though suspect, is correctly copied from its source
significant word	a word in general use, as opposed to a name; includes > appellatives
simplex	a name consisting of one principal > element, e.g. DODD or (THE) WYKE
sing.	singular
specific	see 'compound'
Swed	Swedish
topographic(al)	relating to landscape
transferred name	a name transplanted from one place to another, e.g. HUYTON HILL
W	(as language) Welsh

W	(as direction) West
W.	Western
WS	West Saxon, the OE dialect of Wessex
WYorks	West Yorkshire/West Riding of Yorkshire
Yorks	Yorkshire

BIBLIOGRAPHY, with ABBREVIATIONS

Published and unpublished works, including maps.
For non-bibliographical abbreviations, see list of Symbols, Terms and Abbreviations.

Aitchison, George. 1935. 'Calgarth Hall', CW2 35, 207–12.
Andrews, J. S. & J. A. Andrews. 1991, 1992. 'A Roman road from Kendal to Ambleside. A field survey.' CW2 91, 49–57; CW2 92, 57–65.
Armitt, Mary L. 1906. 'Ambleside town and chapel: some contributions towards their history.' CW2 6, 1–94.
Armitt, M. L. 1908. 'Fullers and freeholders of the parish of Grasmere.' CW2 8, 136–205.
Atkin, Mary. 1988–9. 'Hollin names in north-west England.' Nomina 12, 77–88.
Atkin, M. A. 1991. 'The medieval land use of Kentmere.' CW2 91, 69–78.
Awty, Brian G. 1977. 'Force Forge in the seventeenth century.' CW2 78, 97–112.
Bailey, Richard N. 1980. *Viking Age Sculpture in Northern England*. London: Collins.
Bailey, Richard N. 1985. 'Aspects of Viking-Age sculpture in Cumbria.' In Baldwin & Whyte, 53–64.
Bailey, Richard N. & Rosemary Cramp. 1988. *Corpus of Anglo-Saxon Stone Sculpture 2. Cumberland, Westmorland and Lancashire North-of-the-Sands*. Oxford: OUP/British Academy.
Baines, Sir Edward. 1829. *A Companion to the Lakes of Cumberland, Westmoreland and Lancashire*. London: Hurst, Chance & Co. Also 3rd edn, London: Simpkin & Marshall, 1834.
Baines, Edward. 1870. *The History of the County ... of Lancaster*. 2 vols. London: Routledge.
Baldwin, John R. & Ian D. Whyte (eds). 1985. *The Scandinavians in Cumbria*. Edinburgh: Scottish Society for Northern Studies.
Banks, T. M., C. Nichol & D. G. Bridge. 1994. 'Lead mines in the manor of Kinniside.' CW2 94, 215–47.
Barber, John B. & George Atkinson. 1928. *Lakeland Passes*. Ulverston: James Atkinson.
Barnes, J. A. 1904. 'Ancient corduroy roads near Gilpin Bridge.' CW2 4, 207–10.
Barrow, John. 1886. *Mountain Ascents in Westmoreland and Cumberland*. London [publ. unspec.]
Basden, Eric B. 1997. *Index of Celtic Elements in ... Watson's* The History of Celtic Place-Names of Scotland. Compiled 1978. Edinburgh: Scottish Place-Name Society. [Revised version by A. G. James & S. Taylor online from the Scottish P-N Society website at: <http://www.st-andrews.ac.uk/institutes/sassi/spns/WatsIndex2.htm>]
Bede's Ecclesiastical History of the English People, ed. & trans. Bertram Colgrave & R. A. B. Mynors. Oxford: Clarendon. 1969.
Binns, Alison. 1995. 'Pre-Reformation dedications to St Oswald in England and Scotland: a gazetteer.' In *Oswald, Northumbrian King to European Saint*, ed. Claire Stancliffe & Eric Cambridge. Stamford: Paul Watkins; 241–71.
Birkett, David. 1999. 'And it's still Seldom Seen.' *Conserving Lakeland* 34, 6–7.
Björkman, Erik. 1910. *Nordische Personnamen in England*. Halle: Niemeyer.

Bosworth, Joseph & T. Northcote Toller. 1898. *An Anglo-Saxon Dictionary*. Supplement by T. N. Toller 1921; Enlarged Addenda and Corrigenda by Alistair Campbell 1972. Oxford: OUP.
Bott, George. 1994. *Keswick. The Story of a Lakeland Town*. Keswick: Cumbria County Library.
Bouch, C. M. Lowther. 1936. 'A twelfth century Cumberland surname.' *CW2* 36, 24–33.
Bragg, Melvyn. 1990. *Land of the Lakes*. 2nd edn. London: Hodder & Stoughton.
Brann, M. L. 1985. 'A survey of Walls Castle, Ravenglass, Cumbria.' *CW2* 85, 81–5.
Brearley, Denis. 1974. *Lake District Place-Names explained for the general reader*. Newcastle upon Tyne: Frank Graham.
Breeze, Andrew. 1999. 'The name of the river Mite.' *CW2* 99, 277–8; revised reprint in Coates & Breeze 2000, 70–1.
Breeze, Andrew. 2000: sections by Breeze in Coates & Breeze.
Bridges — List of Bridges, British Library MS Add. 37721.
Brunskill, R. W. 1985. 'Vernacular building traditions in the Lake District.' In Baldwin & Whyte, 135–59.
Brydson, A. P. [1908]. *Some Records of Two Lakeland Townships*. Ulverston: Holmes.
Brydson, A. P. 1911. *Sidelights on Mediæval Windermere*. Kendal: Wilson.
Camden's Britannia 1695. A Facsimile, intr. Stuart Piggott. Newton Abbot: David & Charles. 1971.
Cameron, Kenneth. 1968–9. 'Addenda & corrigenda: vols XXV & XXVI, English Place-Name Elements, Parts I & II'. *JEPNS* 1, 9–42.
Cameron, Kenneth. 1996. *English Place-Names*. 4th edn. London: Batsford.
Census — Index to Censuses, by Roland Grigg, *Cumbrian Genealogy*, <http://sylgrigg.users.btopenworld.com/CensusIndex.htm>, accessed 2004.
Cessford, C. 1994. 'Dinogad's smock.' *CW2* 94, 297–9.
Cherry, J. 1961. 'Cairns in the Birker Fell and Ulpha Fell area.' *CW2* 61, 7–15.
Cherry, J. & P. J. Cherry. 1987. 'Prehistoric habitation sites in West Cumbria. Part V: Eskmeals to Haverigg.' *CW2* 87, 1–10.
Clare, T. 1999. 'The environs of the Castlerigg stone circle: an analysis of the landscape of the Naddle Valley near Keswick.' *CW2* 99, 67–87.
Clark, Keith. 1995. 'Registered archaeological sites.' *MHAS* 2, 15–17.
Clark, Keith. 2000. 'Social and economic changes in Matterdale between 1800 and 2000.' *MHAS* 7, 25–45.
Clark, Keith. 2001. 'The Brownrigg family: ups and downs.' *MHAS* 8, 20–31.
Clarke, James. 1789. *A Survey of the Lakes of Cumberland, Westmorland, and Lancashire*. 2nd edn. London: priv. publ.
Cleasby, Richard & Gudbrand Vigfusson, rev. William A. Craigie. 1957. *An Icelandic-English Dictionary*. Oxford: Clarendon.
CN — *The Notebooks of Samuel Taylor Coleridge*, I *1794–1804. Text & Notes*, ed. Kathleen Coburn. London: Routledge. 1957. [Cited by note number, rather than page]
Coates, Richard. 1984. 'Coldharbour — for the last time?' *Nomina* 4, 73–8.
Coates, Richard. 1988. 'Celtic etymologies of four place-names in England.' In his *Toponymic Topics*. Brighton: Younsmere Press; 24–38, esp. 30–3.
Coates, Richard. 1999. 'New light from old wicks: the progeny of Latin vicus.' *Nomina* 22, 75–116.
Coates, Richard & Andrew Breeze. 2000. *Celtic Voices, English Places. Studies of the Celtic Impact on Place-Names in England*. Stamford: Shaun Tyas.
Coates, Richard 2000: sections by Coates in Coates & Breeze above.
Coates, Richard. Forthcoming. 'The Duddon: a river of Cumbria.' To appear in *JEPNS*.

Coates, Richard. Forthcoming. 'Maiden Castle, Geoffrey of Monmouth and Hārūn al-Rašīd.' To appear in *Nomina*.
Cole, Ann. 1987–8. 'The distribution and usage of the place-name elements *botm*, *bytme* and *botn*.' *JEPNS* 20, 38–46.
Coleridge: see *CN*
Collingwood, R. G. 1924. 'Castle How, Peel Wyke.' *CW2* 24, 74–87.
Collingwood, R. G. 1916. 'The Castle Rock of St. John's Vale.' *CW2* 16, 224–8.
Collingwood, W. G. 1895. *Thorstein of the Mere. A Saga of the Northmen*. London: Arnold.
Collingwood, W. G. 1904. 'The home of the Derwentwater family.' *CW2* 4, 257–87.
Collingwood, W. G. 1907. 'Bampton crosses.' *CW2* 7, 302–4.
Collingwood, W. G. 1912. *Elizabethan Keswick*. Kendal: Wilson.
Collingwood, W. G. 1918. 'Mountain names.' *CW2* 18, 93–104.
Collingwood, W. G. 1920. 'More mountain names.' *CW2* 20, 243–5.
Collingwood, W. G. 1921. 'Thirteenth-century Keswick.' *CW2* 21, 159–73.
Collingwood, W. G. & R. G. Collingwood. 1923. '*Tillesburc*.' *CW2* 23, 138–41.
Collingwood, W. G. 1925. *Lake District History*. Kendal: Wilson.
Collingwood, W. G. 1926. 'An inventory of the ancient monuments of Westmorland and Lancashire-North-of-the-Sands.' *CW2* 16, 1–62.
Collingwood, W. G. 1929. 'Ravenglass, Coniston and Penrith in ancient deeds.' *CW2* 29, 39–48.
Collingwood, W. G. 1931. 'Gleanings from Rydal Muniments.' *CW2* 31, 1–7.
Collingwood, W. G. 1933. *The Lake Counties*. Revised edn. London: Warne.
Corrigan, Linda M. Forthcoming. A Study of Early Recorded Place-Names of the South Cumbria Area. PhD thesis, University of Manchester.
Cowper, H. Swainson. 1891. 'Hawkshead Hall.' *CW* 11, 7–49.
Cowper, H. S. 1895. 'The homes of the Kirkbys of Kirkby Ireleth.' *CW* 13, 269–86.
Cowper, H. S. 1897a. 'Hawkshead folk-lore.' *CW* 14, 371–89.
Cowper, H. S. 1897b. 'Sites of local beacons.' *CW* 14, 139–43.
Cowper, H. S. 1899. *Hawkshead*. London: Bemrose.
Cowper, H. S. 1901. 'Cawmire or Comer Hall.' *CW2* 1, 119–28.
Cowper, H. S. 1904. 'The will of Edward Ridge.' *CW2* 4, 146–53.
Cowper, H. S. 1907. 'The Ambleside "Curates'" Bible.' *CW2* 7, 143–8.
Cox, Barrie. 1994. *English Inn and Tavern Names*. Nottingham: Centre for English Name Studies.
CRO — Cumbria Record Office
CrosthwaiteM — Maps by Peter Crosthwaite: *The Lake of Derwent* 1782, *The Lake of Ullswater* 1783, *The Lake of Windermere* 1783, *The Lake of Broadwater or Bassenthwaite Lake* 1785, *Coniston Lake* 1788. Also 2nd edns of the above, 1809. London [publ. unspec.].
Crowe, C. J. 1984. 'Cartmel, the earliest Christian community.' *CW2* 84, 61–6.
Cumbrian Ancestors. Notes for Genealogical Researchers. 2nd edn. Cumbria Archive Service. Cumbria County Council. 1993.
Curwen, John F. 1907. 'Thornthwaite Hall, Westmorland.' *CW2* 7, 137–42.
CW — *Transactions of the Cumberland & Westmorland Antiquarian & Archaeological Society*. Kendal. [In this Bibliography *CW* = Original Series, *CW2* = New Series, *CW3* = Third Series. Within the Dictionary, articles from *CW* which are used solely as a source of early spellings are cited, for brevity, in the form '*CW* 1926, 78'. Articles used for other content are cited by author, date and page and included in this Bibliography. Throughout the New Series, volume numbers match years, from vol. 1 in 1901 to vol. 100 in 2000, except that vols 44 to 59 appeared one year late, e.g. vol. 44 in 1945; two volumes appeared in 1960.]

CWAAS — Cumberland & Westmorland Antiquarian & Archæological Society
D&HM — Map by Donald & Hodskinson 1774: *The County of Cumberland Surveyed.* [Publ. unspec.]
Danby, Peter B. 1994. *Lake District Grid Squares Explored.* Edinburgh: Pentland.
Davies-Shiel, M. 1971. 'The terminology of early iron-smelting in Lakeland.' CW2 71, 280–4.
Davies-Shiel, M. 1972. 'A little-known late mediaeval industry, Part I. The making of potash for soap in Lakeland.' CW2 72, 85–111.
Davies-Shiel, M. 1974. 'A little-known late medieval industry, Part II. The ash burners.' CW2 74, 33–64.
Davy, John. 1857. *The Angler in the Lake District, or, Piscatory Colloquies.* London: Longman etc.
Denton 1610 — *John Denton: An Accompt of the most considerable Estates and Families in the County of Cumberland*, ed. R. S. Ferguson. Kendal: Wilson. 1887.
Denton 1687–8 — *Thomas Denton: A Perambulation of Cumberland*, ed. Angus J. L. Winchester with Mary Wane. Surtees Society 207/CWAAS Record Series 16. Woodbridge: Boydell. 2003.
DEPN — Eilert Ekwall. 1960. *The Concise Oxford Dictionary of English Place-Names.* 4th edn. Oxford: Clarendon.
Dickinson, J. C. 1980. *The Land of Cartmel.* Kendal: Wilson.
Dilley, Robert S. 1967. 'The Cumberland court leet and the use of the common lands.' CW2 67, 125–51.
Dilley, Robert S. 1970. 'Some words used in the agrarian history of Cumberland.' CW2 70, 192–204.
DOE — Ashley Crandell Amos, Angus Cameron et al. 1986– . *Dictionary of Old English.* Toronto: Pontifical Institute of Medieval Studies.
Domesday Book — *Domesday Book* 30. *Yorkshire*, ed. M. L. Faull & M. Stinson. 2 vols. Chichester: Phillimore. 1986.
DOST — *Dictionary of the Older Scottish Tongue*, ed. from the collections of William A. Craigie. 1937 – 2002. London: OUP etc. See also DSL.
Drummond, Peter. 1991. *Scottish Hill and Mountain Names.* [Place unspec.] Scottish Mountaineering Trust.
DSL — *Dictionary of the Scots Language.* Online edition of DOST and SND, with Supplement (2005), at: <http://www.dsl.ac.uk/dsl/>
Ducatus Lancastriae. III *Calendar of the Pleadings from the Fourteenth Year to the End of the Reign of Queen Elizabeth.* London: Record Commission. 1834.
Dymond, C. W. 1902. 'An exploration of "Sunken Kirk".' CW2 2, 53–76.
EDD — Joseph Wright. 1898–1905. *The English Dialect Dictionary.* 6 vols. London: Frowde.
Ekwall (without further qualification) — PNLancs; see also DEPN, ERN, Scand & Celt.
Elliott, G. 1960. 'The system of cultivation and evidence of enclosure in the Cumberland open fields in the 16th century.' CW2 59, 85–104.
Ellis, Stanley. 1985. 'Scandinavian influences on Cumbrian dialect.' In Baldwin & Whyte, 161–7.
Ellwood, T. 1895. *Lakeland and Iceland, being a Glossary of Words in the Dialect of Cumberland, Westmorland and North Lancashire.* London: English Dialect Society; repr. Felin Fach: Llanerch Publishers, 1995.
Ellwood, T. 1899. 'The mountain sheep: their origin and marking.' CW 15, 1–8.
EPNS — English Place-Name Society/Survey
ERN — Eilert Ekwall. 1928. *English River Names.* Oxford: Clarendon.

Ewbank — see Machell
Fair, M. C. 1922. 'The Eskdale Twentyfour Book.' CW2 22, 73–8.
Fair, M. C. 1929. 'Austhwaite and Dalegarth.' CW2 29, 265–8.
Fair, Mary C. 1931. 'Drigg, Barnscar, Carleton and Benfold.' CW2 31, 63–8.
Fair, Mary C. 1936. 'Loweswater Pele and Parks.' CW2 36, 126–8.
Fair, M. C. 1945. 'An almost forgotten religious house of S. Cumberland: Seaton Priory otherwise Lekeley.' CW2 44, 134–7.
Fair, M. C. 1951. 'Notes on the history of Ulpha.' CW2 50, 99–104.
Fair, M. C., W. C. Calvert & H. Peck. 1954. 'Calder Abbey.' CW2 53, 81–97.
Fell, Alfred. 1908. *The Early Iron Industry of Furness and District*. Ulverston: Hume Kitchin.
Fellows-Jensen, Gillian. 1968. *Scandinavian Personal Names in Lincolnshire and Yorkshire*. Copenhagen: Institut for Navneforskning.
Fellows-Jensen, Gillian. 1980. 'Common Gaelic *áirge*, Old Scandinavian *ærgi* or *erg*?' *Nomina* 4, 67–74.
Fellows-Jensen, Gillian. 1985a. 'Scandinavian settlement in Cumbria and Dumfriesshire: the place-name evidence.' In Baldwin & Whyte, 65–82.
Fellows-Jensen 1985b: see *SSNNW*
Fellows-Jensen, Gillian. 1987. 'To divide the Danes from the Norwegians: on Scandinavian settlement in the British Isles.' *Nomina* 11, 35–60.
Fellows-Jensen, Gillian. 1992. 'Scandinavian place-names of the Irish Sea province.' In *Viking Treasure from the North West: the Cuerdale Hoard in its Context*, ed. J. A. Graham-Campbell. Liverpool: National Museums and Galleries on Merseyside; 31–42.
Ferguson, Robert. 1856. *The Northmen in Cumberland and Westmoreland*. London: Longman.
Ferguson, Robert. 1873. *The Dialect of Cumberland, with a Chapter on its Place-Names*. London: Williams & Northgate.
Field, John. 1972. *English Field-Names. A Dictionary*. Newton Abbot: David & Charles.
Fiennes — *The Illustrated Journals of Celia Fiennes 1685–c. 1712*, ed. Christopher Morris. London: Macdonald. 1982.
Flanagan, Deirdre & Laurence Flanagan. 1994. *Irish Place Names*. Dublin: Gill & Macmillan.
Fleming — *Sir Daniel Fleming's Description of Cumberland, Westmorland and Furness, 1671*. In *Fleming-Senhouse Papers*, ed. Edward Hughes. Cumberland Record Series 2. Carlisle: Cumberland County Council. 1961.
Fleming, *Accounts* — *The Estate and Household Accounts of Sir Daniel Fleming, 1688–1701*, ed. Blake Tyson, transcr. B. G. Fell. CWAAS Record Series 13. Kendal. 2001.
Fletcher, W. & Clare I. Fell. 1987. 'Stone-based huts and other structures at Smithy Beck, Ennerdale.' CW2 87, 27–36.
Ford, William. 1839. *A Description of the Scenery in the Lake District*. Carlisle: Charles Thurnam. Also 3rd edn, 1843.
Forsyth, Jennifer. 1998. 'Jonathan Wilson (1693–1780) — his diary and account book.' CW2 98, 207–31.
Fountains Chartulary — *Abstracts of the Charters and other Documents Contained in the Chartulary of the Cistercian Abbey of Fountains*, ed. William T. Lancaster. 2 vols. Leeds: Whitehead. 1915.
Fritzner — Johan Fritzner. 1886–96. *Ordbog over det gamle norske sprog*. 3 vols. 2nd edn, revised by D. A. Seip & T. Knudsen, 1954. Supplement by Finn Hødnebø, 1972. Oslo: Møller. [Online from the University of Oslo website at: <http://www.dok.hf.uio.no/perl/search/search.cgi?appid=86&tabid=1275>]

Furness Coucher — *The Coucher Book of Furness Abbey*, I, pts 1–3 ed. J. C. Atkinson, 1886–8, II, pts 1–3, ed. John Brownbill, 1915–19. Chetham Society NS 9, 11, 14, 74, 76, 78. [Page numbers continuous through each volume: I: 1–260, 261–536, 537–728; II: 1–288, 289–584, 585–end]

Gambles, Robert. 1989. *Out of the Forest. The Natural World and the Place-names of Cumbria.* Kendal: Laverock Books.

Gambles, R. H. 1993. 'The spa resorts and mineral springs of Cumbria.' *CW2* 93, 183–95.

Gambles, Robert. 1994. *Lake District Place Names*. Revised edn. Skipton: Dalesman.

Geiriadur Prifysgol Cymru/A Dictionary of the Welsh Language, ed. R. J. Thomas et al. 1950–2002. Cardiff: University of Wales.

Gelling, Margaret. 1984. *Place-Names in the Landscape*. London & Melbourne: Dent.

Gelling, Margaret. 1997. *Signposts to the Past*. 3rd edn. Chichester: Phillimore.

Gelling, Margaret & Ann Cole. 2000. *The Landscape of Place-Names*. Stamford: Shaun Tyas.

Geological Survey of Great Britain, 1:50,000 map series.

George, Ron. 1995. *St. Cuthbert's Church Lorton*. [Pamphlet]

Gilpin, William. 1786. *Observations relative chiefly to Picturesque Beauty, made in the Year 1772 on Several Parts of England; particularly the Mountains, and Lakes of Cumberland and Westmorland.* 2 vols. London: Blamire.

Godwin, Jeremy. 1988. 'High Mill, Lorton, from its deeds of 1728–1940.' *CW2* 88, 248–50.

Godwin, Jeremy. 1996. '"Robin Hood", near Langwathby.' *CW2* 96, 236.

Graham, T. H. B. & W. G. Collingwood. 1925. 'Patron saints of the diocese of Carlisle.' *CW2* 25, 1–27.

Grant, Alison. 2002. 'A new approach to the inversion compounds of north-west England.' *Nomina* 25, 65–90.

Grant, Alison. 2004. Scandinavian Place-Names in Northern Britain as Evidence for Language Contact and Interaction. PhD thesis, University of Glasgow.

Gray — *Correspondence of Thomas Gray*, ed. Paget Toynbee & Leonard Whibley. 3 vols. Oxford: Clarendon. 1935. [Page numbers continuous through volumes]

GreenwoodM — C. & I. Greenwood. 1823. *Map of the County of Cumberland*. London: Greenwood, Pringle & Co.

Halliday, Geoffrey. 1997. *A Flora of Cumbria*. Lancaster: University of Lancaster Centre for North-West Regional Studies.

Hamp, Eric P. 1975. '*Alauno, -ā*: Linguistic change and proper names.' *Beiträge zur Namenforschung* (New Series) 10, 173–8.

Hanks, Patrick & Flavia Hodges. 1988. *A Dictionary of Surnames*. Oxford: OUP.

Hanks, Patrick & Flavia Hodges. 1990. *A Dictionary of First Names*. Oxford: OUP.

Haszeldine, Bob. 1998. 'The Three Shire Stones.' *Conserving Lakeland* 31, 8–9.

Hay, David & Joan. 1978. *Ullswater through the Centuries*. [Chesham:] Friends of the Lake District.

Hay, Thomas. 1939. 'Old walls and trackways.' *CW2* 39, 15–18.

Hay, Thomas. 1937. 'Three interesting sites.' *CW2* 37, 52–5.

Heaton Cooper. 1983. *The Tarns of Lakeland*. 3rd edn. Kendal: Peters.

Hervey, G. A. K. & J. A. G. Barnes. 1970. *Natural History of the Lake District*. London: Warne.

Higham, Mary C. 1977–8. 'The "*erg*" place-names of northern England.' *JEPNS* 10, 7–17.

Higham, Mary C. 1995. 'Scandinavian settlement in north-west England, with a special study of *Ireby* names.' In *Scandinavian Settlement in Northern Britain*, ed. B. E. Crawford. London: Leicester University Press; 195–205.

Higham, Nick. 1985. 'The Scandinavians in North Cumbria: raids and settlement in the later ninth to mid tenth centuries'. In Baldwin & Whyte, 37–51.

Higham, N.J. & Barri Jones. 1985. *The Carvetii*. Gloucester: Sutton.
Higham, Nick. 1986a. *The Northern Counties to AD 1000*. London & New York: Longman.
Higham, N. J. 1986b. 'The origins of Inglewood Forest.' *CW2* 86, 85–100.
Hindle, Paul. 1998. *Roads and Tracks of the Lake District*. Milnthorpe: Cicerone.
Hodge, E. G. 1940. 'Belle Isle – 1.' *Country Life* August 3rd, 1940, 98–101.
HodgsonM — T. Hodgson 1828. *Plan of the County of Westmorland*. London.
Hodgson, Plans — Plans of various estates in Cumberland, 1865–6. British Library Maps 136.a.4.
Hogg, Robert. 1972. 'Factors which have affected the spread of early settlement in the Lake Counties.' *CW2* 72, 1–35.
Hough, Carole. 1996. 'Place-name evidence relating to the interpretation of Old English legal terminology.' *Leeds Studies in English* 27, 19–48.
Hough, Carole. 1996–7. 'The ladies of Portinscale.' *JEPNS* 29, 71–8.
Hough, Carole. 2003. 'Lark Hall in Lanarkshire and related place-names.' *Notes and Queries* 248 (NS 50), no. 1, 1–3.
Housman, John. 1817. *A Topographical Description of Cumberland, Westmoreland, Lancashire and a Part of the West Riding of Yorkshire*. 8th edn. Carlisle [publ. unspec.].
Howard-Davis, Christine. 1987. 'The Tannery, Rusland, South Cumbria.' *CW2* 87, 237–50.
Hoyte, Patrick. 1999. 'The story of Seathwaite Tarn Reservoir.' *CW2* 99, 267–76.
Hudleston, Annette. 1995. 'A short history of Hutton and the Huttons of Hutton John.' *MHAS* 2, 22–8.
Hudlestone, C. Roy. 1979. 'Askew of Standing Stones.' *CW2* 79, 57–74.
Hutchinson, William. 1776. *An Excursion to the Lakes in Westmoreland and Cumberland*. London: Wilkie & Charnley.
Hutchinson, William. 1794. *The History of Cumberland*. 2 vols. Carlisle: Jollie.
Insley, John. 1986. 'Toponymy and settlement in the North-West: a review of Gillian Fellows-Jensen, *Scandinavian Settlement-Names in the North-West*.' *Nomina* 10, 169–76.
Insley, John. 1994. *Scandinavian Personal Names in Norfolk*. Acta Academiae Regiae Gustavi Adolphi 62. Stockholm: Almqvist & Wiksell.
Insley, John. 2005. 'Windermere.' *Namn och Bygd* 93, 65–80.
Jackson, K. H. 1951. '"Common Gaelic": the evolution of the Goidelic languages.' *Proceedings of the British Academy* 37, 71–97.
Jackson, K. H. 1953. *Language and History in Early Britain*. Edinburgh: University Press.
Jackson, K. H. 1963. 'Angles and Britons in Northumbria and Cumbria.' In *Angles and Britons: the O'Donnell Lectures*, ed. H. Lewis. Cardiff: University of Wales Press; 60–84.
Jackson, Kenneth. 1968–9. 'Addenda & corrigenda: vols XXV & XXVI, English Place-Name Elements, Parts I & II.' *JEPNS* 1, 43–52.
Jackson, K. H. 1970. Appendix II [place-names], in 'The British section of the Antonine Itinerary', by A. L. F. Rivet, *Britannia* 1, 68–82.
JefferysM — Map by Thomas Jefferys 1770: *The County of Westmoreland surveyed*. Charing Cross: Jefferys.
JEPNS — *Journal of the English Place-Name Society*
Johnson, Susan. 1966. 'Disagreement by the Duddon, 1825–1832.' *CW2* 66, 369–81.
Johnson, Susan. 1981. 'Borrowdale, its land tenure and the records of the Lawson manor.' *CW2* 81, 63–71.
Johnson, Susan. 1988. 'Iron mining on the high fells above Eskdale and Langdale, and miners' roads.' *CW2* 88, 246–8.
Johnston, James B. 1934. *Place-Names of Scotland*. London: Murray.

Jones, G. P. 1973. 'Doubts about the Brittonic derivation of some Westmorland place-names.' *CW2* 73, 356–8.
Jones, Owen Glynne. 1900. *Rock-Climbing in the English Lake District*. Keswick: Abraham.
Kendale — *Records relating to the Barony of Kendale*. I–II ed. William Farrer & John F. Curwen, III ed. Curwen. Kendal: Wilson. 1923–6.
Kipling, Charlotte. 1973. 'The netting sites of Windermere.' *CW2* 73, 111–19.
Kirkby, D. P. 1962. 'Strathclyde and Cumbria: a survey of historical development to 1092.' *CW2* 62, 77–94.
Kitson, P. R. 1996. 'British and European river-names.' *Transactions of the Philological Society* 94, 73–118.
Kitson, P. R. Forthcoming. *Guide to Anglo-Saxon Charter Boundaries*. EPNS.
Landnámabók. In *Íslendingabók, Landnámabók*, ed. Jakob Benediktsson. Íslenzk fornrit 1. 2 vols. Reykjavík: Hið íslenzka fornritafélag. 1968. [Page numbers continuous through volumes]
Lascelles, Venetia. 2003. *Field Names of the Duddon Valley*. Duddon Valley Local History Group.
Later Records — *The Later Records relating to North Westmorland*, ed. John F. Curwen. CWAAS Record Series 8. Kendal.1932.
LDHER — Lake District Historic Environment Record (formerly Sites & Monuments Record), Lake District National Park Authority, Kendal.
Lee, Joan. 1998. *The Place-Names of Cumbria*. Carlisle: Cumbria Heritage Services.
Leland, John. *The Itinerary of John Leland*, ed. Lucy Toulmin Smith. 5 vols. London: Centaur Press. 1964.
Lind, E. H. 1905–15. *Norsk-isländske dopnamn ock fingerade namn från medeltiden*. Uppsala: Lundequistska. [Numbering is by column]
Lind, E. H. 1921. *Norsk-isländske personbinamn från medeltiden*. Uppsala: Lundequistska.
Lindkvist, Harald. 1912. *Middle-English Place-Names of Scandinavian Origin*. Uppsala: [Uppsala University Press].
Lindop, Grevel. 1993. *A Literary Guide to the Lake District*. London: Chatto & Windus.
Loney, Helen L. & Andrew W. Hoaen. 2000. 'Excavations at Baldhowend, Matterdale, 1998: an interim report.' *CW2* 100, 89–103.
Machell c.1692 — Papers of Thomas Machell, CRO Carlisle, Dean & Chapter collection. 2 vols.
Machell, ed. Ewbank — *Antiquary on Horseback. The Collections of the Rev. Thos. Machell*, ed. Jane M. Ewbank. CWAAS Extra Series 19. Kendal. 1963. [Selections in modernised spelling]
Maclean, Hector. 1912. 'Caerthannoc or Maidencastle, Soulby Fell.' *CW2* 12, 143–6.
McKay, Patrick. 1999. *A Dictionary of Ulster Place-Names*. Belfast: Institute of Irish Studies, Queen's University of Belfast.
McKinley, R. A. 1981. *The Surnames of Lancashire*. London: Leopard's Head Press.
Margary, Harry. 1991. *The Old Series O. S. Maps of England and Wales. 8 Northern England and Isle of Man*. Lympne Castle: Margary.
Margary, Ivan D. 1957. *Roman Roads in Britain*. II *North of the Foss Way — Bristol Channel*. London: Phoenix House.
Marshall, J. D. & Michael Davies-Shiel. 1977. *The Industrial Archaeology of the Lake Counties*. 2nd edn. Beckermet: Moon.
Martindale, J. H. 1911. 'Hewthwaite Hall, Cockermouth.' *CW2* 11, 163–6.
Marwick, Hugh. 1952. *Orkney Farm-Names*. Kirkwall: Mackintosh.
Matras, Christian. 1933. *Stednavne paa de færøske norðuroyar*. Copenhagen: Thiele.
MED — Hans Kurath *et al*. 1956– . *Middle English Dictionary*. Ann Arbor: Michigan University Press.

MHAS — *Matterdale Historical & Archaeological Society, Yearbook and Transactions*
Microfiche Concordance — Antonette diPaolo Healey & Richard L. Venezky. 1980. *A Microfiche Concordance to Old English*. Newark: University of Delaware Press.
Miles, John. 1993. *Exploring Lakeland's Wildlife*. Carlisle: Miles & Miles.
Mills, A. D. 1998. *A Dictionary of English Place-Names*. 2nd edn. Oxford: OUP.
Mills, David. 1976. *The Place-Names of Lancashire*. London: Batsford.
Mills, Ken. 1999. *The Cumbrian Yew Book*. Aspatria: Yew Trees for the Millennium/Solway Rural Initiative.
Millward, Roy & Adrian Robinson. 1970. *The Lake District*. London: Eyre & Spottiswood.
Monkhouse, F. J. 1960. *The English Lake District. A Description of the O.S. One-inch Tourist Map: The Lake District*. Sheffield: The Geographical Association.
Moorman, Mary. 1957. *William Wordsworth. A Biography.* I *The Early Years 1770–1803*. London etc: OUP.
MordenM — *The county maps from William Camden's Britannia, 1695* by Robert Morden, intro. J.B. Harley. Newton Abbot: David & Charles. 1972.
NG — O. H. Rygh et al. 1897–1924. *Norske Gaardnavne*. 19 vols. Kristiania: Fabritius. [Online at: <http://www.dokpro.uio.no/rygh_ng/rygh_form.html>]
NGIndledning — O. Rygh. 1898. *Norske Gaardnavne. Forord og Indledning*. Kristiania: Fabritius.
Nicolson, J. & R. Burn. 1777. *The History and Antiquities of the Counties of Westmorland and Cumberland*. 3 vols. London: Strahan.
Nicolaisen, W. F. H. 2001. *Scottish Place-Names*. New edn. Edinburgh: Donald.
Noble, M. E. 1901. *A History of the Parish of Bampton*. Kendal: Wilson.
Noble, Miss. 1903. 'Towtop Kirk, Bampton.' *CW2* 3, 265–8.
Noble, Miss. 1907. 'The stone circle on Knipe Scar.' *CW2* 7, 211–14.
Noreen, Adolf. 1923. *Altnordische Grammatik* I. *Altisländische und altnorwegische Grammatik*. 4th edn. Repr. 1970. Tübingen: Niemeyer.
NSL — Jørn Sandnes & Ola Stemshaug. 1990. *Norsk Stadnamnleksikon*. 3rd edn. Oslo: Det norske samlaget.
OED — J. A. Simpson & E. S. C. Weiner. 1989. *The Oxford English Dictionary*. 2nd edn. Oxford: Clarendon.
One Inch map — *Ordnance Survey Touring Map and Guide 3, Lake District*. 1994. [= *One Inch to One Mile Tourist Map* in previous printings]
ONP — *Ordbog over det norrøne prosasprog/A Dictionary of Old Norse Prose.* 1989–. Copenhagen: Arnamagnæanske Kommission. [List of headwords online from the dictionary's website at: <http://www.onp.hum.ku.dk/onpscript/cgionp.exe/wld>]
Oppenheimer, Lehmann J. 1903. 'Birkness Crag chimney and gully.' *Climbers' Club Journal* 5, 147–54.
OS — Ordnance Survey, 6 inch maps (unless otherwise specified)
OSGazetteer — *Ordnance Survey Gazetteer of Great Britain*. 1989. 2nd edn. London: Ordnance Survey/Macmillan.
OSNB — Ordnance Survey Name Books (or Name Lists), Public Record Office OS 34. [All volumes are indexed, therefore page references are not specified in the Dictionary unless the index is unhelpful]
Cumberland:
/35–6 Crosthwaite 1862 (incorporating Above Derwent, Borrowdale, Keswick, St John's)
Westmorland:
/437 Askham 1860
/438 Bampton 1860
/439–40 Barton 1860 (incorporating Martindale, Patterdale)

/448 Grasmere 1859 (incorporating Langdales, Rydal)
/451-3 Kendal 1858 (incorporating Crook, Hugill, Kentmere, Longsleddale, Nether Staveley, Over Staveley, Strickland Ketil, Strickland Roger, Underbarrow)
/465 Shap 1860
/469 Windermere 1859 (incorporating Bowness, Troutbeck).

O'Sullivan, D. M. 1984. 'Pre-Conquest settlement patterns in Cumbria.' In *Studies in Late Anglo-Saxon Settlement*, ed. M. L. Faull. Oxford: Department of External Studies; 143–54.

O'Sullivan, D. M. 1985. 'Cumbria before the Vikings: a review of some "Dark Age" problems in North-West England.' In Baldwin & Whyte, 17–35.

Otley, Jonathan. 1823. *A Concise Description of the English Lakes*. Keswick: priv. publ.

Owen, Hywel Wyn. 1998. *The Place-Names of Wales*. Cardiff: University of Wales Press.

Padel, O. J. 1980–2. 'Welsh bwlch "bald, hairless".' *Bulletin of the Board of Celtic Studies* 29, 523–6.

Padel, O. J. 1985. *Cornish Place-Name Elements*. EPNS 56–7. Nottingham.

Palmer, J. H. 1944. *Historic Farmhouses in and around Westmorland*. Kendal: Westmorland Gazette.

Palmer, W. T. 1952. *Byways in Lakeland*. London: Hale.

Parker, C. A. 1903. 'Mould from Gill, St. Bees.' CW2 3, 223–5.

Parker, C. A. 1904. 'The so-called "Danish camp" at Gosforth.' CW2 4, 195–8.

Parker, C. A. 1909. 'Some mediæval crosses, cross sites, and cross names in West Cumberland.' CW2 9, 78–119.

Parker, C. A. 1918. 'A pedigree of the family of Docker.' CW2 18, 161–73.

Parker, Charles A. & Mary C. Fair. 1922. 'Bloomery sites in Eskdale and Wasdale (Part 1).' CW2 22, 90–7.

Parker, C. A. 1926. *The Gosforth District. Its Antiquities and Places of Interest*. Revised edn, by W. G. Collingwood. Kendal: Wilson.

Parker, F. H. M. 1907. 'A Calendar of the Feet of Fines for Cumberland.' CW2 7, 215–61.

Parson, W. & W. White. 1829. *History, Directory, and Gazetteer, of the Counties of Cumberland and Westmorland, with ... Furness and Cartmel*. Leeds & Newcastle: White.

Parsons, Margaret A. 1993. 'Pasture farming in Troutbeck, Westmorland, 1550–1750'. CW2 93, 115–30.

Parsons, M. A. 1997. 'The woodland of Troutbeck and its exploitation to 1800.' CW2 97, 79–100.

Pearsall, W. H. 1961. 'Place-names as clues in pursuit of ecological history.' *Namn och Bygd* 49, 72–89.

Pearsall, W. H. & W. Pennington. 1973. *The Lake District. A Landscape History*. London: Bloomsbury Books.

Peascod, W. & G. Rushworth. 1949. *Buttermere and Newlands Area*. Rock-Climbing Guides to the English Lake District. Manchester: Fell & Rock Climbing Club of the English Lake District.

Perriam, D. R. & J. Robinson. 1998. *The Medieval Fortified Buildings of Cumbria*. CWAAS Extra Series 29. Kendal.

Pevsner, Nikolaus. 1967. *The Buildings of England* 33. *Cumberland and Westmorland*. Harmondsworth: Penguin.

Pevsner, Nikolaus. 1969. *The Buildings of England* 37. *Lancashire 2: The Rural North*. Harmondsworth: Penguin.

Phythian-Adams, Charles. 1996. *Land of the Cumbrians*. Aldershot: Scolar Press.

Pipe Rolls of Cumberland and Westmorland, 1222–1260, ed. F. H. M. Parker. CWAAS Extra Series 12. Kendal. 1905.

PNChesh — John McN. Dodgson & Alexander Rumble. 1970–97. *The Place-Names of Cheshire*. EPNS 44–8, 54, 74. Nottingham: EPNS.

PNCumb — A. M. Armstrong, A. Mawer, F. M. Stenton & Bruce Dickins. 1950–2. *The Place-Names of Cumberland*. EPNS 20–22. Cambridge: CUP. [Page numbers continuous through volumes: I: i–vi, 1–258, II: 259–457, III: i–lxxx, 459–565]

PNDerbs — Kenneth Cameron. 1959. *The Place-Names of Derbyshire*. EPNS 27–9. Cambridge: CUP.

PNDu — Victor Watts. Forthcoming. *The Place-Names of County Durham*. Nottingham: EPNS.

PNEssex — P. H. Reaney. 1935. *The Place-Names of Essex*. EPNS 12. Cambridge: CUP.

PNLancs — Eilert Ekwall. 1922. *The Place-Names of Lancashire*. Chetham Society NS 81. Manchester.

PNNotts — J. E. B. Gover, Allen Mawer & F. M. Stenton. 1940. *The Place-Names of Nottinghamshire*. EPNS 17. Cambridge: CUP.

PNNYorks — A. H. Smith. 1928. *The Place-Names of the North Riding of Yorkshire*. EPNS 5. Cambridge: CUP.

PNWestm — A. H. Smith. 1967. *The Place-Names of Westmorland*. EPNS 42–3. Cambridge: CUP.

PNWYorks — A. H. Smith. 1961–3. *The Place-Names of the West Riding of Yorkshire*. EPNS 30–7. Cambridge: CUP.

Postlethwaite, John. 1913. *Mines and Mining in the (English) Lake District*. 3rd edn. Whitehaven: Moss.

PR — Parish Registers [In the Dictionary, the relevant PRs are those of the parish indicated after the Grid Reference, unless otherwise specified. Printed editions, transcripts, and originals have been used as available; for listing of PRs see *Cumbrian Ancestors*]

Prior, Herman. 1865. *Ascents and Passes in the Lake District of England*. London [publ. unspec.]

Ragg, Frederick W. 1924. 'Early Barton.' *CW2* 24, 295–350.

Ramshaw, David & John Adams. 1993. *The English Lakes. The Hills, the People, their History*. Carlisle: P3 Publications.

Ratcliffe, Derek. 2002. *Lakeland. The Wildlife of Cumbria*. London: Harper Collins.

RCHM — Royal Commission on Historic Monuments, *Inventory of Historical Monuments of Westmorland*. London. 1936.

Reaney, P. H. 1976. *A Dictionary of British Surnames*. 2nd edn., by R. M. Wilson. London: Routledge & Kegan Paul.

Richardson, A. *et al*. 1990. 'The Roman road over Kirkstone Pass: Ambleside to Old Penrith.' *CW2* 90, 105–25.

Richardson, Colin. 1996. 'A find of Viking-period silver brooches and fragments from Flusco, Newbiggin, Cumbria.' *CW2* 96, 35–44.

Rivet, A. L. F. & C. Smith. 1979. *The Place-Names of Roman Britain*. London: Batsford.

Roberts, D. F. *et al*. 1981a. 'Genetic structure in Cumbria'. *Journal of Biosocial Science* 13, 317–36.

Roberts, D. F. *et al*. 1981b. 'Genetic variation in Cumbria.' *Annals of Human Biology* 8, 135–44.

Roberts, Gordon. 1999. 'The place name *Naddle*.' *CW2* 86. [Appendix to Clare 1999]

Robinson, J. W. 1904. 'Smuggler's hold on Great Gable'. *CW2* 4, 351–2.

Rollinson, William. 1967. *A History of Man in Lakeland*. London: Dent.

Rollinson, William. 1996. *A History of Cumberland and Westmorland*. 2nd edn. Chichester: Phillimore.

Rollinson, William. 1997. *The Cumbrian Dictionary of Dialect, Tradition and Folklore*. Otley: Smith Settle.
Rygh, O. 1901. *Gamle personnavne i norske stedsnavne*. Kristiania: Fabritius.
Rygh — see also NG and NG Indledning
SaxtonM — *Christopher Saxton's 16th Century Maps. The Counties of England and Wales*, intro. by William Ravenhill. Shrewsbury: Chatsworth Library. 1992.
Scand & Celt — Eilert Ekwall. 1918. *Scandinavians and Celts in the North-West of England*. Lunds Universitets Årsskrift, N. F. Avd. 1, Bd 14, Nr 27.
Scotland — *Calendar of Documents relating to Scotland*. 4 vols. Edinburgh, 1881–8.
Scott, Joe (ed.). 1995. *A Lakeland Valley through Time*. Staveley & District History Society.
Scott, Sir Walter. *Poetical Works*, ed. J. Logie Robertson. London: OUP. 1904; repr. 1957.
Searle, William George. 1897. *Onomasticon Anglo-Saxonicum*. Cambridge: CUP.
Sedgefield, W. J. 1915. *The Place-Names of Cumberland and Westmorland*. Manchester: Manchester UP.
Sephton, John. 1913. *A Handbook of Lancashire Place-Names*. Liverpool: Henry Young & Sons.
'SGJ' [Susan Johnson]. 1961. Ms. notes on Cumbrian place-names, in EPNS library, University of Nottingham.
Shackleton, E. H. 1967. *Lakeland Geology*. Clapham: Dalesman.
Simpson, Gertrude M. 1928. 'Grasmere field-names.' CW2 28, 273–91.
Size, Nicholas. 1929. *The Secret Valley*. Kendal: Wilson.
Smith, Elements — A. H. Smith. 1956. *English Place-Name Elements*. EPNS 25–6. Cambridge: CUP. [See Cameron 1968–9 and Jackson 1968–9 for Addenda & Corrigenda.]
Smith, A. H. 1966–9. 'Whelter.' *Saga-Book of the Viking Society* 17, 61–2.
Smith, A. H. — see also Smith, Elements, PNWestm, PNEYorks, PNNYorks, PNWYorks
Smith, John & F. Merigot. 1791–5. *Twenty Views of the Lake District*. London.
Smith, W. P. Haskett. 1894. *Climbing in the British Isles*. I *England*. London: Longmans, Green & Co.
Smith, W. P. Haskett. 1903. 'Wastdale Head 600 years ago.' *Climbers' Club Journal* 5, 3–15.
Smyth, Alfred P. 1975–9. *Scandinavian York and Dublin*. 2 vols. Dublin: Templekieran.
SND — William Grant & David D. Murchison. 1931–76. *The Scottish National Dictionary*. 10 vols. Edinburgh: SND Association.
Somervell, John. 1930. *Water-Power Mills of South Westmorland*. Kendal: Wilson.
Spence, J. E. 1938. 'Ancient enclosures on Town Bank, Kinniside.' CW2 38, 63–70.
Spence, J. E. 1939. 'Ancient remains in Ennerdale and Kinniside.' CW2 39, 31–4.
Spencer, Brian. 1983. *A Visitor's Guide to the Lake District*. Ashbourne: Moorland.
Spittal, Jeffrey & John Field. 1990. *A Reader's Guide to the Place-Names of the United Kingdom*. Stamford: Watkins. [Supplement forthcoming, ed. Shaun Tyas]
SSNNW — Gillian Fellows-Jensen. 1985. *Scandinavian Settlement Names in the North-West*. Copenhagen: Reitzel.
St Bees — *Register of the Priory of St. Bees*, ed. James Wilson. Surtees Society 126. Kendal. 1915.
Steers, J. A. 1964. *The Coastline of England and Wales*. Cambridge: CUP.
Stockdale, J. 1872. *Annales Caermoelenses*. Ulverston: Kitchin.
Sutcliff, Rosemary. 1956. *The Shield Ring*. London: OUP.
Swift, F. B. & C. G. Bulman. 1959. 'Uldale Church.' CW2 59, 51–64.
Swift, F. B. 1966. 'The oldest parish registers of Bassenthwaite.' CW2 66, 276–92.
Swift, F. B. 1977. 'St Mary's Chapel, Uldale'. CW2 77, 181.

Sykes, W. S. 1926. 'On the identification of some ancient places in South Cumberland.' *CW2* 26, 103–49.
Taylor, M. Waistell. 1886. 'Prehistoric remains on Moordivock.' *CW* 1886, 323–47.
Taylor, S. 1927. 'A Flookburgh Glossary.' *CW2* 27, 152–63.
Taylor, S. 1941. 'The Irtons of Irton Hall.' *CW2* 41, 72–122.
Taylor, Simon. Forthcoming. Elements Glossary, *Place-Names of Fife* IV.
Taylor, Violet. 1995. 'Ulcatrow. An initial review.' *MHAS* 2, 12–15.
Terrier — Gosforth Terrier 1778, extracted by Neville Ramsden in transcription of Gosforth PRs 1690–1839, in CRO Carlisle; original in CRO Whitehaven. 1993.
Thompson, Bruce Logan. 1942. 'Mardale and Haweswater.' *CW2* 42, 13–42.
Threlkeld, Cumbria. 1998. [Pamphlet]
Towneley Second Shepherds' Play = *Secunda Pastorum* — *The Wakefield Pageants in the Towneley Cycle*, ed. A. C. Cawley. Manchester: Manchester UP. 1958; 43–63.
Turner, V. E. 1991. 'Results of survey work carried out between the Shap and Askham Fells.' *CW2* 91, 1–11.
Valentine, H. 1935. 'Ancient Pottery in Whinfell Parish.' *CW2* 35, 37–41.
VCHCumb — *The Victoria County History of the Counties of England. Cumberland*. Vols I–II, ed. H. Arthur Doubleday. 1901–5. Westminster: Constable.
VCHLancs — *The Victoria County History of the Counties of England. A History of Lancashire*. Vol. VIII, ed. William Farrer & J. Brownbill. 1914. London: Constable.
VEPN — *The Vocabulary of English Place-Names. Á–BOX*, by David Parsons & Tania Styles, with Carole Hough, 1997. *BRACE–CÆSTER*, by David N. Parsons & Tania Styles, 2000. *CEAFOR–COCK-PIT*, by David N. Parsons, 2004. Nottingham: Centre for English Name Studies/EPNS.
Wainwright, A. 1955–66. *A Pictorial Guide to the Lakeland Fells*. I *The Eastern Fells*, 1955, II *The Far Eastern Fells*, 1957, III *The Central Fells*, 1958, IV *The Southern Fells*, 1960, V, *The Northern Fells*, 1962, VI *The North Western Fells*, 1964, VII *The Western Fells*, 1966. Kendal: Westmorland Gazette.
Wainwright, A. 1974. *The Outlying Fells of Lakeland*. Kendal: Westmorland Gazette.
Wallenberg, J. K. 1931. *Kentish Place-Names*. Uppsala: Lundequist.
Wallenberg, J. K. 1934. *The Place-Names of Kent*. Uppsala: Appelberg.
Ward, James Clifton. 1876. *The Geology of the North Part of the English Lake District*. London: HMSO.
Warriner, Frank. 1926. 'Some South Cumberland place names.' *CW2* 26, 77–103.
Watson, Angus. 1995. *The Ochils. Placenames, History, Tradition*. Perth & Kinross District Libraries.
Watson, W. J. 1926. *The History of the Celtic Place-Names of Scotland*. Edinburgh: Blackwood; repr. Dublin: Irish Academic Press, 1986. [Indexed in Basden 1997]
Watts, Victor. 2002. *A Dictionary of County Durham Place-Names*. Popular Series 3. Nottingham: EPNS.
Watts, Victor. 2004. *The Cambridge Dictionary of English Place-Names*. Cambridge: CUP.
West 1774, map — *A Map of the Liberty of Furness in the County of Lancaster*, surveyed by Brasier 1745, copied by Richardson 1772, in West 1774.
West, Thomas. 1774. *The Antiquities of Furness*. London: priv. publ.
West, Thomas. 1780. *A Guide to the Lakes in Cumberland, Westmorland and Lancashire*. 2nd edn. London: Richardson & Urquhart. Also 3rd edn, London: Law, 1784.
Whaley, Diana. 1996. 'Anglo-Scandinavian problems in Cumbria, with particular reference to the Derwentwater area.' *Nomina* 19, 91–113.
Whaley, Diana. 2001. 'Trusmadoor and other "pass" words.' *Nomina* 24, 77–95.

Whaley, Diana. Forthcoming. 'The semantics of *stöng, stang*.' In *Cultural Contacts in the North Atlantic Region: The Evidence of Names*, ed. Doreen Waugh, Peder Gammeltoft and Carole Hough.

Whyte, Ian D. 1985. 'Shielings and the pastoral economy of the Lake District in Medieval and Early Modern times.' In Baldwin & Whyte, 103–17.

Wild, Chris *et al*. 2001. 'Evidence for medieval clearance in the Seathwaite Valley, Cumbria.' *CW3* 1, 53–68.

Winchester, Angus J. L. 1984. 'Shielings in Upper Eskdale.' *CW2* 84, 267–8.

Winchester, Angus J. L. 1985. 'The multiple estate: a framework for the evolution of settlement in Anglo-Saxon and Scandinavian Cumbria.' In Baldwin & Whyte, 89–101.

Winchester, Angus J. L. 1987. *Landscape and Society in Medieval Cumbria*. Edinburgh: John Donald.

Winchester, Angus. 1990. *Discovering Parish Boundaries*. Princes Risborough: Shire.

Winchester, Angus J. L. 2000. *The Harvest of the Hills. Rural Life in Northern England and the Scottish Borders, 1400–1700*. Edinburgh: Edinburgh UP.

Wiseman, W. G. 1987. 'The medieval hospitals of Cumbria.' *CW2* 87, 83–100.

Wood, Oliver. 1971. 'The collieries of J. C. Curwen.' *CW2* 71, 199–236.

Withycombe, E. G. 1977. *The Oxford Dictionary of English Christian Names*. 3rd edn. Oxford: OUP.

Wordsworth, Dorothy, 'Excursion on the Banks of Ullswater, 1805.' In *Journals of Dorothy Wordsworth*, ed. E. de Selincourt. 2 vols. London: Macmillan. 1941; I, 418–9.

Wordsworth, Dorothy, *Journals — Dorothy Wordsworth. The Grasmere Journals*, ed. Pamela Woof. Oxford: Clarendon. 1991.

Wordsworth, *Poems — William Wordsworth. Poetical Works*, ed. Thomas Hutchinson, rev. Ernest de Selincourt. London: OUP. 1936.

[Wordsworth] *The Poetical Works of William Wordsworth*, ed. E. de Selincourt. 5 vols. Oxford: Clarendon. 1940–9.

Wordsworth, *Prose — The Prose Works of William Wordsworth*, ed. W. J. B. Owen & J. W. Smyser. 2 vols. Oxford: Clarendon. 1974.

Wyld, Henry Cecil, with T. Oakes Hurst. 1911. *The Place Names of Lancashire*. London: Constable.

YatesM — William Yates. 1786. *A Map of the County of Lancashire*, intro. by J. B. Harley. Birkenhead: Historic Society of Lancashire & Cheshire. 1968.

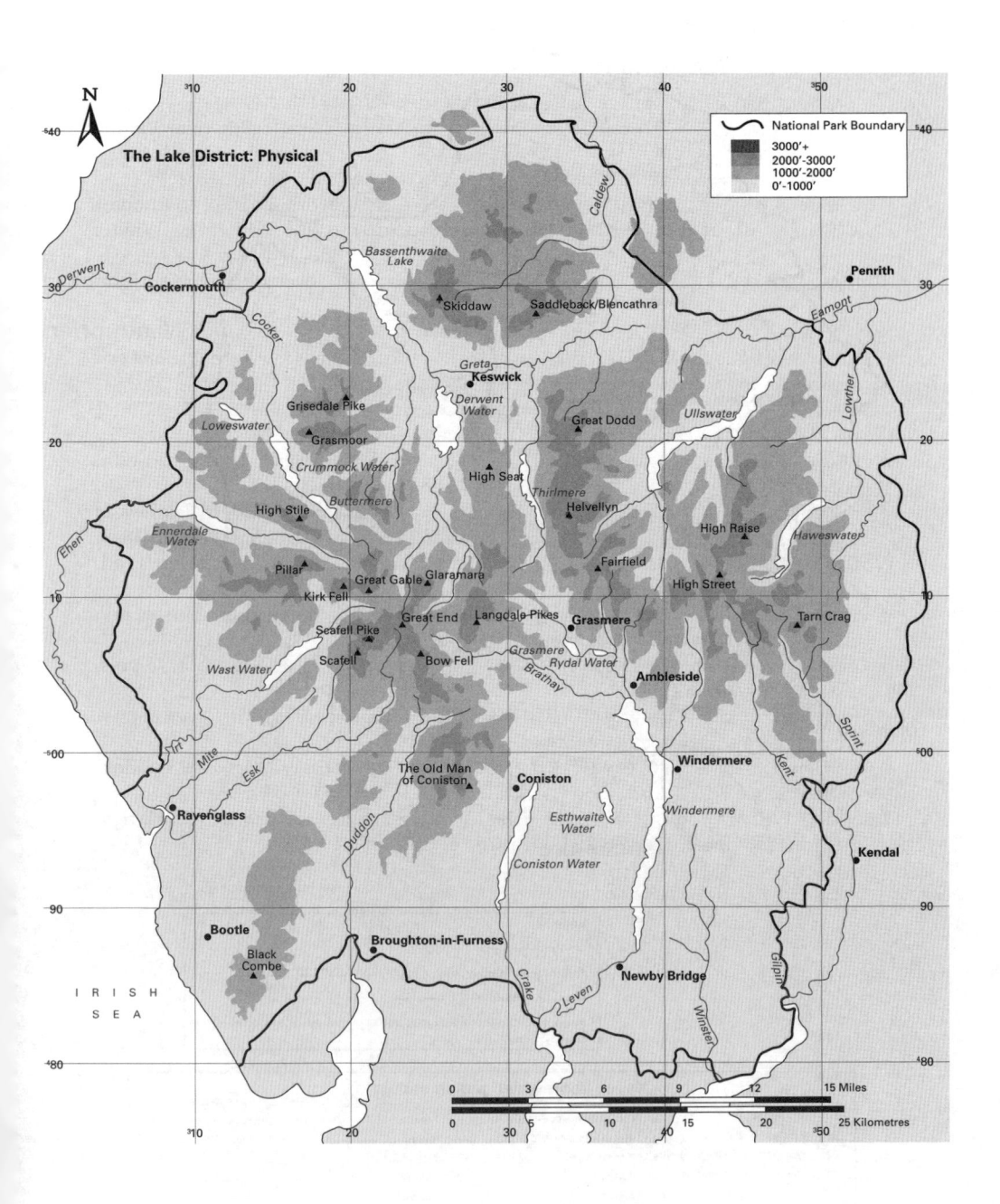

PLACE-NAMES

LAKE DISTRICT

The Lake District: Administrative

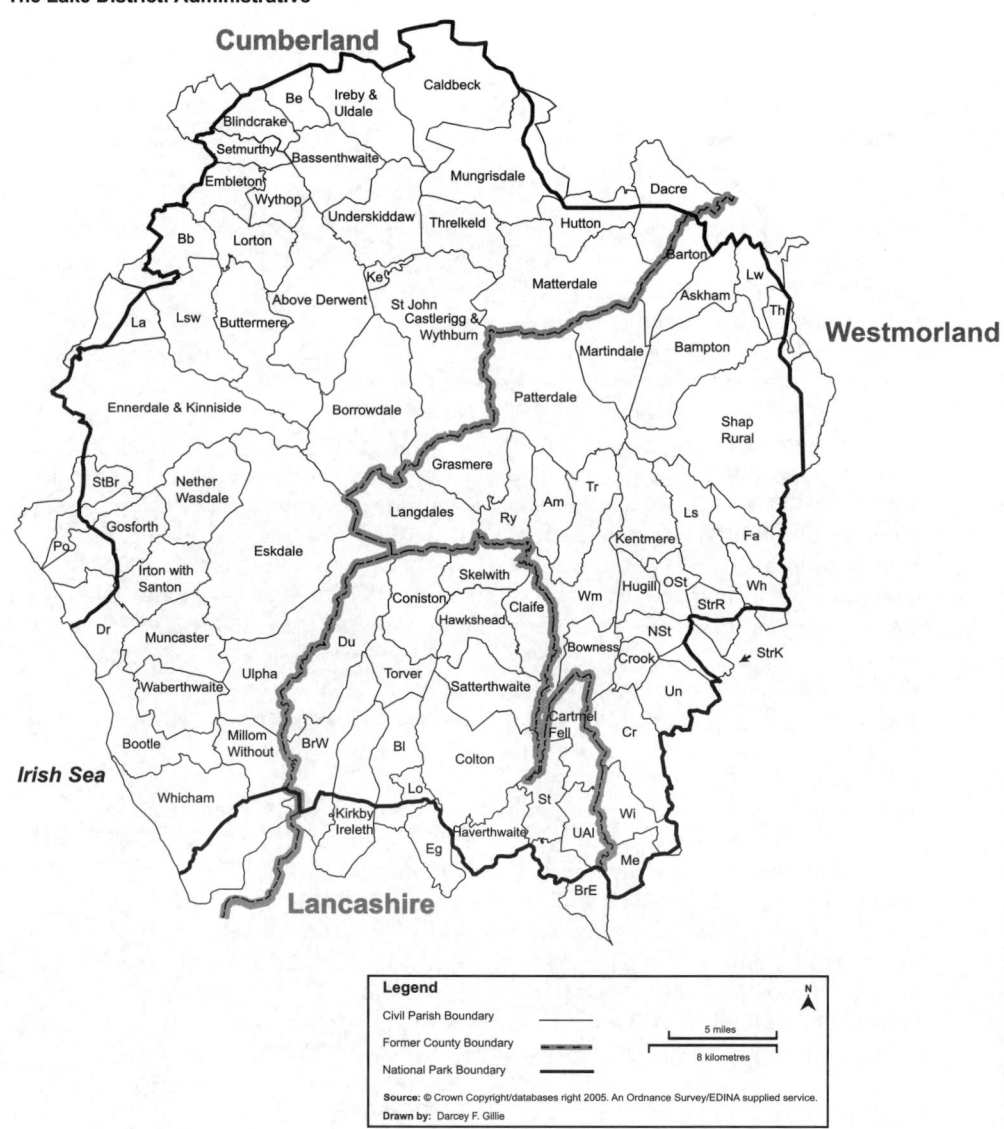

Cumberland

Westmorland

Irish Sea

Lancashire

Legend

Civil Parish Boundary	———
Former County Boundary	– – –
National Park Boundary	━━━

5 miles / 8 kilometres

Source: © Crown Copyright/databases right 2005. An Ordnance Survey/EDINA supplied service.
Drawn by: Darcey F. Gillie

Key to abbreviated parish names

Am	Ambleside	Du	Dunnerdale with Seathwaite	Lw	Lowther	StrK	Strickland Ketel
Bb	Blindbothel	Eg	Egton with Newland	Me	Meathop & Ulpha	StrR	Strickland Roger
Be	Bewaldeth & Snittlegarth	Fa	Fawcett Forest	NSt	Nether Staveley	Tr	Troutbeck
Bl	Blawith & Subberthwaite	Ke	Keswick	OSt	Over Staveley	Th	Thrimby
BrE	Broughton East	La	Lamplugh	Po	Ponsonby	UAl	Upper Allithwaite
BrW	Broughton West	Lo	Lowick	Ry	Rydale & Loughrigg	Un	Underbarrow & Bradleyfield
Cr	Crosthwaite & Lythe	Ls	Longsleddale	StBr	St Beidget Beckermet	Wh	Whitwell & Selside
Dr	Drigg & Carlton	Lsw	Loweswater	St	Staveley-in-Cartmel	Wi	Witherslack
						Wm	Windermere

A DICTIONARY OF LAKE DISTRICT PLACE-NAMES

A

AARON CRAGS NY2210 (1970ft/601m.) Borrowdale.
From 1867 (OS).
♦ Probably 'the rocky heights frequented by eagles'; *erne* 'eagle' plus **crag**. (a) 'Aaron' here may be a corruption of dial. *erne* 'eagle' from OE *earn*, as suggested by Brearley (1974). ERNE CRAG would provide a parallel, and see EAGLE CRAG. (b) The pers.n. *Aaron* is a possibility, but crags named from people are rare, and the form here may be influenced by AARON SLACK, only a mile away.

AARON SLACK NY2110 Borrowdale.
From 1867 (OS).
♦ Apparently 'Aaron's track'; pers.n. plus **slack**, which often means 'small valley', but here refers to a track in a natural groove. The occurrence of the rare name *Aaron* (originally the Old Testament brother of Moses) is striking, and A~ Slack pairs curiously with Moses' Trod, a similarly steep track on the other side of WINDY GAP associated with a famous whisky-distiller and smuggler (Robinson 1904, 352).

ABBEY NY1729 (hab.) Embleton.
The Abbey 1655 (*PNCumb* 384), *Abbey* 1676 (PR), *High, Low Abbey* 1867 (OS).
♦ The reason for *abbey* (an adoption from OFr into ME) here is unknown. I am grateful to Mr Derek Denman of Lorton & Derwentfells Local History Society for noting an Abbey Gate here and in nearby Blindbothel (where it was *Abbey* in 1774 and 1830), both beside ancient commons gates.

ABBEY FLATTS NY0507 (hab.) St Bridget Beckermet.
From 1867 (OS).
♦ 'The level ground belonging to the abbey.' ON *flǫt* 'level area of ground' was adopted into ME as *flat*, and well describes this plateau above the R. CALDER, where the Cistercian CALDER ABBEY held 193 acres at the Dissolution in the late 1530s (Winchester 1987, 155). Whether the specific sense of *flatt*, 'a division of a large open arable field, a furlong' (Dilley 1970, 197) is relevant here is not certain.

ABBOT PARK SD3188 (hab.) Colton.
Abbot park c.1535 (West 1774, 104), *Abbotpke* 1623 (PR).
♦ The word *abbot* was borrowed from Lat *abbas* into OE and re-borrowed into ME. This place was among the possessions of FURNESS Abbey: see **park**.

ABBOT'S BAY NY2519 Borrowdale.
Not on 1867 or 1900 OS maps.
♦ The name appears to be recent and to be formed from *abbot* as in ABBOT PARK, plus **bay**, here an inlet in DERWENTWATER. It may recall the involvement of the pre-Reformation FURNESS Abbey hereabouts (see GRANGE-IN-BORROWDALE).

1

ABBOTS READING FARM SD3485 Haverthwaite.
a Messuage called the Abbott Ryddinge 1597 (*Ducatus Lancastriae* III, 390), *Abbotridding* 1632 (Colton PR); probably = *ye Ridding at Caserend* 1629, *Ridding* 1630 (Colton PR).
♦ 'The abbot's clearing'; *abbot* as in ABBOT PARK, plus **ridding**. 'Abbots' presumably refers to the abbot(s) of FURNESS Abbey, influential hereabouts. **Farm** has, as so often, only recently appeared on maps (it is not on the 1966 One Inch map), but in this case it serves the useful function of distinguishing A~ R~ and A~ R~ Farm, which are separate places.

ABOVE BECK: A~ B~ FELLS SD2999, BELOW BECK FELLS SD2798 Coniston.
abovebeck in Conistone 1714/5 (Torver PR); *Above Beck Fells* 1850 (OS); *Below Beck Fells* 1851 (OS).
♦ 'The high, unenclosed ground belonging to Above and Below Beck, the areas north and south of the stream'; *above* (from OE *on bufan*), *below* (ME), **beck** 'stream', **fell** 'high ground'. Above Beck Fells rise N of LEVERS WATER Beck, and Below B~ F~ rise to the S, and the names presumably refer to these relative positions. Above Beck and Below Beck were the two parts of the ancient manor of Coniston (*VCHLancs* 8, 367); compare, e.g., Above Beck, a quarter of Grasmere township (*CW* 1908, 154).

ADAM-A-COVE NY2404 Eskdale/Ulpha.
Adam a Cove 1867 (OS); cf. *Adam a' Crag* 1802 (CN 1219).
♦ A curious name of unusual construction, which is matched by Adam-a-Crag just to the N. Both places are on the rocky ridge of CRINKLE CRAGS (and Coleridge in *CN* seems to have confused the two). Various Cumbrian p.ns contain the name *Adam*, but this person, like others, is unidentified. The syllable -*a*- sometimes represents *of*, as in CROOKAFLEET, but whether it does so here is unclear. **Cove** here, as elsewhere, is a high recess or corrie.

ADAM SEAT NY4709 (2180ft/664m.) Longsleddale.
Adamsyd 1580–1633 (*PNWestm* I, 162).
♦ 'Adam's hill-side', from the pers.n. *Adam*, common in England from the 12th cent., plus, judging from the 1580 form, the ME reflex of OE **sīde** 'hillside'. This has been influenced by *seat* from ON **sæti**, probably because of the seat-like shape of the hill, shelving away steeply on three sides.

ADDYFIELD SD4089 (hab.) Cartmel Fell.
From 1703 (Stockdale 1872, 164).
♦ Seemingly 'Adam's field', with the common pet-form of *Adam* as specific plus **field**. For examples of OE *feld* and its reflexes with a pers.n. as specific, see *PNWestm* II, 250, **feld** (d).

AIKBANK FARM NY0900 (hab.) Irton.
Aikbanke 1718, *Oakbank* 1775 (*PNCumb* 403).
AIKBANK/AKEBANK: A ~ MILL NY1125 Blindbothel.
Ake Bank 1774 (DHM), *Ake Banks* 1803 (CN 1518); *Akebank Mill* 1867 (OS).
♦ A~ is 'the slope where oak trees grow'; **oak/eik(i)**, **bank(e)**. *Aik* is the local form of the tree name, from ON *eik(i)* 'oak tree(s)', which has held out against the influence of Standard English *oak* (from OE *āc*) seen in the 1775 spelling. On oak trees in Lakeland, see **oak**. **Farm** has been added relatively recently to the Irton name; in the Blindbothel one, **mill** is ultimately from OE **mylen**.

AIK BECK NY4722 Barton.
Aik Beck(e) 1650 (*PNWestm* I, 2).
♦ 'The stream where oak trees grow', with *aik* as in AIKBANK plus **beck**. The stream runs parallel with, and then joins, another whose name contains a tree-name, ELDER BECK.

AIKEN NY1926, A~ BECK NY1826 Lorton. Both from 1867 (OS).
AIKEN CRAG NY3914 Patterdale. From 1920 (OS).
♦ The Lorton A~ is a tract of fell-side. The name possibly originated as a dat. pl. or derivative of ON **eik(i)** 'oak tree(s)' (cf. AIKIN (KNOTT)), but in the absence of early forms the origin is uncertain. **Beck** 'stream' and **crag** 'rocky height' form the secondary names.

AIKIN: A~ KNOTT NY2119 Above Derwent.
Haykin 1569, *Aykinge* 1576, *Akine* 1591, *Ekine in Newlands* 1606 (*PNCumb* 372); *Aikin* 1842 (Newlands PR); *Aikin Knott* 1867 (OS).
♦ The **knott** or rugged height rises above A~, which is of uncertain origin. The main possibilities are: (a) '(The place) at the oak trees' from ON **eik** in the dat. pl. *eikum* (Brearley 1974, Sedgefield 1915 and cf. AIKBANK); (b) 'oak slope' with **eik** compounded with ON *kinn* 'slope' (cf. KINN and *PNWestm* I, 24). The proximity to ancient oak woods would encourage these; see **oak** on oak trees in Lakeland. (c) An OE adj. *ǣcen*, **ācen* 'oaken' existed (see *VEPN* and cf. Aikenheads, *PNWYorks* VI, 188), but the adj. would be unlikely to stand alone in a p.n. (d) A surname or forename, originally the diminutive of *Adam*, is also possible.

AIN HOUSE NY1201 Irton.
White Annis 1774 (DHM), *Whiteanahouse* 1819 (PR), *Whiteain House* 1867 (OS), *Ain House* 1900 (OS), originally Thwaite End but within living memory called White Ain House (*CW* 1926, 78).
♦ Uncertain, in the absence of clear evidence from early spellings, but (a) perhaps 'the lone house', previously 'the white lone house'. **White** appears to have been dropped from the name, which may consist of *ain* as a form for the word *one* meaning 'lone, solitary' (ON *einn* or cf. Scots *ane*), plus **house**, a compound also assumed in AYNSOME. (b) Development of *Thwaite End* 'end of the clearing' (**thwaite, end**) to *Th'White Ain(d)* then *Ain (House)* was suggested by Sedgefield (1915, also Brearley 1974); the early spellings of THWAITE HEAD and THWAITEHILL would confirm that the *th-* in *thwaite* can be mistaken for the definite article and omitted. On the other hand Thwaite End and (White) Ain House may have been completely separate names.

AIRA BECK NY3720 Matterdale, A~ FORCE NY3920, A~ POINT NY4019.
Ayrauhe beke, Ayrauch beke c.1250, *Ayrath'* 1253, *riuulum de Ayra* 1292, *Ayragh* 1316 (*PNCumb* 3); *Airey Force* 1789 (*PNCumb* 254); *Airy Point* 1867 (OS).
♦ AIRA is 'the river at the gravelly spit'. The 1st el. is ON *eyrr* 'gravel spit or bank' — the alluvial A~ **Point** which juts into ULLSWATER and at which the river empties into the lake. The 2nd el. is ON **á** 'river'; the early spellings with *-h-, -ch-, -th* or *-gh* preserve an ancient form of the word, with final spirant consonant. **Bekkr/**

beck 'stream' has been added to the river name, and the famous waterfall A~ **Force** is named from the Beck or from the area, which was *terra de Ayrauhe* c.1250 (*PNCumb* 254). (Pl. 11b)

ALCOCK TARN NY3407 Grasmere.
Allcock Tarn 1899 (OS).
♦ 'Alcock's pool'; surname plus **tarn** from **tjǫrn**. The tarn lies below Grey Crag, and was formerly called Grey Crag Tarn or Butter Crags Tarn. It was renamed in the late 19th cent. after Mr Alcock of Hollins (now HOLLENS FARM) enlarged it by damming and stocked it with trout (Brearley 1974, Heaton Cooper 1983, 133–4; I am grateful to Michelle Kelly of The Armitt, Ambleside, for the latter reference here).

ALDBY FARM NY4627 Dacre.
Aldebi, Aldeby 1203 to 1375 (*PNCumb* 186).
♦ 'The old settlement', from OE *ald* '**old**' and ON **bœr/bý**. Names in *-by* are rare in the sparsely populated uplands of central Lakeland; this one is situated at the N. edge of the National Park. The meaning of the name, and the fact that its specific el. is not ON, could point to coining after the period of Scand settlement, perhaps following a time of settlement desertion (cf. Winchester on other 'old-by' names, 1987, 38). **Farm** has been added later.

ALLAN BANK NY3307 (hab.) Grasmere.
From 1809 (*CN* 1098n.).
♦ Probably 'Allan's hill-side'; pers.n. plus **bank**. 'Allan' is probably the forename or surname of mainly Breton origin, since dial. *allan* 'water-meadow' as possibly in ALLAN TARN would not be apt here. The house was built in 1805 (Simpson 1928, 281) and was home to the Wordsworth family 1808–11.

ALLAN TARN SD2989 Colton.
From 1851 (OS).
♦ This is a small pool in the R. CRAKE, as it flows out of CONISTON WATER. The 1st el. could be the pers.n. (see ALLAN BANK) or a dial. word *allan* meaning 'water-meadow' (*PNWestm* I, xvi and 202, *PNCumb* 460). The 2nd el. is **tarn** 'small pool' from ON **tjǫrn**.

ALLEN CRAGS NY2308 (2575ft/785m.) Borrowdale.
From 1867 (OS).
♦ This is a case of **crags** referring to an independent, rock-strewn mountain. The origin of A~ is uncertain, but presumably it is the forename or surname (cf. ALLAN BANK), since dial. *allan* 'water-meadow' would not fit the topography, nor is there a river large enough to merit the Celtic name Allen (variant of ELLEN).

ALLERDALE: A~ DISTRICT NY1736, A~ RAMBLE NY1432, NY2131, NY2527 etc. Several parishes.
Alnerdall 11th cent. (in 13th-cent. copy), *Alredale* c.1150 to 1316, *Allerdale* c.1190 to present (*PNCumb* 1).
♦ Allerdale is 'the valley of the ELLEN'. The syllable now spelt *-er-* was added to the Brit. river name in order to compound it with ON **dalr** 'valley'; cf. MITERDALE. It may descend from an ON gen. sing. *-ar* inflexion, or from ON *ár*, gen. sing. of *á* 'river' (*SSNNW* 211–12, cf. 241 on Miterdale, and cf. BORROWDALE (HEAD)). A~ was a medieval barony which was the forerunner of A~ Below Derwent (*Allerdall ward beneath the watter of Darwen*, 1580, *PNCumb* 259), one of the five wards of the pre-1974 county of

Cumberland. It lay N and E of the DERWENT valley, while A~ Above Derwent (*Allerdall ward above Derwent 1671, PNCumb* 335) comprised the high fells and coastal area to the S and W. A~ District is currently an administrative division of NW Cumbria. The word *district*, adopted from Lat via Fr, has been used to refer to discrete, especially administrative, units of territory at least since 1664 (*OED*, sense 2). A major local government organisation into urban and rural districts took place in 1894, but the present district councils (including also those of COPELAND and SOUTH LAKELAND) were constituted as part of the 1974 local government reorganisation. (I am grateful to Dr Daniel Bloomer of Cumbria County Council for clarification here.) The A~ Ramble is a 54-mile walk from SEATHWAITE in Borrowdale to the Solway Firth. The word *ramble* has been used of walks or excursions since the 17th cent. (*OED*).

AMBLESIDE NY3704 (hab. & parish).
Amdeseta (in error for *Ameleseta*) 1090–7 (in 1308 copy), *Amelsat(e)* 13th cent., *Amylside* 1379–1403 to 1650, *Ambelsede* 1437 (*PNWestm* I, 182).
♦ 'The shieling at Ámelr, the sandbank by the river.' The name is usually taken to consist of ON **á** 'river', ON *melr* 'sand-bank' and ON **sætr** 'shieling'. A~ thus seems to have developed from a temporary outlying settlement to a major centre.

AMETHYST GREEN (hab.) SD0898 Drigg. From 1867 (OS).
♦ Named from HMS *Amethyst* by William Porter (b. 1782), its master and the builder of this house (Brearley 1974). **Green** 'grassy place, (village) green' is a common generic in p.ns from the Early Modern period onwards.

ANCROW BROW NY4905 Longsleddale.
Arnecrewe, Arncro(o)e 1579 (*PNWestm* I, 160–1), *Iron-craw vulgo Arncraw* c.1692 (Machell II, 104), *Crow Brow* 1770 (JefferysM).
♦ A~ may be 'the eagle's perch' or 'the pen of a man named Qrn or Árni'. Smith suggests origins in OE *earn*/ ON *ǫrn* 'eagle', or possibly the related pers.ns *Qrn* or *Arni*, plus a fairly rare Early ModE *crew, croo* of Celtic origin (cf. early W *creu*, OIr *cró* 'sty, pen', *PNWestm* I, 160–1). The hill-side is a featureless sweep without rocks of the kind attractive to eagles, hence the pers.n. may be more likely. **Brow** 'hill-slope' seems to be a secondary addition.

ANGLER'S CRAG NY0915 Ennerdale.
Anglers Crag 1867 (OS); cf. *Angling Stone* 1802 (*CN* 1208).
♦ The **crag** is a rock buttress dropping steeply into ENNERDALE WATER, a fine fishing lake. The Anglers' Inn used to be just round the lake foot (Davy 1857, 118). The *Angling Stone* referred to by Coleridge in 1802 must be here or close by.

ANGLE TARN NY2407 Borrowdale. From 1867 (OS).
ANGLE TARN NY4114 Martindale, ANGLETARN PIKES NY4114 Martindale/ Patterdale.
Angilterne 13th cent., *Angletarn(e)* 1573 (*PNWestm* I, 15); *The Pikes* 1865 (*PNWestm* II, 218), *Angletarn Pikes* 1920 (OS).
♦ (a) Probably 'the fishing pool' in both cases, from ON *ǫngull* and/or OE *angel* '(fishing) hook' plus **tjǫrn/tarn**. Both offer good trout fishing, though at the price of a strenuous climb, especially to the Borrowdale tarn. Cf. the Norw Fisketjøn 'fishing pool' (*NG* X, 369). Other possibilities are less

likely: (b) that 'Angle' represents comparison with the shape of a fish-hook, since both tarns have small beaked projections (cf. Angeltveit, *NSL*); (c) that it is an ON pers.n. (cf. Angelshaug, *NSL*); (d) it is unlikely to imply 'corner, outlying spot' (cf. *VEPN: angle*), since this would entail exceptionally early influence from OFr. The name of Angletarn **Pikes** refers to twin rocky peaks.

ANNAS, RIVER SD0887 Bootle/Whicham, ANNASIDE SD0986 (hab.) Whicham, ANNASIDE BANKS SD0885 Bootle/Whicham.
(R. Annas:) *Annaside Beck* 1867 (OS); (Annaside:) *Ainresate* c.1145, *Aynerset* c.1150 to 1340 (p), *Annerside* 1503 (*PNCumb* 448); *Annaside Banks* 1867 (OS).
♦ Annaside is probably 'Einar's shieling', from the ON pers.n. *Einarr* and ON **sætr** 'shieling', later replaced by **side**. The river name Annas (not included in *PNCumb*) appears to be a back-formation created once '-side' was taken to refer to the river banks. In Annaside Banks, **bank** refers, unusually in Lakeland, to coastal sand-dunes.

APPLETHWAITE NY2625 (hab.) Underskiddaw, A~ GILL NY2626.
Apelthwayt c.1220 (*PNCumb* 321).
♦ 'The clearing where apple(-trees) grow.' The 1st el. is OE *æppel* 'apple, apple tree', which could possibly have replaced its ON cognate *epli*; both words could refer to fruit in general, as well as apples in particular. The 2nd el. is the common ON **þveit** 'clearing'. A~ **Gill** 'ravine with stream' (not named on OS 1867) is the lower reach of the How Gill which forms HOWGILL TONGUE.

APPLETHWAITE: A~ COMMON NY4202 Troutbeck.
Appelthwayte & variants 1256 to 1379, *Appeltretweyt* 1279 (*PNWestm* I, 194); *the Common at A~* 1702 (*PNWestm* I, 194), *Applethwaite Common* 1770 (JefferysM), 1859 (OSNB).
♦ A~ is 'the clearing where apple trees grow', cf. APPLETHWAITE, Underskiddaw. The 1279 spelling containing the word *tre(e)* may represent a variant, or else the original form of the name. The **Common** was the high, unenclosed grazing attached to the former township of A~. Machell referred to it as *a comon . . . called Aplethwate fell* and attributed the name to plentiful apple trees, 'tho now there are not many' (c.1692, II, 319 and 330).

APPLETREE HOLME SD2788 (hab.) Blawith.
Apple Tree Holme 1729/30 (PR), *Appletree Holme* 1851 (OS).
♦ 'The patch of dry ground where apple trees grow', with *appletree* ultimately from OE *æppel-trēow* (though the name may well be postmedieval), plus **holm(e)**. The place is well named as a *holm(e)*, since the site is still an island of cultivated land amidst the mosses which rise on either side.

APPLETREE WORTH: A~ W~ BECK SD2593 Broughton West.
Appletreewort 1669 (PR); *Appletree Worth Beck* 1851 (OS).
♦ A~ W~ is 'the enclosure with apple tree(s)', with the reflexes of OE *æppel-trēow* and of OE *worð* 'enclosure'. Though *worð, worth* is extremely common in English p.ns, this is its sole occurrence in this volume. The **beck** 'stream' runs past A~ W~, which is now a mainly coniferous plantation.

ARD CRAGS NY2019 Above Derwent/ Buttermere.
From 1867 (OS).
♦ Without early forms the origin of 'Ard' is unknown, but possibilities include (a) a dial. word meaning 'parched, dry' (so Brearley 1974); (b) the word 'hard' (cf. HARD CRAG); and (c) a Celtic word meaning 'height, hill', cf. W *ardd*, Ir *ard*. The **crags** form part of a long, high, rocky ridge above KESKADALE BECK.

ARKLID FARM SD2988 Colton, A~ INTAKE SD3189.
Arkeredyn 1573 (*PNLancs* 218), *Arclid* 1734/5 (PR); *Arklid Intake* 1851 (OS).
♦ A tantalising name. The syllable *ark* could be the outcome of **ærgi/erg** 'shieling', as in Arkholme, Anglesark (*PNLancs* 180, 48) and possibly PAVEY ARK, while the 2nd el. may possibly be (a) a counterpart of W *rhedyn* 'fern, bracken' which seems to be present in GLENRIDDING; or (b) **ridding**, a clearing (so Ekwall, *PNLancs*), common locally and sometimes taking the form *redding*. Without earlier forms this remains highly uncertain, however, and a compound of either (a) or (b) with **ærgi/erg** would be problematic at best. Arclid (Green) in Cheshire has no spellings resembling *Arkeredyn* and is probably of different origin (*PNChesh* II, 264–5). **Farm** has been added to the name, and the **intake** 'piece of land enclosed from moor or waste' is a tract of moor a mile E of A~ F~.

ARMASIDE NY1527 (hab.) Lorton.
Harmondesheved 1368, *Hermudesheved* 1369 (*PNCumb* 408), *Harmaside* 1597, *Armesyde* 1605/6 (PR).
♦ Probably 'Hermund's height', with gen. sing. of the common ON pers.n. *Hermundr* or OE *Heremund* (Searle 1897, 294) plus OE **hēafod**/ ME **heved**, possibly replacing ON **hǫfuð** 'head, high place'. *PNCumb* 408 suggests that the modern form is influenced by the word *hermit*, ME *er(e)mite* (cf. ARMATHWAITE). The site is on a rise above LORTON VALE, and Winchester sees it in the context of clearance and colonisation of the wooded slopes above Lorton in the 12th–13th centuries (1987, 147).

ARMATHWAITE: A~ HALL NY2032 Bassenthwaite.
Ermicetwayth 1278, *Ermitethwayt* 1292, *Ermethwayt* 1303 (*PNCumb* 263); *Armathwaite Hall* 1761 (PR), 1785 (CrosthwaiteM).
♦ A~ is 'the clearing associated with a hermit', from *er(e)mite* 'hermit' (OFr adopted into ME), plus **þveit**. **Hall** here now applies to a 'castellated Victorian Tudor mansion' built in 1881 (Pevsner 1967, 65).

ARMBOTH NY3017 St John's , A~ FELL NY2915.
Armabothe 1530 (*PNCumb* 311–2), *Armboth Fell* 1867 (OS).
♦ The 1st el. is elusive but may, as suggested in *PNCumb* 312, be a shortening of ME *armite*, *er(e)mite* 'hermit' or of an ON pers.n. such as *Arnmóðr* (or else *Árni*, so Sedgefield 1915 and Brearley 1974). The 2nd el. is ODan **bōþ** 'booth, shed', recalling the humble beginnings of a hamlet now lost to THIRLMERE Reservoir. A~ **Fell** is the high ground above and SW of A~.

ARMING HOW, GREAT SD1899 Eskdale.
From 1867 (OS).
♦ **Great** distinguishes this from Little A~ H~ just to the W; but 'Arming' is obscure in the absence of early spellings. **How(e)** is from ON **haugr** 'hill, mound'.

ARMONT/ARMENT HOUSE NY1800 Eskdale.
Armithow 1578, *Harmothow(e)s, Harmethowes* 1587 (*PNCumb* 390), *Armont-, Armanhouse* 1662 (*JEPNS* 2, 58), *Armont House* 1774 (DHM).
♦ Originally 'hermit hill', from *er(e)mite* 'hermit' (OFr adopted into ME) plus **how(e)** (from ON **haugr**) referring to the compact hill on which the house is situated. An *-s* was seemingly added to *how(e)*, which encouraged replacement by **house**.

ARNDALE BECK SD4290 Crosthwaite.
From 1857 (*PNWestm* I, 82).
♦ The 1st syll. may be related to OE *earn* or ON *ǫrn*, both 'eagle', or a pers.n. derived from them, but without earlier forms it is impossible to tell, and the first record of the name is strikingly late. **Dale** and **beck** refer, as usual, to a 'valley' and 'stream' respectively.

ARNISON CRAG NY3914 Patterdale.
Annstone Crag 1860 (*PNWestm* II, 224).
♦ Given the disparity between the 19th-cent. and current forms, it is uncertain whether the 1st el. of the name is (a) the surname *Arnison*; or (b) the forename *Ann(e)* or the reflex of OE *earn* 'eagle', plus ModE *stone*, with influence from the surname. The **crag** is a rocky summit of 1424ft/ 433m., perched above Arnison Gill.

ARNSIDE NY3301 (hab.) Skelwith, A~ INTAKE NY3301 Hawkshead.
Arnesyd c.1535 (*PNLancs* 219), *Arneside* 1535 (West 1774, 105), *Ernesyde* 1537 (*PNLancs* 219); *Arnside Intake* 1850 (OS).
♦ A~ is probably either 'Arni's shieling' or 'Arni's high place' from the ON pers.n. name *Árni* and **sætr** 'shieling' or ON *set* 'shieling' (*PNLancs* 219), though 'side' can also result from OE **hēafod**, ME **heued** 'head, high place' as in another Arnside (*Arnolvesheued* 1246, *PNLancs* 169), or alternatively from **sīde** 'hill-slope, side'. If Arnside was a shieling on outlying pastureland which became a permanent farm, the steep bluff of the **Intake** further extends the development of the waste.

ARTHUR'S PIKE NY4620 (1747ft/532m.) Barton.
From 1859 (*PNWestm* II, 212).
♦ 'Peak of (an unidentified) Arthur'; pers.n., **pike**. As a forename, *Arthur* echoes the name of the legendary British hero King Arthur, and it gave rise to a surname which could also result from the ON pers.n. *Arnþórr* (Hanks & Hodges 1988). The forename was 'never very common until the 19th C[ent.]', though more frequent in medieval Cumberland than many other parts (Withycombe 1977), and Brearley cites a 12th-cent. instance of *Arturus* from a document of St Bees priory in connection with A~ Pike (1974). It is, however, impossible to tell how old this hill-name might be. The Pike is a summit on BARTON FELL.

ARTHUR WOOD SD3799 Claife.
From 1850 (OS).
♦ On the pers.n. *Arthur*, see above. **Wood** is rarely qualified by a pers.n., but cf. HARRY GUARDS WOOD.

ARTLE CRAG NY4710 Shap Rural.
From 1859 (*PNWestm* II, 173).
♦ Uncertain, but perhaps 'Arnkel's or Arkle's rocky height'; pers.n. plus **crag**. Brearley (1974) suggests derivation from the ON forename *Arnketill*, and this would be paralleled by 'Artle' from *Arnketill* or *Arnkell* as assumed in Artlebeck (*Arkelbec* c.1200, *PNLancs*

168) and Artlethorn (*Arkilterne* 1200–25, Wyld 1911, 55; cf. also Artlegarth, *PNWestm* II, 34). There is also a surname *Arkle*, *Arkell* etc. which derives from *Arnketill* or *Arnkell*. Although crag is rarely qualified by a pers.n., there are examples, e.g. JENKIN CRAG, where 'Jenkin' is similarly either a forename or surname.

ASH CRAGS NY2809 St John's.
From 1867 (OS).
♦ Assuming that the 1st el. is the tree-name **ash**, its appearance on **crags** at over 2000ft/600m. is striking.

ASHES Examples at:
SD4697 (hab.) Nether Staveley.
Thesses 1558, *le/the Ashes*, *le Eshes* 1634 (*PNWestm* I, 173).
SD4187 (hab.) Cartmel Fell.
Ashes 1752 (*CW* 1966, 226), 1851 (OS).
♦ 'The ash trees'; see **ash**. The Nether Staveley example is not far from YEWS, and cf. ROWAN TREE HILL for another tree reference in Cartmel Fell. Other p.ns formed from the pl. of names of tree species include BIRKS, ELLERS, ESPS and OAKS.

ASHES BECK SD3489 Colton.
From 1851 (OS).
♦ Apparently 'the stream where ash trees grow'; **ash, beck**. Ash grows along the banks, along with alder, hazel, oak, etc.

ASH GILL BECK SD2695 Torver.
From 1851 (OS).
♦ 'The stream in Ash Gill, the ravine where ash trees grow'; **ash, gil(l), beck**. The sides of gills provide trees with relatively good soil and protection from grazing sheep.

ASH HOUSE SD1887 Millom Without.
Ashouse 1642 (*PNCumb* 418).
♦ Presumably 'the house by the ash tree(s)'; **ash, house**.

ASHLACK: A~ HALL SD2485 Kirkby Ireleth.
Eskeslac 1270–80 (p), 1284, *Esselac* 1325 (p) (*PNLancs* 220); *Ashlack Hall* 1850 (OS).
ASH SLACK SD3289 (hab.) Colton.
Ashslack 1679 (PR).
♦ A~ is 'the valley where ash trees grow'; **eski/ash** (see **askr**), **slakki** 'valley'; also **hall**. The 1st el. in the Kirkby Ireleth name is from ON *eski* 'ash trees, ash copse', influenced by its English counterpart *ash*, while the Colton one may be direct from the English word. Earlier writers took the element 'ash' in the Colton name in a different sense and associated it with old iron bloomeries (Cowper 1899, 364; Brydson 1911, 55). On the Hall, see Cowper 1895, 281–5.

ASHLEY GREEN NY3503 (hab.) Rydal.
From 1919 (OS).
♦ Probably 'the grassy place associated with the Ashley family'. Given the late appearance of the name, A~ is most likely to be a surname derived from the p.n. meaning 'ash wood' (**æsc, lēah**), found in several counties. If so, the p.n. is similar to ASKEW GREEN; Smith also compares it with Ashleygarth (*Ash Leagarth* 1722, *PNWestm* I, 205, 210); see also **green** 'grassy place'.

ASHNESS: A~ BRIDGE NY2719 Borrowdale, A~ FARM NY2719, A~ FELL NY2718.
Eschenes(bec) 1209–10 (*Furness Coucher* II, 570), *Esknese* 1211, *Eshnese* 1588, *Ashnes* 1604 (*PNCumb* 349); *Ashniss Bridge* 1753 (*CW* 1899, 121); *Ashness Fell* 1867 (OS).
♦ 'The headland where ash trees grow, from ON **eski** 'ash trees, ash

copse' (see **askr**) and ON **nes** 'headland, promontory'. The modern form is a result of anglicising. The famous **bridge** (see **brycg**) was an essential link in the old fell route between KESWICK and AMBLESIDE. See also **farm** and **fell**, which here refers to high grazing land rather than a distinct peak.

ASH SLACK: see ASHLACK

ASH SPRING SD4894 Underbarrow.
From 1857 (*PNWestm* I, 103).
♦ 'The plantation of ash trees'; **ash**, **spring**. It is still wooded.

ASKEW GREEN SD4285 (hab.) Witherslack.
Haskew Green 1712 (PR), *Askew Green* 1741, *Hasha Green* 1770, *Asker Green* 1857 (*PNWestm* I, 78).
♦ Probably 'the grassy place associated with the Askew family'; surname plus **green**. (a) There is a p.n. Aiskew, from ON **eik(i)** 'oak' (see **oak**) and ON **skógr** 'wood', with examples in *PNNYorks* 236 and 294, and a lost example in *PNWestm* II, 108. A similar p.n. could have given rise to the name Askew Green. (b) However, the p.n. also gave rise to a surname, and various Cumbrian families of the name of *Askew* are recorded (see HASKEW BECK and TARN; *CW* 1907, 270–1; Hudleston 1979; Hanks & Hodges 1988), and this seems a more likely origin for A~ G~.

ASKEW/HASKEW RIGG FARM NY3727 Mungrisdale.
Haskoerigge 1580, *Harsko rigg* 1605, *Askoerig* 1673, *Askrigg* 1702, *Haskewrigg* 1787 (*PNCumb* 227), *Askew-Rigg* 1814 (PR).
♦ 'The ridge at (H)askew.' The meaning of (H)askew is elusive: (a) possibly 'hare wood' from OE **hara** 'hare', plus OE **sceaga** 'wood' influenced by its relative ON **skógr**, if *PNCumb* 227 is correct to compare it with Haresceugh (*PNCumb* 216, cf. perhaps also Askew (Mire), *PNCumb* 278). (b) Alternatively, since only one *Har-* spelling has been noted, connection with a surname or p.n. Askew is possible (cf. ASKEW GREEN). **Rigg** 'ridge' (see **hryggr**) would be added later, and **farm** later still.

ASKHAM NY5123 (hab. & parish), A~ FELL NY4922.
Ascum, Askum 1232 to 1476, *Aschum* 1278 (p) (*PNWestm* II, 200); *Askham Fell* 1920 (OS).
♦ '(The place at) the ash trees', from *askum*, dat. pl. of ON **askr**, or the OE equivalent *æscum* or *ascum* with later Scand influence. The *-ham* spelling is therefore misleading. The **Fell** is a tract of high ground rather than a distinct summit.

ASKILL NY1222 (hab.), A~ KNOTT NY1222 Loweswater.
Askel 1737 (PR); *Asgill Knot* 1803 (*CN* 1518).
♦ The forms of A~ are too late for interpretation, but possibilities would include derivation from ON **askr** 'ash' and ON **kelda** 'spring', and there are springs and a well in the vicinity. A~ **Knott** is a neat, craggy height.

ASTLEY'S PLANTATION SD3987 Staveley-in-Cartmel.
From 1851 (OS).
♦ 'Astley' has the appearance of a surname derived from a p.n., such as Astley (Green), Greater Manchester (*PNLancs* 101). This is typical of p.ns in **plantation** in that it first appears in the 19th cent.

ATKINSON GROUND SD3297 Coniston.
Atkinsonground Conistone 1790 (Hawkshead PR).
♦ 'The farm of the Atkinson family.' This is one of the many farm-names in High Furness consisting of a surname, especially of the patronymic type, plus **ground**. Atkinson is among the commonest surnames in the Coniston PR. That the buildings are older than the 1790 record of the name is suggested by traces of a 17th-cent. loft-ladder noted by Collingwood (1926, 46).

AUGHERTREE NY2538 (hab.) Ireby and Uldale, A~ FELL NY2637.
Alcotewraye 1540, *Alkatre* 1580, *Ancautre* 1576, *Awhatree* 1777 (*PNCumb* 327), *Auhertree, Awhatree* 1794 (Hutchinson II, 370); *Aughertree Fell* 1900 (OS). Pronounced [æfətri:] (*PNCumb* 327).
♦ Originally 'the nook by Aldcot, the old cottage'. The 1540 spelling suggests derivation from OE *ald* **'old'**, OE **cot** 'cottage' and ON **vrá** 'nook' or their reflexes in ME (cf. 2nd and 3rd elements of ULCAT ROW). However, the *t* of *cot* seems to have been taken with the last syllable, which has been re-interpreted as *tree*. A~ **Fell** is a distinct height rising to some 948ft/300m.

AUSIN FELL WOOD SD3691 Satterthwaite.
From 1851 (OS).
♦ *Ausin* may be a dial. form for 'oxen': see OWSEN FELL; also **fell** and **wood**.

AUTERSTONE NY4521 Barton.
Otherstone 1761, *Otterstone* 1773, *Alter- or Oder Stone* 1839, *Auterstone* 1859 (*PNWestm* II, 212).
♦ 'A great limestone crag and scar, possibly so called from a supposed resemblance to an altar' (*PNWestm*). The 1st el., that is, may be *altar* from OE *altar*, and spellings in *au-* are common 13th–16th cent. See also **stone**.

AYNSOME/AYNESOM(E): A~ MANOR SD3879 Broughton East, BORWICK'S A~ SD3879 (hab.).
Aynson 1491, *Ayneson* 1537, *Aynsam* 1592, *Aynsome* 1597 (*PNLancs* 198); (A~ Manor:) *Aynsome* 1851, 1893 (OS); (Borwick's A~:) *Aynsome Farm* 1851, 1893 (OS).
♦ A~ is '(at the) lonely houses', from ON dat. pl. *einhúsum* (*einn* 'one, sole, lonely' and **hús** 'house, building'), or its OE counterpart *ānhūsum*; cf. AIN HOUSE and NEWSHAM. The word *manor* (ME *maner(e)*, adopted from Anglo-Norman), was used in the sense 'lord's estate' c.1290 and 'mansion' c.1300 (*OED*, senses 1 and 2a). However, this example and the others in this volume, BASSENFELL M~ and MANOR FARM, are considerably more recent. 'Borwick' appears to be a surname, originating in a p.n. such as Borwick (*PNLancs* 188), or possibly Berwick (several instances including Northumberland) or Barwick (including WYorks). It is very common in Hawkshead PRs (Cowper 1899, 365).

AYSIDE SD3983 (hab.) Staveley-in-Cartmel.
Aysshed 1491, *Aysett* 1537, *Ayshead* 1573, 1592, *Aysyde* 1591 (*PNLancs* 199).
♦ Uncertain. The miscellaneous early spellings make the division between 1st and 2nd elements unclear, but they tend to point to OE **hēafod** 'head, height' as 2nd el., rather than ON **sætr** 'shieling'. This means that *-s* belongs with the 1st el. Wyld (1911, 60)

identified *Aykesheued* 1279 (p) with Ayside, and if this were correct, Ayside would mean 'high place where oak grows', from reflexes of ON **eik(i)** (see **oak**) and OE *hēafod*. If *Aykesheued* is not the forerunner of Ayside but rather of OAK HEAD, as Ekwall thought more likely (*PNLancs* 199), the 1st el. may be ON **á** 'river', especially since nearby streams feed the R. Eea or Ay which may originate in *á*. Neither ON *á* nor ON *eik* 'oak' regularly has a gen. sing. in -*s*, so one would have to assume in either case that the name postdated the period when ON grammar remained intact.

B

BACKBARROW SD3584 (hab.) Haverthwaite, OLD B~ SD3685 (hab.) Staveley-in-Cartmel.
(Backbarrow:) *Bakbarowe* (*Mil*) 1537, *Bak(e)baray(fell)* 1538 (*PNLancs* 198), *Backbarrow* c.1692, Machell I, 314, 1698 (Colton PR); *Old Backbarrow* 1851 (OS), presumably = *Backbarowe* 1592 (Cartmel PR), though it is difficult to be certain which early forms belong to B~ and which to Old B~.
♦ B~ is 'the hill with the backlike top', as Ekwall puts it, noting that this describes the ridge above **Old** B~, which he takes as 'no doubt the original B~' (*PNLancs* 198). The elements derive from OE *bæc* 'back' and OE **berg** 'hill'.

BAGGRA: B~ YEAT NY2636 (hab.) Ireby and Uldale.
Bagrawe 1399, *Bagray* 1535 (*PNCumb* 328); *Baggeraw-gate* 1700/1, *Bagerahyate* 1706 (Uldale PR), *Bagrowgate* 1774 (DHM), *Baggrow Gate* 1867 (OS).
♦ An interesting and problematic name. B~ has similar early spellings to the Cumbrian Baggrow and Baggara (*PNCumb* 259), and the 2nd el. of all three seems to be from OE *rāw* '**row** (especially of houses or trees)'. The 1st el. could be (a) a topographical *bagge* referring to a low-lying, wet place (Zachrisson, cited in *PNCumb* 259); or (b) the word 'bag' (OE **bagga*) referring to a dead-end or bag-shaped landscape feature (see *VEPN*). (c) B~ could be a disparaging 'beggar's row' comparable with various English examples of RATTEN ROW 'rat-infested row', applied to poor dwellings. ME *bagge* is used of a beggar's bag (*PNCumb* 259–60), though whether it (as distinct from *baggere* 'hawker') can also refer to a beggar is uncertain (*VEPN*). The site is exposed and quite high, though as seen from the N it sits in a hollow. The land must once have been quite marginal. The final el., *yeat*, is the dial. descendant of OE **geat** 'gate'.

BAKERSTEAD NY1502 Eskdale.
Bakerste(a)d 1570 (*PNCumb* 390).
♦ Presumably 'the Bakers' place or farm', from the occupational surname and **stead**, though I have not found *Baker* in the Eskdale PR.

BAKESTALL NY2630 (2208ft/673m.) Underskiddaw.
Baikstall Pike 1839 (*PNCumb* 264), *Bakestall* 1867 (OS).
♦ Meaning uncertain. Possibilities include: (a) *bake* (from OE *bacan*) in its dial. sense of 'the drying out of ground', plus *stall* (OE *stall* 'place') in its dial. sense of 'shed, temporary hut, sheepfold or shelter' (cf. *EDD*). (b) As a far shot, one might envisage corruption of a word meaning 'baking-stool' — a stool used in bread-making (cf. Scots *bake-stule*, *bake-stuil* etc., *DOST*). Such a name could have been inspired by the striking appearance of B~ when seen from the N, with a massive buttress to the right, and a great rocky recess edged by DEAD CRAGS to the left; cf. perhaps BRANDRETH.

BALD HOWE/BALDHOW NY4023 (hill & hab.) Matterdale.
Bawthow(e) 1592 (frequent), *Balthow* 1650, *Boathow* 1678, *Balldhow* 1725 (*PNCumb* 257), *Boathow* 1774 (DHM).
♦ The 2nd el. is clearly **how**(**e**) from ON **haugr** 'hill, mound', but the 1st is obscure since no single explanation

accounts for the various spellings. They perhaps arise from perceptions of the hill as both 'bald, bare of vegetation' (ME *balled*, see *VEPN*) and as boat-shaped (cf. perhaps some examples of BOAT HOW(E)), since it has an elongated oval shape with a level top and the farm at its NE end. As at two Boat Hows, there are prehistoric remains here, at Baldhowend, including Romano-British hut circles and cairns (Loney & Hoaen 2000, LDHER 18884).

BALDMIRE SD1381 (hab.) Whicham. *Baldmyre* 1651, *Bouldmyre* 1668 (*PNCumb* 445).
♦ Possibly 'the bare marsh', from ME *balled* 'rounded, bald, bare' (see *VEPN*) and the reflex of ON **mýrr**.

BALLA WRAY SD3799 (hab.) Claife. *Balla Wray* 1891 (OS).
♦ Neither the building nor the name appears on the 1850 OS map. This is close to (HIGH and LOW) WRAY and may be named on the model of that or thought of as part of the same *wray* 'nook' (see **vrá**). 'Balla' is obscure to me.

BALL HALL SD2291 (hab.) Dunnerdale. *Bellhaw* 1772, *Bellhall* 1776, *Ballhall* 1833 (Seathwaite PR), *Ball Hall* 1851 (OS).
♦ Perhaps 'the bell-shaped hill', as suggested by the 1772 spelling, from **bell** and **how(e)** (see **haugr**). This could refer to the shapely hill above B~ H~, called The Knott. **Hall** would be a natural adaptation of the name as applied to a residence, and *hall* and *haw(e)*, *how(e)* are often confused in this area. Without earlier evidence, however, the origin of the name remains uncertain.

BAMPTON NY5118 (hab. & parish), B~ COMMON NY4716, B~ GRANGE NY5218. *Bampton(e)* c.1160 to present, *Banton* early 13th cent. to 1699 (*PNWestm* II, 189); *Bampton Common* 1865; *Bampton gra(u)nge* 1540 to 1777 (*PNWestm* II, 193).
♦ 'The settlement by the tree' or possibly 'made of beams', from OE *bēam* 'tree, beam' and OE **tūn** 'farmstead, settlement, village' with *-p-* intruding as a glide. See also **common, grange**. The two parts of B~ parish were known as *B~ Cundal(e)* and *B~ Patri(c)k* in medieval and early modern times, from feudal owners. B~ Grange is an independent village across the R. LOWTHER from B~; the use of *grange* recalls connections with Shap Abbey (*Kendale* II, 205; Noble 1901, 148).

BAND, THE NY2605 Langdales.
From 1865 (*PNWestm* I, 204).
♦ This is a **band** in the sense of a projecting ridge, here the 'well-known walkers' highroad' onto BOWFELL from LANGDALE (Wainwright 1960, Bowfell 5). (Pl. 26)

BANDRAKE HEAD SD3187 (hab.) Colton. *banryghed* c.1535 (*PNLancs* 216), *Baynrig* 1539 (Brearley 1974, source not specified), *Bandrigheade* 1623, *Banrickehead* 1643, *Bandrakehead* 1678 (PR).
♦ B~ may be 'the straight ridge'. The 1st el. seems, if Brearley's identification is correct (1974), to have been from ON *beinn* 'straight, direct', with *-d-* entering as a glide and/or under influence of **band** 'stratum of rock' or 'projecting ridge'. The 2nd el. was originally **rigg** from ON **hryggr** 'ridge', but this has been replaced by

rake 'steep path'. **Head** refers to the hamlet's situation, fairly high and at the end of the ridge.

BANK NY1425 (hab.) Blindbothel.
le Bank' 1293 (*PNCumb* 447), *Bank* 1774 (DHM).
♦ 'The hill-slope'; **bank(e)**. High and Low Bank are distinguished on the 1:25,000 map.

BANK: FAR B~ SD1092 (hab.) Waberthwaite, MIDDLE B~ SD1091 (hab.), NEAR B~ SD1091 (hab.).
Bank (twice) 1774 (DHM); *Farbank* 1786 (PR); *Middle Bank* 1867; *Nearbank* 1811 (Corney PR).
♦ These occupy the low slopes E of ESKMEALS Pool, and may be named according to their relative proximity to CORNEY or to SEATON (HALL); **bank** 'hill-slope', **far, middle, near** (under **far**).

BANK END SD1988 (hab.) Broughton West.
From 1666/7 (PR); *Banken* 1745 (West 1774, map).
BANK END SD2692 (hab.) Torver.
Bankend 1585–91 (CW 1957, 69).
♦ Simply 'the end of the hill-slope'; **bank, end**. The Broughton example lies below the S point of Bleansley Bank, and the Torver one below Cat Bank.

BANK GROUND SD3196 Coniston.
Bank ground in Conistone 1713 (Cowper 1899, 420).
♦ 'The farm of the Bank family'; surname plus **ground**. Persons named *Bank* feature frequently in the PR of Hawkshead, to which this part of Coniston (Monk Coniston) belonged. Collingwood refers to a 'Francis Banke (of Bank Ground)', accused of illicit fishing in a bond of 1589 (1929, 42), and the 1713 record names 'William Bank of Bank Ground', though he was residing in York at the time. The family name 'may be of purely local origin' (Cowper 1899, 366). High and Low B~ G~ are distinguished on larger scale maps.

BANK HEAD NY1834 (hab.) Blindcrake.
Bank-Head 1717 (Isel PR).
BANK HEAD SD4992 (hab.) Underbarrow.
From 1823 (*PNWestm* I, 103).
♦ 'The head of the hill-slope' (**bank, head**), on rising ground in the DERWENT and KENT valleys, respectively.

BANK HOUSE SD1785 Millom Without.
Bankhouse 1591 (PR), *Bank Ho* 1774 (DHM).
BANK HOUSE NY3733 Mungrisdale.
From 1867 (OS).
BANK HOUSE FARM NY0704 Gosforth.
Banck House 1596 (PR), *Bank House* 1774 (DHM).
BANK HOUSE FARM NY3923 Matterdale.
Banckehowse 1589 (*PNCumb* 222).
♦ Simply 'the house on the hill-slope' (**bank, hús/house**), **farm** being added later in two cases. The Millom example is on BAYSTONE BANK, while the others stand on the rising ground above the CALDEW and BLENG rivers, and Matterdale Beck, respectively.

BANNA FELL NY1017 (1348ft/411m.) Ennerdale.
Bennefell c.1599 (CW 1925, 194), *Bannyfell* 1609, *Banafell*, *Benefell* 1610 (*PNCumb* 386), *Bannerfield* 1802 (CN 1207), *Banna Fell* 1867 (OS).
♦ 'Banna' is of uncertain meaning, but could be *banner* in the sense of 'boundary marker' as in JOHN BELL'S BANNER, especially since the **fell** stands at the N extremity of what is now Ennerdale and Kinniside civil

parish. The *ben(n)*- spellings, however, slightly discourage this interpretation.

BANNEL HEAD SD4995 (hab.) Strickland Ketel.
Banalled(-gyll) 1170–84, *Bavelhed* (in error for *Banelhed*) 1518–29, *Bannaley bank* 1836 (*PNWestm* I, 155).
♦ Uncertain. The syllable -*(h)ed*-, later Head, may be from OE **hēafod** 'head, high place', and this forms one of several names referring to land projecting above the KENT valley (*SSNNW* 355), but if so the 1170–84 spelling is suspiciously modern for its date, since *heved* or similar would be usual, and the rest of the name is an enigma.

BANNERDALE NY3429 Mungrisdale, B~ CRAGS NY3329.
Banner-Dale; *Bannerdale Cragg* both 1704 (*CW2* 50, 1951, 126), *Bannerdale Crag* 1800 (*CN* 793).
♦ Without earlier spellings it is impossible to tell whether this is another 'holly valley', like BANNERDALE in Martindale, or whether of some other origin (cf. BANNER RIGG). Bannerdale today is bare of trees and of habitation, and runs, broad and open, down from the amphitheatre of B~ **Crags**.

BANNERDALE NY4215 Martindale, B~ BECK NY4215.
Baynwytdale 1256 to 1279, *Baynwythdale* 1278, *Bannerdale* 1588; *Baynwytdalebek* 1265 (*PNWestm* II, 215).
♦ 'The valley where holly grows', from ON *beinviðr* 'holly' and ON **dalr** 'valley'; see also **bekkr/beck** 'stream'.

BANNER RIGG SD4299 (871ft/265m.) Windermere, BANNERRIGG FARM SD4298.
Baynerigg 1374 (Brearley 1974, source not specified), *Bannerrigge* 1677; cf. *Banerhowe* 1256 (*PNWestm* I, 194).
♦ Various explanations. Though a neat cone, its slightly elongated top just qualifies this as a **hryggr/rigg** 'ridge'. For the 1st el. (a) 'Banner', if authentic, seems most likely to be 'boundary-marker', as in JOHN BELL'S BANNER; the hill is quite close to the boundary with Hugill. (b) 'Holly' (ON *beinviðr*) is also possible, as in BANNERDALE, Martindale and BENNETT HEAD. (c) If Brearley's identification of *Baynerigg* with this hill is correct (1974), the 1st el. could be ON *beinn* 'straight, direct' (cf. BANDRAKE), though it does not obviously suit the topography. (d) A nickname (among the solutions suggested in *PNWestm* I, 194) is less likely since *hryggr/rigg* is rarely or never qualified by pers.ns in some 60 names in this volume, but rather by descriptive terms.

BANNEST: B~ HILL NY3534 (hab.) Caldbeck.
Banehirste, Bainehirste (Mosse) 1560, *the Baniste* 1581, *Banhurste* 1595, *Baynehurste* 1598 (*PNCumb* 278); *Banist-Hill* 1817 (Mungrisdale PR).
♦ Obscure, but possibly 'the straight-growing wood'. (a) The 16th-cent. spellings favour ON *beinn* 'straight' or 'hospitable' as 1st el. (b) *Beinn* is also recorded as a tree-name, but can be ruled out, since it is rare and refers to ebony, unless this is a shortening of *beinviðr* 'holly' (cf. BANNERDALE, BENNETHEAD), in which case Bannest would mean 'Holly Wood'. The 2nd el. is clearly from OE **hyrst** 'wood, wooded hill'. **Hill** has been added later.

BANNISDALE NY5103 Fawcett Forest, B~ BECK NY5103, B~ FELL NY5005, B~ HEAD NY5104 (hab.).

Banenendesdala 1175-84, *Banandesdale* & variants 1251 to 1539, *Bannesdale* 1357; *Bannendisdalebec* 1195 (*PNWestm* I, 137-8); *Bannisdale Fell* 1828 (HodgsonM);' *Bannandesdale hed, -heade* 1542 (*PNWestm* I, 139).
♦ 'Bannand's valley', assumed to be from an otherwise unrecorded ON nickname based on the present participle of *banna* 'to forbid, curse', plus **dalr** (e.g. Ekwall *DEPN*, Smith in *PNWestm*); see also **bekkr/beck** 'stream', **fell** 'hill', **head**.

BARBER GREEN SD3982 (hab.) Staveley-in-Cartmel.
Barber Greene 1594 (Cartmel PR).
♦ 'The green place or farm of the Barber family'; surname plus **green**. The surname *Barber* is common in the Cartmel PRs, and Richard and Hugh Barber are recorded as living at B~ G~ in 1594 and 1603/4 respectively.

BARF NY2126 Above Derwent.
Barth 1800 (*CN* 104), *Barugh, Baragh* 1803 (*CN* 1627-8).
♦ 'The hill.' This is a classic **berg** or **barrow**, with steep contours running down towards BASSENTHWAITE to the E. For a legend relating to Barf, see (THE) BISHOP. (Pl. 7)

BARFIELD SD1087 (hab.) Whicham, B~ TARN SD1086.
Bar Field 1774 (DHM); (B~ Tarn:) *Barlake* 1794 (Hutchinson I, 550), *Barfield Tarn* 1867 (OS), possibly = *Fosseterne* early 13th cent. (*Furness Coucher* II, 528).
♦ B~ has been identified with *Scalgarthbare* 1462 (*PNCumb* 448, where further spellings are given), which seems to be 'hill (**berg/barrow**) by the enclosure (**garðr**) with the shieling (**skáli**)'. If so, Barfield and *Bar-lake* may simply have been formed from the last el., plus **field** and **lake**, while Barfield Tarn is a more recent coinage.

BARKBETH NY2431 (hab.) Bassenthwaite, BARKBETHDALE NY2430.
Barkebethe 1614, *Barkboth* 1646 (*PNCumb* 264); *Barkbeth Dale* 1867 (OS).
♦ B~ may be 'the bark-shed', a store for tanning, with elements derived ultimately from ON *bǫrkr* 'bark' and ON **búð/bōþ** 'hut, shed, booth', as in BARKBOOTH, and cf. BARKHOUSE. The 17th-cent. spellings of the 2nd el. are matched by (*Arm*)*bothe* 1530, (*Arma*)*beth* 1552 for ARMBOTH (*PNCumb* 311). B~ stands at the NW end of the valley of Barkbeth**dale**, surrounded by woodland of oak (whose bark was the favoured agent for tanning), ash and sycamore.

BARKBOOTH SD4190 (hab.) Crosthwaite.
Bark(e)booth 1535, 1633 to 1694, *Bark(e)bough* 1655, *Bark(e)bore* 1789 to 1857 (*PNWestm* I, 82).
♦ 'The bark-shed', presumably bark stored and dried for tanning; elements derived from ON *bǫrkr* 'bark' and ON **búð/bōþ** 'hut, shed, booth'. Cf. BARKHOUSE and probably BARKBETH.

BARKER KNOTT FARM SD4094 Bowness.
Barker Knott 1718 (*PNWestm* I, 186).
♦ 'The compact hill associated with the Barker family'; surname plus **knott**, with **farm** added. This seems likely, though I do not know of evidence of this family name locally. The farm lies below the hill.

BARKHOUSE NY1931 Setmurthy.
the lowe Barkehouse, high Barkhouse 1578 (*PNCumb* 435).
BARK HOUSE WOOD SD3792 Satterthwaite.
From 1851 (OS).

♦ The name refers to storage of bark for tanning (so *PNCumb*), with 1st el. ME, ModE *bark* from ON *bǫrkr* plus **house**; also **wood** in the Satterthwaite example. The compound *bark-house* is recorded as a common noun from 1483 (*OED: bark* n. 1, sense 10). Bark was a valuable commodity: there are, e.g., both 14th- and 19th-cent. records of bark for tanning fetching substantial sums (*CW* 1909, 30, *CW* 1974, 85; see also Howard-Davis 1987, esp. 240, on the tanning process). The Satterthwaite name is, by chance or design, a counterpart to Boat House Wood, under a mile to the S.

BARN FARM SD4795 Strickland Ketel. Building shown but not named 1862, OS.
♦ Self-explanatory. The word *barn* goes back to OE *bere-ærn*, lit. 'barley-building', which appears in names from the OE period onwards, but this and LOW BARN, the only other instance of the word in this volume, appear to be recent. See also **farm**.

BARNSCAR SD1395 Muncaster.
Remains of the City of Barnsea 1774 (DHM), *ruins . . . called by the country people, Barnscar, or Bardskew, in the maps, Barnsea* 1794 (Hutchinson, I, 562).
♦ This is a major Bronze Age site, with cairnfields and traces of settlements, at 558ft/170m. (LDHER 1437, SAM No. 32861), which invites the suspicion that the 1st el. may be from **borran**, but as *PNCumb* 424 puts it, 'the forms so far found for the name are too late and too varied to admit of interpretation'. Barnscar is also the name of a skerry thought to mean 'Barni's reef' (*Barnesker* 1338, *PNCumb* 377), and it is possible that it has influenced the modern name. If *Barnskew*, among the 1794 spellings, has any validity, it might point to the reflex of ON **skógr**/OE **sceaga** 'wood' as 2nd el.

BARROW NY2221 (1493ft/455m.) Above Derwent.
From 1800 (*CN* 778).
♦ 'The hill', showing an outcome of OE **berg** different from nearby BARF.

BARROW: B~ BAY NY2620 St John's.
Barrowe 1578 (as hab., *PNCumb* 316); *Barrow Bay* 1867 (OS).
♦ Barrow is 'the hill' (**berg/barrow**), as in nearby B~ Beck and B~ House, though it is not obvious which feature is meant here. The **bay** is an inlet in DERWENTWATER.

BARROW BECK NY3629 Mungrisdale.
From 1867 (OS).
♦ 'The stream by Barrow, the hill'; **berg/barrow, beck**. As in BARROW, St John's, it is not obvious which hill or mound is referred to.

BARROWFIELD SD4890 (hab.) Underbarrow, B~ WOOD SD4892.
Bar(r)owfeld(e) c.1540, *Barrey Fyld* 1556 (*PNWestm* I, 103); *Barrowfield Wood* 1863 (OS).
♦ 'The open land or field by the hill'; **berg/barrow, field, wood**.

BARROW HOLLIN SD4084 Upper Allithwaite.
From 1851 (OS).
♦ Given the curious order of elements, this could be 'the hill where holly grows', or 'holly tree(s) on the hill'; **berg/barrow, hollin**. The hill rises above the unusually-named settlement of Barrow Wife.

BARTON NY4826 (hab. & parish), B~ FELL NY4621, B~ HALL NY4725, B~ HOUSE NY4825, B~ PARK NY4622 Barton.

Bartun, Barton(a) c.1184 to 1823 (*PNWestm* II, 209); *Barton Fell* 1588 (*PNWestm* II, 212); *Barton hall* c.1692 (Machell I, 381, with hesitation), previously *Barton Manor House* 1650, also *New House* 1750, 1865 (*PNWestm* II, 212); (B~ House:) previously *Fernilee* 1920 (OS, not shown or named OS 1865); (B~ Park:) *parcum de Barton* 1279 (*PNWestm* II, 212).

♦ 'The barley farm or outlying grange', from OE *beretūn* or, more probably, *bærtūn*, which was used of farms, especially outliers of large estates, used for storing crops. This B~ became the centre of a very large parish. See also **fell** (here in the sense of 'upland grazing' rather than a distinct peak), **hall**, **hūs/house** and **park**. (Pl. 17)

BASE BROWN NY2211 (2120 ft/646m.) Borrowdale.
Bess Brown 1774 (*PNCumb* 352).
♦ The spellings are too late to determine whether this might be 'Brúni's cowshed', like BAYSBROWN in Langdale. This would be an odd name for a mountain, and a rugged and remote place for a cowshed, yet the main alternative, of origins in the Cumbrian surname Basebrown, is also unlikely, since ROBINSON is the only certain example in this volume of a mountain name consisting solely of a surname.

BASKELL/BASKILL FARM SD1993 Ulpha.
Baskell in Ulfay 1664, *Baskall* 1774 (*PNCumb* 438); see also *JEPNS* 2, 59.
♦ Uncertain, and *PNCumb* attempts no explanation. One might guess at (a) a 1st el. such as ON *báss* 'cowshed' (as in BAYSBROWN) plus ON **skáli** 'shieling', since the spellings and the location would be compatible with the latter; or (b) ON *bú* 'farmstead, estate' plus the common ON pers.n. *Áskell* (cf. BEWALDETH). Another example of this name, with similar spellings, is in Broughton-in-Furness, at SD2390 (*PNLancs* 222), hence the need to differentiate in 1664.

BASON CRAG NY4514 Bampton.
From 1859 (*PNWestm* II, 194).
♦ Probably 'the bowl-shaped rocky height'; *bason/basin*, **crag**. With WHELTER CRAGS, this forms the rim of the impressive Whelter corrie, hence 'Bason' may simply be *basin*. The word was adopted into ME from OFr *bas(c)in*, and was commonly spelt *bason* from 16th cent. onwards.

BASSENFELL MANOR NY2132 Bassenthwaite.
Bassenfell 1900 (OS).
♦ Neither building nor name is shown on the 1867 OS map. The name is seemingly a modern coinage from the 1st part of BASSENTHWAITE and **fell**; the place lies just W of Bassenthwaite village. On the word *manor*, see AYNSOME M~.

BASSENTHWAITE NY2332 (hab. & parish), B~ COMMON NY2529, B~ LAKE NY2129 Bassenthwaite/Under-skiddaw/Above Derwent/Wythop/ Setmurthy.
(Hab:) *Bistunthweit* c.1160 (p), *Bastunthuait* c.1175 (p), *Basyngthwaite* 1492 (*PNCumb* 263, q.v. for further forms); *Bassenthwaite common* 1794 (Hutchinson II, 155); *Bastunwater* c.1220, *aquam de* (water/lake of) *Bastantheweyt* 1279, *Bassyntwater* 1539, *Bassenthaitlake* 1675, *Broad-Water, commonly called Bassenthwaite-Water* 1789 (*PNCumb* 32).
♦ 'Bastun's clearing'; pers.n. plus **þveit** 'clearing'; also **common**, *lake*. The 1st el. is usually taken to be the Anglo-French nickname or surname

Bastun, originally meaning 'stick', while the 2nd is ON **þveit** 'clearing'. The lake, in early times known as 'Bastun's water', takes its name from the village. On the word *lake*, see LAKE DISTRICT; B~ Lake (often referred to simply as 'Bass') is unique among the sixteen main lakes of the district in containing the word.

BATEMAN FOLD SD4394 Crook.
From 1865 (*PNWestm* I, 178).
♦ 'The pen or farm of the Bateman family', surname plus **fold**. The name *Ba(i)teman* or *Batemond* is recorded locally from the 16th–17th cent. (*PNWestm* I, 178 and Kendal PRs).

BAWD HALL NY2119 Above Derwent.
Boodhole 1575, *Bodehole* 1578 (*PNCumb* 372), *Board Hall* 1823 (GreenwoodM), *Bowed-Hall* 1842 (Newlands PR); Brearley 1974 gives Boad Hole as an alternative modern form.
♦ Uncertain, but possibly 'the hollow with a dwelling'. The 1st el. could reflect OE *bold* 'dwelling' or possibly ON **búð** 'hut'. The 2nd el. seems to be **hole** from OE **holh** or ON **holr** 'hollow', though the site is above the NEWLANDS BECK. Cf. BOADHOLE.

BAYSBROWN NY3104 (hab.) Langdales.
Basebrun 1216–72, *Bays(e)browne* c.1512, *Baysborne* 1536 (*PNWestm* I, 203).
♦ Probably 'Brúni's cowshed'. (a) The name seems to be an inversion compound, with ON *báss* 'cowshed' as the generic in first position, qualified by an ON pers.n. in second position and fully stressed (see *SSNNW* 49, 53, 55; *Brúni* is quite common, Lind 1905–15, 171). This Gaelic-influenced word order suggests an early date, and Millward & Robinson believe that this was the first farm site in Langdale (1970, 166; see also *CW* 1908, 159). Cf.

BASE BROWN. (b) Ekwall suggested ON *brún* 'edge, brink' as 2nd el., hence normal Germanic word order and the meaning 'the slope with the cowshed' (*DEPN*: Baisbrowne).

BAYSTONE BANK SD1785 Millom Without, B~ B~ RESERVOIR SD1785 Whicham/Millom Without.
Bakestone banke 1570, *Basting banke* 1597 (*PNCumb* 445); *Baystone Bank Reservoir* 1900 (OS).
♦ 'Baystone' is from OE *bæcstān* 'baking stone' or its reflex, as in BECKSTONES, hence the place where such stones are found and/or manufactured; see also **bank** and **reservoir**.

BAYSTONES NY4005 Troutbeck.
(*the Cragge of*) *Baystons* 1551 (*PNWestm* I, 189), *Baystones* 1859 (OSNB, citing a perambulation of 1552 and giving current variants as *Bakestones, Back Stones*).
♦ 'The (place where one can obtain) baking stones.' This seems to be of the same origin as BAYSTONE (BANK) above.

BEACON TARN SD2790 Blawith.
From 1823 (Otley 25), 1851 (OS).
♦ 'The small mountain pool by the beacon site'; **beacon, tjǫrn/tarn**. The word *beacon*, of which this is the only occurrence on the One Inch map and hence in this volume, derives from OE *bēcn* 'sign'. The prominent height just N of the tarn is an ancient beacon site (Cowper 1897b, 142; Heaton Cooper 1983, 31–2). (Pl. 29)

BEAUTHORN NY4422 (hab.) Matterdale.
From 1839 (Ford 133), 1867 (OS).
♦ Presumably 'beautiful thorn', from Fr *beau* and **thorn**.

BECKCES NY4127 (hab.) Hutton.
From 1567 (spelling not specified, *MHAS* 1996, 41), *Becks* 1604/5, *the Becks* 1614 (Greystoke PR), *ye Beckeces* 1655, *Bexes* 1728 (*PNCumb* 213).
♦ Originally 'the streams', with **beck** from ON **bekkr** and an English pl. -s. The place is by Skitwath Beck and close to Swinescales Beck. The final '-ces' is either (a) due to a curious doubling of the pl., for which Millses in Matterdale (*Mills* 1680, *PNCumb* 223) might provide a parallel; or (b) derives from dial. *cess* 'peat-bog', as in Buckscess (*PNWestm* I, 190).

BECKCOTE FARM NY0507 St Bridget Beckermet.
Beck Cote 1583 (*PNCumb* 340).
♦ Presumably 'the cottage by the stream'; **bekkr/beck, cot(e)**, and **farm** added later. Winchester believes that BECK COTE and STRUDDA BANK may have resulted from 12th–13th-cent. expansion onto the side of COLD FELL (1987, 156).

BECKFOOT Examples at:
　NY1016 (hab.) Ennerdale. Gill Beck flowing into ENNERDALE WATER.
Becfoot 1647 (PR).
　NY1600 (hab.) Eskdale. WHILLAN BECK at its confluence with the ESK.
Bekkfoote 1578 (*PNCumb* 390).
　SD1989 (hab.) Millom Without. LOGAN BECK near its confluence with the DUDDON.
Beckfout 1669 (*PNCumb* 418).
　NY5020 (hab.) Bampton. HELTONDALE BECK near its confluence with the LOWTHER.
Bekfoote 1564 (*PNWestm* II, 194).
♦ 'The lower end of the stream' (**beck, foot**), referring to the confluences detailed above. Concerning the Millom example, an account of c.1870 speaks of 'a small mountain farm called Beckfoot' standing 'on a mountain torrent, in the low ground near its junction with the river' (*CW* 1964, 350).

BECK HEAD SD4484 (hab.) Witherslack.
Becheheade 1615, *Beck Head* 1681 (*PNWestm* I, 78).
♦ 'The upper end of the stream'; **beck, head**. Smith quotes *The Beetham Repository*, 1770: 'situate where the Beck springs Romantickly out of the Rock' (*PNWestm* I, 78). B~ H~ and B~ H~ Farm are distinguished on larger scale maps.

BECKHOUSE NY1629 Embleton.
From 1658 (*PNCumb* 384).
♦ The place is situated on TOM RUDD BECK; **beck, house**.

BECKSIDE Examples at:
　SD1584 (hab.) Whicham. Whitecombe Beck.
From 1649 (*PNCumb* 445).
　NY3629 (hab.) Mungrisdale. R. GLENDERAMACKIN.
ye Beckside in Grisedale 1693 (*PNCumb* 226).
　SD4595 (hab.) Crook.
Beck Side 1770 (JefferysM).
　SD4693 (hab.) Underbarrow. Chapel Beck.
From 1717 (*PNWestm* I, 103).
BECKSIDE FARM SD3185 Colton. COLTON BECK.
Beckside 1681/2 (PR), *Beck Side* 1851 (OS).
♦ 'The place beside the stream' (**beck, side**) — as listed above. In Mungrisdale, Low and High B~ are distinguished on larger scale maps. **Farm** has been added to the Colton example.

BECKSTONES Examples at:
　SD1890 (hab.) Millom Without.
From 1615 (PR), 1774 (DHM).

SD2585 (hab.) Kirkby Ireleth.
Beckstones 1687/8 (PR).
NY4015 (hab.) Patterdale.
From 1724 (PR).
BECKSTONES: B~ PLANTATION NY2125
Above Derwent.
Beckstones 1574 (*PNCumb* 372); cf.
Bakestanbek (= B~ Gill) c.1215 (*PNCumb* 4).
♦ B~ is probably 'the (place where one can obtain) baking-stones', or in some cases 'the stones in the stream'. The 1215 spelling for Beckstones Gill, Above Derwent, shows that this is OE *bæcstān* 'baking stone, flat stone usable as a baking tray', and indeed large, flat slabs of Skiddaw slate are found in the Gill. The word is quite common in p.ns (see also BAYSTONE(S)), for some places in N. England were centres for the manufacture of such stones (see *VEPN* for further examples). See also **plantation**. Without pre-17th-cent. spellings the other examples are less certain. The 1st el. could have its more obvious sense 'stream', ON **bekkr**/dial. *beck*, and all four places are situated near streams. However, 'stream-stones' would hardly be distinctive or meaningful in a Lakeland context. (Pl. 8)

BECKTHORNS NY3220 (hab.) St John's.
Beckthones 1592, *Beckthornes* 1647 (*PNCumb* 316).
♦ Presumably 'the (haw)thorns beside the stream'; **bekkr/beck** 'stream', here referring to Beckthorns Gill, plus **þorn/thorn**, which might well refer to hawthorn.

BECK WYTHOP: see WYTHOP

BEDA FELL NY4216 Martindale, BEDAFELL KNOTT NY4116, BEDA HEAD NY4217 (1664ft/507m.).
Beda Fell 1860 (*PNWestm* II, 218); *Bedafell Nut* [sic] 1863, *Bedafell Knott* 1899 (OS); *Beda Head* 1860 (*PNWestm* II, 218).
♦ Obscure. 'Beda' could consist of the OE pers.n. *Beda* plus *-a* from ON **haugr** 'hill' as in ULPHA, SD1993; if so, **Fell** would have been added later. Without early spellings, however, the age and true origin of the p.n. is unknown. B~ **Knott** 'craggy height' and **Head** are high points on the ridge of the Fell.

BEECH HILL NY1124 (hab.) Blindbothel.
From 1867 (OS).
BEECH HILL NY4902 (hab.) Longsleddale.
Bitchell 1692, *Bitch Hill* (*Bridge*) 1699 (*PNWestm* I, 162).
BEECH HILL HOTEL SD3892 Cartmel Fell.
Bitch hill 1783 (CrosthwaiteM); cf. *Beech Hill Plantation* 1851 (OS).
♦ Probably 'the beech hill', with the reflex of OE *bēce* plus **hill**. The beech grows widely in Cumbria, including lower-lying parts of Lakeland, though 'most trees have probably been planted' (Halliday 1997, 138). Early spellings for the Longsleddale and Cartmel Fell examples probably show shortening of this word, though the reflex of OE *bicce* 'female dog' cannot be ruled out (see *VEPN: bicce*). The Blindbothel name can certainly be taken at face value, since there are well-established beech trees there (thanks to Mr Chris Harris, present owner, for this information). The word *hotel* was adopted into English from French in the 17th cent., though not used with the sense of 'an inn . . . of a superior kind' until the 18th cent. (*OED*, sense 3). It appears five times in the names in this volume.

BELL, THE SD2897 (1099ft/335m.) Coniston.
(The) Bell 1802 (*CN* 1228), *The Bell* 1851 (OS).
♦ 'The **bell**-shaped hill.' Coleridge wrote of 'the Bell & the Scrow, two black Peaks, perfectly breast-shaped' (*CN* 1228).

BELLART HOW: B~ H~ MOSS SD4583 Witherslack.
Bellart How 1741, *Bellet How* 1770, *Bellert How* 1845 (*PNWestm* I, 78), *Bennet House* 1862, *Bellart How* 1899 (OS); *Bellart Howe Moss* 1862 (OS).
♦ 'Bellart' is of uncertain origin, though the 1770 spelling might encourage association with the surname derived either from Fr *bel* 'beautiful' or from the female name *Isabel* (Reaney 1976: *Bellett*); also **how(e)** 'hill' (see **haugr**) and **moss** 'bog'.

BELL CRAGS/CRAGGS NY2914 St John's.
Bell Craggs 1805 (*PNCumb* 316).
♦ 'The rock outcrops on the rounded hill'; **bell**, **crag**. The crags fringe a small hill rising to 1831ft/558m., which might be the *bell* in the name.

BELLE GRANGE SD3898 Claife, B~ G~ BAY SD3898 Troutbeck/Windermere.
Bellgrange 1789 (Hawkshead PR), *Bella Grange* 1809 (CrosthwaiteM); *Belle Grange Bay* 1859 (OSNB).
BELLE ISLE SD3996 Windermere.
Belleisle (*Lodge*) 1791 (Smith & Merigot), *Bella Island* 1809 (CrosthwaiteM, revised from *Windermere Isld* 1783), *Belle Isle* 1823, previously *le holme* 1324, *isle of Wynandermere* 1347, *the Lange holme* 1566 (*PNWestm* I, 193), '*the Holme, or great island*' 1780 (West 57).
♦ The island, the largest in the lake and a manorial power-base (Brydson 1911, 20, Fiennes 1685–1712, 166) shares its name with the house built there in 1774 by a Mr English (on which see Hodge 1940), and the local historian Cowper believed that 'it was probably under the hands of English that it was grotesquely re-named Bella Island, a new form of which (Belle Isle) is unfortunately still sometimes used' (1899, 47). However, island and house were soon afterwards purchased by Isabella Curwen, and the originally French fem. adj. *belle* seems both to refer to the beauty of island and house and to pun on the name of its owner, who was presumably also the inspiration for the names of Bella Pit and the ill-fated Isabella Pit in the Curwen collieries (mentioned in Wood 1971, 209–10). Belle Grange and Belle Mount were also Curwen properties, again favouring the surmise that the Curwens were the namers of Belle Isle. Either way, modern sensibilities might be struck by the irony that these *belle* residences were funded by the underground labour of children as young as five (Wood 1971, 223–4). See also **grange**, **bay** and **isle**.

BELLES KNOTT NY2908 Grasmere.
From 1859 (OSNB).
♦ 'Belles' is obscure but may be from the pet-form of *Isabella*, or the surname *Bell*. **Knott** is a rocky summit.

BELL GROVE NY4523 (hab.) Matterdale.
Previously *Rumney's Mead* 1867, 1920 (OS), 1966 (OS One Inch).
♦ The meaning of **bell** here is uncertain, but the situation, the wooded bank of Ramps Beck, makes **grove** 'copse' apt here. In connection with the previous name, an Anthony Rumney appears as a tenant hereabouts in a document of c.1474 owned three centuries later by James

Clarke (see OLD CHURCH), and the wood just to the N is still Rumney's Plantation.

BELL HILL SD4693 Underbarrow.
Belle 1220–46, *Bell Hill* 1857 (*PNWestm* I, 101).
♦ B~ in itself probably meant 'the bell-shaped hill' (see **bell**); **hill** has been added later.

BELL HILL FARM SD0898 Drigg.
Belhill 1658, *Bell Hill* 1777 (*PNCumb* 377).
♦ Probably 'the bell-shaped hill' (**bell**, **hill** as above), with **farm** added later. The place is beside a small oval hill — a distinct rise on the low coastal plain.

BELLMAN GROUND SD3994 Bowness.
Bel(l)man ground 1706 (*PNWestm* I, 186).
♦ 'The farm of the Bellman family'; surname plus **ground**. The surname is recorded locally in documents of the 14th, 16th and 17th cent. (*PNWestm* I, 186). Brydson reports a local legend that Bellman was a local carrier and a patron of the parish church of Bowness, named from the bells on his carrying horses (1911, 98).

BELL RIB NY1707 Nether Wasdale.
From 1867 (OS).
♦ This is a perilous rocky summit on the S of YEWBARROW. If to be taken at face value, **bell** presumably refers to its steep, domed shape, and *rib* (from OE *rib(b)*, unique in this volume) to its position as part of a linear arête (illustrated, together with DROPPING CRAG, in Wainwright 1966, Yewbarrow 2). The same name is applied to part of the Wastwater SCREES, at NY1604.

BELMOUNT SD3599 (hab.) Claife.
Bell-mont 1780 (West 55), *Bell-mount* 1784 (West 54), *Belmont* 1802, *Belmount* 1804 (Hawkshead PR).
♦ This is a rarity in being formed from elements of Fr origin. It is a late name, given partly for its style, but the etymological sense of *beau/bel* 'beautiful' and *mont* 'hill' is also apt: 'A handsome modern house, *Bell-mont* is charmingly situated, and commands a delightful view of the lake [Esthwaite]' (West 1780, 55; see also Pevsner 1969, 140). The name did not refer solely to the mansion, however: Hawkshead PR for 1837 refer to Joseph Nicholson 'of Belmont Quarry man'.

BELOW BECK FELLS: see ABOVE BECK FELLS

BELT ASH COPPICE SD3896 Claife.
From 1851 (OS).
♦ 'Belt' is puzzling, as is the structure of the name, though **ash** and **coppice** are unproblematic. (a) It may simply be *belt* 'belt (of woodland)', which occurs in names that are 'late', i.e. after c.1750 (*VEPN* I, 152), and cf. Big Belt at SD4682, as well as the strips of woodland called Belt Plantation(s) in NYorks, at NZ8904 and SE6280. (b) An English surname *Belt* also exists, but is not common.

BENGARTH NY1104 Gosforth.
From 1572 (PR), 1733 (*PNCumb* 395).
♦ 'The enclosure or farm of the Benn family'; surname plus **garth**. As noted in *PNCumb* 395, the PRs contain records of Alice Ben in 1583, and Joseph Benn in 1720. Another Bengarth is recorded in Nether Wasdale (PR for 1760).

BENN, THE NY3019 St John's.
From 1867 (OS).
♦ This is a prominent hill, presently clad in conifers, and its name, like that of THE GLEN, seems likely to be an import from Scotland, of *beinn*, the most common Gaelic term for 'mountain', often anglicised to *ben(n)*.

BENNETHEAD/BENNET HEAD NY4423 (hab.) Matterdale.
Baynwytheued 1285 (p), *Banerhede* 1487, *Benethead* 1580 (*PNCumb* 254).
♦ 'The high place where holly grows', from ON *beinviðr* 'holly', and OE **hēafod** 'head, high place, headland'.

BENTY HOWE NY4713 Bampton.
Benty how 1865 (*PNWestm* II, 194).
♦ 'The hill where bent-grass grows', from an adj. derived from *bent* (see WHITE HORSE BENT; *OED: benty*; *VEPN: *benti*), plus **how(e)** from ON **haugr**.

BESSYBOOT NY2612 (approx. 1640ft/ 500m.) Borrowdale.
From 1867 (OS).
♦ 'Bessy's shelter or fold', with a pet form of *Elizabeth* as 1st el. The 2nd may be either (a) ON **búð** 'hut' (Collingwood 1918, 98); or (b) dial. *bought, bucht* 'sheepfold' (cf. *EDD: bought* sb. 2, mainly Scots and Northumberland). Either would fit the situation, a summit on ROSTHWAITE FELL, and there are comparable names such as NAN BIELD.

BETHECAR: HIGH B~ SD3089 (hab.) Colton, LOW B~ SD3089 (hab.), B~ MOOR SD3090.
Bothaker 1509, *bethokar* c.1535, *Betaker* 1537 (*PNLancs* 218), *Bithecar* 1629, *Bethakar* 1632/3, *Bothaker* 1633/4 (PR); *High Bethecar; Low Bethecar; Bethecar Moor* all 1851 (OS).

♦ B~ is probably 'Bethoc's shieling'. Ekwall suggested as 1st el. a Gaelic female pers.n. *Beathag*, earlier *Bethoc*, plus 2nd el. Gaelic-Norse **ærgi/erg** 'shieling' (*PNLancs* 218), and although earlier and more consistent spellings would be reassuring, this seems reasonable. The farms of **High** and **Low** B~ are likely original shieling sites, on the S. part of B~ **Moor**.

BETWEEN GUARDS NY0905 (hab.) Gosforth.
betwixt Gards 1662, *Between-gards* 1781; cf. *Gards* 1605/6, *Gardsend* 1607, *Abovegards* 1664 (PR).
♦ '(The place) between the enclosures' or 'the middle farm'; *between* from OE *betwēonan*, plus **garðr/g(u)ard**. This is the only appearance of *between* in the p.ns in this volume, but see *VEPN* for further p.ns containing it. The name is elliptical, lacking an explicit generic el. The place lies on a steep slope between Guards Head and Guards End.

BEWALDETH NY2134 (hab. & parish), HIGH B~ NY2234.
Bualdith 1255, *Bowaldef* 1260 (p), *Boaldith* 1278, *Bowaldeth* 1292 (p) (*PNCumb* 264–5); *High Bewaldeth* 1867 (OS). Stressed on 2nd syllable.
♦ 'The farm or estate of a woman named Aldgȳþ/Aldgifu.' Ekwall plausibly derived this from ON *bú* 'farmstead, estate', common in Norway but rare in English p.ns, and an OE female pers.n. (*Scand & Celt* 19, cf. *PNCumb* 265). With the generic el. placed first, this is a so-called inversion compound.

BEWBARROW CRAG NY5114 Shap Rural.
From 1859 (*PNWestm* II, 173).
♦ The 1st el. is obscure, unless it is from ON *bú* 'farmstead, estate' as in

BEWALDETH. The 2nd seems likely to derive from OE/ON **berg** 'hill'; see also **crag** 'rocky height'.

BIELD, THE NY3103 (hab.) Langdales.
The Beald 1697 (Grasmere PR), *Bield* 1839 (*PNWestm* I, 205).
♦ 'The shelter'; **bield**. Bield Crags lie above and are probably named from the farm rather than the reverse.

BIGERT MIRE SD1792 (hab.) Ulpha.
Biggatmire 1646, *Brigetmire* 1664, *Bigertmyre* 1722, *Beggar mire* 1774 (*PNCumb* 438).
♦ B~ is conceivably of similar origin to BIGGARDS 'barley fields', though the 17th-cent. spellings do not point that way; see also **mýrr/mire** 'bog'.

BIGGARDS NY3238 (hab.) Caldbeck.
Biggards 1626, *Bigyards* 1695 (*PNCumb* 278).
♦ 'The barley fields', from dial. *bigg* 'barley' (ON *bygg*) and the reflex of ON **garðr** 'enclosure, field' or its OE cognate *geard*, with English pl. -s.

BIGLAND: B~ HALL SD3583 Haverthwaite, B~ TARN SD3582.
Biglande 1537 (*PNLancs* 198); *Bigland Hall*; *Bigland Tarn* both 1851 (OS).
♦ 'The land where barley grows', with dial. *bigg* as in BIGGARDS, plus **land**; also **hall** and **tjǫrn/tarn**. The Hall is 'mainly of 1809' (Pevsner 1969, 138). The Bigland family arms, *azure two ears of big wheat in pale couped and bearded* or alludes to the etymology of B~.

BINSEY NY2235 (1467ft/447m.) Ireby and Uldale.
Binsey, Binsey Fell, Binsell fell [sic], *Binser fell* 1687–8 (Denton 51, 139, 167), *Binsay (hawse)* 1742, *Binsa* 1777, *Binsay fell, Binsell fell* [sic] 1777 (*PNCumb* 300), *Binsey* 1867 (OS).
♦ The spellings are too late to suggest an interpretation, though to judge from the 18th-cent. forms the final syllable could derive from **how(e)** (ON **haugr** 'hill, mound'). More conjecturally still, the 1st el. could be dial. *bing* (especially Scots, from ON *bingr*) 'a heap'. This would be apt for this smoothly conical hill topped by a cairn of unknown period (LDHER 876).

BIRCH BANK SD2687 (hab.) Blawith.
Birkbank in Blawith 1718 (Lowick PR), *Birk Bank* 1851, 1893, 1919 (OS).
♦ Simply 'the hill-slope where birch grows'; **birk, bank**; cf. BIRK BANK. The evidence of the PRs of Lowick and Blawith, and of the early OS maps, is that the standardisation to *birch* happened fairly recently.

BIRCHCLOSE: HIGH B~ NY4125 Matterdale.
Birkeclousse 1487 (*PNCumb* 257); *High Birk Close* (& *L. Birk Close*) 1774 (DHM), *High Birch-close* 1775 (*PNCumb* 257).
♦ B~ is 'the enclosure or farm with birches', with the English-derived *birch* replacing the Norse-derived **birk**, plus **close**. **High** B~ lies S of Low B~ and slightly higher up the skirts of (LITTLE) MELL FELL.

BIRDHOW NY2001 (hab.) Eskdale.
Bird How 1587 (in 1660 copy, CW 1922, 75), *Birdhows* 1658, *Bird how* 1679 (PR), *Bird House* 1774 (DHM); see also *JEPNS* 2, 58.
♦ 'The hill notable for birds.' In OE, *brid(d)* referred to the young of birds, and sometimes of other animals, while *fugol* (later *fowl*) referred to birds in general. The usage in ME was transitional. Without clearer evidence of date, it is impossible to know

whether the specific or general sense is meant here. **How** is from ON **haugr** 'hill, mound'.

BIRK BANK NY1226 (hab.) Blindbothel.
Birkbank 1610 (Lorton PR).
♦ 'The hill-slope where birch grows'; **birk**, **bank**, cf. BIRCH BANK.

BIRKBY: B~ FELL SD1396 Muncaster.
Bretteby c.1215 to 1323, *Breteby* 1278, *Bertby* 1332 (*PNCumb* 424); *Birkby Fell* 1823 (GreenwoodM), 1867 (OS).
♦ B~ is 'the village of Britons', from *Breta*, gen. pl. of ON *Bretar* 'Britons, Welsh', plus ON **bœr/bý**. The name is therefore an important witness that there were people whose cultural identity was in some way clearly Celtic as late as the Scand-speaking 10th cent., and as far S in Cumbria as this. The fact of the p.n. also suggests that this was unusual enough to be distinctive. The **Fell** is a stretch of high grazing, rich in prehistoric remains, rather than a distinct peak, and Birkby itself is now marked as part of B~ Fell on larger scale maps.

BIRK CRAG NY3113 (1171ft/357m.) St John's.
From 1867 (OS).
BIRK CRAG NY4221 (1047ft/319m.) Matterdale.
From 1867 (OS).
♦ 'The rocky height where birch grows'; **birk**, **crag**.

BIRK DAULT SD3483 (hab.) Haverthwaite.
Birkdalt 1775 (*CW* 1940, 14), *Birk Dault* 1851 (OS).
♦ 'The portion of land where birch grows', from dial. **birk** 'birch' and dial. *da(u)lt* 'portion, share of the common field or some privilege or duty'. This is the sole occurrence of *dault* in the names in this volume, but for examples in field-names from the 17th cent. onwards, see *PNWestm* II, 245–6 and 303.

BIRKER: B~ BECK SD1799 Eskdale, B~ FELL SD1697, B~ FELL NY2100, B~ FORCE NY1800, LOW B~ NY1800 (hab.), LOW B~ POOL SD1999 (stream), BIRKERTHWAITE SD1798 (hab.).
Bircherhe 1215 (Brearley 1974), *Birkergh* 1279 (p), *Byrker* 1432, *Birker* 1560 (*PNCumb* 342); *Bircherhebec* c.1205, *Byrker bek* 1432 (*PNCumb* 4); *Birker Fells* 1802 (*CN* 1222), *Birker Moor* (= Birker Fell, SD1697) 1867 (OS); *Birker Force* 1867 (OS); *Low birker* 1639 (PR); *Low Birker Pool* 1867 (OS), *Birkerthwait* 1741 (*PNCumb* 342).
♦ Birker is 'the shieling by the birch trees', from ON **birki** 'birch trees' and Gaelic-Norse **ærgi/erg** 'shieling', with secondary names containing **bekkr** 'stream', **fell** 'hill, tract of high unenclosed ground', **force** 'waterfall', **low**, **pool** 'slow stream', and **þveit/ thwaite** 'clearing'. The shieling developed into a substantial enough settlement to give its name to a township (absorbed into Eskdale in 1934), whose upland grazing area was its Fell and through which B~ Beck flows. The Birker Fell marked at SD1697 is a wide expanse of unenclosed grazing, while the one at NY2100, which does not appear on the 1867 OS map, is the NW slope of HARTER FELL. B~ Force is a waterfall in Low B~ Pool, which flows through the mossy basin in which the tiny Low B~ Tarn is situated (giving its name to a TARN CRAG). Birkerthwaite lies in a remote situation near the head of B~ Beck.

BIRKET/BIRKETT HOUSES SD4193 Cartmel Fell.
Birkett Houses 1665 (*PNLancs* 200).
♦ The **houses** are named from the B~ family, recorded locally from the 16th

cent. and associated with this place from the 17th (Forsyth 1998, 209). On the origin of the surname, whose meaning is virtually equivalent to Birks Head, just to the NW, see BIRKETT BANK.

BIRKETT BANK NY3123 (hab.) St John's, B~ FIELD (hab.) NY3425 Threlkeld, B~ MIRE (hab.) NY3124 St John's.
Byrketbank 1552, *Birkeheadbancke* 1571 (*PNCumb* 316); *Bürckhet fealde* 1574, *Birket field* 1615 (*PNCumb* 253); *Bryketmyre* 1530, *Birkheadmyre* 1567, *Birkett Myre* 1590 (*PNCumb* 316); cf. *Byrkehevidbek* 1278 (*CW* 1923, 186).
♦ B~ is from ON **birk(i)**/OE **bierce** 'birch' and OE **hēafod** 'head, high place', but the p.n. gave rise to a surname, still very much in evidence in the parishes of St John's, Threlkeld, and further afield, and it may be the surname rather than the p.n. which is incorporated into some of these names, formed from **bank, field** and **mýrr/mire**. The German-type spelling of B~ Field in 1574 reflects the fact that it comes from records of the Augsburg miners in the Keswick area.

BIRKETT FELL NY3619 Matterdale.
Previously *Nameless Fell* 1867, 1920 (OS).
♦ B~ is a local surname (recorded, e.g., at Troutbeck Gills in Matterdale PR 1697), but the name of the **fell**, like the memorial cairn on it, commemorates Lord Birkett of Ulverston, who in the 1960s successfully resisted proposals to make ULLSWATER into a reservoir (Birkett 1999, 7).

BIRK FELL NY2901 Coniston.
From 1850 (OS)
BIRK FELL NY4018 Patterdale (1670ft/509m.).
Byrcfel 1279, *Birk Fell* 1787 (*PNWestm* II, 224).
♦ 'The hill where birches grow'; **birk(i)**, **fell**.

BIRK FIELD SD4899 (hab.) Over Staveley.
Birkfield c.1692 (Machell II, 112), *Birch Field* 1836 (*PNWestm* I, 175).
♦ 'The open ground or field where birch grows'; **birk(i)**, **field**.

BIRKHOUSE MOOR NY3616 Patterdale.
From 1860 (*PNWestm* II, 224).
♦ B~ is probably a p.n., as is often the case with the specifics of 'Moor' names, and its 1st el. presumably **birk(i)** 'birch trees', but the 2nd could be either be **house** or a corruption of **howes** 'hills', from ON **haugr**. This is a high tract of rough grass E of the HELVELLYN massif.

BIRK KNOTT SD2990 Colton.
From 1851 (OS).
♦ 'The rugged height where birches grow'; **birk(i)**, **knott**. This S Lakeland example is far less craggy than the classic **knott**.

BIRK MOSS Examples at:
    NY0715 (hab.) Ennerdale.
Byrkemoss (p) 1534–5 (*CW* 1931, 179), *Birk Moss* 1593 (*PNCumb* 386).
    NY3035 Caldbeck.
From 1867 (OS).
    SD4393 (hab.) Crook.
Birk Mosse 1556, *Birk Moss* 1836 (*PNWestm* I, 179).
♦ 'The bog where birch grows'; **birk(i)**, **moss**. A Birk Gill and Birk Hill lie close to the Caldbeck example, so that it is not clear which name is primary, but it is likely to be B~ Moss, since mosses are a favourite habitat of the birch (Halliday 1997, 140).

BIRKRIGG NY2219 (hab.) Above Derwent.
Birkeryg' 1293 (*PNCumb* 369).

BIRK RIGG NY4602 (1042ft/318m.) Over Staveley/Kentmere.
Birkrigge 1622 (*PNWestm* I, 175).
♦ 'The ridge or hill where birch grows'; **birk(i), hryggr/rigg**.

BIRK ROW/BIRKROW SD2987 (hab.) Blawith.
Byrkerowe 1564, *Birkraye* 1640 (*PNLancs* 214).
♦ Probably 'the row or hamlet where birches grow'; **birk, row**. The local form *birk* has survived in this name but not in nearby BIRCH BANK and CROOKED BIRCH. If the 1640 spelling is significant, the name is equivalent to BIRKWRAY, but the 1564 and modern forms point to *row* as 2nd el.

BIRKS Examples at:
SD2092 (hab.) Dunnerdale, B~ WOOD SD2093.
Birks; *Birks Wood* both 1850 (OS).
SD2399 (hab.) Ulpha.
la Birkis 1398–9 (*Furness Coucher* II, 579), *Birks* 1703 (PR).
NY3814 (2040ft/622m.) Patterdale.
Birks Crag 1860 (*PNWestm* II, 224).
♦ 'The birch trees'; **birk(i)**. One of a number of names formed from the plural of tree names: see ASHES. The **wood** at SD2093 is beside the farm Birks and presumably takes its name from it. In the Patterdale example, formerly *Birks Crag*, it is curious that **crag** appears to have been dropped from the name of this rock-girt height.

BIRKS BROW SD4191 (hab.) Cartmel Fell.
From 1851 (OS).
♦ 'The hill-slope where birches grow'; **birk, brow**, and cf. BIRKET HOUSES for other references to birch in the locality.

BIRKS CRAG NY4612 (1495ft/456m.) Bampton.

From 1860 (OSNB).
♦ 'The rocky height where birch grows'; cf. BIRK CRAG.

BIRK SIDE NY3313 St John's.
From 1862 (OSNB).
♦ 'The hill-slope where birches grow'; **birk, side**. This is a flank of NETHERMOST PIKE.

BIRKWRAY SD3599 (hab.) Hawkshead.
Byrkwray 1600 (*PNLancs* 218).
♦ 'The nook of land with birches' (**birk, vrá/wray**), though the situation is not as secluded as other places named *wray*.

BISHOP, THE NY2126. Above Derwent.
The Bishop's Rock, so called from a fancied resemblance 1886 (Barrow 146), *The Bishop* 1900 (OS).
CLERK, THE NY2126. Above Derwent.
From 1900 (OS).
♦ The Bishop is a fourteen-foot rock pillar part-way up a scree-slope on BARF, whitewashed through the patronage of the landlord of the nearby SWAN HOTEL; the Clerk is its smaller counterpart in the woods below. Barrow, quoted above, gives the commonsense explanation (as also Barber & Atkinson 1928, 67), and indeed it is easy to see the rock as a figure in a pulpit. However, a local story claims that the names and the whitewashing commemorate a bishop-elect of Derry (Londonderry) who, pausing on his way to the port of Whitehaven in 1783 and inspired either by drink or religious fervour, made a wager that he could ride to the top of Barf. The horse charged up the scree slope to the great rock, where horse and rider fell to their death. The bishop was buried, a clerk officiating, in the wood below. The date and source of the story are unknown to

me. *Bishop* is from OE *biscop, clerk* a loan ultimately from Lat *clericus*. (Pl. 7)

BISHOP WOODS SD3794 Claife.
From 1851 (OS).
♦ *Bishop* as above, plus **wood**. According to Cowper, possibly 'the property of Bishop Watson (Llandaff), whose estate lay across Windermere' (1899, 46, 363).

BLACK BECK Examples at:
SD1395 Muncaster.
From 1867 (OS).
NY1710 Nether Wasdale.
(betwixt) Black becks 1664 (Winchester 2000, 169), *Black Beck* (PR, though identification is difficult).
SD3385 (hab.) Colton.
Blackbeck 1632/3 (PR).
SD3597 Hawkshead.
From 1851 (OS).
SD3683 Haverthwaite.
From 1851 (OS).
BLACKBECK: B~ KNOTTS NY1509 Nether Wasdale.
Blakebeck' 1285 (*PNCumb* 5); *Blackbeck Knotts* 1867 (OS).
BLACKBECK TARN NY2013 Buttermere.
From 1867 (OS).
♦ Probably 'the black stream' (**black, bekkr/beck**), for the Haverthwaite instance at least flows through peat mosses. The 1285 spelling would alternatively be compatible with origins in OE *blāc* 'pale (shining, or foaming)'. There are two examples in Nether Wasdale; it is the one that flows into Nether Beck that gives its name to the **knotts** or craggy heights above. B~ **Tarn**, a small pool high on HAYSTACKS, feeds Black Beck, Buttermere.

BLACK BROW NY3810 Patterdale/Ambleside.
From 1920 (OS).
♦ Simply 'the black edge' (**black,**

brow) — a beetling curve of crags, rather more rugged than most other *brows*.

BLACK COMBE SD1385 (1970ft/600m.) Whicham.
Black-coum 1671 (Fleming 61), *Black Comb* 1687–8 (Denton 51).
♦ 'The black crest'; **black, comb(e)**². This 'dark mountain' (Hutchinson 1794, I, 533) and great landmark is heathery (Gambles 1989, 29), and the name may well refer to the dark effect of this compared with the lighter vegetation of WHITE COMBE, but its dark appearance is also due to its crags, among which are BLACK CRAGS. Yet another explanation is Wordsworth's: 'BLACK COMB (dread name | Derived from clouds and storms!)' (*Poems* 174).

BLACK CRAG Examples at:
NY1219 Loweswater.
Black-crag 1780 (West 138).
NY2303 Eskdale/Birker.
Blakrag 1242 (*Furness Coucher* II, 564).
NY2418 Above Derwent.
Black Cragge 1569 (*PNCumb* 372).
NY2703 Langdales.
From 1862 (OS).
NY3618 Patterdale.
From 1863 (OS).
NY3814 Patterdale.
From 1863 (OS).
NY4117 Martindale.
From 1863 (OS).
BLACK CRAGS SD1383 Whicham.
From 1867 (OS).
BLACK CRAGS NY2508 Langdales.
From 1859 (OSNB).
♦ 'The black, rocky height(s) or outcrop(s)'; **blæc/black, crag**. The Whicham crags are on BLACK COMBE.

BLACK FELL NY3302 Skelwith.
From 1850 (OS).

♦ 'The dark hill'; **black, fell**, perhaps named from its rocky summit, Black Crag.

BLACK HALL NY2301 Ulpha.
le Blakehalle 1398–9 (*Furness Coucher* II, 579), *Black Hall* 1610 (*PNCumb* 438).
♦ This is probably from the ME reflex of OE **blæc** 'black', though OE *blāc* 'pale' is possible, plus **hall**. This is an unusually early occurrence of *hall*, and exceptional in having an adj. as 1st el., whereas most later examples have a p.n. On the identification of the 1398–9 spelling with this place, see BORROWDALE NY2515.

BLACKHAZEL BECK NY3130 Mungrisdale.
From 1867 (OS).
♦ Self-explanatory; **blæc/black, hazel, beck** 'stream'.

BLACK MOSS SD2288 Broughton West.
From 1851 (OS).
BLACKMOSS SD4398 (hab.) Windermere.
le Blackemos 1390–4 (*PNWestm* I, 195).
♦ 'The black bog'; **blæc/black, mos/mosi**. The Broughton example is the boggy ground between Galloper Pool and Kirkby Pool.

BLACK POTS NY0913 Ennerdale.
le Blakepottes 1338 (*PNCumb* 386).
♦ 'The black pits or hollows' (**blæc/black, potte**), referring to the area around the headstreams of the R. CALDER.

BLACK SAIL PASS NY1811 Ennerdale, B~ S~ HUT NY1912.
Le Blacksayl 1322 (*PNCumb* 391), *Black Sail Pass* 1865 (Prior 237); cf. *brook of le Blackzol* 1322 (= Sail Beck, *PNCumb* 26).
♦ B~ S~ may be 'the dark mire', with OE **blæc** 'black' or its reflex plus 2nd el. probably from ON *seyla* 'puddle', either referring to Sail Beck or to the boggy terrain, or else from an ON *seila* 'hollow' (*PNWestm* II, xi); see **sail/sale**, also **pass**. *Hut*, an originally Germanic word transmitted through French, first appears in English in the 17th cent. (*OED*).

BLACK SAILS NY2800 (2444ft/745m.) Coniston.
From 1850 (OS).
♦ **Black** is straightforward, but **sail/sale** in hill-names is one of the puzzles of Lakeland p.ns. See List of Common Elements for the range of possible meanings, including 'hollow' and 'mire' which could describe the land below this peak twinned with WETHERLAM, though not the peak itself.

BLACK SIKE SD2194 Dunnerdale.
From 1850 (OS).
♦ 'The black stream'; **black, syke/sike**. The water is clear, but dark moss in places gives a very dark appearance.

BLACK WARS NY2604 Langdales.
From 1862 (OS).
♦ An elusive name for a curved line of crag on the W of PIKE OF BLISCO; **black** plus the unusual 'Wars'. This is impossible to explain in the absence of early spellings, but connections with OE *waru* 'shelter, guard, care' or conceivably with ON *varða* 'cairn' are among the possibilities.

BLACKWELL SD3994 (hab.) Bowness.
From 1751 (*PNWestm* I, 186).
♦ Presumably 'the black spring'; **black, well**.

BLAKEBANK SD4591 (hab.) Underbarrow.
Blaykbank(e) 1615, *Bleakebancke* 1634, *Blake Bank* 1635 to 1865, *Blaikeban(c)ke* 1649, 1657, *Blackbank(e)* 1653 to 1722 (*PNWestm* I, 103).

♦ Probably 'the bleak or pale yellow slope' (**blake, bank**), though **blæc/black** cannot be ruled out as 1st el.

BLAKEBECK NY3627 (hab.) Mungrisdale, BLAKE HILLS FARM NY3628.
Blackbeck (hab.) 1774 (DHM); *Blaikhills in Grisedale* 1677, *Blackehills* 1688 (*PNCumb* 226), *Black hill* 1774 (DHM).
♦ In both cases a natural feature (a **beck** or stream and **hills**) has given its name to a farm, but in the light of the early ModE spellings in *Black*, it is not clear whether the specific is **blake** 'pale' or **black**, nor to which of the features it referred. If the stream, 'dark, black' would describe its smooth and clear-running waters.

BLAKE FELL NY1019 (1878ft/573m.) Lamplugh.
From 1774 (DHM), 1867 (OS).
♦ 'The pale hill'; **blake, fell**. 'Pale' seems the most likely meaning of 'Blake' here, since the screes on the W side give the fell a somewhat pale appearance; 'bleak' is another possibility.

BLAKE HOLME SD3889 Windermere, B~ H~ PLANTATION SD3989 Cartmel Fell.
Blakeholme 1638 (Cartmel PR), *Blake holm Isld* 1783 (CrosthwaiteM); *Blake Holme Plantation* 1851 (OS).
♦ B~ H~ is 'the pale-coloured islet'; **blake, holm(e)**. The **Plantation** is a woodland planting above the E shore opposite, and is named from the island or from nearby Blake Holme Nab.

BLAKELEY: B~ RAISE NY0613 Ennerdale, B~ MOSS NY0614.
(*by the hillsyde called*) *Blake lee unto Blake lee Raies* 1578 (CW 1966, 115); *Blake lee Raids* 1578 (*PNCumb* 386); *Blakeley Moss* 1867 (OS).

♦ B~ may mean 'pale clearing', with **blake** and the reflex of OE **lēah** 'woodland clearing', but the age and meaning of the name are elusive. B~ is a hill, which may formerly have been called Whorl 'rounded one' (*PNCumb* 388). B~ **Raise** 'cairn' (from ON **hreysi**) is the summit, while the **Moss** is the lower ground to the NW.

BLAKE RIGG NY2804 Langdales.
Blakerigg, (perhaps originally Blea Cragg) 1823 (Otley 29).
♦ Probably 'the pale or bleak ridge'; **blake, hryggr/rigg**. Although this is a craggy height above BLEA TARN, and BLEA RIGG gives evidence of confusion between **bleak** and **blea**, one would need evidence for Otley's conjecture that this was once Blea Crag.

BLAWITH SD2888 (hab. & parish), B~ FELLS SD2790, B~ KNOTT SD2688 (806ft/248m.) Kirkby Ireleth.
(*foresta de*) *Blawit* 1276, *Blawith* 1341, *Blathe* 1600 (*PNLancs* 214); *Blawith Fells* 1851 (OS), *Blow Knott or Blawith Knott* 1851 (OS).
♦ 'The dark wood', from ON **blár** 'blue, dark' and ON **viðr** 'wood, timber'; and indeed in medieval times this was **forest** combining a hunting preserve and woodland (*VCHLancs* 363, n. 2). The **Fells** are a stretch of upland grazing, named with reference to the parish of B~ rather than the hamlet. The summit of B~ **Knott** ('compact hill') is very close to the boundary of Kirkby Ireleth with Subberthwaite, which has recently been joined with Blawith and must have been considered part of Blawith at the time when the hill was named.

BLEABERRY FELL NY2819 Borrowdale/St John's.
From 1823 (GreenwoodM), 1867 (OS).

♦ 'The hill where bilberries grow'; dial. *bleaberry* plus **fell**. ON *bláber*, literally 'blue, dark berry' gives local *bleaberry* or *blaeberry* 'bleaberry', known elsewhere as *bilberry* or *blueberry*, i.e. *Vaccinium myrtillus*. This grows as low scrub to high altitudes (including the top of SCAFELL PIKE, Halliday 1997, 225), though berries rarely form due to cropping of young shoots by sheep.

BLEABERRY GILL NY1111 Ennerdale.
From 1867 (OS), seemingly = *Swallow Gill* 1827 (plan, CRO D/Lec/94).
♦ 'The ravine where bilberries grow'; *bleaberry* as above, plus **gil(l)**.

BLEABERRY HAWS SD2694 Torver.
From 1851 (OS).
♦ 'The hills where bilberries grow'; *bleaberry* as in BLEABERRY FELL plus *haws*, a local form of **how(e)** from **haugr**.

BLEABERRY TARN NY1615 Buttermere.
Blebba Tarn 1809 (CrosthwaiteM), *Burtness Tarn, or Bleaberry Tarn* 1823 (Otley 31).
♦ 'The mountain pool where bilberries grow'; *bleaberry* as in BLEABERRY FELL plus **tjǫrn/tarn**. The tarn lies in a corrie where the berries still abound; it is above BURTNESS, which explains the alternative name given in 1823.

BLEA CRAG NY2610 Borrowdale.
From 1867 (OS).
BLEA CRAG NY3007 Grasmere.
From c.1692 (Machell II, 128), 1859 (OSNB).
♦ 'The dark rocky height'; **blár/blea**, **crag**. The example at NY2610 is a compact but extremely prominent rock turret, dark against the hill-side.

BLEAK HOW NY2712 Borrowdale.
From 1867 (OS).
♦ Either 'the bleak, exposed hill' or 'the pale hill'. In the 1st el., ModE *bleak* seems likely, given the exposed and commanding position, but 'pale' (ON *bleikr* 'pale', see **blake**) is also possible, cf. Blake Bank, spelt *Bleak-* in 1634 (*PNWestm* I, 103). The rock is slightly yellow, and possibly paler in some lights than that of the surrounding crags. The 2nd el. is the reflex of ON **haugr** 'hill, mound'.

BLEA MOSS NY5212 Shap Rural.
From 1859 (*PNWestm* II, 173).
♦ 'The dark bog'; **blár/blea**, **moss**.

BLEANSLEY: LOWER B~ SD2089 (hab.) Broughton West.
Blengeslit 1292, *Bleansle* 1570 (*PNLancs* 222).
♦ Apparently 'Blæing's slope', from an ON pers.n. *Blæingr* and ON *hlíð* 'slope', possibly encouraged by OE *hlið*, **hlid* 'slope'; Ireleth as in KIRKBY I~ may provide a parallel. Middle and **Lower** B~ stand on the long slope above the R. LICKLE. Like nearby STENNERLEY, then, this looks like an Anglian name in **-lēah** 'woodland clearing' but is probably not.

BLEA RIGG NY2908 (1778ft/541m.) Langdales/Grasmere.
Blea Rigg or Bleak Rigg 1859 (OSNB).
♦ 'The dark ridge'; **blár/blea**, **hryggr/ rigg**. *Blea* could describe the ridge itself or derive from BLEA CRAG, at its SE end.

BLEA ROCK NY2611 Borrowdale.
Not 1867, 1900 (OS).
♦ 'The dark rock', with dial. **blea** from ON **blár** 'dark, blue', plus ModE *rock*. This is one of only three occurrences of *rock* (which is probably

from OFr *roke, roche*) in this volume, the others being CASTLE R~ and PILLAR R~; none is recorded before the 18th cent. B~ R~, though not massive, is extremely prominent, especially when seen from the N. It is also known as Gash Rock (Wainwright 1958, Sergeant's Crag 2, where B~ R~ is illustrated).

BLEASE: B~ FARM NY3125 Threlkeld, B~ FELL NY3026, B~ GILL NY3126.
Blease 1719 (Lorton PR), *Blaze* 1774 (DHM), *Blees* 1776 (PR); *Bleas Fell* 1789 (*PNCumb* 253); *Blease Gill* 1867 (OS).
♦ B~ is probably from ON *blesi* 'a blaze, white star' (as on a horse's forehead); **farm** has been added, and secondary names contain **fell** 'high ground', **gil(l)** 'ravine with stream'. 'Blease' may refer to pale markings on the fell above the farm (cf. a Blaze Fell, *PNCumb* 220, Blaze Hill, *PNWestm* II, 173, and Blesen, Norway, *NG* IV, 244). Much of it is grassy, but on the W side there is a streak of broken rock on the upper part and a scatter of flattish rocks on the lower slopes; otherwise the 'Blease' could conceivably be the 'strange and unusual canyon' high on B~ Gill (Wainwright 1962, Blencathra 12). Ekwall also mentions the sense 'opening between hills' in some Swed p.ns (*PNLancs* 165–6). B~ and B~ Fell form the farthest W of five pairs of settlement names and fell names which belong to the five buttresses of the majestic S front of BLENCATHRA: see GATEGILL FELL, HALL'S FELL, DODDICK FELL and SCALES FELL. *Fell* in these names may imply both a distinct height and a grazing area designated for the farms in question.

BLEA TARN Examples at:
    NY1601 Eskdale.
Bleaterne 1587 (*PNCumb* 32).

NY2904 Langdales, BLEATARN HOUSE NY2904.
Blatarne 1612 (*PNWestm* I, 15); *Bleatarn House* 1862 (OS); perhaps = *blay-terne* 1690 (as hab., Grasmere PR).
NY2914 Borrowdale, BLEATARN GILL NY2815.
Both from 1867 (OS).
♦ 'The dark mountain pool'; **blár/blea, tjǫrn/tarn**. Although none of these examples appears in medieval records, there was also a *Blaterne*, recorded 1217 in the former Westmor-land (cited *PNCumb* 32, which also gives Scandinavian parallels). Bleatarn **Gill** flows out of the Borrowdale tarn; and B~ **House** is close to the Langdale one.

BLEATHWAITE CRAG NY4409 Kentmere.
From 1857 (*PNWestm* I, 166).
♦ Seemingly 'the rocky height above Bleathwaite, the dark clearing'; **blár/blea, þveit/thwaite, crag**. It is also close to BLEA WATER.

BLEATHWAITE: B~ PASTURE SD2895 Coniston.
Bleathwaite Pasture 1851 (OS).
♦ Wyld cites a spelling *Blythwait* (p) c.1400 from the *Furness Coucher* (Wyld 1911, 72), but the identification with this place is far from sure. The modern form Bleathwaite looks likely to be 'the dark clearing' as in the Kentmere example; **blár/blea, þveit/thwaite**. The **pasture** is a tract of high grazing land.

BLEA WATER NY4410 Shap Rural.
From 1823 (*PNWestm* II, 15).
♦ 'The dark lake'; **blár/blea, water**. Among the highest and smallest of the lakes called *water*, this might equally have been called a **tarn** (see **tjǫrn**).

BLELHAM TARN NY3600 Hawkshead/Claife.
Blalam terne 1537 (*PNLancs* 192).

♦ B~ is probably 'the dark pool', with 1st el. from ON **blár** 'blue, black, dark' and 2nd el. possibly OE *lum(m)* 'pool' (so *PNLancs*), which has been reformed under the influence of settlement names in OE *-hām* 'homestead, settlement'. **Tarn** (from ON **tjǫrn**) would be a later addition.

BLENCATHRA: see SADDLEBACK (Pl. 16)

BLENCATHRA CENTRE NY3025 Threlkeld.
♦ Now an outdoor pursuits centre, this was formerly the Blencathra Hospital (1904–75), for the treatment of tuberculosis, then for geriatric care (*Threlkeld, Cumbria*, 1998, 2 and 29). It stands at nearly 1000ft/300m. on the flanks of BLENCATHRA (see SADDLEBACK).

BLENG, RIVER NY0805, NY1208 Ennerdale/ Nether Wasdale, BLENGDALE NY0805 Gosforth, BLENGDALE FOREST NY0906, BLENG FELL NY0705.
(River:) *Bleng* 1576, *Brenge* 1577, *Blaing*, *Bleng* 1774 (DHM), 1802 (*CN* 1205); cf. *Bleyng[f]it* 1391 (*PNCumb* 5); *Blengdale* 1655 (*PNCumb* 395); *Bleng Fell* 1867 (OS); cf. *Blengfellyeat* 1735, *Blengfell Gate* 1742 (Gosforth PR), *Blaing and Ponsonby Fells, Blaing Fellgate* 1774 (DHM).
♦ B~ is probably 'the dark one', from ON *blæingr*, which derives from the adj. **blár** 'dark, blue' and here characterises a river, though it seems to be a pers.n. in BLEANSLEY. Blean (Beck) (*Blayngbek* 1153, *PNNYorks* 263) seems to be of the same origin (*ERN* 37). The secondary names contain **dale** 'valley', **fell**, here a distinct hill, and **forest** in the modern sense of woodland.

BLENNERHAZEL NY0603 (hab.) Gosforth.
♦ 'According to Dr. Parker the house was built by the Coalbanks of Blennerhasset and therefore is modern' (Warriner 1926, 80). This would also suggest that the name was modelled on that of Blennerhasset, with the substitution of **hazel**. Blennerhasset, N of the National Park, is thought to be from a counterpart to W *blaen-dre* 'hill farm' with ON *heysætr* 'hay shieling' added (*PNCumb* 265–6).

BLINDBOTHEL (parish).
Blendebothel 1278, 1279, *Blyndebothill* (p), *Blyndbothel* (p) 1333 (*PNCumb* 345).
♦ The 1st el. is obscure since the early *-e-* spellings do not support the assumption of OE *blind* (which could have had the sense 'dark', 'hidden' or 'remote, cut off'). The 2nd el. appears to be OE *bōðl* 'house, dwelling', Although the name does not appear on the One Inch map, and the original settlement cannot be located, B~ remains as a parish name.

BLINDCRAKE NY1434 (hab. & parish) Blindcrake.
Blenecreyc later 12th cent., *Blencraic* & variants later 12th cent. to 1610, *Bleyncreyk* 1260 (*PNCumb* 266–7).
♦ 'The top of the rocky hill', from Cumbric **blain* corresponding to W *blaen* 'top, point' plus Cumbric **creig* corresponding to W *craig* 'rock, cliff'. The village is set well above the lower DERWENT valley.

BLIND TARN SD2696 Torver.
From 1745 (West 1774, map).
♦ 'The closed-off pool.' The **tarn** is in a deep hollow, virtually surrounded by steep scree slopes. It has no inlets or outlets, and that is usually presumed to be the sense of *blind* (from OE *blind*) here.

BLINDTARN GILL NY3208 Grasmere.
Blintarn Gill 1572 (*CW* 1928, 276).

♦ 'The stream at Blind Tarn, the filled in pool'; *blind*, **tjǫrn/tarn, gil(l)**. The *gil(l)* or stream runs out of Blindtarn Moss, which seems to represent a tarn now filled with vegetation. *Blind* (from OE *blind*) here may therefore mean 'filled in' or 'hidden' (see *PNWestm* I, 199 for references).

BLOWICK/BLEAWICK NY3917 (hab.) Patterdale.
Blawyk 1256 (p), *Blowyk, Blewyk* 1292; pronounced [bli:wik] (*PNWestm* II, 221).
♦ 'The dark bay', from ON **blár** 'dark, blue' and ON **vík** 'bay, inlet'. The settlement takes its name from what is now called Blowick Bay. The spelling and alternative pronunciation with *Blo-* rather than *Blea-* reflect the influence of standard English, as Fellows-Jensen points out (*SSNNW* 107).

BOADHOLE SD1987 Millom Without.
Boad holle 1626, *Boad-whole* 1680 (*PNCumb* 418), *Boadhole alias Littlehousefeild* 1697 (*JEPNS* 2, 59).
♦ Possibly 'the hollow with a dwelling', cf. BAWD HALL.

BOADLE GROUND SD0898 Drigg.
From 1821 (PR).
♦ 'The farm of the Boadle/Boodle family'; surname plus **ground**. The family name appears as *Boodle* 1647, *Bodle* 1735 etc. in Drigg PR, and is even more common in the PRs of nearby Muncaster.

BOAT HOW NY0810 Ennerdale.
From 1867 (OS), probably = *Balthow* c.1612 (*CW* 1931, 164).
BOAT HOW NY1911 Ennerdale.
Bawthow, Bawethow (= this or one above) 1578 (*PNCumb* 386), *Boat How* 1826 (CRO D/Lec/94), *Boat How* 1867 (OS).

BOATHOW CRAG NY1013 Ennerdale.
Boathow Crag 1867 (OS), probably = *Barter Crag* 1802 (*CN* 1208).
♦ It is striking that 'Boat' co-occurs with **how(e)** (from ON **haugr**) four times in the names on the One Inch map (and in at least one other Boat How at NY4711 on larger scale maps, and cf. Boat How Intake, Grasmere, Simpson 1928, 282, and BALD HOWE), but rarely with any other element (there is a Boat Crag at 1621). In every case the *how(e)* is a minor bump on a broad, steep hill-side, and in two instances (at NY0810 Ennerdale and NY1703 Eskdale, below) there are very significant archaeological remains. Nevertheless, the Boat Hows may or may not be of the same origin, and they present an intriguing puzzle. In BOAT HOW, Eskdale, 'boat' seems to be a corruption of ON **búð, bóþ** 'a hut or shed', and corruption of this or (more likely) some other el. may have taken place in the two Ennerdale examples. This el. may have referred to some man-made structure, and Boat How at NY0810, Ennerdale, lies just above Townbank cairnfield and settlements (LDHER 9342, SAM No. 27825–8; see also Spence 1938, 64 and 1939, 31–2). There is no known archaeology at Boat How NY1911. In Ennerdale (NY1013), Boat How is a miniature oval eminence at 1191ft/363m. on THE SIDE, and the line of the **Crag** 'rocky height' juts above it. Here, 'boat' could be taken at face value, describing the topography, but without early spellings it is unclear whether it is a corruption of some other el. Coleridge's 1802 spelling could be a transcription of a local pronunciation of 'Boathow Crag'. Simpson believed the Grasmere example to be *bought*, a sheep fold (1928, 282) as possibly also in BESSYBOOT.

BOAT HOW NY1703 Eskdale.
the Low Bothhow 1587 (*PNCumb* 391).
♦ Probably 'the hill with the hut', from ON **búð/bóþ** 'booth, hut' and ON **haugr** 'hill' or their reflexes. There are prehistoric remains here — an enclosure containing three hut circles and eight clearance cairns, together with an adjacent hut circle and cairnfield (LDHER 6327, SAM No. 23697), as there are at BOAT HOW, Ennerdale (NY0810), and it is conceivable that the name was motivated by them. The neat oval contours of this minor height, which make it a classic *how(e)*, may subsequently have encouraged the corruption to 'Boat'.

BOG HOUSE NY2324 Above Derwent.
Bogg in Braithwaite 1799, *The Bog* 1845, *Bog house* 1857 (Thornthwaite PR).
♦ This may previously have been *Nordmanthait* 1210–16 (Collingwood 1921, 168; see ORMATHWAITE). 'The **house** by the bog' — the flat, wet alluvial land between DERWENTWATER and BASSENTHWAITE LAKE. This, with WHITE BOG, is the only instance in this volume of ME, ModE *bog*, since **moss** and **mire** are the more common local terms.

BOGLE CRAG SD3393 Satterthwaite.
Bogle Crag 1851 (OS), *Bogley Crag* (Cowper 1899, 326).
♦ 'The **crag** or rocky height haunted by goblins.' A *bogle* (also *boggle, boggart* in Cumbrian dial., Rollinson 1997) is a folklore goblin or phantom, and Cowper regarded the place as 'thoroughly haunted, although we have never heard of any distinctly-recorded apparition' (1899, 326–7, cf. Cowper 1897a, 381–9). The place is currently under the calming influence of the Forestry Commission.

BOLTON: B~ HALL NY0802, HALL B~ NY0803, B~ WOOD NY1004, Gosforth.
Boutonam c.1170, *Bouelton* c.1230, 1279, *Bothelton(a)* 1251 to c.1300, *Bolton* 1294; *Bolton Hall* 1497 (*PNCumb* 394); *Halboulton* 1637 (PR); (*the*) *Bolton wood* 1802 (*CN* 1205–6).
♦ B~ meant 'the settlement with the building', from OE *bōðl* 'building' and OE **tūn** 'settlement, farmstead, village', which co-occur so frequently in N. English p.ns that they may be seen as constituting a compound el. Ekwall suggested that this was used of 'the village proper in contradistinction to the outlying parts' (*DEPN*: Bolton), but this is not certain (see *VEPN*: **bōðl-tūn*). **Hall** Bolton is the old manor house (Parker 1926, 49), and distinct from nearby Bolton Hall, but it was also a township (Denton 1687–8, 84); see **hall** for comment on this and similar names, and see **wood**.

BOMBY NY5217 Bampton.
Bondby 1292 to 1489, *Bonby(e)* 1540, 1626, *Bomby(e)* 1648 to present (*PNWestm* II, 189).
♦ 'The settlement of the freeholder(s) or customary tenant(s)', from ON *bóndi* 'freeholding farmer or peasant' and ON **bœr/bý**, or their ME reflexes. Insley points out that *bóndi* was adopted into the ME of N. England and S. Scotland in the sense of 'customary tenant' and argues that this and similar names may be ME coinages (1986, 171).

BONFIRE HALL SD4793 Underbarrow.
Baynfierhowe 1641, *Bonefire Hall* 1738 (*PNWestm* I, 103), *Bonefire How* 1770 (JefferysM).
♦ Originally 'bonfire hill', with ModE *bonfire* plus **how(e)** from ON **haugr** 'hill'. The building below was then named B~ **Hall**. The hill, though not

the highest in the neighbourhood, is prominent and might possibly have been used for beacons. The 1641 spelling, which refers to a tenement, hints at the origins of the word *bonfire* — OE *bānfȳr* 'fire of bones' (human or animal), here with ON *bein* 'bone' as 1st el. For a glimpse of the national events which occasioned the lighting of bonfires in the late 17th cent., see Fleming *Accounts* 346 (index).

BONNING GATE SD4895 (hab.) Strickland Ketel.
Bonne Yat 1597, *Bonning(e)gate, -yeat(s)* 1599, 1615, 1619, *Bony yate* 1650, *Bunnion yeat* 1836 (*PNWestm* I, 155).
♦ The 1st el. may, as suggested by Smith, be a surname, though the spellings do not clearly indicate which — possibly the Lancs name *Bonney*, or *Bunyan* (*PNWestm*). In the 2nd el., the Y- spellings probably indicate **gate** from OE **geat**.

BONSCALE NY4420 (hab.) Martindale, B~ PIKE NY4520.
Bonscal(l) 1588–1823 (*PNWestm* II, 218), *Bandscale* 1770 (JefferysM); *Bonscale Pike* 1899 (OS).
♦ The 2nd el. is presumably from ON *skáli* 'shieling', but the 1st is elusive — possibly, as suggested in *PNWestm*, ON *bóndi* 'freeholding farmer' (or its ME reflex meaning 'customary tenant', cf. BOMBY). This would be supported by the relative frequency with which **skáli** is qualified by references to persons, whether by name (JOHNSCALES, KESKADALE) or rank (PRIORSCALES). The farm is at the foot of the precipitous NW slopes of B~ **Pike**, which is a summit on SWARTH FELL, and 'sometimes referred to as Swarth Fell; named Toughmoss Pike on Bartholomew's map' (Wainwright 1957, Bonscale Pike 1).

BOONWOOD NY0604 (hab.) Gosforth.
Abovewood(e) 1597, 1598, *Bounwood* 1692 (PR), *Bone Wood* 1774 (DHM), *Bonewood* 1802 (*CN* 1206).
♦ Almost certainly '(the place) above the wood'. (a) 'Boon-' is probably a form of *above* (OE *bufan*), as suggested by Warriner (1926, 81) and supported by *Above Wood* applying to the same place in the PRs. (Cf. ABOVE BECK, also Boon Town, *PNCumb* 446, and *VEPN*: *bufan*.) (b) This makes it unlikely that the 1st el. is the word *boon*, meaning a customary payment or service made by tenants (so Sedgefield 1915 and Brearley 1974, as an alternative), unless *Abovewood* is a secondary development. Low, Middle and High B~ are distinguished on larger scale maps.

BOOT NY1701 (hab.) Eskdale.
Bout, the Bought 1587 (*PNCumb* 389), *Bought* 1632 (PR).
♦ (a) Probably 'the bend', from ME *bouȝt*, which, as pointed out in *PNCumb* 389, aptly describes this angle in the valley where the WHELLAN BECK meets the ESK. *PNCumb* also compares *close called boughte de bekk* (bend in the river) in Nether Wasdale, 1578. (b) Less persuasively, dial. *bucht, bought* 'a sheepfold, shed' has also been suggested (e.g. Fair 1922, 76, adding 'the Parish pinfold is here'; Warriner 1926, 81).

BOOTH HOLME SD1991 (hab.) Ulpha.
From 1867 (OS).
♦ Possibly 'water-meadow with hut', with elements derived from ON **búð/ bōþ** and ON **holmr**. The site is in a curve of the R. DUDDON.

BOOTHWAITE NOOK SD2186 (hab.) Broughton West.
From 1663 (original spelling not

specified, *CW* 1932, 65), 1666 (spelling not specified, Broughton-in-Furness PR).
♦ B~, despite the lack of medieval forms, seems likely to be 'the clearing with the hut', from ON **búð/bóþ** 'hut, booth' and ON **þveit** 'clearing' or their reflexes. **Nook** will have been added later.

BOOTLE SD1088 (hab. & parish), B~ FELL SD1388, B~ STATION SD0989.
Bodele 1086, *Botle* c.1135 (copy) to 1321, *Botel(l)* c.1170 to 1440, *Botyl, Botil(l)* c.1200 to 1513 (*PNCumb* 345); *Bootle Fell*; *Bootle Station* both 1867 (OS).
♦ 'The building or dwelling', from OE *bōðl, bōtl* 'building'. This place in the S of the coastal plain is, with NEWTON and WHICHAM, one of only three in this volume whose names are recorded in Domesday Book, 1086 (though see KENDAL). The **fell** is a stretch of upland grazing belonging to the parish, rather than a prominent summit. The word *station* was first adopted from Lat, possibly via Fr, in the 15th cent.; *OED*'s first citation for (*railway*) *station* is from 1830.

BORDER END NY2201 Eskdale.
From 1867 (OS).
♦ Probably 'the end hill near the boundary'; *border*, **end**. *Border* (adopted from OFr *bordure* into ME) may here refer to the position close to a boundary, nowadays that between Eskdale and Ulpha. The sense of **end** in the name of this rocky height is slightly elusive but the evidence of GREAT END, and the situation, may point to 'hill at the end or edge of a range of hills'.

BORDERSIDE SD4190 (hab.) Crosthwaite. From 1700 (*PNWestm* I, 83).
♦ 'The place beside the boundary'; *border* as in BORDER END, **side**. This is very close to the boundary with Cartmel Fell, and is one of a local cluster of names in **-side** (see List of Common Elements).

BORDRIGGS/BORDERIGGS SD2287 (hab.) Broughton West.
Borderigges 1330 (p), *Bordriggs* 1587 (*PNLancs* 222), *Boardridge* 1786 (YatesM).
♦ B~ may mean 'the flat-topped ridges'. The place is, as Ekwall noted in *PNLancs*, between two ridges (**rigg**, see **hryggr**), whose fairly level tops would suggest OE *bord* 'board, plank, shield' as the 1st el. (or its cognate ON *borð*), though I do not know of evidence for these words in topographical use. Other possibilities such as 'boundary' (cf. BORDER END), or the OE pers.n. *Brorda* (Mills 1976) are less convincing here.

BOREDALE/BOARDALE NY4117 Martindale, B~ BECK NY4217, B~ HAUSE NY4015 Patterdale, B~ HEAD NY4117 (hab.) Martindale.
Burdale 1250 (p) to 1337, *Buredale* 1256 to c.1290 (p), *Bourdal(e), -dall* 1291 to 1377 (*PNWestm* II, 216); *B(o)urdalbek* 1337; *Bowerdalehead* 1588 (*PNWestm* II, 218).
♦ 'The valley with the storehouse', from ON *búr* or possibly OE *būr* 'dwelling' and ON **dalr** 'valley'; also **bekkr/beck** 'stream', **hals/hause** 'neck, col'. Wordsworth's assertion that 'the name of Boardale . . . shews that the wild swine were once numerous in that nook' (*Prose* II, 285) is therefore incorrect. The Hause is a junction of five tracks; the name B~ Hause, though not shown on earlier maps, is implied by names such as *Chapel in the Hause* and *Hause Crag* (e.g. 1920 OS).

BORETREE TARN SD3587 Colton.
Bortree Tarn 1851 (OS), possibly = *Dulas* in medieval times (Cowper 1899, 90).
♦ 'The **tarn** or small lake where elder grows.' *Bor(e)tree*, or more commonly *b(o)urtree*, is first recorded from the 15th cent. (*MED*, *OED*). It is a dial. word for the elder, *Sambucus nigra*, which is common in Lakeland excluding the high fells (Halliday 1997, 406).

BORRANS: HIGH B~ NY4300 (hab.), B~ RESERVOIR NY4201 Windermere.
the Borrance 1597, *Borwaines* 1653 (*CW* 1906, 73), *Borrans*, *Borrans-Ring* c.1692 (Machell II, 327, ed. Ewbank, 130), *Borrans* (= High Borrans) 1862 (OS); cf. *Low Burrans* 1823 (*PNWestm* I, 195).
♦ **Borrans** is 'the cairns', this being the site of Iron Age and Romano-British settlements (LDHER 1907, SAM No. 14105); see also **high**, **reservoir**.

BORROW BECK NY5205 Fawcett Forest/Shap Rural, BORROWDALE HEAD NY5304 (hab.) Fawcett Forest, BORROWDALE MOSS NY5006, HIGH BORROW/BARROW BRIDGE NY5404 (hab.).
Borra watter, *torrentis de Borra* 1170–84, *Borghra* 12th cent., *Burgra* 1195, *Boroudalebek* 1333, *Borrow-Beck* 1836 (*PNWestm* I, 4); (B~ Head:) *caput de Borgherdala* 1175, *Boroudale Hedde* 1539, cf. *Borgher(e)dal* 1154–89 (*PNWestm* I, 138–9); *Borrowdale Moss* 1836 (*PNWestm* I, 139); *High Borrow Bridge* 1712; cf. *Borrowbridge* 1651 (both *PNWestm* I, 139).
♦ **Borrow Beck** was once *Borghra*, *Burgra*, normalised *Borgará*, 'the river by the fortification', from *borgar*, gen. sing. of ON *borg* 'fort', plus ON *á* 'river'; there was a Roman fort at Low Borrow Bridge (*PNWestm* I, 138). The Borrowdale valley is named either from the river *Borgará* (or gen. sing. *Borgarár*) plus **dalr** 'valley' or, less likely, directly from *borgar* 'fortress' plus **dalr**. The river was renamed Borrowdale **Beck**, perhaps because *á* 'river' was no longer understood, and subsequently shortened to Borrow Beck. B~ **Moss** is the damp area where the Borrow Beck rises, and **High** Borrow **Bridge** crosses the beck in the upper part of the dale, giving its name to a habitation. 'Barrow' on the One Inch map seems to be in error (and I am grateful to Mr Brian Nevinson of Borrowdale Head Farm for confirming this).

BORROWDALE NY2515 (parish), B~ FELLS NY2313, NY2813, B~ GATES NY2517 (hab.), B~ YEWS NY2312.
Borgordale c.1170, *Borudale* 1209 to 1337, *Borcheredale* c.1209 to 1292 (*PNCumb* 349); *Borrowdale fells* 1687–8 (Denton 51), *Borradale fells in Cumberland* c.1692 (Machell II, 113), *Borrowdale Fells* 1867 (OS); *Borrowdale Gates* 1867 (OS); *Borrowdale Yews* 1866 (Hodgson Plans 23), 1867 (OS).
♦ B~ is 'the valley of the Borgará, the river by the fortification', or 'the valley with the fortification', cf. BORROWDALE above; also **fell**, **gate** from **geat**, **yew**. The river-name *Borgará*, in the spelling *Borghra*, is recorded in the 13th cent. as a name for the upper Derwent (see ERN 40–1), and the *borg* is the Iron Age hillfort on CASTLE CRAG. B~ Fells is the stretch of highland grazing within B~: 'This name applies to the whole of Borrowdale Township, with the exception of the old enclosed land' (OSNB Crosthwaite, 1862, 2, 179). B~ Gates is a large 19th-cent. house, with fine views of the craggy 'jaws of B~'. B~ Yews, 'Those fraternal Four of Borrowdale, | Joined in one solemn and

capacious grove' (Wordsworth, 'Yew-Trees', *Poems* 146) are now a group of three — one a venerable ruin — with a fourth below them at the beckside. The name Borrowdale also applied, on the evidence of a FURNESS Abbey document of 1398–9, to upper DUNNERDALE, since it names places in 'Borrowdale' that are to be identified with BIRKS (SD2399), BLACK HALL and GAITSCALE (see Collingwood 1918, 96–7).

BORWICK FOLD SD4497 Nether Staveley. From 1836 (*PNWestm* I, 173).
♦ 'The pen or farm of the Borwick family' (surname plus **fold**), who are recorded locally as *Borwicke, Borrick* from the late 16th cent. onwards (*PNWestm*, cf. AYNSOME).

BOUCH HOUSE NY1429 Embleton.
Boutehouse 1701, *Bou(t)ch House* 1707–8 (PR).
♦ Possibly, as suggested in *PNCumb* 384, from a local family name, Percival Bowch being recorded in 1570 and another Bouch in the PR for 1816. If the surname *Bouch* is identical with *Bueth* and variants in medieval documents (as suggested by Bouch, 1936, 24–33), this might help to explain the 1701 and 1707–8 spellings. See also **house**.

BOUTH SD3285 (hab.) Colton, BOUTH FALL STILE SD3385 (hab.).
Bouthe 1336 to 1577, *Bowth* 1577 (*PNLancs* 216), *Bow(e)th* 1623 (PR); *Booth Fall Stile* 1751, *Bouth Fall Stile* 1756 (PR).
♦ B~ is 'the shelter' or 'shieling', from ON **búð/bóþ**, probably in its specifically OWN form *búð* (*SSNNW* 62). Now a reasonable-sized village, this may have originated as a dairy farm belonging to COLTON (so Ekwall, *PNLancs* 216). The 'Fall' in Bouth Fall Stile is 'clearing, felled woodland' (from OE *(ge)fall* 'fall, felling, clearing of timber'), and traditional coppicing is still practised here; cf. FAWE PARK. **Stile/steel** is a steep ascent, presumably referring to the gradient when approached from the SE rather than a wooden stile. The date on the farmhouse, 1751, matches that of the name's first appearance in the PR. (Pl. 30a)

BOWDERDALE NY1607 (hab.) Nether Wasdale.
Boutherdal(beck), Beutherdal(bek) 1322, *Bouthdale* 1338, *Bowderdale (Close)* 1540 (*PNCumb* 440).
♦ 'The valley with a hut'; OWN **búð** plus **dalr**. The *-er-* spellings, reflecting the ON gen. sing. *-ar*, place this name among those definitely from the Scand-speaking period.

BOWDER STONE NY2516 Borrowdale.
bowders stone 1751 (*PNCumb* 353), *Bowthor Stone* 1774 (DHM), *Bowder-Stone* 1776 (Hutchinson 175), *Bootherstone* 1786 (Gilpin I, 194), *Bowder-stone, Powder-Stone or Bounder-Stone* 1789 (*PNCumb* 353), *Bowdar stone, Bowder Stone* 1794 (Hutchinson II, 210–1).
♦ 'The boulder rock', with a dial. version of *boulder* plus **stone**. This is an erratic rock, spectacularly perched and much visited since the 18th cent., 'much the largest stone in England, being at least equal in size to a first rate man of war' (Hutchinson 1794, II, 218, quoting *Gentleman's Magazine* 1751).

BOWERBANK NY4724 (hab.) Barton.
Bovrbank c.1290 (p), *Bourebank(e)* 1292 (p) to 1699, *Bowerebank* 1292 (p) (*PNWestm* II, 209–10).
♦ 'The hill-slope with a cottage or storehouse', from OE *būr* 'cottage' or ON *búr* 'storehouse' plus ODan **banke**

'hill-slope', or more likely their ME reflexes.

BOWES LODGE SD4283 Witherslack.
From 1857 (*PNWestm* I, 78).
♦ 'The **lodge** of the Bowes family', who appear in local records from the 17th cent. (*PNWestm*).

BOW FELL NY2406 (2960ft/902m.) Eskdale/Langdales, BOWFELL LINKS NY2406.
Bowesfel 1242 (*Furness Coucher* II, 562), *Bowfell* 1692 (Machell II, 153), *Bow Fell* 1774 (DHM), 1862 (OS).
♦ Either 'the bowed mountain' or 'Bowe's mountain'. (a) The 1st el. may be 'bow' (ON *bogi* or OE *boga*), either referring to the massive shoulder of THE BAND (Collingwood 1918, 95–6), or else to the hill as a whole (Ekwall, *DEPN*, and cf. BOWSCALE FELL). (b) An alternative possibility is 'Bowe's mountain', since Bowe is recorded as a pers.n., *Robertus filius Bowe* (Robert son of Bowe) in a Cumberland deed of 1333 (Brearley 1974, 17). Nearby BUSCOE, probably *Bowesscard*, *Bouscard* in 1242, seems to share the same enigmatic 1st el., and the fact that both 1242 forms contain a genitive -*s* might favour the assumption of a pers.n. **Fell** here refers to a distinct and majestic mountain, not a tract of grazing. 'Links' is presumably the word derived from OE *hlinc* 'ridge, bank'. It seems a relatively recent innovation here, not appearing on the OS maps of 1863, 1899 or 1920.

BOWKERSTEAD: B~ FARM SD3391 Satterthwaite, LOW B~ SD3391.
From later 17th cent. (*CW* 1981, 103, spellings not specified); *High Bowkerstead* (= present B~ Farm); *Low Bowkerstead* both 1851 (OS).
♦ B~ is 'the Bowkers' place or farm', from **stead** qualified by the occupational surname *Bowker*, which referred originally to those involved in bleaching cloth, or possibly to butchers. It is characteristic of Lancashire, especially the Manchester area (McKinley 1981, 251–3). See also **farm**, **low**.

BOWLAND BRIDGE SD4189 (hab.) Crosthwaite.
From 1632 (*PNWestm* I, 83).
♦ Bowland is a WYorks p.n. with elements from OE *boga* 'bow, bend, curve' and OE *land* 'land' (*PNWYorks* VI, 112). In the present name it could either be a p.n. of similar origin or a surname derived from the p.n. The **bridge** (see **brycg**) and the settlement named from it straddle the R. WINSTER.

BOWMANSTEAD SD3096 Coniston.
Bowmansteed 1620 (PR).
♦ Presumably 'the Bowmans' place or farm', from the occupational surname *Bowman* (though it does not appear in the index to Coniston PR) plus **stead**.

BOWNESS NY1015 (hab.) Ennerdale, B~ KNOTT NY1115.
Bownus 1670 (PR); *Bowness Knot* 1802 (*CN* 1207).
BOWNESS FARM NY2229 Bassenthwaite.
Bawnas 1591 (PR), *Boneas* 1669/70 (PR), *Bowness* 1774 (DHM), 1780 (West 120).
♦ (a) B~ is probably 'the curved headland' in both these cases, from ON *bogi*/OE *boga* 'bow' and ON **nes**/OE **næss** 'headland' — an apt description of these promontories jutting into ENNERDALE WATER and BASSENTHWAITE LAKE. (b) However, given the early evidence for BOWNESS-ON-WINDERMERE, 'bull' cannot be ruled out as the 1st el. B~ **Knott** is a small craggy height above the

Ennerdale B~; **farm** added to the Bassenthwaite B~ identifies it as a habitation.

BOWNESS-ON-WINDERMERE SD4096 (hab. & parish).
Bulebas [*sic*] 1190–1210, *Bulnes* 1282 (p) to 1454, *Bowlnes* 1566, *Bownas* 1675 (*PNWestm* I, 185–6), *Bowness on Windermere* 1899 (OS).
♦ 'The headland where the bull grazes', from OE *bula* 'bull' and OE **næss** 'headland', perhaps referring to the keeping of a parish bull (cf. BULL CLOSE). The affix 'on-Windermere' is relatively recent, distinguishing the place from its Cumbrian namesakes, perhaps especially B~ on Solway.

BOWSCALE NY3531 Mungrisdale, B~ FELL NY3330 (2306ft/702m.), B~ TARN NY3331.
le Bouschale, le Bouscale 1361, *Bowscalez* 1485 (*PNCumb* 181–2); *Bowskale-Fells* 1704 (*CW* 1902, 199), *Bowscale Fell* 1839 (Ford 82); *Bowscale Tarn* 1774 (DHM), *Booth-scale-tarn* (Clarke 1789, ***ix [*sic*]).
♦ 'The shieling on the curving hill', from ON *bogi* (or possibly OE *boga*) 'bow, curve' and ON **skáli** 'shieling '; also **fell, tjǫrn/tarn**. B~ lies below the sweeping curves of B~ Fell, which may in fact have been named Boga or Bogi (cf. *DEPN*). Less likely is that the 1st el. is the rather uncommon ON pers.n. *Bogi* (Lind 1905–15, 151). This is a classic example of a once-temporary shieling site which grew into a hamlet. The tarn nestles in a deep bowl high on the fell.

BOXTREE/BOXTREES SD4495 (hab.) Crook.
Boxtree 1865 (*PNWestm* I, 179).
♦ The box (*Buxus sempervirens*) is currently very rare in Lakeland (Halliday 1997, 317), and one or more specimens would be distinctive. Both *box* and *tree* derive from OE (*box, trēo(w)*).

BRACELET: B~ HALL SD2391 Broughton West.
Bracelet 1614, *Breuslot* 1660, *Braslet* 1663 (*PNLancs* 222); (Bracelet Hall:) *Bracelet* 1851, 1893 (OS).
♦ 'The broad plateau.' Although the spellings of B~ are not maximally helpful, they are compatible with origins in ON *breiðr* 'broad' and ON *slétta* 'level ground' (so *PNLancs*), assuming influence from the word *bracelet*, which was adopted into ME from OFr. Bracelet Moor, though not flat, is a long stretch of high ground which is exceptionally level for this area. See also **hall**, seemingly added in relatively recent times.

BRACK BARROW/BROCK BARROW SD2294 Dunnerdale.
Brock Barrow 1850, 1891 (OS).
♦ This may well be 'the hill frequented by badgers'; **brock, barrow** from **berg**. *Brock B~* also appears on the current 1:25,000 and the 1966 One Inch maps, and may be the more authentic form; there is also a Brock Barrow at SD2989. If *brack* is correct, it could mean 'thicket, patch of brushwood' from OE *bracu* (see *VEPN*).

BRACKENBARROW FARM SD2994 Torver.
Bracanbergh pre-1220 (*CW* 1929, 41), *brakenbarrowe* 1585 (*CW* 1958, 65), *Brackenbarrow* 1599 (PR).
♦ B~ is 'the hill where bracken grows' (**bra(c)ken, barrow** from **berg**). It gives its name to the farm below, now distinguished from the hill by the el. **farm**.

BRACKENCLOSE NY1807 Nether Wasdale.
Brackenclose (Wood) 1867, 1900 (OS).

♦ 'The enclosure where bracken grows'; **bra(c)ken**, **close**; cf. 'John Pattison's Brecken Close' at Watermillock, mentioned in 1627 (*CW* 1884, 32).

BRACKEN RIGGS (hill-side) NY2920 St John's, BRACKENRIGG (hab.) NY2921. (Hill:) *Bracken Riggs* 1867 (OS); (hab.:) previously *Scott How* 1867, 1900 (OS).
♦ 'The ridge(s) where bracken grows'; **bra(c)ken**, **rigg**. Of the five Brackenriggs covered in *PNCumb* (pp. 126, 153, 257, 278, and 290), all but one first appear in the late 16th cent. or early 17th cent., suggesting either an onomastic fashion, or that places so named are in the band of only moderately important settlements which are often first recorded in 16th-cent. documents.

BRACKENTHWAITE NY1522 (hab.) Buttermere, B~ FELL NY1721.
Brakenthwayt 1230 to 1541 (*PNCumb* 354); *Brackenthwaite Fell* 1867 (OS).
♦ 'The clearing where bracken grows'; **bra(c)ken**, **þveit**. B~ **Fell** is a tract of upland grazing quite far from B~ itself; the name preserves the memory of B~ as an independent township, before it was incorporated into Buttermere in 1934.

BRADLEYFIELD SD4991 (parish of Underbarrow and B~), B~ HOUSE SD4992 Underbarrow.
Bradeleyfe(i)ld & variants 1525, cf. *Brathelaf* 1272, *Bradeley*, *Bradelay* 1292 (*PNWestm* I, 101, q.v. for further spellings); (B~ House:) *Bradleyfield* 1863 (OS), *Bradleyfield House* 1899 (OS).
♦ 'Bradley' is 'the broad clearing', from OE **brād** 'broad' (with influence from ON **breiðr** on some spellings) and OE **lēah** 'woodland clearing'. **Field** has been added later, and **house** later still. The medieval evidence for a p.n. 'Bradley' casts doubt on the claim that Bradleyfield 'derived its name from its ancient possessors, the Bradleys, who came from Lancashire' (Parson & White 1829, 646). Bradleyfield on the 1863 map is the name of the present B~ House, while the tract of land called Bradleyfield on the 1994 One Inch map was B~ Allotments in 1863 (and still on 1899 and 1920 maps).

BRAE FELL NY2835 (1920ft/585m.) Caldbeck.
Braythefel 1242 (*PNCumb* 276).
♦ 'The broad hill' (**breiðr**, **fell**) — an apt description.

BRAESTEADS NY3715 Patterdale.
Braysteads 1839, *Braesteads* 1860 (*PNWestm* II, 224).
♦ Probably 'the place or farmstead on the brow'; it lies below the steep slope on the N of GRISEDALE. Smith suggested the reflexes of ON *brá* 'brow (of a hill)' and OE **stede** 'place, site' (*PNWestm* II, 224 and cf. II, 18), but it is uncertain whether ON *brá*, which normally means 'eyelash', can refer to a hill brow, and B~ is more likely to contain **brow**, in a spelling influenced by *brae*, the Anglicisation of Gaelic *bràighe* 'upland, upper part' (cf. GILLBREA and *VEPN*: *bro* ME 'steep slope').

BRAITHWAITE NY2323 (hab.) Above Derwent, LITTLE B~ NY2323 (hab.).
Braithait c.1160, *Braythwayt(e)* 1230, *Braythethwayt* 1286; *Great*, *Mikkel*, *Lytyll Braytwha(i)t* 1560s (*PNCumb* 369).
♦ 'The broad clearing'; **breiðr**, **þveit**; also **little**. The name is apt for this, a thriving settlement in medieval and modern times, with good grazing and

communications, and a situation clear of the wet land between DERWENTWATER and BASSENTHWAITE. (Pl. 9)

BRAM CRAGG/BRAMCRAG (hab.) NY3121 St John's.
Bryrincrag in Fornesyd 1595, *Bryamcrage* 1614, *Brian Cragge* 1616, *Bramecragg* 1657 (*PNCumb* 316).
♦ Perhaps 'the rocky height below which broom or bramble grows'; **broom**, **crag**. The range of spellings for the 1st el. is difficult to account for, but this may be (a) the reflex of OE *brōm* 'broom' (with spellings some of which resemble NYorks *breeam*, cited for *broom* in *EDD*); or possibly (b) ME *brame* 'bramble, blackberry', cf. BRANTHWAITE. Broom grows in the neighbourhood today (Halliday 1997, 305). (c) A further possibility is ME *breme* 'rough, rugged' (*OED*: *breme*, sense 7). The settlement takes its name from the beetling crags which clad the hill-side above.

BRAMLEY NY1023 (hab.) Blindbothel. From 1545 (CRO D/Law/1/239), 1739 (Loweswater PR).
♦ A puzzling name, since it looks like a classic OE p.n., perhaps from OE **brōm** 'broom' or OE *brǣmel* 'bramble' and OE **lēah** 'woodland clearing', yet there are no early spellings, and the situation does not suggest a prime site settled in Anglian times. This may be a transferred p.n., e.g. from one of the Yorkshire Bramleys, or else a surname derived from it.

BRANDELHOW: B~ PARK NY2520 Above Derwent, OLD B~ NY2420.
Brandelaw, *Brandelow* 1569, 1573; *Branley Park* 1787 (*PNCumb* 372), *Brandelow (park)* 1794 (Hutchinson II, 173); *Old Brandelhow* 1867 (OS).
♦ Probably 'Brandulf's hill'. (a) The 2nd el. is probably **how**(e) from ON **haugr**, though the absence of -*h*- in the 16th-cent. spellings is disconcerting. In this case the 1st el. could be the pers.n. *Brandulf* (so Brearley), cf. Seat Sandal from *Sandulf*. (b) The 2nd el. could possibly be a different 'hill' word, *low* from OE *hlāw*, but since THE LOW in Ulpha is the only probable instance of it in this volume, this seems unlikely. In this case the 1st el. could suggest clearing by burning (cf. OE *brand* 'fire, burning'). See also **old**.

BRANDLINGILL NY1226 (hab.) Blindbothel.
Branling Gill 1774 (DHM).
♦ Perhaps 'the salmon stream'. The 1st el. may be *brandling*, referring to young salmon, a very common species in Lakeland. **Gil(l)** is a ravine with a stream, here especially the stream.

BRANDRETH NY2111 (2344ft./715m.) Buttermere/Borrowdale/Ennerdale.
'*the Three-footed Brandreth where Lord Lowther and Lord Egremont and Sir Wilfred Lawson meets*' 1805 (CW 1981, 67), *Brandreth* 1867 (OS).
♦ 'The hill with a beacon or shaped like a gridiron.' *Brandreth* is a local word from ON *brand-reið* 'grate', literally 'flame-holder', referring to a three-legged gridiron, but the motivation for the name is less certain. The hill itself is quite triangular, with ridges running NW, NE and S. On the other hand, a beacon is likely, since B~ is easily visible from far and wide, and is the point where three manors joined (as seen in the 1805 quotation) and three civil parishes still join. 'Brandreth stones' or 'three-footed brandreths' are found elsewhere in Lakeland, especially in association with boundaries, e.g. the one still marked on larger scale maps at NY2818.

BRANDY CRAG SD2299 Ulpha.
From 1867 (OS).
♦ Possibly 'the **crag** or rocky height associated with illicit brandy': M. C. Fair (quoted *PNCumb* 438) suggested that the name commemorates a smugglers' cache on the packhorse route inland from RAVENGLASS. The word *brandy* first appears in the 17th cent., both in its full form *brandwine/ brandewine* (adopted from Dutch *brandewijn* 'burnt' or 'distilled wine') and in the shortened *brandy* (*OED*).

BRANKEN WALL SD0997 (hab.) Muncaster.
Brankin Wall 1679 (*PNCumb* 425).
♦ Obscure. This is the only instance in this volume of 'wall', presumably from OE *wall*, as the generic el., and the meaning of 'branken' is elusive, unless there is some connection with ME *brankand* 'strutting, prancing' (*MED*; cf. *EDD*: *brank* v. 2 and sb. 4 and *SND*: *brank* v. 2, sense 1, also sense 2 for *brankin* 'smartly dressed'). *EDD* also has *brank* sb. 2 'bracken', but only in Suffolk. *Brank(s)* occurs in Scottish p.ns, and is no less problematic there (Dr Simon Taylor forthcoming; I am grateful for a preview of the relevant entry).

BRANSTREE/BRANT STREET NY4709 Shap Rural.
Bransty(e) 1578 (*PNWestm* II, 173), *Branstree* 1828 (HodgsonM), 1839 (Ford 128), *Branstreet* 1860 (OSNB).
♦ 'The steep path', with dial. reflexes of ON/OE *brant* 'steep' and of ON **stígr**/OE **stīg** 'path', influenced by *street* from OE *strēt*. This area of fell is on the track from HAWESWATER to LONGSLEDDALE.

BRANT FELL SD4096 (629ft/192m.) Bowness.
From 1563 (*PNWestm* I, 187).

♦ 'The steep hill', with dial. *brant* 'steep' of ON and/or OE origin, and **fell**. Though not high, the hill is compact and quite sheer.

BRANTHWAITE NY2837 (hab.) Caldbeck.
Braunthwait 1332 (p), *Braunthwayt* 1345 (p), *Brounthwayt* 1357, *Branthwaite* 1560 (*PNCumb* 276).
♦ Possibly 'the clearing where broom grows'; **brōm**, **þveit**. Another Cumbrian Branthwaite has *Braun-* spellings in 13th–14th cent. alongside *Bram-* and *Brom-*, which might suggest OE *brōm* 'broom' as the 1st el. in both cases (*DEPN* and *SSNNW* 218). ME *brame* 'bramble', as suggested in *PNCumb*, is slightly less likely.

BRANTRAKE/BRANT RAKE SD1498 (hab.) Eskdale.
Brantrake 1662 (PR).
♦ 'The steep track', from dial. *brant* 'steep' (of ON and/or OE origin) and **rake**, presumably referring to a track up the steep slope which rises above the settlement.

BRANTWOOD SD3195 (hab.) Coniston.
Brantwood 1851 (OS), possibly = *Brentwode* 1356 (*PNLancs* 219).
♦ 'The burnt wood' or 'steep wood'. (a) If the 1356 spelling belongs here, the name refers to a **wood** that is 'burnt' (ME *brende*). This would be compatible with the prevalence of charcoal-burning in Furness up to the 20th cent., and would be paralleled in names such as Brandwood (*PNLancs* 59) and Brentwood, Essex (see *DEPN*). (b) Alternatively, the 1st el. could be dial. *brant* 'steep', of ON and/or OE origin. Ruskin's famous house (described in Pevsner 1969, 108–9) nestles among woodland on the steep slopes above CONISTON WATER. If the 1356 form refers to this place, this is

the earliest record of the el. **wood** in this volume.

BRATHAY, RIVER NY3503 Skelwith/ Rydal, B~ HALL NY3603 Skelwith.
Braitha 1157–63 (*PNWestm* I, 5); (B~ Hall:) *Brathey* 1577 (SaxtonM), *Brathay Hall* 1920 (OS).
♦ 'The broad river' (**breiðr, á**), a name paralleled in Iceland, the Faeroes and possibly Norway. Machell refers to 'Breathay (w[hi]ch at severall places assumes several names)' (c.1692, II, 153). The present **hall** is late 18th-cent. (Pevsner 1969, 75).

BREASTY HAW SD3492 Satterthwaite.
Breasty Hall 1851 (OS).
♦ Perhaps 'the breast-shaped hill'. 'Breasty', if genuine, may refer to the rising contours now hidden under mainly coniferous plantations; cf. *breast* as in WHITELESS BREAST. *Haw* is the form of **how(e)** 'hill' (see **haugr**) favoured in Furness; corruption to **hall** is common, and encouraged by the presence of a building (the 1851 map shows one at this point).

BRIDGE END Examples at:
 SD2490 (hab.) Kirkby Ireleth.
Bridgend in Woodland 1681 (PR), *Bridge-end* 1748 (Woodland PR), *Bridge End* 1851 (OS).
 NY3914 (hab.) Patterdale.
From 1859 (*PNWestm* II, 224).
 NY5100 (hab.) Longsleddale.
From 1720 (PR), 1836 (*PNWestm* I, 162).
BRIDGE END FARM NY3119 St John's, LOW B~ E~ F~ NY3120.
Brighend 1571, *Bridgend* 1629 (*PNCumb* 316); *Low Bridge End* 1809; cf. *High Bridge End* 1776 (both PR).
♦ Simply **bridge** from **brycg, end**, and **farm** added in one instance. The Kirkby settlement is located beside STEERS POOL, and the bridge there is Bridge End Bridge (1851, OS); the Patterdale farm is beside DEEPDALE Bridge, and the Longsleddale one beside a footbridge across the SPRINT. The two farms in St John's lie beside bridges over ST JOHN'S BECK. Though there is little difference in altitude, Bridge End Farm (*High B~ E~* in the PR) stands on a higher reach of the beck than Low B~ E~ F~.

BRIDGEFIELD SD2986 (hab.) Colton.
Brigfield 1627, *Bridgefield* 1628 (PR), *Bridgefield* 1851 (OS).
♦ This is beside the R. CRAKE; **brycg/ bridge, field**.

BRIDGE/BRIGG HOUSE SD4288 Cartmel Fell.
Brigghouse 1757 (*CW* 1966, 226).
♦ This is close to Lobby Bridge over the R. WINSTER and probably named from it; **brycg/bridge, house**. However, the fact that it was owned by the Briggs family in the early 16th cent. (*VCHLancs* 283) raises the possibility that B~ is the surname here. The 1757 form and one of the current variants show the local form *brigg* which is partly from ON *bryggja*.

BRIDGE HOUSE NY3022 St John's.
Bryghous 1565, *Bridghouse* 1573 (*PNCumb* 316).
♦ Simply 'The house by the bridge', with reflexes of **hūs, brycg/bryggja**; cf. the entry above and BRIGHOUSE. This place is at a crossing over ST JOHN'S BECK.

BRIDGE PETTON NY0702 (hab.) Gosforth.
(*moram de*) *Brigerpetin* c.1285, *Briggepeting* 1303 (*PNCumb* 394).
♦ The 1st el. is clearly **bryggja/bridge** (q.v.), acting as the generic in an

inversion compound, though the *-er* in the c.1285 spelling is puzzling, conceivably from the ON pl. *bryggjur*. The place is near a minor bridging point. The remainder of the name is obscure, although a pers.n. is likely.

BRIERY NY2824 (hab.) Keswick.
de Beriery 1283 (Sedgefield 1915), *Briery* 1867 (OS).
♦ Presumably 'the place with briars'. Although *briery/briary* is normally an adj., *OED* gives two 16th-cent. examples of its use as a noun.

BRIERY CLOSE NY3901 Troutbeck.
From 1649 (spelling not specified, *CW* 1964, 162), *Bryary close* 1721, *Briery Close* 1839 (*PNWestm* I, 189).
♦ 'The enclosure or farm where briars grow', with adj. *briery* (cf. BRIERY above), plus **close**.

BRIGHAM NY2723 (hab.) Keswick.
Brigholm c.1240 (*PNCumb* 302).
♦ 'The waterside land by the bridge.' The name shows a synthesis of English and Scand features. *Brig-* owes its meaning 'bridge' to OE **brycg**, but its form mainly to ON **bryggja**, which most often means 'quay'. The 2nd el. is from ON **holmr** 'land flanked or surrounded by water', but this has been replaced or influenced by *-ham* from OE **hām**. Brigham lies at a bridging point in a gentle curve in the R. GRETA, on land bounded by smaller tributaries to E and S.

BRIGHOUSE SD0895 Muncaster.
Brighouse 1733 (PR), *Brig House* 1774 (DHM).
BRIGHOUSE SD1994 Ulpha.
Brigghouse 1657 (*PNCumb* 438).
♦ 'The house by the bridge', with reflexes of **hūs**, **brycg/bryggja**; cf. BRIDGE HOUSE. B~ in Muncaster is not far from Walls Bridge; a Bridge House is shown on the 1867 map, on the other side of the railway line. The Ulpha example is next to Crosby Gill.

BRIM FELL SD2798 Coniston.
Brimfell 1684 (spelling not specified, *CW* 1910, 379), *Brim Fell* 1850 (OS).
♦ Probably 'the hill with the steep edge'. *VEPN*: *brimme* comments on the difficulty of being sure whether ME *brimme* 'bank, shore, edge' occurs in p.ns, because of the possibility of confusion, e.g. with the reflex of OE **brēme* 'broom, place where broom grows'. However, 'edge' would fit this high **fell** with its precipitous E side, but not 'broom'.

BRIMMER HEAD FARM NY3208 Grasmere.
Brima Head, *Brymar Head* 1616, *Brim(m)erhead* 1719 (*PNWestm* I, 199), *Brimer Head* 1828 (HodgsonM), *Brimmer Head*, *Bremer-head* 1859 (OSNB).
♦ B~ H~ F~ lies at the head of EASEDALE, which might suggest that 'Brimmer' is a valley name. It is also below Brinhowe Crag, and Brimmer is identified with Brinhowe Crag in *PNWestm* I, 199, citing *CW* 1928 (i.e. Simpson 1928, 282), but the relationship between the two and the meaning of Brimmer remains unexplained.

BROAD CRAG NY2207 Eskdale.
Bread Crag, *Broad Crag* 1802 (*CN* 1219–20).
♦ Self-explanatory. **Broad** is from OE **brād**, but the first of Coleridge's 1802 spellings may conceivably be a trace of the ON cognate **breiðr**, found in BRAITHWAITE; see also **crag** 'rocky height'. The name well describes 'the second of the Scafell Pikes, and a worthy mountain in itself', whose W flank presents an 'imposing . . . semi-

circle of crags' (Wainwright 1960, Scafell Pike 9).

BROAD END Examples at:
NY2528 Underskiddaw.
From 1867 (OS).
NY2530 Bassenthwaite.
From 1867 (OS).
NY4007 Troutbeck.
From 1859 (*PNWestm* I, 189).
♦ Simply from **brād/broad, end**. The first two listed are the blunt ends of spurs projecting, respectively, from the N side of SKIDDAW and the NW of (SKIDDAW) LITTLE MAN. The Troutbeck example is similarly the rounded culmination of a long, high ridge.

BROADFOLD SD4797 (hab.) Nether Staveley.
Broadfold c.1692 (Machell II, 114), *Broad fould* 1770 (JefferysM).
♦ 'The wide pen or farm'; **brād/broad, fold**.

BROADGATE SD1886 (hab.) Millom Without.
Brodyett 1609, *Brod-yaitt* 1610 (*PNCumb* 418), *Broadgate alias Roughthwaites, Cooksons alias High Broadgate* both 1696 (*JEPNS* 2, 59).
♦ 'The broad gate', with **broad** from OE **brād** and Standard English *gate* replacing the earlier *y*- pronunciation. This seems to refer to a gate (**geat/gate**) rather than a road (**gata**), despite evidence of an 'ancient hollow-way' just to the W (Dymond 1902, 75).

BROADGATE FARM SD4399 Windermere/Hugill.
Bradgate 1597 (Kendal PR), *Brodgat* 1614, *Broadgate* 1618 (Windermere PR).
♦ B~ is probably 'the wide road', with **broad** from OE **brād**, plus **gate** from ON **gata** 'way, road', since there are no *y*- spellings to suggest the reflex of OE **geat** 'gate'. The name may conceivably refer to the Roman road linking forts at *Alauna* (Watercrook near Kendal) and GALAVA near AMBLESIDE and passing a presumed fort at B~, where Andrews & Andrews believe the road to have temporarily bifurcated (1991 and 1992, esp. 1992, 59). A Roman 'bulla' or ring, of jet, was found at Broadgate in 1808 (LDHER 16528), but the site has not been excavated, and the exact course of the Roman road is not known. **Farm**, as so often, has been added quite recently.

BROADMOOR NY0815 Ennerdale.
Braythemyre 1334, *Braimer* late 16th cent. (*CW* 1931, 165), *Broad Moor* 1867 (OS).
♦ 'The broad bog' (**breiðr, mýrr**), much of it now planted with conifers, the rest rough grass and rushes. If the identification of the 1334 spelling with Broadmoor is correct, the name was originally ON, but has been reformed from English **broad** plus **moor**, of joint OE and ON origin.

BROADNESS FARM NY2229 Bassenthwaite.
Broadnes 1625/6 (PR), *Bradness* 1693, *Bradeness* 1698 (PR), *Braidness* 1774 (DHM), *Bradness* 1780 (West 120), *Broadness* 1787 (*PNCumb* 264), *Braidness* 1865 (Hodgson Plan 14).
♦ B~ is 'the broad headland', with OE **brād** 'broad, wide' and OE **næss** 'headland', their ON cognates **breiðr** and **nes** or their ME or ModE descendants. The spellings with *Braid-* might suggest origins in ON *breiðr*. This headland on the E of BASSENTHWAITE LAKE is exceptionally broad, contrasting strikingly with SCARNESS just to the N. **Farm** has been added relatively recently.

BROAD OAK Examples at:
   SD1194 (hab.) Muncaster.
From 1660 (*PNCumb* 425).
   SD4389 (hab.) Crosthwaite.
Broadoke 1535, in 1669 copy (*PNWestm* I, 83).
   SD4691 (hab.) Underbarrow.
the Brodeake 1582 (*PNWestm* I, 104).
BROAD OAKS NY4001 (hab.) Troutbeck.
Bradoke early 17th cent. (*CW* 1997, 82), *Broad Oak* 1839 (*PNWestm* I, 189).
♦ 'The (place by the) spreading oak(s)'; **brād/broad**, **oak**. Concerning the Troutbeck B~ O~, one of many coppices created within Old Park, Troutbeck at the beginning of the early 17th cent., Parsons remarks that these predominantly oak coppices would have served the growing demand for charcoal from the local iron and tanning industries (1997, 82). The farmhouse was first built in 1565 (Palmer 1944, 20).

BROADRAYNE/BROAD RAIN: HIGH B~ NY3309 (hab.) Grasmere.
Broad roan 1630 (*PNWestm* I, 199), *broad-Rain, Broad-Raine* 1688 (Fleming Accounts 5–6); *Highbroad Rain* 1839 (Ford 38), *High Broadrain* 1859 (OSNB), *High Broadrain, Low Broadrain* both 1863 (OS).
♦ Probably 'the broad strip of land'. ON *rein* and its presumed OE cognate, **rān*, refer to a strip of land, often forming a boundary (*PNWestm* II, 280, and cf. Simpson 1928, 278 on division of land into 'dales' hereabouts); plus **brād/broad**, **high**. OSNB cites *Broadrains, Broadrane* as variants to *Low Broadrain* (1859), and indeed Broadrayne (Farm) and Low Broadrayne are one and the same, constituting a larger holding than High Broadrayne, which appears (though is not named) on a plan of c.1730 in the possession of Mr & Mrs Dennison of Broadrayne Farm. I am grateful to them for a sight of this, and for the suggestion that the name B~ might reflect 'the fact that this site could have been the earliest and easiest to clear for farming in this valley . . . the broadest strip of land between the wet valley floor and the totally uncultivable high fell behind the farm' and 'facing both W and S'. They also note that it forms a strip of land between the fell to the E and the ancient routeway N over DUNMAIL RAISE.

BROAD STAND NY2106 Eskdale.
From 1867 (OS).
♦ This is a platform of rock on the high ridge between SCAFELL and SCAFELL PIKE, which is 'the greatest single obstacle confronting ridge-walkers on the hills of Lakeland' (Wainwright 1960, Scafell 3; Jones 1900, 33–7 for the climb); **broad** from OE **brād**, plus *stand*, which refers at lower altitudes to a hunter's stand used for shooting game.

BROAD TONGUE NY1906 Eskdale.
The Broadtongue 1587 (*PNCumb* 391).
♦ This is the triangular piece of land, broad compared with many tongues, between Long Gill and Oliver Gill; **brād/broad**, **tongue**.

BROADWATER SD0889 (hab.) Bootle.
From 1702 (*PNCumb* 348).
♦ Simply **brād/broad**, **water**, but there is no pool currently or on the 1867 OS map, so either (a) *water* refers to the stream which lies just to the N; or (b) if *water* referred to a pool, drainage has changed the landscape, and reduced the aptness of the name, as also in THE TARN and HYTON.

BROCKA/BROCA SD4180 (hab.) Upper Allithwaite.
Cf. *Broca Hill, Broca Cottage* 1851 (OS).
♦ Probably a reference to badgers, either (a) 'badger hill' (**brock** plus **how(e)** from **haugr**); or (b) 'badger sett' (OE *brocc-hol*), as in the Yorkshire Brocka Beck (*Brocholebec(h)* 1109–14, *PNNYorks* 82).

BROCK CRAG NY2102 Eskdale.
From 1802 (*CN* 1220).
BROCK CRAG NY4519 Martindale/Barton.
Brock Cragg 1752 (*PNWestm* II, 218).
BROCK CRAGS NY4113 (1842ft/561m.) Patterdale.
From 1865 (*PNWestm* II, 224).
♦ 'The rocky height(s) frequented by badgers'; see **brock** on the word and the animal, and **crag**.

BROCKHOLE NATIONAL PARK CENTRE NY3800 Troutbeck.
Brockholes 1859 (*PNWestm* I, 190).
♦ B~ is 'the badger sett' (ultimately from OE *brocc-hol*): on the word and the animal, see **brock**. This is the Visitor Centre of the Lake District National Park, which was founded in 1951.

BROCKLE BECK NY2822 Borrowdale/St John's.
From 1867 (OS).
♦ Probably 'the stream by Brockhole, the badger sett', assuming 'Brockle' to be a shortening of Brockhole (from OE *brocc-hol*); see also **brock** on the animal, and **beck**.

BROCKLEBANK GROUND SD2793 Torver.
From 1718 (PR).
♦ 'The farm of the B~ family'; surname plus **ground**. The surname B~ appears frequently in the PR, though the 1718 record is of the baptism of Benjamin Benson. B~ originated as a p.n. (for instance one near Wigton, Cumbria) meaning 'bank with a badger sett'.

BROCKLECRAG NY2532 (hab.) Ireby and Uldale.
Brockelcrag 1645 (Uldale PR), *Brockle Crag* 1774 (DHM).
♦ The farm lies below Brockle **Crag**, which is probably 'the rocky hill-side with the badger sett', with 'Brockle' contracted from Brockhole (OE *brocc-hol*); and see **brock**.

BROCKSTONES/BROCK STONE NY4605 (hab.) Kentmere.
brockstone 1605 (CW 1988, 123), *Brockstone* 1770 (JefferysM), *Brock Stone*, *Brockstones* 1823 (*PNWestm* I, 166).
♦ 'The badger stone(s)' (**brock**, **stone**), presumably a site of badger setts, as is, certainly, the Brock Stone (alias Badger Rock) at NY4505, and Tongue Scar at NY4507, where Danby reports badgers having lived for generations (1994, 175). The earliest forms suggest that the name was originally in the singular.

BROOM, THE/BROOM FARM SD4592 Underbarrow.
Broom 1745 (*PNWestm* I, 106).
♦ Simply 'the place where broom grows'; **broom**, with **farm** added later.

BROOM FELL NY1927 (1670ft/509m.) Wythop.
From 1867 (OS).
♦ Simply 'the hill where broom grows'; **broom**, **fell**. The plant is currently found in this area (Halliday 1997, 305).

BROOMHILL NY0800 Irton (hab.).
♦ Seemingly = *High House* 1867 (OS), 1966 (OS One Inch).

BROOMHILL SD1187 Bootle (hab.).
Broom Hill 1867 (OS).
♦ 'The hill where broom grows' (**broom, hill**) and hence the inhabited place by it. There are further examples of this name in lowland Cumbria (*PNCumb* 520).

BROTHERILKELD NY2101 Eskdale.
Butherulkul c.1210, *Butherulkil, Brutherulkil* 1242, *Botherulki(l)* c.1260 (p) (*PNCumb* 343), *Butterilket* 1867 (OS).
♦ 'Ulfkell's booths or huts.' In this fascinating name, the 1st el. preserves the pl. *-ir* of OWN **búðir** 'booths, huts' almost intact. The modern form 'Brother-' may well be influenced by the fact that this formed the centre of a large estate managed by FURNESS Abbey (Millward & Robinson 1970, 159, Winchester 1987, 42). The stressed 2nd el. '-ilkeld', from the ON pers.n. *Ulfkell*, is the specific or qualifying el., hence the name is a so-called inversion compound. See BUTTERMERE for a legendary explanation of this name.

BROTHERS WATER NY4012 Patterdale.
Brother-water 1671 (Fleming 27), *Brother water, Broader-water* c.1692 (Machell I, 95, 720), *Broader Water* 1770 (JefferysM), *Broad Water called by some Brotherwater* 1787, *Brothers water* 1802 (Dorothy Wordsworth, *Journals*, Apr. 16).
♦ The name may derive from the ON pers.n. *Bróðir* (so *PNWestm* I, 15), though it is uncertain whether was current in OWN (Insley 1994). Meanwhile, Hutchinson remarked that it was 'called *Broad Water*, by others *Brother Water*, from two brothers being drowned in it; — and what is singular, a similar accident occurred about seven years ago' (1794, I, 429); Dorothy Wordsworth specified that the later brothers fell through the ice in winter ('Excursion on the Banks of Ullswater, 1805'). The story may account for the present form of the name with pl. *-s*. The 2nd el. **water** is here used for a largish tarn.

BROUGHTON: B~ BANK SD3780 (hab.) Broughton East, B~ HOUSE SD3981, B~ LODGE SD3980, FIELD B~ SD3881, WOOD B~ SD3781.
Brocton 1276, *Broghton* 1314 to 1429 (*PNLancs* 198); *Broughton Bank; Broughton House; Broughton Lodge* all 1851 (OS); *Feild Broughton* 1593 (Cartmel PR); *Woodbroughton* 1604 (Cartmel PR).
♦ 'The village on the stream', from OE *brōc* 'stream, brook' and OE *tūn* 'farmstead, settlement, village'. **Field** B~ and **Wood** B~ are on the two arms of the R. Eea. B~ **Bank** is at the edge of the parish. B~ **Lodge**, an 18th-cent. house of five bays and two and a half storeys (Pevsner 1969, 78), is among the grander dwellings named *lodge*.

BROUGHTON IN FURNESS SD2187 Broughton West, B~ MILLS SD2290 (hab.), B~ MOOR SD2493 (hab.).
Brocton 1196, 1235, *Broghtona* c.1300, *Broghton* 1378 (*PNLancs* 222); *Broughton Mill* 1676/7, *Broughton Mills* 1790 (Broughton-in-Furness PR); *Broughton Moor* 1851 (OS).
♦ 'The village on the stream', as above. The place is well enough situated to have been founded and named in the Anglian period. For Furness, see FURNESS FELLS. B~ **Mills** (see **mylen**) and **Moor** are presumably named from the parish of Broughton-in-Furness (now B~ West), for they are not outliers of the village. Walk Mill, close to B~ Mills, may suggest fulling mills rather than corn mills.

BROW NY3103 (hab.) Langdales.
From 1865 (*PNWestm* I, 206).
BROW NY4825 (hab.) Barton.
From 1675 (*PNWestm* II, 212).
♦ 'The hill-slope.' Both places are on or beneath a **brow**, respectively the moderate slope under Bield Crag, and the slopes above BARTON village. Low and High Brow are distinguished at Barton on larger scale maps.

BROW, THE NY2400 Dunnerdale, BROW SIDE FELL NY2500.
Both from 1850 (OS).
♦ The B~ is simply 'the hill-slope' (**brow**), here a steep slope above the DUDDON. The exact sense of **side** is elusive, but may be 'hill-side'; the **fell** is the higher ground around GREY FRIAR. 'The name Browside apparently not used before c.1799. "Troutal" was used for both farms before that' ('Results of research in Parish Registers by Mr Percy Tyson', cited 'SGJ' 1961).

BROW EDGE SD3583 (hab.) Haverthwaite.
Browedge 1751/2, perhaps cf. *Brow* 1680 (Colton PR).
♦ 'The edge of the hill-slope'; **brow, edge**. High and Low B~ E~ are distinguished on larger scale maps.

BROWFOOT NY4500 (hab.) Hugill.
From 1865, cf. *Brow* 1738 (*PNWestm* I, 170).
♦ 'The bottom of the hill-slope'; **brow, foot**.

BROW HEAD FARM NY3604 Rydal.
brow head 1682, *the Brow-head in Loughrigg* 1695 (Grasmere PR), *Brow Head* 1706 (*PNWestm* I, 210).
♦ 'The higher part of the slope'; **brow, head**, with **farm** added later. The slope in question is named Miller Brow (to NW) and Loughrigg Brow (to SE) on larger scale maps.

BROWN BAND NY1310 Nether Wasdale.
From 1867 (OS).
♦ The name applies to the broad, steep SW flank of HAYCOCK; **brown, band.** Another Brown Band, in Eskdale, is recorded from 1587 (*PNCumb* 391).

BROWN BECK NY4720 Askham.
From 1863 (OS).
♦ Simply 'the brown stream'; **brown, beck**. The adj. is more often applied to hills features than to streams.

BROWN COVE NY3415 Patterdale, BROWNCOVE CRAGS NY3315.
Brown Cove 1828 (HodgsonM); *Browncove Crags* 1862 (OSNB Crosthwaite).
♦ 'The brown corrie'; **brown, cove**. This is a great U-shaped corrie; it is not obviously more brown than its surroundings. Browncove **Crags**, forming a high arc of rock outcrops to the W, are, as Wainwright remarks, 'oddly named because Brown Cove is on the other side of the ridge' (1955, Helvellyn 18). The fact that Machell seems to refer to this corrie as *Catstee* might suggest that Brown Cove is a late re-naming (see CATSTYCAM).

BROWN CRAG NY2710 Borrowdale.
Blown [sic] *Crag* 1867, 1900 (OS).
BROWN CRAG NY3217 St John's.
From 1867 (OS).
♦ Simply 'the brown rocky height'; **brown, crag**. A set of rocks part-way up a steep fellside, the Borrowdale example may be named in relation to its darker-coloured neighbour, BLEA CRAG NY2610. The St John's crag is a rocky summit above Brown How and between White Crags and WHITESIDE.

BROWNDALE BECK NY3419 Matterdale.
From 1867 (OS).

♦ Apparently 'the stream in B~, the brown valley'; **brown, dale, beck**.

BROWN DODD NY2617 Borrowdale.
Broundodde 1272 (*PNCumb* 349).
♦ 'The brown rounded hill'; **brúnn/brūn, dodd**. The rocks have a reddish-brown tinge. This is among a number of apparently bland topographical names which might be taken as modern but are in fact several centuries old.

BROWNEY GILL NY2604 Langdales.
Brown Gill 1862 (OS), 1966 (OS One Inch).
♦ 'The brown ravine with stream'; **brown, gil(l)**. It descends beside BROWN HOWE, and one wonders whether the form 'Browney', only recently adopted by the Ordnance Survey, might derive from Brown Howe (cf. POOLEY). The brown colour might result from the presence of iron ore, which was mined here (Johnson 1988, 247).

BROWN HAW SD2293 (1263ft/385m.) Dunnerdale.
From 1850 (OS).
♦ 'The brown hill'; **brown, haugr/how(e)**. This is one of several rocky hillocks in this area named *haw*, a local variant of *how(e)*; cf. BROWN HOW(E).

BROWN HILLS NY3719 Matterdale.
From 1867 (OS).
♦ Self-explanatory; **brown, hill**.

BROWN HOW NY1115 Ennerdale.
From 1867 (OS).
BROWN HOWE Examples at:
  NY2604 Langdales.
From 1862 (OS).
  SD2890 (hab.) Blawith.
Brown Hall 1850 (OS).
  NY4608 Kentmere/Longsleddale.
From 1857 (*PNWestm* I, 166).
  NY4812 (1648ft/502m.) Shap Rural.
From 1860 (OSNB).
  NY5108 Shap Rural.
Brownhow Pike 1828 (HodgsonM), *Brown Howe* 1860 (OSNB).
♦ 'The brown hill'; **brúnn/brown, haugr/how(e)**. Two of the examples are not quite straightforward. In the Blawith name, **hall** and *how(e)*, or perhaps locally *haw(e)*, have alternated, as they frequently do in Furness, so that without earlier spellings it is unclear which is more original. For the Kentmere example, Smith's suggestion that it is named from the family of Richard *Broune*, recorded from 1332 (*PNWestm* I, 166), cannot be ruled out, but proximity to a Brown Crags makes the colour term more likely. Whether by accident or design, the Langdale hill is next to another named from its colour: BLACK WARS. The OSNB description of the more southerly B~ H~ in Shap is 'hill feature with pile of stone on top . . . It is not a *howe*.' Perhaps the surveyors expected a burial mound. Although these names all first appear (to my knowledge) in the 19th cent., it is not impossible that they are several centuries older: cf. BROWN DODD, BROWNRIGG.

BROWN KNOTTS NY2719 Borrowdale.
From 1867 (OS).
♦ This is a rather modest range of crags part-way up BLEABERRY FELL; **brown, knott** 'craggy height'.

BROWN PIKE SD2696 (2237ft/682m.) Dunnerdale.
From 1851 (OS).
♦ Simply 'the brown peak'; **brown, pike**.

BROWN RIGG Examples at:
  SD1896 Ulpha.

From 1867 (OS).
NY3014 St John's.
Brounewrigge 1760 (*PNCumb* 316).
NY4720 Barton/Askham.
From 1859 (*PNWestm* II, 212).
BROWNRIGG NY3040 (hab.) Caldbeck, LOW B~ NY3140.
Brunrigg juxta (next to) *Caudebec* 1209 (*PNCumb* 276).
BROWNRIGG FARM NY4024 Matterdale.
Le Brounrigg 1323 (*PNCumb* 257).
BROWNRIGG WELL NY3315 (spring) St John's.
From 1862 (OSNB).
♦ B~ is 'the brown ridge', with elements from ON **brúnn**, possibly encouraged by OE **brūn** 'brown', and ON **hryggr** 'ridge'; **farm** has been added in one instance, and see also **low**, **well**. Although the classic *rigg* is linear, some examples, including the Caldbeck and St John's ones, are not, and the Ulpha B~ R~, if the maps are correct, is unusually located in a dip. In some cases the name contrasts with other colour references: the Caldbeck place is near GREENRIGG, while the Matterdale B~ is directly opposite GREENROW. Low B~, Caldbeck, is unnamed on the 1867 OS map and appears as B~ Hall Farm on the current 1:25,000 map. B~ Well, also known as Whelpside Gill Spring, is 'a copiously supplied spring on the west top of Helvellyn, well know[n] to tourists' (OSNB). The nearest Brown Rigg is on the other side of THIRLMERE, and this B~ may be a surname. This surmise would be encouraged by the form 'Brownrigg's Well' (Palmer 1952, 236).

BROWNSPRING COPPICE SD4597 Nether Staveley.
From 1858 (OSNB).
♦ (a) Smith suggests that 'Brown' may be the local surname, recorded locally from the late 14th cent. (*PNWestm* I, 173) and common in the Kendal PRs. (b) The colour term **brown** is another possibility. **Spring** is probably 'copse, plantation', with its meaning virtually duplicated by the addition of **coppice** 'coppice wood'.

BROWS, THE SD3789 (hab.) Colton.
From 1851 (OS).
♦ 'The hill-slopes'; **brow**. The site overlooks the S end of WINDERMERE.

BROW SIDE FELL: see BROW, THE

BRUND FELL NY2616 (410m/1363ft) Borrowdale.
Brunt Fell 1782 (CrosthwaiteM), *Brund Fell* 1867 (OS).
♦ Probably 'the burned hill', with 1st el. as in BRUNDHOLME NY2924, which has a *Brunt-* spelling in 1615 (*PNCumb* 321), plus **fell**. In the absence of earlier evidence, however, confusion with *brant* 'steep' could not be ruled out. If this has happened, this would be another BRANT FELL (cf. BRUNT KNOTT).

BRUNDHOLME NY2924 (hab.) Underskiddaw, B~ WOOD NY2824.
Brundholm(e) 1292 (p) (*PNCumb* 321); *Brundholme Wood* 1867 (OS).
♦ 'The waterside land which has been burned.' The 1st el. is probably a NW version of the ME past participle *brend* 'burned', influenced by ON *brunninn* 'burnt' (*SSNNW* 220, *VEPN*: *brend*). The many places distinguished in p.ns as 'burnt' may be literally so — either cleared by burning in order to improve pasture (see Winchester 2000, 136–7) or 'destroyed by fire', or else 'of a burnt colour, brownish' or possibly 'shining', since some stream-names contain the el. (*VEPN*: *brend*). The 2nd el. is ON **holmr** 'land flanked or

surrounded by water', which aptly describes the site, in the angle where the GLENDERATERRA BECK flows into the GRETA. See also **wood**.

BRUNDRIGG SD4895 (hab.) Strickland Ketel.
Brondrig(g) 1344 to 1651, *Brundrig(g)* 1390–5 to present, *Brend(e)rig, -ryg* 1396, 1508 (*PNWestm* I, 152–3, q.v. for further forms).
♦ 'The burned ridge.' Despite the philological complexities (*PNWestm*), the 1st el. seems to be the same as in BRUNDHOLME, while the 2nd is ON **hryggr**/OE **hrycg** 'ridge' or their ME reflex.

BRUNT KNOTT NY4800 (1400ft/427m.) Over Staveley, BRUNT KNOTT FARM NY4700.
Brant Knott 1670, *Burnt Knott* 1692 (*PNWestm* I, 175), *Brunt Knott* 1828 (HodgsonM), 1857 (*PNWestm* I, 175).
♦ Either 'the steep, rocky summit' or 'the rocky summit cleared by burning'. The 1st el. is either (a) from *brant* 'steep', as in BRANT FELL; B~ K~ is very steep on the W side; or (b) a form of 'burnt' (so *PNWestm*, and cf. BRUNDHOLME). The 2nd el. is **knott** 'compact hill, craggy height'.

BRUNT TONGUE NY5009 Shap Rural. From 1860 (OSNB).
♦ This is the **tongue** of land that is either 'steep' or 'burned', with the same two possibilities for 'Brunt' as in B~ KNOTT. This is the steep ridge in the angle between MOSEDALE BECK and Low Mosedale Beck.

BRUTS MOSS NY3521 Matterdale.
Bruts 1867, *Bruts Moss* 1900 (OS).
♦ 'Bruts' of unknown origin, plus **moss** 'bog'.

BRYAN BECK SD4090 (hab.) Cartmel Fell.
Brian beck 1722 (*CW* 1966, 238, as hab.), *Bryan Beck* 1851 (OS).
♦ Seemingly 'Brian's **beck** or stream'. The Celtic name *Brian, Bryan* was brought from Ireland to NW England by Scand settlers, and to the rest of England by Bretons at the Norman Conquest. It gave rise to a patronymic surname current from the ME period onwards.

BRYERS FOLD SD3895 Claife.
Briers 1770 (JefferysM), 1783 (CrosthwaiteM), 1851 (OS).
♦ Probably 'the pen or farm of the Bryers family'; surname plus **fold**. *Bryer(s)* may be a surname as, for instance, in KNIPE FOLD and CARTMELL FOLD, though there is also a local p.n. referring to a place with briars, e.g. *breares* 1597 (Hawkshead PR).

BUCKBARROW NY1305 (hill and hab.) Nether Wasdale.
Buckeborowe 1578 (*PNCumb* 441).
BUCK BARROW SD1591 (799ft/244m.) Waberthwaite, BUCKBARROW BECK SD1390.
Bokkeberghes 1319, *Blakbery* or *Bukbury* 1548 (*PNCumb* 365, q.v. for other eccentric forms); *Buckbarrow Beck* 1867 (OS).
BUCKBARROW CRAG NY4807 Longsleddale.
From 1828 (HodgsonM), cf. *the highte of Buckbarrowe* 1578 (*PNWestm* I, 162).
♦ Buckbarrow is 'the hill where buck are found' (**buck, berg**), though whether *buck* refers to he-goats or stags is uncertain. The Wasdale B~ has also given its name to the farm below. The Waberthwaite B~ is the twin peak to KINMONT B~ B~ (q.v.). See also **beck** 'stream', and **crag** 'rocky height', which is appropriately added to a pre-

existing Buckbarrow in Longsleddale, since the hill is extremely rugged.

BUCK CRAG Examples at:
SD4081 (hab.) Upper Allithwaite.
Buckcragge 1576 (*PNLancs* 199).
NY4214 Martindale.
Buckecrag' 1265 (*PNWestm* II, 218).
NY4304 Windermere.
From 1863 (OS).
BUCK CRAGS NY5007 Fawcett Forest/ Shap Rural.
From 1865 (*PNWestm* II, 174).

♦ 'The rocky height(s) frequented by bucks'; **buck**, referring to stags or he-goats, plus **crag**. The Shap example is near Buck Stone (*Bokeston* 1279), which was 'the great stone where they were wont to stand to watch the deer as they passed' in a source of c.1200 quoted in *PNWestm* II, 174; and the Martindale one is on the W edge of DEER FOREST. Another application of *buck* is, however, suggested when Ekwall writes of the Upper Allithwaite example, 'the place stands at a rocky hill, stated to bear a certain resemblance to a buck' (*PNLancs* 199), though he does not specify deer or goat.

BUCKHOLME ISLAND NY1933 Blindcrake.
Buckholm I. 1785 (CrosthwaiteM), *Buckham Island* 1810 (*PNCumb* 301).
BUCKHOLME: B~ WOOD NY5225 Lowther.
Bucholm 1249, *Buckeholm* 1260; *Buckholme Wood* 1859 (*PNWestm* II, 184).

♦ B~ is 'the island or water-flanked land frequented by male deer or goats'; **holm(e)**, **buck**; also **island**, **wood**. The Blindcrake example is an islet in the R. DERWENT and must have been simply named B~ before the addition of *island*; goats rather than deer are more likely here.

BUCK PIKE SD2696 Coniston/Dunnerdale.
From 1851 (OS).

♦ Presumably 'the peak frequented by the (wild) he-goat' (**buck**, **pike**), cf. nearby GOAT CRAG and GOAT'S WATER, though *buck* can also refer to the stag.

BULL CLOSE NY3402 Skelwith.
Bulclose 1653, *Bull Close* 1654 (Hawkshead PR).

♦ Presumably 'the farm where the bull was kept', perhaps serving the stock of the township, cf. the stipulation 'that every Townshipp within the mannor of Watermillock shall keep a Town bull according to the ancient custom' 1678 (*CW* 1884, 41); *bull* from OE *bula*, **close**.

BULL CRAG NY2613 Borrowdale.
From 1867 (OS).
BULL CRAG NY2710 Borrowdale.
From 1780 (West 95), 1867 (OS).

♦ 'The rocky height associated with the bull'; **bull**, **crag**; cf. BULL CLOSE above.

BULL HAW MOSS SD2794 Torver.
From 1851 (OS).

♦ 'The peat bog by Bull Haw, bull hill'; *bull* is from OE *bula*, while *haw* is a common equivalent in High Furness of **how(e)** from ON **haugr** 'hill'; also **moss** 'bog'.

BULLY HOUSE NY1831 Setmurthy.
Bullyhows 1547 (CRO D/Lec/314/38, pers.comm. Dr Angus Winchester), *Buly house* 1654 (*PNCumb* 435), *Bully House* 1774 (DHM).

♦ The 1st el. is obscure, unless *bully* is a variant of *bullace* 'wild plum', as in some Midlands dialects (*EDD*); plus **house**, unless the 1547 spelling points to derivation from **howe** 'hill'.

BULMAN STRANDS SD4393 (hab.) Crook.
Bulmerstrandes 1525, *Bul(l)manstrands* 1637 (*PNWestm* I, 177).
♦ B~ seems to be 'the pool or marsh where bull(s) graze', with **bull** from OE *bula* 'bull' and the reflex of OE *mere* 'pool' (**mere**[1]) or ON **mýrr** 'bog, marsh'; the name has evidently been influenced by the surname *Bulman*. *Strand* is probably from ON *strǫnd* 'shore'; for an alternative possibility, of 'stream', see under WASDALE NY1204. The place is beside the R. GILPIN.

BUNGALOW, THE NY4316 Martindale. From 1920 (OS).
♦ This now-familiar word for a one-storey building, sometimes for seasonal use, originates in the Hindi adj. meaning 'Bengali', and occurs in English from the 17th cent. onwards, at first mainly in Indian contexts (see *OED*). This B~ was a deer-shooting lodge at which the German Kaiser was entertained by Lord Lonsdale in 1895 (Ramshaw & Adams 1993, 76).

BURBLETHWAITE HALL SD4189 Cartmel Fell.
Burbelthwayt 1351 (*PNLancs* 200).
♦ Possibly 'the clearing where bog-rhubarb grows'. (a) The 1st el. may be ME *bur-blade* (*PNWestm* II, 238), cf. dial. *burblek* 'bog rhubarb'. Fellows-Jensen favours this (*SSNNW* 220–1), and with Ekwall (*PNLancs* 200) compares the field-name *Burbladthwayt* (*PNWYorks* VI, 248); further support comes from *Burbladthwayt* 1318, near Coverham in Yorks (Lindkvist 1912, 106). (b) An alternative for the 1st el. might be ME and dial. *burble, burbel* 'bubbling spring'. The 2nd el. is þveit 'clearing', extremely common in Furness; see also **hall**.

BURNBANK FARM SD4180 Upper Allithwaite.
♦ 'The farm on the stream bank', referring to a small stream in front of the farm; **burn, bank, farm**. I am extremely grateful to the present owners, Mr & Mrs Cochrane, for the following information. 'Burn Bank' was a renaming (with permission from the Post Office) of the farm they bought in 1979 as Skinner Hill no. 2; *burn* rather than *beck* was chosen since Mrs Cochrane's mother was Scottish.

BURNBANK FELL NY1121 (1465ft/ 447m.) Loweswater.
From 1867 (OS).
♦ B~ may have been a p.n. which is now lost from the map and from local usage (I am grateful to Mr K. Bell of HUDSON PLACE for confirming that there is no 'Burnbank' locally). (a) B~ could mean simply 'the bank of the stream(s)' (**burn, bank**), and the grassy dome of the **Fell** rises above streams including Holme Beck. (b) However, *burn* is rare in Lakeland, and corruption of some other el. such as **borran** 'cairn' might be suspected, as also suggested for BURN MOOR. Piles of stones on the fell are of unknown age and purpose, but the presence of Bronze Age cairns close by on CARLING KNOTT could well be significant.

BURNBANKS NY5016 (hab.) Bampton.
Burnbanks 1828 (HodgsonM), *Burn Banks* 1859 (*PNWestm* II, 194).
♦ (a) Perhaps 'the hill-slopes with cairns'; **borran, bank**. It is tempting to take 'Burn' as a corruption of *borran* 'cairn', for there is a Bronze Age round cairn here (LDHER 1563, SAM No. 22516) and several other monuments in the vicinity (see Turner 1991, 6–7;

Hodgson's map of 1828 indicates 'Giants Graves on Burnbanks' here). There would also be a clear parallel in the late 13th-cent. *le Burghanbank* (*PNWestm* II, 102). (b) Alternatively, 'the banks of the stream' (**burn, bank**) would also describe this site above HAWESWATER BECK, though *burn* is not common in Lakeland p.ns.

BURN BARROW: B~ B~ WOOD SD3582 Haverthwaite.
Burnbarow 1611 (Cartmel PR), *Burnbarrow* 1660 (Fell 1908, 200), *Burn Barrow Wood* 1851 (OS).
♦ Apparently 'the hill by the stream' (**burn, berg/barrow**), with the **Wood** secondarily named from it; there is also a B~ B~ Moss.

BURNEY SD2585 (hab.) Kirkby Ireleth, GREAT B~ SD2685 (979ft/298).
Burney 1691, 1742, *Bourney* 1775 (PR), *Burney; Great Burney* both 1851 (OS).
♦ B~ seems to take its name from the hill which rises above it, on which **Great** and Little B~ form the peaks, while B~ End stands on the far side. The lack of early spellings puts an etymology beyond reach, but the presence of prehistoric remains including a Bronze Age ring cairn SE of Great B~ (LDHER 2171, SAM 92) could suggest **borran** 'cairn' from OE *burgæsn* 'tumulus' and **how(e)** from ON **haugr** 'hill'. *Burney* also exists as a surname (see Hanks & Hodges 1988).

BURN KNOTT/KNOTTS SD3286 Colton.
Burn Knotts 1851 (OS).
♦ Presumably 'the rugged height(s) by the stream'; **burn, knott**, unless the 1st el. is of other origin (cf. BURNMOOR, Eskdale). The place is quite near Wear Beck.

BURN MOOR SD1592 Waberthwaite/ Ulpha.
From 1867 (OS).
♦ As with BURNMOOR in Eskdale, the 2nd el. is clearly **moor** 'upland waste', but the identity of 'burn' is uncertain. (a) **Burn** 'stream' from OE *burna* is the most obvious solution, though **beck** would be expected in Lakeland and no one stream is noteworthy in the local drainage system. (b) A reference to cairns (**borran**) would fit the landscape here, for the area is rich in cairns, especially to the W (LDHER 4781) and NE (LDHER 16598, SAM No. 32842); see also **moor**.

BURNMOOR: B~ LODGE NY1804 Eskdale, B~ TARN NY1804.
Burnam moore 1578, *Burmer Moore* 1587 (*PNCumb* 32), probably also = *Burham More* 1674 (CRO D/Lec/94); *Burnmoor Lodge* 1900 (OS); *Burman Tarne, Burneman tarne* 1570, *Burnmoor Tarn* 1784, 'probably identical with *Burmeswater* 1322, *Brumberwater* 1539 ... *Broumeberwater* 1542' (*PNCumb* 32).
♦ *PNCumb* reasonably judges that 'the forms are too inconsistent for any derivation to be suggested'. However, they do seem to rule out the possibility of **burn** 'stream', while on the other hand the presence of MAIDEN CASTLE and other prehistoric monuments might suggest a corruption of **borran**. See also **lodge, moor, tjǫrn/tarn** 'small mountain pool'. The Lodge is a gamekeeper's cottage.

BURNS FARM NY3024 St John's.
Bournes 1530, *Burnys* 1537, *Burnesse* 1533–8, *Burnes* 1552, *Bourns* 1564, (*The*) *Bowrness* 1569 (*PNCumb* 316).
♦ A difficult name, with at least three possible explanations. (a) 'The mounds', with **borrans**, but no archaeological

remains here are recorded in LDHER. (b) 'The stronghold on the headland', from OE *burh* 'fortified place' and OE **næss** 'headland'; but the place is on a modest shelf which would hardly amount to a *burh* site or a *næss* in this area. (c) 'The streams', from **burn**, is unlikely in light of recurrent *-nes-* in the spellings cited, and in others. **Farm** has been added relatively recently.

BURNT HORSE NY2928 (1361ft/416m.) Underskiddaw.
From 1867 (OS).
♦ Probably 'the neck of land where burning has taken place', with 1st el. as in BRUNDHOLME and **hals/hause**. *Hause* would be appropriate to this spur above SALEHOW BECK, between SKIDDAW and BLENCATHRA, and the curious spelling *Horse* is matched by, e.g., a 19th-cent. spelling of DALE HAUSE. Burning would be either deliberate, to encourage regrowth of vegetation, or accidental.

BURNT HOUSE NY1103 Irton.
From 1774 (DHM), 1802 (CN 1205).
♦ Although it is usually tracts of open country rather than buildings that are referred to as 'burnt' in Lakeland p.ns (see BRUNDHOLME), the three dozen examples of Burnt **House** in *OSGazetteer* suggest that the name is to be taken at face value.

BURNTHWAITE NY1909 (hab.) Nether Wasdale.
From 1725/6 (Wasdale Head PR).
♦ Possibly 'the clearing by the stream'. The 1st el. may well be **burn** 'stream', referring to Fogmire Beck, but modern 'burn' could also result from **borran** 'cairn, burial mound', though no archaeological remains are recorded in LDHER. The 2nd el. is from ON **þveit**, later **thwaite** 'clearing'.

BURN TOD NY2832 (hill spur) Ireby and Uldale.
From 1867 (OS).
♦ A baffling name, which applies to a spur of KNOTT, Caldbeck, lying between Burntod Gill and Hause Gill. (a) **Burn** 'stream' could be the 1st el., given the proximity to two streams, while 'Tod' could conceivably be a word meaning 'brushwood' or 'brushy vegetation' (OE *todd*, ON *toddi*, *PNWestm* I, 134, II, 293). (b) Some corruption may have taken place, e.g. of *Burnt **Dodd** 'rounded, compact summit' or *Burnt Odd 'point', as in GREENODD. (c) *Tod* meaning 'fox' seems impossible as a generic 2nd el.

BURROW HOUSE SD3891 Cartmel Fell.
From 1851 (OS).
♦ The **house** is beside Burrow Bridge, but the origin of 'Burrow' is elusive. It may be the surname derived ultimately from OE *be(o)rg* 'hill, mound' or from OE *burh* 'fortified place' (Hanks & Hodges 1988; they indicate that the form Burrow is frequent in Yorks and Lancs).

BURTHWAITE NY1828 (hab.) Wythop.
Burthwait in Wythop 1722 (Embleton PR).
♦ The name probably refers to a building of some kind, and has 2nd el. **þveit/thwaite** 'clearing', but the exact identity of the 1st el. is obscure in the absence of early spellings. Possible etymons include ON *búr* 'store-house' (cf. BOREDALE), OE *būr* 'cottage, dwelling', and ON **birk(i)** 'birch(es)' (cf. Burthwaite, *PNWestm* II, 108).

BURTNESS: B~ COMBE NY1714 Buttermere, B~ WOOD NY1715.
Birkness (*Field*) 16th cent. (spelling not specified, Winchester 1987, 142), *All*

the lands called *Birkmesfield* [sic] *and Gatescath* 1794 (Hutchinson, quoting 'the several licences of alienation', II, 126, n.), *Burtness*; *Burtness Comb*; *Burtness Wood* all 1867 (OS).

♦ B~ is probably 'the headland where birches grow'. The 1st el. is from ON **birk(i)**'birch(es)', as is supported by the local form Birkness (cf. Oppenheimer 1903, 154, who also mentions traces of abundant birches). The 2nd el. is ON **nes** 'headland', loosely used since the long, steep contours flank BUTTERMERE without projecting into it. B~ **Combe** is a classic corrie or cirque (i.e. **comb(e)**¹). B~ **Wood** is planted with conifers, but also contains deciduous trees including birch.

BUSCOE: B~ SIKE NY2505 Langdales.
Probably = *Bowesscard, Bouescarth, Bouscard* 1242 (*Furness Coucher* II, 562–4); *Buscoe Sike* 1862 (OS).

♦ Probably 'the gap by the bow/Bow Fell' or 'Bowe's gap or pass', with 1st el. either *Bow(e)* referring to BOW FELL (q.v.) or a curve in the landscape (cf. BOWSCALE), or a pers.n. *Bowe*, plus **skarð** 'gap, pass'; also **syke/sike**. The identification of *Bouscard, Bowes-scard* with Buscoe seems correct from its place in the 1242 charter bounds. Buscoe is the steep valley down which B~ Sike runs, on the SE flank of Bow Fell. At its head is the col now called Three Tarns Gap, in which THREE TARNS, formerly named *Tarns of Buscoe*, lie, and hence Buscoe probably included this high point and was originally the name of Three Tarns Gap. *Syke/sike* normally refers to more sluggish, less headlong streams than this one.

BUSK HOUSE NY3003 Langdales, B~ PIKE NY3004.

Busk 1862 (OS); *Busk Pike* 1862 (OS); cf. *the Busk* 1696, *the Bush* 1698 (Grasmere PR).

♦ Busk is 'bush, thicket', from a ME *busk* of OE or ON origin (*VEPN: busc, *buskr*); see also **house, pike**.

BUTTERMERE NY1815 (lake), B~ NY1716 (hab. & parish), B~ FELL NY1915, B~ MOSS NY1916.
(Lake:) *water of Buttermere* 1343 (*PNCumb* 33); (hab.:) *Butermere* 1230 (*PNCumb* 355); *Buttermeer fells* 1687–8 (Denton 51), *Buttermere Fell* 1867 (OS); *Buttermere Moss* 1839 (Ford 77).

♦ 'Butter lake, the lake with good pasture-land'; from OE *butere* 'butter', conveying the fertile nature of the flat alluvial land at both ends of the lake, plus **mere**¹ 'lake'; also **fell** 'high ground', **moss** 'bog'. The village and parish are named from the lake. B~ **Fell**, marked N of B~ lake on the 1994 One Inch map, but S on the 1:25,000, refers to the grazing attached to B~ parish. B~ Moss is a high, wet plateau on ROBINSON. Buttermere, together with BROTHERILKELD and BUTTERWICK, has been linked in elaborate legend with one Earl Boethar, leader of Anglo-Scandinavian resistance to the Normans under Ranulf Meschin in the late 11th cent. (e.g. Nicholas Size, *The Secret Valley* 1929, 9, 12, 13; cf. Rosemary Sutcliff, *The Shield Ring*, 1956). Although the p.n. interpretations are incorrect, the legend has some basis insofar as *VCH* deemed it 'not disputed that the family of the Scotic ruler, Bueth or Boet, held its own against the Norman intruder', with possession of the barony of Gillesland, for fifty years after the Norman Conquest (*VCH Cumb* I, 305–6 and 306 n. 1). (Pls 5, 6)

BUTTERWICK NY5019 Bampton.
Buttyrwick, -wyk 1246, *Butterwik* & variants 1279 to 1777 (*PNWestm* II, 189).
◆ 'The farm where butter is made', from OE *butere* and OE **wīc** '(specialised) farm, settlement', a parallel to KESWICK and a likely name for this verdant spot beside the Pow Beck. Five more English examples of this name are cited in *DEPN*, four of them recorded in Domesday Book, 1086. See BUTTERMERE for a legendary explanation of this name.

BYERSTEAD NY1428 Embleton.
Byerstead 1726 (Lorton PR), *Biersteeds* 1814 (PR).
◆ Despite the absence of earlier forms, probably 'the place or farmstead with a hut or shed', from the reflexes of OE *bȳre* 'shed, hut' and OE **stede** 'place', as in Byerstead, recorded from 13th cent. (*PNCumb* 433).

C

CABIN, THE SD3093 (bay) Colton. From 1851 (OS).
♦ This is a bay in CONISTON WATER indicated in blue on the 1994 One Inch map; no building is shown there or on the 1851 map. The word *cabin*, of French origin, first appears in the sense of a small, rough habitation c.1440 (*OED*: *cabin* n., sense 2).

CADDY WELL NY1402 (spring) Irton. From 1867 (OS).
♦ *PNCumb* 403 suggests a connection with a Roger Cady, recorded as parochial chaplain of Irton 1472. The Caddy family were prominent in Gosforth in earlier centuries (Parker 1926, 38–9), and the surname appears quite frequently, e.g. in Irton and Muncaster PRs c.1700; see also **well**.

CAER MOTE NY2036 (antiquity) Bewaldeth.
Carmott 1687–8 (Denton 51), *Carmalt* 1777 (*PNCumb* 326), *Caer-Mot* 1780 (West 121).
♦ Perhaps 'the fort where wethers graze'. The 1st el. seems to be Cumbric *cair (W *caer*) 'fortified place or town', and indeed this is the site of two turf and timber Roman forts: a larger earlier one and a smaller later one (LDHER 882, SAM No. 23794). The 2nd el. may be the Cumbric cognate of W *mollt* 'wether' (*PNCumb* 326, citing Sir Ifor Williams). A hill and habitation of the same name lie just outside the National Park.

CAFFELL SIDE NY2617 Borrowdale. From 1867 (OS).
♦ 'The slope of Caffell' which may be 'calf hill' from *calf* (OE *calf*, ON *kalfr*) and **fell**; cf. CAW FELL. 'Side' probably means 'hill-slope' (from OE **sīde**), given the situation, though other possibilities cannot be ruled out in the absence of early spellings.

CAISTON BECK NY3910 Patterdale.
Keystone Beck 1860 (but 'pronounced as if written Kaystone Beck', OSNB).
♦ Without earlier forms C~ (also in C~ Glen) is obscure. From the comment in OSNB, *Keystone* must be a misleading spelling. It is conceivable that the name contains OE *cǣg* 'key' or its reflex, which seems to be applied in p.ns to certain topographical features, especially hill-spurs, or an OE *cǣg 'stone, gravel' (on both of which, see *VEPN*: *cǣg*). The 2nd el. seems to be **stone**, and the word **beck** is from ON **bekkr** 'stream', but the name need not date from the period of Scand settlement. The valley of C~ B~ is shown as Sandale on Hodgson's map of 1828.

CALDBECK NY3239 (hab. & parish), C~ FELLS NY3135.
Caldebek 11th cent. (in 13th-cent. copy), *Caldbec* c.1175 (*PNCumb* 275); *Caldbeck fells* 1687–8 (Denton 51).
♦ 'The cold stream'; ON *kaldr*, **bekkr**. The village and parish are named from the Cald Beck (*aquam de Caldebc* 1228, *PNCumb* 7). In connection with the name, John Denton noted that 'the brook [is] fed by at least an hundred cold springs flowing into it' from the fells (quoted Hutchinson 1794, II, 376, among several learned conjectures on the name Caldbeck).

CALDER, RIVER NY0506, NY0712 Ennerdale/ St Bridget Beckermet, C~ ABBEY NY0406 St Bridget, C~ BRIDGE NY0406 (hab.).

(River:) *Kalder* c.1200 (*ERN* 60), *aquam de* (river of) *Kalder, de Calder, Caldre* 1292 (*PNCumb* 7); (hab:) *Calder* 1178 (p), *Kaldir* c.1190; (C~ Abbey:) *Caldram* 1152, *Kaldra, Kaldre, Caldir, Caldre* 13th cent. (*PNCumb* 427), *Calder-Abbey* 1671 (Fleming 39), 1774 (DHM), 1867 (OS); (C~ Bridge:) *pons aque de Calder* (bridge of C~ River) 16th cent., *Calder Bridge* 1644 (*PNCumb* 427).
♦ Calder is probably 'the swift stream', from the Cumbric counterparts of W *caled* 'swift, hard' and *dw(f)r* 'river, stream', Brit. **dubro*- (*PNCumb* 7 and *ERN* 59–62); also *abbey* from OFr *abaie*, **bridge** from **brycg**. Calder Abbey, founded in 1134, has left the mark of its ownership in the names of the locality: ABBEY FLATTS, FARMERY, PRIORLING and PRIOR SCALES in this volume. See Winchester 1987, 152–8 on the history of the Abbey lands. C~ Bridge is the name of the bridge and the village beside it — a bridge rather than a ford being necessary given the depth of the river here.

CALDEW, RIVER NY3431 Mungrisdale/Caldbeck.
Caldeu 1189 to 1292 (*PNCumb* 7), *Calde* 1228 (*ERN* 63), *Caldew(e)* 1307 (*PNCumb* 7).
♦ Probably 'the cold river', from OE *cald* and OE *ēa* 'river', influenced by OFr *ewe* 'water'. Another 'cold stream', CALDBECK, is a major tributary. However, the possibility that this is a Brit. name meaning 'swift river' from the same root as CALDER cannot be ruled out altogether (see *ERN* 63). The fast-flowing C~ was a vital power source for mills in the industrial age.

CALEBRECK/CALEBRACK NY3435 (hab.) Caldbeck.
Kelebrycke 1588, *Calebreck* 1679 (*PNCumb* 278).
♦ Uncertain. The spellings of the 1st el. would be compatible with dial. *kale, cale, keal(e)* 'cabbage' or 'pottage' (from OE *cāl* or ON *kál*), but certainty is impossible. The 2nd el. may derive from ON *brekka* 'slope', which would suit the situation, though *brekka* has not been found in the older strata of Lakeland names (see *SSNNW* 82).

CALF COVE NY3510 Rydal.
From 1865 (*PNWestm* I, 210).
♦ This is a compact, rocky **cove** or corrie at 1970ft/600m., and an extremely unlikely place for a calf, as are CALF CRAG and CALFHOW, or the Calf Coves near the summits of GREAT END (NY2208) and GREY FRIAR (NY2500). If this is *calf* from OE *calf* and/or ON *kalfr*, therefore, it must have some transferred or whimsical sense (cf. CALF CRAG).

CALF CRAG NY3010 (1762ft/537m.) Grasmere.
From 1865 (*PNWestm* I, 199).
♦ Smith may be right to group this with other 'late names in this parish ... formed with the names of animals, etc.' (*PNWestm* I, 199), but the **crag** is far from likely pastures and farm sites. *Calf* here may therefore have its occasional figurative sense of 'small outlier' (Smith, *Elements*: *calf* (1)), referring to Calf Crag's position in relation to the bulkier Carrs (see PIKE OF CARRS).

CALFHOW PIKE NY3321 St John's.
Calfhou 1278, *pyke of Cauvey* 1589, *Calevo pyke* 17th cent. (*PNCumb* 312).
♦ C~ appears to be 'calf hill', from ON *kalfr* 'calf' and ON **haugr** 'hill' or its ME reflex; the **pike** is its highest point. The hill is cut off from the

valley by crags, and hence as unlikely a place for calves as the 'Calf' names above.

CALGARTH: C~ HALL SD3999 Windermere.
le Calvegart(rige) 1390–4, *Chalfegarth* 1442, *Calfgarth(e)* 1445, *(the) Calgarth* 1446 to present; *Calgarth Hall* 1777 (*PNWestm* I, 194).
♦ C~ is 'the enclosure for calves', with the reflexes of *kalfa*, gen. pl. of ON *kalfr* (or its OE counterpart *calf* 'calf'), and of ON **garðr** 'enclosure'. The pronunciation 'Co'garth' was noted in 1901 (see COWMIRE HALL). C~ **Hall**, a substantial mansion, predated 1777 considerably (Aitchison 1935).

CALVA, GREAT NY2931 (2264ft/690m.) Underskiddaw/Ireby and Uldale, LITTLE C~ NY2831 (2016ft/642m.)
Calva 1794 (*PNCumb* 320), *Great Calva*; *Little Calva* both 1867 (OS).
♦ Probably 'the hill where calves graze' with the reflexes of ON *kalfr*/OE *calf* 'calf' and ON **haugr** 'hill', cf. another Calva (*Caluowe* 1395, *PNCumb* 367) and CALFHOW PIKE. See also **great**, **little**.

CAM CRAG NY2611 Borrowdale.
From 1867 (OS).
♦ Probably 'the **crag** or rocky height on or near the ridge', with 'Cam' ultimately from ON *kambr* 'ridge, crest' (see **comb(e)**²) and referring either to the Crag itself or to ROSTHWAITE Cam, a rocky summit a little to the NW (illustrated in Wainwright 1960, Rosthwaite Fell 5). 'Crooked crag' is also possible (again see **comb(e)**²).

CAM SPOUT CRAG NY2105 Eskdale.
From 1867 (OS), seemingly *Spout Crag* 1802 (*CN* 1219).
♦ 'The rocky height by Cam Spout, the waterfall on the crest/the crooked waterfall.' Cam Spout is a set of waterfalls on SCAFELL. As with CAM CRAG, 'Cam' could mean either 'ridge, crest' or 'crooked'. The word *spout* appears in later ME as *spoute* 'spout, water-outlet' and is of doubtful, but probably ON, origin. As well as its normal application to a pipe or projecting lip, it is applied in the 16th–17th cent. to a waterspout or cascade (*OED*, sense 6), which is clearly the sense here. The **crag** 'rocky height' is a short way to the S.

CANNY HILL SD3785 (hab.) Staveley-in-Cartmel.
Kanny Hill 1616/7, *Cannyhill* 1618 (Cartmel PR).
♦ Although *canny, conny* is a dial. word meaning 'nice, pleasant', Canny here is the name of a family which appears frequently in association with C~ H~ in the early 17th-cent. Cartmel PRs; see also **hill**.

CAPE, THE NY3613 Patterdale.
From 1860 (*PNWestm* II, 224).
♦ This is the only instance in this volume of the word *cape* 'headland', adopted into ME from OFr *cap*. The Cape projects E from the top of ST SUNDAY CRAG.

CAPELL CRAG NY2411 Borrowdale.
From 1570 (*PNCumb* 353).
♦ Probably 'horse crag' with ME and dial. **capel** 'horse, nag', plus **crag**, here a cluster of outcrops on THORNYTHWAITE FELL. There is a noticeable (though not unique) predilection for references to fauna in the crag names hereabouts: BULL (twice), HIND, EAGLE, HERON and DOVENEST.

CAPPLE BARROW SD4295 (642ft/196m.) Crook.
Caplesberg 1170–84, *Capelsbarghe* 1558–69 (*PNWestm* I, 179).
CAPPLEBARROW NY5003 (1683ft/513m.) Longsleddale/Fawcett Forest.
From 1836 (*PNWestm* I, 162).
♦ 'Horse hill', from ME and dial. **capel** 'horse, nag' and **barrow** from OE **berg** 'hill, mound'.

CAPPLEFALL NY5203 Fawcett Forest.
Capull fall 1542 (*PNWestm* I, 139).
♦ Probably 'horse felling' from ME **capel** 'horse, nag' (cf. nearby CAPPLEBARROW, and CAPPLERIGG) and the reflex of OE (*ge*)*fall* 'fall, felling, clearing of timber'. The steep, rocky location could suggest an alternative sense of *fall*, and that this might be an unusual case of a name commemorating an accident.

CAPPLERIGG SD4794 (hab.) Underbarrow.
Capellrig c.1692 (Machell II, 111).
♦ 'Horse ridge'; **capel**, **hryggr/rigg**. The name applies to a hill and the farm below it. A track, C~ Lane, goes over the hill — perhaps an old packhorse route?

CARHULLAN NY4918 (hab.) Bampton.
Kerholand 1336 (CW 1922, 302), *Carholand* 1420, *Carehullend*, *Carehullan* 1540 (*PNWestm* II, 189–90). Stress is on 2nd syllable.
♦ A historically important but problematic name. The 1st el. (a) appears to be the Cumbric **cair* (cf. W *caer*) 'fortified place' (or, as Nicolaisen suggests for Cumbric territories, 'a stockaded farm or manor-house' (2001, 208). (b) An OE or Gaelic *carr* 'rock' would not match the site. Smith suggests that it could refer to TOWTOP KIRK (*PNWestm* II, 190), but this seems rather forced. The final two syllables are also obscure. (a) I am grateful to Dr Oliver Padel for the suggestion that the likeliest explanation is a Brittonic pers.n. with the common suffix *-an*, though the exact form would be unclear. (b) A pre-existing OE p.n. such as *Hōland*, consisting of OE *hōh* 'heel-shaped hill' or OE **hol(h)** 'hollow' plus **land** is suggested in *PNWestm* II, 190, following Ekwall, but neither of these would fit the topography particularly well. Carhullan has been cited in various arguments concerning British-Anglian relations, and if it did result from addition of a Cumbric el. to a preexisting OE p.n. it might constitute evidence for a Cumbric resurgence (see Introduction, p. xix–xx), but given the uncertainties surrounding its interpretation it is difficult to use as historical evidence.

CARK: HIGH C~ SD3882 (hab.) Staveley-in-Cartmel, C~ HALL ALLOTMENT SD3782.
Karke 1491 (*Lancs*197); (High Cark:) *Ouer Carke* 1606 (*PNLancs* 199), *High Carke* 1747 (Finsthwaite PR); *Cark Hall Allotment* 1851 (OS).
♦ (a) Cark is probably from Cumbric **carreg* 'rock', referring to the ridge on which it stands (Ekwall, *PNLancs* 199, citing OW *carrecc*). (b) Ekwall suggested as an alternative that the name may have been a counterpart to a W *carrog* 'brook, stream', referring to the river Eea (*PNLancs* 197). **High** C~ is distinguished from Cark (*Nether Cark* 1601, Cartmel PR), which lies at SD3676 outside the National Park. **Hall** is presumably from High Cark Hall at SD3782. The use of *allotment* to refer to a small portion of land assigned to a particular person or purpose dates from 17th cent. (*OED*, sense 4); it is

common in the minor names of South Lakeland, though unique among the names in this volume.

CARLETON: C~ HALL SD0898 Drigg, HALL C~ SD0797, C~ HEAD SD0898.
Karlton c.1240, *Carleton* 1334; *Carlton Hall* 1578 (*PNCumb* 377); *Hall Carleton* 1733 (*PNCumb* 377); *Carleton Head* 1819 (PR).
♦ C~ is 'the settlement of churls or peasants', from a compound consisting of OE *ceorl* 'a free peasant' (with Scand influence on its first consonant, which would otherwise be spelt *Ch*) and **tūn** 'farmstead, settlement, village'; also **hall**, **head**, here 'upper part'. 'On 18th-cent. and early 19th-cent. maps the farm now known as Hall Carleton was marked clearly as Carleton Hall, the house . . . now called Carleton Hall . . . being marked as a church or chapelry without name' (Fair 1931, 65). The pattern of '*hall* plus p.n.' seen in Hall Carleton is favoured locally: see **hall**.

CARLING KNOTT NY1120 (1781ft/ 543m.) Loweswater.
Carling-knot, Carline-knot 1780 (West 138), *Carlingknot* 1794 (Hutchinson II, 135).
♦ 'Old woman's craggy height.' 'Carling' is probably from ON *kerling* 'old woman, crone', as also in the Yorkshire hill-names Carling Howe and Carlinghow (*PNNYorks* 151, *PNWYorks* II, 180, respectively), which have 12th- or 13th-cent. spellings in *kerling-*. Ekwall suspected some reference to witches in the Yorkshire names (*DEPN*), but *kerling* is used in Iceland of distinctive rocks and stones (as in Kerlingarfjöll, the name of a mountain range), and *carling* may indicate personification of the hill or a feature on it, cf. CARL (SIDE) and **man**.

Kerling also occurs quite frequently in Norw mountain- and habitation names; it is interpreted in *NG* as possibly referring to ownership by a widow (e.g. vol. XV, 50). **Knott** is appropriate to this distinct and rocky peak. The hill is distinguished by two Bronze Age cairns, one on the summit (LDHER 1068, 1069, SAM Nos. 27655, 27654).

CARL SIDE NY2528 (2448ft/746m.) Underskiddaw.
Karlesheved 1368 (Brearley 1974), *Carl Side (Tarn)* 1823 (GreenwoodM), *Carl Side* 1867 (OS).
♦ As in other N. hill-names containing 'Carl(e)', this may derive from ON *karl* 'man, usually old and low-status', just as *carling* as in CARLING KNOTT is probably from *kerling* '(old) woman'. 'Carl' could refer to a cairn or other summit feature (cf. **man**, **lad** and *PNWestm* II, 13–14). If the 1368 form truly refers to Carl Side, the 2nd el. is 'height, head' (OE **hēafod**/ME **heved**) rather than 'hill-slope' (OE **sīde**) or 'shieling' (ON **sætr**).

CARROCK FELL NY3433 (2174ft/663m.) Caldbeck/Mungrisdale, C~ BECK NY3334 Caldbeck.
Carroc 1208, *verticem de* (summit of) *Carrock* 1568, *the mountain Carrak, Carrick* 1610 (*PNCumb* 305), *Carrock fell* 1687–8 (Denton 51); *Carrock Beck* 1867 (OS).
♦ C~ is 'the rocky height', from Cumbric **carreg* 'rock', and hence of the same origin as CARK. This is a natural fortress of igneous and volcanic rocks, unique among the N. fells of Lakeland, and site of an ancient fort. **Fell** has been added to the mountain-name, and the **beck** 'stream' lies N of the Fell.

CARRS, GREAT NY2701 Coniston/Ulpha, LITTLE C~ NY2701.
Both from 1839 (Ford 31), 1850 (OS).
♦ 'The greater and lesser crags', for these form part of the craggy ridge to the W of TILBERTHWAITE FELLS. The word *carr* is found in Old Northumbrian texts, with the sense 'rock', and is thought to be of Celtic origin, cf. Gaelic *carr* 'rock shelf', W *carreg* 'rock' (Watson 1926, 433); also **great**, **little**.

CARSLEDDAM NY2527 Underskiddaw.
(*Skiddaw Tent is called*) *Carsleddam* 1800 (*CN* 804).
♦ An enigmatic name, which looks as though it might contain Gaelic *carr* 'rock' or Cumbric **cair* 'fortified place' as 1st el., but no convincing explanation is forthcoming at present. This is a prominent, if not quite independent, peak on the SKIDDAW massif.

CARTER GROUND SD2292 Dunnerdale.
From 1666/7 (original spelling not specified, Broughton-in-Furness PR).
♦ 'The farm of the Carter family'; surname plus **ground**, matching a pattern also seen in HARTLEY G~, JACKSON G~, PICKTHALL G~, STAINTON G~ and STEPHENSON G~, all of which are or were farms resulting from cultivation of the fell around the R. LICKLE and its tributaries. The 1666/7 record relates to a family named Carter, and the surname appears frequently in the PRs for Broughton.

CARTMEL SD3878 (hab. & parish), C~ FELL SD4188.
Ceartmel 12th cent., *Cartmel* 12th cent. (frequent), *Kertmell* 1157–63, *Cermel* 1187, *Carmel* 1188, *Caertmel* c.1188, *Karmel* 1190, *Kartemel* 1199 (*PNLancs* 195); *Cartmelefell* 1537, *Carpmanfell* 1577 (*PNLancs* 200).

♦ C~ itself lies just outside the Lake District National Park, but enters into other names (C~ FELL and STAVELEY-IN-C~). The age, linguistic origin and meaning of C~ is something of an enigma. If, as seems linguistically likely, the 2nd el. is ON *melr* 'sandbank', the 1st el. may well also be ON, possibly **kartr* 'rough, poor soil', encouraged by the cognate OE *ceart* (see *PNLancs* 195–6 for detail), but re-modelling or replacement of an earlier Celtic name cannot be ruled out. Cartmel seems to correspond with the place called *Cherchebi* 'village with a church' in Domesday Book, 1086, and J. C. Crowe (1984), pointing out the absence of sand at Cartmel, suggests that *Cherchebi* may have been the original name, while the name Cartmel could have belonged to, and described, an earlier site from which the church was relocated. (I am grateful to Dr Mary Higham for confirming that Cartmel stands on limestone.) C~ **Fell** is the 'high ground belonging to Cartmel', W of the WINSTER, and the village there, hence also the former township and modern civil parish of that name.

CARTMELL FOLD SD4491 Crosthwaite.
From 1535 (in 1669 copy, *PNWestm* I, 83).
♦ 'The pen or farm of the Cartmell family' (surname plus **fold**), who appear in local records of the 16th to 18th cent. (*PNWestm* I, 83); the surname is from the nearby p.n. CARTMEL.

CASTLE, THE NY4720 (1581ft/482m.) Askham.
From 1865 (*PNWestm* II, 201).
♦ This is an area of WHITESTONE MOOR near the boundary with Barton, but there is 'no known archaeological

reason for the name' (LDHER 15433); **castle**.

CASTLE CRAG Examples at:
NY2416 Borrowdale.
From 1769 (Gray 1088).
NY2918 St John's.
From 1867 (OS).
NY3011 St John's.
From 1794 (Hutchinson II, 154).
♦ 'The rocky height with a fortification, or like a fortification'; **castle**, **crag**. These may be named from their formidable appearance, like a natural fortress, from the presence of a man-made fort, or both. The Borrowdale example is a jutting fang in the 'Jaws of Borrowdale' on which perched a univallate Iron Age hillfort (LDHER 1232, SAM No. 23680); this may be the *borg* 'fort' of BORROWDALE (q.v.). In St John's, a fort is marked on OS maps at NY2918 (LDHER 5497, SAM No. 23682), while at NY3011 the name refers to a fortress-like buttress above WYTH BURN. Hutchinson commented on the presence of 'freestone ... which has been got at some distance', a well, and three tiered defensive trenches on the S side (1794, II, 154), but there is no archaeological record on LDHER.

CASTLE HOW Examples at:
NY2030 Wythop.
From 1780 (West 118).
NY2300 Ulpha.
Castle How 1867 (OS), possibly = *Castle How or Hall* (undated, among field-names listed *PNCumb* 438).
NY2534 (1004ft/305m.) Ireby and Uldale.
From 1867 (OS), possibly cf. *John de Kaystre al. Castre* 1411 (*PNCumb* 328).
NY3007 Grasmere.
Castle Howe 1865 (*PNWestm* I, 199).
♦ 'The hill with a fortification, or like a fortification'; **castle**, **how(e)** from

haugr. Hutchinson noted that the Wythop C~ H~ 'has trenches cast up in the most accessible parts' (1794, II, 154; cf. Collingwood 1924). This is an iron age hillfort (LDHER 886, SAM No. 23792). The Ulpha example is a compact height in a strategic position which has traces of a rectangular stone building of unknown date and purpose (LDHER 1360). It is also close to DEMMING CRAG (q.v.), and jointly the two sites and their names inspired Collingwood to ask, 'Have we chanced upon another pre-Norman chieftain and his stronghold?' (1918, 97 and 238). In Ireby and Uldale, a medieval moated site lies one mile NW of the hill (LDHER 885, SAM No. 23796), and this could explain the pers.n. recorded in 1411. However, it may be too far from Castle How to account for the p.n., and no archaeological remains are known there (LDHER N11034). At Grasmere, there is 'a possible circular stone structure, much of which is covered in vegetation' (LDHER 32530), so *castle* may refer to this, though it could also describe the natural rocky fortress formed by Great and Little Castle How (distinguished on larger scale maps) together with Raw Pike.

CASTLEHOWS POINT NY4522 Matterdale.
Castlehow's Point 1867 (OS).
♦ 'Probably to be associated with the family of Thomas *Castelhow*', named in Watermillock PR for 1579 (*PNCumb* 257). The surname presumably derives from a place called CASTLE HOW(E), possibly one of those above. The **point** projects into ULLSWATER.

CASTLE INN NY2132 Bassenthwaite.
From 1838 (PR), 1867 (OS).
♦ The motivation for this name is uncertain, since the nearest known

fortification is at CASTLE HOW, Wythop, a mile away across BASSEN-THWAITE LAKE. **Castle** is known as an inn name from at least the 17th cent., since an example is among the 164 hostelry names recorded in Pepys' diary (Cox 1994, 100). The word *inn* is from OE *inn* 'a dwelling or lodging place', which came to be used in ME of a public house or small hotel. Several Lakeland hostelries have names containing *inn*, but the only other one named on the One Inch map which is the basis for this volume is the PHEASANT INN.

CASTLE NOOK NY2316 Above Derwent. From 1867 (OS).
♦ 'The fortress-like remote spot'; **castle, nook**. This may refer not to a man-made structure, but to the natural bastion formed by crags at this point. This is an old mining site, but no fortifications have been found here.

CASTLERIGG NY2822 St John's (hab.), C~ FELL NY2819 Borrowdale/St John's.
Castlerig 1256 (*DEPN*); *Castle rigge fell* 1610 (*PNCumb* 312).
♦ 'The ridge with/by the fortification'; **castle, hryggr**; also **fell**. Ekwall in *DEPN* specifies Derwentwater Castle, but no conclusive evidence of this has been found (Collingwood 1904, 277–83, Bott 1994, 10–11). The magnificent C~ Stone Circle (SAM No. 22565) is a mile from the present C~, and it is possible that the name refers to it. C~ Fell is the grazing land belonging to what used to be a separate township.

CASTLE ROCK NY3219 St John's.
The Castle Rocks of St. John's 1776 (Hutchinson 122), possibly = *Castel Lindolf/Liadolf/Liudolf* 13th cent. (*CW* 1916, 225); *Castle-rocks* 1784, cited in *PNCumb* 316 in relation to this place, is in fact the one near Derwentwater.
♦ 'The fortress-like crag'; **castle** plus *rock* as in BLEA ROCK. 'A shaken massive pile of rocks, which stand in the midst of this little vale, disunited from the adjoining mountains; and have so much the real form and resemblance of a castle, that they bear the name of The Castle Rocks of St. John's' (Hutchinson 1776, 122). The inspiration for the castle in Scott's *Bridal of Triermain* (*Poetical Works* 553–84), the rock came to be known as Castle Rock of Triermain, and indeed is shown as such on earlier OS maps (e.g. 1966 OS One Inch). Collingwood believed that this was the site of a pre-Norman fortification (1916), and if correct, *castle* may also refer to this. However, no evidence has been found, and there is no LDHER record.

CASTLESTEADS NY5125 Lowther.
Castlesteads (end of word unclear) c.1692 (Machell I, 655A), *Castlestead* 1859 (*PNWestm* II, 184).
♦ 'The place or farmstead on the site of the castle'; **castle, stead**. There are documentary traces of a castle (a *castellum* mentioned 1174 and c.1250 is thought to be this one, *PNWestm* II, 184), and on the ground a rectangular ramparted enclosure of the medieval period (LDHER 3832).

CAT BANK SD3097 (hab.) Coniston.
Catt(e)ban(c)ke 1600/1, 1645, *Cotebancke* 1645 (PR).
♦ 'The slope frequented by wild cats'; **cat(te), bank(e)**.

CAT BELLS NY2419 (1481ft/451m.) Above Derwent.
Catbel(close) 1454 (Winchester 1987, 107), *Catbels* 1794 (Hutchinson II, 163).
♦ 'The bell-shaped hill frequented by

wild cats'; **cat(te)**, **bell(e)**. The two summits (one more prominent) could explain the plural name. (Pl. 12)

CAT BIELDS NY1207 Nether Wasdale. From 1867 (OS).
♦ 'The refuges of wild cats'; **cat(te)**, **bield**. This is a rock-strewn point on the S spur of SEATALLAN. Hutchinson remarked on the abundance of wild cats on the mountain-tops of Wasdale (1794, I, 581).

CATHOW NY0414 (hab.) Ennerdale.
Katthay 1578 (*PNCumb* 386), *Catay* 1643, *Catta* 1677/8 (PR).
♦ Probably 'the enclosure frequented by wild cats', from ME **cat(te)** and ME *hay* 'enclosure', derived from OE (*ge*)*hæg* or *hege*. The present form seems to have been influenced by the more familiar **how(e)** from ON **haugr** 'hill, mound'.

CATSTYE CAM/CATSTYCAM/ CATCHEDICAM NY3415 (2920ft/890m.) Patterdale.
Casticand 1671 (see below), *Catstee-Cam* c.1692 (Machell I, 714), *Catchidecam (called by Camden, Casticand)* 1770 (Gray 1095), *Catchedicam* 1805 (Scott, 'Helvellyn', *Poetical Works* 703), *Cachedecam called locally Catsty Cam* 1828 (HodgsonM), *Catstye Cam* 1860 (*PNWestm* II, 224).
♦ Seemingly 'the ridge by Catsty(e), the cats' steep path', from ME **cat(te)** 'wild-cat', with the reflexes of ON **stígr** or OE **stīg** '(steep) path', and of ON *kambr* 'comb, crest, ridge' (see **comb(e)**²), though the form Catchedicam is puzzling. The place may have been frequented by wild cats, or perhaps it was whimsically thought to take a cat's agility to reach this high summit. Machell used *Catstee* to refer to BROWN COVE, the corrie below C~ C~ where GLENRIDDING BECK rises (c.1692, I, 716). A much-quoted couplet from Camden's *Britannia* contains this name: 'Skiddaw, Lauvellin and Casticand | Are the highest hills in all England' (1695 edn, cf. Fleming 1671, 61; also Hutchinson 1794, II, 166). (Pl. 20)

CAUDALE: C~ BECK NY4011 Patterdale, CAUDALEBECK FARM NY4011, CAUDLE/CAUDALE HEAD NY4110, CAUDALE MOOR NY4110.
Cawdale 1753 (CW 1899, 125); *Cawdale Beck (Bridge)* 1769; (Caudalebeck Farm:) *Caudalebeck* & variants 1860 (OSNB); *caput* (head of) *Cawdell* late 12th cent., *Caldell heade* 1588; (Caudale Moor:) *Caldell mosse* 1588 (*PNWestm* II, 224), *Cawdal more* c.1692 (Machell II, 319).
♦ C~ is 'the cold valley' (ON *kaldr* 'cold' and **dalr**), seemingly from its north-facing situation; also **beck** 'stream', **farm**, **head**, **moor**. C~ Head is a high, enclosed cirque of crags at the head of the valley, and the Moor (previously **Moss**) the high ground above it. The spelling of Caudle (Head), with the weakening of the final element which was already in evidence in the 12th cent., reflects local pronunciation more accurately than Caudale. Cf. CAWDALE.

CAUSEWAY END SD3484 (hab.) Haverthwaite.
Caserend 1626/7, *Cawserend* 1676/7, *Cawseyend* 1679, *Casewayend* 1688, *Casser end* 1696 (Colton PR).
♦ 'The end of the paved track'; **causeway**, **end**. The Causeway crosses RUSLAND POOL and the wet land E of it. Several 17th-cent. spellings are included here to show their variety.

CAUSEWAY FOOT NY2921 (hab.) St John's.
Cassafott 1564, *Caussafoot* 1565 (*PNCumb* 316); *Causeway-foot* 1770 (Gray 1097); cf. *Kawsayheade* 1602 (*PNCumb* 316).
♦ 'The lower end of the paved track'; **causeway**, **foot**. This is close to the present A591, but may be associated with 'an old road, constructed in part as a causeway, [which] exists at NY 289222' (Clare 1999, 74).

CAUSEY PIKE NY2120 Above Derwent.
Cawsey-pike 1776 (Hutchinson 135), *Causeway Pike* 1800 (*CN* 779, but *Causey Pike* in later entries).
♦ 'The peak by the causeway or paved track'; **causey**, **pike**. Though not close to it, the mountain may be named from the former Roman road referred to as (*le*) *Cauce/Chauchey* etc. in 13th-cent. documents, which, aided by a stone bridge, traversed the marsh between DERWENTWATER and BASSENTHWAITE (*CW* 1921, 154, 171, Margary road no. 753, 1957, II, 90, 92). The prominent peak and knuckle-like summit ridge of C~ P~ makes it a distinctive landmark from there.

CAW SD2394 (1735ft/529m.) Dunnerdale, C~ MOSS SD2494.
Calf(heud) 1170–84 (*PNLancs* 194), *Caw* 1851 (OS); *Coe Moss* 1802 (*CN* 1225), *Caw Moss* 1851 (OS).
♦ Probably 'the calf', in the sense of a smaller outlier (to the more massive range including the OLD MAN OF CONISTON); cf. CALF CRAG. The 12th-cent. form means 'calf head', presumably the summit. C~ **Moss** is the low wet ground to the NE.

CAWDALE NY4817 Bampton, C~ BECK NY4717, C~ EDGE NY4717.
Caldell (*grandge*) 1449, *Cawdell* 1564;

Cordale Beck; *Cordale Edge* both 1860 (OSNB).
♦ C~ is 'the cold valley', from the ON adj. *kaldr* also found in CAUDALE and CALDBECK, plus **dalr**. The valley gives its name to the stream (**beck**) running through it and the scarp (**edge**) above it. Other derogatory names in Bampton include HUNGRY HILL and WILLDALE. The beck runs E, with steep slopes on either side which, especially to the S, restrict the sunlight. Noble saw *Cordale* as 'only a corruption of *Cundale*' (1901, 153), but this is misleading.

CAW FELL NY1310 (2405ft/733m.) Ennerdale/Nether Wasdale, CAWFELL BECK Ennerdale NY1009, CAW GILL NY0909.
Cawffelde 1578 (*PNCumb* 387), *Cawfell* 1687–8 (Denton 51), *Caw Fell* 1826 (CRO D/Lec/94); *Cawfell Beck* 1867 (OS), previously *Great Cogill Beck* 1826, *Great Cawgill Beck* 1827 (plans CRO D/Lec/94); *Caw Gill* 1867 (OS), previously *Little Cogill Beck* 1826, *Little Cawgill Beck* 1827 (plans CRO D/Lec/94).
♦ Probably 'calf hill', *calf* from OE *calf*, plus **fell** 'hill'; also **beck** 'stream', **gil(l)** 'stream in a ravine'. The 1578 spelling points to 'calf field', but both elements are puzzling, given the high, remote, situation, which might suggest *fell* 'mountain' rather than *field* as the original form. Literal reference to a calf or calves seems unlikely, and it may be that the hill was thought of as 'calf' to its higher neighbour, HAYCOCK (cf. CAW). Cawfell Beck and Caw Gill both rise on the SW slopes of C~ F~.

CHAPEL NY2331 Bassenthwaite, CH~ BECK NY2331.
Chapell 1670 (PR); *Chapel Beck* 1867 (OS).

♦ The word *chapel*, of OFr and ultimately Lat origin, is first recorded before 1225 (*OED*). The name denotes BASSENTHWAITE chapel and the area around it. The present building, at NY2231, was erected in 1878 to replace earlier buildings nearby (Swift 1966, 277). The **beck** or stream flows close by.

CHAPEL CRAGS NY1615 Buttermere.
From 1867 (OS).
♦ The elements are simply *chapel* as above, and **crag**, but the motivation for the name is less certain. The architecture of this rocky rampart might possibly explain the name, but it also directly faces BUTTERMERE Church.

CHAPEL HILL SD1097 (587ft/179m.) Muncaster.
From 1867 (OS).
♦ *Chapel* as in CHAPEL, **hill**. The hill is above Chapel Wood, and both appear to be named from the 'monument known as "Chapels"' (*CW* 1934, 42) on the flank of the hill.

CHAPEL HOUSE SD3785 Staveley-in-Cartmel.
Chapelhowse 1623/4 (Cartmel PR).
CHAPEL HOUSE SD4692 Underbarrow.
From 1777 (*PNWestm* I, 104).
CHAPEL HOUSE FARM NY2536, CHAPELHOUSE RESERVOIR NY2535 Ireby and Uldale.
Chappel-house 1653/4 (Uldale PR).
♦ For *chapel*, see CHAPEL, plus **house**; also **farm**. The first two examples are close to their respective church or chapel, and, at least in the Underbarrow case, belonged to it (*PNWestm* I, 104). The third is two miles SE of St James' Church, Uldale, which is documented from the 13th cent., and about a mile from the Victorian St John's (on these, see Swift & Bulman 1959, 52). A 14th-cent. chapel of ease dedicated to St Mary also existed, though in decay in the 16th cent., and its location is unknown (Swift 1977). The **reservoir** is 20th-cent.

CHAPEL STILE NY3105 (hab.) Langdales.
Langdalle Chappell steele 1682 (Hawkshead PR), *Chapel stile* 1800 (*CN* 800).
♦ 'The (place by) the chapel stile or steps'; *chapel* as in CHAPEL plus **steel/stile**. This is named from *Langdale Chappell*, recorded 1660 (*PNWestm* I, 206) and drawn by Machell (c.1692, II, 153). Church or chapel stiles, often of stone, were an alternative to gates in the walls of churchyards. The registers of several Lakeland parishes record similar p.ns formed from *Chapel*, *Church* or *Kirk* and *Stile* or *Steel*, but this is the only one to appear on the modern One Inch map. Of the Church Stile in Grasmere (*Kirke Stile* 1580), Simpson notes 'there used to be a stile into the Churchyard opposite this house' (1928, 283).

CHAPPELS FARM SD1684 Whicham.
the Chappell 1661, *Chappells* 1718 (*PNCumb* 445), *Chappels* 1867 (OS).
♦ *Chapel* as in CHAPEL; **farm** has, as frequently, been added to a pre-existing name. The place lies a mile SW of the present chapel at Hallthwaites in Millom, and close to a parish boundary, so the reason for the name is not immediately clear.

CHARLESGROUND SD1192 Waberthwaite.
From 1774 (DHM).
♦ The 1st el. may be, as usual in 'Ground' names, a surname, although there are no persons of that name in the index to the PRs. At 427ft/130m., this is a characteristic site for a farm

named **ground**, though the el. is less common hereabouts than in Furness.

CHARLEY HOUSES SD4193 Crook.
From 1857 (*PNWestm* I, 179).
♦ This is presumably the pers.n., the pet form of *Charles*, plus **house**.

CHESTNUT HILL NY2823 (686ft/209m.) Keswick.
From 1797 (*PNCumb* 303).
♦ 'The **hill** where chestnuts grow.' Though not native to the Lake District (nor to the British Isles), the horse chestnut has been planted quite widely in Lakeland valleys (see Halliday 1997, 32–3). *Chestnut* first appears in the 16th cent., though it had an OE precursor (see *OED*: *chestnut*; *chesteine*, *chesten*).

CHURCH BECK SD2998 Coniston.
From 1788 (CrosthwaiteM), previously *Saincte Martine Beck* c.1620 (*CW* 1929, 274).
♦ This is the lower stretch of the stream that flows from LEVERS WATER, past the church in CONISTON village and into CONISTON LAKE. The word *church* (which alternates with *kirk* in Lakeland p.ns) is from OE *cirice*; see also **beck** 'stream'.

CHURCH HOUSE SD1993 Ulpha.
Kirkhouse 1867 (OS).
♦ This is situated very close to ULPHA church, which has 17th-cent. features (Pevsner 1967, 195). Elements are Standard English *church*, replacing dial. *kirk* from ON *kirkja*, and **house**.

CINDERDALE BECK NY1619 Buttermere.
Cinderrill [*sic*] *Beck* 1867 (OS), *Cinderdale Beck* 1900 (OS).
♦ Slag-heaps near another stream of this name suggest an early bloomery or iron-smelting furnace (Parker 1926, 74). OE *sinder* and ON *sindr* both meant 'slag', but the name is impossible to date. The **dale** is a narrow valley formed by the **beck** 'stream'.

CLAIFE (parish): C~ HEIGHTS SD3797.
Clayf 1272–80 (p) (*PNLancs* 219); *Claife Heights* 1851 (OS).
♦ 'The steep hill-side', from ON *kleif* — an extremely rare p.n. element in England, though paralleled in Norway and Iceland; also **height**. Claife Heights rise steeply to 886ft/270m. from the W shore of WINDERMERE.

CLAPPERSGATE NY3603 (hab.) Rydal.
Clap(p)ergate 1588 (*PNWestm* I, 209).
♦ This has been interpreted as 'the way over the rough bridge', from dial. *clapper* 'a rough bridge, probably one made of planks laid on piles of stones' and the reflex of ON **gata** 'way, road' (so *PNWestm*, which also admits the possibility of **geat** 'gate'); and the position of C~ at a crossing of the R. BRATHAY would certainly be compatible with this. However, doubt has been cast on this explanation since definite cases of *clapper* have only been found S of the Thames, and the name may instead be from *clap(er)-gate*, a swing-gate or a tilting stile (*VEPN*: *clap-gate*); as a further complication some p.ns containing *clapper* may be from ME *clapper* 'rabbit warren' (see *VEPN*).

CLAY GAP NY2938 (hab.) Caldbeck.
Claye Gappe 1594 (*PNCumb* 278).
♦ 'The cleft where the soil is clayey', with 1st el. ultimately from OE *clǣg* 'clay', plus **gap**. I am most grateful to Mr Christopher James of Clay Gap for confirming that 'the soil is an extremely heavy clay here', as witness a former brick and tile works nearby.

The landscape does not present a very obvious *gap*.

CLEABARROW SD4296 (hab.) Bowness.
From 1619 (Windermere PR), *Claybarrow* 1770 (JefferysM).
♦ 'The clayey hill', with 1st el. OE *clǣg* 'clay', as is clear from Jefferys' spelling (though his p.n. forms are often eccentric) and from the local clay soil (I am most grateful to K. R. Moffat, Hon. Secretary of Windermere Golf Club, Cleabarrow, for confirming this). The 2nd el. is **barrow** from OE **berg** 'hill, mound'. There are other p.ns in which *clǣg* qualifies a term for 'hill' (Smith, *Elements*).

CLERK, THE: see BISHOP

CLEWS GILL NY1315 Ennerdale.
From 1867 (OS).
♦ The **gill** is, as usual, a steep rocky ravine with a stream. 'Clews' is problematic, despite a close parallel in *Clewes*, a field-name in Nether Wasdale recorded in 1578 (*PNCumb* 442). (a) It could be a spelling of *clough*, *cleugh* from OE *clōh* 'ravine'. This might be geologically apt, since this is 'the dividing line between the Ennerdale and Buttermere Granophyre and the Skiddaw Slate' (Postlethwaite 1913, 138), though the pl. would be puzzling. (b) A surname derived from this word, or p.ns formed from it (*Clews* occurs especially in WMidlands, Hanks & Hodges 1988: *Clough*). (c) *Clews* also occurs as a NCy variant of a word meaning 'sluice', though the Cumberland form of this is given as *cloor* (*EDD*: *clow* sb. 1).

CLIMB STILE SD2590 (hab.) Kirkby Ireleth.

Climsteele 1649 (Torver PR), *Clim-Steel* 1753, *Climb style* 1780 (Woodland PR), *Climb Stile* 1851 (OS).
♦ **Stile** (ultimately from OE *stigel*) is a steep place (or sometimes a stone or wooden stile in the modern sense), while presumably 'Climb' is the noun meaning 'ascent' (ultimately from the OE verb *climban*) and reinforces *stile*, though it is unusual in p.ns. If 'Climb' were the verb, it would form an imperative p.n., also unusual. The place is reached by an ascent which is steep by local standards.

CLINTS CRAGS NY1635 Blindcrake.
Clints Crags 1867 (OS), cf. *Clints* 1810 (*PNCumb* 268).
♦ 'The rocky heights.' Dial. *clint* comes from ODan *klint* 'rock, crag' (cf. OWN *klettr*); it can refer specifically to hard, flinty rock (*EDD*). **Crag(s)** more or less duplicates the meaning, but distinguishes this place from the adjacent Clints Park.

CLOSE, THE NY1830 (hab.) Embleton.
Close 1705 (PR).
♦ 'The enclosure or farm'; **close.** This was still simply *Close* on the 1966 One Inch map.

CLOUGH HEAD NY3322 (2382ft/726m.) St John's.
From 1867 (OS).
♦ 'The top of the ravine, or hill above the ravine(s)', with ME or ModE *clough* from OE *clōh* 'ravine' plus **head**. The summit rises above a massive wall of rock and scree to its NW 'with one breach only where a walker may safely venture; [the crags] are riven by deep gullies, one of which (Sandbed Gill) is the rockiest and roughest watercourse in the Helvellyn range' (Wainwright 1955, Clough Head 2).

CLOVEN STONE NY3028 Threlkeld/ Mungrisdale.
From 1867 (OS).
♦ This rock, shaped by the elements into a natural sculpture, is one of several old boundary stones of this name in the Lake District and elsewhere (e.g. *the cloven gray stone*, recorded from Eskdale in 1587, *PNCumb* 391 and see *VEPN: clofen*). It is illustrated in Wainwright 1962, Mungrisdale Common 4. *Cloven* is past participle of *cleave* (OE *clēofan*, cf. ON *kljúfa*), and see **stone**.

COALBECK FARM NY2032 Blindcrake.
(Stream:) *Coltbecke* 1629 (*PNCumb* 9), *Cole beck* 1678 (OgilbyM); (hab.:) *Coalbeck* 1705, *Cole-Beck* 1712 (Isel PR).
♦ The place stands on Coal Beck, whose 1st el. is slightly obscure. Derivation from OE *colt* 'colt' (cf. COLTHOUSE) is favoured by the 1629 spelling, but if it were discounted one of the etymologies suggested for COLEDALE might apply. See also **beck** 'stream' and **farm**, which has been added relatively recently.

COCKENSKELL/COCKENSHELL SD2789 (hab.) Blawith.
Cockanscales 1284 (p), *Cokainscalis* 14th cent. (p), *Cockenscale* 1632 (*PNLancs* 214–5), *Cokenshell* 1736 (PR); cf. *Cokayn* 14th cent. (p) (*PNLancs* 203).
♦ 'The shieling(s) belonging to Cocken or Cockayne', a place in Dalton, Low Furness which was established by the monks of FURNESS. The final el. is clearly ON **skáli**, dial. **scale** 'shieling'. As for 'Cocken': (a) It seems to be named from (the Land of) Cockayne, a topsy-turvy realm of plenty and indolence created by medieval wishful imagination. (b) Collingwood, citing *Colganschales*

c.1320, stated that C~ 'must surely mean the "hut of Cogan", an Irish or Manx man ... who came over with the Norse' (1931, 3–4). However, the persistent spellings in *Co(c)k-* are against this.

COCKER, RIVER NY1327 Blindbothel/ Embleton/Lorton.
the Koker c.1170, *Kok'* 1195, *Coker* early13th cent. to 1305 (*PNCumb* 9).
♦ Apparently 'the crooked one', from a Brit. **kukrā* (*ERN* 84, *PNCumb* 9, Padel 1985, 62), though Jackson hints at reservations (1953, 578). 'Crooked' would be an apt description; see also CRUMMOCK (WATER).

COCKERMOUTH NY1230 (hab. & parish).
Cokyrmoth c.1150, *Kokermue* 1194, *Cokirmowth* c.1220 (*PNCumb* 361).
♦ 'The mouth of the COCKER', with the river-name above, plus OE *mūða* 'estuary'. Situated less than two miles outside the Lake District National Park, C~ is included in this volume because of its size and historical importance.

COCK HAG SD4493 (hab.) Crook.
From 1829 (*PNWestm* I, 179).
♦ 'The clearing frequented by woodcock', from the reflexes of OE **cocc** and of ON *hǫgg* 'cutting, place where timber is felled'.

COCK HOW, LOW NY0514 (hab.) Ennerdale.
Low Cockhow 1867 (OS), cf. *Cockhow* 1578 (*PNCumb* 387).
♦ C~ H~ is presumably 'the hill frequented by wild birds' (**cocc**, **haugr/how(e)**), referring to a small compact hill on whose N side Low C~ H~ is situated. A High C~ H~ is shown on the S side on the 1867 map.

COCKLAW: C~ FELL NY4803 Longsleddale.
Cokelaike(moore) 1238–46 (*PNWestm* I, 162); *Cockley fell (wall)* c.1692 (Machell II, 103), *Cockley Fell* 1836, *Cocklaw Fell* 1865 (*PNWestm* I, 162).
♦ 'Cocks' play, the place where blackcocks play', to judge from the 13th-cent. spelling, with the reflexes of OE **cocc** 'cock, male bird, wild bird' (or its gen. pl. *cocca*) and of ON *leikr* 'play, sport'; also **fell**. Names such as Cocklake, Cocklaw and Cockley are common, and seem to refer to the male of the black grouse, 'noisy and theatrical in his "leking"' (Gambles 1989, 185, cf. *PNCumb* 466: *cokelayk*). Grouse butts on the fell nowadays suggest that this is suitable terrain.

COCKLEY BECK NY2401 (hab.) Dunnerdale, C~ B~ FELL NY2501.
Cockley beck 1664 (Eskdale PR), (*Manor of*) *Cockley Beck* 1791 (*VCHLancs*, 407); *Cockley Beck Fells* 1802 (*CN* 1228), *Cockley Beck Fell* 1850 (OS).
♦ C~ is probably 'the place where blackcocks play'. (a) C~ looks characteristically OE, perhaps from OE **cocc** '(wood)cock' and OE **lēah** 'woodland clearing', but a name from the Anglian period would be exceptional hereabouts and medieval spellings are lacking. (b) A ME or hybrid compound meaning 'place where (black)cocks play' is therefore more likely, as in COCKLAW (FELL), which also appears with the spelling *Cockley*. The **beck** 'stream' from which the settlement is named is now C~ B~ Gill. The **Fell** is the high grazing E of C~ B~.

COCKLEY MOOR NY3722 Matterdale. From 1867 (OS).
♦ C~ may be 'the place where blackcocks play'. Like COCKLEY BECK, this is a relatively minor p.n. unrecorded until the modern period, and it too may have similar origins to COCKLAW (FELL). See also **moor**.

COCKPIT, THE NY4822 Barton/Askham. From 1886 (Taylor 1886, 337).
♦ This is a Bronze Age stone circle (LDHER 2944, SAM No. 22538, also described in Taylor 1886, 337–8), whose name points to its similarity to an enclosed arena for cockfighting, a Cumbrian pastime which probably continued long after it became illegal in 1835. The first Ordnance Surveyors remarked 'it is supposed by some to have been a Cock Pit' (OSNB 1860), and cockfighting could have taken place as part of the shepherds' meets hereabouts; however the absence of a deep pit might tell against this literal interpretation. The first citation for *cockpit* in *OED*, from 1587, is also the first in which it is used figuratively to describe a landscape feature — a mountain cirque; the earliest p.n. cited in *VEPN*: *cock-pit* is from 1573. The word is a compound of the reflexes of OE **cocc** and OE *pytt*. (Pl. 17)

COCKRIGG CRAGS NY3016 St John's. From 1867 (OS).
♦ Early spellings are lacking, but this may be *cock* (OE **cocc**) in the sense 'woodcock or blackcock, wild bird', plus **rigg** 'ridge' from **hryggr/hrycg**, here referring to a steep wedge of ground. The **crags** are the rocks cladding the ridge.

COCKS CLOSE SD5299 Strickland Roger. From 1857 (*PNWestm* I, 158).
♦ 'Enclosure or farm of the Cock family'; surname plus **close**. The family of James *Cocke* is on record locally from 1669 (*PNWestm* I, 158). The site is partially enclosed by a bend in the R. SPRINT.

COCKSHOTT/COCKSHOT: C~ POINT SD3996 Bowness.
Cocke shott, the Cockshott 1677, 1746 (*PNWestm* I, 187); *Cockshot Point* 1851 (OS).
♦ C~ is one of several ME and ModE minor names containing *cocksho(o)t*, a glade or path in woodland 'through which woodcocks, etc. might dart or "shoot" so as to be caught by nets stretched across the opening' (*OED*, cf. Cameron 1968–9, 15 and *VEPN*: *coccscyte*; see also **cocc**). The **point** projects into WINDERMERE; there is a C~ beside DERWENTWATER too (Ford 1839, 50 and NY2622 on larger scale maps).

COCKUP NY2531 (1655ft/505m.) Bassenthwaite, GREAT C~ NY2733 (1726ft/526m.) Ireby and Uldale, LITTLE C~ NY2633.
(*an high hill called*) *Coppack* 1687–8 (Denton 171, 172), *Cockup* 1867 (OS); *Great Cockup; Little Cockup* both 1867 (OS).
♦ Perhaps 'the blind valley frequented by wild birds'. If, as often, Denton's 17th-cent. spelling is unreliable, C~ could derive from OE **cocc** 'cock' (perhaps woodcock or black grouse, cf. nearby OVER WATER) and OE **hop** 'blind, rounded valley', with the valley name transferred to the hills, but without more evidence this cannot be proven. Little C~ is a flank of Great C~.

CODALE HEAD NY2908 Grasmere, C~ TARN NY2908.
Codale Head 1865 (*PNWestm* I, 200); *Codale Terne in Little Langdale* c.1692 (Machell II, 128), *Codale Tarn* 1828 (HodgsonM).
♦ The 1st el. of C~ is obscure, perhaps with the same possibilities as COLEDALE, which appears as *Cow-* in early modern times; see also **dalr/dale**, **head**, **tjǫrn/tarn**.

COFA PIKE NY3512 Patterdale.
Cove 1780 (West 50), *Cawkhaw Pike* 1860 (OSNB), 1920 (OS).
♦ 'The peak above the corrie.' (a) C~ may well be the word **cove**, preserved by coincidence in its OE spelling *cofa*. This is favoured by the 1780 spelling and the fact that the **pike** rises above Cawk Cove and the corrie or basin formed by GRISEDALE TARN. (b) *Cawkhaw* may be an alternative name rather than a variant of *Cove, Cofa*. The Ordnance Surveyors commented, 'This name of "Cawk" is the common name given by the miners and country people to sulphate of barytes/baryta, a good deal of which has been obtained here and an inferior kind is visible in the rock' (OSNB 1860).

COGRA MOSS NY0919 Lamplugh.
From 1867 (OS).
♦ C~ is of unknown origin. A reservoir now fills the boggy hollow of the **moss**.

COLDBARROW FELL NY2812 Borrowdale.
From 1867 (OS).
♦ Unless the form is deceptive, C~ is 'the cold, exposed hill', with the reflexes of OE *cald* and OE **berg**. **Fell** must have been added to an already complete name, hence 'the mountain called Coldbarrow'.

COLD FELL NY0509 St Bridget Beckermet.
Caldfell 1687–8 (Denton 87), 1777 (*PNCumb* 340).
♦ 'The cold, exposed **fell** or hill.' The original spelling, with the *-a-* of ON *kaldr* 'cold' and N. dial. *cald*, has, as so often, given way to one influenced by Standard English.

COLD HARBOUR SD4793 (hab.) Underbarrow.
From 1745 (*PNWestm* I, 104).
♦ 'The chilly refuge.' The word *harbour*, from ME *hereberȝe*, is used to refer to a coastal haven from the 13th cent. onwards, but is used more broadly, and over 300 instances of the p.n. C~ H~ or Coldharbour are recorded, mainly from after 1600 and with a particular concentration in SE England (Coates 1984, 73, 75). Coates has shown that, at least in London, 'the name *Coldharbour* was applied from c.1600 till *c.*1666 as a fashionable derogatory term for a miserable house (usually, it appears, at some distance from others)' (p. 75), but it is difficult to know how far a name such as the Underbarrow one arose through influence of this fashion and how far it arose spontaneously as a derogatory name for a farm in an exposed situation (cf. similar names such as Cold Comfort (Farm)).

COLD PIKE NY2603 Ulpha/Langdales.
From 1862 (OS).
♦ 'The cold, exposed peak'; Standard English *cold* plus **pike**. Though exposed, this is not obviously more so than the average.

COLEDALE: C~ BECK NY2122 Above Derwent, C~ HAUSE NY1821, HIGH C~ NY2222.
Coldale 1260 to 1578, *Colledale* 1339 (*PNCumb* 370), *Cowdall* 1569 (Crosthwaite PR); *Coledale Beck* 1867 (OS); *Cowdal Haws* 1800 (CN 805), *Cowdale Halse* 1803 (CN 1616; Coleridge also uses *Coledale* H~); *High Coledale* 1867 (OS).
♦ The 1st el. may be (a) ON *kol*/OE *col* 'coal, charcoal', alluding to charcoal burning; (b) *Kola/Kolla*, gen. case of an ON pers.n. *Koli* or *Kolli*, though the dominant medieval spelling *Coldale* lacks the expected medial vowel. Fellows-Jensen (*SSNNW* 115), like *PNCumb*, prefers (b) and within that option she points out that *Koli* explains the modern spelling better than *Kolli*. The 2nd el. is clearly ON **dalr** 'valley'; see also **beck** 'stream', **hals/hause** 'neck of land', **high**.

COLTHOUSE SD3698 Claife, C~ HEIGHTS SD3697.
Colthowse c.1535 (*PNLancs* 219); *Colthouse Heights* 1851 (OS).
♦ Presumably self-explanatory; *colt*, from OE *colt* 'young horse or ass', **house**, cf. Horsehouse, N. Yorks, and *Coltclose* 1485, *Coulthouse(wood)* 1677 (*PNCumb* 205). Although *colt* is unique in this volume, a Colt Park at Holm Cultram was 'originally the enclosure used by the Cistercian monks as a pasture for their young horses' (Elliott 1960, 98), and since Colthouse is close to HAWKSHEAD, an outpost of the great Cistercian house of FURNESS, it is tempting to wonder whether this might also have been monastic. The use of **Height(s)** matches CLAIFE H~ in the same parish.

COLTON SD3186 (hab. & parish), C~ BECK SD3186.
Coleton 1202, *Colton* 1332 (p) (*PNLancs* 216); '*Coulton* was the usual spelling until about 1850' (*VCHLancs* 383); *Coulton Beck* (as hab.) 1693 (PR).
♦ The 2nd el. is OE *tūn* 'farmstead, settlement, village'. The 1st may be (a) OE *cōl* 'charcoal', appropriate to this part of FURNESS FELLS where charcoal-burning continued well into the 20th cent. (b) An OE pers.n. such as *Cola* was favoured by Ekwall in *PNLancs*. (c) Ekwall, having found evidence for identifying *Cole* 1247 with the present C~ Beck, later suggested derivation

from a Brit. river-name Cole which might have meant 'hazel stream' (*DEPN*, cf. *ERN* 86, also Coates 2000, 318, 382), but if the river-name was Cole it could have been a back-formation, and the spelling would not tally well with OW *coll* 'hazel-trees' (I am indebted to Dr Oliver Padel for these points). See also **beck** 'stream'.

COLWITH NY3303 (hab.) Langdale.
Colleth 1578 (*CW* 1908, 148), *Collwth* 1619/20 (Hawkshead PR), *Collath* 1800 (*PNWestm* I, 206).
♦ (a) Probably 'charcoal wood', from ON *kol* 'charcoal' and ON **viðr** 'wood'; *kolviðr* also occurs as a compound meaning 'wood for charcoal'. The valley is still clothed with deciduous woods. (b) Less likely is the possibility that early spellings point to ON **vað** 'ford', which would have been at the site of the present Colwith Bridge over the BRATHAY.

COMB BECK NY2124 Above Derwent, COMB PLANTATION NY2124.
Both from 1867 (OS).
♦ Without early spellings or clear guidance from the surrounding landscape (where a C~ Gill and C~ Bridge are also found), it is not clear whether C~ here originally denoted a valley or a ridge: see **comb(e)**[1] and **comb(e)**[2]; also **beck** and **plantation**.

COMB CRAGS NY3313 St John's.
From 1862 (OSNB).
♦ 'The rocky heights at Comb, the corrie'; **comb(e)**[1], **crag**. The arc of crags forms part of the rocky wall enclosing a corrie or cirque.

COMBE: C~ DOOR NY2511 Borrowdale, C~ GILL NY2512, C~ HEAD NY2410.
All from 1867 (OS).
♦ The Combe is a high corrie or bowl-shaped valley in the BORROWDALE FELLS; **comb(e)**[1], with secondary names formed from OE **duru** 'door, col', ON **gil** 'ravine with a stream', and OE **hēafod** 'head, top, high place' or their reflexes.

COMMON, THE SD4299 Windermere.
Common 1611 (PR).
♦ This is the high, formerly unenclosed, ground around BANNER RIGG; see **common**. The 1611 spelling presumably referred to the farm here (Palmer 1944, 32), rather than the open ground.

COMMON FELL NY3820 (1811ft/552m.) Matterdale.
From 1589 (*PNCumb* 257).
♦ 'The hill on the common pasture/with common grazing'; **common, fell**. It is on WATERMILLOCK COMMON, which name is used locally in preference to C~ F~, probably to effect a distinction from MATTERDALE COMMON (Roger Connard and Keith Clark, pers.comm. gratefully acknowledged). The careful regulation of rights, for instance the digging of peats, 'upon the said Common-fell or in any other place of our Common', is documented in 17th-cent. materials in *CW* 1884, 30–1.

CONISTON SD3097 (hab. & parish), C~ FELLS SD2999, C~ MOOR SD3099 (1194ft/364m.), MONK C~ SD3198, MONK C~ MOOR SD3296, C~ WATER SD3094.
Coningeston' 1157–63 (*PNLancs* 215); *Coniston fell* 1693 (Hawkshead PR), *Coniston Fells* 1800 (*CN* 801), 1851 (OS); *Coniston Moor* 1850 (OS); *Monke Coneston* 1568 (*PNLancs* 219); *Monk Coniston Moor* 1851 (OS); *Coningston Water and sometimes anciently Thurstonwater* 1671 (Fleming 33), *Coniston Lake* 1788 (CrosthwaiteM),

Coniston Water, called in some old books *Thurston Water* 1823 (Otley 7), previously *turstiniwatra* 1157–63, *Thurstainewater* 1196 (*PNLancs* 192).

♦ 'The king's estate or village.' The 2nd el. is OE **tūn**, and the whole name may, like numerous English Kingstons, be from OE *cyninges-tūn*. Anglian names are rare hereabouts, but the site is low-lying and favourable (*SSNNW* 350, 186), with mineral resources as well as agricultural possibilities. Scand influence is, meanwhile, shown by the *-o-* of early and modern spellings, and Ekwall speculated that this could have been the centre of a 'small Scandinavian mountain kingdom' (*PNLancs* 215). Monk C~ refers to ownership by FURNESS Abbey; the word *monk* goes back to OE *munuc*. This was a division of the township, and the house now named Monk C~ was formerly *Water Head House* (OS 1851). The main village of C~ was until recently often distinguished as Church C~, and prior to that Fleming C~ (*VCHLancs*, 365). See also **fell**, **moor**, **water**. The older name of C~ Water, 'Thurstein water' applied also to the upper reaches of the R. CRAKE, which are still *Thurston Vale* on 19th-cent. maps. It provided W. G. Collingwood with the name of his fictional hero in *Thorstein of the Mere* and seems to be commemorated in THURSTON and THURSTON VILLE. For C~ OLD MAN, see OLD MAN OF C~. (Pl. 29)

COOKSON PLACE SD0899 Irton.
Cookstone Place 1713 (*PNCumb* 403), *Cockstone Place* 1774 (DHM), *Cookson Place* 1833, *Cookstone Place* 1835 (PR).
♦ 'The Cooksons' farm.' Like other names in **place**, this farm name probably has a surname as the specific, though I have not (in a rapid survey) noted the surname in the Irton PRs. *Cookstone* may be a variant of *Cookson* as *Johnston(e)* is of *Johnson*, though it is not listed in the surname dictionaries of Reaney (1976) or Hanks & Hodges (1988).

COOMB NY3132 Mungrisdale, C~ HEIGHT NY3132 (2058ft/627m.).
Comb; *Comb Height* both 1867 (OS).
♦ Despite the lack of early spellings, the topography suggests that Coomb here is 'the ridge or crest' (**comb(e)**^{2}), which rises steeply to C~ **Height**.

COP STONE/TOP STONE NY4921 Askham.
Cop Stone 1859 (*PNWestm* II, 201), *Top Stone* (1st edition One Inch map, cited *PNWestm* II, 201).
♦ The 1st el. seems to be OE *cop(p)* 'top, summit' (*OED*: *cop*, n. 2), as also in HIGH and LOW KOP and WHINCOP; see also **stone**. This is a Bronze Age megalith, possibly part of a stone circle (LDHER 2955, SAM No. 22533) which, with other 'antiquities and oddities of Moor Divock', is illustrated in Wainwright 1957, Loadpot Hill 3.

COPELAND: C~ DISTRICT NY1203 several parishes, C~ FOREST NY1407 Nether Wasdale.
Caupalandia c.1125, c.1135, *Coupland(a)* c.1125 to 1500, *Copelandia* c.1140 (*PNCumb* 2); *foreste de Coupland* c.1282 (in 1594 copy), *Couplandfell* 1321 (*PNCumb* 37).
♦ C~ is 'land that has been bought', from ON *kaup(a)land*. The name may give valuable evidence of a large-scale, organised and rather consensual take-over of territory by Scandinavians, probably in the 10th cent. The western barony of C~ was one of the great feudal fiefdoms established soon after the Norman Conquest, and the **forest** was the upland part of it (see Winchester 1987, 14–22). The barony

then formed the basis for later administrative divisions including the Ward of Allerdale above Derwent and the present Copeland District. On the word *district*, see ALLERDALE DISTRICT.

CORNEY SD1191 (hab.) Waberthwaite, C~ FELL SD1391, C~ HALL SD1190, HIGH C~ SD1292.
Cornai & variants c.1160 (in early copy) and frequently, *duas* (two) *Cornays* 1190–1200; *Corney Fell* 1685; *Corney Hall* 1685; *High Corney* 1657 (*PNCumb* 364).
♦ (a) Apparently 'the island frequented by herons', from OE *corn*, assumed to be a variant of *cran*, *cron* 'heron', plus OE *ēg* 'island', possibly referring to the oval of high ground W of C~ and almost surrounded by streams. There is nowadays a heronry nearby on the ESK estuary (Danby 1994, 182). (b) This solution is preferred to OE *corn* 'grain, corn' since the situation makes a reference to arable farming unlikely (*PNCumb* 364), and because it is doubtful whether OE *corn* was applied to growing crops. (Smith, *Elements*, claims not, and an instance from AD 897 in *OED*: *corn*, n. 1, sense 4a is ambiguous; there are clear ME examples of growing *corn*.) C~ **Fell** is a long, steep tract of upland grazing; see also **hall** and **high**.

CORNHOW NY1422 (hab.) Buttermere. *Cornhowe* 1612 (Lorton PR).
♦ Probably 'the hill where corn grows'; *corn* from OE *corn* 'corn, grain, seed' (see under CORNEY), plus **how**(e). The elevation is extremely modest — suitable for arable farming.

CORRIDOR ROUTE NY2208 Eskdale. Not on 1867 or 1900 OS maps.
♦ The name speaks for itself, and seems to be a modern fellwalkers' name for the 'well-blazoned track' formerly known as the Guides Route which is 'the one and only easy passage' from STY HEAD to the col leading to SCAFELL and SCAFELL PIKE (Wainwright 1960, Scafell Pike 16). *Corridor* and *route* were both adopted into English from French.

COTE FARM NY4418 Martindale. *Coate* 1646/7 (PR).
♦ 'The cottage', with the reflex of OE **cot**(e), and **farm** added later. The relationship to COTEHOW, on the other side of STEEL KNOTTS, is unclear.

COTE HOW NY3605 (hab.) Rydal. *cotehowe* 1597 (*CW* 1906, 82), (*the?*) *Coothow* 1601 (Grasmere PR), *Coat-how* 1631 (*CW* 1908, 190).
COTEHOW/COTHOW NY4318 (hab.) Martindale.
Star Inn or Cote How 1828 (HodgsonM), *Coate How* 1829 (Parson & White), *Coathow* 1838 (*PNWestm* II, 218).
♦ 'The hill by/with the cottage'; **cot**(e), **haugr/how**(e). The Rydal name now refers to a row of cottages, while the Martindale C~ is a farm which was once the Star Inn.

COVE NY4323 (hab.) Matterdale. *Cove* 1581 (*PNCumb* 257).
♦ The name presumably refers to the rather secluded situation, though it is unlike the high, wild sites of most Lakeland places named **cove** (q.v.), and might refer instead to a hut.

COVE, THE SD2696 Torver. From 1851 (OS).
♦ A classic **cove** — a corrie quite high on the OLD MAN OF CONISTON.

COWCOVE BECK NY2103 Eskdale. From 1867 (OS).

♦ The **beck** or stream rises in a damp basin or **cove**. Whether *cow* (from OE *cū*) can also be taken at face value is less certain, since, at over 1000ft/305m., this seems a somewhat remote spot for cattle grazing.

COWMIRE: C~ HALL SD4288 Crosthwaite.
Caluemyer 1332, *Calmire* 1535 to 1748, *Cawmyer* & variants 1589 to c.1722 (*PNWestm* I, 81); *Cawmire or Comer Hall*, locally pronounced 'Co'mer' 1901 (Cowper 120).
♦ 'The bog where calves graze', from *kalfa*, gen. pl. of ON *kalfr* 'calf', and/or its OE counterpart *calfa*, plus ON **mýrr** 'bog, mire'; also **hall**. The form 'Cow-' has come about partly by regular sound change, partly by association. The Hall is a 17th-cent. mansion of six bays built onto a pele tower (Cowper 1901, Palmer 1944, 34, Pevsner 1967, 245).

COWSTY: C~ KNOTTS NY4505 Kentmere.
Cowsty 1836; *Cowsty Knotts* 1857 (*PNWestm* I, 166).
♦ C~ is probably 'the cow-track', with the reflex of OE *cū* plus **sty(e)**; also **knott** 'compact hill, rocky height'. 'Wide greens, lanes and tracks with names like Cowsty and Swinsty and Rake (a track to the pastures) . . . honeycombed the settlements' (Atkin 1991, 74). The Knott(s) constitute a rugged hill.

CRABTREEBECK NY1321 (hab.) Loweswater.
Cabletrebecke 1703/4 (CRO D/Law/1/273), *Crabtree Beck* 1756 (PR).
♦ *Crab* meaning 'crab-apple', 'wild apple' is common in dial. (*EDD*: *crab*, sb.2); *tree* derives from OE *trēo(w)*, and the **beck** 'stream' has given its name to the farm.

CRACOE NY4724 (hab.) Barton.
Crakey 1666, *Cracow* 1697, *Crakah* 1744, *Crakhall* 1765 (*PNWestm* II, 212).
♦ 'The hill frequented by crows', from ON *kráka* 'crow' and **haugr/how(e)** 'hill, mound' or their reflexes.

CRAG Examples at:
NY1931 (hab.) Setmurthy.
Crag 1578 (CRO D/Lec/301/158v–161, pers.comm. Dr Angus Winchester), *Cragg* 1764 (PR), *Crag* 1774 (DHM).
SD3392 (hab.) Satterthwaite.
From 1851 (OS), possibly = *Cragge* 1681 (Colton PR).
SD4694 (hab.) Underbarrow.
ye/the Cragg(e) 1606 to 1634 (*PNWestm* I, 104).
CRAGG FARM Examples at:
SD1099 Irton.
the Cragg 1646 (*PNCumb* 403).
SD1297 Muncaster.
Cragg 1745 (PR).
SD4897 Nether Staveley.
Crag c.1692 (Machell II, 114, ed. Ewbank, 109); cf. *Staveley Crage* earlier 16th cent. (*PNWestm* I, 174).
♦ '(The place below) the rocky height' (**crag(ge)**), e.g. Beacon Crag in Satterthwaite and RAVEN CRAG in Muncaster. The Setmurthy example stands below a height which counts as a *crag* in the relatively gentle local terrain. In Underbarrow, High and Low C~ are distinguished on larger scale maps, as also on Jefferys' map of 1770. In three instances, **farm** has been added to the simplex name Crag(g).

CRAG FARM HOUSE NY0814 Ennerdale,
CRAG FELL NY0914.
Crag 1606 (*PNCumb* 387), 1867 (OS); *Crag Fell* 1867 (OS).
♦ Simply 'the rocky height' (**crag(ge)**), with **farm** and **house** added to the originally simplex name. Crag **Fell** is the summit above Revelin Crag.

CRAGG HALL/CRAG HALL SD1887 Millom Without.
Crag(g) Hall 1599, 1610, *Crag hale* 1607, 1657 (*PNCumb* 418).
♦ 'The hall at Crag, the rocky height'; **crag(ge)**, **hall**. The rocky knoll of KNOTT HILL, which rises just to the W, may be the *crag* in question.

CRAGHALL, LOW SD1891 (hab.) Ulpha. From 1774 (also *High Crag Hall*, DHM).
♦ This is on a small rocky elevation (**low, crag**), as is High Craghall just to the W, and it is possible that **hall** has been substituted for **how(e)** 'hill' (from ON **haugr**): cf. three examples of Hall from *haugr* in *PNCumb* 477, and cf. nearby CASTLE HOW.

CRAG HILL NY1920 (2753ft/839m.) Buttermere/Above Derwent. From 1867 (OS).
♦ 'The hill topped by rock outcrops'; **crag, hill**. Three major crags (EEL C~, Scott C~ and SCAR C~) distinguish this mountain which, with its near neighbour SAND HILL, is a rare case of a significant Lakeland fell named *hill*. It is often known as EEL CRAG. (Pl. 5)

CRAG HOUSE NY1002 Irton.
Santon Craggehowse 1563 (*PNCumb* 403).
CRAG HOUSE SD4396 Nether Staveley.
Crag House 1862 (OS), possibly = *Cragg house* 1618 (Windermere PR).
CRAG HOUSES NY1717 Buttermere. From 1867 (OS); cf. *Crag House Crags* 1799 (*CN* 539).
♦ All three locations are next to rock outcrops: Bull Crag in the Irton example, and High House Crags in the Buttermere one, known to Coleridge as Crag House Crags; **crag(ge), house**.

CRAG WOOD: CRAGWOOD HOUSE NY3900 Troutbeck.
Crag Wood 1859 (OSNB).

♦ Simply from **crag** 'rocky height' and **wood**, with **house** added later. OSNB describes Crag Wood as 'forest wood with scattered fir trees', and the house first appears on OS maps, named *Crag Wood*, in 1920.

CRAG WOOD: HIGH C~ W~ SD4485 Witherslack, LOW C~ W~ SD4485.
Cragg Wood 1804 (*PNWestm* I, 78); *High Crag Wood*; *Low Crag Wood* both 1862 (OS).
♦ 'The woods on the rocky height' (**crag, wood**, also **high, low**) — the steep escarpment forming the S. stretch of WHITBARROW SCAR.

CRAKE, RIVER SD2987 Blawith.
Crayke, Crec 1157–63, *Craic* 1196, *Craik* 13th cent. (*ERN* 102, also *PNLancs*191).
♦ 'The rocky one', from Cumbric **creig* 'rock, cliff', Brit. **crac̦io-*, which would seem to be apt (*ERN* 102; also found in BLINDCRAKE). Lindkvist's hypothetical ON **kreik*, referring to a bend in a stream, is unnecessary (1912, 69).

CRINKLE CRAGS NY2505 (2816ft/815m.) Eskdale/Ulpha/Langdales, CRINKLE GILL NY2504 Langdales.
Crinkle Crags 1862 (OS), *Cringle Craggs* (Ellwood 1895, 14); *Crinkle Gill* 1862 (OS).
♦ These are the **crags** or rocky heights that are either: (a) 'jagged, corrugated' (with 1st el. from OE *cringol* 'wrinkled, twisted'); or (b) 'curving ' (with ON *kringla* 'circle', as assumed by Ellwood, 1895, 14). This is a good example of a whole mountain named from its rock outcrops, and whatever the origin of 'Crinkle' it has become customary to refer to the bumps and gullies of the summit ridge as 'Crinkles'. The **Gil(l)** 'ravine with stream' rises just below the Crags.

CROASDALE/CROSSDALE NY0917 (hab.) Ennerdale, C~ BECK NY0816 Lamplugh/Ennerdale.
Crossedale 1279 (p), *Crozedal* 1294 (p), *Cros(s)dale* 1593 (*PNCumb* 385); *Crossdalebec* 1230 (*PNCumb* 10).
♦ 'Cross valley', perhaps 'valley marked with a cross' (*PNCumb* 385); **kross, dalr** 'valley'; also **bekkr/beck** 'stream'.

CROFT END SD2491 (hab.) Broughton West.
From 1677 (original spelling not specified, Broughton-in-Furness PR), 1851 (OS).
♦ 'The end of the small enclosure'; **croft, end**.

CROFTFOOT NY0916 (hab.) Ennerdale.
From 1643/4 (PR), 1774 (DHM).
♦ 'The lower end of the small enclosure'; **croft, foot**.

CROFT HEAD NY4501 (hab.) Hugill.
From 1738 (*PNWestm* I, 171).
CROFT HEAD FARM NY4325 Matterdale.
ye Crofthead 1700; cf. *Crofte foote* 1589 (*PNCumb* 257).
♦ 'The top of the small enclosure'; **croft, head**, with **farm** added later in the Matterdale example.

CROFT HOUSE FARM NY2538 Ireby and Uldale.
Not on 1867 or 1901 OS, nor 1966 OS One Inch.
♦ Seemingly a recent name formed from **croft, house, farm**.

CROGLINHURST SD2189 (hab.) Broughton West.
Croglinhurst 1671 (original spelling not specified), *Croglynst* 1699, *Croglinest* 1699/1700 (Broughton-in-Furness PR); cf. *Croglin* 1767, *Clogland* 1815, *Croglin* 1831 (Kirkby Ireleth PR).
♦ Uncertain. The 2nd el. 'hurst' may derive from OE **hyrst** 'wooded hill(ock)', while the 1st el. may be comparable to Croglin, a river- and place-name that has been tentatively interpreted as from ON **krókr** 'bend' and OE **hlynn** 'torrent' (*ERN* 105–6, *PNCumb* 9, 183). Croglinhurst lies between the R. LICKLE and the steep slope to the W, and Ekwall suggested that the Lickle had formerly been named Croglin (*PNLancs* 191–2).

CROOK SD4695 (hab. & parish), C~ END SD4595 (hab.), C~ FOOT SD4393 (hab.), C~ HALL SD4594.
Crok(e) 1170–84 to 1647, *Crook(e)* 1220–46 to present (*PNWestm* I, 177); *Crook End* 1706 (*PNWestm* I, 179); (C~ Foot:) cf. *Low end of Crook* 1862 (OS); *Crook(e) hall (als. Chapell Howse)* 1631 (*PNWestm* I, 179), previously *Thwatterden-hall (or Crook-hall)* 1671 (Fleming 15).
♦ C~ is from ON **krókr** 'bend, or corner of land'. The village is in a triangle of land between the two main streams forming the R. POOL. C~ **End** and **Foot** are close to parish boundaries. C~ **Hall** was called *Thwatterden Hall* in the 17th cent. (above and Machell II, 341, c.1692).

CROOK, THE SD2094 (hab.) Ulpha.
Crooke in Ulpha 1655 (*PNCumb* 438).
♦ 'The bend or corner of land' (**krókr/crook**) — at a quite tight bend in the R. DUDDON.

CROOKABECK NY4015 Patterdale.
Crooksbeck 1770 (JefferysM), *Crockey beck* 1839, *Crookedbeck* 1860 (*PNWestm* II, 225).
♦ 'The bend in the stream, the corner by the stream', or possibly 'the crooked stream'. The main elements are from **krókr/crook** 'bend, corner of

land' and **bekkr** 'stream', but the original structure of the name is unclear, and the past participle *-ed* in the 1860 spelling may be a rationalisation of the earlier forms. Cf. CROOK-A-DYKE, CROOKAFLEET. The place is opposite a bend in GOLDRILL BECK.

CROOK-A-DYKE/CROOKDYKE NY4621 (hab.) Barton.
Crookendicke 1676, *Crokeadicke* 1679, *Crookdike* 1686, 1748, *Crooked Dike* 1734, *Crookaydike* 1787 (*PNWestm* II, 212).
♦ Either (a) 'the bend in the dike, the corner by the dike', from the reflexes of **krókr** and **dike,** and *-a-* possibly representing the preposition 'of'; or (b) 'the crooked dike', if the *-a-* represents the past participle ending of the verb *crook* 'to bend', as in CROOKED BIRCH. Cf. CROOKABECK, CROOKAFLEET.

CROOKAFLEET/CROOK-A-FLEET NY3633 (hab.) Mungrisdale.
Crook-a-Fleet 1815, *Crook of Fleet* 1864 (PR).
♦ Probably 'the bend in the stream', referring to a small tributary of the R. CALDEW, although the Caldew itself bends sharply just S of here. The 1st el. is **krókr/crook** 'bend, corner of land'. *Fleet* only occurs here and possibly in FLEETWITH in this volume, but it seems to be the dial. word *fleet*, meaning 'watershed, boggy ground where streams rise' (*EDD*: *fleet* 1, sense 11), which descends from ON *fljót* or OE *flēot* 'river, estuary' (usually referring to quite large rivers). The connecting *-a-* is likely here to represent *o'* or *of* (cf. Crook of Lune in Sedburgh, which is *Crooke of Loyne* 1614, *Crooka Loyne* 1647, *PNWYorks* VI, 267). Cf. the previous two names.

CROOK CRAG SD1998 Eskdale.
From 1867 (OS).

♦ 'Crook' usually derives from ON **krókr** 'hook, bend, corner of land', but the **crag** is not obviously in such a situation, and the reason for the name is not obvious.

CROOKDALE NY5306 Shap Rural, C~ BECK NY5306.
Crookdale 1570 (*PNWestm* II, 174); *Crookdale Beck* 1860 (OSNB).
♦ 'The valley with bends'; **krókr, dalr** or their reflexes, also **beck** 'stream'. There are numerous tight bends in the beck.

CROOKED BIRCH SD2687 (hab.) Blawith.
Crookberk 1720, *Crooked Birch in Subberthwaite* 1796 (Lowick PR), *Crake Birk* 1851, 1893 (OS).
♦ Distinctively crooked trees have been used as landmarks since medieval times, and a famous early occurrence of the word *crooked* (which may be partly due to ON influence) is in the Towneley *Second Shepherds' Play* line 403, where three shepherds pledge to meet *at the crokyd thorne*; see also **birk**. C~ B~ is close to BIRCH BANK FARM.

CROOK FARM/CROOKS SD3386 Colton.
Crooks 1679, *Crook* 1700 (PR).
♦ 'Crook(s)' (from **krókr** 'bend, corner of land') here probably refers to bends in the R. CRAKE, which are still in evidence despite modern embankments and drainage ditches. The farm was named Crooks on earlier editions of the One Inch map (e.g. 1966), **farm** being specified recently.

CROOKWATH NY3821 (hab.) Matterdale.
del Crokwath 1332 (p) (*PNCumb* 221–2).
♦ 'The ford at the bend' — a crossing place on a bend in AIRA BECK; **krókr, vað.**

CROPPLE HOW SD1297 (hab.) Muncaster. From 1753 (PR).
♦ Without earlier forms the meaning of the 1st el. is uncertain, but among the possibilities are derivation from OE *crop(p)*, *croppa* 'sprout, cluster of berries' or from ON *kroppr* 'lump', hence 'hump, hill-top' (cf. Smith, *Elements*: *crop(p)*). 'Cropple' could be a contraction of *Cropp Hill, to which explanatory How has been added. 'How' is clearly the reflex of ON **haugr** 'hill, mound', appropriate to this small, neat hill above the R. ESK.

CROSBY GILL SD1895 Ulpha, CROSBYTHWAITE SD1994 (hab.).
Crosby Gill 1867 (OS); *Crosbythwaite* 1702 (spelling not specified, CW 1966, 316), *Crossby Thwaite* 1705 (PR).
♦ These names and Crosby Bridge SD2093 might suggest a pre-existing p.n. Crosby (ON **kross** 'cross' and ON **bœr/bý** 'farm, village, settlement', which could be linked to the existence of a church in Ulpha). Names in *-by*, however, are virtually non-existent in this part of Cumbria, and it is possible that Crosby is transferred from elsewhere. The final elements are **gil(l)** 'ravine with stream' and **þveit/thwaite** 'clearing'. Crosbythwaite, like BIRKERTHWAITE, commemorates a process of recovery of the high waste between the ESK and DUDDON rivers.

CROSS NY0900 Irton.
From 1803 (Muncaster PR).
♦ This is 'one of the most highly decorated crosses in Cumbria', carved from sandstone in the first half of the 9th cent. (Bailey & Cramp 1988, 115–17). As to whether this should be regarded as a p.n. as well as an antiquity, OS maps have differed, and it has not been used locally as a p.n. within living memory (I am most grateful to Mrs Barbara Wright, churchwarden, and the congregation of St Paul's, Irton, for this information). However, its appearance in the 1803 PR supports its former status as a p.n. The word *cross* is from ON **kross**.

CROSS/THE CROSS (hab.) SD1094 Waberthwaite.
Cross in Waberthwaite 1680 (*PNCumb* 440).
♦ This farm is believed to be named from its situation about a quarter of a mile from a cross beside a ford over the R. ESK. Parker, discussing crosses in the area, found no other explanation (1909, 89); the word *cross* is from ON **kross**.

CROSS DORMONT/DORMANT NY4622 (hab.) Barton.
Trostermod 1202, *Trostormot* 1256 to 1333, *Trostormod(e)* 1275 to 1383, *Trostormond(e)* 1295 to 1401, *Crostormount* 1634 to 1777 (*PNWestm* II, 210), *Trostermont* — *some call it Crossdormont* c.1692 (Machell I, 381).
♦ 'Thormod's brushwood.' This is clearly an inversion compound. The 1st, generic, el. is taken to be ON *tros* 'rubbish, leaves and twigs gathered for fuel' and hence possibly 'undergrowth, brushwood'. The 2nd, specific, el. is the ON pers.n. *Þormóðr* (*PNWestm*, acknowledging Ekwall). Machell thought this was 'Tristram's mount', remarking that the nearby fortified *mount* 'is thought to have taken its denom[ination] fro[m] Tristram, one of King Arthur's knights' (c.1692, I, 690) but this cannot, unfortunately, be upheld. Nor is *Cross*, from **kross**, authentic in this name, but it is a good example of folk etymology.

CROSSES FARM NY4100 Windermere.
Crosses 1622 (PR); cf. *Crosshouse* 1558

(*PNWestm* I, 195).

♦ The cottages called 'The Crosses' may have been linked with the nearby chapel of ST CATHERINE'S (according to Brydson, 1911, 107); **kross**, with **farm** added later.

CROSS GATE NY5018 (hab.) Bampton.
Christ's Cross Gate 1907 (*CW* 1907, 303), 1920 (OS); cf. *Cristcroft* 1361, *Crystcroftes* 1505 (*PNWestm* II, 194).
♦ The 1361 and 1505 spellings imply 'Christian's enclosure or farm', if *Crist* is a shortening of the forename *Christian* which is preserved in nearby Christen Sike (*PNWestm*; see Noble 1901, 148 for another explanation of Christen/Christening Sike). An original **croft** 'enclosure, farm' seems to have been replaced by *cross* (from **kross**) under influence of the apparent reference to Christ, and conceivably also of a local standing cross which Noble (1901, 148) and Collingwood (1907, 303) linked with the name. **Gate** here may have its most obvious meaning (from OE **geat**), though 'road' from ON **gata** is also possible, as in Crossgates, Lorton, a crossing of the ways.

CROSS HOUSE SD1088 Bootle.
Croshouse 1668, *Crosshouse* 1750 (PR); cf. *The Cross* 1664 (*PNCumb* 348).
♦ The **house** is near an ancient crossroads, which may explain 'Cross' (see **kross**), but Parker pointed out that this is also a likely site for a raised cross, perhaps marking the boundary of the former SEATON nunnery, and functioning as a sanctuary cross (1909, 88–9).

CROSSLANDS SD3489 (hab.) Colton.
Crosslands 1630 (PR).
♦ (a) *Crosland* occurs in Domesday Book referring to a place in WYorks (*PNWYorks* II, 265) which is taken to be 'tract of land with a cross', and this could be the same. (b) Alternatively, 'Cross' in the adjectival sense 'transverse' would give the sense 'lands lying crosswise'; or (c) it could refer to crossroads, which Cowper notes is apt (1899, 60); **kross, land**.

CROSTHWAITE SD4491 (hab., parish of C~ and Lyth).
Crosthwaite & variants 1187–1200 to present, *Crossethwaite* & variants 1303 to 1607 (*PNWestm* I, 80).
CROSTHWAITE, GREAT NY2524 (hab. & former parish) Keswick, LITTLE C~ NY2327 (hab.) Underskiddaw.
Crostweit c.1150 (p); *Great Crosthayth* c.1220; (Little Crosthwaite:) *Crosttwait* c.1220, *parva Crostweyt in Bastenthwayt* 1292, *Lytyll Crostwat* 1563 (*PNCumb* 302).
♦ 'The clearing with a cross', from ON **kross,** originally borrowed from Gaelic, and ON **þveit** 'clearing'. C~ at NY2524, though now absorbed into KESWICK town and parish, is a former parish centre, the site of a splendid medieval church and before that, according to Jocelyn of Furness c.1200, the place where the missionary St Kentigern (St Mungo, cf. MUNGRISDALE) raised a cross in the 6th cent.; Jocelyn claimed that it was previously known as *Crosfeld* and that a church had been built there recently (see Whaley 1996, 106 and n. 70). The name therefore probably refers in some way to this Christian history. The example at SD4491 also gave its name to a township or civil parish; the present church there dates from 1878. (Pl. 10)

CROW HOW NY3605 (hab.) Rydal.
From 1865 (*PNWestm* I, 210).
CROWHOW END NY2200 Eskdale.
From 1867 (OS).

♦ 'The hill frequented by crows' (with elements descended from OE *crāwe* and ON **haugr**), and the **end** of such a hill. Though *crow* is reasonably common in English p.ns, these are the only occurrences in this volume. The carrion crow (*Corvus corone corone*) can nest high on fellsides (Hervey & Barnes 1970, 153, Ratcliffe 2002, 189), and in the Rydal example the compact hill presently covered with tall trees is a likely habitat. If to be taken at face value, this is one of several ornithological references in Rydal parish (see ERNE CRAG).

CRUMMOCK WATER NY1518 Buttermere. *lake of Crumbokwatre* 1230, *Crombocwater* 1307 (*PNCumb* 33).
♦ The meaning of C~ seems to be 'Crooked one', from Brit. **crumbāco-* 'crooked' (*ERN* 108, Jackson 1953, 509). This may have been an older name of the R. COCKER which flows out of the lake: its winding course, the apparent meaning of Cocker and the occurrence of *Cromboc* (12th cent.) as a river-name in N. Cumbria (*PNCumb* 10) would all favour this. Alternatively, the 'crooked' feature could be the lake itself, for HAUSE POINT creates a bend, which appears more prominent from certain viewpoints than others. **Water** is the main Lakeland term for 'lake'. (Pl. 6)

CUBBEN NY1301 (hab.) Irton. *Cubboone* 1668, *Cubbon* 1691, *Cubban* 1723 (*PNCumb* 403), *Cubbam* 1774 (DHM).
♦ Of uncertain origin, but conceivably 'the sheds'. Dial. *cub* 'cattle shed' is cited from 1546 in *OED*, while *EDD* lists *cub* sb. 2 as referring, in NCy and elsewhere, to a crib for animal feed, a coop, chest or store, and cites Middle Dutch *cubbe* 'cattle-stall, shed, barn' and a cognate in the dial. of Bremen. The dial. *cub* or an OE pers.n. *Cubba* are suggested as possible origins of Cubley, WYorks (*PNWYorks* I, 337). If the word had an unrecorded antecedent in OE, the *-en* (and the 1774 spelling) could conceivably reflect an OE dat. pl. ending in *-um*. The farm is in a quite high, isolated situation.

CUCKOO BROW WOOD SD3796 Claife. From 1851 (OS).
♦ Probably a recent and self-explanatory name. *Cuckoo*, a ME adoption from OFr, appears only here in this volume; the dial. word for the bird is *gowk*, from ON *gaukr*. See also **brow** 'hill-slope', **wood**.

CUMBLANDS SD0897 (hab.) Drigg. *Cumerlandes* 1578, *Cumlands* 1799 (*PNCumb* 377).
♦ 'Cumb-' is problematic, and no explanation is attempted in *PNCumb*. The other seven compounds with **-lands** in this volume have descriptive words ('Green', 'Long', 'New') as their specific, or else references to ownership ('Kirk-'). (a) This could conceivably be from dial. *cumber* meaning 'trouble, encumbrance', as in 'cumbersome' and in the dial. compound *cumber-ground* 'useless person or thing', applied especially to trees (*EDD*: *cumber* sb. 1; there are NCy examples though not specifically Cumbrian). (b) Alternatively, it is tempting to connect 'Cumb-' with the British inhabitants of Cumbria (see p. xix, and *PNCumb* 1 on the county name Cumberland), but without earlier spellings this is no more than fanciful. The place is on the small rise referred to in the entry for BELL HILL FARM above.

CUMBRIA WAY NY2519, NY2825, NY3134 etc., CUMBRIA CYCLE WAY SD0898, SD2088, SD3482 etc., CUMBRIAN MOUNTAINS NY2005, NY2717 Several parishes.
Cf. *Cumbra land* 945 (in c.960 copy), *to Cumber lande* 11th cent., *comitatu Cumbrie* 1230, *Cumbria* 1231 (*PNCumb* 1).
♦ *Cumbria* originated as a Latinised form of OE *Cumbra land* or *Cumberland* 'land of the *Cwmry*', *Cumbra* being gen. pl. of the OE name, adopted from Brit., for the pre-Conquest Celtic people of NW England and S. Scotland (see p. xix). Since 1974 it has designated the county which resulted from amalgamating most of the former Cumberland and Westmorland, together with the FURNESS district of Lancashire (Lancashire-North-of-the-Sands) and the Sedbergh area of the West Riding of Yorkshire. The word *mountain*, adopted into ME from OFr *montaigne*, is not part of the traditional vocabulary of Lakeland p.ns, occurring only here and in MOUNTAIN VIEW in this volume, though the Gough map, c.1330, shows *Montes Cumbrenses* running N from WINDERMERE (*CW* 1918, 3 and facing). The C~ Way is a 70-mile walking route between Ulverston and Carlisle, while the C~ Cycle Way is a 280-mile route around Cumbria. On this use of *way*, see DALES WAY.

CUNSEY: C~ BECK SD3793 Claife/Hawkshead/Satterthwaite, C~ WOOD SD3793 Satterthwaite, HIGH C~ SD3894 (hab.) Claife, LOW C~ FARM SD3793 Satterthwaite.

Concey (*myll*) 1537, *Consay* 1593, *Consey* (*nabb*) 1649 (*PNLancs* 219–20), 'a common old spelling was *Cunza*' (Cowper 1899, 350); *Cunsey Beck*; *Cunsey Wood*; *High Cunsey* all 1851 (OS); (Low C~:) *Cunsey House* 1851 (OS).
♦ Perhaps 'the king's river', from *konungs*, gen. sing. of ON *konungr*, and *á* 'river', as suggested by Ekwall (*PNLancs* 219–20), though without earlier spellings this can only be accepted with caution. See also **wood**, **high, low, farm**. Ekwall believed that the **beck** 'stream', which delimits parishes, could conceivably have bounded the territory of a petty Scand king based at CONISTON (q.v.; *PNLancs* 215).

CUNSWICK: C~ HALL SD4893 Underbarrow, C~ SCAR SD4994, C~ TARN SD4893.
Coneswic 1189–1201, *Conyngeswyk*(*e*) 1301, *Cunneswyke* & variants 1407 to 1650; *the Haull* 1589 (*PNWestm* I, 101), *Cunswick Hall* 1770 (JefferysM); *Cunswick Scar* 1858 (OSNB); *Cunswick Tarn* 1865 (*PNWestm* I, 104).
♦ 'The king's farm', from *konungs*, gen. sing. of ON *konungr* (or OE *cyninges* influenced by this) and OE **wīc** '(specialised) farm or settlement'; also **hall**, **scar** 'escarpment, crag' and **tjǫrn/tarn** 'small mountain pool'. The remains of the 'ancient hall of the Leyburne family' include the gateway to a pele tower (Palmer 1944, 36). The Scar is a very long, prominent escarpment. The tarn is mentioned but not named by Machell (c.1692, ed. Ewbank, 85).

D

DACRE NY4526 (hab. & parish), DACREBANK NY4527 (hab.), DACRE BECK NY4426 Hutton/Dacre/Matterdale, DACRE LODGE NY4526 Dacre.
Dacor c.1125, *Dacre* c.1200 (p); *Dacrebancke* 1604 (*PNCumb* 186); (D~ Beck:) *amnem Dacore* c.750, *riuulus de Daker, de Dakre* 1292, *Dakerbek* 1323, *Dacrebeck* 1568 (*PNCumb* 10); *Dacre Lodge* 1867 (OS).
♦ 'The trickling one', from a Cumbric **dagr* 'tear-drop' (*ERN* 111, Jackson 1968–9, 46, and cf. *PNWYorks* V, 139 for a Yorks parallel). This referred to a stream which gave its name to the village, and then was itself distinguished by the addition of **bekkr/beck** 'stream'; see also **bank** 'slope', **lodge**. Dacre Bank is a hill-slope N of D~ (not the bank of D~ Beck), and a habitation named from it. D~ Lodge is at the entrance to The Park. Anglian and Viking sculpture in the church as well as the Brittonic origins of the name suggest that settlements flourished at Dacre from early times.

DALE BECK NY3036 Caldbeck.
From 1867 (OS).
♦ 'The stream in the valley'; **dale, beck**. This is one of several reaches, each separately named, of a stream that eventually forms the Cald Beck (see CALDBECK).

DALE BOTTOM NY2921 (hab.) St John's.
From 1605 (*PNCumb* 317).
♦ 'The floor or inner part of the valley' (**dale, bottom**), referring to the wedge of low ground named NADD(A)LE.

DALE END NY3103 (hab.) Langdales.
Dale-End in Little Langdale 1690 (Grasmere PR).
DALE END NY5100 (hab.) Longsleddale/Whitwell.
From 1635 (*PNWestm* I, 162).
♦ The Longsleddale example is probably 'the end of the valley' (**dale; end**), being near the point where the R. SPRINT leaves the enclosed part of the valley, but the description fits the Langdale example less well, and it could contain the local term *dale*, referring to a portion of arable land or meadow (Simpson 1928, 277–8, Atkin 1991, 75, 76 and p. 70 (map), cf. Lindkvist 1912, 30–5).

DALE END NY3306 (hab.) Grasmere.
Tail end 1669 (spelling not specified, *CW* 1928, 283), *Tailend* 1678 (*CW* 1908, 204), *the Tail-end* 1695 (PR), *Tail End or Dale End* 1828 (HodgsonM).
♦ This is an **end**, but not of a valley (**dale**), for the early forms suggest that the 1st el. is from OE *tægl* 'tail, narrow strip of land' (so Smith, *PNWestm* I, 200). Armitt suggested a specifically territorial sense, 'where the township of Grasmere "tails off"' (1908, 193n.). 'Dale' in the modern form could have been influenced either by **dalr/dale** or by dial. *dale* mentioned in connection with DALE END, Little Langdale.

DALEFOOT NY4920 (hab.) Bampton.
From 1682 (*PNWestm* II, 194).
♦ 'The lower end of the valley'; **dale, foot**. The reference is to HELTONDALE, where the name Dalefoot contrasts with Dalehead.

DALEGARTH: D~ HALL NY1600 Eskdale, D~ STATION NY1700.
Da[lega]rthe, Dal[egart]he 1432 (damaged deed, *St Bees* 576), *Dalegarth* 1437 (spelling not specified, *CW* 1929, 265), *Dale garthe* c.1503 (*PNCumb* 343), *Awstwait (now called Dalegarth)* 1671 (Fleming 38); *Dalegarth Hall* 1773 (PR).
♦ D~ is 'the enclosure or farm in the valley'; **dalr/dale, garth. Hall** has been added somewhat later, referring to the seat of the Stanleys (cf. STANLEY FORCE), though it 'degenerated into a farm' after they moved in the late 17th cent. (Fair 1929, 265, who speculates on the development of the site that could have motivated the change of name from Austhwaite to Dalegarth, 266–8). The Station is the E. terminus of the RAVENGLASS & ESKDALE RAILWAY; cf. BOOTLE STATION for the history of the word *station*.

DALE HAUSE NY4719 (1385ft/422m.) Askham.
Dale Horse 1859 (*PNWestm* II, 201).
♦ 'The neck of land above the valley'; **dale, hals/hause**. It rises W of HELTONDALE.

DALE HEAD Examples at:
 NY2215 (2473ft/753m.) Buttermere/Above Derwent, DALEHEAD CRAGS NY2215 Above Derwent.
Dale Head 1774 (DHM); *Dalehead Crags* 1867 (OS).
 NY2400 (hab.) Dunnerdale.
Dalehead 1724 (Seathwaite PR), 1774 (DHM), 1774 (WestM), *Dale Head* 1850 (OS).
 NY4316 Martindale.
Martendaleheved 1363, *Dailehead* 1573 (*PNWestm* II, 218).
DALE HEAD HALL NY3117 St John's.
Daylhead 1567, *Dalehead, at the head of Buresdale* 1777 (*PNCumb* 317); *Dalehead house* 1784 (West 83), *Dale Head Hall* 1837 (Wythburn PR); cf. *Dalehead Parke* 1552.
♦ 'The head of the valley' (**dalr/dale, hēafod/head**), a name with rather different applications. D~ H~ at NY2215 is a grand eminence, topped by **crag**s, at the centre of the horseshoe of fells which closes off the S end of the NEWLANDS valley, while the Martindale name marks the point above which the valley bifurcates into BANNERDALE and RAM(P)SGILL. The Dunnerdale example is a habitation. D~ H~ **Hall** is the former seat of lords of the manor, over half a mile S of Dale Head, St John's. The valley in this case is that of How Beck, which flows into ST JOHN'S BECK, previously known as *Bure*, hence *Buresdale* in the 1777 form. Thomas Denton further gave *Bure* as a name for the GRETA into which St John's Beck flows, as far down as KESWICK (1687–8, 52, 136).

DALEMAIN NY4726 (hab.) Dacre.
Daleman early 12th cent., *Dalman Ald* 1252, *Dalman* 1285, *Dalmayne in Neubynging'* 1292 (*PNCumb* 186–7).
♦ 'Máni's valley', an inversion compound in which the 1st el. is the generic, ON **dalr** 'valley', and the 2nd el. a pers.n. as specific, either ON *Máni* (cf. MANESTY) or possibly Gaelic *Maine* (*Scand & Celt* 23, *SSNNW* 119).

DALE PARK: D~ P~ BECK SD3592 Satterthwaite, HIGH D~ P~ SD3593, LOW D~ P~ SD3591, MIDDLE D~ P~ SD3592.
Dall park c.1535 (West 1774, 104), *Dalepke* 1602 (Hawkshead PR), 1631 (Colton PR); *Dale Park Beck*; *High Dale Park* both 1851 (OS); *Low Dalepark* 1780 (PR); *Middle Dale Park* 1851 (OS).
♦ D~ P~ 'the (deer) park in the valley' is a long stretch of high ground E of D~ P~ Beck. It is 'supposed to have been inclosed for deer by Abbot Banke

of Furness about 1516, and became one of the four customary divisions of Satterthwaite township' (*VCHLancs* 370, 380–1, cf. Cowper 1899, 49–50, 91, who is more cautious). See **dalr/dale**, **park**, **beck** 'stream', **high**, **low**, **middle**.

DALES WAY SD4296 Bowness/Crook/Nether Staveley.
♦ Devised in 1968, the D~ W~ covers some 80 miles between Ilkley and BOWNESS. Its name refers primarily to the Yorkshire Dales (**dale** from ON **dalr** 'valley'). The use of *Way* (ultimately from OE *weg* 'path') to refer to long-distance recreational routes is a 20th-cent. development; cf. CUMBRIA WAY.

DAN BECKS NY3500 Hawkshead.
From 1850 (OS), which also shows *Dan Beck Coppice*; both are areas of woodland.
♦ It is uncertain whether Dan Beck is a stream-name or pers.n.; it appears to consist of the short form of *Daniel*, plus **beck** 'stream', but *Beck(s)* also exists as a surname.

DARLING FELL NY1222 (1282ft/391m.) Loweswater.
From 1867 (OS).
DARLING HOW NY1825 Lorton.
Derlingow 1541 (*PNCumb* 457).
♦ It is striking that D~ co-occurs with two 'hill' words (**fell**, and **how(e)** from **haugr**), and its meaning is unclear. Since pers.ns rarely qualify hill names, D~ is unlikely to be the surname, which is especially common in Scotland, still less the OE pers.n. *Derling/Dirling*.

DASH BECK NY2532 Bassenthwaite/Ireby and Uldale/Underskiddaw, D~ FARM NY2632 Ireby and Uldale.

Dash Beck 1867 (OS); *Dash* 1660 (*PNCumb* 328), 1867 (OS).
♦ D~ B~ is 'the stream with the waterfall', for *dash* is 'the violent throwing and breaking of water . . . upon or against anything; a splash' (*OED*, n. 1, sense 4a), here referring to WHITEWATER DASH, the spectacular waterfall on Dash **Beck**. D~ **Farm**, formerly just Dash, may be named more from the beck than the falls, which are half a mile away.

DAWSON FOLD SD4588 Crosthwaite.
From 1535 (*PNWestm* I, 83).
♦ 'The pen or farm of the Dawson family' (surname plus **fold**), who are documented locally from late 14th cent. to 17th cent. (*PNWestm* I, 83).

DEAD CRAGS NY2631 Bassenthwaite/Underskiddaw.
From 1867 (OS).
♦ The **crags** stand above Dead Beck, and it can only be conjecture which name came first, and what 'dead' means here. It could refer to the forbidding appearance of D~ C~, a 'silent and strange crater', 'a dark . . . rampart of buttresses and jutting aretes' (Wainwright 1962, Bakestall 2), or conceivably to some fatal accident (cf. the suggestion for OE *dēad* 'dead' in Smith, *Elements*). Other conceivable meanings of 'dead' do not fit the landscape today: 'dried up' (cf. *PNWestm* II, 109) would not describe Dead Beck, nor would 'lacking in vegetation' describe Dead Crags.

DEEPDALE NY3520 Matterdale, D~ CRAG NY3419.
depedell 1589, *Deepdale* 1749; *depedell cragg* 1589 (*PNCumb* 222).
DEEPDALE NY3812 Patterdale, D~ BECK NY3812, D~ COMMON NY3612, D~ HALL NY3914, D~ HAUSE NY3612.

Duppedale pre-1184, *Depedal(e)* c.1197 (p) to 1648; cf. *Diupedalhed* 13th cent. (*PNWestm* II, 221–2); *Deepdale Beck* c.1692 (Machell I, 726), *Deepdale Common* 1865; *Deepdale Hall* 1865; *Deepdale Hause, Hoss* 1860 (all *PNWestm* II, 225).
♦ 'The deep valley', with elements going back to OE *dēop* or (clearly in the Patterdale spellings) ON *djúpr* 'deep', plus **dalr**. The Matterdale name refers to the sheer valley sides in the upper reaches of the AIRA BECK, the Patterdale one to the course of D~ Beck. See also **beck**, **common** 'unenclosed pasture for common use', **hall** and **hals/hause** 'neck of land'. The Hall is a 17th-cent. farmhouse.

DEEP GILL NY1312 Ennerdale.
From 1867 (OS).
♦ 'The deep ravine with a stream', with 1st el. as in DEEPDALE, plus **gil(l)**. This is a classic *gill*, fed by many streams.

DEER BIELDS NY3009 Grasmere.
From 1859 (OSNB).
♦ 'The deer shelters'; *deer* from OE *dēor* and/or ON *dýr* 'wild animal, deer', plus dial. **bield** 'shelter'. This is a long rank of tightly-packed crags, 'a very craggy feature supposed to have been a resort & shelter for wild deer, formerly' (OSNB). The word *deer* occurs in five Lakeland names in this volume, all recorded in the 16th cent. or later.

DEER DYKE/DIKE MOSS SD3382 Haverthwaite.
Deer Dike Moss 1851 (OS).
♦ On the word *deer* see DEER BIELDS; **dyke/dike** could be an embankment or more likely a stream, as it is in the p. n. Otter Dike, very close by. D~ D~ **Moss** 'bog', with FISH HOUSE MOSS, forms part of a long stretch of level ground S of the lower LEVEN, partitioned into separately named Mosses and now meshed by drainage ditches. It is still frequented by deer (pers.comm. Rob Petley-Jones of English Nature, gratefully acknowledged).

DEER FOREST NY4314 Martindale.
♦ Self-explanatory; *deer* as in DEER BIELDS, plus **forest**. The uplands SE of BANNERDALE are still a refuge of deer. *Forest* therefore has its ancient meaning here. D~ F~ seems only to have appeared on recent OS maps, however (not appearing on 1863 or 1899 Six Inch maps), and it is debatable whether it should be regarded as a p.n.

DEER HILLS NY3136 Caldbeck.
From 1867 (OS).
♦ Now grouse moor, but a suitable habitat for deer, which are also commemorated in nearby DEER RUDDING and indirectly in PARKEND; *deer* as in DEER BIELDS plus **hill**.

DEER RUDDING NY3637 (hab.) Mungrisdale.
Deyruddinge, Dayriddyng 1559, *Dearuddinge* 1569, *ye Deareruddeinge* 1649 (*PNCumb* 211), *Ruddin* 1774 (DHM).
♦ Probably 'the clearing frequented by deer'; *deer* as in DEER BIELDS plus **ridding**.

DEMMING CRAG NY2200 Eskdale/Ulpha.
Deming Crag 1823 (GreenwoodM), *Demming Crag* 1867 (OS).
♦ The 2nd el. is **crag** 'rocky height', but the 1st el. is uncertain. Collingwood thought that the name may have signalled a place of judgement and

assembly (1918, 97 and 238; cf. *OED*: *deeming*, sense 1). *Dem(ming)* can also refer in ME and dial. to the damming of water (*EDD*, *OED*, Smith, *Elements*), but this can hardly apply here.

DENNYHILL NY5117 (hab.) Bampton.
Denny Hill 1743 (*PNWestm* II, 194).
♦ 'Denny' is obscure, but is perhaps the surname, which is of various origins (Hanks & Hodges 1988); see also **hill**.

DERWENT, RIVER NY1832, NY2425 Borrowdale/Underskiddaw/Blindcrake, D~ BANK NY2522 (hab.) Above Derwent, D~ BAY (hab.) NY2521, DERWENT FELLS NY2117, DERWENT ISLE NY2622 Keswick, DERWENT WATER NY2520 Keswick/Borrowdale/Above Derwent.
(River:) *Derventione* (applied to Roman fort at Papcastle, on the D~), *Deruventionis fluvii*, 8th cent. (*PNCumb* 11, *ERN* 122–3, Rivet & Smith 1979, 333–5), possibly = *Derwennydd*, late 6th–early 7th cent. (Cessford 1994); *Derwent Bank* 1823 (Otley 77); *Derwent Bay* 1867 (OS); *Derwentfelles* 1256 (*PNCumb* 370); *Darwen Isle* 1695 (MordenM); *Derewentewatre*, *Derwentwater* early 13th cent. (*PNCumb* 33).
♦ '(River) with oak trees', traditionally explained from Brit. *derwā 'oak' plus suffixes, hence of the same origin as other English rivers named Derwent, Darwen, Darent and Dart (*ERN* 122–3); the names may be still more ancient, ascribable to an 'Old European' stratum (Kitson 1996, 77–9); also **bank, bay, fell, isle, water**. The river gave its name to Derwent Water (which was also known as the *Lake of Derwent, Keswick Lake,* or *Keswick Water* in the 18th–19th centuries) and thence to D~ Bank and D~ Bay (the bay and house so named, see Bott 1994, 61). D~ Fells now applies to the hills SW of the lake, but previously designated a much larger medieval forest (see Winchester 1987, 39–40). For the fascinating range of earlier names for D~ Isle, see *PNCumb* 302. (Pls 12, 13, 15)

DERWENTFOLDS NY2925 (hab.) Threlkeld.
Darrawnfealdt 1574 (Collingwood 1912, 160), *Derwent Fold* 1774 (DHM).
♦ 'The pens or farm by the Derwent'; river-name plus **fold**. The place is three miles from DERWENTWATER and the R. DERWENT, though close to its tributary the GRETA. The strange 1574 spelling is from the accounts of the German miners based at KESWICK.

DEVIL'S GALLOP SD3694 Hawkshead.
From 1851 (OS).
♦ The word *devil* goes back to OE *dēofol*, while *gallop* is of Fr origin, appearing as an English noun and verb first in the 16th cent. The name of a mile-long hill-slope S of ESTHWAITE WATER, D~ G~ is perhaps just a piece of 'local superstition' (Cowper 1899, 363). It is tempting to connect it with the local legend of a stupendous gallop undertaken during the Civil War by 'Robin the Devil' — Robert Philipson (e.g. Machell c.1692, ed. Ewbank, 88, Gilpin 1786, I, 139–42); but the route in the story would not have taken in Devil's Gallop.

DEVOKE WATER SD1596 Eskdale.
Duvokeswater c.1205, *Duuokwat'* 1279, *Duffockiswatir* c.1280, *Devoke* 1626 (*PNCumb* 33). Stress is on the first syllable.
♦ 'Devoke' seems to be 'the black one', from a Brit. root **Dubāco-* or *Dubācā (*PNCumb* 33, acknowledging

Jackson). D~ could originally have been a pers.n., as suggested as one possibility in ERN 255 and PNCumb, and this would be encouraged by the gen. sing. -es in the c.1205 spelling. Alternatively, it could have been the name of D~ W~ itself, or an earlier name of LINBECK (GILL), from which the tarn drains into the R. ESK. The same el. may well appear in MOOR DIVOCK. That the names belong to the earliest identifiable layer of language, the Brittonic, is fitting, for both these tracts of high moor are extremely rich in cairns and other prehistoric remains (Cherry 1961 for Devoke area; Taylor 1886 for Moor Divock). The word **water** 'lake' may have been felt suitable since this is the largest of the Lakeland tarns.

DOBBIN WOOD NY4120 Matterdale.
From 1867 (OS).
♦ Presumably 'the **wood** belonging to Dobbin', a hypocoristic form of *Robert*.

DOB GILL NY3113 St John's.
From 1823 (Otley 27), 1867 (OS).
♦ Probably 'the ravine/stream with pools'. The 2nd el. **gill** is a ravine with a stream. (a) The 1st el. seems likely to be a spelling of **dub(b)**, a pool in a stream; this would be paralleled by *Dobwra* 1292 for DUBWATH, if that identification is correct. The short stretch of D~ G~, unlike others in this part of the valley, is characterised by several deep, black pools. (I am very grateful to Rod Grimshaw of Green Farm, St John's in the Vale, and local informants consulted by him, for this information; he adds that *dob/dub* might alternatively refer to Harrop Tarn, out of which the Gill flows.) Other explanations seem much less likely: (b) 'Dob' could be a shortening of *Robert* (cf. the surname *Dobson*), but a forename with **gill** would be rare at best. (c) It might be compared with the 1st el. of Dobbshall, which 'could be a variant of dial. *dobby* "a sprite"' (*PNWestm* I, xiv). (d) Gaelic *dubh* 'dark' is unlikely in a name recorded so late.

DOCKER NOOK NY5001 (hab.) Longsleddale.
Docker/Dockar nook(e) 1622 to 1700 (*PNWestm* I, 162).
♦ A **nook** is a secluded place, but D~ is elusive. Smith in *PNWestm* suggests that it is either a p.n. or a surname derived from a p.n. If a p.n., it could have shared origins with Docker township (*PNWestm* I, 129–30), which may mean 'the shieling at the hollow' (with ON *dǫkk* plus Gaelic-Norse **ærgi/erg** 'shieling') or 'the shieling where dock grows' (OE *docce* plus **ærgi/erg**); cf. the same problematic syllable in DOCKRAY and, strikingly since it shares the el. **nook**, DOCKRAY NOOK. If a surname, there may have been a link with the family traced in Parker 1918.

DOCKRAY NY3921 (hab.) Matterdale.
Dokwra 1278 to 1487, *Dockewra* 1279 (p), *Dockray* 1577, *Dockera* 1589 (*PNCumb* 333).
♦ This may be 'the nook in a hollow' or alternatively 'the nook where dock or sorrel grows', with 2nd el. **vrá** 'nook' in either case. The co-occurrence of 'Dock-' with a word meaning 'nook' (**vrá/wray** or **nook** itself) in several Cumbrian names is striking (*PNCumb* 333). (a) It could favour interpretation of the 1st el. as from ON *dǫkk* 'hollow' (so Ekwall, *Scand & Celt* 77), and a compound of two ON elements is generally more likely than an OE-ON hybrid. It has been objected that ON *ǫ* is unlikely to

develop into *o* (*PNCumb* 334), but there are examples: ON *hǫfuð* 'head' gives ME *hoved*, and ON **knǫttr** is a possible source of *knott*, while for several words evidence is lacking since *ǫ* alternates in the ON paradigm with *a*, and it is the *a* that survives in Lakeland names (e.g. STANGER, STANGRAH; on the sound-change *a > ǫ* and problems concerning it, see Noreen 1923, 69–80). (b) The 1st el. could alternatively be the plant-name OE *docce* (so *PNCumb* 333, *SSNNW* 225–6), referring to damp-loving plants of the *Rumex* genus: broad-leaved dock or common sorrel; sheep's sorrel prefers dry conditions (Halliday 1997, 172 and 169; cf. *DOE*: *docce*). This plant-name appears in DOCK TARN but is not otherwise common and the repeated collocation with *vrá/wray* would be surprising.

DOCKRAY NOOK NY0820 (hab.) Lamplugh.
Dockaranooke 1583, *Dockerer-nook* 1586 (*PNCumb* 333).
♦ D~ is a p.n., either 'the nook in the hollow', or possibly 'the nook where dock or sorrel grows', with **vrá/wray** 'nook' as generic in either case, cf. DOCKRAY. In D~ Nook, Dockray could either be the p.n. or the surname derived from it (e.g. *Dockeray* in PR 1751) — the same problem as encountered in DOCKER NOOK. **Nook** 'secluded place' duplicates part of the meaning of the original name.

DOCK TARN NY2714 Borrowdale.
Docketerne c.1209–10 (*Furness Coucher* II, 2, 570).
♦ 'The mountain pool where water-plants grow.' The exact botanic significance of OE *docce* is elusive (cf. DOCKRAY), but 'water-lily' is considered likely in p.ns (*DOE*: *docce*), and indeed *Nymphaea alba*, the white water-lily, is presently abundant at D~ T~, its highest location in Cumbria (Halliday 1997, 116). The **tarn** was also described in 1862 as 'a small sheet of water, very full of weeds' (OSNB Crosthwaite). Vegetation tends to be rather sparse in tarns, which are typically high, shallow and stony.

DODD Examples at:
    NY1615 Loweswater.
Dodd 1867 (OS); *le Dod de Gillenfinchor . . . the other high Dod of Gillefinchor* 1230 may = Little Dodd and Dodd (*PNCumb* 411–12), though identification is very difficult, not least because there is a Little Dodd at both NY1415 and NY1319.
    NY1623 Buttermere.
From 1799 (*CN* 537).
    NY2427 (1612ft/491m.) Underskiddaw, D~ WOOD NY2427.
the Dod 1834 (Baines 119), (*lofty Skiddaw with his*) *Dodd* 1839 (Ford 74); *Dodd Wood* 1867 (OS).
    NY2718 Borrowdale.
From 1867 (OS).
DODD, THE NY4619 (1752ft/534m.) Askham.
The Dod 1859 (*PNWestm* II, 200–1).
♦ A **dodd** is normally a steep, compact, rounded summit, though it can also be a spur of a larger hill. The Loweswater example is a distinct, rocky knob rising to 1490ft/454m., while the Buttermere peak is connected to RED PIKE, by THE SADDLE. The Underskiddaw example, though a compact and distinct summit, is in the shadow of a higher mountain, here SKIDDAW, and hence is also known as Skiddaw D~, Little D~ and D~ Fell (Wainwright 1962, Dodd 1). The mainly coniferous D~ **Wood** now clothes the entire hill. The Borrowdale D~ is a stony spur on

ASHNESS FELL and below HIGH SEAT, and The D~, Askham is also a spur, this time of LOADPOT HILL.

DODD/DOD CRAG NY2920 (1497ft/456m.) St John's.
dodcrage 1615 (*PNCumb* 317).
♦ 'The rocky summit.' This is a craggy spur of CASTLERIGG FELL; **dodd, crag**.

DODDICK NY3326 (hab.) Threlkeld, D~ FELL NY3327.
Doddick 1678 (*PNCumb* 253), *Doddack* 1720 (PR), *Duddock* 1774 (DHM), *Doddock* 1787 (*PNCumb* 253); *Doddock-Pikes* 1704 (CW 1951, 127), *Doddick Fell* 1867 (OS).
♦ D~ is obscure. A connection with **dodd** 'compact, rounded summit' is possible, referring to D~ **Fell**, a distinct summit on BLENCATHRA. For the 2nd syllable a diminutive ending has been suggested (Brearley 1974), or a connection with *dike* (also Brearley 1974, cf. Sedgefield 1915), but the evidence of the spellings is inconclusive. Whether or not D~ originally referred to D~ **Fell**, it is likely that the name D~ Fell derives from D~ as an established farm-name. The pair of names forms part of an interesting set referring to the settlements and buttresses of BLENCATHRA: see BLEASE.

DODDS HOWE SD4391 (hab.) Crosthwaite.
Dodg(e)how 1535 (in 1669 copy), 1716, *Dogehow* 1714, *Dodds How* 1865 (*PNWestm* I, 83).
♦ The name, designating both a hill (**how(e)** from **haugr**) and the nearby settlement, probably refers to the Dodd family, who appear in local PRs from the 18th cent.

DODD WOOD NY1324 Blindbothel.
Dod Wood 1867 (OS).

♦ 'The wood at the rounded summit'; **dodd, wood**.

DODGSON WOOD SD3092 Colton.
From 1851 (OS).
♦ Persons named *Dodgson* appear frequently in the Colton PRs; see also **wood**.

DOD HILL NY4005 Troutbeck.
Dod Hill 1859 (*PNWestm* I, 190); cf. *a Peat moss called Dod in Troutbeck forest* c.1692 (Machell II, 323, ed. Ewbank p. 128).
♦ 'The hill called Dod', from **dod(d)** 'compact, rounded hill', with **hill** added later.

DOE GREEN/DAWGREEN NY4219 (hab.) Martindale.
Daw Green 1749 (PR), *Doe Green* 1828 (HodgsonM), *Daw Green* 1851 (Census), *Dawgreen* 1863 (OS); cf. *Jo: Dawes of Greene* 1637 (PR). Local pronunciation as *daw*.
♦ 'The grassy place associated with the Dawes'; surname plus **green**. From the reference to the Dawes family in the PR for 1637 and subsequent years, this prime site on level ground locally known as 'the Green' seems likely to be named from them (I am most grateful to Mr Brian Humphrey of Doe Green for this information. He believes that the form Doe Green may have been adopted by a Mr Buxton who rented the house during the 1930s; if so the form was, on Hodgson's evidence, a long-established variant.)

DOLLYWAGGON PIKE NY3413 St John's/ Patterdale.
Dolly Waggon Pike 1839 (Ford 39), 1860 (OSBN Barton).
♦ An impressive peak, the southernmost summit of HELVELLYN, this is well named as a **Pike**, but

'Dollywaggon' continues to defy explanation. (a) It could be a corruption of an older name that is now unrecoverable. Like Helvellyn, Dollywaggon may have originated as a Brittonic name, but no satisfactory explanation along these lines presents itself. (b) If the name is more recent, *Dolly* as a pet form of *Dorothy* is a possibility, for there are references to named women in Lakeland p.ns, many of them associated with minor, high places (e.g. NAN BIELD, WILLIE WIFE MOOR). *Dolly* can also refer to 'various contrivances', including a 'small platform on wheels or rollers' (*OED*: *dolly* n. 1, sense 4h), hence something not unlike a *waggon*, but this may be sheer coincidence. *Waggon* is an early Modern English adoption from Dutch *wagen*, while the cognate OE *wægn* developed into *wain* (see *OED*: *wag(g)on* n.).

DORE HEAD NY1709 Nether Wasdale.
Previously *le Mikeldor de Yowberg* (the great door on YEWBARROW) 1332 (*PNCumb* 441).
♦ 'The top of the door or col', with elements derived from OE **duru** and **hēafod**. There is also a Great Door at the other, S, side of YEWBARROW, but this does not correspond with the *Mikeldore* of the document.

DOUP CRAG NY4109 Troutbeck.
From 1859 (*PNWestm* I, 190).
♦ Uncertain, but perhaps 'the **crag** or rocky height above the declivity'. (a) The 1st el. may be dial. *doup*, which can mean 'a cavity, hollow' (Taylor 1927, 155, Lascelles 2003, 12), as probably in DOUPS below. This is a linear scarp high above treacherous slopes, especially to the E. (b) Smith assumed what seems to be the same el., but in a different sense, citing NCy *doup* 'posterior, buttocks' from ON **daup* 'rounded cavity' (*PNWestm*; the asterisk is unnecessary since the word is attested — see below). (c) A further possibility is dial. *dowp, doup* 'carrion crow' (*EDD*: *dowp*).

DOUPS NY2526 Underskiddaw.
From 1867 (OS).
♦ 'The hollows.' Dial. *doup* 'hollow' may be connected with ON *daup*, which refers to a long, narrow, steep-sided declivity, difficult to cross without a bridge (Fritzner), and with *djúpr*, adj. 'deep' and *djúp* 'a deep channel, natural or excavated'. The place is framed on one side by forbidding cliffs.

DOVE COTTAGE NY3407 Grasmere.
From 1920 (OS).
♦ Simply **dove**, **cottage**. This only became the customary name of the cottage after the Wordsworth Trust acquired it in 1890 (Lindop 1993, 67). Home to Wordsworth at the height of his poetic powers (1799–1808), and then of de Quincey, it had previously been an inn:
'For at the bottom of the brow,
Where once the DOVE and OLIVE-BOUGH
Offered a greeting of good ale
To all who entered Grasmere Vale . . .
a Poet harbours now,
A simple water-drinking Bard.'
(Wordsworth, *The Waggoner* I, 52–60, *Poems* 138). 'The house had no name when the Wordsworths took it over, and they never gave it one . . . [their] address was simply "Town End, Grasmere"' (Moorman 1957, 459–60; I am grateful to my colleague John Saunders for this reference).

DOVE CRAG NY3710 Patterdale/Rydal.
Douekrag 1275 (*PNWestm* II, 225).

♦ 'The rocky height frequented by doves'; **dove**, perhaps referring to rock doves, plus **crag(ge)**, and cf. DOW CRAG.

DOVEDALE NY3811 Patterdale, D~ BECK NY3811.
(D~ Beck:) Previously *Hartsop Beck* 1863, 1900, 1919 (OS).
♦ Dovedale would mean 'the valley frequented by doves or pigeons', but since early forms seem to be lacking it appears to have been named secondarily from DOVE CRAG NY3710, using the ubiquitous **dale** 'valley'. D~ **Beck** 'stream' is also still known as Hartsop Beck (Wainwright 1955, Dove Crag 3; OS One Inch 1966).

DOVE NEST NY3802 (hab.) Ambleside.
From 1787 (*PNWestm* I, 183).
♦ This is, as Smith remarks, 'a frequent type of house-name' (*PNWestm* I, 183), so **dove** need not have any particular reference. NEST is the only other habitation name of the sort in this volume. Crosthwaite's map of 1809 has an illustration of the house D~ N~.

DOVENEST CRAG NY2511 Borrowdale.
Dove Nest Crag 1867 (OS); cf. *Dove's-nest* 1769 (Gray 1088).
♦ This is a remarkable rock wall whose caves and gullies make it a suitable nesting place, real or imagined, for doves or pigeons, perhaps rock doves; **dove**, *nest* from OE *nest*, and **crag**.

DOW BANK NY3305 Grasmere.
From 1920 (OS).
♦ Seemingly 'the steep slope frequented by pigeons or doves'; **dow**, **bank**. There are no early forms but cf. nearby *Dovekrag* 1273 (*PNWestm* I, 200).

DOW CRAG Examples at:
SD2099 (1325ft/404m.) Eskdale/Ulpha.
From 1867 (OS).
NY2206 Eskdale.
(= This or the example at SD2099:) *Dove cragg* 1587, *Doe-Cragg* 1794 (Hutchinson I, 580).
SD2697 (2555ft/779m.) Dunnerdale/Torver.
Dow Cragg; [*Dove Cragg;*] 1823 (Otley 24).
♦ 'The rocky height frequented by pigeons or doves'; **dow**, perhaps referring to rock doves, plus **crag(ge)**; cf. DOVE CRAG. The rock wall at NY2206 is 'Scafell Pike's grandest crag' (Wainwright 1960, Scafell Pike 4, who illustrates the crag and notes that it is also 'known to climbers as Esk Buttress'). The Dunnerdale/Torver example too is spectacular, a fine rock-climb which gives its name to the whole mountain.

DOWN IN THE DALE NY1808 (hab.) Nether Wasdale.
Downey dale 1729, *Downy in the Dale* 1815, *Down in the Dale* 1818 (Wasdale Head PR).
♦ '(The place) down in the **dale** or valley.' This unusual farm-name may be just what it seems. The situation is the level, damp ground near the head of WAST WATER.

DOWTHWAITE: DOWTHWAITEHEAD NY3720 (hab.) Matterdale.
Dowethweyt 1285 (p); *Dowthate Head* 1563 (*PNCumb* 222).
♦ 'Dow-' is difficult to explain. (a) It could be from an ON pers.n. or nickname such as *Dúfa*. This is assumed in another Dowthwaite (*PNEYorks* 47–8), and is common in Denmark (*SSNNW* 120; see also Lind 1905–15, 115). Other possibilities are:

(b) ON *dúfa*/OE *dūfe* 'dove, pigeon' (*SSNNW* 120 again), or (c) ME *dow(e)* 'dough', perhaps characterising the soil (Lindkvist 1912, 106), though this is not an obviously apt description of the ground here (Keith Clark, pers. comm.). The remaining elements are clearly ON **þveit** 'clearing' and ME or ModE **head**. Dowthwaite is part of the glacial valley through which the upper-mid reach of the AIRA BECK flows, with Dowthwaitehead at its upper end.

DRIGG (hab., parish of D~ and Carleton): D~ POINT SD0795.
Dreg 1175–1199 to 1363, *Dregg(e)* 1279 to 1514, *Drigg* 1572 (*PNCumb* 376–7); *Drigg Point* 1867 (OS).
♦ Probably 'the place of dragging.' Drigg lies N of Drigg **Point**, a sandy spit of land which deflects the R. IRT two miles S to share an estuary with the MITE and ESK. The name D~ has usually been assumed to refer to portage (e.g. *DEPN*, *PNCumb*), since it may constitute a mutated counterpart of ON *drag*, which often has the sense 'portage' in Norw p.ns (*NG Indledning* 47, *NSL*: Drag). However, the early cartographic evidence suggests that the deflection of the Irt has taken place since the mid 17th cent. (Steers 1964, 81–3). I am indebted to Ann Cole for drawing my attention to this, and for suggesting that some other kind of dragging may be meant — perhaps of boats onto the shore.

DROPPING CRAG NY1607 Nether Wasdale.
Le Droppingcrag 1322 (*PNCumb* 440).
♦ *PNCumb* suggests 'the dripping **crag(ge)**', but this is a rocky wall with a steep drop, so 'falling away steeply' seems a more likely meaning, as it is for another Dropping Crag, close to the summit of Scafell Pike. See also BELL RIB.

DRYBARROWS NY4916 Bampton.
Dribarows, *Driberijs* 1366 (*PNWestm* II, 190).
♦ 'The dry hills', with reflexes of OE *drȳge* 'dry' and OE **berg** 'hill'. There is a cluster of minor summits hereabouts.

DRYGROVE GILL NY4508 Kentmere. From 1858 (OSNB).
♦ D~ is 'the dry stream-bed or groove', with reflex of OE *drȳge* 'dry' and **gro(o)ve** 'stream-bed, hollow'. It is dry in its upper course, but a stream gathers momentum lower down; **gil(l)** could refer to ravine, stream or both. The Ordnance Surveyors described it as 'a deep ravine, well known by this name' (OSNB).

DRY HALL SD2291 Dunnerdale.
Dryhall 1677 (original spelling not specified, Broughton-in-Furness PR), 1750/51 (Seathwaite PR).
♦ Somewhat uncertain. The 1st el. may be as it seems, from OE *drȳge* 'dry' (notwithstanding a spring close by), cf. DRYBARROWS, DRYHOWE. 'Hall' could be **hall**, or perhaps more likely **how(e)** 'hill' (see **haugr**), referring to The Knott which rises above. For local confusion of *hall* and *haw(e)*, *how(e)*, see BALL HALL.

DRYHOWE NY5202 (hab.) Whitwell, D~ PASTURE NY5102 (1202ft/366m.).
Ryehow als. *Dryhow* 1612, *Dryhowe* 1836 (*PNWestm* I, 150).
♦ D~ is 'the dry hill', with *dry* from OE *drȳge* and **how(e)** 'hill' from ON **haugr**. The area is not free of streams and bog, but it was perhaps perceived as 'dry' compared with the other, E, side of BANNISDALE, which is richly

veined with streams. D~ **Pasture** is a high rocky slope, not verdant lowland.

DUBBS: D~ RESERVOIR NY4201 Windermere.
(*aquam de* [river of]) *Dubbes(mor)* 1292, (*water* [river] *of*) *le(z) Dubbes* 1324, *Dubbys* 1377 (*PNWestm* I, 194); *Dubbs Reservoir* 1919 (OS).
♦ A **dub(b)** is a pool in a river. The **reservoir** was formed, and named, from the stream called Dubbs, now Dubbs Beck.

DUB HOW FARM SD3694 Claife.
Dub How 1851 (OS).
♦ 'The hill by the pool' (**dub(b)**, **haugr/how(e)**), with **farm** added later. The place lies between OUT DUBS TARN and a height rising to 427ft/ 130m.

DUBWATH NY1931 (hab.) Setmurthy.
Dubwath (*Beck*), *Dubwath* (*Bridge*) both 1867 (OS), possibly = *Dobwra* 1292 (*PNCumb* 435), cf. *Dubbes* 1195 (*CW* 1921, 157).
♦ If the identification with the 1292 spelling is correct, the original meaning may have been 'the nook by the pool', from **dub(b)** 'pool' and ON **vrá** 'nook, secluded place'. The situation by a stream would have encouraged replacement of the 2nd el. by *wath*, from ON **vað** 'ford'.

DUDDERWICK NY4610 Shap Rural.
From 1859 (*PNWestm* II, 174).
♦ A puzzling name, with the ring of an Anglo-Saxon village, yet this is no village, but a steep slope SW of HAWESWATER, and there are no early forms. The *-wick* is more likely to be ON **vík** 'bay, creek', which is sometimes applied to inland sites (cf. FROSWICK), than OE **wīc** 'settlement, specialised place'. The name is conceivably linked with a lost *Duddrigg(bank)* (17th cent., mentioned in Thompson 1942, 24).

DUDDON, RIVER SD2093, NY2502 etc. Ulpha/Millom Without/Dunnerdale/ Broughton, D~ BRIDGE SD1988 (hab.) Millom Without, D~ HALL SD1989.
Dudun pre-1140 (in later copy, *ERN* 137, *DEPN*), *aqua descendit . . . per Dudenam usque mare* (the river descends through *Dudena* to the sea) 1157–63 (in later copy, *PNLancs* 191), *Dudene(e)* c.1180 to 1500 (*PNCumb* 11), *Duthen* 1196 (*PNLancs* 191), *Doden* 1279 to 1459, *Dodyn* c.1280 to 1535, *Dudden* early 16th cent. (*PNCumb* 11); *pontem de* (bridge of) *Duden, Doden* 1292, *Dodenbrig* 1332 (p), *Duddonbridge* 1610; *Duddon Hall* 1703 (*PNCumb* 415). See also forms for DUNNERDALE.
♦ The name is obscure. The 2nd el. may possibly be OE *denu* 'valley', which would tally with most of the early records and with the context of the 1157–63 form, which suggests that *Duden(a)* was a valley-name (here with -(*a*)*m* added to fit Lat syntax). Ekwall in *ERN* 137 hesitated about *denu* on the basis of the *Dudun* spelling, but *PNCumb* 11 cautions again attaching too much significance to this spelling. For the 1st el., suggestions include a river-name based on a root meaning 'black' (Cumbric *duß, **dū̄*, cf. DEVOKE) or an OE pers.n. *Dudd(a)* (Ekwall in *PNLancs* 191 gave both but in *ERN* 137 considered the name 'obscure'; *PNCumb* favours the pers.n. A new, wholly Brittonic, interpretation is forthcoming from Richard Coates).The river flows through DUNNERDALE (q.v.). See also **brycg/bridge** and **hall**. The Georgian D~ Hall was also known as D~ Grove in the 19th cent. (*CW* 1966, 375).

DUNGEON GHYLL FORCE NY2806 Langdales.
Dungeon Gill Force 1800 (*CN* 753).
♦ 'The waterfall at D~ Ghyll, the cavernous ravine.' *Dungeon*, an adoption from OFr, was used in English from the 14th cent., originally to mean either a castle keep or a prison cell, often subterranean (*OED*). Dorothy Wordsworth refers to 'those fissures or caverns, which in the language of the country are called dungeons' (*Journal*, cited *PNWestm* I, xix n.2). See **gil(l)** 'ravine with stream' for the spelling *ghyll*, and **force** 'waterfall'.

DUNMAIL RAISE NY3211 (782ft/238m.) St John's/Grasmere.
Dunbalraise stone 1576, *Dunbalraise stones* 1610, *Dunnimail or Dunmail-raise* 1610 (*PNCumb* 312).
♦ 'Dunmail's cairn', apparently referring to one of the last Cumbrian kings, who flourished in the mid 10th cent. (*PNCumb* 312, xxvi, Phythian-Adams 1996, 112), plus *raise* from **hreysi**. Though now disturbed by modern roadworks, this at the end of the 17th cent. was a heap of stones some 72 yards in circumference with a wall over the top (according to Machell's drawing and description, c.1692, II, 132B, ed. Ewbank p. 145; cf. LDHER 1239, SAM No. 22556). The place is an important watershed, which divided the ancient counties of Cumberland and Westmorland, the medieval dioceses of Carlisle and York and possibly before that the territory of the Cumbrian Celts and Yorkshire (Phythian-Adams 1996, 74, 119). There is an unsubstantiated tradition that, as well as being a boundary marker, the Raise is the burial cairn of Dunmail.

DUNNERDALE SD2093, NY2502 etc. (parish of D~ with Seathwaite), D~ FELLS SD2091, SD2393, HALL D~ SD2195.
Dunerdale 1293 (p), *Dunerdale, Donerdale* 1300, *Donesdale* 1412 (*PNLancs* 223); *Dunnerdale Fells* 1851 (OS), cf. *Donnerdale Fell* 1802 (*CN* 1225); *Hall Dunerdale* 1666/7 (Broughton-in-Furness PR), *Hall Dunnerdale* 1745 (West 1774, map), 1850 (OS).
♦ Probably 'the valley of the DUDDON', with ON **dalr** added to an original river-name, as it is in ALLERDALE, MITERDALE etc. The origin of DUDDON itself is obscure (see the entry), but the valley name could well be derived from it through a Scandinavianised *Duðnardal(r)* or *Duðnárdal(r)* (cf. the 1196 spelling *Duthen* for the river), with early assimilation of *-ð-*. The medial syllable *-er-* could go back to *ár*, gen. sing. of ON **á** 'river', possibly as part of a compound element *árdalr* 'river-valley' (*SSNNW* 325), or simply to a gen. sing. *-ar* added to the river-name itself. Earlier scholars took the simpler line that D~ contained the OE pers.n. *Dunnere* (Wyld 1911, Sephton 1913, 36), and Watts (2004) mentions the possibility of *Dunhere*, though he thinks a Scandinavianised version of the river-name more likely. The two D~ FELLS illustrate the two main applications of **fell**: the ones at SD2091 are rugged heights (see STICKLE PIKE, GREAT STICKLE), while those at SD2393 comprise a tract of upland grazing. The pattern of '*hall* plus p.n.' seen in Hall D~ is favoured locally: see **hall**.

DUNTHWAITE NY1732 (hab.) Setmurthy.
Dunthuaytt 1385 (Winchester 1987, 158), *Dunthwayte* 1570 (*PNCumb* 435).
♦ This may be 'the clearing with light soil'. The 2nd el. is the common **þveit**

'clearing'. The 1st is obscure, but (a) an ON-derived solution would be most likely, and the nearest match is to *dúnn* 'down (on birds)', which may apply to light earth in the Norw p.n. Dun, especially since two examples stand on gravel terraces (*NSL*: Dun). The fact that Dunthwaite is next to a small patch of alluvium in a bend of the R. DERWENT (Geological Survey) favours this interpretation. The following are rare locally and unlikely to compound with *þveit*: (b) Gaelic *dún* 'fort', which may occur, e.g., in Dunmallard, near POOLEY BRIDGE (*PNCumb* 187); (c) OE *dūn* 'hill', which is absent from the names in this volume.

DURHAM BRIDGE FARM SD4489 Crosthwaite.

Durham brigge 1535, *Durham Bridge* 1634 (*PNWestm* I, 83).

♦ 'Durham' may be a surname, although it does not appear in the Crosthwaite PRs. The **bridge** (see **brycg**) crosses the R. GILPIN; **farm** has been added in modern times.

DYKE SD1195 (hab.) Muncaster.
From 1769 (*PNCumb* 425–6).
♦ The name must refer to a noteworthy wall, embankment or ditch (**dyke**).

DYKE NOOK NY2329 (hab.) Bassenthwaite.
From 1789 (*PNCumb* 264), seemingly named *Far House* 1867 (OS), but *Dyke Nook* 1900 (OS).
♦ 'The secluded place by the embankment or wall'; **dyke**, **nook**.

E

EAGLE CRAG NY2712 Borrowdale. *Eagle Cragg* 1774 (DHM).
EAGLE CRAG NY3514 Patterdale. From c.1692 (Machell I, 718).
♦ The word *eagle* was adopted into ME from OFr *egle, aigle, OED*'s first citation being from c.1380; see also **crag(ge)**. West, one of several 18th-cent. writers to mention the eagles on the imposing rock citadel in Borrowdale, says that on its front '*the bird of Jove* has his annual nest, which the dalesmen are careful to rob' — a dangerous occupation, but worth it since 'the devastation made on the fold, in the breeding season, by one eyrie, is computed at a lamb a day' (1780, 95–6). Machell tells a story about an eagle in Patterdale (c.1692, I, 718 and 724), and Hutchinson reported golden eagles in nearby Martindale in 1794 (I, 449). Some at least of the records may refer to sea eagles rather than goldens (Ratcliffe 2002, 237). Eagles are also commemorated, using the dial. word, in ERNE CRAG and conceivably in AARON C~, HERON C~ and IRON C~, St John's. Compare also RAVEN CRAG.

EALINGHEARTH/ELINGHEARTH SD3485 (hab.) Haverthwaite.
Elinath 1688, *Eleing Hearth* 1694, *E(l)linarth* 1698 (Colton PR), *Eelinharth* 1746 (Finsthwaite PR).
♦ An *ealing hearth* was a kiln for producing dried wood or, albeit to some extent by accident, ash, often seemingly for lead smelting (see Davies-Shiel on kilnwood kilns, 1974, 37–45). They were plentiful in this area. Miles Sawrey of Greythwaite, some four miles from E~, was c.1545 granted 'licence to make a little house and hearth called the Ealing hearth ... in ... Graythwayte ... and to use such broken wood & sticks there' (Fell 1908, 105), and again in 1546, Miles and William Sawrey were permitted two more (Davies-Shiel takes these to refer to the place now named Ealinghearth, 1974, 43, but this does not seem proven, nor consistent with his association of a large pit at Graythwaite with the 16th-cent. records). The origin and meaning of the word *e(a)ling* is obscure, though it seems to denote the process of burning in a kiln and this might be compatible with origins in OE *ǣling* 'burning' (*DOE*). A document of 1538 refers to 'the art called elying of asshes' (Davies-Shiel 1972, 102) and one of c.1565 referring to hearths also mentions 'the ealing ashes there to be made' (Fell 1908, 105–6, 428). The word *hearth* goes back to OE *heorð*.

EAMONT, RIVER NY4827 Dacre/Barton.
æt Eamotum 10th cent., *Amont* 12th cent., *Emot(t)* 1220–47 (in 15th-cent. copy), *Amod* 1231, *Eamont* 1558 (*PNCumb* 12).
♦ 'The river confluences', from OE *ēa(ge)mōt* 'river-meet', in the pl. to judge from the 10th-cent. dat.pl. form. The DACRE BECK meets the E~ near BARTON, while the E~ and LOWTHER join at Brougham. The name must originally have referred to the confluences or a place nearby, then transferred to the river itself. Early *A*-spellings are influenced by ON *á* 'river', and the later *-mont* spellings by association with OFr *mont* 'hill'.

EASEDALE NY3208 Grasmere, E~ HOUSE NY3208, E~ TARN NY3008, FAR E~ GILL NY3109.

Asedale 1332 (p), 1375 (p), *Aisedale* 1706 (*PNWestm* I, 199); *Easedale House* 1920 (OS); *Easdale Terne* c.1692 (Machell II, 128), *Easedale Tarn* 1802 (*PNWestm* I, 200); *Far Easedale Gill* 1920 (OS); cf. *Far-Easedale* 1857 (Davy 38), *Easedale Gill* 1865 (*PNWestm* I, 200).

♦ 'Ási's valley', with ON pers.n. *Ási* and ON **dalr**; also **house, tjǫrn/tarn** 'small mountain pool', **far, gil(l)** 'ravine with stream'. Far E~ is presumably so called for its distance from GRASMERE.

EAST HOUSE NY1830 Embleton.
the Esthouse 1578 (*PNCumb* 384), *East House* 1677 (PR).
EAST HOUSE NY2320 Above Derwent.
From 1833 (Newlands PR).
♦ 'The house on the east side'; *east* from OE *ēast*, **hūs/house**. The Embleton example may be so named from its position on the E boundary of the parish, where it abuts WYTHOP (cf. WESTRAY), while the Above Derwent one is on the E side of the NEWLANDS valley.

EASTHWAITE NY1303 (hab.) Irton.
Ustethayt 1292 (p), *Ouesthwait in the town of Santon* 1338, *Ewsthwate* 1578, *Easthwaite* 1731 (*PNCumb* 402).
♦ 'The eastern clearing', from ON *austr* 'east' (anglicised in the modern form) and ON **þveit** 'clearing'.

EASTWARD NY5116 (hab.) Bampton.
Esforth c.1240, *Eastward* 1727 (*PNWestm* II, 194).
♦ A puzzling name, since the spellings could point to 'the eastern ford', with **ford** as 2nd el., developing into *-ward* under influence of *eastward* and perhaps encouraged by the Norse-derived word for 'ford', **wath**. However, there is no ford here, nor any watercourse of sufficient size to require a ford. 'East' (OE *ēast*) could perhaps refer to the location near the E boundary of Bampton township or to the general situation, facing E with a fine view over the LOWTHER valley. (I am indebted to Mr J. Thompson of Eastward for information about the site.)

ECCLERIGG: E~ HOUSE NY3801 Troutbeck.
Ecklrigg 1756 (*CW* 1984, 171), *Ecclerigg* 1787 (*PNWestm* I, 190); *Ecclerigg House* 1859 (OSNB).
♦ English p.ns in *Eccles-* frequently derive from Brit. **eglēs* 'church', adopted from Lat *ecclesia*. This is possibly the el. joined with **rigg** 'ridge, hill' (see **hryggr/hrycg**) here, since E~ Farm at NY3900 is next to a church, and ecclesiastical sites often show great continuity. The lack of spellings containing *-s-*, however, makes origins in *eccles* far from certain. This may have been the site of one of the numerous chapels-of-ease in the huge ancient parish of Kendal (Mary Atkin, pers.comm.).

EEL CRAG NY1820 Buttermere/Above Derwent.
Ill-crags, Ill Crags 1800 (*CN* 779, 783).
EEL CRAGS NY2316 Above Derwent.
From 1867 (OS).
♦ A tantalising pair of names, probably meaning 'treacherous **crags** or rocky heights'. 'Eel' is most likely a corruption of **ill** meaning 'treacherous, rough, steep', as in ILL CRAG and ILL BELL, and as suggested by Coleridge's version of Eel Crag, and his explanation, 'Scar Crags with the hollow bason so scored called Ill-crags' (*CN* 779). In the case of Eel Crags a fanciful reference to the sinuous terrain on the W end of the long ridge of HIGH SPY as being eel-like is not impossible, and

might at least explain the corruption of an original 'Ill Crag(s)'. Eel Crag is a buttress of CRAG HILL, and the name is often applied to the whole mountain (e.g. Wainwright 1964, Eel Crag 1).

EEL TARN NY1801 Eskdale.
Eiletarne 1660 (*PNCumb* 34).
♦ (a) Presumably 'the mountain pool where eels abound', from the reflex of OE *ēl* 'eel', plus **tjǫrn/tarn**. The large and long-lived eel was an important food source in medieval times, and is still found widely in Lake District lakes and tarns (Hervey & Barnes 1970, 108). (b) Heaton Cooper suggested derivation from the ON **ill(r)** 'evil' (which he represents *ee*), perhaps referring to the occasional appearance of will o' the wisp at the tarn (1983, 44).

EHEN, RIVER NY0515 Ennerdale.
Egre c.1125 (in 15th-cent. copy), *Eggre* 1152, *Eghena, Eghene* c.1160 (in 15th-cent. copy, *PNCumb* 13, Watts 2004; c.1160 form given as *Ehgena* ERN 144 and *DEPN*).
♦ Obscure. The early spellings, variously in *-r-* and *-n-*, cannot be reconciled, and *PNCumb* deems this 'a pre-English name of which no satisfactory explanation can be offered', evidently sceptical of Ekwall's suggestion in *ERN* 144–5 of 'cold one'. Ekwall, probably rightly, considered the *Ehgena/Eghen* spellings to be the more authentic while the *Eg(g)re* type seem to be influenced by Egremont. Suggesting similar origins to the Herefordshire river Eign, he proposed a Scandinavianised version of a name based on an OW (strictly Brit.) **i̯agin-* or **i̯egin-* 'cold', derived from a root meaning 'ice'. In its upper reaches, the Ehen is named LIZA.

ELDER BECK NY4723 (stream) Barton, ELDERBECK NY4723 (hab.).
(Hab:) *Ellerbeck(e)* 1578, *Elderbeck* 1676 (*PNWestm* II, 212).
♦ 'The stream where alder or elder grows.' The 1st el. is (a) probably **eller** 'alder' (as Smith assumes in *PNWestm*); though (b) *elder*, from OE *ellærn*, which occurs in ME with *eller* and *eldre* spellings, cannot be ruled out. The geographical distribution of alder (see **eller**) and elder, *Sambucus nigra*, is similar, though the elder is more strongly associated with hedge-rows and habitations (Halliday 1997, 406). The 2nd el. is the common **bekkr/beck** 'stream'.

ELFHOWE SD4699 (hab.) Over Staveley.
Elfhow 1694, *Elph howe* 1836 (*PNWestm* I, 175–6).
♦ 'The hill associated with elves', with the reflexes of OE *ælf* 'elf', pl. *ylfe* and ON **haugr** 'hill, mound'. The date of this name and the exact nature of the sprites — whether threatening or playful — is elusive. Though supernatural beings are often associated with burial mounds or other antiquities, no such archaeology is recorded here in LDHER. The hill Elf Howe gives its name to settlements distinguished as Middle and Low Elfhowe (1:25,000 map). Another Elfhow is recorded from Hutton in 1484 (*CW* 1930, 72), and compare ELVA (HILL and PLAIN).

ELLEN, RIVER NY2536 Ireby and Uldale.
(*ad*) *aquam Alne* 1171–5 (in 1333 copy) to 1578, *aquam de Al(l)en* late 13th cent. (*PNCumb* 13).
♦ A Brit. river-name **Alauno-*, also believed to result in the Northumberland Aln. Its meaning is obscure. Ekwall pointed out the resemblance of

the stem to the names of some Continental Celtic deities and suggested some common meaning such as 'holy' or 'mighty' in *ERN* 6–7, but did not repeat the suggestion in *DEPN*. More recently, Hamp has argued for a root meaning 'speckled' and hence 'trout (river)' (1975, 174).

ELLERAY SD4198 Windermere.
Ellerie 1640, *Elleray* 1648/9 (PR).
♦ Probably 'the alder nook' (**eller, vrá**). Elleray also appears as a surname, presumably derived from the p.n., in the PRs.

ELLERBECK SD1196 (hab.) Muncaster.
Ellerbank c.1215, *Ellerbecke in Birkby* 1678 (*PNCumb* 424).
♦ 'The hill-slope where alder grows', unless the c.1215 form is erroneous. From ON *elri(r)* 'alder(-copse)' (see **eller**) and **bank(e)**, which has been replaced by **beck**, encouraged by the situation by a stream.

ELLERBECK FARM SD4694 Crook.
Ellerbecke 1636 (*PNWestm* I, 179).
♦ E~ is 'the stream where alder grows'; **eller, beck**. Farm has, as so often, been added relatively recently.

ELLERGARTH NY3106 (hab.) Langdales.
Ellers 1692 (Grasmere PR), 1865 (*PNWestm* I, 206), 1966 (OS One Inch).
♦ Originally 'the alders, the place where alders grow' (**eller**, cf. ELLERS below). The modification to Eller**garth** appears to be recent.

ELLER HOW SD4181 (hab.) Upper Allithwaite.
From 1851 (OS).
♦ 'The hill where alder grows', and the settlement below, named from it; **eller, haugr/how(e)**.

ELLER MIRE SD2385 (hab.) Kirkby Ireleth.
Ellermire 1683/4 (Kirkby Ireleth PR).
♦ Presumably 'the bog where alder grows'; **eller, mýrr**.

ELLERS NY2417 (hab.) Borrowdale.
Ellers or *Ellersfielde* 1578, *Ellars in Borradaill* 1703 (*PNCumb* 353).
♦ 'The alders, the place where alder grows'; **eller**. See ELLERGARTH above, and ASHES for similar names.

ELLERSIDE SD3586 Colton.
Ellersyde 1617 (Cartmel PR).
♦ Evidently 'the slope where alder grows', with **eller**, and **side** probably in the sense of 'hill-slope'. Shown as Great Ellerside on 1:25,000 map.

ELLERWOOD SD3994 (hab.) Bowness.
From 1920 (OS), seemingly *Middle Farm* 1862 (OS).
♦ Apparently a quite recent name meaning 'alder-wood'; **eller, wood**.

ELMHOW NY3715 (hab.) Patterdale.
Elm How 1828 (HodgsonM).
♦ 'The hill where elms grow', with elements derived from OE *elm* (cf. ON *almr*) and ON **haugr** 'hill'. The wych elm, *Ulmus glabra*, was extremely widespread in Lakeland until devastated by Dutch elm disease in recent decades (Halliday 1997, 135), yet this is the only reference to the tree in the names in this volume.

ELTER WATER NY3304 (lake) Langdales/Rydal, ELTERWATER NY3204 (hab.) Langdales, ELTERMERE NY3204 (hab.).
(Lake:) *Elterwatter, Helterwatra* 1154–89, *Heltewatra, Elterwat'* 1157–63 (*PNWestm* I, 16); (village:) *Elterwater* 1684 (Grasmere PR); *Eltermere* 1920 (OS).

♦ 'The lake frequented by swans', from ON *elptr/alpt* 'swan', in the gen. sing. form with *-ar*, and **water**, probably replacing ON *vatn* 'lake'. Whooper swans (*cygnus cygnus*) still winter on the lake. The village is named from the lake. Eltermere seems to be a relatively recent name containing **mere**[1] 'lake'.

ELVA HILL NY1731 Setmurthy, E~ PLAIN NY1731 (hab.).
Both from 1867 (OS).
♦ E~ is probably 'elf hill', with the reflex of OE *ælfe* 'elf', pl. *ylfe*, plus **haugr/how(e)**, cf. *Elfhow* 1488, *Elfa-hills* 1848, Hutton-in-the-Forest (*PNCumb* 209), and ELFHOWE above. **Hill** was presumably added after the 'hill' component in Elva was no longer recognised. As often, a prehistoric site seems to have been connected with supernatural beings, for there is a stone circle on the SE side (LDHER 874, SAM No. 23793); cf. Elf Howe, Kentmere (NY4600), a burial mound according to Danby (1994, 176; I have not been able to verify this from LDHER). There is another Elfa Hill in Setmurthy (1867 OS). Elva P~ lies below Elva H~. The word *plain* was adopted into ME from OFr and used, as now, to designate relatively level stretches of open ground. FELL P~ and LOW P~ are the only other occurrences in this corpus.

EMBLETON NY1730 (hab. & parish), E~ HIGH COMMON NY1627.
Emelton 1195 to 1407, *Embelton* 1233 (p) to 1322, *Embleton* 1243 to present (*PNCumb* 383); *Embleton High Common* 1867 (OS).
♦ E~ is perhaps 'Eanbald's settlement'. The 1st el. is either (a) an OE pers.n. such as *Eanbald* (or *Æmele*, cited by Sedgefield, 1915), or (b) OE *emel* 'caterpillar', which is understandably rare in p.ns (see Smith, *Elements*). The 2nd el. is OE **tūn** 'farmstead, settlement, village', a sign of relative antiquity and of local importance, which continues in its status as a parish centre. See also **high**, **common** 'unenclosed pasture for common use'.

EMERALD BANK NY2320 (hab.) Above Derwent.
Emerald 1820, *Emerald Bank* 1842 (Newlands PR).
♦ Presumably 'the bright green slope', with *emerald* — not a standard p.n. element — from *emeraude*, adopted from OFr into ME, plus **bank**. This would be an apt description, and cf. GREEN BANK.

ENNERDALE (parish): E~ BRIDGE NY0715 (hab.), E~ FELL NY1113, E~ FOREST NY1314, E~ WATER NY1015.
Anenderdale c.1135 (in early copy), *Ananderdale* 1190s (in 1308 copy), *Enderdale* 1303, *Eghennerdale* 1323; *Bridge in Enerdall* 1656 (*PNCumb* 385), *Enerdale-bridge* 1687–8 (Denton 98); *Ennerdale and Wast dale Fells* 1802 (*CN* 1212), *Ennerdale Fell was E~ Common* in 1826 (CRO D/Lec/94), *Ennerdale Fell* 1867 (OS); *the forest of Einerdale* c.1539–45 (Leland V, 52), *Ennerdale Forest* 1687–8 (Denton); *the water of Ennerdale* 1799 (*CN* 540), *Ennerdale Water* 1867 (OS), previously *Eyneswater* 1322 (*PNCumb* 34), *(ye) Broadwater* 1557–8, 1612 (*CW* 1931, 166, 164).
♦ 'Anund's valley.' The name Ennerdale seems originally to have derived from *Anundar*, gen. sing. of the ON pers.n. *Anundr/Qnundr*, and ON **dalr** 'valley', but there has been cross-influence between this p.n. and EHEN, the name of the river which flows through the valley. The **bridge** (see **brycg**) in E~ gives its name to the

village beside it. E~ **Fell** is the high grazing land belonging to the parish, not a distinct peak. Though the modern E~ **Forest** is a large coniferous plantation, Denton wrote in 1687–8, 'this Enner is a Forest full of red-deer' (p. 98); see also **water** 'lake'. (Pl. 4)

ERNE CRAG NY3608 Rydal.
le Ernekrag, Ernecrag 13th cent. (*PNWestm* I, 210), *Earing Crag* 1863 (OS), *Erne Crag (or Earing Crag)* 1955 (Wainwright, Heron Pike 2).
♦ 'The rocky height frequented by eagles', with elements from OE *earn* 'eagle' and **crag(ge)**. The name is equivalent to EAGLE CRAG (q.v., on the threat posed by eagles). *Erne* is equated specifically with *Falco albiulla* by Hutchinson (1794, I, 450), but with the white-tailed sea-eagle *Haliaeetus albicilla* by Ratcliffe (2002, 237); perhaps this is what Hutchinson also meant. Machell drew an eagle, with caption, 'A sort of Eagle called an Iron here, In Scotland a Naron, being of a Blackish brown colour'; 'or Earn or Erne' is added in a different ink (c.1692, I, 724A). Other ornithological references in Rydal parish or on its boundaries include CROW HOW, DOVE CRAG, ELTERWATER and HERON PIKE.

ERNE NEST CRAG NY3712 Patterdale.
Earnest Crag 1860 (OSNB).
♦ 'Eagle's nest crag', with elements derived from OE *earn* 'eagle', OE *nest* and **crag(ge)**. There are no early forms, but cf. ERNE CRAG.

ESK, RIVER SD1297 Eskdale/Muncaster/ Drigg/Bootle, ESK HAUSE NY2308 Borrowdale/Eskdale, ESK PIKE NY2307 (2907ft/885m.).
Esc c.1140, *Esk* c.1180 (*PNCumb* 14); *Eskhals* 1242 (*PNCumb* 389); *Esk Pike ... is the hitherto nameless mountain lying between Esk Hause and Ore Gap* 1876 (Ward 46, n.), seemingly = *Tung(e)* 13th cent. (Collingwood 1918, 96).

♦ Like other rivers named Esk, Exe or Axe, this derives from a Brit. name **Iscā* (Cumbric **esc*) which may have meant simply 'water' (see references in Coates 2000, 365). It is sometimes known as South Esk, to distinguish it from the Esk which empties into the Solway Firth. It rises at the pass Esk Hause (ON **hals** 'neck, col'). The name Esk Pike does not appear on the OS maps of 1867 or 1900. As seen above, it was coined by the geologist James Clifton Ward for the summit formerly known as Tongue. TONGUE HEAD is still shown on the map at NY2408 as a distinct point NE of Esk Pike.

ESKDALE SD1699 (parish), E~ FELL NY1803, E~ GREEN NY1300 (hab.), E~ MILL NY1701, E~ MOOR NY1703, E~ NEEDLE NY2202.
Eskedal(e) 1285 to 1368, *Asshdale* 1461 (*PNCumb* 388); *Eskdale Fell* 1867 (OS); *Eskdale Green* 1867 (OS), 'formerly known as Yeat House Green' (*CW* 1922, 98); *Eskdale Moor* 1867 (OS); (Eskdale Needle:) previously *The Steeple* 1867 (OS).
♦ 'The valley (**dalr**) of the ESK', though spellings such as the 1461 one show that the river-name was associated with ON **eski** 'ash trees, ash copse' and hence ME *ash(e), esh(e)* (see **askr**). E~ **Fell** is the tract of upland grazing in the centre of the parish, while E~ **Moor** is unenclosed land at a lower level. The village of E~ **Green** lies on the flat land between Mere Beck and the Esk. E~ **Mill** has been a cornmill since medieval times and boasts a still functioning waterwheel. E~ Needle, a 50 ft rock column, seems only recently to have acquired

its name, cf. NAPES NEEDLE.

ESKHOLME: HIGH E~ SD1197 (hab.), LOW E~ SD1197 (hab.) Muncaster.
High Hestholm 1737 (PR), 1774 (DHM), *High Hestholme* 1826, *High Eskholm* 1827 (PR); *Low Hestholm* 1757 (PR), 1774 (DHM), *Low Hestholm* 1825, *Low Eskholme* 1826 (PR).
♦ The present name clearly and aptly means 'land beside the ESK', with 2nd el. from ON **holmr** 'land beside water, water-meadow', but it has apparently been reformed from *Hestholm*, in which the 1st el. may be from ON *hestr* 'horse, stallion' (cf. Eastholme and Hestham seemingly of the same origin, *PNCumb* 120 and 415). On the evidence of the PRs, the change took place about the 1820s.

ESKIN NY1829 (hab.) Wythop.
Eskyn 1578 (*PNCumb* 457).
♦ Uncertain. (a) One line of interpretation is suggested by Heskin in Lancs, which has *Eskin* 1260 alongside early spellings in *H*-, and for which Ekwall suggests an etymon related to W *hesgen* 'sedge, rush', or 'marsh' in p.ns (*PNLancs* 130–1, refined in Breeze 2000, 225–6). (b) It is tempting to link this with ON **eski** 'ash trees, ash copse' (see **ash**), postulating a noun **eskin* of the same meaning and parallel with ON *espin(i)*, *eikini* (Fritzner), which refer to aspens and oaks respectively, but no direct evidence for this is known to me. (c) *Esk-* 'ash' plus *kinn* 'slope' as perhaps in KINN is another possibility.

ESKMEALS SD0893 Bootle, E~ HOUSE SD0993.
Eskmeal 1610, *Eskmeals* 1777, *Esk-Meols* 1842 (*PNCumb* 345), *Eskmeals* (= Eskmeals House) 1867, 1900 (OS).
♦ 'The dunes beside the ESK'— the extensive sands around the estuary; from the river-name and ON *melr* 'sand-bank'. 'Esk-meold is plain dry low ground near ye River Esk between ye mountaines and ye sea. Every such place ye old Inhabitants termed Mule or Meold' (Fleming 1671, 38).

ESP FORD SD4490 (hab.) Crosthwaite.
Esp foord 1694 (*PNWestm* I, 83).
♦ 'The ford where aspen grows', with elements from OE *æsp(e)*/ON *espi* and OE **ford**. The aspen, *Populus tremula*, is found in all parts of the Lake District, though not in abundance (see Halliday 1997, 187). The place is at a crossing over the R. GILPIN.

ESPS FARM NY1429 Embleton.
Esps 1723, *The Esps* 1726 (PR), 1774 (DHM).
ESPS FARM SD2985 Lowick.
Esps 1722 (PR).
♦ E~ is presumably 'the aspen trees', with the reflex of OE *æsp(e)*/ON *espi*, and see ASHES for similar names. **Farm** has been added later in both cases.

ESTHWAITE: E~ HALL SD3595 Hawkshead, E~ LODGE SD3596, E~ WATER SD3696.
Estwyth 1539 (*PNLancs* 218), *Easthwaite* 1625 (PR); *Estead Hall* [sic] 1770 (JefferysM), *Esthwaite Hall* 1850 (OS); *Esthwaite Lodge* 1851 (OS); *the Mere of Hawkshed Estwater* 1537 (*Furness Coucher* II, 617), *Estthwaite Water* 1565 (West 1774, Appx IX; spelling possibly normalised), *Eastwait-water* 1671 (Fleming 29), *Esthwaite Water* 1851 (OS). (*Estwayt* 1326, identified with E~ in Wyld 1911, 120, in fact belongs to Nottinghamshire.)
♦ Probably either (a) 'the eastern clearing', with ME *e(a)st*, probably replacing ON *aust(r)* 'east', and ON **þveit** 'clearing', in which case the

name may conceivably be another territorial marker interacting with CUNSEY and CONISTON; or (b) 'the clearing where ash trees grow', from ON *eski* 'ash trees, ash copse' (see **askr**) and again *þveit*. Fellows-Jensen notes the use of ash boughs for winter sheep fodder (*SSNNW* 352). (c) Breeze suggests a Brittonic solution: that CUNSEY BECK, which flows out of E~ Water, may previously have been called *Ystwyth* 'agile one'; cf. W Aberystwyth 'mouth of the Ystwyth' (Breeze 2000, 64–6). However, Breeze's premiss that *-wyth* is 'anomalous and difficult to explain' as a spelling for *þveit, thwaite*, does not hold, since the spelling is paralleled, e.g. in *Kyrkwythe* c.1535 for KIRKTHWAITE, some six miles away (*PNLancs* 216), and in *Estwyt* 1483, *Estwith* 1563 for Eastwood, Notts, whose numerous medieval spellings clearly point to *þveit, thwaite* as 2nd el. (*PNNotts* 144). E~ Hall was the 'birthplace 1519 of Archbishop Sandys; later buildings modernized into a farmhouse' (Collingwood 1926, 42). E~ **Lodge** is a Regency house, now Hawkshead Youth Hostel; see also **water**.

ETHER KNOTT NY2617 Borrowdale.
Ether Knott 1862 (OSNB).
♦ This **knott** is a rocky height on GRANGE FELL. The origin and meaning of 'Ether' is obscure. The spelling might encourage the sense 'adder', since *EDD* has evidence from Scotland, Cheshire and Leicestershire of *ether* in this sense, but on the other hand the place is known locally as *Heather Knott* (I am very grateful to R. W. Richardson of Fold Head Farm, Watendlath for this), which matches Wainwright's description of Ether Knott as 'behind a barricade of long heather' (1958, Grange Fell 1).

EUSEMERE NY4624 (hab.) Barton.
Osemire, Osemyre 1278, 1279, *Ewes mire als. Ults mire* 1650 (*PNWestm* II, 210–1), *Eusemere* 1801 (*CN* 1020).
♦ Probably 'the bog near the outlet', from ON *óss, ósi* 'estuary, outlet of a lake', adopted into ME, and ON **mýrr** 'bog', since ON *ó* often produces *eu, ew* in the local dial. from the 16th cent. (*PNWestm* I, liii–liv) and there is a Norw parallel (*PNWestm* II, 211). The place is close to the point where ULLSWATER flows out into the R. EAMONT. The alternative form *Ults* is puzzling, however, especially since *Ults Rigg* occurs in 1650 as a variant for the field-name Useriggs, near to Eusemere.

EWE CRAGS NY4404 Kentmere.
From 1865 (*PNWestm* I, 166).
♦ This may be taken at face value as 'rocky heights frequented by ewes' (*ewe* from OE *eowu*, plus **crag**), cf. the more numerous YEW CRAG(S).

EYCOTT HILL NY3829 Mungrisdale.
Great Aiket Pike 1823 (GreenwoodM), *Eycott Hill* 1867 (OS).
♦ The spellings are too late and disparate to give a clear indication of origins. (a) *Aiket* could point to a name meaning 'high place where oak grows', cf. OAK HEAD and other names containing tree-names plus OE **hēafod**, ME **heved**: BIRKETT (BANK) and HESKETT. (b) *Eycott* (which could be heard as *Aiket*) has the appearance of a p.n. from OE *ēg* 'island' and OE *cot* 'cottage', and a similar name exists — Eycote in Gloucestershire (*DEPN*); but since it only appears late and in a **hill** name here, E~ may be a p.n. transferred from elsewhere, or more likely a surname derived from such a p.n.

F

FAIRBANK FARM SD4597 Nether Staveley, HIGH F~ SD4497.
Faverbancke & variants 1297 to 1570, *Fairebancke* & variants 1332 to 1823 (*PNWestm* I, 173); *Troutron (alias Higher Fair Bank)* c.1692 (Machell ed. Ewbank, 111), *High Fairbank, Middle Fairbank, Low Fairbank* all 1862 (OS).
♦ 'The fair hill-slope', with 1st el. from ON *fagr* 'fair, beautiful', giving way to the cognate ME *fair(e)* from OE *fæger*. **Bank** here refers to the slopes S of the R. GOWAN, and *Middle Fairbank* (1862 OS) is now GOWAN BANK FARM, while *Low Fairbank* is Fairbank **Farm**.

FAIRFIELD NY1526 (hab.) Lorton.
From 1867 (OS).
FAIRFIELD NY3511 (2863ft/873m.) Patterdale/Grasmere.
ye Fair-Field (Fleming 1671, 16).
♦ 'Beautiful field or open country', with 1st el. from OE *fæger* 'fair, lovely' and **field**. The Lorton example is on low ground E of the R. COCKER, while the mountain Fairfield doubtless owes its name to its grassy top. Fleming wrote of 'a very high Mountaine called Ridale-Head, on top whereof is a very fair and levell ground called ye Fair-Field' (1671, 16).

FAIR RIGG SD3884 (hab.) Staveley-in-Cartmel.
Fayrigg 1604 (Cartmel PR).
♦ 'The fair, fine ridge', with the reflex of OE *fæger* 'fair, lovely' plus **hryggr/rigg** — whether viewed with the eye of a farmer or an aesthete is impossible to say. This, like LINE RIGGS, is the kind of modest height which qualifies as a *rigg* in the gentle landscape of Cartmel.

FALCON CRAG NY2720 Borrowdale.
Falcon and Wallow Crags 1776 (Hutchinson 145).
FALCON CRAG NY3512 Patterdale.
Not on 1863, 1900 (OS).
♦ The Borrowdale **crag**, and presumably the Patterdale one, was 'once the home of peregrine falcons', according to Danby (1994, 61). The word *falcon* resulted from the adoption of OFr *faucon* into ME.

FALLEN YEW SD4692 (hab.) Underbarrow.
From 1745 (*PNWestm* I, 104).
♦ One of several Lakeland places named from the **yew** tree. *Fallen* derives from OE past participle *(ge)fallen*. The original farmhouse of this name dates from 1565 (Palmer 1944, 20).

FAR as affix: see main name, e.g. for FAR SAWREY see SAWREY

FAR END SD1187 (hab.) Whicham.
From 1774 (DHM).
FAR END SD3098 (hab.) Coniston.
From 1606 (PR).
♦ 'The outer edge of the village or parish'; **far, end**. The Coniston example is simply at the end of CONISTON village; cf. the various hamlets named TOWN END. The Whicham one outlies BOOTLE, but the name may refer to its position at the N edge of Whicham parish, far from WHICHAM itself.

FAR HOUSES SD2486 Kirkby Ireleth.
Farhouses 1688/9 (PR).
♦ It is not immediately clear from what point the **houses** were perceived

as being **far** — perhaps from the former township centre of WOODLAND.

FARMERY NY0707 (hab.) St Bridget Beckermet.
Furmery, del Fermerie (p) 1279, *Farmery* 1754 (*PNCumb* 338–9).
♦ 'The infirmary', from OFr *(en)fermerie*, in this case a hospital for the sick of CALDER ABBEY. The remote situation, two miles from the abbey, is not untypical of monastic infirmaries (so *PNCumb*).

FARTHWAITE NY0610 (hab.) Ennerdale.
farrethwate 1578 (*PNCumb* 388, among field-names).
♦ Possibly 'the remote clearing'; **far**, **þveit/thwaite**. The place is at the western extremity of Kinniside, which came to be merged with Ennerdale township.

FAULDS NY2939 (hab.) Caldbeck, F~ BROW NY2940 (1125ft/343m.).
Faldes 1560, *Folds* 1754, *The Faulds* 1777, possibly = *de Falda* (p) 1285 (*PNCumb* 278).
♦ 'The animal pens'; **fold** and cf. FOLDS. F~ **Brow** is the high ground above Faulds.

FAWE PARK NY2522 (hab.) Above Derwent.
(ate) Fal 1260 (p), *Falpark* 1369 (*PNCumb* 370).
♦ F~ is probably 'the clearing, felled woodland', from OE *(ge)fall* or ON *fall* in these senses. The site, next to DERWENTWATER, would be a very likely candidate for clearance. **Park** here probably has the sense 'large enclosure for beasts of the chase'.

FELLBARROW NY1324 (1365ft/416m.) Blindbothel/Loweswater.
From 1867 (OS).
♦ Presumably 'the summit on the upland'; **fell**, **barrow** from **berg**. This is the highest point on a range of grassy heights.

FELLBOROUGH SD3894 (hab.) Claife.
From 1892 (OS).
♦ Perhaps 'the summit on the upland'; **fell**, **barrow** from **berg**. This is one of two farms in a clearing by WINDERMERE, the other, Waterbarrow, being closer to the shore. This pairing and the unlikelihood of a name containing *-borough* from OE *burh* 'fortified place' hereabouts would encourage the surmise that this was originally another FELLBARROW.

FELL COTTAGE SD1185 Whicham.
From 1867 (OS).
♦ The place is below the W slopes of BLACK COMBE; **fell**, **cottage**.

FELLDYKE NY0819 (hab.) Lamplugh.
Fell Dike 1581 (*PNCumb* 406).
♦ (a) This may simply be *felldyke*, one of several terms which denote the 'head-dyke', the 'hill-side boundary between improved land and rough grazing' (Winchester 2000, 53); **fell**, **dyke**. (b) *PNCumb* suggests Scots *fail* 'turf' as 1st el.

FELL EDGE SD4388 (hab.) Crosthwaite.
Edge 1535, *Fell Edge* 1638 (*PNWestm* I, 83), 1862 (OS).
FELL EDGE SD4591 (hab.) Crosthwaite.
From 1862 (OS).
♦ The elements are clearly **fell, edge**, but the application of these is slightly elusive. *Fell* could either refer to the unenclosed grazing of Crosthwaite (as suggested in *PNWestm*) or to a specific hill. *Edge* in the example at SD4388 could refer to its location on the fringe of WHITBARROW SCAR or to the

escarpment itself. F~ E~ at SD4591 lies below a distinct height rising to 617ft/188m., but is also next to the boundary with Underbarrow and is in that sense at the edge of the Crosthwaite grazing. The early modern spellings probably belong to the example at SD4388.

FELL END/FELLEND Examples at:
   NY0614 (hab.) Ennerdale.
From 1593 (*PNCumb* 387).
   NY2234 (hab.) Bewaldeth.
Fell-end 1750 (Uldale PR), *Fellend* 1774 (DHM), *Fell end* 1803 (*CN* 1518).
   SD2388 (hab.) Kirkby Ireleth.
Fellend in Woodland 1685 (PR), *Fell End* 1851 (OS).
HIGH F~ E~ SD4383 (hab.), LOW F~ E~ SD4584 (hab.) Witherslack.
Fell End 1741 (*PNWestm* I, 78); *High Fell End* 1730, 1753 (PR), *Low Fell End* 1745 (PR). Cf. *Upper Fell End* 1676; *Over Fell End* 1701; *Far Fell End* 1715 (all PR).
♦ Simply 'the edge of the high ground'; **fell, end**; also **high, low**. The Ennerdale place stands where the contours of HECKBARLEY begin to level out; the Bewaldeth one is situated at about 750ft/230m. on the flank of BINSEY; and the Kirkby Ireleth one at the S end of a distinct stretch of high ground rising out of level, wet land. The Witherslack sites lie below WHITBARROW.

FELL FOOT NY2903 (hab.) Langdales.
ffell=foote in Langdalle 1682/3 (Hawkshead PR); *Fell Foot* 1704 (*PNWestm* I, 206).
FELL FOOT SD4699 (hab.) Hugill.
Cf. *Fellfoot (Mill)* 1857 (*PNWestm* I, 171).
FELL FOOT PARK SD3886 Staveley-in-Cartmel.
Fell Foot in Staveley 1782 (Finsthwaite PR), *Fell Foot* 1783 (CrosthwaiteM), 1859 (Stockdale 1872, 511).

♦ 'The place below the high ground'; **fell, foot**; also **park**. The Langdales and Hugill farms are not below prominent peaks (though the Langdale one is at the foot of a small knoll, and of WRYNOSE PASS), so that *fell* probably has the sense 'high unenclosed grazing land'. Fell Foot Park, still simply Fell Foot on earlier editions of the One Inch map (e.g. 1966), is a caravan park at the foot of a steep slope at the SE tip of WINDERMERE.

FELL GATE SD2487 (hab.) Colton.
From 1850 (OS).
FELL GATE SD4794 (hab.) Underbarrow.
Fellyatt 1717, *Fellyeat* 1725 (*PNWestm* I, 104).
♦ Presumably 'the gate onto the high ground'; **fell, geat/gate**. Fellgates marked the division between open pastures (fell) and enclosures, and controlled the movement of livestock between the two.

FELLGREEN SD1188 (hab.) Bootle.
Fell Green 1867 (OS).
♦ 'The grassy place by the high ground'; **green, fell**. It is on the fringe of BOOTLE FELL.

FELL PLAIN SD4596 (hab.) Crook.
From 1836 (*PNWestm* I, 179).
♦ As the name suggests, this is a high but relatively level tract of land (cf. ELVA PLAIN). Smith takes the **fell** to be specifically Crook Fell (*Crockefell* 1570, *PNWestm* I, 179).

FELLSIDE/FELL SIDE Examples at:
   SD1188 (hab.) Bootle.
Felsyde 1579 (*PNCumb* 348).
   NY3037 (hab.) Caldbeck, FELLSIDE BROW NY3036.
Felside 1560 (*PNCumb* 278-9); *Great Fellside, Little Fellside* 1774 (DHM);

Fellside Brow 1867 (OS).
   SD3584 (hab.) Haverthwaite.
From 1851 (OS).
   SD4490 Crosthwaite.
Fellside & variants 1535 to 1738; cf. *Crosthwaite Fell(s)* 1631, 1754 (*PNWestm* I, 83).
FELLSIDE FARM NY1224 Blindbothel.
Fell Side 1653 (*PNCumb* 423).
♦ 'The slope or side of the hill'; **fell**, **side**; also **brow** 'hill-side', **farm**. The Bootle example is by BOOTLE FELL, and the Blindbothel one by WHINFELL; *farm* has been added to the latter. The Caldbeck Fellside relates to CALDBECK FELLS, with F~ Brow on higher, steeper ground above it. The Haverthwaite place is on the slope of BACKBARROW above the R. LEVEN, while the Crosthwaite one lies below rising ground, its name grouping with other local p.ns formed from **side** (q.v.).

FENWICK SD1689 Millom Without.
Fenacke 1630, *Fenwicke* 1654 (*PNCumb* 418), *Phoenix* 1774 (DHM).
♦ The spellings are too late for certainty, but if this is of the same origin as the Northumberland and Yorkshire Fenwicks, the meaning is 'settlement by the marsh', and the elements derive from OE *fen* 'marsh' and OE **wīc** 'settlement, specialised place'. That the land has been reclaimed from waste is clear on the 1867 OS map.

FERNEY GREEN SD4096 (hab.) Bowness.
Furneagreen 1614, *Furnellgreene* 1622, *Farnellgreene* 1622 (Windermere PR), *Forney Gr.* 1823, *Ferney Green* 1865 (*PNWestm* I, 187).
♦ **Green** designates a 'grassy place', but the 1st el. is uncertain: either derived from OE *fearnig* 'ferny' or possibly, judging from the 1622 spellings, a surname such as *Farnell, Furnell*.

FERNGILL CRAG NY2909 Grasmere.
From 1859 (OSNB).
♦ 'The rocky height by Ferngill, the ravine where fern grows'; *fern*, **gil(l)**; also **crag**. *Fern*, derived from OE *fearn*, appears less often than **bracken** in Lakeland p.ns, and in Cumbrian p.ns generally (according to Gambles, who takes *bracken* in p.ns to mean bracken, the female fern, and *fern* the male, 1989, 12–13 and 22–3).

FERRY HOUSE, THE SD3995 Claife.
the Ferry c.1692 (Machell II, 314), *The Ferry* 1783 (CrosthwaiteM).
FERRY NAB SD3995 Bowness.
From 1851 (OS).
♦ The word *ferry* derives from ON *ferja*; see also **house** and **nab** 'promontory'. The ferry crossed (and still crosses) the narrow 'waist' of WINDERMERE, an ancient and vital link in the route between Kendal to the E and HAWKSHEAD and points W, documented as early as 1454 (*Kendale*, II, 73). The Ferry House is on the site of the old Ferry Inn (shown on 1851 OS map) on the W side, while the Nab is on the E.

FEWLING STONES NY5111 (1667ft/ 508m.) Shap Rural.
Fewlingstone 1859 (*PNWestm* II, 174), *Fewling Stones* 1863 (OS).
♦ 'Rocks where wild-fowling took place', with dial. *fule* for *fowl* (so *PNWestm* II, xiv), plus **stone**.

FIDDLER HALL SD3884 Staveley-in-Cartmel.
Fidler Hawe 1589 (*PNLancs* 199), *Fidlerhall* 1625 (Cartmel PR).
♦ Probably 'Fiddler's hill'. *Fiddler* and variants occur as a common noun (player on the fiddle) and as a surname from ME onwards. The 16th-cent. spelling *hawe* is the form of

DICTIONARY

how(e) 'hill' (from ON **haugr**) favoured in Furness and often interchangeable there with **hall**.

FIELD CLOSE SD4697 Nether Staveley.
Seemingly = *Demesne* 1862 (OS), *South View* 1899, 1920 (OS).
♦ 'The enclosure or farm by the field'; **field**, **close**.

FIELD END NY0802 (hab.) Gosforth.
From 1778 (original spelling not specified, Terrier).
FIELD END SD3783 (hab.) Staveley-in-Cartmel.
Feildend 1592 (Cartmel PR).
♦ Self-explanatory; **field**, **end**. The Staveley site is the meeting of level ground with the slope of Knotts Hill.

FIELD FOOT NY3605 (hab.) Rydal.
the Field-Foot in Loughrdge [sic] *side* 1697/8 (Grasmere PR).
♦ 'The lower end of the field' (**field**, **foot**), now a short row of houses beneath a wooded slope.

FIELD GATE NY5218 (hab.) Bampton.
Field Yate or Gate 1712 (*PNWestm* II, 194).
♦ 'The gate to the field'; **field**, specifically *Bampton field*, recorded from 1683 (so *PNWestm*), plus the reflex of OE **geat** 'gate'. The Y-spelling of 1712 suggests this rather than the reflex of ON **gata** 'road'. On earlier editions of the One Inch map (e.g. 1966), the name appears as Fieldgate House.

FIELD HEAD Examples at:
   SD1598 (hab.) Eskdale.
Fieldhead 1753 (PR).
   SD3691 (hab.) Satterthwaite.
Fieldhead 1781 (PR).
   NY3727 (hab.) Mungrisdale.
Feildhead 1714 (*PNCumb* 227).

FIELD HEAD HOUSE SD3499 Hawkshead.
ffeyldehed c.1535, *Feldhed* 1539 (*PNLancs* 218), *Haukeshead field head* 1659 (PR).
♦ Simply 'the upper end of the field'; **field**, **head**; also **house**. The Eskdale F~ H~, for example, is at the N edge of the enclosed, level land along Fisher Beck (1867 OS map), and the Mungrisdale one stands on what is still the boundary between cultivated land and rough grassland. In Hawkshead, Ekwall suggests a specific reference to 'the upper end of Hawkshead townfield' (*PNLancs* 218); cf. *Hauxhead field* 1598 (PR).

FIELDSIDE NY2823 (hab.) St John's.
Fieldsid 1571, *Feldside* 1585 (*PNCumb* 317).
♦ 'The **side** of the open, or enclosed, ground', depending on the meaning of **field** at the time when the name was given.

FINSTHWAITE SD3687 (hab.) Colton, F~ HEIGHTS SD3688, F~ HOUSE SD3687.
Fynnesthwayt 1336 (*PNLancs* 217); *Finsthwaite Heights* 1851 (OS); *FinsthwaiteHouse* 1826 (Finsthwaite PR), previously *Ploome Greene* (according to Cowper 1899, 61; occurs in Hawkshead PR e.g. 1605/6).
♦ 'Finn's clearing', from the ON pers.n. *Finnr* and ON **þveit** 'clearing'; also **height**, **house**.

FISHER CRAG NY3016 St John's.
From 1823 (GreenwoodM), 1867 (OS).
FISHER PLACE NY3118 St John's, F~ GILL NY3218.
Fisher Place 1774 (DHM), *Fishers Place* 1787 (*PNCumb* 317), *Fisher Place* 1867 (OS); (Fisher Gill:) *Fisherplace Gill* 1867 (OS), and present 1:25,000 map; *Fisher Gill* 1867 (OS) is on the other, W, side of Thirlmere.

♦ 'The rocky height, farm and stream associated with the Fisher family'; surname plus **crag, place, gil(l)**. 'Fisher' is presumably a surname, like the 1st el. in other p.ns containing *place*. Persons named Fisher appear in the PRs in the early 1800s, and other 'Fisher' p.ns are found in the vicinity, such as Fisher's Wife's Rake at NY3222. F~ Crag is on the W side of THIRLMERE while F~ Place is on the NE. Stream names such as Fisher Gill, with surname qualifying *gil(l)*, are rare but not unparalleled: cf. TAYLOR G~ and probably WILEY G~.

FISHERGROUND FARM NY1400 Eskdale.
Fisher Ground 1716 (PR).
♦ F~ is 'the farm of the Fisher family'; surname plus **ground**; also **farm**. *Fisher* is a common Lakeland surname, which occurs in the Eskdale PRs, though I have not found it associated with Fisher Ground. *Ground* is applied to farms, especially those founded about the 16th cent. *Farm* seems to have been added recently for clarification.

FISH HOUSE: F~ H~ MOSS SD3382 Haverthwaite.
Fish House 1804 (Colton PR); *Fish House Moss* 1851 (OS).
♦ Fish House stands on the R. LEVEN, and its name can be taken at face value (*fish* ultimately from OE *fisc*, and **house**). 'At Fish-house' occurs in 1711 and in 1713 in reference to part of a fishery on the Leven rented by the BACKBARROW iron company, partly to supply its workforce (Fell 1908, 306, 287). On the **Moss** 'bog', see DEER DYKE MOSS.

FLASKA/FLASKEW NY3625 Matterdale.
in bosco de Flatscogh in vasto (in the woodlands of F~, in the waste) 1278 (*CW* 1923, 186), *Flatschow, Flatscough* c.1278 in later copy (*CW* 1918, 102).
♦ 'The level wood', from ON *flatr* 'flat, level' and ON **skógr** 'wood' (cf. *PNCumb* 222 and 187 for local and Norw parallels). Though not flat, the gradient of this long incline is gentle by local standards. The name, and its context in the 1278 document, gives a glimpse of ecological change, for in 1794 the place was commended by Hutchinson for its 'excellent turbary [turf-cutting], and luxuriant herbage' (I, 411), and it is only partially wooded now.

FLASS NY1203 Irton.
From 1867 (OS).
♦ Despite the lack of early forms, this probably goes back to ON *flask* 'swamp, pool', which occurs in ME p.ns spelt *flasshe*, and in Flass, Durham (Watts 2002). This a low slope around F~ Tarn.

FLAT FELL NY0513 (869ft/265m.) Ennerdale.
Flatt fell (ende) 1578 (*PNCumb* 387).
♦ The unusually broad, level top of the **fell** suggests that the name is to be taken at face value. The ON adj. *flatr* 'flat' was adopted into ME as *flat*. It also seems to occur in FLASKO, while the related noun appears in ABBEY FLATTS, FLATTS and THORNFLAT(T).

FLATTS SD0990 (hab.) Bootle.
the Flatts 1542 (*PNCumb* 348).
♦ 'The level ground', with ON noun *flǫt* adopted into ME as *flat* and given an English pl. *-s*. F~ Farm and Cottage are distinguished on larger scale maps.

FLEETWITH NY2113 Buttermere, F~ PIKE NY2014 (2126ft/648m.).
Fleetwith 1780 (West 131), *Fleetwath*

1783, *Fleetwith* 1784 (*PNCumb* 356).
♦ The puzzling mountain name Fleetwith seems to refer to nearby features. The 1st el. may be *fleet* 'stream or boggy ground' as in CROOKAFLEET, while the 2nd is either a ford (*wath* from ON **vað**) at one of the two becks below, or a wood (*with* from ON **viðr**). However, the spellings are too late and inconsistent for certainty. F~ **Pike** is the higher summit nearby.

FLODDER: F~ HALL SD4587 Crosthwaite. *Flodder* 1535 to 1777, *Floder* 1663 (*PNWestm* I, 83); *Flodder Hall* 1862 (OS).
♦ (a) Smith suggests origins in a dial. word *flother, fludder*, a form of the postulated OE *flōdor* 'channel' (*PNWestm* I, 83 and cf. I, 7). (b) A dial. word *flodder* 'froth, brown scum' also exists, at least in S. Cumbria (Taylor 1927, 156), but it is difficult to make sense of this as a potential p.n. About the **Hall**, Palmer wrote, 'although now a farm, it can be seen from its size to have been of considerable importance, and it is the finest house in the [Lyth] valley'; 'largely rebuilt early in the 17th century, it has retained parts of an earlier building' (1944, 42).

FLOSHGATE NY4523 (hab.) Dacre.
Cf. (*del*) *Floshe* (p), *Floche* (p) 1278 (*PNCumb* 187).
♦ Probably 'the gate onto the lakeside', from ME *flasshe* 'pool, watery place' and **gate**. The place is close to the W shore of ULLSWATER.

FLOUTERN TARN NY1217 Loweswater. *water of Flutern* 1343 (*PNCumb* 34), *Floutern Tarn* 1774 (DHM), previously *Blutterne* 1230 (*PNCumb* 34).
♦ The 2nd el. *-tern(e)* is clearly from ON **tjǫrn** 'small mountain pool', duplicated by the later addition of Tarn. The *Flu-/Flou-* el. is obscure.

FOLD GATE SD3596 Hawkshead (hab.). *foulyeat* 1598/9, *fouldeyate* 1634 (PR), *Fold Yeat* 1851 (OS); cf. *Fould* 1770 (JeffreysM).
FOLDGATE FARM SD1191 Waberthwaite. *Foulyate* 1702 (*PNCumb* 365), *Foulgate* 1774 (DHM), *Fold gate* 1819 (PR).
♦ Originally 'foul, dirty gate', with reflexes of OE *fūl* 'foul' and OE **geat**, but the name has been 'improved' by a change to **fold**, and Standard English *gate* has replaced dial. *yeat/ yate*; **farm** has subsequently been added in the Waberthwaite example. (In the Hawkshead PR, continuity of occupants establishes that *foulyeat* and *fouldeyate* are the same place, while in Waberthwaite the identification of *Foulyate* 1702 with Foldgate tentatively suggested in *PNCumb* is proved by the 1774 map.)

FOLDS SD1791 (hab.) Ulpha.
Foulds 1717 (*PNCumb* 438).
♦ 'The animal pens'; **fold**, and cf. FAULDS.

FORCE: F~ FORGE SD3390 Satterthwaite, F~ MILLS SD3491.
Fosse 1577 (SaxtonM), *forse* 1631 (Hawkshead PR); *Forse Forge* 1668; *Force Myln* 1537 (*PNLancs* 220).
♦ The **force** or waterfall is Force Falls in RUSLAND POOL, whose cascading upper reaches are known as Force Beck, while on Saxton's map of 1577 the stream as a whole is designated *Fosse*. On the word *forge* see (THE) FORGE; also **mylen/mill**. F~ Forge was established to process iron ore in the early 17th cent., and may originally have been located at F~ Mills (Awty 1977, 97, 110), where the falls are particularly powerful.

FORCE CRAG NY1921 Above Derwent.
From 1800 (*CN* 804).
♦ 'The rocky height beside the waterfall'; **force**, **crag**. This is the precipice flanking Low Force in Pudding Beck.

FORD HOUSE SD1089 Bootle.
the Ford House 1542 (*PNCumb* 348).
♦ **Ford** may here refer to a track which passes over ESKMEALS Pool just W of the **house**.

FORE SLACK SD1685 (hab.) Whicham.
Foreslack 1646, *Forslack* 1651, *Forceslacke* 1686 (*PNCumb* 445).
♦ Uncertain. The 2nd el. appears to be from ON **slakki** 'valley', but the 1st is obscure. The 1686 spelling would suggest **force** 'waterfall' but there do not appear to be falls nearby.

FOREST, THE NY5303 Fawcett Forest.
(*ye*/*the*) *Forrest* 1581 to 1618 (*PNWestm* I, 138).
♦ This is the **forest** (presumably in the sense of a hunting preserve) that has given its name to the parish of Fawcett Forest. Today it is a treeless grouse moor.

FOREST HALL NY5401 Fawcett Forest.
forrest Hall c.1692 (Machell II, 102). Machell refers to F~ H~ as a manor house 'w[hi]ch has halfe of the forrest lying in Demane'; **forest** (see THE FOREST above), **hall**.

FOREST SIDE NY3408 Grasmere.
the Forrest syd 1610 (*CW* 1908, 154), cf. (*de*) *Foresta* 1332 (p), *the common waste ... called* The Forest 1777 (*PNWestm* I, 200).
♦ Presumably 'the place beside the hunting area'; **forest**, **side**. Although F~ S~ happens to be on the edge of a plantation, *forest* here probably has its older sense of a protected hunting area. Wordsworth wrote 'Upon the forest-side in Grasmere Vale | There dwelt a shepherd, Michael was his name ... ' (*Poems* 104).

FORGE, THE SD3793 Satterthwaite.
From 1851 (OS), possibly = *Fourge* 1628, *fforge* 1630 (Colton PR).
♦ This, situated by CUNSEY BECK, is a reminder of the days when charcoal-burning and iron bloomeries flourished in the woodlands of High Furness. An iron-furnace is known to have existed in 1711, later a bobbin mill (Danby 1994, 204). The word *forge* was adopted into ME from OFr *forge*; *OED*'s first citation for sense 4, a 'hearth or furnace for melting or refining metals' is 1601. It 'has been used indiscriminately ... to indicate bloomery, bloom-smithy, furnace and forge' (Davies-Shiel 1971, 283-4).

FORGE HOUSE SD1499 Eskdale.
From 1867 (OS), previously *Howe Howe* or *Howe Powe* (Parker & Fair 1922, 94).
♦ This is an old bloomery site (Parker & Fair 1922, 94), and possibly the location of a forge for which permission was granted in 1354 (Sykes 1926, 113). On the word *forge*, see (THE) FORGE. This is one of a cluster, with F~ Bridge and F~ Hill, of names referring to a nearby forge; it is known locally as Forge **Farm** rather than F~ **House** (Parker & Fair 1922, 94 and the present owners, to whom I am grateful for the information).

FORNSIDE NY3220 (hab.) St John's.
Fornesate 1303, *Fornset*(*t*) 1530s, *Furnesyde* 1530 (*PNCumb* 312-13), *Fornside* 1695 (MordenM).
♦ 'Forni's shieling' or 'the old (former) shieling', with 1st el. either

the ON pers.n. *Forni* or ON adj. *forn* 'old', plus ON **sætr** 'shieling, summer-pasture'. If this is an old or former shieling site it finds a parallel in OLD SCALES.

FORTH HOUSE NY1332 Setmurthy.
Frith house 1786, 1858 (PR), *Forth House* 1865 (OS).
♦ Uncertain, but possibly 'the **house** by the wood or scrubland'. The PR spellings might suggest ME and ModE *frith* from OE *fyrhþ(e)* 'a wood' as in FRITH WOOD, rather than **ford/forth**. The lower DERWENT valley is amply wooded today.

FOULE CRAG NY3228 Mungrisdale.
Foul-cragg 1794 (Hutchinson I, 424), *Fools crags* 1867 (OS).
♦ (a) This is probably the **crag** or rocky height that is *foul* in the sense of difficult or dangerous (from OE *fūl*, EDD: *foul*, sense 9): the place is 'a perpendicular rocky precipice' (Hutchinson) which terrified early visitors and moved A. Wainwright to use the adj. 'unpleasant' (1962, Blencathra 25, and 26 for illustrations). (b) 'Frequented by birds' (*fowl* from OE *fugol*), hence a place for fowling (cf. FEWLING STONES) seems much less likely.

FOULSHAW: HIGH F~ SD4683 (hab.) Witherslack, LOW F~ SD4781 (hab.), MIDDLE F~ SD4682 (hab.), F~ MOSS SD4682 Witherslack/Meathop/Ulpha.
Fowleshawe 1612 to 1652 (*PNWestm* I, 79); *High Fowlshaw (house)* 1779 (Witherslack PR), *High Foulshaw* 1862 (OS); *Low Fowlshaw* 1759 (Witherslack PR); *Middle Foulshaw*; *Foulshaw Moss* both 1862 (OS).
♦ F~ is probably 'bird wood', from OE *fugol* 'bird' and OE **sceaga** 'wood' (see **skógr**), though, as Smith remarks,

OE *fūl* 'foul, dirty' cannot be ruled out, especially since the place was described in 1770 as 'a large Moss or Turvery [peat-cutting area]' (*PNWestm* I, 79). See also **high, low, middle, moss** 'bog'.

FOULSYKE NY1421 (hab.) Loweswater.
Fulsike 1545 (CRO D/Law/1/239), *Foulsyke* 1740 (PR).
♦ 'The muddy, sluggish stream', with the reflex of OE *fūl* 'foul' and **syke**. For other Cumbrian instances of this name see *PNCumb* 529.

FOX CRAGS SD1593 Muncaster/Ulpha/Waberthwaite.
From 1867 (OS).
♦ This rock-strewn height is a natural and characteristic retreat of foxes. This persistent enemy of the farmer gives rise to many p.ns containing *fox* (OE *fox*) or the dial. *tod(d)* as in TODD CRAG. Gambles notes that older p.ns refer to the now extinct grey fell fox, larger than the small red fox (1989, 117).

FOXFIELD SD4087 (hab.) Cartmel Fell.
Foxfeild 1605/6 (Cartmel PR).
♦ Short of undetectable corruption, simply 'the **field** frequented by foxes', on which, see FOX CRAGS.

FOX GHYLL/GILL NY3605 (hab.) Rydal.
fox(e)gill 1645, 1647 (Grasmere PR).
♦ 'Fox ravine/stream'; *fox* as in FOX CRAGS, plus **gil(l)**. Cf. nearby Fox How (from 1613, *PNWestm* I, 210) and Fox How Farm.

FOXHOLE BANK SD4392 (hab.) Crosthwaite.
From 1865 (*PNWestm* I, 84).
♦ OE *fox-hol* 'fox's earth' is found in p.ns as early as *Foxehola* 1086 (Foxhall, Suffolk), but this seems to be a much

more recent example; *fox* as in FOX CRAGS, **hol(r)** 'hollow', **bank** 'slope'.

FRIAR'S CRAG NY2622 Keswick.
Fryer Cragg 1771 (*CW* 1975, 295), *Friarcrag* 1784 (*PNCumb* 303), *Friar's Crag, Friar Crag* 1839 (Ford 46, 50).
♦ Clearly *friar* (from ME, OFr *frere* 'brother (in a religious order)') plus **crag** 'rocky height or outcrop', but the motivation for the name of this magnificent viewpoint jutting into DERWENTWATER is uncertain. It is associated in local folklore with St Herbert (see ST HERBERT'S ISLAND), or sometimes with the Cistercian monks whose GRANGE was at the S end of the lake.

FRITH HALL SD1891 Ulpha.
Recorded 1770 (spelling not specified, *CW* 1966, 374), *Froth Hall* 1774 (DHM), *Frith Hall* 1830 (Seathwaite PR).
♦ Probably 'the hall by the wood or scrubland'. 'Frith' may be the ME and ModE *frith* from OE *fyrhþ(e)* 'a wood': see FRITH WOOD. This is 'said to have been a hunting lodge' and subsequently a farm, but by 1950 was 'the tottering shell of a once fine building with its great hall and massive chimney stack' (Fair 1951, 102, 104). The word **hall** may therefore be taken at face value; otherwise the hill-top situation might have led one to suspect that it could have replaced **how(e)** 'hill' from ON **haugr**.

FRITH WOOD NY5115 Shap Rural.
From 1859 (*PNWestm* II, 174).
♦ 'The wood called or at Frith.' *Frith* originates in OE *fyrhþ(e)*, often translated 'a wood' but more recently explained as 'land overgrown with brushwood, scrubland on the edge of forest' (Gelling & Cole 2000, 224). In Cumbria it can also refer to 'unused pasture-land' (*EDD*: *frith*, sense 2). **Wood** may have been added later, when the word was no longer current.

FROST HOLE SD4898 (hab.) Over Staveley.
Frosthole 1629, *Frosty holes* 1836, *Forest Hall* 1865 (*PNWestm* I, 176).
♦ 'The hollow prone to frost', from the reflexes of OE *frost* and OE **hol(h)** 'hollow'. The place is at about 787ft/ 240m., beside a stream. This is a good example of the politer kind of derogatory p.n.; the 1865 form also illustrates the treacherous ease with which even self-evident names can become mangled in transmission.

FROSWICK NY4308 (2359ft/719m.) Troutbeck/Kentmere.
From 1828 (HodgsonM).
♦ According to Smith, this name originally designated a cove near the Roman Road [High Street], before being used of a mountain. It is, in any case, strange as a hill-name, since *-wick* most often derives from OE **wīc** 'settlement, specialised place, etc.' or from ON **vík** 'bay, inlet'. This may be a parallel to the occasional use of ON *vík* in Norw p.ns for other topographical curves, such as bends in rivers or in hill contours (*NGIndledning* 85: *vík, NSL*: Vik). The meaning of the 1st el. is obscure.

FROZEN FELL NY2833 Ireby and Uldale.
From 1867 (OS).
♦ If to be taken at face value, the name may refer to the mainly N or NW-facing slopes of the **fell**. *Frozen* is from OE past participle (*ge*)*frosen*.

FURNESS SD2690, F~ FELLS SD3289. Several parishes.
Futhþernessa c.1150 (in 13th-cent. copy), *Fudernesium* 1127–33, *Fodernesio*

1127 (p), *ffurnesio* 1153–60, *Fornesio* 1158, *Furnes* 1170, *Furneis* 1169 to 1246; *Montanis de Furnesio* 1196, *Fournes-fell'* 1338 (*PNLancs* 200–1), *Furness Fells* 1774 (West 104), *Heigh Furnes* 1584; cf. *Lowfurnes* 1546, *Playne Furneys* 1582 (*PNLancs* 201).

♦ Possibly 'headland by Fuð', where Fuð is an island name meaning 'cleft, cunt'. The name Furness is a celebrated problem, to which Ekwall's solution is probably the best, though not ultimately provable (*PNLancs* 200–1). The 2nd el. is probably ON **nes** 'headland', which may be the present Rampside, at or near the SE point of the Furness peninsula. The 1st el. may be the same as in Fouldray (*Fotherey* c.1327), now Peel Island, off Rampside, and may be *fuð*, also found in the names of Norw islands and skerries. This seems to mean 'cleft, split', and is applied mainly to female genitalia or the cleft between the buttocks, but may belong to a set of words which have both anatomical and topographical applications (see further Fritzner). The island is rounded, with a N–S cleft. Other suggestions, including OE *fodor* or ON *fóðr* 'fodder' (so Wyld 1911, 130), account for the early *-u-* spellings less well. The 12th-cent. founders of the great F~ Abbey seem to have unwittingly bowdlerised the name, adopting a spelling influenced by OFr *fornais* 'furnace'. The abbey was a major landowner in Cumbria until its dissolution in 1537, and hence its charters are a major source of early p.n. forms; see also GRANGE-IN-BORROWDALE and MONK CONISTON (under C~). F~ is the name given to Low or Plain F~, the peninsula between the DUDDON and LEVEN estuaries (mainly outside the National Park), together with High F~ or F~ **Fell**s, the equally large tract of higher hills to the N (within the National Park).

FUSEDALE NY4418 Martindale, F~ BECK NY4418.
Fe(e)usdale 1220–47, *Fehusdale* 1278, *Fewesdale* 1279 (*PNWestm* II, 216); *Fusedale Beck* 1863 (OS).
♦ 'The valley with the cowshed', from ON *fé-hús*, lit. 'cattle-house' and ON **dalr** 'valley'; and its **beck** 'stream'.

FUSETHWAITE YEAT NY4101 Windermere.
Fusethwaite Yeat c.1729 (*CW* 1982, 152).
♦ F~ is probably 'the clearing with/by the cattle-shed', from ON *féhús* 'cattle-shed' as in FUSEDALE plus ON **þveit** 'clearing'. **Yeat**, from **geat**, is a gate, either man-made or a natural gap.

G

GAIT CRAGS NY2205 Eskdale.
From 1867 (OS).
♦ 'The rocky heights frequented by goats', with the reflex of ON **geit** 'she-goat', plus **crag**; cf. GOAT CRAG(S) and probably GATE CRAG. High and Low G~ Crags are distinguished on larger scale maps.

GAITKINS NY2503 Ulpha.
From 1867 (OS).
♦ From the topography, a craggy brow high above WRYNOSE BOTTOM, one might make a guess — though it is only that — at origins in ON **geit** 'she-goat' and ON *kinn* 'cheek, hill-side' (see KINN). It is impossible to tell whether this name bears any relation to nearby GAITSCALE.

GAITSCALE: G~ CLOSE NY2402 Ulpha, G~ GILL NY2503.
Gaytschale 1398–9 (*Furness Coucher* II, 579), *Gaitscale* 1723 (Seathwaite PR), *Gate Scale* 1774 (DHM); *Gaitscale Close*; *Gaitscale Gill* both 1867 (OS).
♦ Gaitscale is presumably 'the shieling where goats are kept', with the reflexes of ON **geit** 'she-goat' and ON **skáli** 'shieling'. From this the nearby 'enclosed fell' (**close**) and 'ravine with stream' (**gil(l)**) were then named. On the 1398–9 spelling, see BIRKS, Ulpha.

GALAVA ROMAN FORT NY3703 Ambleside.
Galava 3rd cent. (Rivet & Smith 1979, 365).
♦ The name G~ appears to be of Brit. origin, but its meaning is obscure. Transfer of a river-name meaning 'vigorous stream' (**gala* plus **-aṷa* suffix) was suggested by Jackson (1970, 74). The original turf and timber fort is dated to AD 79, while a second, of the Hadrianic period, AD 117–38, had stone defences (LDHER 1877, SAM No. 13567). There was an associated *vicus* or civilian settlement. See also under GLANNOVENTA.

GALE FELL NY1316 Loweswater.
From 1867 (OS).
♦ Probably 'the hill where bog-myrtle grows'. (a) 'Gale' is likely to be the Northern word for bog-myrtle (OE *gagel*), *Myrica gale*, which grows especially in valley bottoms and lowland raised mosses. Its modern altitude limit is 1509ft/460m. (Halliday 1997, 137), similar to that of the damp Gale Fell (1699ft/518m). (b) Alternatively, 'Gale' could be from ON *geil* 'ravine', hence a narrow track or valley. The **fell** is not an independent summit but a flank of STARLING DODD.

GALESYKE NY1303 (hab.) Nether Wasdale.
From 1570 (*PNCumb* 441).
♦ Probably 'the slow stream where bog-myrtle grows', with the same specific as GALE FELL, plus **syke** 'slow stream or ditch'. Hutchinson mentions 'an aromatic shrub, called *Gale*' in Wasdale (1794, I, 582). The low-lying situation beside the R. IRT more or less excludes ON *geil* 'ravine, narrow track' as the origin of the specific.

GAP, THE NY1204 Nether Wasdale.
del Gap 1539, *Gape in Neither Wasdaill* 1659 (*PNCumb* 441).
♦ The word **gap** is adopted from ON; *OED*'s first record is c.1380. This gap passes through rising ground above NETHER WASDALE, though it looks far from spectacular on the map.

GARBURN/GARBOURN: G~ PASS NY4304 Windermere/Kentmere.
(Top of) Garburne (fell), *(Top of the Hight of) Garburne-(mosse)* c.1692 (Machell II, 106); *Garburne Pass* 1719 (*PNWestm* I, 196).
♦ G~ is believed to be 'the stream in the triangle of land', from OE *gāra* 'gore, triangle', referring to a point of land projecting into Kentmere parish (*PNWestm*), plus OE *burna* 'stream' (**burn**), presumably Hall Gill which rises close by. The **pass** carries an ancient trackway and drove-road between the TROUTBECK and KENTMERE valleys (Hindle 1998, 113, 127, Atkin 1991, 72).

GARGILL NY5406 Shap Rural.
Gargyll (yate) 1535, *Gargill* 1865 (*PNWestm* II, 174).
♦ 'The ravine/stream in a corner of land.' OE *gāra* 'gore' often refers to a corner of land, but since Gargill is (at least according to the map) a tract of fell, the feature referred to here is unclear. Similarly, it is uncertain which of the small nearby **gills** is meant.

GARNER BANK SD1598 Eskdale.
From 1867 (OS).
♦ Garner is presumably the surname, of various origin, although I have not been able to find it in the Eskdale PRs (Garnet occurs 1701, 1708). The site is higher and more desolate than the average **bank** 'hill-slope'.

GARNETT BRIDGE SD5299 (hab.) Strickland Roger.
Garnet [bridge] 1651 (*PNWestm* I, 158), *Garnet-Bridge* 1797 (Longsleddale PR).
♦ The **bridge** (see **brycg**) over the R. SPRINT, and the village of the same name, are named from the local family of Garnett.

GARTH HEADS NY4218 (hab.) Martindale.
Garth heade 1588 (*PNWestm* II, 218).
♦ Presumably 'the upper end of the enclosure'; **garth, head**. The phrase *garth head* was used to refer to part of a farm (17th-cent. examples *CW* 1884, 42, *CW* 1903, 154).

GARTHROW SD4791 (hab.) Underbarrow.
From 1626 (*PNWestm* I, 104).
♦ 'The row of houses by/with the enclosure'; **garth, row**.

GASGALE CRAGS NY1721 Buttermere, G~ GILL NY1721.
River Gasgale (= G~ Gill) 1803 (*CN* 1518); *Gasgale Crags*; *Gasgale Gill* both 1867 (OS).
♦ The 2nd el. of Gasgale could be ON *geil* 'ravine, narrow track' (see under **gil(l)**), in which case the 1st is obscure. More likely, however, is shortening of a name such as GAITSCALE 'goat shieling': cf. the *Gas-* spellings for GATESGARTH, Buttermere. G~ **Gill** 'ravine with stream' is the rocky gorge through which the LIZA BECK flows; the **crags** 'rocky heights' tower above.

GASKETH: see GATESGARTH

GATE CRAG SD1899 Eskdale.
From 1867 (OS), possibly = *Gaytecrag* 1290 (Brearley 1974, location not specified).
♦ 'The rocky height frequented by she-goats'; **geit, crag**; cf. GAIT CRAGS and GOAT CRAG(S). This is suggested by the 1290 spelling and the likelihood that a crag would be associated with goats rather than a 'gate' (from OE **geat**) or 'way, road' (from ON **gata**).

GATEGILL NY3226 (hab.) Threlkeld, G~ FELL NY3126.
Gategill 1774 (DHM), *Gait-Gill* 1775

(PR); *Gategill Fell* 1867 (OS).

♦ The origin and sense of 'Gate' (whether **geat** 'gate', **gata** 'road', 'pasturage' or **geit** 'she-goat') cannot be determined in the absence of earlier spellings, though goats would certainly not be out of place. **Gil(l)** is here classically applied to a steep, rocky ravine and watercourse, and Gate Gill has given its name to a hamlet and **Fell**. Gategill and G~ Fell form part of an interesting set of names referring to the buttresses of BLENCATHRA and the settlements at their feet: see BLEASE.

GATE HOUSE SD1184 Whicham.
From 1867 (OS).
♦ Probably to be taken at face value. Other examples of Gate House have early spellings in Y- (*Yatehouse* 1570 in Eskdale, *PNCumb* 391; *Yeathouse* 1674 in Torver, PR), which might favour 'gate' (from OE **geat**) rather than 'road' (from ON **gata**) as the 1st el. here; see also **house**.

GATERIGGHOW/GATTERRIGGHOW NY1003 (hab.) Irton.
Garterhow 1681, *Gaiterhow* 1690, *Gatrighow* 1728, *Gaithrigghow* 1736 (*PNCumb* 403).
♦ Partly obscure. The 1681 and 1690 spellings might suggest a surname as 1st el., but the elements appear to derive from ON **geit** 'she-goat', **hryggr/rigg** 'ridge' and **haugr/how(e)** 'hill'.

GASKETH SD0998 (hab.) Irton.
From 1867 (OS), and see GATESGARTH NY1001.
GATESCARTH PASS NY4709 Shap Rural/Longsleddale.
Gaitscarthe, Gaytskarthe 1578 (*PNWestm* II, 174), *Gate Scarth* c.1692 (Machell II, 103).

GATESGARTH NY1001 (hab.) Irton. *Gascarth* 1778 (PR).
Probably = *Catescart* 1278 (p) (*PNCumb* 402, which gives Gatesgarth and Gasketh as variants of each other, but they both exist, two miles apart, and it is not certain to which the early spellings belong).
GATESGARTH NY1915 (hab.) Buttermere, GATESGARTHDALE BECK NY2014.
Gatescarthe(heved) c.1211, *Gatescarth(e)* 1260 to 1562, *Gascard* 1268, *Gascarth* 1273, *Gaitescarth* 1318 (*PNCumb* 356), *Gaskath* 1696/7 (Lorton PR); *Gatesgarthdale Beck* 1867 (OS), cf. *Gascadale* 1786 (= Little Gatesgarthdale, *PNCumb* 356).
♦ Probably 'the mountain pass frequented by goats', with the reflex of ON **geita**, gen. pl. of **geit** 'she-goat' (perhaps merging with the cognate OE **gāt(a)**) plus ON **skarð** 'gap, notch, mountain pass'. The 1st el. could alternatively be ON **gata** 'way, road'. Through a misdivision of the name, the -s was (except in the Shap example) taken as belonging to the 1st el., and the 2nd el. understood as **garðr/garth** rather than *skarð/scarth*. That Gasketh is of the same origin as Gatescarth or Gatesgarth is suggested by the spellings in *Gasc-* or *Gask-* for the Buttermere Gatesgarth. The mountain pass referred to in the Buttermere name is HONISTER PASS, for G~ lies at the foot of the pass, which comprises Gatesgarthdale and Little Gatesgarthdale, and *Gatescartheheved* 'Head of Gatescarth' seems from its position in the charter bounds of c.1211 to refer to Honister Hause. Explanatory **pass** has been added later to the Shap name; and see **dale** 'valley' and **beck** 'stream'.

GATESIDE SD1483 (hab.) Whicham.
Gate Side 1651 (*PNCumb* 445).

♦ The 1651 spelling, and the location next to a major routeway (the present A595), suggests that the 1st el. is **gate** 'road' from ON **gata** rather than **gate** 'gate' from OE **geat**; see also **side**.

GAVEL CRAG NY4309 Kentmere.
From 1857 (*PNWestm* I, 166).
♦ 'The gable-like rocky height', with dial. *gavel* from ON *gafl* 'gable, end of a building or natural feature', perhaps alluding to the way the hill projects E over the upper reaches of the KENT, plus **crag**. Cf. the following two names and GREAT GABLE, GREEN GABLE.

GAVEL FELL NY1118 (1720ft/526m.) Loweswater/Ennerdale.
Gavelfell c.1599 (*CW* 1925, 194).
♦ 'The gable-like hill', with *gavel* as in GAVEL CRAG, plus **fell**.

GAVEL PIKE NY3713 (2577ft/785m.) Patterdale.
From 1860 (OSNB).
♦ 'The gable-shaped peak', from *gavel* as in GAVEL CRAG, plus **pike**. This is 'a very steep and rugged point on the end of a small ridge that juts out from The Cape' (OSNB).

GAWTHWAITE/GOATHWAITE SD2784 (hab.) Blawith, GAWTHWAITE LAND(S) SD2786.
Golderswatt 1552 (*PNLancs* 214), *Golderwith* 16th cent. (*VCHLancs* 354, n.), *Goathwait* 1720, *Gawthwaite* 1755; *Gathwt Lands* 1749, *Gawthwaite Lands* 1791 (Lowick PR), *Goathwaite Land* 1851 (OS).
♦ The 1st el. is uncertain; the 2nd from ON þveit 'clearing'. G~ **Land(s)** is a northerly extension from Gawthwaite; there are also a G~ Moor and G~ Moss.

GHYLL(-): see also GILL(-).

GHYLL NY2625 (hab.) Underskiddaw.
Gill 1867 (OS).
GHYLL FARM NY0611 Ennerdale.
Gill 1605 (*PNCumb* 387).
GILL NY1305 (hab.) Nether Wasdale.
le gyll 1570 (*PNCumb* 441), *Gill* 1714 (PR).
♦ 'The ravine with a stream', from ON **gil** or the dial. word descended from it. The spelling has varied between *gill* and *ghyll* (see **gill**). The Underskiddaw example is situated where How Gill turns into APPLETHWAITE GILL. The Ennerdale name refers to the narrow valley through which Sleven Beck flows; **farm** has been added later here. The Nether Wasdale place stands on Gill Beck.

GHYLL/GILL BANK NY2320 (hab.) Above Derwent.
Gylbanke 1564 (*PNCumb* 372).
GHYLL BANK/GILLBANK NY4700 (hab.) Over Staveley.
Gillbank 1836, cf. *Gill* 1714 (*PNWestm* I, 176).
GILL BANK NY1801 (hab.) Eskdale.
Gilbank 1570 (*PNCumb* 391).
GILLBANK SD3698 (hab.) Claife.
Gill Bank 1891 (OS).
Neither building nor name appears on the 1850 OS map.
♦ 'The bank of the ravine/stream', or 'slope by the stream', with the reflexes of **gil(l)** and **bank(e)**. *Gill* is applied rather loosely in the Above Derwent name, since the place is on the E bank of NEWLANDS BECK, which is not steep by the normal standards of a gill. The Eskdale place is by WHILLAN BECK.

GHYLL/GILL FOOT (hab.) NY3309 Grasmere.
Gill foot 1687 (PR).
♦ 'The lower end of the ravine'; **gil(l)**, **foot**. The hamlet is the site of a packhorse bridge over Green Burn,

and at the foot of the ravine known as the 'Fairy Glen' (1920 OS).

GHYLL HEAD/GILL HEAD SD3992 (hab.) Cartmel Fell.
Gill head 1783 (CrosthwaiteM), *Gill Head* 1851 (OS).
♦ As the name implies, the site is high on a narrow valley with a stream, though not at its very top; **gil(l)**, **head**.

GIBSON KNOTT NY3110 Grasmere.
♦ 'Gibson's rocky height'; surname plus **knott**. 'Gibson' is presumably the surname, ultimately derived from a pet-form of *Gilbert*, plus *son*.

GILL(-): see also GHYLL(-)

GILL BECK NY5019 Bampton, GILLHEAD NY5019.
Gill Beck 1863 (OS); *Gillhead* 1673 (*PNWestm* II, 195).
♦ 'The stream in the ravine' and 'the top of the ravine'; **gil(l)**, **beck**, **hēafod/head**.

GILLBREA/GILLBRAE/GILLBROW: HIGH G~ NY1526 (hab.) Lorton.
Gylbancke 1570 (Winchester 1987, 148), *Gilbrowe* 1597, *Gilbray* 1608 (PR), *Gillbrow* 1867 (OS).
♦ Probably 'the steep hill-side with a ravine and stream'. The 1st el. is clearly **gil(l)** 'ravine with stream', while the 2nd seems to have been **bank(e)**, early replaced by **brow**, which may also have been influenced by **brae**, the Anglicisation of Gaelic *bràighe* 'upland, upper part'. **High** G~ lies above Gillbrae Farm, and H~ G~ and Gillbrae were distinguished on the 1900 OS map (Gillbrae Farm and High Gillbrea appear with those spellings on the current 1:25,000 map). Winchester believes that 'the two hamlets of Gillbrow and Highside [see HIGH SIDE] probably represent twelfth- or thirteenth-century grants of land carved out of the forest along the margin of the settled area at Lorton', and that High Gillbrow represents a further stage of intaking at the end of the medieval period. 'High Gillbrow, a ring-fenced holding on the fellside above the freehold land at Gillbrow, existed by 1547' (1987, 148–9).

GILLBROW NY2219 (hab.) Above Derwent.
gilbrowe 1578 (*PNCumb* 372).
♦ 'The slope above the ravine with stream'; **gil(l)**, **brow**. It is on a tributary of KESKADALE BECK.

GILL CRAG NY3712 Patterdale.
From 1860 (OSNB).
♦ 'The rocky height above the ravine with stream'; **gil(l)**, **crag**. This is a rocky ridge above DOVEDALE BECK; it is the crest of Hartsop above How, and G~ C~ can refer to the fell as a whole (Wainwright 1955, Hartsop above How 1, cf. 1:25,000 map).

GILLERCOMB NY2111 Borrowdale, G~ HEAD NY2111 Ennerdale/Borrowdale.
Gillercombe 1759 (*CW* 1976, 106), *Giller Coom* 1839 (Ford 60); *Gillercomb Head* 1867 (OS), previously *Le Bradscarth* 1338 (Collingwood 1920, 244).
♦ Perhaps 'the valley of many streams', from ON *gilra*, gen. pl. of **gil** 'ravine with stream' and the originally Brit. word **comb(e)**[1] 'corrie, rounded valley'. This would be an apt description of G~, a marshy basin backed by a rocky wall rising to G~ **Head**, through which many streamlets flow into SOUR MILK GILL. In the absence of early spellings the 1st el. must be regarded as uncertain, but the main alternative interpretation, 'snares',

available for GILLERTHWAITE, seems implausible here given the high situation, at about 1300ft or 400m.

GILLERTHWAITE NY1414 (hab.) Ennerdale.
Gillerthwait 1604 (*PNCumb* 387).
GILLERTHWAITE NY1421 (hab.) Loweswater.
Gillerthw[ai]t 1740 (PR), *Gillerthwat* 1774 (DHM).
♦ Uncertain, perhaps 'clearing by the streams' or 'clearing with snares'. The 2nd el. is clearly the reflex of ON þveit 'clearing', but the 1st is difficult. (a) It may possibly be from ON gen. pl. *gilra* 'of streams' (see **gil**), since several streams flow into the R. LIZA near G~ in Ennerdale, though streams are less abundant near the Loweswater G~. (b) It may alternatively be derived from ON *gildra*, gen. pl. of *gildra* or *gildri* 'snare', which may occur in *Gilderbeckescove* and *Gilderbeckhow*, recorded from Eskdale in 1587 (*PNCumb* 393). Without *Gild-* spellings for either Gillerthwaite, however, this remains doubtful, since elsewhere such spellings survive well into the modern period (*PNCumb* 141, 175, 393).

GILL FORCE NY1700 Eskdale.
From 1867 (OS).
♦ 'The waterfall in the ravine/stream'; **gil(l)**, **force**. The falls are on the River ESK, hence this is an unusual application of *gill*, which usually refers to tributary streams in narrow ravines.

GILL HOUSE NY1102 Irton.
From 1692 (*PNCumb* 403), 1774 (DHM), 1802 (*CN* 1205).
♦ 'The house by the stream in the ravine' (**gil(l)**, **house**), here MECKLIN BECK.

GILLSIDE NY3716 (hab.) Patterdale.
From 1860 (*PNWestm* II, 225).
♦ 'The place beside the ravine with stream'; **gil(l)**, **side**. It is next to GLENRIDDING BECK.

GILLSROW NY3726 Matterdale.
Gilles 1589 (*PNCumb* 223), *Gills* 1774 (DHM), *Gillsrow* 1867 (OS).
♦ Originally 'the ravines with streams' (**gill** in pl.), to which **row** (a row of houses?), an el. common in this area, has been added later.

GILPIN, RIVER SD4392 Crook/Crosthwaite, G~ BANK SD4687 Crosthwaite, G~ COTTAGE SD4686 Crook, G~ FARM SD4685, G~ LODGE SD4295, G~ MILL SD4394, GILPINPARK PLANTATION SD4395.
(River:) *the watt' of Gylpyne* 16th cent., *Gilpin* 1614 (*PNWestm* I, 7); *Gilping Bank* 1857 (*PNWestm* I, 84); (G~ Cottage:) *Gilpin Cottage* 1899 (OS); (Gilpin Farm:) *Gilpins* [sic] *Cottage* 1862, *Gilpin Farm* 1899 (OS); *Gilpin Lodge* 1920 (OS); *Gilpin Mill* 1863 (OS); *Gilpinpark Plantation* 1862 (OS).
♦ G~ is a renowned Westmorland surname, appearing from the 13th cent., and the river may take its name from the family (as Ekwall thought, *ERN* 172), perhaps by back-formation via names such as Gilpin Beck and G~ Bank; the meaning of Gilpin is unclear (see *PNWestm* I, 7–8). (Cf. GOWAN, RIVER for a similar process.) G~ **Bank** is a small hillock beside the river. G~ **Lodge** is next to Gilpinpark **Plantation**, which is beside the upper reaches of the river but may be named from the river or the family. G~ **Mill** (see **mylen**) is somewhat lower down; a fulling mill is recorded here in a will of c.1700 (Somervell 1930, 113). See also **cottage** and **farm**.

GIMMER CRAG NY2707 Langdales.
Ghimmer-crag 1805 (publ. 1819, Wordsworth, *Poems* 143).
♦ 'The rocky height frequented by yearling ewes', with *gimmer* from ON *gymbr*, referring to a ewe that has not yet lambed, plus **crag**. Cf. EWE/YEW CRAG(S). (Wordsworth refers to *Ghimmer-crag*, as 'tall twin brother' of Raven Crag, which fits the Langdale situation, though the passage is set much further N in 'St John's Vale'.)

GLADE HOW NY1306 Nether Wasdale. From 1867 (OS).
GLEDE HOWE NY5212 Shap Rural. From 1859 (*PNWestm* II, 174).
♦ Both may mean 'kite hill', with variant spellings of dial. *glead*, referring to a kite or other bird of prey, plus **how(e)** from ON **haugr** 'hill, mound'. The kite is extinct in England (Gambles 1989, 161), though red kites are seen in Lakeland, at nearby NADDLE FOREST and elsewhere (Miles 1993, 43).

GLADSTONE KNOTT NY2504 Langdales.
♦ This **knott** or cluster of crags, close to the former Cumberland/Westmorland boundary, is not named (nor accurately mapped) on the 1867 OS map, so that it is tempting to speculate that the name commemorates the famous 19th-cent. Prime Minister, William Ewart Gladstone. The surname derives from a p.n. in Scotland.

GLANNOVENTA ROMAN FORT SD0895 Muncaster.
Clanoventa 3rd cent., *Glannibanta* c.400 (Rivet & Smith 1979, 367).
♦ 'The place by the shore', a Latinised form of a Brit. name composed of **glanno-* 'bank, shore' and *-venta*, an el. of uncertain meaning which may be pre-Celtic (Rivet & Smith 1979, 367, cf. 262–5). *PNCumb*

512 identifies the 2nd el. with W *-went* 'place, field', and Coates suggests 'central place, market' as possible senses of *went* (2000, 356). The fort, with its *vicus* (civilian settlement), provided coastal defence by the sands at RAVENGLASS, at the mouth of the R. ESK; it dates from the Hadrianic period, c.AD 122, with later rebuilding (LDHER 1378, SAM No. 13569). For the nearby bath-house see WALLS; and other forts from the same defensive chain across the S of Lakeland are named on modern maps: MEDIOBOGDUM (HARDKNOTT) and GALAVA (AMBLESIDE). (Pl. 28)

GLARAMARA NY2410 (2560ft/781m.) Borrowdale.
Hovedgleuermerhe (head of Glaramara) 1209–10, *Gleuermerghe* 1211, *Glaramara* 1784 (*PNCumb* 350).
♦ A fascinating and difficult name for a rugged mountain, possibly meaning 'the shieling at the ravines'. The 13th-cent. spellings for the final el. match those of names including BIRKER, LANGLEY and MOSSER and point to **ærgi/erg** 'shieling'. The syllables *-gleuerm-* are more elusive. (a) They may, as assumed in *PNCumb*, reflect dat. pl. *gljúfrum* '(at the) ravines', though despite the parallel offered there the resulting construction would be extremely rare. (b) Ekwall suggested instead *gljúf-* 'ravine' plus a word such as *rúm* 'clearing, space' (*DEPN*). (c) Collingwood's suggestion of ON *merki* '(?boundary-)marker' for the final el. raises more problems than it solves (1918, 99).

GLEBE FARM NY1934 Blindcrake.
Previously *Hodge's Folly* 1867 (OS), then *Glebe Cottage* 1900 (OS).
♦ Presumably *glebe* here, as usually, refers to land which is part of the

benefice attached to a church, the nearest being the Norman church at ISEL, two miles SW. The Isel PRs for 1724 refer to a terrier (land inventory) 'of ye Glebe'.

GLEBE HOUSE SD1094 Waberthwaite.
From 1900 (OS), seemingly = *The Parsonage* 1867 (OS).
♦ The **house** and its land presumably belonged to WABERTHWAITE church, which had been 'augmented with Queen Anne's bounty and [had] a small farm belonging to it' by 1829 (Parson & White 280. I am grateful to Mary Atkin for this reference). Cf. GLEBE FARM above.

GLEDE HOWE: see under GLADE HOW

GLEN, THE SD4596 (hab.) Nether Staveley.
Glen 1836 (*PNWestm* I, 174).
♦ 'The valley' or 'the clearing'. (a) This is probably the Gaelic word for 'valley', *gle(a)nn* (see **glen**). I am most grateful to the present owners for the information that 'the farm is slightly up the valley, but the original land belonging to it is mostly in the valley.' The name is likely to be a romantic and relatively recent one, in imitation of Scottish valley-names. It could be much older, either belonging with names such as GLENCOYNE, or with names thought to contain ME *glen* from Gaelic *glenn* (*PNWestm*). However, the lack of early records tells against this. (b) Smith's alternative suggestion, also in *PNWestm*, is of ON **glenna* 'open place in a wood, grassy place among rocks'. Although it is difficult to prove its use in the Lake District, the word is preserved in Norw farm-names, especially around the Oslofjord (*NSL*: *glenne*) and in modern Norw *glenne*.

GLENAMARA/GLEMARA PARK NY3815 Patterdale.
Glemorye parke 1588, *Glenmer park* 1589, *Glenner park* 1629, *Glemara park* 1777, *Blemara park* 1860 (*PNWestm* II, 222).
♦ As the spellings show, G~ occurs mainly with **Park**, but also occasionally as an independent p.n. (pers.comm. Keith Clark). There are two main explanations of G~ (*PNWestm*): (a) It is a Celtic name, whose 1st, generic, el. is from Gaelic *gle(a)nn* or Cumbric **glinn* 'valley' (cf. **glen** and nearby GLENCOYNE and GLENRIDDING) and whose 2nd el. may be 'some unidentified pers.n.'. (b) It is a surname derived from GLARAMARA but influenced by the nearby 'Glen-' names. This seems less likely, in the light of the early spellings, though the longer, more recent form Glenamara may be influenced by Glaramara.

GLENCOYNE NY3718 (valley & hab.) Patterdale, G~ PARK NY3919 Matterdale, G~ WOOD NY3818 Patterdale.
Glencaine 1212, *Glenekone* 1255, *Glenkun* 1425, *Glencoyne* 1443 to 1823 (*PNWestm* II, 222); (G~ Park:) *Glencoine* 1589 (*PNCumb* 254), *Glencoin Park* 1867 (OS); (G~ Wood:) *forest of Glencuyn* 1459, *Glencoyn woods* 1787 (*PNWestm* II, 225). Stressed on 2nd syllable and pronounced [glenki:un] (*PNCumb* 254) or [glenkɔin, glenkiun] (*PNWestm* II, 222); *Linkin* is also recorded from 1862 (letter in OSNB Barton).
♦ Possibly 'the reedy valley' or 'beautiful valley'. A difficult name, of Celtic origin, and usually assumed to be specifically Brittonic, with Cumbric **glinn* 'valley' as its 1st, generic, el. and the Cumbric equivalent of W *cawn* 'reed-grass' as its 2nd (*PNWestm* II, 222, reviving a suggestion made by Ekwall in *Scand & Celt* 112 but rejected by him in *ERN* 178 because the stream

is swift and stony). However, since such a structure would be untypical of Brittonic, the 1st el. at least could be Gaelic rather than Brittonic: see **glen**. Meanwhile, Ekwall's suggestion of a stream-name from a Brit. base *kaino of uncertain meaning, possibly 'beautiful', was refuted (Ekwall *ERN* 178 and *DEPN*; Jackson 1953, 328), yet a Gaelic *cáin*, Ir. *caoin* 'pleasant, fair' appears as early as c.830 in the Ulster p.n. Inishkeen (*Insi Cáin* c.830, McKay 1999, 83, cf. Ardkeen, Drumquin pp. 6, 62) and might explain the 2nd el. of Glencoyne. G~ refers both to the valley of the G~ Beck (G~ dale on 1966 OS One Inch map) and to the fine manor farm. See also **park** (G~ was a medieval hunting park) and **wood**.

GLENDERAMACKIN, RIVER NY3627 St John's/Threlkeld/Mungrisdale.
Glenermachan 1278 (*PNCumb* 15), *Glenermakan* 1278 (*CW* 1923, 186), *Glendermaking(e)* 1589 (*PNCumb* 15), *Glenderamacken* 1774 (*DHM*); = *Grise* 1687–8 (Denton 52).
GLENDERATERRA BECK NY2926 Underskiddaw/Threlkeld.
Glenderterray 1729 (*ERN* 179), *Glenderaterra* 1774 (*DHM*), *Glendoweratera* 1789 (*PNCumb* 15), *Glenduratara* 1803 (*CN* 1517).
♦ The two rivers feed into the R. GRETA, and their names need to be taken together. The only reasonable certainty is that the 1st el. is the Celtic 'valley' term, Cumbric *glinn* or Gaelic *gle(a)nn* (see **glen**). The remainder of both names is obscure, though I am grateful to Dr Oliver Padel for the observation that the 2nd syllable *-er-* is reminiscent of the Brittonic definite article as preserved in Triermain 'homestead of the rock or stone' (*PNCumb* 116). Ekwall held that 'Glender- is very likely identical with *Glunduuar'*, recorded in the *Pipe Rolls of Cumberland* for 1247, and he interpreted the river-names in the light of it, as containing W *glyn* 'a valley' and *dwfr* 'a river' (*ERN* 179). However, *Glunduuar* 1247 is a locational surname (*Michaelis de Glunduuar*) whose identification is very uncertain, and if this spelling is set aside the early forms give no warrant for *dwfr* or similar (and *-d-* in the modern forms could be a glide). Any solution proposed for Glenderamackin and Glenderaterra would need to take account of the fact that they are river-names, rather than valley names; to consider both Gaelic and Brittonic options; to interpret their distinguishing parts (modern '-mackin' and '-terra', see *ERN* 179); and to account for the medial *-a-* in both names, which Ekwall in *DEPN* suggested is from ON *á* 'river', but which does not appear in the medieval forms. Dial. **beck** 'stream' from ON **bekkr** has been added to Glenderaterra.

GLENRIDDING NY3817 (hab.) Patterdale, G~ COMMON NY3516, G~ DODD/DOD NY3817.
Glenredyn 1292, *Glenridding(e)*, *Glenriddyng(e)* 1426 to 1787, *Glen Roden* 1577 (*PNWestm* II, 222), *Glenkrhodden* (*Beck*) 1610 (Denton 1); *Glenridding Common* 1865; *Glenridding Dodd* both 1859 (*PNWestm* II, 225).
♦ Probably 'bracken valley'. As in nearby GLENCOYNE and possibly GLENAMARA, the 1st, generic, el. is a Celtic word for 'valley', either Cumbric *glinn* or Gaelic *gleann* (see further **glen**). The 2nd is usually taken as a Cumbric equivalent of W *rhedyn* 'fern, bracken' (Ferguson 1873, 197–8, *PNWestm* 222–3), with influence from ME **ridding(e)** 'clearing' producing the modern form. This 2nd el. could

have described the valley or could have been a stream-name, referring to what is now Glenridding Beck, and Machell took it as such, writing of 'a Little Gill called Glenridden from the Scottish word glen, w[hi]ch signifies gill or Hollow and Ridden, the name of the River which rises at Catstee' (c.1692, I, 716), though it is also possible that Ridden as a stream-name was a back-formation extrapolated from the valley-name.

GOAT CRAG NY1916 Buttermere.
From 1867 (OS).
GOAT CRAG NY2919 St John's.
From 1867 (OS).
GOAT CRAGS NY2716 Borrowdale.
Gate-Cragg 1789 (*PNCumb* 353), *Goat Crag* 1867 (OS).
♦ 'The rocky height(s) frequented by goats', with Standard English *goat* from OE **gāt**, plus **crag**. At least in the Borrowdale name, *goat* may have replaced a N. form originating in ON **geit**, which is represented in the 1789 spelling above, and which still survives in GAIT CRAGS and GAITSCALE.

GOAT SCAR NY4706 Longsleddale.
Codka [sic] 1770 (JefferysM), *Goatscar* 1857 (*PNWestm* I, 161).
♦ Probably 'the craggy scarp frequented by goats'; **geit/gāt**, **scar**. Smith connects this with a neighbouring *Galtecoue*, early 13th cent., 'wild boar cove' (*PNWestm* I, 161), but the goat seems a more likely denizen of these extremely treacherous crags.

GOATS WATER SD2697 Coniston/Torver.
Geiteswater pre-1220 (*CW* 1929, 41), *Goats Tarn* 1774, *Goats Water* 1786 (*PNLancs* 192, n.1), 1851 (OS).
♦ (a) Probably 'the lake of the goat'; it lies below Goat Crag (and nearby Goat's Hawse), high on the OLD MAN OF CONISTON, one of the last bastions of the Lakeland wild goat (as observed by Palmer 1952, 93). *Goat* is from OE **gāt** (see **geit**). This pool is among the smaller lakes to be designated **water** rather than **tjǫrn/tarn**. (b) Ekwall gives this as Gaits Water and considers *Gait* 'very likely a pers.n. of ON origin' (*PNLancs* 192), but there is little comparative evidence to support this.

GODFERHEAD NY1421 (hab.) Loweswater.
Godfrid 1723 (Lorton PR), *Godfrayhead* 1747 (Loweswater PR), *Godfred* 1774 (DHM), *Godfahead* 1828 (PR).
♦ Uncertain. The spellings may point to a pers.n. or surname *Godfer*, a variant of the name *Godfrey* (an adoption from Norman French), plus **head** 'high place', but such a combination would be rare, at best.

GOLDRILL BECK NY3915 Patterdale.
Goldrill-beck 1780 (West 152); cf. *Goldrelbrigge* 1573 (*PNWestm* II, 225).
♦ 'The stream where marigolds grow.' Early ModE *rille* 'stream', unusual in Lakeland, is duplicated in **beck**, which seems to have been added later. The 1st el. probably goes back to OE *golde* 'marigold', since the colour-term *gold* from OE *gold* would be hard to account for. The marsh marigold or kingcup, *Caltha palustris*, is currently common throughout Lakeland, though the habitats listed by Halliday are mainly near standing water, in wet woodlands or on fells (1997, 117).

GOODCROFT NY5316 Shap Rural.
Gothescrofft 1473, *Goodcroft(e)* 1589 (*PNWestm* II, 174).
♦ The 2nd el. is **croft** 'small enclosure, farm', and the 1st appears

to be based on the adj. *good* — OE *gōd* or ON *góðr*. The -*s*- would suggest that it was used as a pers.n., perhaps a nickname.

GOODY BRIDGE (hab.) NY3307 Grasmere.
Guddy Brig 1586 (*CW* 1928, 284).
♦ The bridge, which crosses EASEDALE Beck, also gives its name to a farm. 'Goody' may be a surname (see Reaney 1976: *Goodiff*); a rustic title (from *goodwife*); or a forename, as seemingly in Wordsworth's poem 'Goody Blake and Harry Gill' (*Poems* 420–1), where Goody occurs alone as well as with Blake.

GOOSEGREEN NY4123 (hab.) Matterdale.
Goose Green 1693 (*CW* 1884, 43), *Goosgreen in Watermillock* 1742 (Greystoke PR).
♦ Presumably 'the green, grassy place with a goose or geese', with the reflex of OE *gōs* and **green**, though there are no early forms to confirm this. Geese were commonly kept, often travelling distances to graze on unenclosed land (*CW* 1981, 69, *CW* 1876, 363).

GOOSEWELL FARM NY2923 St John's.
Gousewell 1567, *Goosewell* 1590 (*PNCumb* 317).
♦ This can probably be taken at face value: *goose* from OE *gōs*, **well**, and **farm**. It is perhaps mere coincidence that the immediate area is rich in ornithological allusions: NEST, Snipes How, TEWET TARN and Woodcock Stone.

GOSFORTH NY0603 (hab. & parish).
Goseford c.1150 to 1396, *Gosford* c.1170 to 1440 (*PNCumb* 393–4).
♦ 'The ford of the geese'; *gōsa*, gen. pl. of OE *gōs* 'goose', and OE **ford**. The Anglian settlement and its name must predate the Scand influx which produced the fine carved cross and hogbacks for which G~ is famous (Bailey & Cramp 1998, 100–4). (Pl. 27)

GOSFORTH CRAG NY1609 Nether Wasdale. From 1867 (OS).
♦ Presumably the **crag** 'rocky height' has some connection with GOSFORTH village or parish, either directly or through a surname.

GOUTHER CRAG: see under GOWDER CRAG

GOWAN, RIVER SD4598 Hugill/Nether Staveley.
Gawen 1828, *Gowan* 1857 (*PNWestm* I, 8); cf. *Gowan br.* 1651 (British Library MS Add 37721, 9r), *Gawen Bridge* 1748 (*PNWestm* I, 175).
♦ (a) The view of Ekwall has normally been accepted, that this is a back-formation, i.e. that the river-name Gowan has been (erroneously) extrapolated from GOWAN BRIDGE. This in turn contains a pers.n. *Gowan* or *Gawain*; a *Gawen* Brathwaite is recorded locally from 1671 (*ERN* 182), and the forename *Gawen* and variants appears occasionally in the Kendal PRs. Cf. GILPIN, RIVER for a similar process. (b) *Gowan* is also used in N. dial. and Scots for flowers including the daisy, buttercup and marsh marigold (*EDD, DSL*).

GOWAN BANK FARM SD4498 Nether Staveley.
Formerly *Middle Fairbank* 1862 (OS).
♦ The **farm** stands on the slopes (**bank**) above the River GOWAN.

GOWBARROW: G~ FELL NY4021 (1578ft/ 481m.) Matterdale, G~ HALL NY4321, G~ PARK NY4021.
Golbery c.1250, *Golebergh* 1294

(*PNCumb* 255); *Gowbarrow Fell* 1867 (OS); *Gowbarrow Hall* 1774 (DHM), *Gobray hall* 1786 (Gilpin II, 79); *Parcus vocat[us] Golbergh* 1436, *Gowberay parke* 1589 (*PNCumb* 255).

♦ Probably 'the breezy hill', from ON *gol, gola* 'breeze, gust of wind' and ON **berg** 'hill, rock' encouraged by OE **berg** 'hill'; also **fell, hall, park**. This is where the Wordsworths saw the daffodils which 'tossed and reeled and danced . . . with the wind' (Dorothy Wordsworth, *Journals*, Apr. 15, 1802). The area of the Park roughly corresponds with the Fell; this was a medieval deer park, often mentioned by visitors in early modern times. In Hutchinson's day it amounted to 'about 2000 acres . . . stocked with six or seven hundred head of fallow deer' (1794, I, 435). According to Clarke, the hall was mentioned in a document of 1474 (see OLDCHURCH); it was the seat of the Swinburn(e) family.

GOUTHER CRAG NY5112 Shap Rural.
From 1859 (*PNWestm* II, 174–5).
GOWDER CRAG NY1410 Nether Wasdale, LITTLE G~ C~ NY1411 Ennerdale.
Gouthcrag 1338 (*PNCumb* 441); *Little Gowder Crag* 1867 (OS).
GOWDER CRAG NY2618 Borrowdale.
Gowder-crag 1769 (Gray 1080), *Gowder Crag* (Baines 1829, Errata slip correcting p. 121).

♦ Probably 'the rocky height with an echo', from ON *gauð* 'barking', in its gen. sing. form *gauðar*, plus **crag**; also **little**. Baines remarked that 'an extremely fine Echo is to be heard' in the vicinity of the Borrowdale example. 'The Colonel, George and myself hollaed loud and long' (1829, 121).

GOWK HILL NY4416 (1539ft/469m.) Martindale.
From 1865 (*PNWestm* II, 218).

♦ Apparently 'cuckoo hill', with dial. *gowk* from ON *gaukr* plus **hill**.

GRAINS GILL/GRAIN GILL NY2310 Borrowdale.
Cf. *Grains* 1867, 1900 (OS).

♦ 'The stream/ravine with forks or confluences'; **grain, gil(l)**. The *grains*, V-shaped when seen from above as on a map, are especially noticeable because of the unusually straight course of this stream. The valley it has created is known as Grains. (Pl. 14)

GRAINSGILL BECK NY3033 Mungrisdale.
Graynscalebeck 1569 (*PNCumb* 15), *Grainsgill Beck* 1867 (OS).

♦ G~ was originally 'the shieling at the confluence', judging from the 1569 spelling, with **grain** plus **scale** from ON **skáli**, later replaced by **gil(l)** 'ravine with stream'. The **Beck** is hence 'the stream at G~' or 'the stream by the shieling at the confluence', and the duplication of the notion of 'stream' in *gill* and *beck* is in this case only apparent. G~ B~ meets the R. CALDEW at NY3232.

GRANGE SD1293 (hab.) Waberthwaite.
From 1784 (PR).

♦ It is not clear whether this is a **grange** in the medieval sense of an outlying farm belonging to a monastery, possibly the priory three miles away at SEATON HALL, or in a more general sense.

GRANGE/GRANGE-IN-BORROWDALE NY2517 (hab.) Borrowdale, GRANGE FELL NY2616.
grangia nostra de Boroudale (our grange in Borrowdale) 1396 (*PNCumb* 350); *Grange Fell* 1867 (OS).

♦ 'The outlying farm or store.' This **Grange** was a focal point for produce from the Borrowdale estates (vaccaries or cattle farms, and probably sheep farms) managed by the monks of FURNESS Abbey. 'In BORROWDALE' distinguishes it from the similar holding of Cartmel Priory at Grange-over-Sands. A place called *Bredinebrigge* 'woven, wattle bridge' in a charter of 1209–10 (*Furness Coucher II*, 570) must have been located at or near Grange, and that may have been the original name of the settlement at this important bridging point over the DERWENT. G~ **Fell** is a tract of high grazing rather than a distinct peak. (Pl. 13)

GRANGE HOUSE NY5218 Bampton.
From 1859 (*PNWestm* II, 195).
◆ The **house** is just N of BAMPTON GRANGE.

GRASMERE NY3306 (lake), NY3307 (hab. & parish), G~ COMMON NY2909.
(Lake:) *Grysemere* 1374 (*DEPN*), *(pool of) Grissemere* 1375 to 1580, *Gresmyer* 1570, *Gresmier Tarne* 1615–30 (*PNWestm* I, 16); (hab:) *Ceresmere* 1203, *Gressemer(e)* 1245 to 1530, *Gresmer(e)* 1274 to 1777, *Grasmere* 1247 to present (*PNWestm* I, 198); *Gresmere Comon* 1689 (Fleming Accounts 7), *Grasmere Common* 1863 (OS).
◆ 'The lake flanked by grass'; **gres, mere**[1]. Early spellings in *Grys-, Gris(s)-* might suggest ON *gríss* 'young pig' as 1st el., but the weight of the evidence points to OE/ON *gres* 'grass', with the modern form influenced by Standard English. 'Here is a great lake called Grass-meerwater, w[hi]ch being well stored with grass about it probably gave name to this water & Parish' (Fleming 1671, 16; Machell also links the name to the fact that 'the soyle is sufficiently fertil for Hay Corn Meadow Pasture', c.1692, II, 129C, also II, 432). The medial *-s(s)e-* may, as suggested by Ekwall in *DEPN*, point to ON *gres-sær* 'grass lake' as the original name, perhaps with encouragement from OE *gres-sǣ* since, as Fellows-Jensen notes, ON *sær* is poorly attested in English p.ns (*SSNNW* 422). **Mere** would then be a secondary addition. The lake has given its name to the village and parish, and hence to the **Common** 'unenclosed pasture for common use'.

GRASMOOR NY1720 (2791ft/851m.) Buttermere.
awful Grasmere (the Skiddaw of the vale) 1780 (West 137), changed to ... *Grasmire* ... 1784 (West 138), *Grasmer* 1786 (Gilpin I, 169), *Grasmere* 1800 (*CN* 1207).
◆ 'The grassy upland waste.' **Gras(s)** can be taken at its face value, referring to the ample grazing on top of this otherwise sheer and rocky mountain, as can **moor**. The 18th-cent. forms in *-mire, -mere* are inappropriate, probably influenced by GRASMERE. (Pl. 5)

GRASSGARTH Examples at:
    SD3582 (hab.) Haverthwaite.
From 1851 (OS).
    SD4499 (hab.) Hugill.
Gresgarth 1602, *Grasgarth* 1611 (*PNWestm* I, 171).
    SD4586 (hab.) Crosthwaite.
Girsgarth 1535 to 1771, *Gras(s)garth* 1711 (*PNWestm* I, 84).
GRASSGUARDS SD2298 (hab.) Ulpha.
Grasgarth 1646 (*PNCumb* 438).
◆ 'The grassy enclosure(s)'; **gres/gras** and **garth**. Grassguards contains the variant form *guard* often favoured in the pl. It is 'a farm high on the fell surrounded by magnificent walls' (Lascelles 2003, 8).

GRASS HOLME SD3892 Windermere.
From 1780 (West 56).
◆ 'The grassy islet'; **grass, holm(e)**.

GRASSOMS: GREAT G~ SD1387 Bootle, LITTLE G~ SD1488.
Gres(s)holmes 1252 (*PNCumb* 346); *Great Grasson*; *Little Grasson* both 1867 (OS).
♦ 'The grassy expanses near water', from ON/OE **gres** 'grass' and ON **holmr** 'land beside water, island'; also **great, little**. Each of these is a tongue of fellside formed by two streams.

GRASS PADDOCKS SD2991 Colton.
From 1851 (OS).
♦ This was a wood in 1851, and is still, but both **grass** and *paddock* 'a small field' (a modified form of *parroc* from OE *pearroc*) are suggestive of former meadow.

GRASSTHWAITE: G~ HOWE/ GRASTHWAITEHOW NY3816 (hab.) Patterdale.
Greswayt 1350-60 (p); *Grassthwaitehow* 1860 (*PNWestm* II, 225).
♦ G~ is 'the grassy clearing' (**gres, þveit**) while **how(e)** derives from ON **haugr** 'hill'.

GRAY BULL NY5309 Shap Rural.
From 1860 (OSNB).
♦ A large erratic granite boulder; **grey**, and reflex of OE *bula*.

GRAY CRAG: see GREY CRAG(S)

GRAYMAINS/GREYMAINS SD1195 (hab.) Muncaster.
Graymaines 1698 (*PNCumb* 426).
♦ 'The grey home farm'; **grey**, and *mains* as in MAINS FARM.

GRAYSTONE HOUSE SD1887 Millom Without.
Grayston Ho 1629 (*PNCumb* 418).
♦ Apparently **grey, stone, house**, though whether the name is to be taken at face value is uncertain. Although in the same parish as GRAY STONES, G~ House is probably too far from the hill to derive its name from there. *Grayston(e)* exists as a surname.

GRAY STONES SD1687 (1283ft/396m.) Millom Without.
From 1867 (OS).
GRAYSTONES NY1726 Lorton.
From 1867 (OS).
♦ These hill-names may refer simply to rock outcrops. However, grey stones may elsewhere be associated with boundaries (see **grey, stone**), and although these are usually strategically placed monoliths, which are not evident on either hill, association with a boundary cannot be ruled out since the Millom example is flanked by a rocky ravine which forms the boundary with Whicham, while the Lorton one, at 1476ft/450m., is the highest point on KIRK FELL and stands close to a (modern but doubtless also ancient) parish boundary, with Embleton and Wythop.

GRAYTHWAITE NY1123 (hab.) Blindbothel.
Magna (Great) *Grathwayte, Parva* (Little) *Grathwayte* 1563, *Grathwat* 1631 (*PNCumb* 423).
GRAYTHWAITE: G~ HALL SD3791, G~ OLD HALL SD3790 Satterthwaite.
Grathpeyte, Grathwayt 13th cent. (Wyld 1911, 136), *Graythwayt* 1336 (*PNLancs* 220); *Graythwaite Hall* 1801 (Hawkshead PR); cf. *Graythwaite high Hall* 1766, *Low Graythwaite* 1809 (CrosthwaiteM), *Graythwaite Low Hall* 1829 (Finsthwaite PR); (G~ Old Hall:) *Graythwaite* 1851 (OS; where G~ Hall is *Graythwaite Hall*).
♦ G~ is probably 'the grey clearing'; **grey, þveit**; also **hall, old**. In the 1st el., ME *gray* from OE *græg* may have replaced the ON cognate *grár* (as

Fellows-Jensen suggests for the Satterthwaite example, *SSNNW* 230). Descriptive words are the most common type of specifics in compounds with *þveit/thwaite*, and *grey* seems to be paralleled in GREYMAINS, so that doubts about the probability of *grey* expressed by Ekwall seem unfounded. His alternative solutions for the Satterthwaite example, of a pers.n., or the adj. *greiðr* 'ready, free', are no more plausible (*PNLancs* 220). G~ Hall (also known as G~ High Hall) is 'an Elizabethan house . . . comfortable Tudor' with later alterations (Pevsner 1969, 119; also Cowper 1899, 50–1). G~ Old Hall (or G~ or G~ Low Hall) is a three-storey 'gentry' house of uncertain date (Cowper 1899, 51, 163–4).

GREAT as affix: see main name, e.g. for GREAT ARMING HOW see ARMING HOW

GREAT BANK NY1401 Irton.
From 1867 (OS).
♦ This is a craggy brow overlooking MITERDALE; **great, bank**.

GREAT BARROW NY1801 Eskdale.
From 1867 (OS); cf. *the Barrow* 1587 (*PNCumb* 390).
♦ 'The big hill'; **great, berg**. This rises to approx. 787ft/240m. Little B~ is the lower, southerly, part of the same mound.

GREAT BAY NY2519 Borrowdale.
From 1867 (OS).
♦ A large inlet at the S end of DERWENTWATER; **great, bay**.

GREAT BORNE/BOURNE NY1216 (2020ft/ 616m.) Ennerdale/Loweswater.
Great Borne 1867 (OS); probably *Hardecnut* 1230 (*Scotland* I, 203), cf. another HARD KNOTT.

♦ 'Bourne' is probably a boundary cairn here: see *OED*: *bourne, bourn*, n. 2 for Early Modern examples of the sense 'boundary'. Great B~, which is strictly speaking the summit of HERDUS (Wainwright 1966, Great Borne 2), lies on a parish boundary and is presently furnished with a very large cairn, and either the hill or the cairn could be described as **great**.

GREAT COCKUP: see COCKUP

GREAT COPPICE SD3596 Hawkshead.
From 1851 (OS).
♦ 'The large coppice wood' (**great, coppice**), currently referring to a small plantation outlying an extensively afforested area.

GREAT CRAG NY2614 Borrowdale.
From 1867 (OS).
♦ An apt description of a massive if not spectacular rocky summit; **great, crag**.

GREAT CROSTHWAITE: see CROSTHWAITE NY2524

GREAT DODD/DOD NY3420 (2808ft/ 856m.) Matterdale.
Dod Fell 1783 (*PNCumb* 222–3), *the great Dod* 1794 (Hutchinson II, 197, quoting Naughley via Clarke), *Great Dod* 1800 (*CN* 777).
♦ 'The large, compact rounded summit' (**great, dod(d)**), highest of those named *dodd*. Little Dodd is to the W.

GREAT END NY2208 (2984ft/910m.) Eskdale/Borrowdale.
the Wastall great ende 1578 (*PNCumb* 353), *Great-End* 1802 (*CN* 1219), 1823 (Otley 85).
♦ This mountain forms the massive, precipitous bulwark at the N end of the SCAFELL range; **great, end**. This is

unusual among names containing *end*, since *end* is here qualified by an adjectival 1st el., whereas most others refer to the end of a feature specified in the 1st el.

GREATEND CRAG NY2517 Borrowdale.
From 1867 (OS).
♦ 'Greatend' here is presumably the NW spur of Grange Fell, faced with the precipitous rocks of the crag; **great, end, crag**.

GREAT GABLE NY2110 (2949ft/899m.) Nether Wasdale/Ennerdale/Borrowdale.
(*le Heye del* [the enclosure or hunting-ground of]) *Mykelgavel* 1338, *Great Gavel* 1783 (*PNCumb* 389).
♦ The name is from **great** and ON *gafl* 'gable', also found in GREEN GABLE and GAVEL CRAG, FELL and PIKE. It well describes the massive, angled shape of this famous mountain, at least as viewed from some directions. (Pl. 23)

GREAT GREEN HOWS: see GREEN HOW(S)

GREAT HAGG SD3586 Colton.
From 1851 (OS).
♦ Probably 'the large plantation for felling'; **great,** and *hagg* as in HAGG. This is a stretch of sloping, wooded land above the R. LEVEN. A 'charcoal fireplace' is shown on the 1851 map.

GREATHALL GILL NY1403 Irton/Eskdale.
Hollegill 1294 (*St Bees* 159, n. 2), *Great Hall Gill* 1867 (OS).
♦ This is a classic **gil(l)** — a rocky ravine with a stream. **Great** differentiates this from nearby Little Hall Gill, while the syllable 'hall', somewhat incongruous in this rugged setting, seems to be a corruption, perhaps of **hol(r)** 'hollow' (which qualifies **gil(l)** in HOLEGILL, HOWGILL). Although there is a hall, WASDALE HALL, under a mile away, it is on the other side of the R. IRT.

GREAT HILL SD3092 Colton.
From 1851 (OS).
♦ The **hill** is substantial, if not spectacularly **great**.

GREAT HOW NY1904 (1699ft/522m.) Eskdale.
the Greathow 1587 (*PNCumb* 391).
GREAT HOW NY3118 (1092ft/333m.) St John's.
From 1789 (*PNCumb* 317).
GREAT HOWE NY4806 Longsleddale.
the Greit(e) howe 1578 (*PNWestm* I, 162–3).
♦ 'The big hill'; **great, haugr/how(e)**. Steep-sided, with a broad, rounded top, the Eskdale G~ H~ is a fairly classic example of a *how(e)*, as are the others, which are **great** by local standards if not in absolute terms.

GREAT INTAKE NY3002 (1339ft/408m.) Coniston.
From 1850 (OS).
GREAT INTAKE SD3493 Satterthwaite.
From 1851 (OS).
♦ 'The large area recently taken into cultivation'; **great, intake**. The Satterthwaite example is one of several intakes now swallowed up in GRIZEDALE FOREST.

GREAT KNOTT NY2504 Langdales.
From 1867 (OS)
GREAT KNOTT SD3391 Satterthwaite.
From 1851 (OS).
♦ 'The large, rugged summit' (**great, knott**). The Satterthwaite example is a wooded height S of Little Knott. It is distinct, and lumpy though not notably rocky, especially compared with other hills called *knott*.

GREAT LINDETH: see LINDETH

GREAT LINGY HILL NY3133 (2021ft/ 616m.) Mungrisdale.
From 1867 (OS).
♦ 'The large heathery hill'; **great, lyng/ling(e)y, hill**. One of the higher hills in the vicinity, topped with a broad, heathery grouse moor. Little L~ H~ is to the SW.

GREAT MOSS NY2205 Birker/Eskdale.
From 1867 (OS).
♦ 'The large bog'; **great, moss**.

GREAT RIGG NY3510 Rydal/Grasmere.
greatrig (head) c.1692 (Machell I, 655A), *Great Rigg* 1828 (HodgsonM).
♦ 'The big ridge' (**great, hryggr/rigg**), appropriate for the N part of the long ridge which rises from RYDAL WATER to FAIRFIELD, and formed part of the parish boundary.

GREAT SCA FELL: see SCA FELL

GREAT SLAB NY2406 Langdales.
♦ This is a massive, tilted tranche of striated rock, and the name simply comprises **great** plus the only occurrence in this volume of *slab*, which is of uncertain origin and first appears c.1290 (*OED*).

GREAT STICKLE: see STICKLE

GREAT TONGUE NY3410 Grasmere.
From 1859 (OSNB).
♦ This is the long, narrow, tongue-shaped ridge formed by TONGUE GILL and Little Tongue Gill (**great, tongue**); Little Tongue lies to the W.

GREAT TOWER PLANTATION: see TOWER PLANTATION, GREAT

GREAT WOOD Examples at:
NY2721 Borrowdale.
From 1867 (OS).
SD3593 Satterthwaite.
From 1851 (OS).
SD3685 Staveley-in-Cartmel.
From 1851 (OS).
♦ A large tract of mixed woodland still stands in Borrowdale E of DERWENTWATER. The Satterthwaite G~ W~ is one of the individual woods now forming part of GRIZEDALE FOREST. Like the Staveley one, it is a reminder that Furness and Cartmel have always been famous for their woodland. See **great, wood**.

GREAT WORM CRAG: see WORM CRAG, GREAT

GREAVES NY4425 (hab.) Matterdale.
Grove 1774 (DHM), *Greeves(-Beck)* 1789 (Clarke 25), *Greaves* 1867 (OS).
GREAVES FARM SD3982 Staveley-in-Cartmel.
Greave 1576/7, *Greaves* 1594, *Greeves* 1605, *the Greaves* 1630, *Groves* 1634 (Cartmel PR).
♦ Probably both 'the grove(s) or copse(s)'; **greave**. The Matterdale place is not presently wooded, though the name of nearby HESKET FARM also refers to (ash) trees. **Farm** has been added relatively recently to the Staveley name.

GREAVES GROUND SD2692 Torver.
Greues Gound [sic] 1693/4, *Gravs Ground* 1698, *Graves Ground* 1702, *Grooves Ground* 1711, *Groves Ground* 1712, *Greaves Ground* 1718 (PR).
♦ 'The farm of the Greaves family'; surname plus **ground**. The surname Greaves appears frequently, with a range of spellings similar to the above, in the earlier Torver PRs.

GREENAH CRAG FARM NY3928 Mungrisdale.
Grenecragge 1555 (*PNCumb* 181); cf. *Grenehowe Mosse* 1589 (*PNCumb* 227).
♦ G~ is 'the green hill'; **green, haugr/how(e)**, also **crag(ge)** 'rock outcrop', **farm**. Derivation of the 2nd el. from *haugr/how(e)* is favoured by the 1589 spelling and by other Greenahs with spellings pointing to this (*PNCumb* 255, 304). *Farm* has been added later. The farm nestles below the Crag from which it is named.

GREEN BANK SD2190 (hab.) Dunnerdale.
Greenbank 1760 (Kirkby Ireleth PR).
GREEN BANK SD3880 (hab.) Broughton East.
Green Banck 1599/1600 (Cartmel PR).
GREENBANK FARM NY3914 Patterdale.
Greenbank 1839 (*PNWestm* II, 225).
GREENBANK FARM SD4292 Crosthwaite.
Green bank 1809 (CrosthwaiteM).
♦ These sites are on moderate, verdant slopes; **green, bank**; also **farm**. The age of the names is not certain, but the building at Patterdale, for example, dates from 1677, and another instance of this name is recorded from 1323 (*Grenebonk, PNLancs* 172). *Farm* is a relatively recent addition in the last two cases.

GREENBURN NY2902 Coniston, G~ BECK NY2902.
Grenebotne pre-1220 (CW 1929, 40), *Greenburn; Greenburn Beck* both 1850 (OS).
♦ If the identification with *Grenebotn* is correct, G~ is 'the green inner part of the valley', from ON **grœnn**/OE **grēne** 'green' and ON **botn** 'head or inner part of valley' rather than **burn** 'stream'. G~ is the high valley above Little Langdale, and G~ **Beck** the stream in it.

GREENBURN BOTTOM NY3110 Grasmere. From 1859 (OSNB); cf. *Greenburn* 1847 (*PNWestm* I, 8).
♦ 'The valley through which Green Burn flows'; **green, burn, bottom**.

GREENCLOSE NY2133 (hab.) Bassenthwaite.
From 1767 (CRO D/Law/1/205,8), 1779 (PR).
GREEN CLOSE NY4226 (hab.) Hutton.
From 1685 (*PNCumb* 213).
♦ 'The green enclosure or farm'; **green, close**.

GREEN COMBE/COMB NY2714 Borrowdale.
Green Comb 1867 (OS).
GREEN COMBE NY2911 St John's.
Green Comb 1867 (OS).
♦ 'The green rounded valley'; **green, comb(e)**[1]. The Borrowdale example is the natural basin in which DOCK TARN lies, and *green* contrasts with nearby Black Knott and White Crag. The St John's one is a high, bowl-shaped corrie.

GREEN CRAG SD2098 Eskdale/Ulpha.
the Green Cragg 1587 (*PNCumb* 391).
♦ The name presumably refers to a greenish tinge of grey in the rock, since this rocky crest is far from grassy; **green, crag(ge)**.

GREENDALE NY1405 Nether Wasdale, G~ TARN NY1407.
Greendale 1867 (OS, as hab.), 'probably' = *Grindale* 1332 (p), *Gryndell* 1570 (*PNCumb* 442); *Greendale Tarn* 1867 (OS).
♦ 'The green valley' (**grēne/grœnn, dalr**) and the small lake (**tjǫrn/tarn**) on the fell above it. If the spellings in *Grin-, Gryn-* truly belong to this name, they represent the natural development of weakened pronunciations

which have later been 'corrected', at least in the written form. G~ is a verdant tract NW of WAST WATER which contrasts strongly with the forbidding screes S of the lake.

GREEN FARM NY5512 Shap Rural.
th' Green in Shapp 1754 (*PNWestm* II, 175).
♦ 'The grassy place or **green**', with **farm** added later.

GREEN GABLE NY2110 (2628 ft/801m.) Ennerdale/Borrowdale.
Green Gavel 1799 (*CN* 544).
♦ 'The grassy, gable-shaped hill'; **green** plus *gavel/gable* as in GREAT GABLE, the larger and stonier neighbour of Green Gable.

GREEN HEAD NY2837 (hab.) Caldbeck.
Greenhead 1867 (OS).
GREENHEAD: G~ GILL/GHYLL NY3508 Grasmere.
Greenhead 1687/8 (Grasmere PR); *Green-head Ghyll* 1800 (Wordsworth, *Poems*104).
♦ 'The green high place'; **green, head**; also **gil(l)** 'ravine with stream'. The Grasmere example is the 'hidden valley' or 'Dell' immortalised as the haunt of the pastoral hero of Wordsworth's *Michael* (cf. MICHAEL'S NOOK). Greenhead was 'a farm 1619, to about 1885. The barn only remains at what is now called Michael's Fold' (Simpson 1928, 284).

GREEN HILL NY2229 (hab.) Bassenthwaite.
From 1831 (PR).
GREEN HILL NY4020 Matterdale.
From 1900 (OS).
♦ G~ H~, Bassenthwaite, is, like neighbouring SAND HILL, a point on the gentle slopes above BASSENTHWAITE LAKE, and it may be more than coincidence that the two names pair and contrast so neatly; **green, hill**. The Matterdale example is among a few local names signalling verdant places (see GREENROW).

GREEN HOLE NY2305 Eskdale.
Green Hall 1802 (*CN* 1220), *Green Hole* 1867 (OS).
♦ 'The green hollow'; **green, hol(r)**. This is a low, wet tract beside LINGCOVE BECK, 'a great hollow Dell' (*CN* 1220).

GREENHOLME FARM SD2889 Blawith.
Green Holme 1735 (PR), 1851 (OS), *Green Holme Farm* 1893 (OS), possibly = *Greenholm* 1689 (Fleming *Accounts* 9).
♦ G~ is presumably 'green, grassy pasture next to water', with ME **grene** or ModE **green** and the reflex of ON **holmr** 'island, land beside water'. **Farm** has been added to the pre-existing name.

GREEN HOW Examples at:
SD1799 (654ft/199m.) Eskdale.
From 1867 (OS).
SD1898 Eskdale.
From 1867 (OS).
NY1907 Eskdale.
From 1867 (OS); *the greene howes* 1578, *the Greenhow* 1587 (*PNCumb* 391, or = SD1898), *Greenehowe* 1664 (Winchester 2000, 169).
SD2095 (940 ft/287m.) Ulpha.
From 1867 (OS).
NY2537 (1053ft/321m.) Ireby and Uldale.
From 1867 (OS).
GREAT GREEN HOWS SD3589 Colton, GREEN HOWS TARN SD3690 Colton/Satterthwaite.
Great Green Hows 1851 (OS).
♦ 'The green hill'; **green, haugr/how(e)**; also **great, tjǫrn/tarn**. Great Green Hows in Colton are modest

rounded heights, now forested. Little G~ H~ is to the N, as is G~ H~ Tarn, which appears on the 1851 OS map as a small, unnamed oval pool on Little Green How; it has been artificially enhanced.

GREENHURST SD3681 (hab.) Haverthwaite.
Green Hurst 1851 (OS), possibly = *Greenhurst* 1630 (Cartmell PR).
♦ Seemingly 'the green wooded hillock'; **grēne/green, hyrst**.

GREENLANDS NY0901 (hab.) Irton.
Greenland 1821 (PR).
GREENLANDS NY1431 (hab.) Setmurthy.
Green Lands 1838 (PR).
♦ The Irton example is on a gentle slope above the R. IRT, the Setmurthy one on relatively level ground above Bitter Beck; **green, land**.

GREEN MOOR SD2589 (hab.) Kirkby Ireleth.
Greenmore 1681 (PR), *Green Moor* 1851 (OS).
♦ The site, an area of springs and streams, illustrates the fact that a Cumbrian **moor** is frequently wet; see also **green**.

GREENODD SD3182 (hab.) Egton with Newland.
Green Odd 1774 (*PNLancs* 213).
♦ 'The green point', referring to the triangle of land at the confluence of the CRAKE and LEVEN. The elements go back to ON **grœnn** 'green' and ON *oddi* 'point of land', and the name may be old, as W. G. Collingwood implies in his use of it in his Viking tale *Thorstein of the Mere*, but as Ekwall pointed out, the elements continue in dial. use, so that the name 'need not be old' (*PNLancs*). The village of G~ is outside the National Park, but the point of land within.

GREEN QUARTER NY4604 (hab.) Kentmere, G~ Q~ FELL NY4603.
greine quarter 1605 (*CW* 1988, 123); *Gre quarter fell* [*sic*] c.1692 (Machell II, 106), *Green Quarter Fell* 1858 (OSNB).
♦ G~ Q~ is the name both of a historic subdivision of Kentmere and of the hamlet in it. **Green** may, as Smith suggests, be the noun 'grassy place' here rather than the colour adj. (*PNWestm* I, 167), for the farms cluster around a small green, but the hamlet also has 'possibly the best land in the township', with 'far less stony land and less severe gradients' than the aptly-name Cragg Quarter (Atkin 1991, 69, 72), so the adj. is also possible. The word *quarter* was adopted from OFr into early ME. G~ Q~ **Fell** is 'a large tract of moor' (OSNB).

GREEN RIGG SD2693 (hab.) Torver.
Grenrigge 1591 (*CW* 1958, 66), *Grenerig* 1603, *Greinrige* 1612 (PR), *Green Rigg* (hill & hab.) 1851 (OS).
GREENRIGG: G~ HOUSE NY2938 Caldbeck, HIGH G~ HOUSE NY2837.
Greneric 1163, *Grenerig* 1232 (p) to 1505 (*PNCumb* 276).
GREENRIGGS SD4691 (hab.) Underbarrow.
Grenerigge & variants, 1256 to 1619, *Greneriges* 1332 (*PNWestm* I, 101–2).
♦ 'The green ridge(s); **grēne, hryggr/rigg**; also **house, high**. G~ in Underbarrow is shown as Low G~ on the 1:25,000 map. G~ House, and High G~ House, Caldbeck, stand at either end of the height called Greenrigg; the naming of these on recent maps is inconsistent.

GREENROW NY4124 (hab.) Matterdale.
Greneyrowe 1597 (*PNCumb* 255).

♦ 'The row at Greenah, the green hill'; **green, haugr/how(e), row**. As indicated in *PNCumb*, this is probably connected with Greenah (*Grenehou* 1295 (p)) just to the N. The settlement as shown on 1867 OS is not strictly a *row*, though it forms part of a string of settlements about the 900ft contour (Mr Keith Clark, pers.comm. and *MHAS* 2001, 20), and this is probably the slightly looser use of the term characteristic of this area. I am also indebted to Mr Clark for the suggestion that the 'Green' names in Matterdale and neighbouring parishes reflect the underlying geology, since they stand on the Devonian slates just N of the point where these give way to Borrowdale volcanics. The names concerned include GREEN CLOSE NY4226, Greenbank NY3825, Greenhead NY4123, Green How NY3821 and GREEN HILL NY4120.

GREEN SIDE NY3518 Patterdale.
Greenside c.1692 (Machell II, 155).
♦ This steep 'green hill-slope' gave its name to the lead mines below, now disused; **green, side**.

GREENSLACK SD2087 (hab.) Broughton West.
Green Slack 1703 (Broughton-in-Furness PR).
♦ 'The green valley'; **green, slakki/slack**.

GREEN TONGUE NY2506 Langdale.
From 1862 (OS).
♦ This is the tongue of land between ROSSETT GILL and Grunting Gill — to Wainwright an 'uninteresting and lengthy grass shoulder' (1966, Bowfell 7); **green, tongue**.

GREEN TREES NY1227 (hab.) Blindbothel.
From 1774 (DHM).

♦ Self-explanatory; **green** plus *tree* ultimately from OE *trēo(w)*. A small mixed plantation stands close by at the present time.

GREENUP: G~ EDGE NY2810 Borrowdale/St John's, G~ GILL NY2712 Borrowdale.
Greenup Hedge 1805 (CW 1981, 67), *Greenup Edge*; *Greenup Gill* both 1867 (OS).
♦ Greenup is 'the green blind valley' (**grēne, hop**), a fair description of the site and, as a presumed Anglian name, a rarity in highland Cumbria. A *Grenehope, Parva Grenehope*, mentioned in a FURNESS charter of c.1209–10, is identified in *PNCumb* 350 with Greenup Edge, but is more likely, from the documentary context, to be the Greenup at NY2417 (Pl. 13). The **edge** is a high scarp, and the **gill**, as usual, a narrow valley with a stream.

GREEN YEW SD4193 (hab.) Crook.
From 1836 (*PNWestm* I, 180).
♦ One of a number of p.ns commemorating a distinctive tree; **green, yew**.

GREETY GATE SD2086 (hab.) Broughton West.
Greety Gate 1667 (original spelling not specified, Broughton-in-Furness PR), *Gredy Gate* 1775 (Kirkby Ireleth PR), *Greedy Gate* 1745 (West 1774, Map), *Gretagate* 1822, *Greetygate in Broughton* 1824 (Kirkby Ireleth PR).
♦ Possibly 'the stony or gravel road'. *Greety* may be the dial. variant of *gritty*, an adj. derived from OE *grēot* 'gravel, small stones'. This interpretation is encouraged by the existence of a lane called *Gretigate*, recorded in Gosforth in the 13th cent. and interpreted as 'gravel road' in *CW* 1961, 1. This hamlet is also on the edge of the marsh beside DUDDON Channel,

DICTIONARY

by the ancient road across the Duddon Sands. The 2nd el., if ultimately from ON **gata** 'road', might therefore refer to this track, though the sense 'gate' (from OE **geat**) is also possible.

GREGG HALL/GRIGGHALL SD4691 Underbarrow.
Grygge Hawle 1556, *Gregghall (Laine)* 1718 (*PNWestm* I, 105).
♦ This seems to be from the local family name *Grigg* (recorded in Rydal and Kendal from 16th cent., *PNWestm* I, 105), plus **hall**. The spellings Gregg Hall, Grigg Hall and Grigghall all appear on the 1:25,000 map for this and related names.

GRETA, RIVER NY2924 Underskiddaw/Keswick/St John's/Threlkeld.
Greta 1278 (*PNCumb* 16, *ERN* 185).
♦ 'The rocky river' — as it visibly is in its upper and middle reaches —, from ON *grjót* 'rock(s), stone(s)' and ON *á* 'river'. There are rivers of the same name in Yorks, Lancs, Norway and Iceland. For an alternative name *Bure*, see entry on DALE HEAD.

GRAY CRAG NY4211 Patterdale.
From 1860 (OSNB).
GREY CRAG NY4907 Longsleddale, GREYCRAG TARN NY4907.
Graycragg, Graycrage 1579; *Grey Crag Tarn* 1828 (HodgsonM), previously *Braban tarne (stagnum)* 1580 (*PNWestm* I, 163).
GREY CRAGS/GREY CRAG NY1714 Buttermere.
From 1867 (OS).
GREY CRAGS NY2627 Underskiddaw.
From 1867 (OS).
♦ 'The grey rocky height(s), or jutting rock(s).' The names may have no more significance than the obvious one, though see **grey**; also **crag**. The Patterdale and Longsleddale examples are independent summits, while the Underskidddaw and Buttermere Crags are on (SKIDDAW) LITTLE MAN and HIGH STILE respectively. The Buttermere example is shown as Grey Crag (singular) on the 1: 25,000 map. Greycrag **Tarn** (see **tjǫrn**) is now a boggy area rather than a pool.

GREY FRIAR NY2500 (hill) Dunnerdale.
From 1786 (*PNLancs* 193, n. 2).
♦ 'Self-explaining' according to Ekwall (*PNLancs*), but still unusual and rather puzzling, since the hill is not notably **grey**, despite several rock outcrops. If 'Friar' is what it seems: *friar* from ME, OFr *frere* 'brother (in a religious order)', it could refer to the slightly hooded appearance of the domed top, or to shrouds of mist. The original 'Greyfriars' were the mendicant order of Friars Minor established by St Francis in the early 13th cent. and so named in English from the colour of their tunics. Further Cumbrian names compare hill-tops with friar's hoods: see *PNWestm* I, 125.

GREY KNOTTS NY2112 (2287ft/697m.) Buttermere/Borrowdale.
From 1867 (OS).
♦ 'The grey rocky heights'; **grey**, **knott**. The name well describes the two summits and other rock outcrops on the broad top of this fell, which produce a very distinctive skyline.

GRICE CROFT SD1786 Whicham.
Groscrofte 1571, *Grisecroft* 1585, *Gricecroft* 1684 (*PNCumb* 445).
♦ **Croft** is a small enclosure, but 'Grice' is uncertain. The 1571 spelling *Gros-* is out of line and may be erroneous. If so, (a) the 1st el. may be dial. *grice* from ON **gríss** 'young pig', hence of the same origin as the 1st el. of GRISE- or GRIZEDALE. (b) *Grice* is

also recorded as a surname, as recorded in Whicham PR 1843 and 1855, and in Bootle PR 1742 etc.

GRIKE NY0814 (1596ft/486m.) Ennerdale. From 1802 (*CN* 1207).
♦ A *grike* is a crack or ravine on a hillside in local dial. (*EDD*), and the name may well refer to the ravines on the N side of Grike.

GRISEDALE: G~ GILL NY2023 Above Derwent, G~ PIKE NY1922 (2593ft/790m).
Grisedal 1323 (*DEPN*); *Grisedale Gill* 1867 (OS); *Grysdale Pike* 1800 (*CN* 805).
GRISEDALE NY3715 Patterdale, G~ BECK NY3614, G~ BRIDGE NY3916, G~ BROW NY3615, G~ FOREST NY3513, G~ HAUSE NY3411, G~ TARN NY3412.
Crisdale 1291, *Gris(e)dal(e)*, *Grys(e)dal(e)* 1292 to 1823 (*PNWestm* II, 223); *grisedale Beck* [flowing from the W into Goldrill Beck] ... *is called by the name of I (or e as wee comonly use to pronounce it here in these partes) from the streightness of it here* c.1692 (Machell I, 726); *Grisedale Bridge* 1920 (OS); *Grisedale Brow* 1865 (*PNWestm* II, 225); *forrest de Greisdale* 1588, *Grisedale Forest* 1865 (*PNWestm* II, 225); *Greisdale Halse* 1799 (*CN* 515), *the Hawes* 1801 (*PNWestm* II, 225); *Grisedale Tarn* c.1692 (Machell I, 655A).
GRIZEDALE SD3394 (valley & hab.) Satterthwaite, G~ BECK SD3393, G~ FOREST SD3295, G~ TARN SD3494.
Grysdale 1336, 1537 (*PNLancs* 220); *Grizedale Beck* 1851 (OS); *Grysdale Tarn* 1802 (*CN* 1204), *Grizedale Tarn* 1851 (OS).
♦ Grisedale or Grizedale (also in MUNGRISDALE) is 'the valley where young pigs graze'; **gríss, dalr**. Oaks, a good source of swine fodder, still grow among the newer, mainly coniferous plantings of the vast G~ Forest, Satterthwaite, and must have been abundant elsewhere in former times. This **Forest** appeared on the 1851 map as a series of individual woods, plantations and coppices, not a named entity. It was acquired by Forestry Commission in 1934. Grisedale Forest, Patterdale, is by contrast virtually treeless today, its name commemorating its former status as a domain for hunting. See also **gil(l)** 'ravine with stream', **pike** 'peak', **beck** 'stream', **brycg/bridge**, **brow** 'hillslope', **hause** 'neck of land' and **tjǫrn/tarn** 'small mountain pool'. (Pl. 22)

GRIZEBECK SD2485 (hab.) Kirkby Ireleth. (*piscarium de* [the fishery at]) *Grisebek* 13th cent. (*PNLancs* 221).
♦ 'The stream where young pigs graze'; **gríss, bekkr/beck**.

GRIZEDALE: see GRISEDALE

GROOVE BECK NY3622 Matterdale. From 1900 (OS).
♦ 'The stream in the hollow or by the copse.' The 1st el. could be either (a) **groove**, reflex of ON *gróf* 'pit' in the sense 'stream-bed, hollow', which would refer to the short, deep course of the beck; or (b) **grove** (OE *grāf(-)*) in the sense 'copse', especially since the stream is known locally as Grave Beck and a remnant of ancient woodland stands, unusually high, above it (information from Roger Connard gratefully acknowledged). The 2nd el. is **beck** 'stream'.

GROVE FARM SD3882 Staveley-in-Cartmel.
Ramp Greaves 1851 (OS).
GROVE, THE SD0896 (hab.) Muncaster. From 1803 (PR).
GROVE, THE SD4283 (hab.) Witherslack. From 1899 (OS).

♦ These are probably instances of **grove** 'grove, copse', ultimately from OE *grāf*. Only a small scrap of woodland is currently present at the Staveley and Witherslack places, while the Muncaster one is surrounded by mixed woodland. The 1803 record refers to 'Nicholas Middleton, gentleman, of the Grove, housekeeper'. The names seem to be relatively recent, and two other examples are first recorded in the 19th cent. (*PNWestm* I, 167 and 196). The Staveley name has been modified from *Ramp* (wild garlic?) *Greaves*, and **farm** added.

GROVEFOOT FARM NY4325 Matterdale.
Grovefoot 1867 (OS).
♦ Whether the 1st el. is **grove** 'copse' or **groove** 'hollow' is unclear here; see also **foot**, **farm**. It lies below LITTLE MELL FELL.

GRUBBINS POINT SD3791 Satterthwaite.
The Grubbins Point 1851 (OS); cf. *Grubbins Wood* 1851 (OS).
♦ 'The headland by Grubbing/ Grubbins, the place where trees have been uprooted.' The name of this **point** projecting into WINDERMERE may refer to a place hereabouts recorded in 1915 as Grubbing (*below Grubbing*, CW 1973, 116). Grubbing was also the name of a coppice woodland at Troutbeck (*CW* 1997, 84). *Grubbing* refers to clearing land by digging out tree roots, as when a payment 'for Grubbing of the Orchard and Cutting it up' was made in 1684 (*CW* 1989, 193, and see *PNWestm* II, 257, 305).

GUARDHOUSE NY3325 Threlkeld.
the garthus 1332 (p), *The Gardhowse* 1571, *Gardess (or Guards)* 1703 (*PNCumb* 252).

♦ 'The house at the enclosure'; **garðr**, **hús**, possibly encouraged by the OE cognates *geard* and *hūs*. Clarke, noting the situation by a narrow pass, with 'remains of a very strong wall', thought that this must have been a defensive tower of the Threlkelds of Threlkeld Hall (1789, 57–8), but was probably misled by the place-name.

GUBBERGILL SD0899 (hab.) Irton.
Gubbergill(heade) 1657 (*PNCumb* 403).
♦ 'Gubber' is unidentified. 'Gill' is presumably from ON **gil** 'ravine with stream', though as with other places on the lowland fringes of Lakeland, the topography is subtle rather than dramatic.

GUERNESS: G~ GILL NY4813 Shap Rural, G~ WOOD NY4813.
Girnishe 1593, *Girnes(s)* 1604, *Girnas* 1619 (*PNWestm* II, 167) ; *Guerness Gill* 1860 (OSNB); *Guerness Wood* 1859 (*PNWestm* II, 167).
♦ The 1st el. may be OE *grin* '(fishing) trap'(so *PNWestm*); the 2nd appears to be OE **næss** 'headland' (reinforced by ON *nes*) — less prominent now since the creation of HAWESWATER RESERVOIR. See also **gill** 'ravine with stream', **wood**.

GUMMER'S HOW SD3988 (1054ft/321m.) Staveley-in-Cartmel.
Gomershow 1818 (Watson, quoted *CW* 1974, 86), *Gunner's How* 1839 (Ford 22), *Gummer's How* 1851 (OS).
♦ (a) Possibly 'Gummer's hill', with *Gummer*, a surname originating in an OE pers.n. such as *Gummar* or *Guthmær/Guðmær* (Searle 1897, 270, 273). The 1818 spelling is from the pen of the then owner, Bishop Watson. (b) the 1839 spelling and the presence of a Ginners Beck a mile to the E might instead point to a derivative of the

common ON pers.n. *Gunnarr*. The generic is **how(e)** 'hill' (see **haugr**); this is the highest point in the Cartmel area.

GURNAL DUBBS/DUBS SD5099 Strickland Roger.
Gurnell Dubs 1828 (HodgsonM), *Gurnal Dubbs* 1858 (OSNB).
♦ 'Gurnal pools', from the local surname *Gurnal/Gurnel* and **dub**. There is one main tarn and outliers, High, Middle and Low Taggleshaw.

GUTHERSCALE NY2421 Above Derwent.
Godrichesskales 1256 (*Scotland* I, 399), *Goderyscales* 1293, *Goderikeschales* 1318 (*PNCumb* 370), *Gudder Scale* 1794 (Hutchinson II, map facing 153).
♦ 'Godric's shieling(s)', from the pers.n. *Gōdrīc*, common in OE (cf. GUTTERBY), and ON **skáli** 'shieling'. The ME pl. *-s* in the early forms has been dropped. (Pl. 12)

GUTTERBY SD1084 Whicham, G~ SPA SD0984.
Godrikeby 1209, *Guderbye*, *Gudderbie* 1526 (*PNCumb* 448).
♦ 'Godric's settlement', from the common OE pers.n. *Gōdrīc* (cf. GUTHERSCALE) and ON **bœr/bý**. *Spa*, originally the name of a Belgian watering-place, was used as an English common noun from the 17th cent. and hence added to English p.ns. This, the sole instance in this volume, refers to an area of saline springs on the coastal sandbanks, which provided 'a sovereign remedy for the scurvy and the gravel' (Hutchinson, quoted in Gambles 1993, 188) from the 17th to 19th centuries.

H

HACKET NY3203 (hab.) Langdales.
hackeath 1612 (*CW* 1986, 143), *Hackart* 1634/5 (Johnson 1988, 247), *the Hackert, Hackert Forge* 1654 (*CW* 1908, 168), *Hacket* 1691 (Fleming *Accounts* 126), *Hackett* 1694, *the Hacket* 1695 (Grasmere PR), *Hackerd in litle Landale* 1695 (Hawkshead PR).

♦ H~ may derive from a rare ME **hacket* 'cleared ground', also found in Gloucestershire (so *PNWestm* I, 206). This does not, however, account for the *-ert, -erd* spellings, which could possibly be influenced by *haggard, haggart* 'stack-yard' from ON *heygarðr*, though examples in *EDD* are mainly Irish, Manx and Scots. The 17th-cent. forge was a bloom-smithy, i.e. a combination of bloomery and forge. Low and High H~ are distinguished on larger scale maps.

HAG END SD4397 (hab.) Nether Staveley.
From 1836 (*PNWestm* I, 174).
♦ Probably 'the end of the plantation', with *hag(g)* as in HAGG, (a), plus **end**.

HAGG SD2189 (hab.) Broughton West.
From 1669 (Broughton-in-Furness PR).
♦ (a) Probably 'the plantation', ultimately from ON *hǫgg* 'felling, place where trees have been cut down or are marked for felling' (cf. *OED*: *hag, n.* 3, and *EDD*: *hag*). (b) 'A piece of soft bog', or firmer ground rising out of a peat bog (*OED*: *hag, n.* 4) does not fit the sloping, partially wooded site.

HAGG GILL NY4206 Troutbeck.
Hag Gill 1859 (*PNWestm* I, 190).
♦ 'The ravine/stream where timber has been cut', with elements derived from ON *hǫgg* 'felling' as in HAGG, (a), plus ON **gil** 'ravine with a stream'. Early Modern woodland management in Troutbeck is discussed in Parsons 1997. There are few trees along the valley's steep sides nowadays.

HALECAT/HALECOTE: HALECAT HOUSE SD4383 Witherslack.
Hale Cat 1256, *Halkat* 1280 (*PNWestm* I, 77); *Halecat House* 1862 (OS).
♦ Possibly 'Kát's nook' or 'wild cat nook'. In this tantalising name, the 1st el. seems to be the generic. (a) It may be OE *halh* 'nook of land', qualified by a pers.n. such as Scand *Kátr* or *Káti*, or by OE *cat*/ON *kǫttr* '(wild) cat'. (b) Alternatively, Halecat could be 'Cat's tail', from ON *hali* 'animal's tail' and OE *catt* '(wild) cat', referring to a tail-like strip of cultivated land; cf. the field-name *Cat tails* (*PNWestm* II, 49 and Scand parallels, I, 77). Fellows-Jensen favours either version of (a) (*SSNNW* 230–1 and references there).

HALHEAD: H~ HALL SD4894 Strickland Ketel.
Halleheued(e) c.1320, *Hal(l)he(a)d(e)* 1516, 1616; *Haul(e)he(a)d(e) Ha(u)ll* 1527 to 1616 (*PNWestm* I, 154).
♦ Halhead is 'the high ground by the hall' (**hall, hēafod**). This gave rise to a surname *Halled* which may be the origin of the 1st el. (so *PNWestm*). The addition of *Hall* to a name originally containing the same word is coincidental.

HALL BECK NY4600 Over Staveley.
From 1857 (*PNWestm* I, 176).
♦ 'The stream by the hall'; **hall, beck**. Hall Beck flows alongside Hall Lane, and both may be named from the

former Staveley Hall (*PNWestm* I, 176).

HALLBECK HOUSE SD1484 Whicham.
Holebecke 1655, *Hoalbeck* 1671, *Wholebeck* 1798 (*PNCumb* 445).
♦ H~ is 'the stream in a hollow'; **hol(r)/hol(h)/hole, bekkr/beck**, cf. HOLBECK GHYLL and nearby HOLEGILL; see also **house**.

HALL GARTH, LOW NY3002 Coniston.
Low Hall Garth (also *High H~ G~*) 1850 (OS); cf. *Hallgarth* 1678/9 (PR); *Low Garth, High Garth* approximately here 1770 (JeffreysM).
♦ 'The hall enclosure or farm' (**hall, garth**); then distinguished as **low** and **high**.

HALLIN FELL NY4319 Martindale, HALLIN BANK NY4319.
Haylin, Haylin 1266 to 1279, *Halin, le Hayling* 1279, *Hawling Fell* 1770, *the mountain Hallin* 1787, *Holling Fell* 1793 (*PNWestm* II, 217), *Hallin Fell* 1809 (CrosthwaiteM); *Hellenbancke* 1588, *Hollin Bank* 1823, *Hallinbank* 1865 (*PNWestm* II, 218).
♦ 13th-cent. documents speak of *Haylin* as an area from which oak trees can be harvested or three acres of pasture exploited (*CW* 10, 446–8), but the name is of uncertain origin and meaning. (a) The early forms could point to ON *heill* 'good fortune', which may occur in Norw mountain-names (*NSL*: Heilhornet), but these seem to be so named from their role as sailors' landmarks, and the *-in(g)* suffix in the Cumbrian name would be unexplained. (b) A *hallen, hallan* is a passage or partition, hence Clarke's explanation: 'From that part of the entrance to a house which is called Hallen, a narrow turn of the Lake Ulles-Water has its name' (1789, xxix); but this is evidently a re-interpretation of an earlier form. (c) *Hollin(g)* spellings would suggest association with holly, but again this seems to be a re-interpretation. See also **fell** 'hill', **bank(e)** 'hill-slope'.

HALLOWBANK NY4605 (hab.) Kentmere.
Hallowbanke 1630 (*PNWestm* I, 167).
♦ (a) 'Hallow' may be a version of *hollow*, cf. *Hallow-stone crag* 1784, now Hollow Stone (*PNCumb* 353), and cf. other names such as HOLLOW MOOR, HOLLOW STONES. (b) Smith suggests 'hall hill' from **hall** and **haugr**, later *how(e)*, comparing *Hallow feet* 1836 (*PNWestm*). The present KENTMERE HALL is not, however, near Hallowbank or in H~ Quarter (cf. GREEN QUARTER on Quarters). The final el. **bank** 'hill-slope' seems unproblematic.

HALL'S FELL NY3226 Threlkeld, HALLSFELL TOP NY3227.
Both from 1867 (OS).
♦ (a) 'Hall's' may, as Brearley (1974) suggests, refer to Threlkeld Hall, below at NY3225 (*Thelkeld Hall* [sic] 1675, *PNCumb* 252), especially since the other great buttresses of BLENCATHRA are named from the places below: DODDICK FELL, GATEGILL F~, SCALES F~, and probably BLEASE F~. (b) Otherwise, the genitive 's would seem more likely to point to a surname. See also **fell** 'mountain' and **top**.

HALL WOOD SD3393 Satterthwaite.
From 1919 (OS), previously *Low Coppice* 1851, 1892 (OS).
HALL WOOD NY4503 Kentmere.
From 1865; cf. *Kentmore Haul* c.1540, *Kentmer(e) Hall(e)* 1617 (*PNWestm* I, 167).
♦ Simply **hall, wood**. In Satterthwaite, the hall in question was presumably nearby Grizedale Hall,

now demolished (Pevsner 1969, 217). The Kentmere wood is attached to KENTMERE Hall, a castellated farmhouse built onto a 14th-cent. pele tower.

HALTCLIFF: H~ BRIDGE NY3636 Caldbeck, H~ HALL NY3637, H~ HOUSE NY3536, H~ VIEW NY3535.
Halteclo 1208 to 1394, *Halteclove* 1211, *Hauteclo* 1252 (p), *Hauteclo(u)ch* 1285 (p), *Hawtecliffe* 1523; *Haltclobrig'* 1363 (*PNCumb* 276–7); *Haltcliffe hall* 1680 (PR); *Haltcliff View* 1900 (OS), previously *Haltcliff* 1867 (OS). H~ House does not appear on 1867, 1900 OS, though it or the buildings just to the N are named *Low Row*.
♦ Haltcliff appears to derive from OFr *halt*, *haut* 'high' and OE *clōh* 'ravine', later replaced by *cliff*. (Another variant ending *-ley* appeared in the 16th to 18th centuries.) Ekwall in *DEPN* suggested *halhiht* 'angular, winding' as the 1st el., on the grounds that Fr *haut* does not occur in English until c.1450, but it occurs, albeit in the French, originally monastic, name Haltemprice, EYorks, (*Hautemprise* 1324) and possibly, compounded with an OE el., in Haltwhistle, Northumberland (*Hautwisel* 1240). The Haltcliff sites are on a broad hill W of the R. Caldew, with TOWNEND as the S extremity. The word *view* was an adoption into late ME from OFr/Anglo-Norman *vewe*, *v(i)eue* etc., but the name H~ View is recent, as is the only other instance of *view* in this volume, MOUNTAIN VIEW.

HAMPSFIELD (hab.) SD4080 Broughton East, H~ FELL SD3979, H~ HALL SD3980. *Hamesfell* 1292–9 (p), *Hampesfell* 1537 (*PNLancs* 198); *Hampsfield Fell* 1851 (OS); *Hamfeldhall* 1577 (*PNLancs* 198); *Hamsfell-Hall* 1671 (Fleming 29).

♦ 'Ham's hill', from ON pers.n. *Hamr* or, since this is rare at best, a short form of *Hamall* (Fellows-Jensen *SSNNW* 130), plus ON **fell/fjall**, ME **fell**. The hamlet was apparently named from the hill to the S, previously Hampsfell but now called Hampsfield Fell; the element **field** then replaced *fell*. H~ **Hall** dates from shortly before 1636 (*VCHLancs* 279–80).

HANGING KNOTTS NY2407 Borrowdale. *Hanging Knot* 1802 (*CN* 1219), *Hanging Knotts* 1867 (OS).
♦ 'The **knotts** or rocky heights on the steep slope.' The verb *hang* is from OE *hangian*; this sense of *hanging* is illustrated, for instance, in: 'On the banks are villages and scattered houses, sweetly situated under woods and hanging grounds' (West 1780, 55).

HANNAKIN SD3597 (hab.) Hawkshead. From 1793; cf. *Anykinsyke* 1659, *Annykin syke* 1668, *Hanikinsicke* 1678, *Anykin Syke* 1697 (PR).
♦ (a) The early forms suggest that the place was named from a nearby stream or **sike**, qualified by the diminutive of a forename: probably *Annikin* from *Ann*. Female names commonly qualify names of minor natural features, and it is striking (though not conclusive) that the baptism of Ann, daughter of Richard Ashburner who lived here until his death in 1697, occurred in 1633 and the p.n. is first recorded in the entry for her mother's burial in 1659 (all PR). (A son John was baptised in 1635, and *Han(e)kin* is also a possible pet-form of *John*, but the *H*-less spellings make this a less likely explanation.) It seems that the presumed pers.n. was then reinterpreted as a p.n. which could stand alone, hence Hannakin (Ekwall,

PNLancs 218). (b) *Anikin, any kin* is a ME and Yorks dial. phrase for 'any kind of' (*MED, EDD*); but it would be hard to account for in a p.n.

HARD CRAG SD3583 (hab.) Haverthwaite.
Possibly = *Hard Cragg, Cartmel* 1745 (Finsthwaite PR).
♦ Seemingly 'the solid rocky height', with *hard* as in HARD KNOTT plus **crag**.

HARD KNOTT NY2302 (1803ft/549m.) Eskdale/Ulpha, HARDKNOTT FORT NY2101 Eskdale, HARDKNOTT PASS NY2201 Eskdale/Ulpha.
Hardecnuut c.1210, (*summitatem del*) *Ardechnut* 1242 (*PNCumb* 343-4); (fort:) *Hardknott Castle* 1867, 1900 (OS); *the Hard Knot pass* 1865 (Prior 57), previously *Wainscarth* (wagon-pass) 1242 (*PNCumb* 343-4).
♦ 'The hard, rugged hill', from ON *harðr* 'hard', probably influenced by the English cognate, ME *hard*, plus ON *knútr* 'knot, hard lump, rugged peak'. For another *Hardecnut*, see GREAT BOURNE. The **Pass** takes its name from the hill, while the fort, some distance from the hill summit, may be named from the pass. The fort, whose Romano-British name was MEDIOBOGDUM, forms a large square covering some three acres, and is quite well preserved (LDHER 3019, SAM No. 13568).

HARD RIGG NY1906 Eskdale, HARDRIGG GILL NY1906.
(*the*) *Hardrigge* 1587 (*PNCumb* 391); *Hardrigg Gill* 1867 (OS).
♦ 'The hard, rugged ridge', with *hard* as in HARDKNOTT, plus **hryggr/rigg**. The **gill** 'ravine with a stream' runs down its E side.

HARE CRAG NY2729 Underskiddaw.

From 1867 (OS).
HARE CRAGS SD2795 Torver.
From 1851 (OS).
♦ (a) 'Hare' here may possibly result from OE **hær* 'rock, cairn', only recorded in p.ns (cf. HARROP and Smith, *Elements*). This could register the presence of ancient remains, at least in the case of the Torver Hare Crags, site of a mound and ring bank of possibly Bronze Age origin (LDHER 1504). (b) 'Hare' may alternatively be from OE *hār* 'hoary, grey'. (c) Since the main habitat of the native brown hare is lowland (e.g. Ratcliffe 2002, 103), a reference to the mammal (**hare** from OE **hara**) seems less likely. See also **crag**.

HARECROFT HALL NY0702 Gosforth.
Hare Croft 1867 (OS).
♦ The meaning of 'Hare', though not of **croft** and **hall**, is elusive. If to be taken at face value, it may be the mammal **hare** or alternatively be from OE *hār* in its postulated sense 'boundary' (cf. HARROP), since nearby Hare Beck may be 'boundary beck', having formed the boundary between the manors of Bolton and Gosforth (Parker 1909, 84). However, Harecroft may instead be a corruption of the name appearing in the 17th-cent. PRs as *Hal(l)ecroft(e), Hal(l)croft* or (in 1637) *Howcroft*. The present H~ Hall, now a boarding school, was built in 1881.

HARE GILL SD1698 Eskdale.
From 1867 (OS).
♦ 'Hare' may be the mammal **hare**, for the comparatively gentle gradients around the **gill** would make a suitable habitat for the brown hare (by contrast with HARE CRAG(S)). A *gil(l)* is a ravine, normally with a stream, and the OS cartographers unusually give priority to the sense 'ravine' for *gill*

DICTIONARY

here, by printing the name in black, though there is a stream.

HARE HALL SD2192 Dunnerdale.
Harehall 1669 (original spelling not specified), *Hair-hall Dunerdale* 1696 (Broughton-in-Furness PR).
♦ 'Hare' is probably the mammal here (see **hare**), and **hall** may also be taken at face value, though confusion with **how**(e) 'hill' (from ON **haugr**) is always possible in this area.

HARE HILL SD4085 (hab.) Upper Allithwaite.
From 1851 (OS).
♦ 'Hare' is probably the mammal here, though reference to grey, hoary colour or a boundary (this is near the boundary with Staveley-in-Cartmel) cannot be ruled out: see **hara/hare**. The place lies below a **hill**, which is not named Hare Hill, but rather Barrow Wife Hill, on the 1851 OS map and on modern maps.

HARE SHAW NY4913 (1639ft/500m.) Shap Rural.
From 1860 (OSNB).
♦ (a) Probably 'hare wood or copse' (**hare, skógr/shaw**), though this is a hill, not presently wooded. (b) Derivation of 'Hare' from OE *hār* 'grey, hoary' cannot be ruled out.

HARPER HILLS NY5114 (1358ft/414m.) Shap Rural.
From 1859 (*PNWestm* II, 175).
♦ H~ may be the occupational surname, derived ultimately from OE *hearpere* or OFr *harpeor* (Reaney 1976), though it is not found in Shap PRs. The pl. **hill**s may be due to the small subsidiary summits.

HARRISON STICKLE NY2807 (2403ft/732m.) Langdales.

From 1828 (HodgsonM).
♦ 'The peak associated with the Harrison family', who are known from local records of the 16th–17th cent. (*PNWestm* I, 206); cf. HARRY PLACE FARM. For the generic *stickle* see STICKLE PIKE.

HARRIS SIDE NY0918 Lamplugh.
From 1867 (OS).
♦ Probably 'the hill-slope associated with the Harris family'. The name refers to the SW flank of MURTON FELL, and *Harris* is presumably the surname, of which there are some instances in the 18th cent. and early 19th cent. Lamplugh PRs.

HARROP: H~ PIKE NY5007 Shap Rural/Longsleddale/Fawcett Forest.
Haropes 12th cent., *Harhopes* 1235; *Harrop Pike* 1823 (*PNWestm* II, 168).
♦ There are multiple possibilities for H~. (a) Smith favours 'craggy valley', from OE *hær* 'rock, cairn' and OE **hop** 'rounded, blind valley' (*Westm* II, 168, suggesting the head of Little Mosedale, and cf. Smith, *Elements*: *hær*). (b) OE **hara** 'hare' is also linguistically possible, as assumed by Ekwall in *DEPN* for Harehope in Northumberland (*Harop* 1185) and Harrop WYorks (*Harrop* 1274), but this high, remote place is unlikely hare habitat. (c) A further possibility is OE *hār* 'hoary, grey' or the same with the sense 'boundary', though such a usage is uncertain (Smith, *Elements*). The **pike** is on a boundary which is probably ancient, and which was marked by *Harrope-raise* 'H~ cairn' (1577, *PNWestm* II, 168).

HARROP: H~ TARN NY3113 St John's.
Harhop c.1280 (*PNCumb* 34); *Harrop Tarn* 1839 (Ford 41).
♦ The 1st el. is uncertain, with the same range of possibilities as HARROP

above (*PNCumb* mentions (b) and (c)), though this place is not on a recent boundary. The 2nd el. is OE **hop** 'rounded, blind valley', apt for the hanging valley in which the **tarn** 'small mountain pool' nestles.

HARROT NY1527 (hill) Lorton.
Harroth Fell 1774 (DHM).
♦ A puzzling name. The hill rises steeply to 958ft/292m., and may be of the same unknown origin as Harrath in Millom (SD1581, *Harroth* 1597, *Harrats* 1794, *PNCumb* 418), another compact vantage point on the edge of a cluster of hills and above a valley. (a) They may possibly be of the same origin as Heriot at NT3652 in Midlothian (*Herth* 1164, 1247, *Hereget* 1198, from OE *heregeat* 'gap (in hills) through which an army might pass', Nicolaisen 2001, 16, 24), but the linguistic differences are perhaps too great. (b) An OE **harað*, **harad* is postulated in Smith, *Elements*, but the only examples given are Hartridge and Hardres, both in Kent, for which the spellings are diverse and far from unambiguous (Wallenberg 1931, 67–8; 1934, 321–2).

HARROW SLACK SD3896 (hab.) Claife.
Harrowslack 1783 (CrosthwaiteM), 1851 (OS).
♦ H~ is of uncertain origin, but may be (a) Gaelic-Norse **ærgi/erg** 'shieling', cf. LITTLE ARROW. (b) Less likely is OE *hearg* 'heathen temple' found in Harrow, former Middlesex, and implausibly assumed in Harrow Head, Nether Wasdale (*PNCumb* 442, 477). The 2nd el. is the reflex of ON **slakki**, here in the sense 'slope' rather than 'valley'.

HARRY GUARDS WOOD NY3100 Coniston.
From 1850 (OS).
♦ The **wood** is W of Guards Beck (1850 OS), and 'Guards' in both may be either the variant of **garths** 'enclosures' or perhaps the name of a onetime owner, though the surname *Guard(s)* does not appear in the index to Coniston PRs. It is presumably coincidental that *Harry Guards* is among field-names recorded from Whicham in 1840 (*PNCumb* 446).

HARRY PLACE FARM NY3106 Langdales.
Harry Place 1671 (*PNWestm* I, 206), *Harrison-place* 1692 (Grasmere PR), *Harrison place* 1706 (*PNWestm* I, 206).
♦ 'The Harrisons' farm', with the same surname as in HARRISON STICKLE, plus **place**, cf. nearby ROBINSON PLACE and WILSON PLACE. The Grasmere PRs refer in 1694 to 'Anthonie Harrison of Harry-place'. **Farm** has, as so often, been added to a pre-existing name.

HARTBARROW SD4090 (hab.) Cartmel Fell.
Hertbergh 1332 (p), *Hertbarrowe* 1537 (*PNLancs* 200).
♦ 'The hill frequented by the hart or stag', with *hart* originating in OE *heorot* or ON *hjǫrtr* and **barrow** also of OE or ON origin (see **berg**).

HART CRAG NY3711 Patterdale/Rydal (2698ft/822m.), LITTLE H~ C~ NY3910 Patterdale.
Hart Crag 1860 (*PNWestm* II, 225); *Little Hart Crag* 1860 (OSNB).
HART CRAG NY4017 Martindale.
From 1860 (*PNWestm* II, 219).
HART CRAG NY4108 Troutbeck.
Heart Cragg 1764 (*PNWestm* I, 190).
♦ 'The rocky height frequented by the hart or stag', with *hart* from OE *heorot* or ON *hjǫrtr*, and **crag**; also **little**. Deer have frequented MARTINDALE through to the present day (cf. Hutchinson 1794, I, [1]).

HARTER FELL SD2199 (2129ft/649m.) Eskdale/Ulpha.
Herter fel c.1210 (*Furness Coucher* II, 565).
HARTER FELL NY4609 (2539ft/774m.) Shap Rural/Longsleddale.
Herterfell, Harterfell 1578 (*PNWestm* II, 168).
◆ 'The mountain of the hart or stag', from ON *hjartar*, gen. sing. of *hjǫrtr* 'stag, deer', and **fell**. The survival of the *-ar* inflectional ending as *-er* shows that these names were given when Scand was a living language locally.

HART HILL NY4618 (2057ft/627m.) Bampton.
From 1859 (*PNWestm* II, 195).
◆ 'Hill frequented by stags', with *hart* from OE *heorot* and **hill**. This is not a distinct summit, but the NE shoulder of LOADPOT HILL, cf. comment on HOWE NY4919.

HARTLEY GROUND SD2189 Broughton West.
From 1666/7 (original spelling not specified, Broughton-in-Furness PR).
◆ 'The farm of the Hartley family'; surname plus **ground**. The place is associated with the Hartleys in the Broughton PRs for 1666/7, 1677/8 and 1719. The surname is frequent in the Broughton PRs.

HARTRIGG NY4506 (hab.) Kentmere.
Hathrigg [sic] 1770 (JefferysM), *Hearth Rigg* 1828 (HodgsonM), *Harthrigg* 1920 (OS).
◆ Perhaps 'the hard ridge'. Despite the modern form suggesting 'hart, stag', the *-th-* in the earlier forms points to a different 1st el., perhaps the ON adj. *harðr* 'hard'. The 2nd el. is **rigg** (see **hryggr**).

HART SIDE NY3519 (2480ft/756m.) Matterdale.
From 1867 (OS).
◆ Presumably 'the hill-slope frequented by harts or stags'; *hart* from OE *heorot* plus **side**, though if the name were older than it appears, ON **sætr** 'shieling' would be an alternative possibility.

HARTSOP NY4013 (hab.) Patterdale, H~ DODD NY4111, HIGH H~ DODD NY3910 (1703ft/519m.), H~ HALL NY3912.
Herteshop(e) & variants, pre-1184 to 1481, *Hartsop(p)* & variants 13th cent. to 1777 (*PNWestm* II, 223); *Hartsop Dod* 1860 (*PNWestm* II, 225); *High Hartsop Dod* 1860 (OSNB); *Harteshop hall* 1577 (*PNWestm* II, 225). Machell refers to *Upper Hartsop, Nether Harsop* [sic] and *Great Dod*, which may = Hartsop Dodd (c.1692, I, 725A).
◆ 'The valley of the stag', from gen. sing. of OE *heorot* 'hart, stag' and OE **hop** 'rounded, blind valley'. There are still deer in this area. Of the ten names in this volume referring to 'hart', this is the only one whose generic is a valley, rather than hill, term. The two **Dodd**s — steep, compact summits —, and the **Hall** lie S of Hartsop.

HASKEW BECK NY5112 Shap Rural, H~ TARN NY5211.
Hasker Beck 1828 (HodgsonM), *Haskew Beck* 1860 (OSNB, as variant to H~ *Gill*); *Hasker Tarn* 1828 (HodgsonM), *Haskew Tarn* 1860 (OSNB).
◆ Probably 'the stream and pool associated with the Askew family'; surname plus **beck, tjǫrn/tarn**. A *John Askew* is named in the PR for 1792; variation between forms with and without *H-* is common.

HASSNESS NY1815 (hab.) Buttermere.
Hassenesse 1578 (Winchester 1987, 141), *Hosenesse, (gate of) hessnesse* 1578 (*PNCumb* 356), *Hesseness (How)* 1809 (CrosthwaiteM), *Haseness* 1815 (PR).
♦ Possibly, as suggested in *PNCumb*, from ON **hals** 'neck of land' and ON **nes** 'headland', referring to a rounded projection into BUTTERMERE. If this is correct, the name is virtually synonymous with HAUSE POINT, which is situated at the corresponding place on Buttermere's sister lake CRUMMOCK WATER. The spellings *hess*- and *Hass*- are unusual, however, and the *hause* is not a very distinct landform.

HATTERINGILL HEAD NY1324 Blindbothel.
From 1867 (OS).
♦ The 1st el. is obscure, unless it is connected with the Scots and N. dial. *hatter* 'to batter, wear out, erode' (*OED*) or 'shake, throw into disorder' (*EDD*); cf. *Hattering Stone* (*PNWYorks* III, 165). **Gill** refers to a modest valley, streamless at first, which runs E below H~ **Head**, a minor summit on WHIN FELL.

HAUSE NY1109 Ennerdale/Nether Wasdale.
Hawse 1826 (CRO D/Lec/94).
HAUSE FARM NY4319 Martindale.
Hawse in Martindale 1631 (cited PR edition, p. 112), *Hause* 1656 (PR), *Horse* 1860 ('corrected' from *Hause*, OSNB).
HAUSE, THE NY4817 Bampton.
The Hause 1899 (OS); cf. *Horse End* 1860 (OSNB).
♦ 'The neck of land, col, ridge'; **hals/hause**. The Ennerdale example is a long, raised neck of land between CAWFELL BECK and R. BLENG and the Bampton one a tongue of high ground between CAWDALE and WILLDALE. The Matterdale place is situated on the *hause* or pass between HALLIN FELL and the hills to the S; **farm** added relatively recently.

HAUSE FOOT NY5505 (hab.) Fawcett Forest.
From 1777; cf. *Crook(e)dale hawse* 1671, *Horse house* 1675 (*PNWestm* II, 139).
♦ 'The lower end of the pass or neck of land' (**hals, foot**), referring to CROOKDALE HAUSE over SHAP FELL.

HAUSE GILL NY2313 Borrowdale.
Horsey Gill [sic] 1867 (OS).
♦ 'The stream in the ravine flowing from the pass'; **hals** 'neck of land', **gill** 'ravine with stream'. The name relates to HONISTER Hause.

HAUSE POINT NY1618 Buttermere.
From 1867 (OS).
HAUSE POINT NY3115 St John's.
From 1867 (OS).
♦ 'The promontory at the neck of land', with **hause** from ON **hals** 'neck, pass', plus **point**. These jut into CRUMMOCK WATER and THIRLMERE respectively. The making of the latter into a reservoir has obscured the landforms, but not eliminated the promontory. The Buttermere example is a *hause* in the sense that it provides a narrow way between the lake and the steep ground above.

HAVERIGG: H~ HOLME SD2691 (hab.) Kirkby Ireleth.
Haverrigge 13th cent. (*PNLancs* 221); *Heverik Holme* 1728 (PR), *Haverig-Holm* 1757 (Woodland PR), *Haverigg Holme* 1851 (OS).
♦ Haverigg is probably 'the ridge where oats are grown' from ON *hafri*, pl. *hafrar* 'oat(s)' (though ON *hafr* 'he-goat', in which -*r* belongs to the root, is also possible) and ON **hryggr**

'ridge'. A **holm(e)** is 'land beside water', here STEERS POOL.

HAVERTHWAITE SD3484 (hab. & parish). *Haverthwayt* 1336 (*PNLancs* 217).
♦ (a) 'The clearing where oats are grown', from ON *hafri* 'oats' and ON **þveit** 'clearing'. (b) Origins in ON *hafr* 'buck, he-goat' or the man's name *Hafr* (in both of which *-r* belongs to the root), or even *Hávarðr* (favoured by Wyld 1911, 150) cannot be ruled out. See also LAKESIDE.

HAWES SD2190 (hab.) Broughton West. Possibly = *Hows* 1671, *Howse* 1692 (Broughton-in-Furness PR).
♦ Probably 'the hills', with **how(e)** from ON **haugr**. Despite the lack of medieval forms, this fits the landscape and the onomastic profile of High Furness better than assuming a variant of **hause/hawse**.

HAWES, THE SD0986 Whicham. From 1867 (OS).
♦ Without early forms it is impossible to determine whether this is (a) 'the neck of land' (with a variant of **hause** from ON **hals**) referring to a passage between two low hills; (b) 'the hills' (**how(e)** from ON **haugr**, with pl. *-s*); or (c) 'the enclosures, pastures', with the reflex of OE *haga* or ON *hagi*, again with pl. *-s*.

HAWESWATER: H~ BECK NY5216 Bampton/Shap Rural, H~ RESERVOIR NY4713.
Havereswater 1199, *Hawse-water* 1671 (*PNWestm* I, 16); *Hawes-Water Beck* 1829 (Parson & White 575), formerly *Halfa* (Noble 1901, 1, Thompson 1942, 15).
♦ Possibly 'Hafr's lake', from a pers.n. such as ON *Hafr* or a postulated OE **Hæfer*, and **water**; also **beck** 'stream' and **reservoir**, on which see Thompson 1942.

HAWK, THE SD2492 Broughton West. From 1851 (OS).
♦ This is a small but craggy eminence on the triangle of land between the River LICKLE and APPLETREE WORTH BECK. This makes it unlikely that it is of the same origin as HOWK. If this is the bird-name from OE *hafoc* or ON *haukr*, it is striking, but cf. (THE) PEWITS.

HAWKBARROW FARM NY0904 Gosforth. *Hawkebarrowe* 1598/9 (PR).
♦ H~ is a hill (**barrow** from OE/ON **berg**) which is either (a) frequented by hawks (OE *hafoc* or ON *haukr*), perhaps meaning sparrowhawks or goshawks (though see HAWK CRAG), or (b) named from a man called *Hafoc* or *Haukr*. As so often, without evidence about the age of the name it is impossible to know whether it was given when these pers.ns were still current or not. The hill gave its name to the farm, **farm** being added later.

HAWK CRAG NY4116 Martindale. From 1860 (*PNWestm* II, 219).
♦ If the **crag**-name indeed contains the bird-name *hawk*, it is rather rare among Lake District names (cf. the above two entries). Both sparrowhawk and goshawk are mainly woodland breeders (Ratcliffe 2002, 189), and it seems more likely that *hawk* here refers to the peregrine falcon, as it does in minor names of the Scottish Borders (I am grateful to Dr Chris Cameron for this).

HAWKSHEAD SD3598 (hab. & parish), H~ HALL PARK SD3397, H~ HILL SD3398 (hab.), H~ MOOR SD3396.
Hovkesete 1198–1208, *Haukesset* c.1220,

Haukesheved 1336 (*PNLancs* 218); *Hawkshead Hall Park* 1851 (OS); cf. *Hauxhead hall* 1659 (PR); (H~ Hill:) *Hyll* c.1535 (*PNLancs* 218), *Haukeshead hill* 1657 (PR); *Hawkshead Moor* 1851 (OS).
♦ 'Hauk's shieling.' The 1st el. is from the ON pers.n. *Haukr*. The 2nd el. was ON **sætr** 'shieling' (so *DEPN*), OE *set* 'fold' or ON *setr* 'settlement' (so Fellows-Jensen, *SSNNW* 64). This was replaced by ME **heved** 'head, hill' (OE **hēafod**), possibly under influence of another Hawkshead (*PNLancs* 186). From seemingly humble beginnings, H~ became a chapelry in the 13th cent., and a parish from 1578 (*PNLancs* 218). H~ **Hill** is a hamlet above, and quite separate from, H~. See also **hall**, **park**, and **moor**. H~ Hall was the manor house of FURNESS Abbey (Cowper 1891, 24).

HAWS: H~ WOOD SD3890 Cartmel Fell. *haws* 1723 (CW 1966, 225); *Haws Wood* 1851 (OS).
♦ 'Haw(s)' is probably the local form of **how(e)** 'hill(ock)' (see **haugr**). The **wood** is on the fairly steep slopes SE of WINDERMERE.

HAWS BANK SD3096 (hab.) Coniston. *Howhousebancke* 1645 (*PNLancs* 216), *Hows Bank* 1851 (OS).
♦ The 17th-cent. spelling suggests 'the house by the hill' (**haugr/how(e)** 'hill', **hús/house**) plus **bank**, here referring to the sloping NW shore of CONISTON WATER. The telescoping of *Howhouse* to *Hows*, *Haws* may have been encouraged by the local use of *haws* to refer to fellsides, e.g. Low Haws SD2896.

HAWSE END NY2421 (hab.) Above Derwent.
Hausend 1567, *Hosend* 1649 (*PNCumb* 372), *Hawse End* (and *Hawse*, to its S) 1774 (DHM).
♦ 'The end of the projecting ridge'; **hals/hause, end**. Situated at the N end of CAT BELLS, the landform is similar to that of THE HAUSE and Hawes End NY4817.

HAWTHWAITE: LOWER H~ SD2189 (hab.) Broughton West.
Hauthwayt 1509–47 (*PNLancs* 222).
♦ Probably 'the clearing by the hill', with reflexes of ON **haugr** 'hill, mound' and ON **þveit** 'clearing'. This would suit the elevated situation. **Lower** H~ is distinguished from Upper H~, a little to the N.

HAY BRIDGE: HIGH H~ B~ SD3387 (hab.) Colton, LOW H~ B~ SD3387 (hab.).
Haybrigge c.1535 (West 1774, 104), *Haybri(d)g* 1628 (PR); *High Hay Bridge*; *Low Hay Bridge* both 1851 (OS).
♦ The bridges span streams which feed indirectly into RUSLAND POOL; see **brycg**. Whether the 1st el. means 'hay' (OE *hēg*) or 'enclosure' (OE *(ge)hæg*, ME *haye*), or is of some other origin is unclear. The **high** settlement is some 25ft/8m. above the **low** one.

HAYCOCK NY1410 (2618ft/797m.) Ennerdale.
Hay Cock 1774 (DHM), *Hay Cocks* 1783 (*PNCumb* 442), *the Hay-Cock* 1802 (CN 1205).
♦ The word *haycock* derives from OE *hēg* 'hay' and OE *cocc* 'a heap' and is first recorded as a compound in 1470 (*OED*). The stony dome which tops the hill resembles a haycock; cf. HAY STACKS.

HAYESWATER NY4312 Patterdale, H~ GILL NY4212.
Haiswater, Hay water c.1692 (Machell I, 726, II, 467), *Haiswater* 1777, *Hays or East Water* 1836 (*PNWestm* I, 16);

Hayeswater Gill 1860 (*PNWestm* II, 225).
♦ The 1st el. is obscure, but (a) a possible connection with a lost *Aisdale* (12th cent. *Aidesdale, PNWestm* II, 221) might suggest derivation from a rare ON pers.n. *Eiðr*, or appellative *eið* 'neck of land'. (b) If the names of Hayeswater and *Aisdale* are unconnected, or *Aidesdale* erroneous, other explanations are possible, e.g. from OE (*ge*)*hæg* or *hege* 'fence, enclosure', as in *les hayes de Askum* 13th cent. (*CW* 1918, 155). This is one of the smaller lakes to be called **water**; it gives its name to the **gill** 'ravine with stream'.

HAY KNOTT NY3036 Caldbeck.
From 1867 (OS).
♦ 'Hay' has the same range of possibilities as in HAY BRIDGE. This is a steep, compact, though not quite independent hill (see **knott**); it stands above Hay Gill.

HAY STACKS/HAYSTACKS NY1913 Buttermere/Ennerdale.
Hay Stacks 1751 (Smith map, *CW* 1918, 50), *by the dalesmen, from its form called Hay-rick* (West 1780, 129), *the Hay Stacks* 1803 (*CN* 1518).
♦ West's evidence above, and the similarity of the mountain's multiple summits to piles of hay in a field (before mechanisation) favour taking this name at face value, as do the parallel hill-names HAYCOCK and the Norw *Stakken* (*NSL*: Stakken, also Hjelmen 'stack, haystack'). *Hay* goes back to OE *hēg*, and *haystack* is recorded from the 15th cent. (*OED*). ON *stakkr* and its reflex *stack* are also used of rock columns, which could apply to Big Stack on H~, but it seems unnecessary to assume that the hill-name itself is a corruption of 'high stacks' (a view attributed to 'learned authorities' by Wainwright, 1966, Haystacks 2).

HAZEL BANK NY2615 (hab.) Borrowdale.
From 1852 (PR), 1867 (OS).
♦ 'The slope where hazel grows'; **hazel, bank**. The place is situated on the long contours on the E side of BORROWDALE.

HAZEL HEAD SD1994 (hab.) Ulpha.
hazlehead 1729 (PR).
♦ 'The high place where hazel grows'; **hazel, head**.

HAZEL HOLME NY0314 (hab.) Ennerdale.
Hazelholme 1867 (OS).
♦ 'The water-meadow where **hazel** grows.' The place is a classic **holm(e)** 'land beside water', set in the angle formed by the EHEN and its tributary Mere Beck. The name seems to be of 19th-cent. origin, since 'Hazelholme was built as a gentleman's residence at Low Mere-beck about 1840' (Banks *et al.* 1994, 233).

HAZELHURST NY3628 (hab.) Mungrisdale.
From 1867 (OS).
♦ Presumably 'the wooded hill where hazel grows'; **hazel, hyrst**. The place is on the lower slopes of SOUTHER FELL, beside a stream which would favour hazels. For another H~, see *PNCumb* 279.

HAZELRIGG SD3784 (hab.) Staveley-in-Cartmel.
Recorded in 1508–9 (*VCHLancs* 281, spelling not specified), *Heselrigg* 1575 (Cartmel PR).
♦ 'The ridge where hazel grows'; **hazel, hryggr/rigg**. For earlier examples of H~, see *SSNNW* 132.

HAZELSEAT SD3692 (hab.) Satterthwaite.
Hazel Seat (Wood) 1851 (OS).
♦ Without earlier spellings it is impossible to judge how old this name is, and whether '-seat' is more likely to be 'shieling, summer pasture' from ON **sætr**, or 'seat, seat-shaped eminence' from ON **sæti**; see also **hazel**. The site is below modest heights typical of Satterthwaite.

HEAD HOUSE SD3982 Upper Allithwaite.
Headhouse 1576/7 (Cartmel PR).
♦ The **house** is on a rise which is presumably the **head** or high place; cf. nearby Newton Heads.

HEALD BROW: H~ B~ PASTURE SD3194 Satterthwaite.
♦ '*The HIELD BROW covered with WOOD also*' 1809 (CrosthwaiteM); *Heald Brow Pasture* 1851 (OS).
♦ **Heald**, from OE *helde* 'hill-slope', is typical of this area, here applying to the long, rampart-like slopes rising steeply from the E shore of CONISTON WATER. H~ **Brow** is the upper part of the slope, and the **Pasture** has long ceased to be the 'grassland for grazing' suggested by the name.

HEALD WOOD SD3898 Claife.
From 1850 (OS).
♦ 'The wood on the hill-slope', with *heald* as in HEALD BROW, plus **wood**. The Heald, the long contours rising above the W shore of WINDERMERE, is named as such at SD3897 on larger scale maps.

HEANING SD4399 (hab.) Windermere.
Heaninge 1619/20 (PR).
♦ 'The enclosure', with the reflex of ON *hegning*, as also in HINNING HOUSE.

HEATHWAITE SD2996 (hab.) Coniston.
From 1851 (OS).

♦ Possibly 'clearing for hay', of the same origin as the HEATHWAITE in Kirkby Ireleth, but without earlier forms no certainty is possible.

HEATHWAITE SD4197 (hab.) Bowness.
Hothwayt 1256 (p), *Heathwaite* 1751 (*PNWestm* I, 187).
♦ Probably 'the high clearing'; ON *hǫ́r, hár* 'high' (cf. 13th cent. *Ho-* spellings for KNIPE, HIGH) plus **þveit**.

HEATHWAITE: H~ FARM SD2486, H~ MOSS SD2487 Kirkby Ireleth.
Heittheuuot 1273 (*PNLancs* 221); (H~ Farm:) previously *High Heathwaite Gate* 1851, 1893, 1919 (OS); *Heathwaite Moss* 1851 (OS).
♦ H~ is possibly 'the clearing where hay is grown'. ON *hey*, OE *hēg* 'hay' has been suggested as 1st el. (Lindkvist 1912, 110, n. 4, *PNLancs* 221, *SSNNW* 129), while 2nd el. is ON **þveit** 'clearing'; also **farm**, **moss** 'bog'.

HECKBARLEY NY0714 (1280ft/390m.) Ennerdale.
Heck-barley 1802 (*CN* 1207).
♦ Obscure. This hill is an outlier of GRIKE. Dial. *heck* derives from OE *hæcc*, and refers to doors, gates and partitions of various sorts, especially wicket-gates in boundaries of parks, forests and parishes, or gates and sluices across streams (see *EDD*: *hack* sb. 2, Smith, *Elements*: *hæc(c)*), but even if this is present in this name, 'barley' remains unexplained.

HECK CRAG NY4214 Martindale.
Heck Crag 1828 (HodgsonM), *Haig Crag* 1863 (OS), cf. *Eck Beck* 1859 (*PNWestm* II, 219).
♦ Partly uncertain. If 'heck' is the dial. word for a gate (see HECK-BARLEY), it could refer to a nearby gate or a barrier across Heck Beck (which is

Haig Beck 1863 OS), or else to the **crag**'s function as a forbidding barrier above MARTINDALE; but certainty is impossible.

HEGDALE NY5317 (hab.) Shap Rural. *Hegdal* pre-1201, *Hekedal* 1279, *Heggedale* 1292 (p), *Heghdale* 1332 (p), *Hegdaile* 1473 (*PNWestm* II, 169).
♦ Probably 'the valley where birdcherry grows', from ON *heggr* 'birdcherry, hagberry or heckberry' (so *PNWestm*) plus **dalr** 'valley. This plant is *Prunus padus*, common in Lakeland, especially in hedgerows (Halliday 1997, 281–2). Alternatively, the 1st element could be the ON pers.n. *Heggr* (see Lund 1905–15, 502 for this name).

HEGGLEHEAD NY3734 (hab.) Mungrisdale, HEGGLE LANE NY3635 (hab.). *Heygill-head* 1581, *Heggill Head* 1600; cf. *Heggill* 1487; *Heggil Loning* 1722 (*PNCumb* 211), *Heggle Lane* 1815, *Eggle Lane* 1827, *Hegle Lane* 1850 (PR).
♦ Heggle is 'the high stream', perhaps coined in the ME period from the reflexes of OE *hēah* 'high' (see **hegh**), and ON **gil** 'ravine with a stream' (contrasting with *Lowgill (felde)* 1581, *PNCumb* 211); also **head, lane**. Hegglehead and Hegglefoot, at NY3635 on larger scale maps, stand on Heggle Sike, which must have been the original *gil*. Heggle Lane lies W of the Sike.

HEIGHT SD4084 (hab.) Upper Allithwaite.
ye Height 1614 (Cartmel PR), possibly = *The Hee* 1604.
HEIGHT, THE NY3141 (hab.) Caldbeck. *Height* 1652 (*PNCumb* 279).
HEIGHTS, THE SD4088 Cartmel Fell. From 1851 (OS).
HEIGHTS, THE SD4599 (hab.) Hugill. *the Height* 1670 (*PNWestm* I, 171).

♦ The el. **height** is especially characteristic of Furness and Cartmel, referring to tracts of high ground; in the Cartmel Fell instance this is dotted with tarns. The Upper Allithwaite example is at about 620ft./190m., which is high for the immediate vicinity. The northerly Caldbeck example is named from the slope on which it stands above Bowten Beck; Low and High Heights were distinguished in the early 18th cent. PRs of Caldbeck.

HELL GILL NY2505 Langdales. From 1862 (OS).
♦ Probably 'the recessed **gill** or ravine with stream'. (a) Despite the menacing appearance of this and other features with names containing 'Hell', they may derive from ON *hellir* 'cave', as in dial. *hell-beck* 'a stream issuing from a cave-like recess' (cited by Smith in *PNWestm* I, 62, discussing a different Hell Gill). (b) An alternative is derivation from ON *hella* 'flat stone', thought to be present in a Yorks *Hell Gill*, recorded from the 13th cent. (*PNNYorks* 259). (c) Meanwhile, a reference to the inferno cannot be entirely ruled out (cf. HELPOT (FARM)).

HELM CRAG NY3209 (1299ft/396m.) Grasmere.
a Mountain . . . called the Helm, on the Top of w[hi]ch there is a Crag which Lookes Like Atlas supporting the Hevens, c.1692 (Machell II, 432, ed. Ewbank, 143), *Helm-Crag* 1770 (Gray 1098), *The Helm* 1859 (OSNB); cf. *Underhelme* 1569 (*PNWestm* I, 202).
♦ The Helm is a rocky scar below the proud top of Helm **Crag**; there are various possible explanations of the name. OE *helm* and ON *hjalmr* meant 'helmet, protection', with the additional sense 'barn' in ON, so that

helm can refer to helmet-shaped hill-tops, but more usually to natural or man-made shelters or perhaps to defensive positions (cf. The Helm, near Kendal, site of Castlesteads, *CW* 1908, 108–12). Cf. HELMSIDE.

HELMSIDE NY3309 (hab.) Grasmere.
From 1865 (*PNWestm* I, 201).
♦ 'The place beside HELM CRAG or The Helm'; see above and **side**.

HELPOT FARM/HELP POT SD4792 Underbarrow.
Helpott 1589, *Hellpot* 1745 (*PNWestm* I, 105), *Help Pot* 1790 (PR).
♦ H~ may be 'hell pit, the infernal pool'. *Hell-pot* is listed in *OED* among the compounds formed from *hell* (under sense 11), and may be comparable with *hell-kettle*, which is used as a local name of chasms and deep pools. The usage of the word in a p.n. may be paralleled in the 14th-cent. *helpots* (*PNDu* I, 37). The modern form *Help Pot* appears to reflect a desire to sanitise the name; **farm** has been added relatively recently.

HELSFELL SD4993 (hab.) Strickland Ketel.
Hellesfel 1272, *Helsfell* 1446 to 1690 (*PNWestm* I, 154).
♦ 'The hill with a cave', from ON *hellir* 'cave', referring to a cave in the limestone at Helsfell Nab (so *PNWestm*), plus ON **fell** 'mountain, hill'.

HELTON NY5122 Askham, HELTONDALE NY4920, HELTONDALE BECK NY4819 Askham/Bampton, HELTON FELL NY4720 Askham, HELTONHEAD NY5021.
♦ *Helton* c.1160 (p) to present; *Heltondale* c.1160 to present; *Heltondale Beck* 1859 (*PNWestm* II, 200, 202), seemingly = *Helton-Beck* c.1692 (Machell I, 89); *Helton Fell* 1920 (OS); *Heltonhead* 1795 (*PNWestm* II, 202).
♦ 'The settlement on the slope', from OE *helde* 'slope' as in HEALD (BROW) or possibly ON *hjalli* 'shelf on a hill-side', plus OE **tūn** 'farmstead, settlement, village'; the site stands proud above the LOWTHER valley. Other explanations of the 1st el. are possible but less likely, e.g. OE *helde* 'tansy' or ON *hjallr* 'shed' (mentioned by Ekwall, *DEPN*). For much of its history H~ was distinguished by the affix *Flechan*, a family name, later *Flecket* (*PNWestm* II, 200). See also **dalr** 'valley', **beck** 'stream', **fell** (here referring to high grazing land rather than a distinct peak) and **hēafod/head**.

HELTON TARN SD4184 Upper Allithwaite/Witherslack.
Heltontern(e) 1278 to 1354 (*PNWestm* I, 77–8).
♦ H~ may be 'the settlement on the slope' though there are other possibilities (see HELTON above). There is seemingly no settlement named H~ today, so that the etymology cannot be checked against the topography. The **Tarn** is unusual, being a pool in the R. WINSTER, at a height of about 33ft/10m. I am indebted to Mr M. J. Clifton of HIGH TARN GREEN for the information that the tarn must have been larger in former times, since the previous edge to the tarn is still visible in the fields; the river was lowered by at least three feet in the early 1950s.

HELVELLYN NY3415 (3116ft/949m.) St John's/Patterdale, H~ GILL NY3216 St John's, H~ SCREES NY3215.
Helvillon 1577, *Lauuellin* 1600 (*DEPN*), *Helvelon* or *Hell Belyn* 1610 (Denton 1), *the man on Hellvellin head* c.1692 (Machell II, 155), *Elvel-in* c.1692

(Machell I, 714), *Helvellin* 18th cent. (Collingwood 1918, 102), *Helvellyn* 1769 (Gray 1078); *Helvellyn Gill*; *H~ Screes* both 1867 (OS).
♦ 'The forms are too late for an etymology to be suggested' (*DEPN*). For the last two syllables, however, a Cumbric equivalent of W *melyn* 'yellow' has been proposed, and Coates suggests 'a Cumbric **hal velyn* or the like, "yellow (upland) moor"', perhaps referring to vegetation on the level summit of the mountain (Coates 1988, 30–3). Although vegetation is extremely sparse on the stony tops of H~, the name might have originally applied to the lower parts, and a Brittonic (Cumbric) solution is attractive for this venerable mountain. Older suggestions such as Collingwood's, of an inversion compound with OE *hlāw* 'mound' or ON *hváll* 'hill' as 1st el., and pers.n. *Willan* as 2nd (1918, 103) are less carefully argued and less plausible. Although — or perhaps because — H~ straddles the old Cumberland–Westmorland border, this famous mountain appears in neither EPNS county survey, though forms are given in *JEPNS* 3, 50. See also **gill** and SCREES. (Pl. 20)

HEN COMB NY1318 (1661ft/509m.) Loweswater.
Hencomb 1803 (*CN* 1518).
♦ 'Ridge frequented by (female) wild birds'; **hen, comb(e)**². The hill is a steep, elongated oval.

HEN CRAG NY2900 Coniston.
Hencrag 1684 (spelling not specified *CW* 1910, 379), *Hen Crag* 1850 (OS).
♦ 'The rocky height frequented by (female) wild birds'; **hen, crag**. The crag is immediately next to Hen Tor, while Henfoot Beck flows below both; the puzzling Henfoot might suggest that some other word is now concealed by 'Hen'.

HEN CROFT SD2390 Broughton West.
From 1675 (original spelling not specified, Broughton-in-Furness PR).
♦ **Croft** is a small enclosure, and hence **hen** could refer to domestic fowl; otherwise to game birds, or the name could be a whimsical allusion to the smallness of the farm.

HEN HOLME SD3997 Windermere.
Hen-holme c.1692 (Machell II, 342, ed. Ewbank, 87), *Hen Holme* 1780 (West 58).
♦ 'Hen island' (**hen, holme**), 'named after its colony of moorhens' (Danby 1994, 209).

HENHOW NY4317 (hab.) Martindale.
Henhowe 1588 (*PNWestm* II, 219).
♦ Presumably 'the hill frequented by (female) fowl'; **hen, haugr**.

HERDUS NY1116 Ennerdale.
Herdhouse 1774 (DHM), 1802 (*CN* 1207).
♦ Seemingly 'herd-house'; *herd* from OE *hi(e)rde, heorde* 'herdsman', plus **house**. H~ is marked on maps W of the summit of GREAT BORNE, but is also applied to G~ B~ as the name of a whole mountain (Wainwright 1966, Great Borne 1).

HERDWICK CROFT NY1932 Blindcrake.
Seemingly *West Lodge* 1867 (OS), together with *Bridge End Farm* 1900, 1920 (OS).
♦ A *herdwick* (OE *heordewic*) was a sheep farm or 'herd farm' as opposed to an arable farm, and the English p.n. Hardwick (several counties) results from it. More locally, *herdwicks* were tracts of land in the charge of a shepherd employed by the owner,

especially those belonging to FURNESS Abbey such as LAWSON PARK and WATER(SIDE) PARK. H~ Croft may, however, refer to the famously hardy H~ mountain sheep which have been variously believed to originate on the Furness herdwicks (*OED*: *herdwick*, sense 2), to be of Norwegian origin (Ellwood 1899, 1–8), or to have swum ashore or been recovered from a wrecked ship: a Spanish galleon at Drigg (Parker 1926, 35) or 'a Danish East Indiaman' (Clarke 1789, 98n.). See also **croft**.

HERON CRAG NY2203 Eskdale.
Earn Crag 1802 (*CN* 1220), *Heron Crag* 1867 (OS).
HERON CRAG NY2712 Borrowdale. 1867 (OS).
♦ The present name is simply 'the rocky height frequented by herons'; heron (ME *heiroun*, *heyron*, from OFr *hairon*, *OED*) plus **crag**. In favour of this is the proximity of the crags to becks which would supply fish, and the present author has seen herons within one or two miles of both. However, if herons have been sighted on the crags at all, it would be as a rarity, since these two lofty rock perches are unlikely habitats for the grey herons which are generally common in Lakeland, but frequently nest in trees (Ratcliffe 2002, 122). A corruption of another specific such as dial. *erne* 'eagle' has been suggested (Gambles 1989, 149, 156, Ratcliffe 2002, 237), and this seems plausible, encouraged by Coleridge's 1802 form; cf. also AARON CRAGS.

HERON PIKE NY3508 (2008ft/612m.) Rydal.
Unnamed 1863, 1899, 1919 (OS).
HERON PIKE NY3717 Patterdale.
From 1863 (OS).

♦ Seemingly 'the peak frequented by herons'; *heron* as in HERON CRAG, **pike**. As with the examples of Heron Crag, the high peak in Rydal is an unusual habitat for the grey heron, while Wainwright remarks on the Patterdale H~ P~: 'from no direction does it look like a pike nor will herons be found there' (1955, Heron Pike 1). Corruption of *erne* 'eagle' may therefore be suspected. However, there is a Heron Island in RYDAL WATER, and there is a heronry in its twin, GRASMERE (Miles 1993, 41), while the Patterdale peak is close to the fishing grounds of ULLSWATER, and Hutchinson in 1794 reported herons breeding in nearby GOWBARROW (I, 455). The names could therefore conceivably refer to infrequent sightings of herons in high places.

HESKET FARM NY4426 Matterdale.
Eskheved 1253 to 1353, *Heskett* 1588 (*PNCumb* 255).
♦ 'The high ground where ash trees grow', from ON **eski** 'ash trees, ash copse' (found in several early Norw p.ns; see **askr**) and OE *hēafod* 'head, high ground', possibly replacing ON *hǫfuð* (see **hēafod/head**). Nearby SPARKET also has -*et* from *hēafod*. **Farm** has been added to a pre-existing name.

HESKETH HALL SD2290 Dunnerdale.
From 1688 (original spelling not specified, Broughton-in-Furness PR), *Heskethaw in Dunnerdale* 1818 (Kirkby Ireleth PR).
♦ In the absence of a nearby place named Hesketh, this appears likely to be a surname, derived from the Cumbrian p.n., or from one of the counterparts in the present Lancs or Yorks (Hanks & Hodges 1988), plus **hall**.

HESKET NEWMARKET NY3338 Caldbeck. *Eskhevid* c.1230, *Eskeheued* c.1250 to 1399, *Hesket* 1523, *Hesket New Market* 1751 (*PNCumb* 277).
♦ H~ is 'the high place where ashes grow', from ON **eski** 'ash trees, ash copse' (see **askr**) and OE **hēafod** or ME **heved** 'head, high place', with initial H- influenced by Hesket (-in-the-Forest, *PNCumb* 199–200). 'It is now called *Hesket-New-Market*, from a market lately set up there, and in contradistinction to another Hesket in the Forest of Englewood' (Hutchinson 1794, II, 376, quoting Gilpin).

HESK FELL SD1794 (1566ft/478m.) Ulpha. From 1867 (OS).
♦ 'Hesk' is obscure; conceivably a Brittonic word meaning 'sedge, coarse grass' or 'wet ground' (W *hesgen*, Breton *hesk*, Cornish *heschen*, listed Padel 1985, 130). The **Fell** is a discrete, almost conical, peak.

HEUGHSCAR HILL NY4823 (1231ft/375m.) Barton/Askham.
Cf. *Heugh Scar* 1859 (*PNWestm* II, 212).
♦ Heughscar is 'the rocky scarp on the heel-shaped hill', with *heugh* from OE *hōh* 'heel, hill-spur, (heel-shaped) hill', dial. **scar** 'escarpment, crag, rock outcrop', plus **hill**. Heugh Scar itself is the outcrop on the N side of the hill, which has the typical profile of a *hōh*, like an inverted human heel, with a long, gradual slope on one side and a more abrupt one on the other (cf. Gelling & Cole 2000, 186).

HEWRIGG FARM NY0801 Irton.
Hewrig 1647, *Hurigg* 1686, *Hawrigg* 1701, *Hewrigg* 1747 (*PNCumb* 403).
♦ In the 1st el., 17th-cent. and current spellings in Hew- could result from OE *hōh* 'heel, hill-spur', cf. HEWTHWAITE below, or from ON **haugr** 'hill, mound', cf. Hewer Hill (*PNCumb* 245). **Rigg** (see **hryggr**) is used here, as frequently in the low-lying fringe of Lakeland, for relatively high ground, not necessarily a distinct ridge. **Farm** has been added to the name in recent times.

HEWTHWAITE HALL NY1532 Setmurthy. *Hotweyt* 1260, *Huthweyt* 1266 (p) (*PNCumb* 434), *Hughthwait* 1705 (Isel PR); *Hewthwaite Hall* 1724 (Isel PR).
♦ H~ is probably 'the clearing by/on the hill-spur', from OE *hōh* 'heel-shaped hill, hill-spur' and ON **þveit** 'clearing'. The place is on rising ground. Such a hybrid of OE and ON is not unlikely here in the lower DERWENT valley. The **Hall** 'has a three-bay front with an inscription of 1581' (Pevsner 1967, 145; see further Martindale 1911).

HIGH as affix: see main name, e.g. for HIGH BETHECAR see BETHECAR

HIGHAM HALL NY1831 Setmurthy.
the Highe 1578 (CRO D/Lec/310/158v–161, pers.comm. Dr Angus Winchester), *High* 1823 (GreenwoodM), *High (The)* 1829 (Parson & White 8), *Higham* 1862 (PR), 1867, 1900, 1926 (OS).
♦ 'Higham' has the appearance of a place-name formed from OE *hēah* 'high' or possibly 'chief' plus *hām* 'village, homestead', like the Highams in several English counties. However, a name in OE -*hām* would be unparalleled in this locality and there is no early evidence for it. It seems that the place was called The High (and was an esquire's residence, among the 'Seats in Cumberland' in Parson & White; I am grateful to Mary Atkin for pointing this out). A simplex p.n. is perfectly possible (cf. HIGH FARM/THE HIGH), but was perhaps felt

to be incomplete and therefore expanded to Higham under influence from places of that name or from the surname *Higham*. The **hall** is a grand building of c.1800 (Pevsner 1977, 37 and 188), now a residential college for continuing education.

HIGH BARROW BRIDGE: see BORROW BECK

HIGH BECK NY1512 Ennerdale, LOW BECK NY1412.
Both from 1867 (OS).
♦ Self-explanatory; **high**, **low**, **beck** 'stream'. The becks flow into the R. LIZA above ENNERDALE WATER, at a higher and lower point respectively.

HIGH BIRK SD3193 Satterthwaite.
From 1851 (OS).
HIGH BIRKS SD4290 (hab.) Crosthwaite.
From 1535 (*PNWestm* I, 82).
♦ 'The upper place where birch grows' (**high**, **birk**). The Satterthwaite example is a stretch of upland, now part of GRIZEDALE FOREST. In Crosthwaite, Low Birks is at SD4290. The selectivity of small-scale OS maps often leaves names such as these stranded without their partners (cf. LOW YEWS, under Y~).

HIGH BORROW BRIDGE: see BORROW BECK

HIGH BROADRAYNE/BROAD RAIN: see BROADRAYNE

HIGH BROW NY3621 (1886ft/575m.) Matterdale.
From 1867 (OS).
♦ 'The high hill-slope'; **high**, **brow**. A distinct if modest summit, this hill does not have the linearity of a classic *brow*. The name may have referred to the straighter contours below, or may conceivably originate in **barrow** from *berg* 'hill' (cf. another High Brow, *PNCumb* 418).

HIGH CLOSE NY3305 (hab.) Rydal.
the Hei close c.1575 (*CW* 1908, 179), *the highe close* 1638, *High Close* 1688 (Grasmere PR).
♦ Simply 'the high enclosure'; **high**, **close**. This is presently a Youth Hostel.

HIGH CRAG Examples at:
  NY1814 Buttermere (2443ft/744m).
High-crag 1780 (West 129).
  NY2713 Borrowdale.
From 1867 (OS).
  NY3413 St John's/Patterdale.
From 1863 (OS).
  NY3500 Hawkshead (514ft/157m.).
From 1850 (OS).
HIGH CRAGS Examples at:
  NY2117 Above Derwent.
From 1867 (OS).
  NY2319 Above Derwent.
High Crag 1839 (Ford 81).
  NY3314 St John's.
From 1867 (OS).
♦ 'The high rocky summit(s) or outcrop(s)'; **high**, **crag**. They are high in absolute terms or relative to the locality. Some are distinct summits, such as the Buttermere H~ C~, the towering culmination of a rock wall which warrants a separate chapter in Wainwright 1966, and the one at NY3413, a worthy partner to the other peaks in the HELVELLYN range. The others are not distinct summits but are on or near the top of a ridge or brow.

HIGH CROSS SD3398 Hawkshead.
From 1850 (OS), possibly cf. *Crosse* 1603 (PR); appears as *Cross High* on 1991 printing of the One Inch map.
♦ 'The high cross-roads'; **high**, **kross/cross**. 'High Cross at Coniston is well known as the junction of cross roads on the watershed between the

Coniston and Hawkshead valleys' (Parker 1909, 98), hence *cros(s)* seems to refer to the crossroads (so Cowper 1899, 363). Another crossroads named High Cross is found just W of Broughton-in-Furness, while two Roman roads cross at High Cross, Leicestershire.

HIGHCROSS NY1321 (hab.) Loweswater. *High Cross* 1725 (PR), *Cross* 1794 (Hutchinson II, map facing 153).
♦ Simply **high** and **kross/cross**, though the meaning here is uncertain. 'One is tempted to believe there may have been a cross here; but there are at least three sets of cross roads within a mile, all on lower ground, one of which might be "Low Cross"' (Parker 1909, 98).

HIGH DAM SD3688 Colton.
From 1851 (OS).
♦ 'The high reservoir', an apt description; **high** plus ME *damme*, ModE *dam*, which can be applied to a pool as well as the artificial bank which contains it.

HIGH DODD NY4118 (1640ft/501m.) Martindale.
The Dod 1865 (*PNWestm* II, 218).
♦ 'The high, compact, rounded hill'; **high**, **dodd**. Low Dodd is the craggy N part of the same hill.

HIGH DYKE NY1026 (hab.) Blindbothel.
High Dike 1774 (DHM).
♦ 'The high embankment or wall'; **high**, **dyke**.

HIGH FALL NY3606 Rydal.
From 1865, also *Low Fall* 1865 (*PNWestm* I, 210); cf. *Rydale Falls* 1802 (Dorothy Wordsworth, *Journals*, June 7), *Rydal Waterfalls* 1829 (Parson & White 617).

♦ 'The upper waterfall'; **high**, *fall*. The use of sing. *fall* to mean *waterfall* is unusual, but *OED*: *fall*, sense II, 7 has an instance from 1832 (Martineau) along with several examples of the more usual pl. High and Low Fall are incorrectly linked with OE (*ge-*)*fall* 'clearing, felled woodland' in *PNWestm* I, 210.

HIGH FARM, THE/THE HIGH SD4390 Crosthwaite.
High (*house*) 1535 (in 1669 copy), *High* 1547 to 1738, *the Highe* 1547 (*PNWestm* I, 84).
♦ The farm is on a pronounced bluff, and it is presumably this to which **high**, here a noun, refers. The element **farm** has been added relatively recently.

HIGH FELL NY1508 Nether Wasdale.
From 1867 (OS).
HIGH FELL SD2899 Coniston.
The High Fell 1850 (OS).
HIGH FELLS NY3319 St John's.
From 1867 (OS).
♦ There is no reason to suspect anything other than **high**, **fell** in these names. They refer to high points but not towering summits. The Nether Wasdale example is on the flanks of the Wasdale RED PIKE, and the Coniston one just below BLACK SAILS. The St John's name refers to the stretch of high ground on the flanks of WATSON'S DODD.

HIGHFIELD HOUSE SD3498 Hawkshead.
Not shown on 1851, 1892 OS maps.
♦ Presumably as it seems: **high**, **field**, **house**.

HIGH FORCE NY4020 Matterdale.
From 1867 (OS).
♦ 'The high waterfall'; **high**, **force**. These are the falls above AIRA FORCE.

HIGHFORD BECK SD1897 Eskdale.
From 1867 (OS).
♦ Seemingly 'the stream with the ford in its upper reaches'; **high, ford, beck**.

HIGHGATE: HIGHGATECLOSE/HIGH GATE CLOSE NY3525 Matterdale.
Highyeat 1697 (*PNCumb* 223); *Highgateclose* 1867 (OS).
HIGHGATE FARM NY4427 Hutton.
the High Yaite 1599, *hyghe Yaite of Penruddock* 1604 (*PNCumb* 210).
♦ H~ is 'the high gate'; **high, gate**; also **close, farm**. The *yeat, yaite* spellings point to the reflex of OE **geat** 'gate' rather than of ON **gata** 'way, road'. If correct, this puts in doubt Hindle's implication, by placing the Matterdale name on a map of the Roman Road from Old Penrith to Lorton, that *gate* here might refer to an ancient routeway (1998, 27). H~ Close lies just S of Highgate, and is slightly larger than the parent settlement. The Hutton example is on a notable rise, and here **farm** has been added to the pre-existing name.

HIGH GREEN NY4103 (hab.) Troutbeck.
From 1716 (*PNWestm* I, 190).
♦ 'The high grassy place'; **high, green**.

HIGH GROUND SD1798 (hab.) Eskdale, LOW G~ SD1798 (hab.).
High Ground 1769; *Low Ground* 1741 (*PNCumb* 344).
♦ 'The upper farm and lower farm'; **high, low, ground**. High G~ lies above Low G~ on the slopes to the S of ESKDALE. Although *ground* could simply refer to 'land, terrain' (see next entry), it more likely has its specific Lakeland sense of 'farm'.

HIGH GROUND SD2995 (hab.) Coniston.
From 1851 (OS).

♦ This may be 'high ground' in the ordinary sense; **high, ground**. *Ground* commonly forms farm-names in High Furness, qualified by surnames, referring to farms resulting from cultivation of former waste. However, another Highground at NY5527, just beyond the NE border of the National Park, is well out of the territory where *ground* occurs in farm-names, and suggests that a more general sense 'land' is also possible here; the late first record would favour this.

HIGH HEADS SD4687 Crosthwaite.
From 1815 (*PNWestm* I, 84).
♦ Situated at 79ft/24m., well above the flat floodplain of the GILPIN, the place qualifies locally as a **head** or 'high place'. **High** 'upper' distinguishes it from Low Heads just to the S.

HIGH HILL NY2523 Keswick.
Hey Hyll 1563, *Hee Hyll* 1564 (*PNCumb* 303), *High Hill* 1782 (CrosthwaiteM), "High Hill," or "How Hill" (CW 1904, 254).
♦ This part of KESWICK lies above its immediate surroundings; **high, hill**. The 16th-cent. spellings are unusual, but not unparalleled (see HIGH HOUSE TARN and HIGH PIKE).

HIGH HOLLOWS: see HOLLOWS

HIGH HOUSE Examples at:
    NY0800 Irton.
From 1774 (DHM).
    NY2837 Caldbeck.
High house (meadows) 1686 (*PNCumb* 279).
    NY4300 Hugill.
From 1670 (*PNWestm* I, 171).
    SD4494 Crook.
From 1836; cf. *the Hee in Crook* 1637 (*PNWestm* I, 180).

NY4700 Over Staveley.
From 1622 (*PNWestm* I, 176).
NY5001 Longsleddale.
From 1770 (JeffreysM).
HIGH HOUSE FARM SD4193 Cartmel Fell.
High House 1851 (OS).
HIGH HOUSES NY2137 Bewaldeth.
High House 1679, *High-Houses* 1708 (Torpenhow PR), *High House* 1774 (DHM).
♦ Self-explanatory; **high, house**. The Hugill example has a counterpart in LOW HOUSE, but the altitude difference is perceptual, since both are set close to the 200m. (646ft) contour; it is a 16th-cent. mansion (Danby 1994, 173; not in Pevsner). The H~ H~ in Longsleddale is marginally lower than LOW HOUSE, but is somewhat farther up the valley. The 1637 form from Crook suggests that at least in that instance *high* may be a noun rather than an adj. In Cartmel Fell, **farm** has been added to the name. The Bewaldeth place is at 820ft/250m.

HIGH HOUSE NY5304 Fawcett Forest, H~ H~ BANK NY5405 Shap Rural (1624ft/495m.), H~ H~ FELL NY5106.
High House 1770 (JeffreysM); *High House Bank*; *High House Fell* both 1860 (OSNB Shap).
♦ H~ H~ is in High BORROWDALE; **high, house, bank, fell**. The Bank is a wide, steep slope rising to a distinct point, while the Fell is an area of rough grazing nearly two miles from the farm.

HIGH HOUSE TARN NY2409 Borrowdale. Cf. *Heehouse* 1565, *Hiehouse* 1575 (Crosthwaite PR), *High House, Seathwaite* 1866 (Borrowdale PR).
♦ 'The pool near High House'; **high, house, tjǫrn/tarn**. The 'High House' may be the one mentioned in the PR, and may have been at the top of BORROWDALE, for John Braithwaite of *Hiehouse* married 'Janet Birkhead of Seatwhait' (nearby SEATHWAITE) in 1575. The tarn (the largest of Lincomb Tarns) is, however, extremely remote, and it is hard to imagine any but a temporary shelter in the near vicinity.

HIGH HURST SD2094 (hab.) Ulpha.
High Herst 1737 (PR, cf. *Low-Herst* 1711/12), *High Hirst* (and *Low Hirst*) 1774 (DHM); cf. *Hirst in Ulfay* 1660 (*PNCumb* 438).
♦ The 2nd el. is probably ME, ModE *hirst/hurst* from OE **hyrst**, normally applied to a small wooded hill. This name, and Hurstside a short way to the SW, suggest that the woodland may have been more extensive in the past than at present. As seen above, the distinction between **High** and **Low** Hurst is at least as old as the 18th cent.

HIGH HARTSOP DODD: see HARTSOP

HIGH INTACK: see INTACK

HIGH KNOTT NY2512 Borrowdale.
From 1867 (OS).
♦ 'The high rugged height'; **high, knott**.

HIGH KOP NY4515 (2179ft/665m.) Bampton, LOW KOP NY4716 (1875ft/572m.).
High Kop; *Low Kop* both 1859 (*PNWestm* II, 195).
♦ 'The higher and lower peaks', with the reflex of OE *copp* 'summit, top', **high, low**. These are points on BAMPTON COMMON.

HIGH LEYS SD4593 (hab.) Crook.
High Leays 1706 (*PNWestm* I, 180).
♦ 'The high clearings or pastures'; **high** plus reflex of **lēah**.

HIGH LOFT WOOD: see LOFT WOOD, HIGH

HIGH MAN SD3296 Coniston/Satterthwaite.
From 1851 (OS).
♦ 'The high peak', a summit on MONK CONISTON MOOR; **high, man**. Now mainly submerged in forestry plantation, it has no detectable cairns (a likely sense of *man*), only telecommunications masts, but it is a fine viewpoint.

HIGH MEREBECK: see MEREBECK, HIGH

HIGH MERE GREAVE: see MERE GREAVE, HIGH

HIGH MILL NY1525 Lorton.
From 1867 (OS); cf. *Overcornemylne* 1478 (Godwin 1988, 248).
HIGH MILL SD4194 Crook.
From 1823 (*PNWestm* I, 180).
♦ 'The upper mill'; **high, mill** from **mylen**. The Lorton site is on the WHIT BECK, a tributary of the COCKER. A mill is associated with Lorton in the earliest (mid 12th-cent.) documentary record, which may refer to High Mill (Winchester 1987, 146). It may also be identical with a fulling mill which was in poor repair in 1478, and a cornmill from 1728 until 1883 when it ceased working (Godwin 1988, Winchester 1987, 118, 146); *Lorton Low Mill* is N of it on the 1867 map. The Crook example is on the R. WINSTER, also close to a Low Mill (1863 OS).

HIGH MOSS NY2021 Above Derwent.
High Moss Hill, (*the*) *High Moss* 1800 (*CN* 778, 781, 783); cf. *the Mosse* 1570 (*PNCumb* 373).
♦ 'The high tract of bog' (**high, moss**), lying between SAIL and OUTERSIDE. (Coleridge may have meant one of these heights by H~ M~ before he became clearer about their names; see *CN* entries and Coburn's notes to them.)

HIGH NOOK FARM NY1220 Loweswater.
High Nook 1738 (PR).
♦ The name describes the situation, on a rising corner of land between two streams, and 'truly Arcadian' (Wainwright 1966, Gavel Fell 5); **high, nook**, with **farm** added.

HIGHPARK NY1420 (hab.) Loweswater, LOWPARK NY1420 (hab.).
the high parke 1578 (*PNCumb* 412); *Low Park* 1749 (PR).
♦ 'The upper and lower park'; **high, low, park**. The names, like nearby Park Beck and Park Bridge, refer to a former hunting park created by the de Multons, Lords of Egremont (*St Bees* 145). It appears to have been divided into separate holdings in the late medieval period (Winchester 1987, 51).

HIGH PIKE NY3135 (2189ft/658m.) Caldbeck, LOW P~ NY3235.
High Pike 1774 (DHM), 1803 (*CN* 1518); *High Pike, Low Pike* 1823 (GreenwoodM).
HIGH PIKE NY3708 (2155ft/656m.) Rydal/Ambleside, LOW P~ NY3707.
the Hey Pyke 16th cent., *the high pike of Rydall* 1614, *High Pike* 1655 (*PNWestm* I, 211); *the Lawe Pyke* 16th cent., *the low pike of Rydall* 1614 (*PNWestm* I, 211).
♦ 'The higher and lower summits'; **high, low, pike**. H~ P~ at NY3135 is high by Caldbeck standards, and a true *pike*, a steep, shapely oval. L~ P~ is its NE spur, not an independent peak. The Rydal/Ambleside H~ P~ is a prominent point on the old boundary ridge, to the N of another Low Pike, which in this case is a distinct peak, though considerably lower than its partner.

HIGH PIKEHOW NY1409 Nether Wasdale.
From 1867 (OS).
♦ 'The high, peaked hill'; **high, pike, haugr/how(e)**. As a small eminence on the S spur of HAYCOCK, this is not as impressive as its name would suggest.

HIGH RAISE NY2809 (2500ft/762m.) Borrowdale/St John's/Langdales.
the Hee-Rase c.1692 (Machell II, 153), *High Raise* 1867 (OS).
HIGH RAISE NY4413 (2634ft/802m.) Martindale/Bampton, LOW R~ NY4513 (2465ft/754m.) Bampton.
High Raise 1860 (*PNWestm* II, 219); *Low Raise* 1859 (*PNWestm* II, 195).
♦ 'The high cairn or tumulus'; **high, hreysi/raise**; also **low**. Both fells called H~ R~ are vantage-points on ancient boundaries (parish and, at NY2809, county), suggesting that 'cairn' rather than 'cairn-shaped mountain' is the meaning. This is further supported by the fact that the Martindale H~ R~ has a Bronze Age round cairn on the summit (LDHER 1604, SAM No. 22543), while there are remains of a prehistoric cairn of uncertain date close to the summit of Low R~ (LDHER 1512).

HIGH RIGG NY3021 St John's, LOW R~ NY3022.
Both from 1867 (OS).
♦ The two hills jointly form a spine down the W side of St John's in the Vale, with the church in the intervening dip; **hryggr/rigg, high, low**.

HIGH ROW NY3535 (hab.) Caldbeck.
From 1723 (*PNCumb* 279).
HIGH ROW NY3821 (hab.) Matterdale.
From 1712 (PR).
HIGH ROW FARM NY3125 Threlkeld.
the Hoighe Rawe 1574, *High Row* 1678 (*PNCumb* 253).

♦ 'The high hamlet'; **high, row**; also **farm**. The Caldbeck example is named in relation to *Low Row* (1660 PR). The Matterdale one is high, on the slope above AIRA BECK, and a *row* in the sense of a settlement, though not the linear hamlet often indicated by *row* elsewhere (see Clark 2001, 23, 25 (photo)). High Row, Threlkeld, is so called in relation to Middle Row and Low Row, the three together making a linear settlement, though not a continuous row, on the contours at the foot of BLENCATHRA. *Farm* has been added relatively recently.

HIGH SADDLE NY2912 (2215ft/675m.) Borrowdale, LOW SADDLE NY2813 (2152ft/656m.).
Both from 1867 (OS).
♦ These two points, with a dip in between, form the saddle-shaped top of COLDBARROW FELL; **saddle, high, low**.

HIGH SCARTH CRAG NY2104 Eskdale.
From 1867 (OS).
♦ S~ C~ is presumably 'the rocky height beside the notch or col' (**skarð, crag**), though it is not obvious which particular 'notch' is meant. **High** may qualify Scarth or Crag.

HIGH SEAT NY2818 (1996ft/618m.) Borrowdale/St John's.
High Seatt 1569 (*PNCumb* 353).
♦ Probably 'the lofty seat'; **high, sæti**. 'Seat' is either from ON **sæti** 'seat' or possibly from ON **sætr** 'shieling'. This is the highest point on a long spine-like ridge.

HIGH SIDE Examples at:
NY1626 (hab.) Lorton.
Highside 1570 (Winchester 1987, 148), 1597 (PR), *High Side* 1774 (DHM).
NY1628 (hab.) Embleton.

Higheside 1578 (*PNCumb* 384).
NY2330 (hab.) Bassenthwaite.
Highside 1581 (PR), 1774 (DHM), *the High Side* 1777 (*PNCumb* 264).
♦ 'The high slope'; **high, side**. The Lorton place is situated on the slopes E of the COCKER (see further GILLBREA for its history), and the Embleton one on a quite steep slope at about 650ft/ 200m. Although 'the high hill-slope', would describe the Bassenthwaite situation well, *side* may be a more general name-forming element here, used under influence of nearby MIRE SIDE (q.v.) and Moss Side or forming a pair with Low Side, since High and Low Side were the two constablewicks of Bassenthwaite parish (*PNCumb* 264).

HIGH SNAB: see SNAB

HIGH SPY NY2316 (2143ft/653m.) Above Derwent.
From 1867 (OS).
♦ Probably 'the high look-out'; **high, spy**. As well as associations with secret agents, the verb *spy* (originally from OFr *espie* n., *espier* vb.) can refer to keeping watch (*OED*: *spy*, vb, sense II, 6a), and this hill is well situated for that. Machell in the 1690s mentions a Spying How, one mile W of TROUTBECK church, and the possibility that a cairn there 'may have been a monument to some scouts or watchmen slain here' (II, 323; ed. Ewbank, p. 128). High Spy is 'also variously known as Eel Crags, Lobstone Band and Scawdel Fell' (Wainwright 1964, High Spy 1).

HIGH STILE NY1614 (2644ft/806m.) Buttermere.
From 1774 (DHM), *High-steel* 1780 (West 129).
♦ 'The high, steep place' (**high, steel/ stile**), an apt description of this grand peak on the ridge between Buttermere and Ennerdale.

HIGH STREET NY4208, NY4517 & NY4823 (track) several parishes, NY4411 (hill, 2718ft/828m.) Patterdale/ Shap Rural. (Track:) *Brethstrette, Brestrett, magnam viam que venit de* (great road which comes from) *Brethstrede* 1220–47, *the Streete* 1650, *High Street* 1793 (*PNWestm* I, 21; cf. I, 167); (hill:) *High Street* 1770 (JefferysM).
♦ 'The high paved road', from the reflexes of OE *hēah* 'high' (see **hegh**) and OE *strēt, strǣt* 'street', adopted from Lat *strata*. A Roman road went from the fort at Brocavum (Brougham) S across the fell-tops to GALAVA (AMBLESIDE) or a point between Ambleside and Watercrook (see Hindle 1998, 17–21); some believe it to have been 'a native track intermittently Romanised' (Richardson *et al.* 1990, 118, citing Hay and Hindle). The name recorded in the 13th cent. seems to have originated as 'street of the Britons', but to have been influenced by ON *breiðr* 'broad'. H~ S~ has also become the name of the mountain — the highest in Far Eastern Lakeland — whose level, grassy top carries the Roman Road.

HIGH SWEDEN BRIDGE: see SWEDEN

HIGH TONGUE SD2397 Dunnerdale.
High Tongue 1851 (OS); cf. *the tongue* 1681 (*CW* 1908, 353).
♦ 'The high, tongue-shaped ridge'; **high, tongue**. The undulating ridge which forms a tongue between the R. DUDDON and TARN BECK contains three heights named (N to S): TROUTAL TONGUE, High Tongue, and HOLLIN HOUSE TONGUE. H~ T~ is indeed the highest point.

HIGH TOVE NY2816 Borrowdale.
High Toue 1805 (*PNCumb* 353), *High Tooves* 1805 (*CW* 1981, 66).
♦ 'Tove' may be the dial. word meaning 'tuft' (*PNCumb*), which might be a suitable description of this damp, heathery, featureless height; also **high**.

HIGH WETHER HOWE: see WETHER HOWE, HIGH

HIGH WHITE STONES NY2709 Borrowdale.
From 1867 (OS).
♦ This is the summit of HIGH RAISE, strewn with light grey stones; **high, white, stone**. Low W~ S~ is just to the NE.

HIGH WOOD NY1204 Nether Wasdale.
From 1867 (OS).
HIGH WOOD SD2989 Colton.
From 1851 (OS).
♦ 'The upper wood'; **high, wood**. The Wasdale example is named in distinction from Low Wood at NY1404 (also on 1867 OS map). The Colton one rises E of the upper CRAKE valley.

HIGH WRAY: see WRAY

HILL NY4023 (hab.) Matterdale.
From 1588 (*PNCumb* 257).

HILL, THE NY1424 (hab.) Loweswater.
Hill 1723 (PR).
HILL, THE SD2389 (hab.) Broughton West.
Hill 1768 (Broughton-in-Furness PR).
HILL FARM NY1226 Blindbothel.
Hill 1774 (DHM), possibly = *the Hill* 1597 (Lorton PR).
HILL FARM/HILL SD4498 Hugill.
ye hill in Hugill 1623 (*PNWestm* I, 171).
♦ Though it might have been thought too bland to be useful, **hill** from OE *hyll* is sufficiently rare as a p.n. element in the Lake District to be fairly distinctive. The Matterdale name refers either to LITTLE MELL FELL to the E, or to its own situation, slightly raised above flat, boggy land. The Loweswater farm is on the slope W of LORTON VALE, while the Broughton place is on a slope W of KIRKBY POOL. In Blindbothel and Hugill, **farm** has been added to a pre-existing name. The farms stand, respectively, on high ground between Little Sandy Beck and Sandy Beck, and on a small knoll.

HILL END SD1483 (hab.) Whicham.
the Hillend 1654 (*PNCumb* 445).
♦ This is below the S spur of BLACK COMBE; **hill, end**.

HILL FARM: see HILL

HILL FELL SD3299 Coniston.
From 1850 (OS).
♦ A rather strange juxtaposition of **hill** and **fell**: perhaps 'the mountain called Hill', or *fell* in the sense of grazing land, or else this is a corruption of Ill Fell (cf. ILL BELL, MARDALE ILL BELL), though the terrain is probably not forbidding enough for this to be likely.

HILL PARK FARM SD3087 Colton.
Hell parke c.1535, *Helparke* 1537, *Hellpark* 1539 (*PNLancs* 218), *Helpke* 1627/8, *Hilpke* 1637, *Hellparke* 1677, *Hillparke* 1689 (PR), *Hill Park* 1851 (OS).
♦ This is a more puzzling name than would first appear, but perhaps 'estate on the shelving hill-side'. The 1st el., if **hill**, would suit the high position, but the 16th-cent. spellings in *Hel(l)*- tell against that and might instead suggest ON *hella* 'stone, flat, rocky height', or rather, since the place is not notable for stones, ON *hjalli* 'shelf (on hill-side)' (mentioned by Ekwall in

PNLancs). Ekwall's other suggestion, ON *hellir* 'cave', can be excluded since there are none in the vicinity. The el. **park** (q.v.) reflects its history as an outlying farm of FURNESS Abbey. *Farm* is a relatively recent addition by officials and cartographers, Hill Park being the name on the farm sign and used by the present owners (to whom I am most grateful for the local information above).

HILL TOP Examples at:
　SD3183 (hab.) Colton.
Hilltop 1722 (PR).
　SD3486 (hab.) Colton.
From 1851 (OS).
　SD3695 (hab.) Claife.
From 1851 (OS).
　SD4083 (hab.) Upper Allithwaite.
Hill Top House 1851 (OS).
HILL TOP FARM NY3122 St John's.
Hill Toppe 1595 (*PNCumb* 317).
♦ Self-explanatory; **hill, top**. H~ T~ at FAR SAWREY, Claife, is renowned as one of the farms once owned by Beatrix Potter. H~ T~, Upper Allithwaite, is a hamlet above HIGH NEWTON, now absorbed into the main village. **Farm** has been added to the St John's name.

HIND CRAG NY2311 Borrowdale.
From 1867 (OS).
♦ Presumably 'the rocky height frequented by (female) red deer', with the reflex of OE/ON *hind* plus **crag**. In 1251 Henry III ordered two hundred hinds to be sent from Cumberland as part of his requisitioning for Christmas (*Scotland* I, 341–2).

HINDSCARTH NY2116 (2385ft/727m.) Above Derwent.
From 1774 (DHM).
♦ An attractive but enigmatic name, which would appear to mean 'the pass of the hind or female deer', from OE/ON *hind* and ON **skarð**, though there is no obvious gap in the landscape.

HINNING HOUSE Examples at:
　SD1188 Bootle.
(*del*) *Heyning in Bothill* 1357 (p), *the he . . . ynghous* 1462, *Heninghouse* c.1500 (*PNCumb* 346).
　SD1297 Muncaster.
Heninghouse 1500 (CW 1926, 113), *Hinning House* 1769 (*PNCumb* 426), *Hening House* 1774 (DHM).
　SD2499 Dunnerdale, H~ H~ CLOSE SD2399, H~ H~ FELL NY2500 (2536ft/ 773m.).
Hinninhouse 1737/8 (Seathwaite PR), *Henning House* 1774 (DHM), *Hinning House* 1850 (OS); *H~ H~ Close*; *H~ H~ Fell* both 1850 (OS).
♦ 'House by H~, the enclosure or enclosed land', with 1st el. from OE *hegning*, ME *haining*, also preserved in HEANING, plus **house**; also **close** 'enclosure, farm', **fell** 'high unenclosed ground'. In Dunnerdale, H~ H~ Close seems to be a secondary settlement, and H~ H~ Fell is a rocky area N of the summit of GREY FRIAR.

HIRD WOOD NY4106 Troutbeck.
From 1859 (*PNWestm* I, 190).
♦ Probably 'the **wood** associated with the Hird family', since the surname is recorded locally from the late 14th cent. onwards; cf. also HOLEHIRD, and Hird House at NY4105 (recorded as *Herds* in the mid 18th cent., CW 1984, 168, 178).

HOATHWAITE FARM SD2994 Torver.
Holtwayt 1272–80 (*PNLancs* 215), *Howthait* 1612 (PR), *Hoathwaite* 1851 (OS).
♦ 'The clearing in the hollow'; **hol(r)**, **þveit**, with **farm** added in recent times.

HOBCARTON NY1923 Lorton, H~ CRAG NY1922, H~ END NY1923 (2010ft/613m.), H~ GILL NY1823.
Hopecartan 1260, *Hobbecartan* 1290 (*PNCumb* 408–9); *Hobcarton Crag; H~ End; H~ Gill* all 1867 (OS).
♦ Possibly 'Cartan's tussocky hill'. An intriguing name, which seems to be an inversion compound. The 1st el. was seemingly already causing confusion in the 13th cent., producing a spelling alternation similar to that in HOPE BECK, and it is difficult to determine which form is the more original. The most promising explanation is (a) a ME **hobb(e)*, assumed from Swed and English dial. words relating to tussocks of thick grass (Smith, *Elements*). This would supply the expected el. referring to a hill. (b) That the 1st el. is, instead, OE **hop** 'valley, secluded place' could be suggested by the proximity and similar spellings of HOPE BECK, and would be supported by the striking parallel of Hopecarton (habitation and valley), Peeblesshire (NT1430, *Hopecarthan* c.1240; *PNCumb* III, xx), which, in an area abounding in 'Hope' names referring to enclosed valleys, must originally have referred to the inner end of the valley (I am grateful to Doreen & Willie Waugh for thoughts on this). However, the Cumbrian Hobcarton is not close to a valley to which *hop* could refer, and it is more likely that an unusual word would be re-interpreted as a common word than the reverse, making explanation (a) somewhat more likely. The 2nd, specific, el. may be the Gaelic pers.n. *Cartán* or its ON equivalent *Kjartan*. See also **crag** 'rocky height', **end,** and **gil(l)** 'ravine with stream'.

HOBGRUMBLE GILL NY4911 Shap Rural.
Hobgrumgill 1828 (HodgsonM), *Hobgrumble Gill* 1859 (*PNWestm* II, 175).
♦ The **gill** 'ravine with stream' is a spectacular rocky groove with cascades, descending several hundred feet to the head of SWINDALE. *Hob* in the sense of 'sprite', 'hobgoblin' would therefore seem the most fitting of the numerous possible interpretations of 'hob' (see *EDD*), but it remains elusive, and 'grum(ble)' is still more so.

HOBKIN GROUND SD2290 Broughton West.
Hobkinsgrd 1706, *Hobkinground* 1716/7 (Broughton-in-Furness PR).
♦ 'The farm of the Hobkin family'; surname plus **ground**. The surname *Hobkin* or *Hopkin* occurs in the Broughton PR, but not in association with H~ G~. Meanwhile, a family named Pritt is recorded at H~ G~ in 1725/6, 1764 and the 1780s, but at *Hopkins* or *Hobkins* in 1723, 1728/9 and 1731. It is not clear whether *Hobkins* and *Hobkin Ground* refer to the same place. Unlike many *grounds* which are clearly on secondary sites reclaimed from upland waste, this is close to the R. LICKLE.

HODGE CLOSE NY3101 Skelwith.
Hodgclose 1623 (Hawkshead PR), *Hodge Close* 1850 (OS).
♦ Presumably 'the farm of the Hodge family', one of three or four instances of a surname qualifying **close** in this volume. The surname *Hodge*, originally a diminutive of *Roger*, is not indexed in Cowper's 1897 edition of the Hawkshead PR, though *Hodgson* (and variants) is very common.

HODGE HILL SD4188 (hab.) Cartmel Fell.
Hodghill 1756 (CW 1966, 226).
♦ Presumably the surname *Hodge*, plus **hill**, though a rapid survey has

not revealed the surname in the Cartmel PRs (where there are several Hodgsons), and this was the residence of the Philipsons in the 17th cent. (*CW* 1895, 300). H~ Hill slopes up W of the WINSTER.

HOGS EARTH NY2618 Borrowdale.
From 1867 (OS).
♦ Uncertain. 'Hog(g)' could be (a) the surname (from ME *hog(g)* 'pig', hence a swineherd); or (b) the dial. word for a yearling sheep. 'Earth' could go back to (a) OE *erð* 'ploughed land' or (b) OE *eorðe* 'ground, soil, animal's lair' which is rarer in p.ns (Smith, *Elements*), but is used to refer to animals' lairs from the 16th cent. onwards (Ekwall, *DEPN*: Foxearth).

HOLBECK: H~ GHYLL NY3902 (hab.) Troutbeck.
holbeck (bridge) 1675 (*CW* 1906, 31), *Hollow Beck* 1787 (*PNWestm* I, 8); *Holbeck-gill* c.1692 (Machell II, 151, ed. Ewbank, 131).
♦ H~ is 'the stream in the hollow' (**hol(r)**, **bekkr/beck**), with *ghyll*, ultimately from ON **gil** 'ravine with stream', added to form the name of the settlement.

HOLE BECK SD2591 (hab.) Broughton West.
Holebeck 1669 (original spelling not specified), *Holbeck* c.1701 (Broughton-in-Furness PR).
♦ 'The stream in a hollow or valley' (**hol(r)/hole**, **bekkr/beck**), and hence the farm beside it.

HOLEGILL SD1186 (hab.) Whicham.
Holgill 1571 (*PNCumb* 449).
♦ 'The ravine/stream in a hollow'; **hol(r)**, **gil(l)**, and cf. HOLE BECK. Holegill Beck forms quite a deep cleft here.

HOLEHIRD NY4100 (hab.) Windermere.
wholl hird 1613, *Holehird* 1624 (PR).
♦ 'The hollow place associated with the Hird family', with **hole** from **hol(r)** and the surname *Hird/Hyrde*, which is recorded from 1390–4 onwards (*PNWestm* I, 196) and is also preserved in HIRD WOOD. Holehird is an instance of a later ME name following the pattern of early inversion compounds, with the generic first and the owner's name second.

HOLE HOUSE SD1893 Ulpha, HOLEHOUSE GILL SD1793.
Whole House 1689 (*PNCumb* 438), *Holehouse* 1749 (Seathwaite PR), *Hole House* 1774 (DHM); *Holehouse Gill* 1867 (OS).
HOLE HOUSE NY2431 Bassenthwaite.
Hole House 1867 (OS), possibly = *Horlose* 1669, *Hourlose* 1670 (PR).
HOLE HOUSE SD3699 Claife.
From 1850 (OS).
HOLE HOUSE FARM NY4725 Barton.
Holehouse 1741 (*PNWestm* II, 212).
♦ H~ H~ is simply 'the house in the hollow'; **hole**, **house**; also **gil(l)** 'ravine with stream', **farm**. All the examples are in low-lying or secluded situations, usually beside water. The 1689 spelling from Ulpha is misleading. In Barton, *farm* has been added to a pre-existing name.

HOLLENS FARM NY3407 Grasmere.
Hollins 1644 (*CW* 1928, 285), *Hollins* 1685 (PR).
♦ H~ is 'the hollies, the place where holly grows'; **hollin** (cf. HOLLINS), with **farm** added later. Machell, describing GRASMERE, reported a 'great store of large holly [trees] all over' (c.1692, ed. Ewbank, p. 133).

HOLLIN BANK SD3298 (hab.) Coniston.
Holingbank Conistone 1800 (Hawkshead

PR), *Hollin Bank* 1851 (OS), probably = *Hollinbanke* 1603.
♦ 'The slope where holly grows'; **hollin, bank**.

HOLLIN CRAG NY2903 Langdales.
From 1862 (OS).
♦ 'The rocky height where holly grows'; **hollin, crag**.

HOLLIN HALL SD4696 Crook.
Hollinghall(e) 1630 (*PNWestm* I, 180), *Holin Hough* 1730 (*CW* 1999, 227), *Hollin Howe* 1865 (*PNWestm* I, 180), probably = *Hollinhall* 1564–5 (*CW* 1909, 41), *Hollin How . . . a Hall of that name* c.1692 (Machell II, 114, ed. Ewbank, 109).
♦ 'The hall where holly grows'; **hollin, hall**. There has been confusion between *hall* and **how(e)** from **haugr** here, as elsewhere; in this case, *hall* seems to be the earlier.

HOLLIN HOUSE: H~ H~ TONGUE SD2296 Dunnerdale.
Holling house 1681 (*CW* 1908, 353); *Tongue* 1765, *Hollinhouse Tongue* c.1781 (spellings not specified, *CW* 1961, 238–9), *Holling House Tongue* 1851 (OS).
♦ Simply 'the tongue of land at Hollin House, the house where holly grows'; **hollin, house, tongue**, cf. HIGH TONGUE. It seems to be called simply *Tongue* in a local deed of 1765.

HOLLIN ROOT NY3023 (hab.) St John's.
Hollinge Roote 1573 (*PNCumb* 317).
HOLLIN ROOT NY4903 (hab.) Longsleddale.
Hollin-Root 1766 (PR).
♦ 'Holly root'; **hollin**, plus *root*, which derives from ON *rót* 'root' or occasionally 'stump'. It is rare in p.ns, but there are at least two other occurrences of Hollin(g) Root in Cumbria: see *PNCumb* 334, *PNWestm* I, 147.

HOLLINS (hab.) Examples at:
    NY1003 Irton.
Hollings 1741 (*PNCumb* 403).
    NY1522 Buttermere.
the Hollings in Brackenthwaitt 1599 (Lorton PR), *Hollins* 1774 (DHM).
    NY1701 Eskdale.
Hollings 1587 (in 1660 copy, *CW* 1922, 75), *Holins* 1665, *Hollins* 1674 (PR).
    SD4189 Cartmel Fell.
the hollings 1701 (*CW* 1966, 240).
♦ 'The hollies, the place where holly grows'; **hollin**, cf. HOLLENS. In the Buttermere instance, Low and High Hollins are distinguished on larger scale maps. For other 'Hollin(g)' names in Eskdale parish, see *PNCumb* 391.

HOLLOWGATE NY5403 (hab.) Fawcett Forest.
Hollow Gate 1770 (JefferysM).
♦ 'The sunken track', with reflexes of **hol(h), gata**, named from what is thought to be an ancient hollow way (*PNWestm* I, 139).

HOLLOW MOOR NY1006 Gosforth.
From 1867 (OS).
♦ The site is on a relatively gradual slope, but not in a 'hollow' as such; **hollow, moor**.

HOLLOW OAK SD3484 (hab.) Haverthwaite.
Hollowoake 1676 (Colton PR).
♦ Self-explanatory; **hollow, oak**.

HOLLOWS NY3922 (hab.) Matterdale.
the Hollesse 1580, *Holleys* 1589, *Holles* 1643, *Holehouse* 1656, *ye Holless* 1668, *the Hollas* 1722; cf. *Lowholhouse* 1698 (*PNCumb* 223).
HOLLOWS FARM NY2417 Borrowdale.
Hollas 1565 to 1610, *Howlas* 1571 (*PNCumb* 353), *Hollows* 1774 (DHM).
HOLLOWS, HIGH NY3524 (hab.) Threlkeld.

From 1867 (OS); cf. *Hollows* 1774 (DHM), *Hollas* 1775 (PR).
♦ Either 'the hollows' or 'the house by the hollow', from **hollow** plus either an *-s* pl. or **house**; also **high**. At both Matterdale and Borrowdale, the situation is a single, sheltered hollow. The common affix *high* in High Hollows distinguishes the place from Hollows rather than producing a self-contradictory description.

HOLLOW STONES SD1399 (hab.) Muncaster.
From 1739 (PR), 1774 (DHM).
♦ This is a row of low outcrops; **hollow, stone**.

HOLLOW STONES NY2007 Eskdale.
Hollow Stones 1802 (CN 1218).
♦ This spot on the flank of SCAFELL PIKE was to Coleridge 'the frightfullest Cove, with huge Precipice Walls'; to Wainwright 'an excellent place for a bivouac' (1966, Scafell Pike 11); **hollow, stone**.

HOLME SD4183 Upper Allithwaite.
the Home 1605/6, *Holme* 1606 (PR).
♦ 'The land flanked by water' — a classic **holm(e)**, being a rise above a bend in the R. WINSTER.

HOLM/HOLME FELL NY3100 (994ft/303m.) Coniston, HOLME GROUND NY3101 Skelwith.
Holme Fell 1850 (OS); *Holme ground* 1657 (Hawkshead PR), *Holme Ground* 1689 (Coniston PR), *Holm Gd* 1745 (West 1774, map).
♦ Holme Ground is 'the farm of the Holme family'; surname plus **ground**. The burial of 'Cuthbert Holme de Holme ground' is recorded in 1657, and the surname *Holm(e)* (from **holmr/holm(e)**) is extremely common in Hawkshead PRs. H~ **Fell** may be named from the family or possibly from a place called Holme.

HOLMEGATE SD0986 (hab.) Whicham.
Holmgate 1867 (OS).
♦ Probably 'the gate onto Holme, the land beside water'; **holm(e), geat/gate**. *Holm(e)* is apt for this site between the R. ANNAS and its tributaries, and the supposition that 'gate' is from OE *geat* 'gate' rather than ON **gata** 'track, road' is supported by another Cumbrian Holmegate, recorded as *Holmeyeat* in 1641 (*PNCumb* 158).

HOLMESHEAD FARM NY3402 Skelwith.
houmes head 1665/6 (Hawkeshead PR).
♦ H~ is 'the upper end of Holme, the place surrounded or flanked by water' (**holm(e), head**), with **farm**, as so often, added later. The site is on high ground in an area of many springs, miniature tarns and becks.

HOLME WELL SD3892 (hab.) Satterthwaite.
From 1795 (PR), 1851 (OS).
♦ 'The spring by the islet'; **holme, well**. There is a spring hereabouts, and *holme* seems likely to refer to nearby Grass Holme in WINDERMERE.

HOLME WOOD NY1221 Loweswater.
From 1867 (OS), perhaps cf. *Common Moor or Waste called the Holme* 1597 (in 18th-cent. copy, CRO D/Law/1/258).
♦ 'The wood at Holme, or on the land beside water'; **holm(e), wood**. The *holm(e)* in question is a discrete tract of land bounded by small streams, fells to the SW and Loweswater to the NE.

HONISTER CRAG NY2114 (2126ft/648m.) Buttermere, H~ PASS NY2114.
'The whole mountain is called *Unnesterre*, or, as I suppose, *Finisterre* [edge of the land], for such it appears to be' 1751 (*Gentleman's Magazine*,

cited Hutchinson 1794, II, 219), *Honistar Crag* 1774 (DHM); *Honister Crag Pass* 1865 (Prior 177).
♦ H~ is possibly 'Húni's farmstead', though this name is a good example of the frustration arising from the lack of early evidence. The 1st el. may be a pers.n. such as *Húni* or *Húnn* (and *Húni* appears in Domesday Book, Björkman 1910, 70), though their gen. would have been *Húna*, *Húns* respectively. (a) In the 2nd el. the *-ster*, *sterre* forms could reflect ON *staðir*, a pl. noun with the sing. meaning 'place, farmstead', and indeed *Húni* and *Húnn* are compounded with *staðir* in Norw p.ns (Lind 1905–15, 599–600). Though rare in England and seemingly unparalleled in the Lake District, *staðir* is common in Norway, Iceland, the N. and W. Isles of Scotland and the Isle of Man. The sites denoted by *staðir* names elsewhere vary, but the Norw and Icel names are associated with secondary settlements rather than prime sites (cf. Fellows-Jensen 1987, 44), and if Honister originally referred to the present Honister Hause, at over 1000ft/300m. and occupied by a slate quarry and Youth Hostel, it would match them, or indeed outdo them in bleakness. (b) *Sætr* 'shieling' as suggested by Brearley (1974) is unpromising since there is no other example of the *-r* being retained except when the el. is a specific (e.g. SATTERTHWAITE). See also **crag** and **pass**.

HOOKER CRAG SD1198 (858ft/231m.) Muncaster.
Hooker Crag 1900 (OS); cf. *Hooker* 1867 (OS).
♦ The **crag** is a small rocky summit on MUNCASTER FELL, next to Hooker Moss. Hooker may be the surname, of various origins but usually based on the word *hook* 'hook, bend'.

HOPE BECK NY1623 (stream) Buttermere, HOPEBECK NY1624 (hab.), HOPE GILL NY1723, HOPEGILL HEAD NY1822.
(Stream:) *Hobbecke* 1578 (*PNCumb* 17); (hab.:) *Hobbeck* 1607, *Hopebecke* 1637/8; *Hope Gill*; *Hopegill Head* both1867 (OS); cf. *the Hope* 1618, *Hope* 1623 (Lorton PR).
♦ 'Hope' may refer to The Hope, a secluded declivity and settlement at NY1623, presumably from OE/ME **hop(e)** 'rounded valley, remote place'. The *hob-* spellings make this explanation somewhat uncertain, though they may simply be influenced by nearby HOBCARTON (q.v.). Hope **Beck** flows through Hope Gill (as shown on the 1:25,000 map), which therefore is a clear case of **gil(l)** meaning 'ravine' rather than 'stream'. Hopegill Head is not merely the top of Hope Gill, but also a noble peak, known locally as Hobcarton Pike (Wainwright 1964, Hopegill Head 2).

HORROCKWOOD NY4422 (hab.) Matterdale.
Hurrock Wodd [sic] 1598, *Horrock Wood* 1793 (*PNCumb* 257).
♦ Probably from N. dial. *hurrock* 'heap of stones' (so *PNCumb*), plus **wood**. The place is not far from ULLSWATER and quite shingly (Keith Clark, pers.comm.).

HORSE CRAGS/CRAG NY2803 Langdales.
Horse Crags 1850 (OS).
♦ If 'Horse' is to be taken at face value, it is from OE *hors*, possibly encouraged by ON *hross* 'horse, mare', which appears in two of the ROSTHWAITES. 'Great' and 'Little' Horse **Crag**s are distinguished on larger scale maps. Perhaps coincidentally, these are close to WRYNOSE, which seems to contain another 'horse' allusion. Cf. also CAPPEL CRAG, KEPPLE CRAG.

HORSEMOOR HILLS NY2533 (hab.) Ireby and Uldale.
Horsemore-hills 1670 (Uldale PR), *Horsemoor-Hills* 1794 (Hutchinson II, 374).
♦ Without earlier forms, there is nothing to contradict the impression that this is 'the sloping moor where horses graze', with *horse* as in HORSE CRAGS, **moor**, **hill**.

HOSES SD2192 (hab.) Dunnerdale.
From 1850 (OS), possibly = *Hosehouse* 1783 (Seathwaite PR).
♦ If the 1783 spelling belongs here, this may be 'the house by the neck of land' (**hals/hause**, **house**). *Hose* is a familiar spelling of *hause*, e.g. in NEWLANDS HAUSE, HAWSE END, and possibly HASSNESS, while *house* is reduced to the indeterminate vowel schwa, [ə], plus -*s*, as, for example, in WHITTAS.

HOSPICE SD3979 Upper Allithwaite.
Hospice of Hampsfell 1851 (OS).
♦ This is an ornamental 'small tower or hospice for the accommodation of visitors' (*VCHLancs* 278), at a viewpoint on Hampsfield Fell. It was built in 1846 by Thomas Remington, vicar of Cartmel (Pevsner 1969, 78).

HOSPITAL PLANTATION NY2123 Above Derwent.
From 1867 (OS).
♦ 'Why *Hospital* Plantation?', asks Wainwright, answering, 'because the solitary dwelling on the Whinlatter Pass road, now named Lakeland View [at NY2124], was once a Fever Hospital' (1964, Grisedale Pike 6). The word *hospital* entered ME from OFr, initially in the sense 'hospice, hostel'. See also **plantation**.

HOUKLER HALL SD2888 Blawith.
Hoglerhowe 1609, *Houghler Hall* 1637 (*PNLancs* 215), *Owler Hall* 1851 (OS), *Houkler Hall* 1893 (OS).
♦ Partly obscure. 'Houkler' may contain **ærgi/erg** 'shieling' (as Ekwall remarks, *PNLancs* 215), but the 1st syllable is obscure. The final el. was evidently **how(e)** from ON **haugr** 'hill' but, as often in Furness, it has given way to **hall**. The hill has consequentially become Houlker [*sic*] Hall Bank (on the modern 1:25,000 map, *Owlerhall Bank* 1851 OS).

HOW: see HOWE

HOWBANK SD1196 (hab.) Muncaster.
From 1763 (PR).
♦ Presumably 'the slope on the hill' (**haugr/how(e)**, **bank**). The place is on the slope above the R. ESK.

HOWBURN NY2836 (hab.) Caldbeck.
Holborn 1867 (OS).
♦ Probably 'the hollow stream, stream in a deep channel', with elements derived from OE **holh** 'hollow' and OE *burna* 'stream', cf. various Holbecks, Holbrooks and Holborns, including the London Holborn. The name is spelt Holburn on the 1:25,000 map (1998 printing).

HOW CRAGS, GREAT and LITTLE SD2799 Coniston.
Both from 1850 (OS).
♦ (a) Seemingly 'the rock outcrops on How, the hill'; **haugr/how(e)**, **crag**; also **great**, **little**. *How* would refer here, unusually, to a long escarpment. (b) However, Collingwood noted that 'the name is locally pronounced Hookrigg ... The two Hookriggs are arêtes with cavities in shattered rocks,

and the O.E. *holc*, used about A.D. 1000, means a hollow' (1920, 243), hence these could conceivably be the larger and smaller 'ridge with cavities'. (c) If an OE origin for Hookrigg is sought, another possibility would be OE *hōc* 'hook, bend, angle', which forms a compound with OE *hrycg* 'ridge' in Hawkridge, Kent (Smith, *Elements*), and the crags do form a crescent shape.

HOW NY2424 (hab.) Above Derwent.
'Probably (*del*) *hou, la howe* 1292 (p)' (*PNCumb* 370), *How-hill* 1769 (Gray 1094).

HOWE NY4919 (hab.) Bampton.
How 1865 (*PNWestm* II, 195); cf. *Low How, High How* 1770 (JefferysM, unless these = The Howes, NY4917, mislocated).

HOWE, THE NY4102 (hab.) Windermere.
How 1770 (JefferysM), possibly = *le Howis, le Howys* 1275–85 to 1401 (*PNWestm* I, 195).

HOWE, THE SD4395 (hab.) Crook.
Howe 1836 (*PNWestm* I, 180).

HOWE, THE SD4588 (hab.) Crosthwaite.
How(e) 1535 (in 1669 copy, *PNWestm* I, 84).

HOWE FARM SD3597 Hawkshead.
Howe in haukeshead field 1658 (PR), *How* 1851 (OS).

HOWE FARM SD4193 Crook.
How 1770 (JefferysM).

HOWE: NEAR H~ NY3728 (hab.) Mungrisdale, FAR H~ (hab.) NY3728.
Cf. *Grysdell How* 1578 (*PNCumb* 227), *Howe* 1813 (PR) possibly = *Grisdale-Howe* 1823 (PR); *Near how* 1900 (OS; = *Underhow* 1867 presumably in error); *Farhow* 1867 (OS).

♦ All these are settlements named from their proximity to hills or mounds (ON **haugr**, later **how(e)**). Most are of modest size; the Above Derwent example is a 66ft/20m. rise above the flat marsh — the smallest *how(e)* on the map. Most are also discrete heights, but Howe in Bampton is on the E slopes of THE PEN, and it is typical of this area of long, sprawling contours that the elements *how(e)* and *hill* are often used of tracts of hill-side rather than self-contained hills with distinct summits. **Farm** is a recent addition in the Hawkshead and Crook Howe Farms. In Mungrisdale, **Far** H~ is situated at How Top (shown on larger scale maps), Near H~ somewhat lower down and closer to MUNGRISDALE itself, and this seems to be the force of *Near*. Cf. also UNDER HOWE.

HOWE GRAIN NY4318 Martindale.
How grane c.1692 (Machell I, 700), *Howe Grain* (*PNWestm* II, 219).

♦ 'The confluence by the hill'; **haugr/ how(e), grain**. The *grain* is the meeting of BANNERDALE BECK and RAMPSGILL BECK, but it is unclear which nearby hill is meant — cf. HOWTOWN.

HOWES NY4910 (1930ft/588m.) Shap Rural.
From 1865 (*PNWestm* II, 175).

HOWES, THE NY5017 Bampton, H~ BECK NY5017.
Howes 1859; *Howes Beck* 1865 (*PNWestm* II, 195).

♦ These are names of upland features, from **how(e)**, the reflex of ON **haugr** 'hill, mound'. Despite the pl. form, neither place has distinct mounds comparable to those at TARN HOWS. The Shap example is a rocky height whose name is more likely to contain the word for 'hill' than the local surname *Howe* as proposed in *PNWestm* II, 175. Cf. also HOW(E) and the habitation name HOWS. See also **beck** 'stream'.

HOWGATE FOOT NY5119 (hab.) Bampton.
From 1704 (*PNWestm* II, 195), possibly
= *Hogarthfoot* 1829 (Parson & White).
♦ 'The lower end of Howgate'; p.n.
plus **foot**. H~ is probably 'the road by
the hill' (**haugr/how(e)**, **gata**), though
hol(r) 'hollow' is also possible as
specific.

HOWGILL: H~ WOOD NY1933 Blindcrake.
Howgill 1689 (a tenement, CRO
D/Law/1/209, 3); *Howgill Wood* 1810
(*PNCumb* 301).
HOWGILL TONGUE NY2726 Under-skiddaw.
From 1867 (OS).
♦ A **gil(l)** is a ravine with a stream,
but 'How' is slightly elusive. (a) It may
be from ON **hol(r)** 'hollow' as in HOLE
GILL or HOWBURN, referring to deep
ravines; this is supported by other
Cumbrian How Gills which have
medieval spellings in *hol-* (*PNCumb*
247, 336, 453). (b) Alternatively, it may
be the reflex of ON **haugr** 'hill,
mound'. A How Gill is shown on the
1:25,000 map in Underskiddaw but
not Blindcrake. See also **wood** and
tunga.

HOW HALL FARM NY0916 Ennerdale.
Carswelhowe in Ennerdale 1523,
Castlehowe 1578, *Caswellhow* 1623
(*PNCumb* 385-6), *the How* 1681/2 (PR),
How- or *Castwell How* 1687-8 (Denton
98), *How Hall* 1774 (DHM).
♦ The present name comprises
how(e) from **haugr** 'hill' plus **hall** (one
of the more modest examples), with
farm added later. However, the hill
appears previously to have been
named from the 'spring where cress
grows' (reflex of OE *cærse* 'cress' plus
well). The farm lies below a distinct,
rounded hill, and there are springs in
the vicinity.

HOW HEAD SD3197 (hab.) Coniston.
How-head Conistone 1798 (Hawkshead
PR).
♦ Presumably 'the top of the hill'
(**haugr/how(e)**, **head**), though the site
is only part way up MONK CONISTON
MOOR, whose broad slopes do not
form a typical *haugr/how(e)*.

HOWK NY3139 Caldbeck.
the Howk 1687-8 (Denton 177), *The
Hawk* 1777 (*PNCumb* 279), *the Howk*
1800 (*CN* 828), *the Hough or fairy Breaks*
1803 (*CN* 1426); cf. *Howkbanke* 1652
(*PNCumb* 279).
♦ This is a great limestone cave, with
waterfalls. The name must be from the
dial. verb *howk*, which Hutchinson
explained as 'the common term in the
north for *scooping out earth*, or any
thing else, and digging *an hole*' (1794,
II, 388-9). Hutchinson also vouches for
the cavern's other name, *The Fairy Kirk*,
which matches Denton's marginal
comment, 'Here have fairies been seen
formerly & of late' (1687-8, 177).

HOWS NY1701 (hab.) Eskdale.
le Howes 1570 (*PNCumb* 391), *Hows*
1790 (PR).
♦ 'The hills', ultimately from ON
haugr. This is a settlement name of the
same origin as HOWES, HOW(E).

HOWTOWN NY4419 Martindale.
How(e)town(e) 1676 to 1770 (*PNWestm*
II, 219).
♦ 'The village by the hill'; **haugr/
how(e)**, **tūn/town**. The *how* referred to
is probably HALLIN FELL, which stands
proudly beside ULLSWATER, the main
alternative being BEDA FELL.

HUBBERSTY HEAD SD4291 (hab.)
Crosthwaite.
Ub(b)erstede 1283, 1411, *Ubbersteade*
16th cent., *Hubberstye head* & variants

1535 to 1823 (*PNWestm* I, 81).

♦ This hamlet name is slightly puzzling. The 1st el. is thought to be an OE pers.n. such as the male name *Hūnbeorht* or the female *Hūnburg* (*PNWestm*), and 'Head' is presumably 'high place', ultimately from OE **hēafod** 'head'. The el. now appearing as *-sty-* has at different times been the reflex of OE **stede** 'place' or of OE **stīg**/ON **stígr** 'steep path' (Smith believed **stīg** to be original, *PNWestm*).

HUDSCALES NY3337 Caldbeck.
Hudscaille 1560 (*PNCumb* 279), *Hudscales* 1774 (DHM), possibly = *hotonscal* 1285, *Hoton Scales* 1292 (p) (*PNCumb* 279).
♦ Probably 'the shielings outlying Hutton'. The 2nd el. suggests that the hamlet developed out of summer pastures with shieling huts (ON **skáli**, dial. **scale**), and if the 13th-cent. spellings refer to this place, they must have operated from a place called Hutton — either H~ SCEUGH or H~ ROOF. If not, the 1st el. could be a pers.n. such as *Hud(d)*, a pet-form of *Hugh* (cf. ROGERSCALES).

HUDSON PLACE NY1122 Loweswater.
Place 1774 (DHM), *Places* 1809 (CrosthwaiteM), *Hudson Place* 1867 (OS).
♦ 'The Hudsons' farm'; surname plus **place**. Standing on a rise above LOWESWATER, this forms a cluster with IREDALE PLACE and Jenkinson Place (NY1122 on larger scale maps). The surname *Hudson* occurs frequently in the Loweswater PRs, and 18th-cent. gravestones commemorating Hudsons in Loweswater churchyard seem to register a move from Kirkhead to *Place*. A Peter Hudson of *Place* died in 1807, but when *Place* appears in the PR in the 1820s and 1830s it is associated with families named Smith and Cook. Hudson Place is presumably a renaming, since 'Iohn Woodhall 1667. I.W.' is inscribed on a stone window mullion at the rear of the house (I am most grateful to Mr Kenneth W. Bell of Hudson Place for the latter information).

HUGH'S LAITHES PIKE NY5015 Shap Rural.
From 1859 (*PNWestm* II, 175).
♦ 'The peak above Hugh's barns', 'said to be named after *Hugh Holme*, the first of the family to live at Chapel Hill in Mardale' (*PNWestm*). The 2nd el. is from ON *hlaða* 'barn'; see also **pike**.

HUGILL (parish): H~ FELL SD4599, H~ HALL SD4599.
Hogail(l), *Hogayl* 1256 to 1375, *Hugayl* 1274, 1383, *Hogill*, *Hogyll* 1341, *Hugill*, *Hugyll* 1459 to 1823 (*PNWestm* I, 169–70); *Hugill Fell* 1828 (HodgsonM); *Hugill Hall* 1738 (*PNWestm* I, 171).
♦ Possibly 'the high ravine'. The 1st el. is uncertain: the main possibilities are (a) 'high', ON *hór*, a by-form of *hǫ́r*, *hár* (*PNWestm* I, 170, citing Lindkvist) or (b) 'bay, bend in river', an ON **hō*, which Fellows-Jensen suggests citing Scand parallels (*SSNNW* 137). As she points out, ON **haugr** 'hill, mound' as suggested in *DEPN* would not result in *-o-* spellings. (c) A hybrid with OE *hōh* 'heel-shaped hill' as in HUTTON is also conceivable (cf. HEWTHWAITE). For the 2nd el., the earliest spellings point to origins in ON *geil* 'ravine or narrow track', which has been replaced, as often, by the reflex of ON **gil** 'ravine with a stream'. Interpretation as 'the high ravine' is supported by the topography, for a narrow valley runs N from R. GOWAN to H~ HALL, and 'the farms and what

was almost certainly the ancient arable' stand 100m. above the river and the present A591 (Mary Atkin, pers.comm. gratefully acknowledged). H~ **Fell** is 'a large district of arable, pasture, and moorland in the Township of Hugill' (OSNB Kendal, 1858, 2, 6); see also **hall**.

HULLETER SD3387 (hab.) Colton.
Hullater 1538 (*PNLancs* 216), *Hulleter* 1626 (PR).
♦ Perhaps 'the lair or shelter on the hill', from ON *hóll* 'hill' and ON *látr* 'lair' (*PNLancs* 194 and see **latter**); this would fit the situation on H~ Scar. Other conjectures might include dial. *hullet* 'owl' plus some generic such as **how**(e) 'hill, mound'; cf. *Hullethole* in Hawkshead (Finsthwaite PR, 1823).

HULLOCKHOWE NY5018 (hab.) Bampton.
Ullockhowe 1619, *Hullockhow*(e) 1638 (*PNWestm* II, 190), possibly cf. *Hueloc* end 13th cent., *Hueloc (skale)* 1324–7 (*CW* 1922, 294, 301).
♦ The 1st two syllables are puzzling, and if the *Hueloc* spellings belong here they are difficult to reconcile with any of the possible explanations. (a) The (*H*)*ullock* spellings may point to the p.n. ULLOCK (cf. the one at NY2423) or a surname derived from it, though no such surname appears in the index to the PRs 1637–1812. (b) Smith suggested alternatively some connection with the lost p.n. *Clattercollackhowe*, which may incorporate the OIr pers.n. *Collachtach* or similar (*PNWestm*). (c) The *-oc* termination could suggest Brittonic (Cumbric) origins (cf. PENRUDDOCK, WATERMILLOCK, etc.), but no obvious Brittonic explanation of the root presents itself. The final el. appears to be **how**(e) from **haugr** 'hill, mound'; the hamlet is on the slopes of a fairly distinct hill.

HUNDHOWE SD4997 (hab.) Strickland Roger.
Hound howe 1578 (*PNWestm* I, 159).
♦ 'The hill associated with dogs or hounds', with 1st el. from ON *hundr*/ OE *hund* 'dog, hound', plus ON **haugr** 'hill' or its reflex **how**(e). This is the steep slope rising to POTTER TARN.

HUNDITH HILL NY1428 (hab.) Embleton.
Hundeth Hill 1710, *Hundath Hill* 1774 (DHM).
♦ Obscure. 'Hundith' may conceivably have meant 'high place associated with dogs', with **hill** added later, though without earlier spellings this is highly conjectural. The 1st el. looks most like the word 'hound, dog' (OE *hund*, ON *hundr*), and a hill-name referring to dogs would be paralleled in HUNDHOWE, WHELPO, WHELP SIDE and conceivably BEECH HILL. In the 2nd syllable, (a) the 18th-cent. spellings of *-eth, -ath* could point to 'head, high place' (OE **hēafod**, ON **hǫfuð**), as in LINDETH, and the landscape would support this. (b) Alternatively, the reflex of ON **viðr** 'wood' is a possibility (cf. early forms of COLWITH). (c) ON **vað** 'ford' can be ruled out, in the absence of a suitable stream.

HUNDREDS, THE NY4004 Troutbeck.
From 1865; cf. *the lowest Hundred, Middle Hundred* 1839 (*PNWestm* I, 190).
♦ The three Hundreds (*hundred* from OE *hundred*) were divisions of the high common grazing land of Old Park (see under TROUT BECK NY4104). In existence before 1604, possibly long before, they became important units not only for regulation of grazing but also for organisation of communal responsibilities such as maintenance of gates and the local millrace (Parsons 1993, esp. 118–23).

HUNGERHILL NY5017 (hab.) Bampton.
Hungerhill 1648 to 1754 (*PNWestm* II, 195).
♦ Simply 'the hungry hill', with the reflexes of OE *hungor* and **hyll**, referring to poor land; cf. CAWDALE for comment on derogatory p.ns in this area. Noble suggested a surname *Hunger* or *Unger*, but without giving evidence (1901, 148).

HUNTING STILE NY3306 (hab.) Grasmere. From 1847 (*PNWestm* I, 201).
♦ The verb *hunt* goes back to OE *huntian*, but what kind of hunting is unclear, as is the age of the name. **Stile** here presumably has its normal sense of a steep ascent, or a track up one; this was a packhorse route and corpse road between Langdale and Grasmere (Simpson 1928, 275, 285–6).

HURLBARROW NY0705 (hab.) Gosforth. *Hurlebarrow* 1638 (PR), *Hurrel Barrow* 1774 (DHM).
♦ Possibly 'the rounded hill', from OE *hwerfel, hwyrfel* 'circle, circular feature', used especially of round-topped hills (*PNCumb* 396, Smith, *Elements*), plus OE **berg** 'hill'. H~ is on the W. slopes of BLENG FELL, whose contours are rounded, or at least oval.

HURSTHOLE POINT NY2227 Wythop. From 1867 (OS).
♦ Uncertain, but probably 'the promontory at Hursthole, the hollow by the wooded hill'; **hyrst, hole, point**. This is a small promontory or point projecting into BASSENTHWAITE LAKE by H~ Bay. A small pool lies — in the *hole*? — just to the W, with a wooded hill-side beyond, which may be the *hurst*. However, without early forms the identity of the elements is uncertain, as is their exact reference.

HUTTON SD2090 Dunnerdale.
From 1850 (OS), possibly cf. *Old Hutton* 1676 (original spelling not specified, Broughton-in-Furness PR).
♦ A puzzling name. 'Huttons' elsewhere in England, including the examples below, are of OE origin, meaning 'settlement at the hill(-spur)', but this one is not recorded early, and hardly occupies a prime site of the sort that attracted Anglian settlers. It is perhaps a later transferred name, or a surname.

HUTTON NY4326 (parish).
Hutton Soylle 1578 (*PNCumb* 212).
HUTTON JOHN NY4326 (parish).
Hoton John 1279 to 1487 (*PNCumb* 210).
HUTTON ROOF NY3734 Mungrisdale, H~ MOOR END NY3627 (hab.).
Hotunerof, Hotonerof 1278, *Hotonrof* 1279, 'Roof' appears as *Ruff, -ruf, roof, roef, rogf*' and *Rof* in 14th cent. (*PNCumb* 210); *Hutton Moor End* 1813 (PR), cf. *Hutton-moor* 1780 (West 139).
HUTTON SCEUGH NY3537 Caldbeck.
Huntone Skewghe al. (or) *Heskethe pasture* 1560 (*PNCumb* 279).
♦ H~ is 'the settlement by the hill(-spur)', from OE *hōh* 'heel-shaped hill' as in HEUGHSCAR plus OE **tūn** 'farmstead, settlement, village'. The distinguishing affixes are, however, problematic: 'Soil' is unexplained, and the identity of 'John' uncertain, though it has been suggested that this refers to the lord of the manor of Hutton c.1250 (Hudleston 1995, 22). H~ (Soil) and H~ John amalgamated into one parish in 1934. For 'Roof' (Hutton Roof being formerly independent) there are two competing explanations: (a) A pers.n. such as ON *Hrólfr, Rolf* or *Riulf, Riolf*, of early Germanic origin (as suggested for another H~ R~, *PNWestm* II, 36); (b) OE *hrōf*/ON *hróf* 'roof', which would

be apt for this airy settlement on a limestone platform at 984ft/300m., albeit dwarfed by neighbouring CARROCK FELL. See also THWAITE. 'Sceugh', pronounced [skiuf] (*PNCumb* 279), goes back to ON **skógr** 'wood'. Higham inclines to see the Hutton names as a group which, like others in the vicinity of Inglewood Forest, 'are likely to represent communities in existence before 900 [and] characterised by association with a wide territory lacking any single nucleated settlement' (1986b, 92). H~ Moor End is at the SE corner of Mungrisdale parish, not far from the boundary with Hutton and hence presumably at the edge of its unenclosed land; **moor, end**. (There are medieval references to the moor of Hutton Roof, *PNCumb* 210, but it would have been separated from H~ M~ End by the parish of Mungrisdale.)

HUYTON HILL NY3601 (hab.) Skelwith.
♦ This name was transferred when the preparatory school H~ H~ was evacuated from Liverpool in 1939/40 and moved into the then PULL WOODS House (built 1891). The school continued to take pupils until 1969. The Liverpool Huyton is assumed to derive, like HYTON below, from OE *hȳð-tūn* 'village by a landing stage' (*PNLancs* 113 cf. 45), and it is fortuitous that the situation of H~ H~, on a small rise above WINDERMERE, was one where the name's historic reference to a landing stage and a **hill** were quite fitting. (I am most grateful to Mrs Muriel Shuttleworth for information about Huyton Hill school.)

HYCEMOOR SD0989 (hab.) Bootle, H~ SIDE SD0888 (hab.), H~ SIDE FARM SD0890. *Hysemore in Seton* 1391, *Haysmore* 1551–3, *Hisemore* 1571 (*PNCumb* 346); *Hise Moor Side* 1774 (DHM); *Hisemoorside* 1671 (PR).
♦ OE *hys(s)e*, referring to a type of water plant, may be the specific here, as tentatively suggested in *PNCumb*. See also **moor, side,** and **farm**, which distinguishes two settlements that were both shown as H~ Side on earlier maps.

HYTON SD0987 Bootle, HYTON SD1087, OLD~ SD0987.
Hytona c.1210 (p), *Hitun* c.1230 (p), *Hyton* 1251 to 1588, *Hiton and Oldhiton* 1358, *Hitton* 1531–3 (*PNCumb* 346).
♦ H~ is assumed to be 'the settlement by the landing-stage' from OE *hȳð* 'landing-stage' and OE **tūn** 'farmstead, settlement', and hence to commemorate a time when the RIVER ANNAS (formerly Annaside Beck) was navigable and flowed more directly into the sea (*PNCumb* 346, Steers 1964, 84; Hyton SD1087 and Old H~ are close to the river; Hyton SD0987 is not.). **Old** in Old H~ is the earliest occurrence of the adj. in the p.ns in this volume. There is some uncertainty as to which Hyton the early spellings refer to, and some inconsistency in the naming of these places on 19th- and 20th-cent. maps.

I

ICKENTHWAITE SD3289 (hab.) Colton.
Yccornewayt c.1535, *Yckorntwayte* 1538 (*PNLancs* 218).
♦ 'Clearing frequented by squirrels', from ON *íkorni* 'squirrel' and ON *þveit* 'clearing' — a rare allusion to the indigenous red squirrel, and still apt, since I~ lies at the S end of GRIZEDALE FOREST, 'the best place to see this mammal' (Miles 1993, 15).

IDLE HILL NY4005 Troutbeck.
From 1859 (*PNWestm* I, 190).
♦ 'The **hill** associated with the Idle family', for whom records exist from the 14th and 16th cents (*PNWestm* I, 190).

ILL BELL NY4307 Troutbeck/Kentmere.
Hill Bell 1770 (JefferysM), *Hill-bell* 1780 (West 59), *Ill Bell* 1857 (*PNWestm* I, 190).
♦ Probably 'the treacherous bell-shaped hill'; **ill, bell**. The 2nd el. is clearly *bell*, which suits the shape (described and illustrated in Wainwright 1957, Ill Bell 2). For the 1st el. the 18th-cent. spelling *Hill* also survived on earlier versions of the 1:25,000 map, and might be taken seriously, as by Smith, who compares Bell Hill and MARDALE ILL BELL (*PNWestm* I, 190–1). But Hill Bell is an implausible construction, and Mardale Ill Bell and Ill Bell are more likely to contain *ill* than *hill*. Both are compact, forbidding, craggy hills with a pile of stones at the summit (and both are on parish boundaries), and *ill* is paralleled in the following names.

ILL CRAG NY2207 Eskdale.
Ill Crags 1802 (*CN* 1219), *Ill Crag* 1867 (OS).
♦ 'The treacherous rocky height'; **ill, crag**. This point on the SCAFELL range is furnished with spectacular and forbidding crags. Cf., probably, EEL CRAG(S).

ILLGILL HEAD NY1604 (1998ft/609m.) Eskdale.
head del Ilgill 1338 (*PNCumb* 391).
♦ 'The height above Ill Gill, the treacherous ravine'; **ill, gil(l), hēafod/head**. The terrain — the stupendous screes S of Wast Water — is truly *ill*, but *gill* is puzzling. There is no 'Ill Gill' on the maps (including OS 1867) and no stream or distinct ravine which could be so called immediately below I~ H~.

ING BRIDGE NY4104 Troutbeck.
Ings bridge 1659 (Bridges), *Ing Bridge* 1859 (OSNB); cf. *Ynggarth* 1560 (*PNWestm* I, 191).
♦ Ing, deriving from ON *eng* 'meadow', names the stretch of pasture along the E of the TROUT BECK, which is crossed by the **bridge** (see **brycg**); cf. next entry.

INGS SD4498 (hab.) Hugill.
the Inges 1546, *Ings* 1625 (*PNWestm* I, 171).
♦ 'The (water-) meadows', from ON *eng* with a ME pl. -(*e*)*s*, and hence the village beside them. As often, the reflex of ON *eng* 'meadow' is here applied to water-meadows, those along the R. GOWAN.

INMANS SD0988 (hab.) Bootle.
Inmans 1778, *Inmonts* 1780 (PR).
♦ This place seems to be named from the surname Inman (originally meaning 'inn-keeper') which is recorded

from 1665 in Bootle PRs. By the late 18th cent., when the p.n. appears in PR entries, however, it is associated with other surnames: Kendal, Jackson, Askew and Poole. Other p.ns of this form are RAWSONS and JENKIN (*Jenkens* 1718).

INNOMINATE TARN NY1912 Buttermere.
Unnamed 1867 (OS); previously *Loaf Tarn* 1864 (*PNCumb* 34).
♦ The **tarn** is one of the pools on the summit of HAYSTACKS. The highly Latinate 'Innominate' (a word first recorded in *OED* from 1638) conveys the same meaning as NAMELESS (FELL). It was first suggested in 'an early *Rock and Fell Climbing Club Journal*' (so *PNCumb*; the *Journal* was first issued in 1907) but is seemingly a misnomer, since the older name was Loaf Tarn, thought to refer to hummocks of peat or vegetation in the tarn.

IN SCAR: see SCAR

INTACK, HIGH NY2737 (hab.) Caldbeck.
the new Intacke 1652, *Intacks* 1680 (*PNCumb* 279), *Intack* 1774 (DHM), *Intacks* 1867 (OS), *High Intack*, also *Middle* and *Low Intack* all 1900 (OS).
♦ 'The upper piece of land enclosed from the waste'; **high**, **intake**. If *new* was still apt in the 17th cent., this was land taken into cultivation then (see **intake**). Low, Middle and High I~ are distinguished on the 1:25,000 map (1998 printing).

INTAKE, THE SD1394 Muncaster.
From 1867 (OS).
♦ 'The land enclosed from the waste' (**intake**), referring to a tract of quite high, sloping terrain. *Intake* is unusual here, outside Furness.

IREBY: HIGH I~ NY2337 (parish of I~ and Uldale).
Ireby c.1150 (p); *Heghireby* 1279 (*PNCumb* 299).
♦ 'The village or farmstead of the Irishmen, Irishman, or of a man called Íri.' The 1st el. is the genitive of ON *Íri* 'Irishman', either in sing. or pl., or else of a pers.n. derived from this; the 2nd el. is ON **bœr/býˊ**. Several other places named Ir(e)by are found in N. England (*PNCumb* 300), and they are important if tantalising nuggets of evidence for the settlement history of the region. They seem to refer to distinct communities of settlers from Ireland, either Irish or of Scand origin. They could alternatively have come from the Isle of Man, as suggested by Fellows-Jensen (1992, 39). High I~ lies at 804ft/245m. and, as Fleming remarked (1671, 47), is so named 'because it stands higher on ye Hillside' than Ireby, which lies a mile N and just outside the National Park. Mary Higham points out that High I~ 'to this day exhibit[s] features which would suggest it may have developed from a dependent shieling site' (1995, 201–2).

IREDALE PLACE NY1122 Loweswater.
Ardale Place [sic] 1867 (OS); cf. *High Iredale* 1738, *Iredell Lands* 1744 (PR).
♦ 'The farm of the Iredale family'; surname plus **place**. I~ could be (a) a p.n., suggested by the 18th-cent. PR entries above; or (b) the surname, recorded from Cumbria in the 16th cent. onwards (Hanks & Hodges 1988), and found in the Loweswater PRs as *Iredale* or *Iredell*. Since the neighbouring farms of HUDSON P~ and Jenkinson P~ are named from a surname plus *place*, (b) seems more likely here.

IRELAND MOSS SD3384 Haverthwaite.
From 1851 (OS).
♦ Now artificially drained, this **moss** is the low, wet, peaty land beside RUSLAND POOL. The reason for 'Ireland' is unknown to me, though (a) one might guess at a connection with the Irishmen who came to work in the furnaces and forges of Furness in the 18th cent., and who may have sailed into nearby GREENODD, then a busy port (Fell 1908, 286, 314–5). (b) Ireland also exists as a surname (e.g. in the Broughton-in-Furness PRs), though the many Mosses in this area are not normally named from surnames.

IRON CRAG NY1212 Ennerdale.
From 1802 (*CN* 1208).
IRON CRAG NY3034 Caldbeck.
From 1867 (OS).
♦ 'The rocky height where iron ore is found'; *iron* ultimately from OE *īren*, plus **crag**. In Ennerdale, iron ore haematite was mined below the linear outcrop of Iron Crag in the 19th cent. (Danby 1994, 88); and perhaps cf. nearby SMITHY BECK. Lead, copper and other minerals were similarly mined from Roughtongill Mine, near the Caldbeck Iron Crag, until 1894 (Danby 1994, 25).

IRON CRAG NY2919 St John's.
From 1794 (Hutchinson, II, 154).
♦ This may be 'the rocky height where eagles are seen', with a version of dial. *erne* 'eagle' plus **crag**. This is a prominent rock wall on BLEABERRY FELL, and in the absence of evidence for iron ore hereabouts, it is possible that this is another ERNE CRAG. The presence of a RAVEN CRAG nearby at NY3018 might also support this interpretation (I am grateful to Mr Ian Ferriday of Grasmere for this point).

IRON KELD NY3301 Hawkshead.
From 1850 (OS).
♦ Apparently 'iron spring', with *iron* (ultimately from OE *īren*) plus **keld**(a). This is a tract of rocky ground, boggy in the hollows. Local historian Cowper, however, did not take 'iron' at face value since he assumed a corruption of 'Arni's spring' (1899, 357).

IRT, RIVER NY1002 Gosforth/Nether Wasdale/Irton.
aquam de (river of) *Irt* 1279, *water of Irte* 1322 (*PNCumb* 17).
♦ Obscure. Ekwall suggested origins in a Brit. word meaning 'fresh, green' or a Gaelic one meaning 'death' (*ERN* 211–2), and Coates accepts Brittonic origins (2000, 362), but in *DEPN* Ekwall instead mentions the possibility of derivation from OE *gyr* 'mud'. The river gives its name to IRTON.

IRTON: I~ FELL NY1302, I~ HALL NY1000, I~ PARK NY1100, I~ PIKE NY1201, I~ ROAD STATION SD1499.
Irton, Yrton 1225 to present, *Hyrton, Hirton* 1225 (p) to 1496 (*PNCumb* 402); *Irton Fell* 1802 (*CN* 1213), 1867 (OS); *Irton Hall* 1675 (*PNCumb* 402); *Irton Park*; *Irton Pike* both 1867 (OS); *Irton Road Station* 1900 (OS).
♦ I~ is 'the settlement on the IRT'; river-name plus OE **tūn**. The name is rather emblematic since the place stands where the river emerges from the Cumbrian dome into the lower, more fertile land where Anglian p.ns are more numerous. I~ **Fell** is a stretch of upland grazing, whereas I~ **Pike** rises to a distinct peak; Coleridge's reference of 1802 might in fact be to the Pike. The present **Hall**, shown as 'I~ Hall School' on 1:25,000 map, was

built 'in the Elizabethan style' in 1873 (Pevsner 1967, 143); the previous Irton Hall had housed the Irton family for some five centuries (Taylor 1941, 72). The Station is on the RAVENGLASS & ESKDALE narrow gauge railway. See OLD COACH ROAD for a note on *road*, BOOTLE STATION on *station*, and **park**.

IRTON HOUSE NY2034 Blindcrake.
From 1865 (Hodgson, Plan 15),1867 (OS), previously *Old Park Farm* 1855 (plan in CRO D/Law/5/35; altered in a later hand to *Irton House Farm*); cf. *Irton Wood* on 1855 plan.
♦ The **House** (and I~ Cottages just to the W) bear the name of IRTON some 20 miles SW, or more probably of the surname derived from it. Marriages of one Maud Redman to Christopher Irton of Irton (1535–62) and then to Sir Thomas Leigh of ISEL seem to have been the route by which the Irton family established an interest in this area (Taylor 1941, 94; I am most grateful to Mrs M. Almond of Irton House for putting me onto this trail).

ISEL NY1533 (hab.) Blindcrake, ISELGATE NY1633, I~ HALL NY1533, ISEL OLD PARK NY1934 (hab.).
Ysala 1195, *Ishale*, *Yshale* c.1235 to 1344, *Isall* 1269, *Issell(e)* 1397; *Isellgait* 1623; *Isill Hall* 1675 (*PNCumb* 267); *Old Parke* 1555, *Isell-old-park* 1718 (*PNCumb* 301).
♦ I~ is thought to be 'Ise's nook', from an OE pers.n. *Īse* plus OE *halh* 'nook of land (often beside water)' (*PNCumb* 267); see also **gata** 'road', **hall**, **old**, **park**. The village is beside the meandering R. DERWENT. 'Gate' in Iselgate presumably means 'road'

(from ON *gata*) rather than 'gate' (from OE **geat**). Hall has its full sense, for Isel Hall is 'quite a spectacular building', mainly 16th-cent. (Pevsner 1967, 143–4). I~ Old Park was formerly a separate township; the farm of that name appears as *Harrisongate* 1867 (OS), *Isel Old Park* 1900 (OS).

IVING HOWE NY3203 (hab.) Langdales.
Ivin How 1693 (Grasmere PR), *Iving how* 1823, *Ivy Howe* 1865 (*PNWestm* I, 206).
♦ 'The hill where ivy grows.' In the 1st el., the modern spelling preserves a N. dial. form of the plant-name from OE *īfig* which more usually appears as *ivin* (*EDD*). Ivy, *Hedera helix*, is widespread in Cumbria but 'generally absent from the higher fells and moorlands' (Halliday 1997, 332). The 2nd el. is **how(e)** from ON **haugr**. Cf. IVY CRAG.

IVY CRAG NY3100 Coniston.
From 1850 (OS).
IVY CRAG NY3504 Rydal.
From 1863 (OS).
♦ Simply 'the rocky height where ivy grows'; *ivy* as in IVING HOWE, plus **crag**.

IVY TREE SD2887 (hab.) Blawith.
Ivintree 1722, *Ivy tree* 1729 (Lowick PR), *Ivy Tree* 1851 (OS).
♦ Presumably 'the ivy-covered tree', though an 'ivy tree' is also possible since a mature ivy growing against a wall can develop a thick, woody, trunk. For the word and plant *ivy*, see IVING HOWE; *tree* is ultimately from OE *trēo(w)*.

J

JACK GAP PLANTATION SD3395 Satterthwaite.
From 1851 (OS).
♦ *Jack* is presumably the forename, and *gap* either an opening in the landscape or (since none is obvious from the map), possibly a way through onto a stretch of grazing (see **gap**). This is one of several **plantation**s that make up the modern GRIZEDALE FOREST.

JACKSON GROUND SD2392 Dunnerdale.
From 1678 (original spelling not specified, Broughton-in-Furness PR).
♦ 'The farm of the Jackson family.' This is one of a local cluster of farm-names consisting of a surname and **ground** (cf. CARTER GROUND). There are numerous entries for *Jackson* in the relevant PRs, for Seathwaite.

JENKIN NY1528 (hab.) Embleton.
Jenkens 1718, *Jenkin* 1814 (PR).
♦ 'The Jenkins' (place).' *Jenkin* is a common Lakeland surname, originating in a pet-form of *John*. This and the following two places were presumably owned by or associated with a person or family of this name, who cannot now be identified. I have not found *Jenkin(s)* in the Embleton PRs, though *Jenkinson* occurs in the early 18th cent.

JENKIN CRAG NY3802 Ambleside.
Jenkyn's Crag 1920 (OS).
♦ 'The **crag** or rocky height associated with an individual or family named Jenkin' — either as a forename, a diminutive of *John*, or as the surname derived from it; cf. JENKIN.

JENKIN HILL NY2727 Underskiddaw.
Genkin-hill in Thorntwhate 1646, *Jenkin Hill* 1648 (*PNCumb* 372), *Jenkin Hill* 1867 (OS).
♦ 'The hill associated with an individual or family named Jenkin.' Cf. JENKIN above, plus **hill**. The hill is high and remote, and one wonders whether the 17th-cent. spellings refer to a settlement some distance from it. I have not found the surname *Jenkin* in the early local PRs, though *Jenkinson* occurs in Crosthwaite PRs.

JOHN BELL'S BANNER NY4109 Patterdale/Troutbeck.
John Bell Banner 1614, *John Bell's Banner* 1865 (*PNWestm* II, 226).
♦ The name refers primarily to a boundary mark, but is sometimes applied to CAUDALE MOOR on which it stands (Wainwright 1957, Caudale Moor 1). *Banner* is from *baner* 'standard, banner' (ME from OFr). The number of John Bells in local records makes it difficult to identify this one. Armitt mentions a John Bell known as 'Old', who witnessed a deed of 1589 though he could not write his name. A further John Bell, 'curate', witnessed a deed of 1617; this was probably the schoolmaster of AMBLESIDE who, with the help of his pupils, made a paved causeway in the neighbourhood (Armitt 1906, 61). Cowper, taking *banner* in the sense 'bounder, boundary', suggests that this 'spirited road-paving curate' may have been involved in riding the boundaries, and may be the person named in J~ B~ B~ (1907, 145).

JOHNSCALES SD4686 Crosthwaite.
Johnskal(l)es 1526, 1535, *Johnscales* 1694 (*PNWestm* I, 84).

♦ 'John's shielings or shieling huts.' Smith notes that the man in question may be John Garnet, its occupant in 1526 (*PNWestm*). *Scale(s)*, from ON **skáli**, refers to an original shieling, which in this case, as often, has become a permanent settlement.

JULIAN HOLME NY0903 (hab.) Gosforth. *Julianholm* 1365, *Gillianholme* 1600 (*PNCumb* 396), *Gilling Holm* 1774 (DHM).

♦ 'The waterside land associated with Julian'; forename plus ON **holmr** or its reflex. *Julian*, originally from *Julianus*, a derivative of Lat *Julius*, is well recorded as a male forename in England from the 13th cent., but also as a variant of the female *Juliana*, alongside *Gillian* (Withycombe 1977, 183–4).

K

KEEN GROUND SD3498 Hawkshead.
Keenground 1656 (PR).
♦ 'The farm of the Keen family'; surname plus **ground**. The surname *Keen* appears frequently in the PRs of Hawkshead and neighbouring parishes. The 1656 entry is the baptism of 'Margarett Keene fil. (daughter of) Lenord de Keenground', and a William Keene was there in 1674 (*CW* 1973, 194).

KELBARROW NY3307 Grasmere.
Keldbergh 1375, *Kellbarro (Hall)* 1578, *Kellbarrow* 1719 (*PNWestm* I, 199).
♦ 'The hill with a spring or springs'; **keld**(a), **berg**. 'Rises', i.e. water emerging from underground, are shown nearby (1920 OS).

KELDAS NY3816 Patterdale.
♦ The name conceivably derives from dial. **keld** (ON **kelda**) 'spring', plus **house** or its ON or OE ancestor **hús/hūs** (cf. WHITTAS), hence 'house at the spring', but without early forms this is pure speculation. This is a minor summit at the E end of BIRKHOUSE MOOR, and there is no building here now or on the 1863 OS map.

KELD GILL NY5413 Shap Rural.
From 1859 (*PNWestm* II, 175).
♦ 'The stream at Keld, the (place by a) spring'; **keld**(a), **gil**(l). The village of Keld (*Keld(e)* 1540, *PNWestm* II, 165) lies just E of the confluence of K~ G~ and the R. LOWTHER, and K~ Gill quite likely takes its name from the village rather than directly from a spring.

KELDHEAD NY4819 (hab.) Bampton.
Keld 1839 (*PNWestm* II, 195), *Keldhead* 1863 (OS).
♦ 'The head of Keld, the (place by a) spring'; **keld**(a), **hēafod**. There are streams and 'rises' nearby.

KELSWICK NY1929 (hab.) Wythop.
Kelsick 1602 (in later copy, Lorton PR), *Kelswick* 1774 (DHM).
♦ Possibly 'the stream issuing from a spring' (**keld, syke**). The interpretation of the 2nd el. depends whether the *-w-* is authentic and significant or not. (a) The translation above assumes that it is not authentic, but has been introduced, perhaps under influence of the *w* in KESWICK, so that Kelswick is of the same origin as Kelsick in Dundraw (*Keldesike* 1292, *PNCumb* 140; cf. also three Keld Sikes, *PNWestm* II, 17, 115, 118). (b) If on the other hand the *-w-* is ancient and significant, it could point to ON *vík* referring to a curve in the landscape (cf. FROSWICK) or to OE *wīc* 'settlement'. On Kelswick chapel, see WIDOW HAUSE.

KELTON: K~ FELL NY0918 Lamplugh.
Keltona c.1150 to c.1280 (*PNCumb* 405); *Kelton fell* 1687–8 (Denton 51).
♦ K~ is probably 'the settlement by the spring'; **kelda, tūn**; also **fell**. The 1st el. may well be ON *kelda* 'spring', for there are springs below the Fell. The 2nd el., as in nearby MURTON (FELL), is OE *tūn* 'settlement, farmstead, village' — often a sign of Anglian settlement in the lower-lying W fringe of the Lake District. K~ F~ has a distinct summit, but since Keltonfell Top lies one mile W across a dip, *fell* here seems instead to imply a stretch of unenclosed grazing land.

KEMPLERIGG NY0802 (hab.) Gosforth.
From 1731 (*PNCumb* 396).

♦ The 1st el. is obscure; the 2nd is presumably **rigg** 'ridge' (see **hryggr**) referring to the situation on a modest bank.

KENDAL SD5192 (hab. & parish).
(Hab:) *Cherchebi* 1086, *Kircabikendala* 1090–7 (in 1308 copy), *Kirkebi-in-Kendal* & variants 1189–1210 to 1609 (*PNWestm* I, 114); (valley, Barony or Ward:) *Kendal(l)* & variants later 12th cent. to present (*PNWestm* I, 61).
♦ 'The valley of the KENT'; river-name plus **dalr**. The town of K~ was known as Kir(k)by Kendal up to the 18th cent., with the affix *Kendal* distinguishing it from Kirkby Lonsdale and K~ Stephen; however, already in the 15th cent. the affix alone was being used as the town name (*PNWestm* I, 115). K~ is little more than a mile outside the Lake District National Park boundary (and is home to the Park's headquarters); it is included in this volume because of its size and historic importance.

KENDALL GROUND SD2786 Lowick.
Kendallground 1720/1 (PR), 1851 (OS).
♦ 'The farm of the Kendall family'; surname plus **ground**. The surname *Kendal(l)* appears frequently in the PRs of Lowick, Blawith and Colton.

KENT, RIVER NY4506 Kentmere.
(*aqua de* [river of]) *Kent(e)* c.1170 to 1777, *Kenet* 1246, 1256 (*PNWestm* I, 8–9; *ERN* 225–8).
♦ The name is Brit., but of uncertain meaning, and its interpretation depends partly on the weight assigned to the majority spellings in *Kent(-)* or the minority type *Kenet*. (a) Ekwall took the *Kenet* type as primary, pointing out that an original *Kenet* could easily syncopate to *Kent* in late OE, especially in compounds such as KENTMERE and KENDAL(E), and hence associated the Kent with two rivers named Kennet(t) and derived from a Brit. *Cunētiū (*ERN* 227, followed by *PNWestm*). Ekwall explained this from a now discredited root **kuno-* 'high, exalted' (see Smith, *Elements*, under the ghost-word **cuno-*, and Breeze 2000, 126). Breeze suggests **cuno-* 'dog' or a pers.n. derived from it in names including Kent and Kennet(t) (126–8). (b) However, *Kent* spellings strongly predominate for the Cumbrian river-name, in contrast with the two river Kennet(t)s, whose early forms show next to no syncope even in compounds (*ERN* 225–6, *Kenteford* 1275 is the exception). A possible derivation which does not assume syncope would be from Brit. **canto-*, **canti-* (Smith, *Elements*, Padel 1985, 37), which seems to mean 'edge, boundary' and may appear in that sense in Pen-y-Ghent, NYorks (Higham 1999, 65). I am grateful to Dr Oliver Padel for this suggestion.

KENTMERE NY4504 (hab. & parish), K~ COMMON NY4408, K~ PARK NY4303, K~ PIKE NY4607 Kentmere/Longsleddale; K~ RESERVOIR NY4408 Kentmere.
Kent(e)mer(e) 1247–60 (p) to 1836 (*PNWestm* I, 165); *Kentmere Common* 1899 (OS); *le Parke* 1617 (*PNWestm* I, 167), *Kent-mere parke* c.1692 (Machell II, 306); *Kentmere Pike*; *Kentmere Reservoir* both 1858 (OSNB).
♦ K~ is 'the pool by the KENT' (river-name plus **mere**[1]), referring to K~ Tarn, which was partially drained in the 1840s, and hence to the settlement and township of K~. 'Kendal from Kent derives its name, | And Kent from Kentmere's springing fountain came' (Machell, quoting from 'Poet Braithwaite', c.1692, ed. Ewbank, 93). See also **common**, **hall**, **park** (here

referring to a tract of enclosed pasture), **pike** 'peak' and **reservoir**. K~ Reservoir was created in 1848 'for supplying the Bobbin Mills in the vicinity of Staveley' (OSNB).

KEPPEL COVE NY3416 Patterdale.
Kepple-Cove (Tarn) 1787 (*PNWestm* II, 226), *Keppel Cove* 1828 (HodgsonM), *Kepple-Cove* 1829 (Parson & White 579).
♦ Probably 'horse corrie'; **capel**, **cove**. The 1st el. may well be ME *capel* 'horse', since an old pony-track runs past, and the tarn, now dry, but formerly dammed to serve GREENSIDE mines, would have provided a watering place. The 2nd el. is simply *cove*, a corrie scooped out by glaciation.

KEPPLE CRAG SD2198 Ulpha.
From 1867 (OS).
KEPPLE CRAG SD1999 Eskdale (1062ft/328m.).
From 1867 (OS).
♦ There are no early spellings, but these rocky heights are probably named from horses, perhaps because near pony-tracks; **capel**, **crag**, and cf. CAPELL CRAG. The example in Eskdale is just above the Woolpack Inn and next to Horse Gill.

KESKADALE: K~ BECK NY2018 Above Derwent, K~ FARM NY2119.
Keskeldale 1260, *Ketelschaledal* 1268; *Keskadale Beck* 1578 (*PNCumb* 370, 18).
♦ The starting point is *Ketils **skáli** 'Ketill's shieling', which commemorates the use of this inner recess of the NEWLANDS valley for summer pasture before it became a permanent settlement; it was so used by the tenant of FAWE, four miles away, c.1270 (Winchester 1987, 95). This gave its name to the valley (ON **dalr**), which in turn gave rise to the **beck** 'stream' and **farm** names.

KESTREL LODGE NY2432 Bassenthwaite. Previously *High Close* 1867, 1900 (OS), 1966 (OS One Inch).
♦ The word *kestrel* appears in ME as *castrel(l)*, and is of uncertain origin; it is rare in p.ns. The kestrel is a compact and adaptable bird of the falcon family, found widely throughout Lakeland (Hervey & Barnes 1970, 153), including here, though buzzards are more common. This is one of the larger places bearing a name in **lodge**, being a Christian residential centre. It was named by the previous owner in 1974, but High Close is still used locally. (I am most grateful to Mrs Marion Beckham, the present owner, for local information.)

KESWICK NY2523 (hab. & parish).
Kesewic c.1240, *Kesewyk* 1266 to 1403, *Chesewyk'* 1285 (p) (*PNCumb* 301).
♦ 'The cheese (dairy) farm', OE *cēsewīc* (see **wīc**), the same name as the London Chiswick, but here with K- due to Scand influence. Now the largest centre of population in the Lake District, K~ enjoys a favourable situation which fits both the Anglian origins and the specific meaning of its name. (Pl. 9)

KETTLE CRAG NY2704 (1263ft/385m.) Langdales.
From 1865 (*PNWestm* I, 206).
♦ Possibly 'the **crag** or rocky height by the hollow'. 'Kettle' may go back to ON *ketill* with the sense 'deep hollow', which is attested for OE *cetel*, while *ketill* occasionally seems to refer to a deep inlet in Norw p.ns (*NG* XI, 131; XIII, 214). This is a knoll with rocks partially enclosing a small dip which might be seen as bowl-like. Elsewhere,

but not here, 'cauldron, bubbling spring or stream', might be appropriate and *Ketill* is also a common ON pers.n., but such names are not normally coupled with *crag*.

KEYHOW NY1300 (hab.) Irton.
Keyhow in Irton 1677 (Eskdale PR); cf. *Highkay Hall, Lowkay Hall* 1774 (DHM).
♦ The 1st el. is obscure, and *PNCumb* attempts no explanation. (a) 'Key' may be a surname, as in Key Scar (*PNWestm* II, 68, and Key Croft and Hill (*PNWYorks* III, 193 and VI, 179), though I have not noted it in the Irton PRs. (b) Some such topographical reference as suggested for CAISTON (BECK) is conceivable. (c) ON *kví* 'animal pen' is rare in Cumbria and the clearest instance, Wheyrigg (*PNCumb* 140), shows a different development. (d) Derivation from OFr *kay*, ME *key* 'quay, wharf' also seems unlikely, despite the proximity of the R. MITE. The 2nd el. appears to be **how(e)** from **haugr** 'hill, mound'.

KEY MOSS SD4284 (hab.) Witherslack.
Key Moss 1741 (*PNWestm* I, 79), *Kay Moss* 1901 (*CW* 1901, 192).
♦ A **moss** is a boggy area, but 'Key' is obscure — possibly a surname or topographical term, as in other minor names (see KEYHOW).

KIDBECK FARM NY1104 Nether Wasdale.
Kydbek in Wassedale 1397 (*PNCumb* 442; cf. *PNCumb* 18 for the stream-name).
♦ 'Kid stream'; ON *kið* 'young goat' or ME *kid(e)* adopted from it, plus **bekkr/beck** 'stream'. K~ already referred to a farm as well as a stream before the recent addition of **farm**.

KID MOOR NY4819 Bampton.
From 1859 (*PNWestm* II, 195).
♦ Presumably 'the upland waste where kids graze'; *kid* as in KIDBECK plus **moor**.

KIDSTY HOWES NY4612 Bampton, K~ PIKE NY4412 (2560ft/846m.) Martindale/ Bampton.
Kidsty Howes 1860 (OSNB); *Kidsey-pike* c.1692 (Machell II, 467), *Kidsway Pyke* 1770 (JefferysM), *Kidstey Pike* 1777 (*PNWestm* II, 195), *Kidstow-pike* 1780 (West 160), *Kidsty Pike* 1823 (*PNWestm* II, 195).
♦ K~ may be 'the path frequented by kids', from *kid(e)* as in KIDBECK, and **sty(e)** 'steep path'. The Howes 'hills, mounds' (from **haugr/how(e)**) constitute a subsidiary summit E of K~ Pike 'peak'.

KILBERT HOW NY4018 Martindale.
From 1920 (OS).
♦ The hill may have been named from a man bearing the ME (and originally ON) name *Ketilbert* (so *PNWestm* II, 219), though the lack of early forms makes this uncertain. **How(e)** is from ON **haugr** 'hill, mound'.

KILN BANK: FAR K~ B~ SD2193 (hab.) Dunnerdale, HIGH K~ B~ FARM SD2194.
Kilnbank 1737 (Seathwaite PR), *Killbank* 1745 (map in West 1774); *Far Kiln Bank, High Kiln Bank* (also *Low Kiln Bank*) 1850 (OS).
KILNBANK/KILN BANK SD2787 (hab.) Blawith.
Killbancke in Blawith 1674 (Torver PR).
♦ Self-explanatory; *kiln* from OE *cyln* 'kiln, for firing or baking', **bank** 'hill-slope', **far, high**, and **farm** added later. The Dunnerdale K~ B~ is the slope E of the R. DUDDON. Davies-Shiel counts this among kiln sites, though whether a potash or kilnwood kiln is difficult to tell (1974, 35). At Blawith, no remains of a kiln are visible today (I

1. Trusmadoor, looking north to the Solway Firth

2. Truss Gap and Swindale Beck

3. Mosedale, Loweswater

4. Scarth Gap (to R) from Ennerdale

5. Buttermere lake and village, with (L to R) Low Bank, Grasmoor, Whiteless Pike, Wandope and Crag Hill

6. Buttermere and Crummock Water

7. The Bishop, Barf

8. Beckstones Gill, Above Derwent, showing potential 'bakestones'

9. Braithwaite, 'the broad clearing', with Keswick in distance

10. Crosthwaite Church, Keswick

19. Thirlmere from above Wythburn

20. Helvellyn: Red Tarn, with part of Swirral Edge leading to Catstycam

21. Patterdale Church

22. Grisedale, Patterdale

23. Yewbarrow, Great Gable and Lingmell, fron Wastwater

24. The Screes, Wastwater

25. Mickeldore (L), between Scafell Pike and Scafell

26. Langdale Pikes from The Band, with Pike O' Stickle to L

27. Viking Age hogbacks at Gosforth Church

28. Walls, Roman bath-house at Glannoventa, Ravenglass

29. Beacon Tarn, with Coniston fells in background

30a. Coppicing at Bouth Fall Stile

30b. Charcoal-burner's hut near Satterthwaite

31. Windermere at dawn, from Low Wray

32. Whitbarrow Scar, with wooded valley of Witherslack to L

am grateful to Mr Livesey of Lowick Bridge Farm, owner of Kiln Bank, for this information), though again the name is suggestive. For late 17th-cent. references to expenditure on lime-kilns, and for the spelling *kill*, see, e.g., Fleming *Accounts* 132, 151.

KILNHILL NY2132 (hab.) Bassenthwaite. *Kiln-hill* 1757, *Killhill* 1760 (PR), *Kill Hill* 1809 (CrosthwaiteM).
♦ On *kiln*, see above. The reference here may have been to a lime-kiln which served a local fulling industry: the fulling of cloth is commemorated in Walk Mill Bridge (cf. WALL-THWAITE), and stretching of cloth in Tenter Hill (cf. the word *tenterhooks*), both two miles to the E. The **hill** here is a modest rise above BASSENTHWAITE.

KILNSTONES NY5002 (hab.) Longsleddale. *Kilstone* c.1692 (Machell II, 104), *Kilne stone* 1700 (*PNWestm* I, 163), possibly = *Revegyll* 1260, *Revegill* 1263 (*Kendale*, II, 367, I, 300).
♦ Apparently 'stony place where a kiln is sited'; *kiln* as in KILN BANK, plus **stone**. The site was identified with *Revegill*, a corn mill of Shap Abbey, by Somervell (1930, 27–9), who quoted the opinion of Mr. Withers that there are traces of corn and fulling mills at K~, and oral tradition of a mill there. Remains of kilns are not in evidence at K~ today (I am grateful to Mr J. Mercer for this information).

KING'S HOW NY2516 Borrowdale.
♦ 'The king's hill', with ModE *king* (from OE *cyning*) plus **how(e)** (from ON *haugr*). This is a modest height, whose name is implicitly explained on a rock below the summit bearing the inscription, 'In loving memory of King Edward VII/Grange Fell is dedicated by his sister Louise/As a sanctuary of rest and peace' (Danby 1994, 103). The King died in 1910.

KINMONT: K~ BECK SD1390 Waberthwaite/Millom Without, K~ BUCK BARROW SD1490 (1754ft/535m.) Waberthwaite, LOW K~ SD1189.
Kinemund 1201–16, *Kenemund* 1236–52, *Kynmont* 1526 (*PNCumb* 364–5); *Kinmont Beck* 1867 (OS); *Kynnimont or Kydment . . . and . . . Klerkburre* alias *Lukberry* 1571 (*CW* 1911, 170), *Kinmont Buckbarrow* 1867 (OS); *Low Kinmont* 1867 (OS).
♦ K~ is believed to be Gaelic *ceann monaidh* 'head of the moor or mountain', cf. Scottish examples (*PNCumb* 364–5) and possibly Kinmond in Westmorland (*PNWestm* II, 47, Coates 2000, 339). The Gaelic *ceann* may also occur in KINNISIDE and conceivably in KINN. See also **beck** 'stream', **low**. One wonders whether K~ originally referred to the hill whose twin peaks are now called K~ Buck Barrow and BUCK BARROW and hence to the territory below the hill. The farms of High, Low and Middle K~ are well down on the W skirts of the hill. Meanwhile, K~ B~ B~ is part of PRIOR PARK (q.v.).

KINN NY2223 (1167ft/356m.) Above Derwent.
Cf. *Keenbrow* 1800 (*CN* 805).
♦ (a) Possibly from ON *kinn* 'cheek', used of steep hill-slopes in early Norw p.ns (Fritzner). The name applies to the steep but long and even contours above COLEDALE BECK, cf. ROOKING. (b) A Gaelic hill-name *ceann* may also be possible (cf. KINMONT, KINNISIDE). It is not common in Scottish hill-names (Drummond 1991, 27), and I have not found examples of *ceann, cionn* as a simplex p.n. in Irish or Scottish p.ns (Flanagan & Flanagan

1994, 47; McKay 1999, Watson 1928), for it typically occurs in a compound or phrasal name meaning 'head/end of' another feature; but it can be qualified by an adjective, functioning as a term for 'height' (e.g. Kinallen and Kinbane, McKay 1999, 90; Kinglass, Watson 1926, 147), and if it can also stand alone with this meaning it is paralleled by its Brittonic cognate *pen(n)* (see PENN).

KINNISIDE: K~ COMMON NY0711 Ennerdale.
Kynisheved 1321, *Kynysyde* 1541, *two little hilles called Keniside and Wasaborowe* 1578 (*PNCumb* 386); *Kinniside Common* 1867 (OS).
♦ K~ appears originally to have been a hill-name. Its 2nd el. is clearly OE **hēafod**, ME **heved** 'head, high place', while the 1st may have been the pre-existing name of a peak — conceivably the present LANK RIGG which is the highest point on K~ **Common**. This 1st el. (a) could have been from Gaelic *ceann* 'head, end' as in KINMONT (so *PNCumb* 386, Coates 2000, 288, with caution, and see KINN above), or (b) it could be from ON *kinn*, again as in KINN above. In either case one would have to assume that a ME gen. sing. *-es, -is* had been added. (c) Ekwall suggested instead OE *Cyne*, a pers.n. (*DEPN*). K~ forms a civil parish with Ennerdale.

KIRKBANK SD1382 (hab.) Whicham.
Kirkebaunk 1278 (p), *Kirkbank(s)* 1653 (*PNCumb* 444).
♦ 'The slope by the church'; **kirk**, **bank(e)**. *Kirk* in Lakeland names is from ON *kirkja* 'church', which was adopted from OE *cirice*, and became widespread throughout N. England and parts of Scotland. The location is the top of the lane leading up from WHICHAM church.

KIRKBARROW NY4926 (hab.) Barton.
Kirk(e)barrow 1677 (*PNWestm* II, 213), *Church Barrow* (Clarke 1789, map facing p. 24).
♦ 'Church hill', with *kirk* as in KIRKBANK plus **berg/barrow**. This modest hill is immediately above St Paul's Church.

KIRKBY (parish of K~ Ireleth): K~ POOL SD2386 (river) Broughton West/Kirkby Ireleth, K~ PARK WOOD SD2386 Kirkby Ireleth.
Kirkebi 1191–8, *Kirkebi Irlid* 1180–99, *Kirkeby Irlith* 1278 (*PNLancs* 220); *Kirkby Pool or Steers Pool* 1851 (OS); *Kirkby Park Wood* 1851 (OS), cf. *Kirkby Park* 1745 (West 1774, map).
♦ K~ is 'the village with a church', from ON *kirkja* 'church' and ON **bý/ bœr**; also **pool** 'river', **park**, **wood**. The parish is distinguished as K~ Ireleth (with I~ meaning 'Íri's slope' or 'slope of the Irishman/men', *PNLancs* 205, *SSNNW* 138, cf. 319). The village, with the modern postal name of Kirkby-in-Furness, is at SD2382, outside the National Park. K~ Pool denotes the lower reaches of STEERS POOL. K~ Park Wood is in the SW corner of Kirkby Ireleth parish.

KIRKBY HOUSE SD4692 Underbarrow.
Kirkbyhouse 1745; cf. *Kirkebybank* 1334 (*PNWestm* I, 105).
♦ 'The house of the Kirkby family'; surname plus **house**. Two families of this name are documented locally. They probably, as Smith suggests, took their name from Kirkby Kendal, now KENDAL (*PNWestm*).

KIRK FELL NY1726 Lorton.
From 1867 (OS).
KIRK FELL NY1910 (2630ft/802m.)

Ennerdale/Nether Wasdale, KIRKFELL CRAGS NY1910 Ennerdale.
Kerkefell, le egge of Kirkefell 1338 (*PNCumb* 392), *Kirk fell* 1802 (*CN* 1208); *Kirkfell Crags* 1867 (OS).
♦ 'Church mountain', with *kirk* as in KIRKBANK, plus **fell**; also **crag** 'rocky height'. The Lorton K~ F~ lies E of LORTON church, and dominates the view from there. A church in Lorton is recorded from at least 1198 (George 1995, 4), though whether it was on the same site is unknown. The Ennerdale K~ F~ forms a massive backdrop for the hamlet of WASDALE HEAD and its church. K~ F~ Crags fringe the N of the top of K~ F~.

KIRKHEAD NY1320 (hab.) Loweswater.
Kirkeheued 1279 (p) (*PNCumb* 410), *Kirkeheved* 1286 (in late medieval copy, *St Bees* 146).
♦ 'The high place above the church', from OE *cirice*, influenced by ON *kirkja* (see KIRKBANK) plus OE **hēafod** 'head, high place, headland', or their ME reflexes. K~ lies above the present, 19th-cent., LOWESWATER church, suggesting continuity of site since the granting of a chapel c.1125 (*St Bees* 2).

KIRKLANDS FARM/KIRKLAND NY0800 Irton.
Kyrkeland 1472 (*PNCumb* 403), *Church Land* 1774 (DHM).
♦ 'Land belonging to the church', probably from ON *kirkju-land* with that meaning, with **farm** added in modern times. The place is very close to IRTON church. The present building is 19th-cent., but the antiquity of the site is shown by the 9th-cent. CROSS there.

KIRKSTONE NY4008 Patterdale, KIRKSTONE PASS NY4009.
Kirkestain pre-1184; *magna via Kirkestain* (great K~ road) 13th cent., *a pass called Kirkstone* 1671 (*PNWestm* II, 223), *Kirkstone Pass* 1860 (OSNB Barton, 2, 91).
♦ 'The church stone', with *kirk* as in KIRKBANK, plus **stone**; also **pass**. The name was probably inspired by the shape of one of the massive boulders which strew the pass, though modern writers disagree about its location. Wordsworth envisaged the Roman legionaries beholding 'This block — and yon, whose church-like frame | Gives to this savage Pass its name' ('The Pass of Kirkstone', *Poems* 171; cf. Machell c.1692, ed. Ewbank, p. 148). K~ Pass links PATTERDALE and ULLSWATER with AMBLESIDE and WINDERMERE and is the course of a Roman Road (Richardson *et al.* 1990).

KIRKTHWAITE COTTAGE SD3287 Colton.
Kyrkwythe c.1535, *Kyrkthwayte* 1537 (*PNLancs* 216).
♦ K~ is 'church clearing', from ON *kirkja* 'church' and ON **þveit** 'clearing' (or their ME reflexes); also **cottage**. K~ is over a mile N of the nearest existing church, at COLTON.

KISKIN SD0986 (hab.) Bootle.
Kiskinge 1678 (PR), *Kiskin* 1702 (*PNCumb* 348), *Kischin* 1772 (PR).
♦ Obscure. *PNCumb* suggests comparison with *le Kiskane* 1368 (*PNCumb* 308), and the field-name *Kiskin Spring*, a field-name (*PNCumb* 268), but they too are unexplained.

KITCHEN GROUND SD1099 Irton.
Kitchin Ground 1688 (*PNCumb* 403).
♦ 'The farm of the Kitchen family'; surname plus **ground**. The 1688 document shows the farm in the possession of a John Kitchin, though

the surname is recorded as early as 1466 in Gosforth parish (*PNCumb* 403), and appears frequently in Irton PRs from 1691 onwards. *Ground* is much less common in this W area of Lakeland than in Furness.

KIT CRAG NY4914 Shap Rural.
From 1859 (*PNWestm* II, 175).
♦ Probably 'Christopher's rocky height'; pers.n. plus **crag**. 'Kit' may be the abbreviation of *Christopher* which flourished especially 16th–18th cent. (Withycombe 1977, 66). The common noun meaning 'bucket, tub' etc., seems less likely. A habitation of the same name is recorded as *Kitcrag* in Cartmel PR for 1625.

KNIGHT, THE NY4017 Patterdale/Martindale.
From 1859 (*PNWestm* II, 226).
♦ This is a pile of stones marking the parish boundary. The word *knight* is from OE *cniht* 'boy, youth', but gained its present sense in the ME period. Given the mainly S. and SW. distribution of OE *cniht* in p.ns, The Knight is regarded as a 'modern oddity' (*VEPN: cniht*).

KNIPE: HIGH K~ NY5219 (hab.) Bampton, LOW K~ NY5119 (hab.), K~ MOOR NY5219.
Gnip(e), Gnyp(e) c.1160 to 1479, *Knipe, Knyp(e)* 1246 to present; *Hognip(p), Hognyp(e)* 1241, *Over G~* 1279, *High Knipe* 1699; *Nether G~* 1279, *Low Knipe* 1676 (*PNWestm* II, 191); *Knipe Moor* 1863 (OS).
♦ K~ is from ON *gnípa* 'jutting crag, rocky summit', here denoting the limestone escarpment now called Knipe Scar, which dominates this part of the LOWTHER valley. The farms of **High** and **Low** K~ lie well below; see also **moor**. The suffix *Patrik* is added to the name K~ in some medieval records, commemorating its 13th-cent. ownership (*PNWestm* II, 191).

KNIPE FOLD SD3499 Hawkshead.
Knipe fold 1713 (Torver PR).
♦ Probably 'the pen or farm of the Knipe family'; surname plus **fold**. *Knipe* derives ultimately from ON *gnípa* 'jutting crag' (cf. KNIPE above), but it does not describe the landscape here, and is probably a surname derived from a p.n., surnames being a common type of specific with *fold*. A family named Knipe or Gnype is well recorded in Furness from the 14th cent. to 18th cent. (*VCHLancs* 280 and n. 32, 282 etc.), and is very frequent in the Hawkshead PRs.

KNIPE TARN SD4294 Crook.
From 1836 (*PNWestm* I, 177).
♦ ON *gnípa* 'jutting crag, rocky summit' has given rise to local p.ns and hence to a surname, recorded locally from the early 14th cent. as *Gnipe, Gnype* and later *Knype* (*PNWestm* I, 177–8 and Kendal PRs). It is not clear whether the **tarn** or small pool (from ON **tjǫrn**) is named from a person or a place.

KNITTLETON SD2586 Kirkby Ireleth.
From 1689 (PR).
♦ This presents a puzzle: the el. *-ton* could point to Anglian origins (OE **tūn** 'settlement'), yet no medieval forms have as yet been found, and the place is not in a main valley and never attained great importance, as would be characteristic of Anglian settlements. Some connection with nearby Knott (End) is conceivable, but this, like Brearley's suggestion of 'nettle ton' (1974), is a shot in the dark.

KNOCK MURTON: see MURTON

KNOTT NY2933 (2329ft/710m.) Caldbeck.
Knot 1794 (*PNCumb* 279).
KNOTT SD4792 (hab.) Underbarrow.
Netherknote, Overknote 1574, *Knot* 1745 (*PNWestm* I, 105).
KNOTT, THE Examples at:
SD1495 (1086ft/331m.) Muncaster.
Knott 1867 (OS); cf. *Knotland* 1570 (*PNCumb* 426).
SD2493 (1093ft/332m.) Broughton West.
From 1851 (OS).
NY4312 Patterdale/Martindale.
From 1865; cf. *Medilknott* 1220–47, and possibly *Thome Knott* 13th cent. (*PNWestm* II, 219).
NY5011 Shap Rural.
From 1859 (*PNWestm* II, 175).
KNOTTS NY2614 Borrowdale.
From 1867 (OS).
KNOTTS NY4321 (hill & hab.) Matterdale.
le Knott 1487, *Knotte* 1589 (*PNCumb* 257), *Knot* 1774 (DHM).
♦ 'The craggy, compact hill'; **knott** probably from ON **knǫttr**. The Muncaster one, for instance, is steep and craggy on three sides. At Underbarrow, the hill above and S of the settlement is the rocky Knott Hill. The 16th-cent. spellings may suggest that there were previously two farms here. The two places named Knotts refer to single, rugged heights and, in the Matterdale case, the settlement below. The alternation of sing. and pl. forms is not unusual, and the pl. is appropriate for 'a hill with numerous isolated outcrops separated by turf or heather' as at Matterdale (Keith Clark, pers.comm. gratefully acknowledged).

KNOTT END Examples at:
SD1397 (hab.) Muncaster.
From 1754 (PR).
SD2291 (hab.) Dunnerdale.

Knotend 1793 (Seathwaite PR).
SD2586 (hab.) Kirkby Ireleth.
From 1699 (PR), 1851 (OS).
KNOTT ENDS NY1608 Nether Wasdale.
From 1867 (OS).
♦ 'The end of (The) Knott, the craggy height'; **knott, end**. K~ Ends, Wasdale, differs in being the name not of a habitation but of the craggy apex of the S. triangle of HIGH FELL, the other points being BLACKBECK KNOTTS and Great Knott.

KNOTT HILL SD1787 (921ft/281m.) Millom Without.
From 1867 (OS); cf. *Knotend* 1608 (*PNCumb* 418).
♦ This is a classic **knott** — craggy and compact, and Knott may well have been the original name, with **hill** added later.

KNOTT RIGG NY1918 Buttermere/Above Derwent.
From 1867 (OS).
♦ 'The ridge with rock outcrops' (**knott**; **hryggr/rigg**), which describes the hill well.

KNOTTS: see KNOTT

KNOWE, THE NY4508 (2509ft/765m.) Kentmere/Longsleddale.
Knowe(tarne) 1574, *The Knowe* 1857 (*PNWestm* I, 163).
♦ 'The hill-top', with *knowe* from OE *cnoll* 'small hill, hill-top'. The K~ is a shapely, elongated summit.

KNOWE CRAGS NY3126 Threlkeld.
From 1867 (OS).
♦ 'The rock outcrops at the summit' (of BLEASE FELL, part of the BLENCATHRA massif), with *knowe* as in THE KNOWE, plus **crag**.

L

LACONBY NY0902 Gosforth.
Lakynbye, Lakenby 1548 (*PNCumb* 396); pronounced [leikənbi].
♦ This is ON **bœr/by** 'farmstead, settlement', probably qualified by a pers.n. since over half of the p.ns in -*by* in the former Cumberland have a pers.n. as specific el. (*PNCumb* 500), but what the pers.n. could have been is uncertain. Although comparison has been made with Lackenby in NYorks (*PNCumb* 396), the OIr pers.n. *Lochān* or the postulated ON nickname suggested in that name (Fellows-Jensen 1968, 354–5) would not account for the pronunciation of Laconby.

LAD CRAGS NY4715 Bampton.
From 1860 (OSNB).
♦ The meaning of **lad** here and elsewhere is obscure (see List of Common Elements), but may be either 'pile' or 'boy, youth'; see also **crag**. There is another Lad Crag below the summit of LOUGHRIGG FELL at NY3404, and a Great and Little Lad Crag well below HAYCOCK, at NY1410 and NY1511.

LAD HOWS NY1719 Buttermere.
Ladhow 1803 (*CN* 1518), *Ladhouse* 1823 (GreenwoodM), *Ladhows* 1867 (OS).
♦ A puzzling name. The pl. Hows (from ON **haugr** 'hill, mound') may seem curious in connection with this impressive single buttress of GRASMOOR, but the pl. also seems to be used rather accidentally in THE HOWES NY5017. Greenwood's form *Ladhouse*, and Wainwright's reference to 'Lad House, now known as Lad Hows' (1964, Grasmoor 2), might point to a building now lost, but *house* may merely have been an attempt to make sense of the pl. *howes*. **Lad** co-occurs with other 'hill' terms, and may mean either 'pile' or 'boy', probably the former; cf. LAD CRAGS above.

LADIES TABLE NY2028 Wythop.
From 1867 (OS).
♦ *Lady* derives from OE *hlǣfdige*, while *table* — rare in p.ns — is an adoption from OFr into ME. This is a small, shapely peak, whose paths and view are now obscured by a plantation so that 'the place is forgotten and only the name remains', as Wainwright puts it. In the absence of early evidence his guesses as to the motivation of the name cannot be confirmed or denied — or bettered: 'A flat boulder, probably used by Victorian picnickers, may have given the place its name, but more likely it is a gentle parody on Lord's Seat nearby' (Wainwright 1964, Sale Fell 5; a drawing is included).

LAD STONES NY2900 Coniston.
From 1850 (OS).
LADSTONES, GREAT NY5312 (1439ft/ 439m.) Shap Rural.
Ladstones 1859 (*PNWestm* II, 175), *Great Ladstones* 1860 (OSNB).
♦ Probably 'piled up stones' (see **lad**, **stone**), though 'boy' may have developed as a further connotation of **lad**; cf. LAD CRAGS, LAD HOWS.

LADY HOLME SD3997 Windermere.
Lady-holme c.1692 (Machell II, 342, ed. Ewbank, 87), 1780 (West 58); cf. *Isle of the Blessed Mary of Winandermere* 1271–2 (translated, *Scotland* I, 537), *Seyntemarieholm* 1334, *Marieholm* 1340 (*CW* 1987, 93) & variants to 1530 (*PNWestm* I, 193) *our ladie chappel of Tholme* 1546 (*PNWestm* I, 193).

♦ 'The islet of Our Lady.' *Lady* is ultimately from OE *hlǣfdige*, while ON **holmr** or its reflex is used in the names of numerous Lakeland islets. This one is named from the medieval chapel to the Virgin there; a chantry and hospital are also recorded (Wiseman 1987, 92–4, Brydson 1911, 75–99).

LADYSIDE PIKE NY1822 Buttermere/Lorton.
From 1867 (OS).
♦ A somewhat obscure name. Wainwright remarks that the hill was 'formerly Lady's Seat' and compares LORD'S SEAT on the other side of the WHINLATTER PASS (1964, Hopegill Head 2 and 6). *Lady*, if to be taken at face value, is ultimately from OE *hlǣfdige*. 'Side' could be from ON **síða**/OE **sīde** '(hill-)side', or could result from the common interchange between **seat** (ON **sæti**) and **side**, but without earlier spellings the original form is uncertain. The **pike** 'peak' is a subsidiary summit of HOPEGILL HEAD.

LADY SYKE/SIKE SD3283 (hab.) Haverthwaite.
Lady Sikes in Cartmel 1728/9 (Colton PR), *Lady-syke* 1799 (Egton-cum-Newland PR).
♦ There are **sykes** or small streams in the vicinity. Why *lady* (from OE *hlǣfdige*) is unclear.

LADY'S RAKE NY2721 Borrowdale.
Lady's-rake 1786 (Gilpin I, 187).
♦ 'The lady's steep track', with *lady* ultimately from OE *hlǣfdige*, plus **rake**. This is a gully leading from the shore of DERWENTWATER over WALLA CRAG. It was claimed to be the escape route from LORD'S ISLAND (q.v.) of Lady Derwentwater after the failure of the Jacobite rebellion and arrest of the third Earl of Derwentwater in 1715, 'by climbing the horrid and stupendous heights with such jewels and valuables as she could secure', as Hutchinson put it in 1776 (p. 158). By 1794, however, he was describing this as a 'traditional tale' (II, 191), and already in 1789 Clarke (who calls the place *Ladies Walk*) classed this among the unlikely tales pedalled to tourists by the boatmen of KESWICK (p. 69, see also Collingwood 1904, 272–5). The scepticism was well based since the island seems to have been abandoned before this time. A different story, of a virtuous Lady Derwentwater escaping from her robber husband after destroying his loot, is reported in *JEPNS* 6, 44.

LAG BANK SD2494 Broughton West.
From 1851 (OS).
♦ This is a standard **bank** 'hill-slope', but the meaning of 'Lag' is elusive. The postulated OE **lagge* 'marsh' in Smith, *Elements*, is rare and southern, and there is no means of knowing which, if any, of the senses of *lag* as recorded in *OED* is relevant: 'last, hindmost' (*lag*, n. 1 and adj.); a stave or lath on a barrel (*lag*, n. 2); or a cleft or mark in a tree or timber (*lag*, n. 3). *EDD* adds *lag* as a term for the wild grey (greylag) goose (int. and sb. 9, sense 2). Brearley's suggestion of Gaelic *lag* 'hollow' (cf. LOGAN BECK) seems unlikely.

LAGGET NY0512 (hab.) Ennerdale.
Lagart 1695/6, *Leggat* 1721/2, *Laggarth* 1733, *Lagget* 1784 (PR).
♦ Obscure. 'The low garth or enclosure', from ON **lágr** and **garðr**, is a possibility, but the forms are too late and inconsistent for this to be more than a tentative conjecture.

LAKE BANK SD2889 (hab.) Blawith.
Lakebank (Hotel) 1893 (OS), *Lakebank Lodge* 1919 (OS).
LAKE BANK SD3696 (hab.) Claife.
From 1851 (OS).
♦ These appear to be quite recent names, from *lake* as in LAKE DISTRICT and **bank**. The Blawith example is on the shore of CONISTON WATER, the Claife one beside ESTHWAITE WATER.

LAKE DISTRICT.
From 1829 (Parson & White, title).
♦ Despite its prominence in the names Lake District, Lakeland or simply 'the Lakes', the word *lake* (adopted into ME from OFr *lac*) only appears in a few, relatively modern, p.ns, **mere**[1] or **water** being the main words for larger lakes. On *district*, see ALLERDALE D~. The occurrence of 'Lake District' in the title of Parson & White's *Gazetteer* of 1829 slightly predates *OED*'s first citation, from Wordsworth in 1835 (*lake*, n. 4, 6a); there may be still earlier occurrences.

LAKESIDE SD3787 (hab.) Colton, LAKESIDE & HAVERTHWAITE RAILWAY SD3585.
Lake Side 1748 (Kitchin map in *CW* 1918, 51).
♦ Simply 'the place beside the lake'; *lake* as in LAKE DISTRICT, **side**. 'The old name was Landing' (*VCHLancs* 383n.): see LANDING HOW. Lakeside is at the foot of WINDERMERE, and forms one terminus of the Lakeside & Haverthwaite steam railway, which preserves a branch line of the former FURNESS Railway.

LAMBFOOT NY1630 (hab.) Embleton.
Langfite c.1210, *Langfit* 1363 (*PNCumb* 384), *Lamfoot* 1676, *Lamthwaite* 1702 (PR).
♦ 'The long meadow'; ON **langr** 'long' and ON *fit* 'meadow, especially by a river bank'. The present form results from regular phonetic processes together with substitution of elements (see *PNCumb*).

LAMB HOWE SD4291 (hab.) Crosthwaite, LAMBHOWE PLANTATION SD4191.
Lamb Howe 1857 (*PNWestm* I, 84); *Lambhowe Plantation* 1862 (OS).
♦ 'The hill frequented by lambs', with the reflexes of OE/ON *lamb* — rare in this volume — and of ON **haugr** 'hill, mound' (assuming that there is no corruption, for instance of *lang-*, as in LAMBFOOT); also **plantation**. The hill is now known as Lambhowe Hill.

LAMB PASTURE NY5302 Fawcett Forest.
From 1836 (*PNWestm* I, 139).
♦ Presumably from the reflex of OE/ON *lamb* and **pasture**, but whether the name is to be taken at face value is not certain. It seems to match nearby Kids Howe, and perhaps Wolf Howe, but as a hill of over 1000ft/300m. it is not a hospitable place for young lambs. Corruption of **lang(r)** as in LAMBFOOT is possible.

LAMPLUGH NY0820 (hab. & parish), L~ FELL NY1019.
Lamplou c.1150 to 1279, *Lamplogh* c.1160, *Landplo(h)* c.1200 (p) to 1266 (*PNCumb* 404–5); *Lamplugh fells* 1687–8 (Denton 51), *Lamplugh Fell* 1867 (OS).
♦ L~ may be 'the bare valley', from a Brittonic *Nant bluch* 'bare valley', as argued by Padel (1980–2, developing a suggestion of Quentel). L- spellings for W *nant* 'valley' and its equivalents are early and widespread in Brittonic dialects, especially in areas of English influence; *blwch* 'hairless, bare' is relatively rare, and Lamplugh constitutes important evidence for its

use in Cumbric. (b) The editors of *PNCumb* proposed Brit. **landā* 'enclosure' as the 1st el., but had no solution for the 2nd el. (p. 405). L~ **Fell** is a tract of grazing rather than a distinct peak.

LANCRIGG NY3308 (hab.) Grasmere.
Langridge 1577 (*PNWestm* I, 201), *Lankrigg* 1804 (*CN* 1812).
♦ 'The long ridge' (**langr, hryggr/ hrycg/rigg**), presumably the one extending NW from the settlement of this name. The *ridge* spelling reflects OE *hrycg* and its Standard English descendant, while the *rigg* spelling reflects ON *hryggr* and its dial. descendant. The name is equivalent to LANK RIGG.

LAND ENDS NY4324 (hab.) Matterdale.
Landends 1587 (*PNCumb* 257).
♦ Self-explanatory; **land, end**. The motivation of the name is uncertain, though the site is just below, and E of, LITTLE MELL FELL.

LANDING: L~ HOW SD3787 Colton.
Lendine 1598/9, *lending* 1621/2 (Hawkshead PR), *Lending* 1726/7, (*the*) *Landing* 1767 (Finsthwaite PR), *Linden* 1745 (West 1774, map), *Landing* 1783 (CrosthwaiteM); (L~ How:) previously *Landing Knott* (*Wood*) 1851 (OS).
♦ 'The hill above Landing, the landing-place', with p.n. Landing plus **how(e)** from ON **haugr** 'hill'. The name Landing is now partly displaced by LAKESIDE, though there is still a Landing Cottage. *OED* has citations from 1609 onwards for *landing* in the concrete sense of a 'landing-place', and this seems the likeliest meaning here, though Cowper regarded it as referring to a ford across the LEVEN as it flows out of WINDERMERE, which was 'eighty yards across, and about two feet deep when the lake was normal' (Cowper 1899, 250). L~ How rises above the S. tip of WINDERMERE.

LANDS POINT SD3096 Coniston.
From 1851 (OS).
♦ **Land** presumably has its obvious sense; the **point** projects into CONISTON WATER.

LANE END SD1093 (hab.) Waberthwaite.
Loning End 1774 (DHM), *Lane end* 1790, *Lonning end* 1796 (PR).
LANE END SD2190 (hab.) Broughton West.
From 1666 (Broughton-in-Furness PR).
LANE ENDS SD3484 (hab.) Haverthwaite.
Possibly = *Lane End* 1727 (Colton PR).
♦ 'The end of the lane, or junction of lanes'; with Standard ModE **lane** (replacing dial. *lon(n)ing* in the Waberthwaite example), plus **end**. L~ E~ SD1093 is an outlier of WABERTHWAITE; the Broughton and Haverthwaite ones are at junctions. The Haverthwaite place is not shown or named on the OS maps of 1851 or 1893, where Gateside appears, so that identification with *Lane End* 1727 is very uncertain. The place is close to CAUSEWAY END.

LANEFOOT NY0615 (hab.) Ennerdale.
From 1867 (OS).
LANEFOOT FARM NY1326 Blindbothel.
Lonenfoot 1734, *Lanefoot* 1736 (*PNCumb* 407).
LANEFOOT FARM NY2224 Above Derwent.
Lane-foot 1731 (*PNCumb* 373).
♦ 'The foot of the lane'; **lane, foot**; also **farm**. As with other places named from lanes, it is often difficult to know precisely which routeway is meant. The Ennerdale farm is situated beside a road running S over the fell. The Above Derwent example is close to

the WHINLATTER PASS as it descends to BRAITHWAITE, and on a route following the W side of BASSEN-THWAITE LAKE. *Farm* has been added relatively recently in two cases.

LANE HEAD SD3184 (hab.) Colton.
Lanehead 1698 (PR).
LANE HEAD NY3727 (hab.) Mungrisdale.
Lanehead 1774 (DHM).
LANE HEAD/LANEHEAD NY3914 (hab.) Patterdale.
From 1860 (*PNWestm* II, 226).
♦ 'The top of the lane' (**lane, head**), though it is not always easy to see which 'lane' is referred to. The Mungrisdale example, for instance (Lane Head Farm on its sign), could reflect its situation near the course of the old Roman Road just W of Troutbeck camps (near the present A66, Hindle 1998, 27), where it rises as it heads E and where, on the 1774 map, turnpike road gives way to open road. The name could alternatively refer to the junction at this point between a small track and the main road.

LANGDALE: GREAT L~ NY3006 (Langdales), GREAT L~ BECK NY3006, LITTLE L~ NY3103, LITTLE L~ TARN NY3003, L~ COMBE NY2608, L~ FELL NY2606, L~ PIKES NY2707.
Lang(e)den(e) 1179 (p) to 1630, *Langdal(e)* 1578 (*PNWestm* I, 203); (Great L~:) *micklelangdaile* 1564 (*CW* 1908, 145), *Great Langdale alias Mickledale* c.1692 (*Machell* II, 134), *Great Langdale* 1706 (*PNWestm* I, 203); (Great L~ Beck:) *Langdale-beck* c.1692 (*Machell* I, 99), *Great Langdale Beck* 1862 (OS); (Little L~:) *Langedenelit(t)le* 1157–63 (*PNWestm* I, 203); (Little L~ Tarn:) *Langdale Tarn* 1745 (map in West 1774), *Little Langdale Tarn* 1823 (Otley 28); *Langdale Combe* 1859 (OSNB); *Langdale Fell* 1738, cf. *del Fell* 1375 (p) (*PNWestm* I, 206); *Langdale Pykes* 1770 (JefferysM), *Langdale Pikes* 1787 (*PNWestm* I, 206).
♦ 'The long valley'; **langr, dalr** (evidently replacing OE *denu* 'valley'); also **great, little, beck** 'stream', **tarn** 'pool', **comb**(e)[1] 'rounded valley', **fell** 'open grazing land', **pike** 'peak'. Little L~, with its tarn, lies S of Great L~. L~ Combe is a classically shaped boggy basin, some distance from Great L~ valley but at the NW corner of the former parish. L~ Fell is not an individual peak but the high grazing land at the head of Great L~. The L~ Pikes are, W to E, PIKE OF STICKLE, HARRISON STICKLE and PAVEY ARK. With their spectacular elevation when seen from S or E, they well deserve the name of *pike*. (Pl. 26)

LANG HOW NY3106 Grasmere.
From 1862 (OS).
♦ 'The long hill'; **langr/long, haugr/how**(e). It is an elongated oval, and a classic *how*(e).

LANGHOWE PIKE NY5213 (1313ft/400m.) Shap Rural.
Langhowe Pike 1860 (OSNB), *Longhowe Pike* 1865 (*PNWestm* II, 175–6).
♦ 'The summit on Langhowe, the long hill'; **langr/long, haugr/how**(e), **pike**.

LANGLEY SD0991 (hab.) Waberthwaite, L~ PARK SD0992 (hab.).
Langliuerh', Lang(e)livere c.1225 (p); *Langley Park* 1702 (*PNCumb* 365).
♦ 'Langlíf's shieling', with ON female pers.n. *Langlíf* and Gaelic-Norse **ærgi/erg** 'shieling'; also **park**.

LANGSTRATH/LONGSTRATH NY2609 Borrowdale, L~ BECK NY2611.
Langestrothe 1189–99, *Langestrode* 1211

(*PNCumb* 351); *Langestrothebec* 1209–10 (*PNCumb* 20).
♦ 'The long, marshy area overgrown with brushwood'; **lang(r)** plus OE *strōd* 'marshy ground overgrown with brushwood' and/or ON *storð* 'brushwood'. The 2nd el. is not, as often assumed, the Gaelic word for 'valley' found in Strathclyde etc., despite the aptness of 'long valley' here. In the *lang/long* variation we see competing influences: the Scand and dial. *-a-* form as against the Standard English *-o-*. See also **beck** 'stream'.

LANK RIGG NY0711 (1775ft/541m.) Ennerdale, LANKRIGG MOSS NY0812.
Lankrig 1826, *Lankrigg* 1827 (CRO D/Lec/94); *Lankrigg Moss* 1867 (OS).
♦ Probably 'the long ridge' (**langr, hryggr/rigg**, cf. LANCRIGG), this being the highest part of KINNISIDE COMMON, and a long watershed from which numerous becks and gills flow. L~ **Moss** is on the NW skirts of the hill.

LANTHWAITE NY1520 (hab.) Buttermere.
Langthwate 1505 (*PNCumb* 354), *Longthwaite* 1774 (DHM).
♦ 'The long clearing'; **langr, þveit**.

LATRIGG NY2724 (1204ft/367m.) Underskiddaw.
Laterhayheved 1220, *Latterigg* 1666 (*PNCumb* 321–2).
♦ The linguistic origins, especially of the 1st el., are doubtful. The main possibilities are (a) ON *látr* 'lair, shelter' or (b) Gaelic *lettir* 'hill, slope' (see **latter**), but neither of these would form a plausible compound with the remaining elements, or is likely to have formed a simplex name to which the remaining elements were added later. The *-hay-* in the 13th-cent. spellings is unexplained, but could point to a word such as OE (*ge*)*hæg* or *hege* 'fence, enclosure'. The final el. is **rigg** 'ridge' from ON **hryggr**, replacing OE **hēafod** 'top, head, high place' (cf. LATTERHEAD).

LATRIGG NY4101 Windermere.
Laterigg 1865 (*PNWestm* I, 196).
♦ Probably 'the ridge with the lair or shelter', from ON *látr* 'lair or shelter' plus **hryggr/rigg** 'ridge' (so *PNWestm*). A Gaelic *letter* 'hill, slope' (see **latter**) is unlikely, given the ON origins of the 2nd el.

LATTER BARROW NY1711 Ennerdale, LATTERBARROW BECK NY0711, LATTERBARROW MOSS NY0711.
Latterbarrow 1823 (Otley 60); *L~ Beck*; *L~ Moss* both 1867 (OS).
LATTERBARROW SD3699 (804ft/245m.) Claife.
From 1850 (OS).
LATTERBARROW: L~ FARM SD4383 Witherslack.
Lat'ber(*brigg*) 1280 (*PNWestm* I, 78).
♦ Probably 'the hill with a lair or shelter'. The 1st el. in the Witherslack p.n. seems to be ON *látr* 'lair or shelter' (see **latter**), while 2nd el. is probably ON **berg** 'hill'. The Ennerdale and Claife hill-names may be of the same origin, but without earlier spellings this is uncertain. See also **beck** 'stream', **moss** 'bog' and **farm**.

LATTERHAW CRAG NY3813 Patterdale.
From 1860 (*PNWestm* II, 226).
♦ L~ is 'the hill with a lair or shelter', if this is from ON *látr* 'lair, shelter' and ON **haugr** 'hill' (so *PNWestm*). The Gaelic alternative mentioned under **latter** is much less likely in such a compound. **Crag** 'rocky height' will be a secondary addition.

LATTERHEAD NY1422 (hab.) Loweswater.
Laterheued 1260 (*PNCumb* 410), *Laterd* 1774 (DHM).
♦ The 2nd el. is clearly OE **hēafod**, ME **heved** 'head, top of, high place', but this could replace the cognate ON *hǫfuð*, with ON *látr* 'lair' or possibly 'shed' as 1st el. (b) Alternatively, the 1st el. could be Gaelic *letter* 'hill-slope' (see **latter**), in which case *hēafod, heved* could have been added later.

LAUNCHY GILL NY3015 St John's.
From 1867 (OS); cf. *Lanshewbray* 1821 (*PNCumb* 19).
♦ 'Launchy' is obscure, but (a) the 1821 spelling is suggestive of origins in a compound such as 'long-shaw' (**long, sceaga/shaw**) i.e. 'long wood'. (b) Otherwise, one might postulate an adj. from the verb 'launch' used in a NCy dial. sense of 'to bound, take great strides' (*EDD: launch*, sense 7), hence 'the leaping stream'. L~ **Gill** 'stream in a ravine' descends from the vestigial L~ Tarn, with waterfalls in its lower reach.

LAVEROCK HOW NY0606 (hab.) Ponsonby.
Laverick Howe 1828, possibly = *Lauerdeshow* c.1215 (*PNCumb* 428 = *St Bees* 304).
♦ 'The hill frequented by larks', with dial. *laverock*, which is closer to OE *lāferce* (cf. ON *lævirki*) than Standard English **lark**, plus **haugr/how(e)** 'hill'. If the identification with *Lauerdeshow* 'Lord's How' is correct, the original specific was from OE *hlāford* 'lord' and has been replaced rather than gradually evolving.

LAWNS/LAWNS HOUSE SD4385 Witherslack.
the Launds, Laund 1691 (PR); cf. *Laundsfield* 1681 (*PNWestm* I, 79); *Lawns House* 1899 (OS).
♦ ME *launde*, of OFr origin, means 'a glade'. The spelling *lawn* became current in the 16th cent., and the sense 'grassed part of a garden' in the 18th cent. (*OED: lawn*); also **house**.

LAWSON PARK SD3195 (hab.) Satterthwaite.
Lawson Park 1535 (West 1774, 104), *Lowsonpke* 1659 (Hawkshead PR), *Lowson Park* 1783 to 1797 (Satterthwaite PR).
♦ L~ is an originally patronymic surname based on *Law*, a diminutive of *Lawrence*; the particular Lawson has not, to my knowledge, been identified. With other Furness **park**s, this was an outlying farm of FURNESS Abbey.

LAYTHWAITE CRAGS NY4714 Bampton.
From 1865; cf. *Laythalt* 1651, *Laythalt als.* (or) *Lathhead* 1716, *Laithwaite* 1774, *Laithald* 1839 (*PNWestm* II, 191).
♦ L~ is obscure. The common el. *thwaite* from *þveit* seems to have replaced a different original. Smith suggests origins in **leið-hólf* 'track-bridge' (*PNWestm*), but although *leið* 'track' would match the known spellings, **hólf* would not; while more obvious solutions such as OE *hald* 'shelter, protection' assume an unusual (though not impossible) hybrid. The name therefore remains opaque, and the original L~ is now submerged under HAWESWATER. See also **crag** 'rocky height'.

LEAGATE NY5300 (hab.) Whitwell.
Lee yeat 1836, *Leagate* 1857 (*PNWestm* I, 150).
♦ Probably 'the gate onto the pasture'; reflexes of **lēah, geat**.

LEAPS BECK NY0818 Lamplugh.
From 1867 (OS).
♦ **Beck** 'stream' is unproblematic, but there is an embarrassment of possibilities for the explanation of 'Leaps': (a) Small waterfalls (*EDD*: *leap* sb. 2, sense 16); (b) Salmon leaps (*OED*: *leap* n. 1, sense 2b); (c) a mining term for a fault, recorded since 18th cent., apt since L~ B~ runs among disused mineshafts and iron-ore tips; (d) reflex of OE *lēap* 'basket', especially for trapping fish (Smith, *Elements*); (e) reflex of OE **hlēp* 'chasm, defile, place to leap across' (Smith, *Elements*).

LEEMING/LEAMING HOUSE NY4421 Matterdale.
Leming House 1811 (*PNCumb* 257), *Lemon Hall* 1839 (Ford 133).
♦ L~ (also in L~ Farm and Cottages) is of uncertain meaning. (a) Situated on the main route along ULLSWATER, it could be the NCy *leam(ing)*, a 'term applied to ancient roads or places situated on such roads' (*EDD*: *leam*, sb. 4). (b) Leeming occurs as a place- or river-name, especially in Yorkshire, where it has been associated with OE *lēoma* 'light', ME *le(e)ming* 'shining' (e.g. by Ekwall, *ERN* 247, *DEPN*), or with a Brittonic root cognate with OIr *leamh* 'elm tree' (*PNNYorks* 227, though Ekwall in *ERN* 247 rejects OW *Lēmein* 'elm river' for Leeming Beck, Yorks). (c) More likely, perhaps, is the surname derived from one of the p.ns. See also **house**.

LEGBARROW: L~ POINT SD3182 Colton.
Legbarro 1577 (*PNLancs* 217), *Legbarrow* 1630, *Leckbarrow* 1676 (Colton PR); *Legbarrow Point* 1850 (OS).
♦ The 2nd el. is **barrow** 'hill' from **berg**, the 1st el. more doubtful, but possibly a pers.n. *Leggr*, in which case the name might be equivalent to the 1st two syllables of LEGBURTHWAITE. The **point** is the triangle of rising land at the confluence of the CRAKE with the LEVEN.

LEGBURTHWAITE NY3119 (hab.) St John's.
Legberthwait 1303, *Legburgthwayte* 1530, *Legbo(u)rtwhat* 1563, *Legberthwayte* 1573 (*PNCumb* 313).
♦ The first two syllables are obscure, but may be a p.n. comprising ON pers.n. *Leggr* (cf. Fellows-Jensen 1968, 184–5), plus ON **berg** 'mountain, rock', or ON *borg* '(natural?) fortress'. The final el. is clearly ON **þveit** 'clearing'.

LEVEN, RIVER SD3583 Colton/Staveley/Haverthwaite.
Levena, Leuena c.1160, *Levene, Leuene* 1246, *Leven* 13th cent. (*ERN* 250–1).
♦ As with many ancient river-names, the meaning is elusive. Ekwall in *PNLancs* 191 conjectured a Brit. root meaning 'elm' (cf. OIr *lem*, W *llwyf*), but six years later he suggested a different Indo-European root connected with notions of pouring or gliding (*ERN* 250–2). Coates considers it Brittonic, 'possibly with late phonology' (2000, 362), but does not elaborate.

LEVERS WATER SD2799 Coniston, L~ HAWSE SD2699.
Lauereswater pre-1220 (*CW* 1929, 41), *Levers water (Damm)* 1713–4 (*CW* 1989, 202), *Levers Tarn* 1786, *Levers Water* 1830 (*PNLancs* 192); *Levers hawse* 1850 (OS), possibly = *Lauerescart* before 1220 (*CW* 1929, 41).
♦ The 1st el. may, as Brearley (1974) suggested, be a pers.n. such as OE *Lāfhere*. Like nearby LOW WATER and GOAT'S WATER, L~ Water is more of a **tarn** than a typical **water**, as the 1786

record (Yates's map) seems to acknowledge. L~ Hause (see **hals**) is the narrow col between SWIRL HOW and the OLD MAN OF CONISTON.

LICKBARROW: HIGH L~ SD4197 (hab.) Windermere.
Licheberg 1220–46, *Lickeberg* 1277, *Licbergh(e)* & variants 1301 to 1411, *Lyckebarowe* 1560 (*PNWestm* I, 186); *High Lickbarrow* 1859 (OSNB).
♦ 'Corpse hill', from OE *līcbe(o)rg*, which is recorded as a compound. The reference is presumably to a burial mound, though none has been found. High L~ and L~ are distinguished, both farms, in OSNB.

LICKLE, RIVER SD2190, SD2393 Dunnerdale/Broughton.
Licul pre-1140, *Likyl* (as p.n.) 1246 (p) (*ERN* 254), *River Lickle* 1745 (West 1774, map).
♦ Uncertain. Ekwall in *PNLancs* 191 suggested ON *lykkja* 'loop' and ON *hylr* (as in TROUTAL), 'pool' or possibly 'slow-moving stream', hence 'stream with bends', but although some terms apply both to pools and streams (see **pool**), there is no evidence that ON *hylr* does, and its normal application to deep pools and etymological relationship with **holr** (de Vries 1977) would discourage the idea. Ekwall later had doubts, tentatively suggesting OE or Brit. etymons (*ERN* 254) and finally labelling the name 'unexplained' (*DEPN*). See CROGLINHURST for a possible alternative name of the river.

LIGHTBECK SD4792 (hab.) Underbarrow. From 1857 (*PNWestm* I, 105).
♦ 'The light or bright stream', and hence the place beside it, with *light* going back to OE adj. *līht*, *lē(o)ht*, plus **beck**. In OSNB (1858) the name refers to the stream, while *Lightbeck Cottage* is a habitation name.

LIGHTWOOD SD4089 (hab.) Cartmel Fell. *Lightwoode* 1656 (PR).
♦ 'The light or bright wood'; *light*, **wood**. The adj. *light* is from OE *līht*, *lē(o)ht*, which occurs quite frequently in names referring to woods or trees, where it is usually assumed to mean 'light-coloured'. The p.n. Lightwood is widespread (see *OSGazetteer*); for a contrasting 'dark wood' see BLAWITH.

LIMEFITT: L~ PARK NY4103 Windermere. *Lynfit* 1560 (*PNWestm* I, 196), *Limefit* 1657 (PR), *Line-fit* c.1692 (Machell II, 330, ed. Ewbank, 117); Limefitt Park not 1863, 1919 (OS).
♦ L~ is probably 'the water-meadow where flax grows', from ON *lín* 'flax' and ON *fit* 'water-meadow'; also **park**. The place is beside TROUT BECK.

LINBECK SD1398 (hab.) Muncaster, L~ GILL SD1497 Muncaster/Eskdale. *Lindebeck(e)* c.1280, 1350–60, *Linbeck* 1660 (*PNCumb* 424, cf. 20); *Linbeck Gill* 1867 (OS).
♦ 'The stream where lime trees grow', from ON (also OE) *lind* 'lime tree, linden' plus **bekkr** 'stream', and hence the habitation beside it. The addition of **gill** 'ravine with stream' has then distinguished the stream and the habitation. The small-leaved lime (*Tilia cordata*) is 'typically a tree of wooded limestone scars and of steep Lake District gills' which afford it some protection from sheep. It grows in S. Lakeland up to c.600m., and the bases can live for 1000 or 2000 years (Halliday 1997, 178–9). Linbeck may have replaced a Celtic stream-name preserved as DEVOKE.

LIN CRAG FARM SD2787 Blawith.
Lincrag 1637 (Torver PR), *Lincragge* 1724 (Lowick PR).
♦ Without earlier forms the el. 'Lin' is enigmatic, since (a) *lind* 'lime(tree)' as in LINBECK; or (b) ON *lín*/OE *līn* 'flax' would be curious neighbours of a crag. Perhaps (c) **ling** 'heather' is the starting-point, cf. LING COVE; see also **crag, farm**.

LINDALE SD4180 (hab.) Upper Allithwaite.
Lindal 1191 (*VCHLancs* 269 n. 10), *Lindale* 1246 (*PNLancs* 199).
♦ Probably 'the valley where lime trees grow'; ON/OE *lind*, **dalr**. Ekwall remarked in 1922 that lime trees still grew in the upper part of the valley (*PNLancs*). Otherwise, ON *lín* or OE *līn* 'flax' might have been considered possible.

LIND END SD2391 (hab.) Broughton West.
Leanend 1703, *Lin-end* 1720/1, *Lindend* 1734 (Broughton-in-Furness PR), *Lind End* 1851 (OS).
♦ Uncertain. The 1st el. may, as in LINDALE, be either 'flax' or 'lime tree'. Either way, **end** would not mean 'end of', as is usual in names such as LANE E~ and TOWN E~, but would mean 'end place' as in BROAD E~, FAR E~, hence 'the end place where flax grows, or with lime trees'.

LINDETH, GREAT SD3385 (hab.) Colton.
Lindeth Wood 1851 (OS).
LINDETH SD4195 (hab.) Bowness, LOW L~ SD4195 (hab.) Crook.
Lintheued 1220–50, *Lyndeheved* & variants 1292 (p) to 1452, *Lindeth, Lyndeth* 1501 to 1777 (*PNWestm* I, 186); *Low Lindeth* 1859 (*PNWestm* I, 180).
♦ L~ is 'the high place where lime trees grow'; OE *lind* 'lime tree', **hēafod**; also **great, low**. Without early forms the Colton example must be regarded as somewhat doubtful, but as a modest hill of 220ft/67m. it is suitable for description as a *hēafod*. There is also a Lancs example (*Lyndeheved* 1344, *PNLancs* 188), and cf. LINDRETH BROW.

LINDRETH BROW SD4693 (hab.) Underbarrow.
Lindeth/Lyndeth Brow(e) 1616, 1836, *Lindreth Brow* 1859 (*PNWestm* I, 105).
♦ The identity of the 1616 spellings to those of the same period for LINDETH in Crook (q.v.) suggests a connection. Either L~ (Brow) is of the same origin, or contains a surname derived from one of the places named Lindeth. See also **brow**.

LINE RIGGS (hab.) SD3886 Staveley-in-Cartmel.
From 1851 (OS).
♦ This may be 'flax ridges', with reflex of OE *līn*/ON *lín*, plus **rigg** (see **hryggr**), cf. comment on FAIR RIGG.

LINEWATH NY3534 (hab.) Caldbeck.
Linewathe 1560 (*PNCumb* 279).
♦ 'The ford where flax is grown', from ON *lín* 'flax' (possibly encouraged by the cognate OE *līn*) and ON **vað** 'ford'. The place is close to a crossing over the R. CALDEW.

LING COMB NY1515 Loweswater.
From 1867 (OS).
♦ 'Heathery valley or corrie' (**lyng/ling, combe**[1]) — a spectacular basin, still distinctive for its heather today.

LING COVE: LINGCOVE BECK NY2304 Eskdale.
Luncoue 1242, *Lyncoue* 1284–90 (*PNCumb* 20), *an huge Rock w[hi]ch is called Ling-Coue* c.1692 (Machell I, 718);

Lingcove Beck 1867 (OS).
♦ L~ is either 'the cove by the torrents' or 'heathery cove'. The 1st el. is either OE *hlynn* 'torrent' (cf. CROGLIN(HURST)) or **lyng/ling** 'heather' (as suggested in *PNCumb* 20). Both would match the landscape. The 2nd el. is OE *cofa* 'cove' (see **cove**). The **beck** 'stream' name is secondary.

LING CRAGS NY1518 Loweswater.
Ling Crag 1809 (CrosthwaiteM), *High Lingcrag, Low Lingcrag* both 1867 (OS).
♦ 'The jutting rocks where heather grows'; **lyng/ling, crag**. Dashes of heather grow amongst the rock.

LINGEYBANK NY2033 (hab.) Blindcrake.
Lingebanckh 1569, *Lingey Bank* 1810 (*PNCumb* 301).
♦ 'The heathery slope'; **ling(e)y, bank**.

LING FELL NY1728 (1243ft/373m.) Wythop.
From 1867 (OS).
♦ 'Heathery hill' (**lyng/ling, fell**), on whose top 'a sea of heather' is still to be found (Wainwright 1964, Ling Fell 3).

LING HOLME SD3893 Windermere.
Lingholm c.1692 (Machell II, 342, ed. Ewbank, 87).
♦ 'The islet where heather grows'; **lyng/ling, holmr**.

LINGHOLM(E): L~ GARDENS NY2522 Above Derwent.
Ling Holm 1787 (*PNCumb* 373).
♦ L~ is 'the islet where heather grows' (**lyng/ling, holmr**). It referred originally to L~ Islands in DERWENTWATER (or presumably, since sing., to the largest of them), and hence to the settlement on the shore nearby. This is the sole occurrence of *garden*, an adoption from Norman Fr into ME, in this volume, though it occurs elsewhere in ME and modern p.ns.

LINGMELL NY1413 (1410ft/435m.) Ennerdale.
From 1867 (OS).
LINGMELL NY2008 Nether Wasdale, L~ BECK NY1908, L~ CRAG NY2008, L~ GILL NY1907 Nether Wasdale/Eskdale.
Lingmale 1578 (*PNCumb* 392), *Lingmell-(side)*, *Lingmall* 1644 (Winchester 2000, 169–70), *Ling mel* 1802 (*CN* 1213); *Lingmell Beck* 1867 (OS), seemingly = *Ederlangebeck* 1294 (*CW* 1926, 106 = *St Bees* 159 n.); *Lingmell Crags* 1839 (Ford 62), *Lingmell Crag* 1867 (OS); *Lingmell Gill* 1867 (OS).

LINGMELL END NY4409 (2183ft/665m.) Kentmere.
Lingmill End 1857; cf. *Ling* 1836 (*PNWestm* I, 167).
♦ Probably 'the heathery hill', with *ling* from ON **lyng** 'heather' and a Celtic word for 'bare hill' (see **mell**). The Ennerdale example is heathery above the line of the recent conifer plantation while the Wasdale one cannot be described as heathery. Without earlier spellings it is impossible to tell when, and in what linguistic situation, these names were formed, but the combination of an originally ON el. with an originally Celtic one might suggest a Norse-Gaelic context. A further possibility would be that they are ME constructions, *mell* 'hill' having become part of the local toponymic vocabulary. The secondary names are formed from **beck** 'stream', **crag** 'rocky height', **gil(l)** 'ravine with stream', and **end**. (Pl. 23)

LINGMOOR FELL NY3004 (1540ft/469m.), L~ TARN NY3005 Langdales.
Lingmoor-Fell 1829 (Parson & White 616), probably = *Lingemouthe* 13th cent.

(bounds of the Baysbrown estate, *CW* 1908, 159); *Lingmer Tarne* c.1692 (Machell II, 128).
♦ 'The heathery upland waste' (**lyng, moor**), still an apt description of the high plateau of L~, and the **fell** 'high ground' and **tarn** 'mountain pool' taking their names from it. The 13th-cent. spelling appears to be corrupt.

LINGY CRAG NY4113 Patterdale.
From c.1692 (Machell I, 726).
♦ 'The heathery rock outcrop'; **lyng, crag**.

LINING CRAG NY2508 Borrowdale/ Langdale (and former Cumberland/ Westmorland boundary).
From 1867 (OS).
LINING CRAG NY2811 Borrowdale, nr St John's boundary.
From 1867 (OS).
♦ Both **crag**s 'rocky heights' are situated on boundaries, which might suggest some connection with surveying or drawing of boundaries (*OED*: *lining* vbl. n. 2), but this cannot be proven.

LINK COVE NY3611 Patterdale.
From 1860 (OSNB Barton, 2, 107).
♦ Perhaps 'the corrie by the ridge', with *link* 'ridge, bank' as in BOWFELL LINKS, **cove**. This is a cove or corrie among the crags at the head of DEEPDALE. In the absence of earlier spellings one cannot rule out corruption of **lyng/ling** 'heather', cf. perhaps LING COVE.

LINSKELDFIELD NY1734 (hab.) Blindcrake.
Linskeldfield 1803 (Isel PR), *Linskill Field* 1810; possibly cf. *Linescales* 1300 (*PNCumb* 268).
♦ If the 1300 spelling denotes this place, it is probably 'the hut where flax grows', from ON *lín*/OE *līn* 'flax', ON **skáli** 'shieling, hut', influenced by ON **kelda** 'spring', and **field** added subsequently.

LISCO/LISCOW FARM NY3627 Mungrisdale.
Lyskay (*Head*) 1577, *Lyskoe*(*head*) 1587, *Lyscoo* 1610 (*PNCumb* 227).
♦ (a) Quite possibly 'the light wood', ON *ljóss* 'light' and **skógr**, though without clearer evidence from early spellings the name remains uncertain. (b) Brearley 1974 suggested a 1st el. from ON *hlíð* slope, which is not impossible, especially since it combines with *skógr* in Litherskew (*PNNYorks* 259). **Farm** has been added to the name.

LITTLE as affix: see main name, e.g. for LITTLE BRAITHWAITE see BRAITHWAITE

LITTLE ARROW SD2895 (hab.), L~ A~ MOOR SD2796.
Litelherga pre-1220 (*CW* 1929, 41), *Little Array* 1610, *Little Harrow* 1671 (*PNLancs* 215); cf. *Neithermore litle Aray*, *Overmorelitle Aray* 1645 (PR); *Little Arrow Moor* 1851 (OS).
♦ 'The small shieling'; **lítill, ærgi/erg**; also **moor**. Spellings and situation confirm that Arrow originates in the Gaelic-Norse *ærgi/erg* 'shieling'; the place then became a permanent settlement, and nearby L~ A~ Intake suggests continuing extension of pasturage up onto the fell.

LITTLECELL BOTTOM SD1491 Waberthwaite.
From 1867 (OS).
♦ This is a corrie just W across the watershed from SELE BOTTOM (q.v.). Without early spellings, both names are obscure, but it is possible that '-cell' and 'Sele' are equivalent, with

the sense 'shieling' or 'willow'; see also **little, bottom**.

LITTLE COCKUP: see COCKUP

LITTLE COVE SD2598 Dunnerdale.
From 1851 (OS).
♦ This is a small recess high above SEATHWAITE TARN; **little, cove**.

LITTLE DALE NY2016 Above Derwent, LITTLEDALE EDGE NY2016.
Liteldale 1332 (p) (*PNCumb* 373); *Littledale Edge* 1867 (OS).
♦ 'The small valley' (**lítill, dalr**), a minor dip between the massive slopes of ROBINSON and HINDSCARTH. L~ **Edge** runs at right angles to the head of the valley.

LITTLE DODD NY1319 (1165ft/355m.) Loweswater.
From 1867 (OS).
LITTLE DODD NY1415 Loweswater/ Ennerdale.
From 1867 (OS).
♦ 'The small, compact summit'; **little, dodd**. The example at NY1319 is the summit of the steep-sided N spur of the much higher HEN COMB. The one at NY1415 lies E of the slightly higher STARLING DODD. See also DODD NY1615.

LITTLE FELL SD1286 Whicham.
LITTLE FELL NY3202 Skelwith.
From 1850 (OS).
♦ 'The small hill'; **little, fell**. The Whicham name refers to the lower slopes to the W of BLACK COMBE. In Skelwith, L~ F~ appears on the 1994 One Inch map as the sole name of an eminence of 692ft/211m., whereas on larger scale maps it is the W. part only, distinguished from the mightier Great How.

LITTLE MAN/SKIDDAW LITTLE MAN/ LOW MAN NY2627 (2837ft/865m.) Underskiddaw.
Little Man 1867 (OS).
♦ 'The minor peak or cairn'; **little, man**. This is a substantial double peak, but lower than SKIDDAW, a mile to the NW, whose main summit is Skiddaw Man. Nicholson & Burn in the 18th cent. described Skiddaw Man as a blue slate stone 'about a man's height' (cited *PNCumb* 320), and *man* in Little Man could refer to a summit cairn, though the present ones are not notable, and reference to the whole mountain seems more likely.

LITTLE NARROWCOVE NY2206 Eskdale.
From 1867 (OS).
♦ The name is an apt description of this rocky recess; **little**, *narrow* as in NARROW MOOR, **cove**.

LITTLE STAND NY2403 (2426ft/739m.) Ulpha.
From 1867 (OS).
♦ For the meaning of *stand*, see BROAD STAND, though this is a different landform — a small summit with rocks and diminutive tarns, perhaps named **little** in relation to its higher neighbour CRINKLE CRAGS.

LITTLE TARN NY2433 Ireby and Uldale.
From 1794 (Hutchinson II, 373), 1900 (OS), also previously *Tarn Nevin* 1867 (OS).
♦ 'The (extra-) small mountain pool'; **little, tjǫrn/tarn**. Tarns are generally small, this one particularly so in relation to nearby OVER WATER.

LITTLETHWAITE NY1424 (hab.) Blindbothel.
Littlethat 1600 (Lorton PR), *Littlethwaite* 1728 (Loweswater PR).

♦ Presumably 'the small clearing'; **lítill/little, þveit/thwaite**.

LITTLE TOWN NY2319 Above Derwent.
Litleton 1578, *Litletowne* 1595 (*PNCumb* 373).
♦ This is **little** by urban standards, indeed, but **town** was applied to small settlements, and L~ T~ was already a hamlet of eight landholdings by 1578 (Winchester 1987, 70–1). The 1578 spelling in *-ton* rather than *-town* in the spelling may also indicate that the name had existed for some time; if so the modern form is a rather artificial 'improvement'.

LITTLEWATER NY5017 (hab.) Bampton.
Lytelwater 1289 (*PNWestm* II, 191).
♦ 'The small pool' (**lȳtel, wæter**), hence the settlement named from it. The pool is now called Littlewater Tarn.

LITTLEWOOD FARM SD4899 Over Staveley.
Littlewood 1836 (*PNWestm* I, 176).
♦ Self-explanatory; **little, wood**, with **farm** added.

LIZA BECK NY1621 Buttermere.
Lissa 1786 (Gilpin II, 4).
LIZA, RIVER NY1713 Ennerdale.
aquam de (river of) *Lesar* 1292, *stream of Lesagh* 1294, 1322 (*PNCumb* 20).
♦ 'The bright, light river', from ON *ljóss* 'light' and *á* 'river'. The name is paralleled in Norway and Iceland. In L~ Beck, **beck** may have been added to a pre-existing name of the same origin, and if so, this is the smallest watercourse to be dignified by a name in *á*. Without earlier spellings, however, certainty is impossible.

LOADPOT HILL/LODEPOT HILL NY4518 (2201ft/671m.) Barton/Askham/Bampton.
From 1860 (OSNB Bampton), 1863 (OS); cf. *Loadpot* 1823 (*PNWestm* II, 202), *Lade Pot* 1828 (HodgsonM).
♦ 'The hill by the Load Pot, the hollow with ore'; *load/lode*, **potte, hill**. The Load Pot is 'a deep hole situate on the north side of Loadpot Hill' (OSNB Bampton, 1860, 27). *Pot* is from ME *potte* 'deep hollow', in this case an old iron-working, and *load/lode*, from OE *lād*, has 'a vein of metal ore' among its senses (*OED*: *lode*, sense 5; see also Hay 1937, 52–3).

LOANTHWAITE/LONETHWAITE SD3599 (hab.) Claife.
Lonethwayt 1537, *Lounthwaite* 1613 (*PNLancs* 219).
♦ 'Calm, sheltered clearing' or 'clearing by the lane', with 1st el. either: (a) dial. *loun(d)*, from ON *logn* 'a calm' (so Fellows-Jensen, *SSNNW* 351); cf. Clarke, 'a *calm day* is said to be *lown*' (1789, xxvi) and *lownded* meaning 'sheltered' (*CN* 1218, 1802); or (b) ME *lone*, a variant of **lane**. The 2nd el. is from **þveit/thwaite**.

LOBBS NY3524 (hab.) Matterdale.
From 1750; cf. nearby *Lobwath* 1787 (*PNCumb* 223), or *Lobthwaite* (Clarke, cited in Hutchinson 1794, II, 196).
♦ An enigmatic name. (a) A dial. word *lobb* means 'lump' (*EDD*), and Smith, *Elements* lists OE **lobb* 'something heavy or clumsy', with examples from S. England, but it is difficult to see how it would apply. (b) Another possibility is derivation from a known surname *Lobb*. Simplex names ending in inflectional *-s* are most likely to be topographical terms such as SYKES or HOW(E)S, tree-names such as ASHES, BIRKS, or surnames, as in (THE) MARSHALLS or RAWSONS.

LOBSTONE BAND NY2316 Above Derwent/Borrowdale.

From 1867 (OS).
♦ Obscure. L~ B~ is the SE buttress of HIGH SPY, and sometimes gives its name to that hill; it is neither a very clear ridge nor a horizontal stratum, so the sense of **band** is uncertain. So too is 'Lob' or 'Lobstone'. (a) Brearley 1974 suggests a mining term *lob*, meaning a step or a step-like vein of mineral; he notes a long history of quarrying hereabouts. (b) Alternatively, *lob* could have the dial. sense of 'lump' (see LOBBS); see also **stān/stone**.

LODGE, THE NY4318 Martindale.
Keepers Lodge 1881 (Census), *The Lodge* 1886 (PR), 1891 (Census).
♦ This was the **lodge** of John Jackson, 'gamekeeper' in 1881, occupied by him, his daughter and 'deer-keeper' son-in-law in 1891.

LODORE: L~ FALLS NY2618 Borrowdale, HIGH L~ NY2618.
Laghedure 1209–10, cf. *Heghedure* 1209–10 (*PNCumb* 350); *Lowdore waterfall* 1769 (*PNCumb* 350), *Fall of Lowdore* 1829 (Baines 121), *Lowdore Fall* 1839 (Ford 165), *Lowdore Cascade* 1867, *Lodore Cascade* 1901 (OS).
♦ L~ is 'the low door', with ON *lágr* '**low**' and OE **duru** either forming a rather rare hybrid or coming together through their ME reflexes; also *fall* 'waterfall' as in HIGH FALL, and **high**. The 'door' is probably the gorge between GOWDER CRAG and Shepherd's Crag through which WATENDLATH BECK flows, forming the famous Falls, though the original L~ is difficult to locate precisely on the basis of the 13th-cent. documents, as is *Heghedure*, which may not correspond with the present High L~.

LOFSHAW HILL NY3827 Hutton.
Cf. *Loftshow Cross* 1787, *hill called Lofts-Cross* 1789 (*PNCumb* 213).
♦ Somewhat uncertain, but the 1st el. seems to be from ON *lopt, loft* 'loft, two-storey building', or possibly 'high place' (cf. LO(W)THWAITE), and the 2nd el. **shaw** from OE **sceaga** 'wood', perhaps replacing the related ON **skógr** 'wood'; also **hill**. The editors of *PNCumb* suggest origins in an ON phrase *loptískógi* 'loft-house in the wood', preserved in Loskay (*PNNYorks* 62, cf. *PNWYorks* I, 82, II, 85).

LOFT WOOD, HIGH SD4086 Cartmel Fell.
High Loft Plantation 1851 (OS).
♦ *Loft*, from ON *lopt, loft* presumably has some such meaning as 'a building with two storeys' (cf. dial. *loft* meaning the upper of two storeys). The sense of 'air, an elevated, airy situation' (cf. LOWTHWAITE) appears scarcely to have outlived the Middle Ages. The **wood** is part way up a long slope, and **high** in relation to Low L~ W~.

LOGAN BECK SD1790 (stream) Muncaster/Ulpha, LOGANBECK SD1890 (hab.) Ulpha.
Loggan becke 1646, *Logan-beck in Thwaits* 1730, *Logganbeck* 1802 (*PNCumb* 20).
♦ Possibly 'the stream in the hollow', from Gaelic *lagán* 'little hollow' (so *PNCumb*, comparing Watson 1926, 140), plus **bekkr/beck**.

LONG BAND NY2812 Borrowdale.
From 1867 (OS).
♦ This refers to the rocks edging a steep escarpment; **langr/long** plus **band**, here 'stratum of rock'.

LONG BROW NY3029 Mungrisdale.
From 1867 (OS); cf. *the Browe ende* 1589 (*PNCumb* 227).
♦ The name denotes the long, even

contours on the W side of MUN-GRISDALE COMMON; **langr/long, brow**.

LONG CLOSE FARM NY1733 Blindcrake.
Long Close 1669 (Isel PR).
LONG CLOSE FARM NY2326 Underskiddaw.
Longclose, Langclose 1563, 1566 (*PNCumb* 323).
♦ 'The long enclosure' (**langr/long; close**) with **farm** added more recently.

LONG CRAG Examples at:
 NY1506 Nether Wasdale.
From 1867 (OS).
 NY1511 Ennerdale.
From 1867 (OS).
 SD2098 Eskdale/Ulpha.
From 1867 (OS).
 NY2305 Eskdale.
From 1802 (*CN* 1219).
 SD2398 Ulpha/Dunnerdale.
From 1867 (OS).
 NY2710 Borrowdale.
From 1867 (OS).
 SD3098 Coniston.
From 1851 (OS).
 NY4019 Martindale.
From 1863 (OS), previously *New Crag* 1770 (*PNWestm* II, 219).
 NY4621 Barton.
From 1859 (*PNWestm* II, 213).
 NY5105 Fawcett Forest.
From 1863 (OS).
LONG CRAGS SD3191 Colton/Satterthwaite.
From 1851 (OS).
♦ Simply 'the long rock outcrops'; **langr/long, crag**.

LONGDALE NY3133 Mungrisdale.
Long Dale 1867 (OS).
♦ Apparently 'the long valley' (**langr/long, dalr/dale**), which could refer to the course of GRAINSGILL BECK and R. CALDEW, though the maps, strangely, show L~ on a short ridge N of this.

LONG FELL NY1628 Embleton.
From 1867 (OS).
LONG FELL NY5509 Shap Rural.
From 1865 (*PNWestm* II, 176).
♦ Self-explanatory; **langr/long; fell**. The Embleton example is a long spur on the fringe of KIRK FELL. The Shap one is a distinct hill and the easternmost point of the Lake District National Park.

LONGFIELD: L~ WOOD SD2488 Kirkby Ireleth.
Longfield (*House*) 1829 (PR); *Longfield Wood* 1851 (OS).
♦ Presumably self-explanatory; **langr/long, field, wood**.

LONG GARTH SD1892 (hab.) Ulpha.
Langarth 1646, *Longegarth* 1661 (*PNCumb* 438).
♦ 'The long enclosure'; **langr/long, garth**.

LONG GRAIN NY1011 Ennerdale.
From 1826 (CRO D/Lec/94).
♦ 'The long branch of a stream'; **langr/long, grain**. This, along with the scarcely shorter Short Grain to the N, is among the feeder streams of WORM GILL.

LONG GRAIN NY4514 Bampton, LONGGRAIN BECK NY4514.
Both from 1863 (OS).
♦ **Grain** here is unusually applied to a hill — the steep spur between Longgrain Beck (MEASAND BECK in its lower reach) and HAWESWATER, though it normally means 'fork in a river, confluence'. Either (a) the hill-name is secondary, perhaps extrapolated from Longgrain Beck or its confluence with Haweswater; or (b) *grain* has a different origin or meaning here. See also **langr/long** and **beck** 'stream'.

LONG GREEN HEAD NY4204 (hab.) Troutbeck/Windermere.
From 1770 (JefferysM).
♦ The elements are **langr/long, green, head**. The place is situated where the ground starts to rise above the fairly wide and verdant valley of the TROUT BECK, hence either 'head of the long green' (with *green* as a noun) or 'long green hill' (with *green* as adj.) is possible.

LONG HEIGHT SD3798 Claife.
♦ This is part of CLAIFE HEIGHTS; **langr/long, height**.

LONG HOUSE SD2396 Dunnerdale, LONGHOUSE GILL SD2496.
longhouse; longhousegill 1681 (*CW* 1908, 353).
LONG HOUSE NY3006 Langdales.
Longhouse 1683, *the Langhouse* 1695 (Grasmere PR).
LONG HOUSES NY4503 Kentmere.
Long houses c.1692 (Machell II, 106).
♦ Self-explanatory; **langr/long, house**, also **gil(l)**.

LONGHOWE: L~ END SD4483 (hab.) Witherslack.
Long How 1762 (PR); *Longhowe End* 1862 (OS).
♦ L~ H~ is 'the long hill', with L~ H~ End just to the S of it; **langr/long, haugr/how(e), end**.

LONGLANDS NY2635 (hab.) Ireby and Uldale, L~ FELL NY2735 (1581ft/482m.).
Langelandes 1399 (*PNCumb* 328); *Langlands* 1687-8 (Denton 172), *Longlands Fell* 1867 (OS).
LONGLANDS SD3879 (hab.) Broughton East, L~S FARM SD3980.
Longelondes 13th cent. (Wyld 1911, 182), *Langlands* 1614/5 (Cartmel PR); *Longlands; Longlands Farm* both 1851 (OS).

♦ Self-explanatory; **langr/long, land, fell, farm**. Although the exact reference of *land* cannot be recaptured, *VCHLancs* states that 'Langlands and Fell Close are field-names' (278 n.7), referring to a rental of 1508-9 for AYNSOME, Broughton East. This together with the qualifier *long* could suggest a specialised sense such as 'strip of arable land in a common-field' (Smith, *Elements: land* (iv)). In Uldale, L~ Fell is a distinct peak rising above the hamlet of L~.

LONG LEA NY3038 (hab.) Caldbeck.
Longe Lee 1652 (*PNCumb* 279).
♦ Probably 'the long meadow'; **langr/long, lēah**.

LONGMIRE NY4101 (hab.) Windermere.
Longmire (Yeat) 1656 (spelling not specified, *CW* 1997, 84), *Longmire* 1823 (*PNWestm* I, 196); cf. *Lang-, Longmyre* 1560 (as derived surname, *PNWestm* I 191).
LONGMIRE: HIGH~ SD3287 (hab.) Colton, LOW L~ (hab.) SD3287.
Longmire 1628 (PR); *High Longmire; Low Longmire* both 1851 (OS).
♦ 'The extensive tract of wet ground'; **langr/long, mýrr/mire**. The Windermere example is a stretch of rough ground E of the TROUT BECK; the Colton one is an area of springs and marsh.

LONGMOOR NY0615 (hab.) Ennerdale.
Lange Moore 1578, *Longmoor* 1716; cf. *Longmoreheade* 1578 (*PNCumb* 387).
♦ Self-explanatory; **langr/long; moor**. Ennerdale also contains a BROADMOOR.

LONG MOSS NY2814 Borrowdale/St John's.
From 1867 (OS).
♦ This is indeed an 'extended tract of bog' (**langr/long, moss**) on

WATENDLATH FELL. 'After rain, wear thigh-length gumboots', advises Wainwright, '... this is one of the wettest walks in Lakeland, and not one to be undertaken with pleasure' (1958, Ullscarf 13).

LONG PIKE NY2208 Eskdale.
From 1867 (OS).
♦ This is the SW spur of GREAT END; **long, pike**.

LONG RIGG NY5013 Shap Rural.
Longrigge, Langrigge 1581, 1590 (*PNWestm* II, 176).
LONGRIGG: LOW L~ NY1702 Eskdale.
Langrigge, the Longrigge 1587 (*PNCumb* 392); *Low Longrigg* 1867 (OS).
♦ 'The long ridge'; **langr/long, hryggr/rigg**, also **low**. Low L~ is the lower, S. part of the ridge which also comprises BOAT HOW and ESKDALE MOOR.

LONG SCAR NY2703 Ulpha/Langdales.
From 1867 (OS).
♦ 'The long outcrop'; **langr/long, scar**.

LONG SIDE NY2428 (2405ft/733m.) Underskiddaw, L~ S~ EDGE NY2428 Bassenthwaite.
Both from 1867 (OS).
♦ 'The long hill-slope'; **langr/long, side**; also **edge**. The contours are broad but steeply rising and are topped by a rocky scarp or edge.

LONGSLEDDALE: see SLEDDALE

LONG STILE NY4411 Shap Rural, SHORT S~ NY4411.
Long Stile 1828 (HodgsonM), *Long Stile, Short Stile* 1858 (OSNB).
♦ 'The long/short steep ascent'; **langr/longr**, *short*, **stile**. These are steep rock ridges ascending at right-angles onto the HIGH STREET ridge. *Long* is common in p.ns, but *short*, ultimately from OE *sc(e)ort*, is not (though cf. SCANDALE).

LONGTHWAITE NY2514 (hab.) Borrowdale.
Longtwhayte, earlier 16th cent. (*PNCumb* 353).
LONGTHWAITE NY4322 (hab.) Matterdale.
Lonktwayt 1253, *Langthweyt* 1285 (p) (*PNCumb* 255).
♦ 'The long clearing'; **langr/long, þveit**.

LONG TONGUE SD3790 Satterthwaite.
From 1851 (OS).
♦ 'The long promontory'; **langr/long, tongue**. Unusually for a *tongue*, this one juts out into a lake — WINDERMERE.

LONG TOP NY2404 Eskdale.
From 1867 (OS), possibly = *Midelfel* 1242 (*Furness Coucher* II, 564, Collingwood 1918, 95), cf. MIDDLEFELL FARM.
♦ This is part of the summit ridge of CRINKLE CRAGS, an extension of the highest of the five rocky hummocks; **langr/long, top**.

LONSCALE/LONGSCALE NY2925 (hab.) Underskiddaw, L~ FELL NY2826.
Lonskell 1566, *Longskell* 1572 (PR), *Longscales* 1774 (DHM); *Lonscale Fell* 1867 (OS).
♦ 'The long shieling or summer pastures' (**langr/long, skáli**), which must have stretched up onto the **fell**. A document of Fountains Abbey, 1256, mentions a clearing or *essart* called *Le Scales* hereabouts (*PNCumb* 323).

LOOKING STEAD NY1811 (2058ft/627m.) Ennerdale/Nether Wasdale.
From 1867 (OS).

♦ 'The look-out place.' Despite the lack of early forms, the situation at a high point on the PILLAR massif, on a boundary and close to a 'Cloven Stone' (on larger scale maps), may suggest that this is indeed a 'look-out place'. The verb *look* is from OE *lōcian*, and see **stead**. Another Looking Steads is found at NY2410 on larger scale maps.

LORD CRAG NY3507 Rydal/Grasmere. *Laverdkrag* 13th cent. (*PNWestm* I, 210).
♦ This rocky height appears in the 13th-cent. bounds of Rydal, and its name may specifically reflect its role as a marker of the boundary of a lord's estate (as suggested by Brearley, 1974 and cf. Winchester 2000, 29); **lord** from OE *hlāford*/ME **laverd**, **crag(ge)**.

LORD'S HOW NY2813 Borrowdale. From 1867 (OS).
♦ 'The lord's hill'; **lord** from OE *hlāford*/ME **laverd** and **how(e)** from ON **haugr**. It is unknown when or why the name was coined.

LORD'S ISLAND NY2621 Borrowdale. *Lords Island* 1771 (*CW* 1975, 295), *Lord's Island* 1776 (Hutchinson 151), previously *island of Derwentwatre* 1230 (*PNCumb* 313).
♦ Self-explanatory; **lord** (see **laverd**), **island**. 'Lord's' since the site of the manor-house of the Radcliffes: 'Yn the one [isle] ys the hedd places of M. Radclyf' (Leland c.1539-45, V, 54), though depending when the name arose it could also refer to the fact that a Radcliffe was created Earl of Derwentwater in 1687. Hutchinson claimed that the island was 'formerly ... only a peninsula, but ... was severed from the land by a ditch' when the Radcliffes took over (1776, 158), but this is 'impossible' (Collingwood 1904, 259). It was seemingly abandoned at some time between 1675 and 1709, certainly before James, the third and last Earl, was executed for his part in the 1715 rebellion (Collingwood 1904, 272; see also LADY'S RAKE). Nevertheless, as early as 1759, John Crofts of Bristol reported, 'You will be shewn one island in the Derwent Water, which was the seat of the unfortunate Earl who took his title from thence' (*CW* 1961, 291).

LORD'S LOT SD4492 Crosthwaite. From 1899 (OS).
♦ **Lord** (see **laverd**) plus *lot*, from OE *hlot*, which has been used in the sense of 'portion of property or land' at least since the late OE period, and though not widespread is found in p.ns and field-names. The exact motivation for the name is uncertain. (a) 'Lord' could be the surname recorded in 1332 in the form *Adam le Lauerd* and connected by Smith with LORD'S SEAT and Bridge in the same parish (*PNWestm* I, 84). A Robert Lord appears in the Crosthwaite PR for 1584. (b) 'Lord's Lot' could alternatively be simply a share of land belonging to the lord of the manor, or, since this is a stony tract of hill-side, an ironic, derogatory use of the same phrase. There is a Lord's High Allotment at SD2595.

LORD'S RAKE NY2006 Eskdale. From 1867 (OS).
♦ 'Lord's steep track'; **lord** (see **laverd**), **rake**. This is a famous gully, straight but extremely steep, whose chute of scree is 'now considered too dangerous for most walkers' (*Conserving Lakeland* 2004). Why 'Lord's' is unknown to me. The location is probably too far from LADY'S RAKE for the names to be intentional counterparts.

LORD'S SEAT NY2026 (1811ft/552m.) Wythop/Lorton/Above Derwent.
Lauerdesate 1247, *the Lordseatt* 16th cent., *Lord's Seat* 1821 (*PNCumb* 409).
♦ 'The seat or summer pastures of the lord.' The 1st el. is clearly from ME **laverd** or its OE antecedent *hlāford*, but even with the relative luxury of a 13th-cent. record, the origins and meaning of the 2nd el. are uncertain. (a) *Seat* from ON **sæti** is suggested by the spellings, by other Cumbrian examples (see below) and the commanding position of this hill. If a 'seat', it is probably the hill as a whole, and as Wainwright pointed out a quest for a seat-shaped rock near the summit is vain (1964, Lord's Seat 10). (b) The reflex of ON **sætr** 'shieling, summer pasture' could be indicated by the location on the S edge of WYTHOP, an area used as pasturage until its colonisation in the 13th cent. (Winchester 1987, 40, and see OLD SCALES).

LORD'S SEAT Examples at:
NY3713 Patterdale.
From 1859 (*PNWestm* II, 226).
SD4487 (706ft/215m.) Crosthwaite.
From 1857 (*PNWestm* I, 84).
NY5106 (1719ft/524m.) Shap Rural.
From 1865 (*PNWestm* II, 176).
♦ 'The lord's seat' (**laverd/lord, sæti**), or possibly 'the lord's shieling' (**laverd/lord, sætr**), since these late-recorded examples are as uncertain as the one above. The Shap hill is close to others whose names may be variations on the same theme: YARLSIDE CRAG and GREAT Y~, and SEAT ROBERT. The Crosthwaite hill is the highest point on WHITBARROW SCAR. For the possibility that 'Lord' in Crosthwaite is a local surname, see LORD'S LOT. Machell mentions (and roughly sketches) a Lord's Seat in Bannerdale, 'a chair of Stone on the Top of an Hill', and reports a tradition that lords used to sit and watch their greyhounds coursing red deer (c.1692, I, 699).

LORD'S WOOD SD2692 Broughton West. *Lords Wood* 1851 (OS).
♦ Self-explanatory; **lord** (see **laverd**), **wood**. This is beside Lord's Gill, while Lord's High Allotment and Lord's Low Allotment stretch away to the N (on larger scale maps). The area was owned by the Broughton family until the late 15th cent., then by the Stanleys, Earls of Derby, and later to the (Gilpin-)Sawreys. I am grateful to Mary Atkin for this information.

LORTON (hab. & parish): HIGH L~ NY1625, LOW L~ NY1525, L~ FELLS NY1825, L~ VALE NY1525.
Loretona c.1150 (in early copy), *Lortun(a)* c.1160, *Lorton(e)* 1197 to 1316; *Overlorton* 1365 (*PNCumb* 408), *High Lorton* 1867 (OS); *Lowe Lurton'* 1570 (*PNCumb* 408); *Lorton fells* 1687–8 (Denton 51).
♦ The 2nd el. is clearly OE **tūn** 'farmstead, village'. The 1st is enigmatic, but Ekwall's suggestion in *DEPN* of a river-name ON *Hlóra* 'roaring', paralleled in the Norw Lora, is attractive. This could have been an earlier name for the lively WHIT BECK on which High Lorton is situated, or of the COCKER which flows through Low L~. Coleridge described it 'roaring' and 'fling[ing] itself down a small chasm of rock' (*CN* 537, 1799). Reference to the Cocker would be more likely especially if, as Winchester thinks possible, **High** and **Low** L~ resulted from the division of a single settlement close to the present church, but the location of the earliest settlement(s) is not certain (Winchester 1987, 146–7). **Fell** in L~ Fells means

'high unenclosed land'. L~ Vale does not appear on 1867 or 1900 OS. **Dale** being the standard Lakeland word for a valley, the word *vale* appears only twice in this volume: here and in ST JOHN'S IN THE VALE, while *valley* is only found in LYTH V~, RUSLAND V~ and WHICHAM V~. Both words were adopted into ME from OFr.

LOTHWAITE: L~ SIDE (hab.) NY2029 Wythop.
Loftweic, Loftwic 1195 (*PNCumb* 457); *Lowthwaite side* 1601, *Lothetsyde* 1637 (Lorton PR), *Lothwaite Side* 1774 (DHM).
♦ L~ is 'the airy or two-storey settlement'. The 1st el. seems to be ON *lopt* 'air, loft', hence either 'hill, airy place' or 'two-storey building', as in LOWTHWAITE, Ireby and Uldale. (The ON pers.n. *Loptr* seems less likely.) From the 12th-cent. spellings it seems that the original 2nd el. was ON **vík**, referring to a curve in the landscape, or OE **wīc** 'settlement, specialised place', but that it was supplanted by **thwaite** 'clearing, settlement', more common locally. L~ now refers to no more than a small hill, and L~ Side, once a gamekeeper's house, is deserted. **Side** could have the sense 'hill-slope' or simply 'place beside (L~)'.

LOUGHRIGG: L~ FELL NY3404 Rydal, L~ TARN NY3404, L~ TERRACE NY3405.
Loghrigg(e) & variants 1274 to 16th cent. (*PNWestm* I, 209); *Loughrigg-fell* c.1692 (Machell II, 430); *Loughrigge Tarne* 1671 (Fleming 15); *Loughrigg Terrace* 1920 (OS).
♦ (a) 'The ridge above the lake', from OE *luh*, an adoption from Brittonic (W *llwch*), and ON **hryggr** 'ridge'; also **fell** 'hill', **tjǫrn/tarn** 'small mountain pool', *terrace*. The lake in question is L~ Tarn (Machell referred to 'the Lough at Loughrigg at the foot of Little Langdale', c.1692, ed. Ewbank, 137), and the 'ridge' is presumably L~ Fell, which is still often known simply as L~. Already in 1274, L~ referred not to a ridge but to an area, since the source translates as 'the said part of *Amelsate* [Ambleside] and *Loghrigg*', and for centuries it referred to a township. (b) An alternative derivation of the 1st el., from ON *laukr* 'leak, garlic', based on identification of the spelling *Loukrig* with this place, was proposed by Ekwall (*DEPN*), and adopted by Smith in the *Elements* entry for *laukr*, but rejected by him in *PNWestm* I, 209–10. L~ Terrace is the only occurrence of *terrace* (an Early Modern adoption from Fr) in this volume. The name is anticipated by Ford: 'The finest views are from the terrace-road under Loughrigg' (1839, 36).

LOW as affix: see main name, e.g. for LOW BROWNRIGG see BROWNRIGG (though some names in LOW have a HIGH counterpart and are treated under H-; these are noted below)

LOW, THE SD2094 (hab.) Ulpha.
Low 1867 (OS), possibly = *ye lawe* 1664, among field-names *PNCumb* 438.
♦ (a) This may be the word *law* from OE *hlāw* 'hill, mound', though this is far from common in the Lake District. This farm lies below a minor hill. (b) The use of the adj. **low** as if a noun to form a p.n. is also conceivable: cf. LOW FARMS.

LOW BANK NY1717 Buttermere.
From 1867 (OS).
♦ At about 1000ft/330m., this is a fine viewpoint, yet low in the sense of being overshadowed by the grander fells behind it; **low, bank** 'hill'. (Pl. 5)

LOW BARN NY1833 (hab.) Blindcrake. From 1867 (OS).
♦ Self explanatory; **low**, *barn* as in BARN FARM.

LOW BECK: see HIGH BECK

LOW BIRKER, LOW BIRKER POOL: see BIRKER

LOWBRIDGE HOUSE NY5301 Whitwell. From 1857; cf. *Bryghouse in Banisdale* 1668 (*PNWestm* I, 150).
♦ Self-explanatory; **low, brycg/ bridge, house**. The house actually appears to be upstream of Bannisdale Low Bridge, closer to the present Bannisdale High Bridge.

LOW CLOSE/LOWCLOSE NY5125 Askham. *Lowclose* 1764 (*PNWestm* II, 202).
♦ Either 'farm of the Low family' or 'the low farm or enclosure'. (a) Smith suggests a possible connection with the Low family, since a John Lowe or Lawe is recorded locally in 1672 (*PNWestm*). (b) The adj. **low** cannot, however, be ruled out, for the generic **close** 'enclosure, farm' is more often qualified by a descriptive term than a surname.

LOW COCK HOW: see COCK HOW, LOW

LOW CRAG SD2296 (947ft/289m.) Ulpha. From 1867 (OS).
♦ A rocky height which seems to be named 'low' in relation to WALLOWBARROW Crag; **low, crag**.

LOW CRAGHALL: see CRAGHALL, LOW

LOWCRAY (LOWCHEY on 1994 One Inch map) NY0805 (hab.) Gosforth. *Lawkroe* 1598, *Lacrey* 1604, *Locrye* 1635 (PR), *Lockeay* 1774 (DHM).
♦ *PNCumb* 396 wisely offers no explanation of this enigmatic name, borne by a farm on a small tributary of the R. BLENG. However, (a) it may possibly be of the same origin as Lacra in Millom, which has spellings including *Lawcray, Lowcray, Loucray* 1403 and a suggested etymology of ON *lauk(v)rá* 'garlic nook' (*PNCumb* 416). (b) Spellings of *crey, crei, cray* and modern Cray are found elsewhere for stream-names from OW *crei* 'fresh, strong' (*PNWYorks* III, 50, VI, 116 and VII, 123–4), but there is no means of knowing whether this is present here.

LOWER BLEANSLEY, L~ HAWTHWAITE: see BLEANSLEY, HAWTHWAITE

LOWER MAN NY3315 (3033ft/925m.) St John's.
From 1823 (Otley 44).
♦ 'The subsidiary summit' (**low, man**) of HELVELLYN, NW of the main summit. Otley, one of the most intelligent and well-informed early writers on Lakeland, says 'that which was anciently called the top of Helvellyn, or Helvellyn Man, is ... now called Lower Man'. He also interestingly (though again without citing evidence) suggests that the name H~ originally applied to part of the mountain other than the summit, which 'was formerly known by another name' (1823, 43–4).

LOWESWATER (lake) NY1420, (hab. & parish) NY1420, L~ FELL NY1219.
(Lake:) *Laweswatre, Lausewatre* c.1203, *lake of Loweswatre* 1230 (*PNCumb* 34); (hab.:) *Lawswater* c.1125 (in late medieval copy, *St Bees* 2), *Lousewater* c.1160 to 1385, *Loswater* 1186, *Laweswater, -watre* 1188 to 1397 (*PNCumb* 410); *Loweswater fells* 1687–8 (Denton 51), *Loswater Fell* 1802 (*CN* 1208).

♦ This is usually assumed (e.g. *PNCumb*) to be 'the leafy lake' (ON *laufsær*, preserved in various instances of Swed Lövsjö(n)), with the explanatory **wæter/water** 'lake' added later; also **fell**. West described the extremities of the lake as 'rivals in beauty of hanging woods, little groves, and waving inclosures' (quoted Hutchinson 1794, II, 135); and more than half of the perimeter is still overhung with trees. The village is secondarily named from the lake. L~ Fell is the high grazing land S of the lake and village.

LOW FARMS SD4588 Crosthwaite.
(the) Low(e) 1535, *(the) Law(e)* 1587 to 1709 (*PNWestm* I, 84).
♦ (a) L~ was probably 'the hill', from the reflex of OE *hlāw* 'hill, tumulus', with **farms** subsequently added to refer to Low Farm and South Low Farm. The hill is rather indistinct (as also at LOW HOUSE, which may also be 'hill' rather than 'low'), cf. also (THE) LOW. (b) A further possibility is that these names have the adj. **low** used as noun, as a counterpart to the nominal use of **high** (see HIGH FARM).

LOW FELL Examples at:
　NY1322 Loweswater.
Low-fell 1780 (West 133).
　SD4290 (hab.) Crosthwaite.
= *Low Fell Cottage* 1863 (OS).
　NY5510 Shap Rural.
From 1859 (*PNWestm* II, 176).
♦ Self-explanatory; **low, fell**. At 1388ft/423m., L~ F~ in Loweswater is in fact the highest point in the immediate vicinity. It is a *fell* in the sense of a distinct hill, though it also supplies ample rough pasture. The Crosthwaite name refers to a lower part of Crosthwaite Fell, and the Shap one is a minor height on SHAP FELLS.

LOWFIELD NY1832 (hab.) Setmurthy.
Lowefielde 1578 (*PNCumb* 435).
♦ Presumably to be taken at its face value; **low, field**.

LOW FOLD SD4493 Crook.
Low Fold 1738, *Lawfald* 1746 (*PNWestm* I, 180).
LOW FOLD NY4600 Over Staveley.
From 1836 (*PNWestm* I, 176).
♦ Presumably 'the low-lying fold or farm'; **low, fold**.

LOW GREEN SD4282 (hab.) Upper Allithwaite.
From 1851 (OS).
♦ 'The low-lying grassy place'; **low, green**. This occupies a similar position to SUNNY GREEN, but is right down beside the R. WINSTER.

LOW GROUND SD1798: see HIGH GROUND

LOW GROVE FARM NY2525 Underskiddaw.
Lowgrave 1567, *Lowgrove* 1600 (*PNCumb* 323).
♦ 'The low copse'; **low, grove**, with **farm** added in recent times.

LOW HALL NY1227 Blindbothel.
From 1789 (Mosser PR).
LOW HALL SD2195 Dunnerdale.
Lowhall 1737 (Seathwaite PR), *Low Hall* 1850 (OS).
♦ Self-explanatory; **low, hall**. The Dunnerdale example is just downriver from HALL DUNNERDALE.

LOW HALL GARTH: see HALL GARTH, LOW

LOW HOLME NY1400 (hab.) Eskdale.
Laweholme 1570, *Lowholme* 1587 (*PNCumb* 392).
♦ 'The low place beside water' (**low,**

holmr/holme) — a modest 100ft/30m. above the river MITE.

LOW HOUSE Examples at:
　NY1522 Buttermere.
Lowehouse 1624/5 (Lorton PR), *Low House* (DHM); cf. *High House* 1578 (*PNCumb* 356).
　NY2132 Bassenthwaite.
From 1828 (PR).
　SD4389 Crosthwaite.
From 1863 (OS).
　SD4599 Hugill.
From 1670 (*PNWestm* I, 171).
　NY5101 Longsleddale.
From 1770 (JefferysM).
LOWHOUSE: L~ BECK (hab.) SD4092 Cartmel Fell.
Lowhouse 1792 (*CW* 1966, 231); cf. *High Old House Beck, Low Old House Beck* both 1851 (OS).
LOW HOUSE FARM NY2219 Above Derwent.
Lawehouse 1599, *Lowhouse* 1739 (*PNCumb* 373).
LOW HOUSE FARM SD4196 Bowness.
Low House 1706 (*PNWestm* I, 187).
♦ Self-explanatory; **low, house**, with **farm** added later in two cases; also **beck** 'stream'. In some instances *low* from OE *hlāw* 'hill' could not be completely ruled out. The first two examples occupy low-lying situations close to, respectively, the R. COCKER and BASSENTHWAITE LAKE. The Hugill and Longsleddale ones are on much the same level as HIGH HOUSE, q.v., but slightly farther down their respective valleys.

LOWICK SD2986 (hab. & parish), L~ BRIDGE SD2986 (hab.), L~ GREEN (hab.) SD2985, L~ HALL SD2885.
Lofwic 1202, *Lowyk* 1246, *Laufwik, Louwyk, Lofwyk* (undated, from *Furness Coucher* I, 435, in *PNLancs* 213);
Lowickbridge 1742/3 (Colton PR), 1851 (OS); *Lowick Green* 1635/6 (Colton PR); *Lowick Hall* 1631/2 (Colton PR), 1851 (OS).
♦ Probably 'the leafy nook'. The spellings of the 1st el. could point to ON *lauf* 'leaf, foliage', which enters into LOWESWATER and various Icel and Norw p.ns. (a) The 2nd el. seems to be ON **vík**, normally 'bay, inlet' but occasionally a bend in a river (as suggested by Lindkvist 1912, 147, n. 4, noting that L~ Green is in a bend of the R. CRAKE; so also *PNLancs* 213–4). (b) OE **wīc** 'settlement, specialised place' is a less attractive solution since it would entail assuming an unusual hybrid compound, or replacement of OE *lēaf* 'leaf', also unusual, as specific. See also **brycg/bridge, green** 'grassy place', **hall**.

LOW KOP: see HIGH KOP

LOW MILL NY3632 (hab.) Mungrisdale.
Lowe Myll 1608 (*PNCumb* 181).
♦ The site is on GRAINSGILL BECK; **low, mill** from **mylen**.

LOW MOSS SD2088 (hab.) Broughton West.
From 1671 (Broughton-in-Furness PR).
LOW MOSS NY2820 St John's/Borrowdale.
From 1867 (OS).
♦ 'The low, boggy tract'; **low, moss**. The Broughton habitation takes its name from a stretch of wet ground by the R. LICKLE. L~ M~, St John's, lies at around 1250ft/380m., but is 'low' in relation to BLEABERRY FELL to its S.

LOWPARK: see HIGHPARK

LOW PIKE NY3235, LOW PIKE NY3707: see HIGH PIKE

LOW PLACE NY1501 (hab.) Eskdale. *Lowe place* 1570 (*PNCumb* 392).
♦ This lies beside the R. MITE and below steep slopes; **low**, **place**.

LOW PLAIN SD4790 (hab.) Underbarrow. From 1858 (OSNB).
♦ The situation is indeed a low-lying, level tract of land; **low**, and *plain* as in ELVA PLAIN.

LOW RAISE: see HIGH RAISE

LOW RIGG: see HIGH RIGG

LOW SADDLE: see HIGH SADDLE

LOWSIDE NY3527 (hab.) Mungrisdale. *Lawside* 1756, *Low-side* 1769 (Gray 1078), 1817 (PR).
♦ 'The low hill-slope' (**low**, **side**) is suggested by the spellings and their dates, and the situation, on the lower slopes of SOUTHER FELL quite close to the R. GLENDERAMACKIN.

LOW SNAB: see SNAB

LOW TARN NY1609 Nether Wasdale. From 1867 (OS).
♦ 'The low-lying mountain pool' (**low, tjǫrn/tarn**) — low in comparison with nearby SCOAT TARN.

LOWTHER, RIVER NY5124 Askham/ Lowther/Bampton, LOWTHER NY5323 (hab. & parish), L~ CASTLE NY5223, L~ PARK NY5223, L~ LEISURE PARK NY5322. (River:) *Lauther* 1157–86, *Lowthre* 12th cent., (*aqua de* 'river of') *Louther* early 13th cent. to 1408, (*aqua de*) *Lowther* from 1249 (*PNWestm* I, 9); (hab:) *Lauuedra* 1174, *Lauder* 1170–80, *Loudr(e)*, *Louder* 1195 to early 13th cent., *Louther* c.1190 to 1476 (*PNWestm* II, 182); *Lowther Castle* 1811 (*PNWestm* II, 184); *Louthre park* 1546 (*PNWestm* II, 185).

♦ The two main explanations of L~ are (a) derivation from a Brit. root *loṷ-* 'to wash' plus Brit. **dubro-* 'water, river', cf. the Scottish Lauder (so *ERN* 266, also Breeze 2000, 67); or (b) 'foaming river', from ON *lauðr* 'foam' and ON *á* 'river' (so *PNWestm* I, 9, *ERN* 267). Against the latter Fellows-Jensen notes the lack of Scand parallels and the different development of ON *au* in the river-name ROTHAY (*SSNNW* 423). The original L~ settlement was named from the river, while L~ NEW TOWN is a short way E of the river, and the present L~ village still further. The present L~ **Castle** is a ruined castellated mansion, 18th-cent. replacing a 17th-cent. predecessor which burned down (Pevsner 1967, 274). There was a medieval fortification: see CASTLESTEADS. A licence for a **park** was granted in 1283 (Perriam & Robinson 1998, 290).

LOWTHER BROW NY4205 Windermere. From 1865 (*PNWestm* I, 196).
♦ The surname *Lowther* is known locally from the 1390s, and the name of this **brow** or steep hill-side may well derive from that (so *PNWestm*).

LOWTHWAITE NY2635 Ireby and Uldale, L~ FELL NY2734 (787ft/240m.). *Louthweit* 1229 (*PNCumb* 327), *Lofthweit* 1245 (*DEPN*), *Lost(t)thawyt* 1295, *Loftthwayt*, *Loftethayt* 1399, *Lowthwate* 1539 (*PNCumb* 327).
♦ The 2nd el. of L~ is from ON **þveit** 'clearing'. The 1st el. seems to be ON *loft, lopt* 'air, sky, loft', and hence also 'two-storey building' — a rarity in the Middle Ages and a meaning favoured here by *PNCumb*. However, another possible sense is 'on a hill', which would suit the reasonably elevated site (Ekwall, *DEPN*: Lothwaite and Fellows-Jensen, citing Scand parallels,

SSNNW 145). The **fell** rises to a distinct peak, though the name presumably also indicates that it is the high grazing available to L~.

LOWTHWAITE NY4123 (hab.) Matterdale.
Lowthwayte 1540 (*PNCumb* 257).
LOWTHWAITE FARM NY3122 St John's.
Lothwait 1571, *Lowthwayt* 1591 (*PNCumb* 317).
♦ (a) Probably 'the low clearing'; ON *lágr* or ME **low**, plus **þveit/thwaite**. (b) 'Clearing on a hill or with a two-storey building', with 1st el. *lopt, loft* as in LOWTHWAITE, Ireby and Uldale, is less likely since there are no comparable early spellings to support this, and the Matterdale hamlet is in a sheltered situation near the foot of a hill.

LOW TODRIGG: see TODRIGG

LOW WATER SD2798 Coniston.
From 1745 (West 1774, map).
♦ Like nearby GOATS WATER and LEVERS WATER, this is a small high-altitude lake enclosed by steep slopes, of the sort more often named **tjǫrn/tarn** than **water**. At 1800ft/549m., it lies higher than the other two, leaving the 1st el. 'Low' an enigma, unless it is a corruption of a word for 'pool', either the originally Celtic *lough* 'pool', assumed in LOUGHRIGG, or possibly ON *lǫgr* 'sea, lake'.

LOW WOOD SD3483 (hab.) Haverthwaite.
Low wood (bridge end) 1768 (Colton PR), *Low Wood* 1851 (OS).
LOW WOOD NY3913 Patterdale.
From 1860 (*PNWestm* II, 226).
♦ Self-explanatory; **low, wood**. L~ W~ in Haverthwaite lies by the R. LEVEN and is one of only a few Lakeland examples of a settlement named from a wood. The Patterdale wood clothes the slopes W of BROTHERS WATER.

LOW WOOD: MIDDLE L~ W~ SD4286 Witherslack.
Low Wood 1685 (*PNWestm* I, 79); *Middle Low Wood* 1862 (OS).
♦ Self-explanatory; **middle, low, wood**. High, Middle and Low Low Wood are distinguished on the 1862 Six Inch map and the current 1:25,000 map.

LOW WRAY: see WRAY (Pl. 31)

LUDDERBURN/LUDDER BURN SD4091 (hab.) Cartmel Fell.
Litterburne 1537 (*PNLancs* 200), *Luderburne* 1605 (Cartmel PR), *Ludderburne* 1619 (*PNLancs* 200).
♦ Probably 'the clear stream', from OE *hlūt(t)or*, ME *lutter*, often used of water, and **burn** from OE *burna* 'stream'. Ekwall favoured this (*PNLancs*) but had previously suggested a connection with the same Celtic hill term as is found in LATRIGG (*Scand & Celt* 91). Low and High L~ are distinguished on the 1:25,000 map.

LUMHOLME SD2190 (hab.) Broughton West.
Lumholme 1671, *Lomholm* 1699 (Broughton-in-Furness PR).
♦ Possibly 'the raised ground above a pool', from the reflexes of OE *lum* 'pool, especially in a river' (cf. BLELHAM and *PNLancs* 62) and of ON **holmr** 'island, land beside water'. The site is above the R. LICKLE.

LYTH VALLEY SD4688 Crosthwaite.
(le) Lyth(e), Lith(e) 1247 to 1777 (*PNWestm* I, 81–2).
♦ L~ is 'the slope', probably from ON *hlíð* 'slope, hill-side'. The valley is the wide flat course of the R. GILPIN. On the word *valley*, see LORTON VALE.

LYULPH'S/LYULF'S TOWER NY4020 Matterdale.
From 1783 (*PNCumb* 255).
◆ A castellated house built c.1780 by Charles Howard of Greystoke, who became 11th Duke of Norfolk, this was named after *Ligulf*, a presumed ancestor of the barons of Greystoke (*PNCumb* 255, Pevsner 1967, 160). The word *tower* is an adoption into ME from OFr; in this volume it only occurs here and in GREAT T~ PLANTATION.

LYZZICK: L~ HALL NY2526 Underskiddaw.
Losaikes c.1220, *Lesake(s)* 1343, 1563 (*PNCumb* 322), *Lissick* 1774 (DHM); *Lyzzick Hall* 1867 (OS).
◆ L~ is 'the bright oaks', from ON *ljóss* 'bright, shining' and ON *eik(i)* (see **oak**). The **hall** is a substantial building, currently a hotel.

MAIDEN CASTLE NY1805 Eskdale. From 1587 (*PNCumb* 392).
MAIDEN CASTLE NY4424 (hab.) Matterdale, NY4524 (antiquity). *Maydencastel* 1285, *syke* [stream] called *Carthanacke, Carthonock* 1589 (*PNCumb* 255).
♦ The Matterdale **castle** is an Iron-Age hill fort (LDHER 1165, SAM No. 23687, and see Maclean 1912), while at Eskdale there is a Bronze Age round cairn of turf and stone (LDHER 1329, SAM No. 23693). The meaning of *maiden*, from OE *mægden* 'girl, unmarried woman', when used in p.ns is a still-unsolved problem. It could imply ownership by young women or nuns, a haunt of girls or courting spot, a stronghold which is 'virgin', unconquered or one so strong that it could be defended by women (see *PNCumb* 255–6). One of the last two meanings seems most apt for the numerous English Maiden Castles, usually prehistoric earthworks, including the spectacular one in Dorset, and for names such as Mayburgh and Mawbray, both in Cumbria and meaning 'maidens' fortification'; similar names are also found in Europe (see Hough 1996 for listing and discussion of English p.ns containing *mægden, maiden*). A Celtic origin is sometimes claimed for the p.n. Maiden Castle (e.g. *maidun* 'Celtic for "fort"', Heaton Cooper 1983, 19), presumably with W *dun* 'fort' and cognates in mind, but this fails to explain the 1st syllable. A wide-ranging investigation of the name-type Maiden Castle is forthcoming from Richard Coates. However, the older name of the Matterdale hill fort, *Carthanacke*, is Celtic, with Cumbric *Cair* 'fortified place', and *Thannock* possibly a pers.n. The place was still known as *Caer-Thanock* to Hutchinson in 1776, who remarked that 'the country people give it the name of *Maiden's Castle*' (p. 81; Maiden Castle in Hutchinson 1794, I, 442).

MAIDEN MOOR NY2318 Above Derwent. From 1839 (Ford 81), 1867 (OS).
♦ The significance of *maiden* in this hill-name is unknown, though see MAIDEN CASTLE for possibilities; also **moor**. Coleridge called it *Maiden Bower* (*CN* 804, 1800), perhaps recalling prehistoric sites of that name in Oxfordshire and Bedfordshire.

MAINS FARM NY4724 Barton. Previously *Maynes New House* 1714, *Maines Hall* 1717, *Demesne(s) House* 1747, 1812, *Mains House* 1735 (*PNWestm* II, 213), *Mainshouse* 1863 (OS).
♦ 'Home farm, demesne farm.' The 1747 spelling is a useful reminder of the origin of *main(s)*, in Anglo-French *demesne* (*OED*). *Main(s)* is used in Scotland and N. England for 'the farm attached to the mansion-house on an estate, the home farm, the chief farm of an estate or township, demesne lands' (*EDD: main* sb. 1). The demesne in question here was recorded in 1578 as *Pooley demaynes* (*PNWestm* II, 213). As seen above, **farm** has replaced **house** in this name.

MAINSGATE SD0999 (hab.) Irton. *Maynesgate* 1626 (*PNCumb* 404).
♦ 'The gate or road to the home farm', with 1st el. *mains* as in MAINS FARM. The 2nd el. may be the reflex either of OE **geat** 'gate' or of ON **gata** 'road, way'.

MALLEN DODD NY2725 Underskiddaw. From 1867 (OS).
♦ 'Mallen's hill-top'; pers.n. plus **dodd**. This is the N. part of LATRIGG, a rounded and quite steep-sided spur. (a) M~ may be a variant of *Mal(l)in*, a pet-form of *Mary*; or the surname derived from it; such a name is suggested by records of two persons surnamed Malleson in the (Crosthwaite) PR. (b) I am grateful to Dr Oliver Padel for the alternative suggestion that this could contain a Gaelic diminutive which appears in Scotland as *maoilinn* 'bare round hillock' (Watson 1926, 146, and see **mell**).

MANESTY NY2518 (hab.) Borrowdale.
Manistie, Maynister 1564, *Manistae* 1565, *Manistye* 1601 (*PNCumb* 353).
♦ Possibly 'Máni's steep path', with ON pers.n. *Máni*, gen. sing. *Mána*, plus **stígr/sty(e)** 'steep path' (cf. THORPHINSTY). M~ lies at the foot of steep contours.

MANOR FARM SD2188 Broughton West. From 1892 (OS; not shown or named on 1850 map).
♦ *Manor* as in AYNSOME MANOR, plus **farm**. This was built late in the 19th century, as the newest farm on the estate of the lord of the manor, Robert Rankin (this information gratefully received from Mrs E. Garnett, tenant of Manor Farm).

MARDALE: M~ BANKS NY4812 Shap Rural, M~ COMMON NY4811, M~ WATERS NY4410.
Merdale 1278 (*PNWestm* II, 169); *Mardale Banks*; *Mardale Common*; *Mardale Waters* all 1858 (OSNB).
♦ M~ is 'the valley with the lake'; **mere**[1], referring to HAWESWATER, plus **dalr**; also **bank, common, water**. The name has taken on a certain poignancy, since the valley and village of M~ were engulfed by the lake when it was expanded into a reservoir to serve Manchester in the 1930s. On Mardale and Haweswater, see Thompson 1942. M~ Waters refers to BLEA WATER, SMALL WATER and the streams draining from them, and more broadly to 'a large tract of pasture & rocky land' (OSNB, 1858, 83).

MARDALE ILL BELL NY4410 Kentmere/Shap.
Hill Bell 1836, *Mardale Ill Bell* 1857 (*PNWestm* I, 165). OSNB shows hesitation between *Hill* and *Ill*, but cites *Ill Bell* in Otley's Lakes Guide, p. 5, adding, 'This is considered by the best authorities the correct spelling' (1858, Kendal 1, 7).
♦ I~ B~ is probably 'the bad, treacherous bell-shaped hill', like the ILL BELL at NY4307 (q.v.), and with MARDALE distinguishing it from its namesake.

MARSHALLS, THE NY2336 (hab.) Ireby and Uldale.
Cf. *Marshall's Hall* 1867 (OS), *Marshall's Cottage* 1900 (OS), both apparently on the same site as The M~.
♦ This appears to contain the surname *Marshall*, which derived from a medieval occupational term referring either to high officials or to farriers. A John Marshal, innkeeper, appears in the Uldale PR 1791, and James Marshall was assistant minister (1781, 1830).

MARSHSIDE COTTAGES SD0890 Bootle.
Marshside 1867 (OS), *Marshside Cottages* 1900 (OS).
♦ The site is on low-lying ground close to the coast, and is one of two places in close proximity named

Marshside on the 1867 OS map. *Marsh* goes back to OE *mersc* but is probably not old here; its only other occurrence in this volume is in (LOW) MEATHOP MARSH. See also **side**, here in the sense 'place beside', plus **cottage**.

MART CRAG NY3004 Langdales.
From 1823 (*PNWestm* I, 207).
MARTCRAG MOOR NY2608 Langdales.
From 1859 (OSNB).
♦ M~ C~ is 'the rocky height frequented by pine martens'; *mart*, **crag**; also **moor**. The *mart* (from OE *mearð*) is probably the pine marten or sweet mart rather than the polecat or foul mart (Gambles 1989, 128–31). The polecat is now extinct, and the pine marten a rarity, in the Lake District (Hervey & Barnes 1970, 193). Speaking of the Langdales and Elterwater c.1692, however, Machell wrote, 'Marts and wild cats are found all over' (ed. Ewbank, 139), and his comment on the Hartsop area reveals their significance to farmers: 'They haue great store of Marts hereabouts, som of wh[ich] are very perniciouse not only to their Lambs but old sheep too' (I, 726). Martcrag Moor is 'a tract of high moory land' (OSNB) close to a different Mart Crag, at NY2607.

MARTINDALE NY4419 (parish), M~ COMMON NY4317.
Martindale, Martyndale 1220–47 (*PNWestm* II, 215); *Martindale Common* 1860 (OSNB).
♦ 'Martin's valley', pers.n. plus **dalr**, also **common**. Possibly 'from an Old Chaple they haue in that Place dedicate to St. Martyn' (Machell c.1692, I, 698; his other suggestion was marts or martens). As in the case of PATTERDALE, however, it is not certain whether the p.n. derives from a dedication, here to the 4th-cent. Bishop of Tours, or vice versa. Both the valley name and the chapel of St Martin were in place at least by the 13th cent. The Common is an area of high grazing that elsewhere might have been called **fell**.

MARY MOUNT NY2619 (hab.) Borrowdale.
♦ The use of *mount* is generally rare and late in Lake District names (though see MOUNTJOY). I am most grateful to Susan Q. Mawdsley of Mary Mount Hotel for the information that the name was given in the 1950s by the then owners, the Braithwaite family, echoing that of the sons' school, Mount St Mary's College in Derbyshire.

MATSON GROUND SD4196 Bowness.
Matsons 1661, 1706 (Windermere PR), *Matson Ground* 1751 (*PNWestm* I, 187).
♦ 'The farm of the Matson family'; surname plus **ground**. *Matson* is recorded locally from the late 14th cent. to 17th cent. Matson Shoal in WINDER-MERE is also named from them (*PNWestm* I, 187, 193).

MATTERDALE (hab. & parish): M~ COMMON NY3521, M~ END NY3923 (hab.).
Mayerdale (in error for *Maþerdale*) c.1250, *Matheresdal'*, *Maderesdale* 1285, *Matherdal(e)* 1323 to 1419, *Madredale* 1380 (p), *Maderdale* 1487, *Matterdall* 1559; *le Common* 1487; *Matterdail End* 1663 (*PNCumb* 221).
♦ M~ is probably 'the valley where bedstraw grows' from ON *maðra* (the cognate of OE *mæddre*) and ON **dalr**; also **common**, **end**. (a) This seems the most likely explanation of the spellings, and ON *maðra* occurs in several Norw names and in the exactly parallel Möðrudalur, Iceland. The plant is thought to be *Galium boreale*,

Northern Bedstraw (Fritzner: *maðra*), which Hutchinson lists (as *Galium boreale, Crosswort madder*) at 'Usemire and lower part of the lake [Ullswater]' (1794, I, 465; also Halliday 1997, 403). (b) An alternative explanation of the 1st el., mentioned in *PNCumb* 221 and favoured by Sedgefield (1915) is the OE pers.n. *Mæðhere*. The main evidence for this would be the *-es-* spelling in 1285, but overall it is less likely. M~ End is at the N end of the valley of M~ Beck, and of the parish of M~, close to the boundaries with Hutton and Watermillock (which were separate from M~ until 1934). It is 'the second hamlet of the valley and is of a later settlement than many other parts, first mentioned in the C16'; it appears to be referred to as Matterdale Hamlet in 1590 and 1599 (Keith Clark 2000, 44 and pers.comm.).

MAY CRAG NY2117 Above Derwent. From 1867 (OS).
♦ 'The rocky height with hawthorn or associated with maidens'; from *may*, an adoption into ME from Anglo-Norman or OFr *mai* (*OED: may* n. 2) in the sense of 'hawthorn blossom', plus **crag**. The month-name from which the flower-name *may* originates seems less likely, as is *May* as a female forename, since this is mainly 19th-cent. (Withycombe 1977) or 'at its height in the early 20th century' (Hanks & Hodges 1990). A further possibility is *may* 'maiden' from OE *mǣge*.

MEADLEY: M~ RESERVOIR NY0514 Ennerdale.
Medlay 1578, *Meedley* 1712 (*PNCumb* 387); *Meadley Reservoir* 1900 (OS).
♦ M~ may derive from OE *mēd, mǣd* 'meadow' and OE **lēah** 'woodland clearing', but without earlier spellings the age and meaning of the name are uncertain. See also **reservoir**.

MEAL FELL NY2833 Ireby and Uldale. From 1867 (OS).
♦ Probably 'the hill called Meal or Mell', 'the bald, bare hill', from Cumbric **mēl* 'bare hill' (see **mell**), with **fell** 'hill' added later . The hill has a stony top, which contrasts with its own grassy sides and with its neighbours. Thus the name may be equivalent to MELL FELL (as suggested in *PNCumb* 328), and this is supported by the spelling *Mealfell* c.1690 for a Mell Fell in former Westmorland (*PNCumb* 212).

MEARNESS FARM SD3281 Haverthwaite, M~ POINT SD3281.
Mean House; Mearness or *Mean House Point* both 1851 (OS).
♦ Despite the lack of early forms, M~ could mean 'boundary headland' from OE (*ge*)*mǣre* (**mere**[2]) plus **næss** 'headland'. For the 1st el., OE *mere* 'pool' (**mere**[1]) is also a possibility. The form *Mean House* seems to misconstrue the name. M~ projects into the channel of the R. LEVEN at GREENODD Sands, very close to the boundary with Egton township. **Farm** distinguishes the habitation from the **Point**, whose name reinforces the meaning 'headland'.

MEASAND: M~ BECK NY4715 Bampton, M~ END NY4715.
Mussaund 1265, *Mesand(e)* 1308 to 1612, *Measand(e)* 1564 to present; *M~ becks* 1714 (*PNWestm* II, 192); *Measand End* 1860 (OSNB).
♦ M~ probably means 'the sandy area by the marsh' from OE *mēos* 'marsh' and OE *sand* 'sand, sandy shore or bank'; also **beck, end**. Formerly a delta between the two main reaches of Haweswater, the area is now submerged under HAWESWATER

RESERVOIR. A former interpretation was 'narrow channel' (ON *mjósund*; Sedgefield 1915). M~ End is a high, rocky scar rising to the S of the beck.

MEATHOP SD4380 (hab. & parish), LOW M~ SD4379 (hab.), LOW M~ MARSH SD4279, M~ MOSS SD4481.
Midhop(e), Midhopp 1184–90 to 1256, *Mithehop* 1190–1210, *Methop(e), Methopp(e)* 1278 to 1650; *Meathop Moss* 1845 (*PNWestm* I, 76); *Low Meathop* 1751 (PR), cf. *Nether Meathop* 1754 (PR, referring to same family); *Low Meathop Marsh; Meathop Moss* both 1862 (OS).
♦ M~ is 'the middle enclave', from OE *midd* 'middle', influenced by ON *miðr*, and OE **hop**, which often means 'blind, secluded valley', but here it refers to a plot of habitable land next to the salt-marshes of the KENT estuary, including M~ Marsh. The element *marsh*, though originating in OE *mersc*, may be fairly recent here; see also SUNNYSIDE FARM, and **low**, **moss** 'bog'.

MECKLIN BECK NY1202 Irton, M~ PARK NY1302.
Both from 1867 (OS).
♦ 'Mecklin' is also the specific for a Bridge and Wood nearby but is unidentified, unless it is a corruption of the plant-name *meckin*, applied to the yellow flag iris and various ferns (*EDD*: *mekkin*). The **Park** is the hillside, rock-strewn and rich in prehistoric earthworks, S of the **beck** 'stream'.

MEDIOBOGDUM NY2101 Eskdale.
Medibogdo, Medebogdo c.700 (Rivet & Smith 1979, 415).
♦ Probably '(The place) in the middle of the bend', from Brit. *mĕdi̯o̯-* 'middle' and **beugh-* 'bend, curve' (Rivet & Smith; Coates considers it of unknown origin, 2000, 2). This is the older name of HARDKNOTT FORT; if Rivet and Smith are correct, it must refer to some aspect of its dramatic position on a platform backed by a semi-circle of crags and between the upper reaches of the R. ESK and Hardknott Gill.

MELBECKS NY2432 (hab.) Bassenthwaite. From 1664 (PR); *Mellbeck* 1774 (DHM), *High Mellbecks, Low Mellbecks* 1823 (GreenwoodM), *Millbecks, Low Millbecks* 1867 (OS).
♦ (a) Probably 'the streams by the sand-bank', with the reflex of ON *melr* 'sand-bank' and **beck**. A gravel-pit known as 'the Sand Quarry' lies close by, and another Melbecks (*PNNYorks* 271), and a Mell Beck (*PNWestm* I, 11) are of the same assumed origin. (b) Despite a history of small mills on nearby streams, and the names Mill Beck, and Walk Mill Bridge (NY2431 on larger scale maps), the 1st el. is less likely to be **mill** from **mylen**, especially since none of the certain 'mill' names in this volume have spellings in *mel-*. (Local information kindly supplied by the Trafford family of PETER HOUSE FARM.)

MELLBREAK/MELBREAK NY1419 (1668ft/ 508m.) Loweswater.
From 1780 (West 133).
♦ This is a very prominent oval mountain which dominates the W. shore of CRUMMOCK WATER, but its name is somewhat obscure. (a) It may, unusually, be of Gaelic origin, from *Meall breac* 'hill that is dappled' (as suggested in Brearley 1974); this is assumed in Maulbrack, Co. Cork (Flanagan & Flanagan 1994, 120, and see **mell**). (b) A wholly ON solution would be *melr* 'sandbank', perhaps referring figuratively to shape, plus ON *brekka* 'hill-slope', a combination

which may occur in the Norw Merbrekke (*NG* VII, 438) and which would be paralleled in Sandbrekken (*NG* XIII, 384; VII, 97). ON *brekka* is, however, surprisingly rare in the Lake District (see CALEBRECK) and is not an obvious term for so dramatic a hill.

MELL FELL, GREAT NY3925 (1759ft/ 536m.) Hutton, LITTLE MELL FELL NY4224 (1657ft/505m.) Matterdale, MELLFELL HOUSE NY4323.
(Great Mell Fell:) *Melfel* 1279 (p), *Great mele Fell* 1487; (Little Mell Fell:) *Mele Fell* 1487, *Little Mellfell* 1789 (*PNCumb* 212); (Mellfell House:) *Mell Fell* 1774 (DHM), *Mellfell* 1867 (OS), *Mellfell House* 1900 (OS).
♦ 'The hill called Mell'; probably from Cumbric **mēl* 'bare hill' (see **mell**), with **fell** 'hill' added later, and see **great, little, house**. These distinct, conical hills are less than two miles apart. Great M~ F~, the higher, though not the larger in extent, lies W of Little M~ F~, and hence was *Westermellfell* in 1589 (and the two hills are shown as *West* and *East Mell Fell* on Greenwood's 1823 map). Mellfell House takes its name from Little M~ F~, as may nearby WATERMILLOCK, which appears to contain a derivative of Brit. **mēl*.

MELLGUARDS NY4419 (hab.) Martindale.
Mel Gards 1758, 1763, *Melguards* 1767 (PR).
♦ Possibly 'enclosures by the sand-bank', with the reflex of ON *melr* 'sandbank' and **garth/g(u)ard** 'enclosure'. However, there is no feature nearby that would obvious qualify as a sandbank. I am indebted for this information to Mr Brian Cook, the present owner, who tentatively suggests a connection with the bobbin **mill** 200 yards away.

MEOLBANK NY0802 (hab.) Gosforth.
Mealbanck 1600 (PR), 1605 (*PNCumb* 396).
♦ Probably 'the slope by the sand-bank', with the reflex of ON *melr* 'sand-bank' and **bank** 'slope, bank'. The place is near Hare Beck.

MERE BECK NY4416 Martindale.
From 1865 (*PNWestm* I, 11).
♦ 'The boundary stream'; **mere**², **beck**. It rises just below the Martindale/Bampton boundary.

MEREBECK, HIGH (hab.) NY0314 Ennerdale.
Merebek 1540 (*PNCumb* 21); cf. *Lowmeerbeck* 1696 (*PNCumb* 387).
♦ M~ is 'the boundary stream'; **mere**², **beck**; also **high** 'upper'. The stream forms the present boundary between Cleator Moor and Ennerdale parishes (and coincidentally of the National Park at this point).

MERE CRAG NY5006 Longsleddale/ Fawcett Forest.
From 1836 (*PNWestm* I, 164).
MERE CRAGS SD1789 Millom Without.
From 1867 (OS).
♦ Probably 'the rock outcrop(s) on the boundary'; **mere**², **crag**. The position on the parish boundary strongly suggests this for the example at NY5006. The other case is less certain, the crags at SD1789 being half a mile S of the present boundary between Ulpha and Millom Without.

MEREGARTH SD3999 (hab.) Windermere.
From 1920 (OS).
♦ 'The enclosure or farm by the lake' — here WINDERMERE; **mere**¹, **garth**.

MEREGILL BECK NY1021 Loweswater.
Cf. *Mear Gill* 1752 (PR).
♦ 'The stream in Meregill, the

boundary ravine'; **mere**[2], **gil(l)**, **beck**. This still marks a County Constabulary and Civil Parish boundary.

MERE GREAVE, HIGH NY4307 Troutbeck. From 1865 (*PNWestm* I, 191).
♦ Probably 'the upper boundary copse'; **high, mere**[2], **greave**. High and Low M~ G~ form part of the steep W. flanks of YOKE close to the boundary with Windermere and may have formed the 'boundary copse', though no woodland is shown on the map now.

MESSENGERMIRE WOOD NY2033 Blindcrake.
Messenger's Mire Wood 1810 (*PNCumb* 301), *Messengermire Wood* 1867 (OS).
♦ 'The **wood** at Messengermire, the marsh or **mire** associated with the Messenger family.' The surname Messenger is recorded locally from 1699 (*PNCumb* 301, citing *CW*), and occurs frequently in PRs of the 18th and 19th cent.

MICHAEL'S NOOK NY3408 (hab.) Grasmere.
Seemingly *Nook* 1859 (OSNB), 1863 (OS), *St Michael's* 1899 (OS).
♦ At least two Michaels are associated with this immediate area: Michael Knott, owner of BROADRAYNE in 1692, who may have given his name to *Michael-place in Grasmere* (1688, Grasmere PR, and see Simpson 1928, 277 and 287), and the shepherd hero of Wordsworth's poem *Michael*. One might conjecture that the name of Wordsworth's Michael was inspired by *Michael-place*, then that the poem influenced the naming of M~'s **Nook** and perhaps nearby Michael's Fold, a renaming of GREENHEAD.

MICKLEDEN NY2606 Langdales.
Mickleden 1862 (OS); Machell's '*Great Langdale alias Mickledale*', c.1692 (II, 134) may result from confusion with this name.
♦ 'The great valley', from OE *micel* 'large' (cf. ON *mikill*) and OE *denu* 'valley', which has not been replaced by **dale** as happened in LANGDALE. The seeming lack of early forms for this name is rather striking. Together with OXENDALE, M~ forms the spacious head of GREAT LANGDALE.

MICKLEDORE NY2106 Eskdale.
the yawning chasm of Mickle Door 1823 (Otley 89).
♦ 'The large door or col', from the reflex of ON *mikill*, OE *micel* 'large', plus **duru/door**. Well known to walkers and geologists, this is the bite-like gap in the crags between SCAFELL and SCAFELL PIKE. There are no early forms, but the name may be old, judging from the occurrence of *le Mikeldor de Yowberg*, now DOOR HEAD on Yewbarrow, in 1332 (*PNCumb* 441). (Pl. 25)

MICKLE RIGG NY2736 (1030ft/314m.) Ireby and Uldale.
From 1867 (OS).
♦ 'The great ridge', from dial. *mickle* (the reflex of ON *mikill*/OE *micel* 'large'), plus **rigg** (see **hryggr**). This hill is fairly elongated, though not so spine-like as many *riggs*.

MIDDALE NY4903 (hab.) Longsleddale.
Midd-Dale 1755 (PR), *Middale* 1836 (*PNWestm* I, 163).
♦ 'The mid-point or middle stretch of the valley' of (LONG)SLEDDALE, with a shortened reflex of OE **middel**, plus **dale**.

MIDDLE BANK: see BANK, Waberthwaite

MIDDLE CRAG NY2815 (1587ft/484m.) Borrowdale/St John's boundary.

Middle Cragg 1805 (*CW* 1981, 67), *Middle Crag* 1867 (OS).
♦ 'The rocky height in the middle'; **middle**, **crag**. This is a modest rocky knoll, which is mentioned in the 1805 bounds of the manor of Borrowdale between HIGH TOVE and 'a standing Cragg' (*CW* 1981, 66–7) — perhaps SHIVERY KNOTT. This may (or may not) have been the intended sense of *middle*.

MIDDLE DODD/DOD NY3909 Patterdale.
Midledod (*syde*) 1588, *Middle Dod* 1860 (*PNWestm* II, 226).
♦ 'The compact hill in the middle'; **middle**, **dodd**. Shaped like 'a gigantic upturned boat' (Wainwright 1955, Middle Dodd 1), it lies roughly between HARTSOP DODD and HIGH HARTSOP DODD.

MIDDLE FELL NY1018 Lamplugh.
Midlefell 1599 (*CW* 1925, 194).
MIDDLE FELL NY1407 Nether Wasdale.
Melfell 1802 (*CN* 1213), *Middle Fell* 1839 (Ford 62).
♦ Simply 'the hill in the middle' (**middle**, **fell**), assuming that Coleridge's 1802 form is (as often) mistaken. The Lamplugh example is a stretch of high grazing land flanked by more distinct peaks. The Nether Wasdale one is a peak between BUCK-BARROW and YEWBARROW.

MIDDLEFELL FARM NY2806 Langdales.
Midlefell place 1654 (Winchester 2000, 112), 1687/8 (Grasmere PR), *Middlefell Place* 1966 (OS One Inch).
♦ 'The farm of the Middlefell family'; surname plus **place**, apparently replaced by **farm** in recent decades. As with most of the other Lakeland names which are, or were, formed from *place*, the specific is probably a family name, since the surname *Midelfell* and variants is recorded from 14th to 16th cent. (*PNWestm* I, 207), and is specifically associated with Great Langdale in the Grasmere PR for 1653 etc. The surname, in turn, may derive either from a stretch of high grazing or **fell** in the **middle** of Langdales township, or specifically to *Midelfel*, a point between BLACK CRAG and BOWFELL in charter bounds of 1242, which Collingwood identified with LONG TOP (q.v.).

MIDDLE GROVE NY3905 (hab.) Ambleside.
midle graue 1707, *Middle Grove* 1838 (*PNWestm* I, 184), cf. above *Grove* 1597 (*CW* 1906, 82), *Groves* 1770 (JeffreysM), *The Groves* 1780 (West, 75).
♦ This is probably 'the middle copse'; **middle**, **grove**. 'The Grove seems to have extended from Ambleside to the head of the Stock valley. The names of High, Middle and Low Grove designate three farmhouses within it, two of which have been in ruins for many years. Formerly, and probably down to the time of the enclosure of Wansfell, the inhabitants of Ambleside had the right of cutting fuel wood in the Grove' (Hills, quoted by Armitt in *CW* 1906, 85).

MIDDLE LOW WOOD: see LOW WOOD, MIDDLE

MIDDLE SWAN BECK: see SWAN BECK, MIDDLE

MIDDLE TONGUE NY3227 Threlkeld.
(*Scorte*)*middeltunge* c.1260 (*PNCumb* 252), *Middle Tongue* 1867 (OS).
♦ 'The middle tongue of land'; **middle**, **tongue**. This is *middle* from its position between GATEGILL FELL and HALL'S FELL on the massive southern rampart of BLENCATHRA. It is a classic

tongue, formed by two streams which converge as Gate Gill part way down the mountainside (hence the 'short' tongue recorded c.1260).

MIDDLETON PLACE SD0992 Waberthwaite.
Midletonplace 1657 (*PNCumb* 365).
♦ 'The Middletons' farm'; with 1st el. a surname, as frequently in names in **place**. Denton comments 'this was anciently the place and habitation of the Middletons' (1687–8, 75), and persons of that name appear in Waberthwaite PRs from 1694 to the 19th cent.

MILKINGSTEAD SD1599 (hab.) Eskdale. From 1728 (PR).
♦ 'Dairy place or farm'; the verb *milk* goes back to OE *milcian*, plus **stead**. The situation by the R. ESK is compatible with taking this at face value.

MILLBECK Examples at:
  NY1623 (hab.) Buttermere.
Milnebeck 1279 (p) (*PNCumb* 354).
  NY2526 (hab.) Underskiddaw.
Milnebek (p) 1260 (*PNCumb* 322).
  NY2906 (hab.) Langdales.
Mill-Beck 1693 (Grasmere PR).
♦ 'The mill-stream', from OE **mylen** 'mill' and ON **bekkr** 'stream', or their reflexes. The Langdale example, situated on STICKLE BECK, may have been a fulling mill during the 16th-cent. heyday of the industry (Armitt 1908, 148), though a corn-mill is also possible. The Underskiddaw example was once a woollen mill (*CW* 1957, 158).

MILLBROW SD1892 Ulpha.
From 1867 (OS).
MILL BROW NY3403 (hab.) Rydal.
Millbrow in Loughrigg 1655 (Grasmere PR).

♦ 'The hill-slope by the mill'; **mylen/mill**, **brow**. The Ulpha example is above HOLEHOUSE GILL, where New Mill, a bobbin mill, is shown on the 1867 OS map. The Rydal mill seems to have been a walk-mill or fulling mill: a nearby field called Tenter Close must have been used for stretching and drying cloth (*CW* 1908, 150).

MILLDAM SD4494 (hab.) Crook.
Milldamm in Crook 1659 (*PNWestm* I, 180).
♦ 'The millpond', **mylen/mill** and *dam*, the term used in Lakeland for a millpond (Davies-Shiel 1971, 281). Somervell noted that 'the fields [here] have every indication of having at one time contained a large mill pond', but believed the former mill to be obscured by modern buildings or the road (1930, 112). This, and the separation by at least a mile from CROOK make it implausible that 'the dam served Crook mill' (*PNWestm* I, 180).

MILLER BRIDGE HOUSE NY3604 Rydal.
Milnerbrigg 1613, *Miller Bridge* 1632 (*PNWestm* I, 210–11).
♦ 'The house by the miller's, or Miller's, bridge'; *miller* from OE *mylnere* (though *mylenweard* 'millward' was more usual) or the surname derived from it, plus **brycg/bridge**, **house**. As in other p.ns containing 'Miller', including MILLER PLACE, there is some doubt whether this is the appellative *milner, miller*, or the surname derived from it. The appellative would be favoured by evidence of a mill at the site in the Early Modern period which was probably used at various times for both corn and fulling (Armitt 1908, 182–3, 205), though nearby Miller Brow might encourage the assumption of a surname.

MILLERGILL BECK SD1085 Whicham.
Not named on 1867 OS map.
♦ The **beck** flows through the ravine of Miller **Gill** and through Whitbeck Mill. *Miller* therefore seems likely to be the appellative here (cf. MILLER BRIDGE HOUSE), though the surname derived from it cannot be excluded.

MILLERGROUND: M~ LANDING SD4098 Windermere.
From 1719 (spelling not specified, *CW* 1964, 18), *Milner ground (miln)* 1738 (*PNWestm* I, 196); *Millerground Landing* 1859 (OSNB).
♦ M~ is 'the farm of the Milner or Miller family'; surname plus **ground**; also *landing* as in LANDING. The Milners of UNDERMILLBECK are traceable in documents of the 14th cent. onwards (*PNWestm* I, 196; *Miller* appears in PRs). That Millerground was a milling site, not merely the onetime residence of people named Miller, is seen in the 1738 spelling. M~ Landing is one of several past and present names of landings on the shores of WINDERMERE.

MILLER HILL SD0890 (hab.) Bootle.
From 1867 (OS).
♦ The **hill** is a slight rise in the low coastal plain. *Miller* may be an occupational description (see MILLER BRIDGE HOUSE) or the surname derived from it, though a rapid survey has not revealed the surname *Miller/Milner* in Bootle PRs. Another Millerhill is recorded from 1589 (*PNCumb* 116).

MILLER MOSS NY3033 Mungrisdale.
From 1867 (OS).
♦ Presumably 'the **moss** associated with the Miller family', though I do not know of records of the family or individual concerned. This is the bog at the headwaters of GRAINSGILL BECK.

MILLER PLACE NY1623 Buttermere.
Milner-place 1610/11, *Millerplace* 1638 (Lorton PR), *Millar Place* 1774 (DHM), *Millerplace* 1814 (PR).
♦ Probably 'the farm of the Miller family'; surname plus **place**. Names in *place* often have a surname as specific element, and this may be the case here, though given the situation, close to MILLBECK, the occupational term *milner/miller* is also possible; cf. MILL PLACE.

MILL GILL NY3219 St John's.
From 1867 (OS); cf. *Legburth'wt Mill* 1813 (PR).
♦ Self-explanatory; **mylen/mill**, **gil(l)** 'ravine with stream'. The water of the *gill* powered LEGBURTHWAITE Mill, still traceable at NY3119 (see 1:25,000 map). It is shown as a corn mill on OS 1867, though there was a fulling mill in the parish (*Walk Mill* 1665, *PNCumb* 318) whose site is unknown to me.

MILL HOUSE FARM SD1199 Irton.
Mill House 1780 (PR).
♦ Self-explanatory; **mylen/mill**, **house**, and **farm** added later.

MILL PLACE NY1001 Irton.
Millplace 1599 (Muncaster PR).
♦ This is beside the R. IRT, close to a number of weirs; **mylen/mill**, **place**. It was the site of the manorial corn mill (Parker 1926, 152).

MILLRIGGS/MILLRIG NY4502 (hab.) Over Staveley, MILLRIGG KNOTT NY4601.
Milrig c.1692 (Machell II, 112, ed. Ewbank, 109); *Millrigg Knott* 1858 (OSNB).
♦ 'The ridge by the mill'; **mylen/mill**, **hryggr/rigg**; also **knott** 'craggy height'. There was at one time a fulling mill here.

MILL SIDE SD4484 (hab.) Witherslack. From 1862 (OS).
◆ 'The place beside the mill'; **mylen/mill**, **side**. It appears next to 'Witherslack Mill (corn)' on the 1862 map.

MIREHOUSE NY2328 Bassenthwaite. *Miras* 1717 (PR), *Myrehouse* 1736 (*PNCumb* 264).
MIRE HOUSE NY3123 St John's. *Mirehouse* 1568 (*PNCumb* 317).
◆ 'The house by the marsh'; **mýrr/mire**, **house**. The Bassenthwaite example is a large, late Georgian mansion (Pevsner 1967, 64), built on a low-lying site that was drained and planted when the common land was divided (Clarke 1789, 99). The St John's situation is also low-lying and damp, beside ST JOHN'S BECK.

MIRESIDE Examples at:
  NY1016 (hab.) Ennerdale.
From 1608 (*PNCumb* 387).
  NY2230 Bassenthwaite.
From 1617 (PR).
  SD2288 Broughton West.
From 1668 (original spelling not specified, Broughton-in-Furness PR).
  SD4390 (hab.) Crosthwaite.
From 1535 (in 1669 copy, *PNWestm* I, 85).
◆ 'The place beside the swampy ground'; **mýrr/mire**; **side**. This aptly describes all four situations. The sense of *side* could shade into 'hill-slope' in some cases. The Bassenthwaite M~ S~, nearby Moss Side (*Mossyd* 1578, *PNCumb* 264) and possibly HIGH SIDE may have been named in response to each other, and the Crosthwaite M~ is also close to other names formed from *side* (see List of Common Elements). The Broughton site is beside an area, now drained, around the stream called Galloper Pool and fed by springs.

MIRKHOLME NY2532 (hab.) Ireby and Uldale.
Myrkeholme 1594 (*PNCumb* 328).
◆ 'The dark place by the river', with 1st el. from ON *myrkr* 'dark' plus **holm(e)** 'land flanked by water, island'; the farm is next to DASH BECK. 'Dark' may refer to the somewhat enclosed situation, where sunlight hours are relatively short (I am very grateful to Judith Cubby for this).

MIRK HOWE SD4391 (hab.) Crosthwaite. *Merkshowe* 1374 (*Kendale*, II, 94), *Mirkhow(e)* 1535 to 1738 (*PNWestm* I, 85), *Mirk house* (hab.), *Mirk Hill* (hill) both 1863 (OS).
◆ 'Darkness hill' or perhaps 'fog hill'. If the 1374 spelling can be relied on, the name is composed of the ON noun *myrkr* 'darkness' (gen. *myrkrs*; cf. adj. *myrkr*) and ON **haugr** 'hill'. Why the place is described as dark is uncertain, but I am most grateful to Dr Peter Bracewell of Mirk Howe for the suggestion that it could refer to the unusual conditions which occur in summer when 'sea fret comes up the Winster and Lyth valleys and surrounds the farm for a few minutes'. Since ON *myrkr* can apply to enclosing sea fogs, this seems quite possible. Mirk Howe remains the accepted name and is shown as such on the 1994 One Inch map, but Mirk Hill is used of the hill, and appears on the 1863 map and the current 1:25,000 map; note also *Mirk House* 1863. These seem to be attempts to differentiate the hill from the farm.

MISLET SD4399 (hab.) Windermere. *Micheleselet* [sic] 1256, *Micheleset* 1256, *Mystlehedd* 1545, *Mis(s)led* 1694, *Misslett* 1702; cf. *Mikeleswyk* 1256 (*PNWestm* I, 195).
◆ Probably 'Michael's shieling', from the ME pers.n. *Michel*, recorded locally

as a surname (*PNWestm*), and ON **sætr** 'shieling', though see that entry for other possibilities.

MITCHELLAND SD4395 (hab.) Crook.
Mitchall Land 1706, *Mitchelland* 1836 (*PNWestm* I, 180).
♦ 'The **land** or estate of the Mitchell family', which is recorded from the Crook area from the 16th cent. (*PNWestm*), and is common in the Kendal PRs.

MITE, RIVER SD1098, NY1602 Irton/Eskdale/Muncaster, MITE HOUSES SD0897 Drigg, MITERDALE NY1501 Eskdale, MITERDALE FOREST NY1301 Irton, MITESIDE SD1098 (hab.).
Mighet 1209, *Mite* 1292 (*PNCumb* 22); *Mitehouse* 1765 (Muncaster PR), *Mite Houses* 1867 (OS); *Meterdal* 1294, *Miterdale* 1334 (*PNCumb* 389); *Mile Side* [sic] 1774 (DHM), *Miteside* 1867 (OS).
♦ M~ is probably 'the trickling one'. Like many river names, this seems to be of Brit. origin. (a) Ekwall suggested that the stem descended from Indo-European *meiĝh-* 'urinate' (cf. some Norw stream-names based on this idea) or *meigh-* 'to drizzle', while the *-t-* reflected a Brit. suffix *-ētion* or *-eto* (ERN 294–5; the name is left unexplained in his later *DEPN*). (b) Breeze suggests derivation from the Cumbric equivalent of W *muchudd* 'jet', 'if local knowledge confirms that the Mite flows over black rocks, or is blackened by the rocks it flows over' (1999). However, local information suggests that the river is not notably black, but that it is 'a "spate" river which only has significant flow immediately after rainfall. In a very dry summer it just becomes a trickle' (pers.comm. from Mr William Paul, gratefully acknowledged); it was described as *a little rill or beck* in the 17th cent. (Denton 1687–8, 78, following John Denton).This would favour one of the earlier explanations suggested by Ekwall. The *-er-* syllable in the valley-name Miterdale derives from an ON gen. sing. *-ar*, rather exceptionally added to a non-Scand name, or from the gen. sing. of ON *á* 'river', cf. ALLERDALE); see also **house**, **dalr** 'valley', **forest** (a product of modern afforestation), **side**.

MITON HILL NY3334 Caldbeck/Mungrisdale.
From 1867 (OS).
♦ M~ is of uncertain origin. It has the appearance of a p.n., but there is none such, to my knowledge, in the vicinity, so that a surname derived from a p.n. elsewhere seems possible, cf. *Mitton, Mutton* (Reaney 1976) or *Mitton, Myton* (Hanks & Hodges 1988). **Hill** usually occurs in relatively recent names.

MONK CONISTON: see CONISTON

MONK FOSS FARM SD1185 Whicham.
Fossa(m) 1135–54, *Munkfossam* c.1200, *Munkeforse* c.1220 (*PNCumb* 448–9).
♦ Either 'the ditch of the monks' or 'the waterfall of the monks'; *monk* from OE *munuc* (an adoption from Lat *monachus*), *foss* 'ditch' or possibly 'rampart', or **force** 'waterfall', **farm**. This is *quandam curracatam terre* 'a certain carrucate or ploughland' granted to FURNESS Abbey in the 12th cent. (*Furness Coucher* II, 522–3). In the view of the *PNCumb* editors, Lat *fossa* 'ditch' was the original el., but since *foss* locally means 'waterfall' (or more usually *force*, from ON *fors*), and since there is a waterfall, Hall Foss, less than a mile to the N, the name has been reinterpreted. Fellows-Jensen however, acknowledging Michael Copeland, suggests that this was part of a larger

estate known originally as *Fors* 'waterfall' (*SSNNW* 147–8).

MONK MOORS SD0892 (hab.) Bootle.
Munkemore in Cauplande c.1290 (*PNCumb* 347).
♦ 'The moors owned by the monks', with *monk* as in MONK FOSS FARM plus **moor**. This is a reference to FURNESS Abbey; see Sykes 1926, 116 and, e.g., *Furness Coucher* II, 526–8 for grants of land in Bootle to the monks.

MONKS BRIDGE NY0610 Ennerdale.
This is marked as a Foot Bridge, but not named, on the 1867 OS map.
♦ From *monk* as in MONK FOSS FARM, plus **brycg/bridge**. The present structure, a packhorse bridge and hence probably later 17th- or early 18th-cent., crosses the R. CALDER just N of Friar Gill. The crossing is presumed to have been a routeway through the lands of CALDER ABBEY (Hindle 1998, 47 and photograph there); it is known locally as Matty Benn's Bridge (Wainwright 1966, Lank Rigg 3).

MOOR NY2822 (hab.) St John's.
Moure 1569 (*PNCumb* 317).
♦ 'The upland waste'; **moor**. Use of an unqualified topographical term as a farm or hamlet name is not uncommon, and the p.n. Moor occurs, e.g., twice in each of Dumfries & Galloway, Dyfed and Devon (*OSGazetteer*); cf. also MOOR FARM. Clare remarks of this p.n., 'a name usually used of common land, and it is probably the latter to which the name recorded in 1569 ... applies' (1999, 71).

MOORAHILL FARM NY4918 Bampton.
Moorah hill 1690, 1723, *Murrah hill* 1698; cf. *Morhowe* 1240–70 (*PNWestm* II, 196).
♦ 'Moora' is evidently 'the hill on the boggy waste' (**mór/moor, haugr/howe**), to which **hill** and much later **farm** have been added. The farm is below the same slope as CARHULLAN, and next to a very small knoll. The place may also have been known as Lodge Hill, 'which seems to have been the early name of Moorah Hill' (Noble 1901, 148, facing 153).

MOORCOCK HALL SD4583 Witherslack.
From 1899 (OS).
♦ The moorcock is the male of the red grouse, *Lagopus lagopus*. The word *moorcock* is first recorded from 1329–30 (*OED*). This is one of the newer and smaller places to bear the name **hall**, and comparison of the 1862 and 1899 maps suggests that the building came into being after drainage of a portion of BELLART HOW MOSS. Place-names consisting of a bird name plus *hall* are quite frequent, especially in N. England and S. Scotland, arising in the Modern period (Hough 2003); cf. THROSTLE HALL below.

MOOR DIVOCK NY4822 Barton.
Moredvuock 1278 (*PNWestm* II, 201), *Moor-dovack* 1784 (West 158, corrected from *Dovack-moor* 1780, West 159).
♦ 'Dufoc's moor' or 'the moor of the stream called *Dufoc, the dark one'. This is a fascinating name, with OE **mōr**/ON **mór** 'upland waste, often boggy' as 1st el., while the 2nd el. and the overall structure are Celtic. The 2nd el. is probably from Brit. *dubāco-* meaning 'dark, black', cf. DEVOKE WATER. (a) This could be a pers.n. comparable to Middle W *Dyfog*. Ekwall, writing in 1928, counted a pers.n. 'very likely' (*ERN* 255; also Smith in *PNWestm* II, 201), and the spelling *Duvokeswater* for Devoke Water might support the idea of a pers.n. (b) It could alternatively be a stream-name, again with a possible

parallel in Devoke (Water) (so Ekwall in 1918, *Scand & Celt* 41–2). The word order resembles that of the Gaelic-influenced inversion compounds (see Introduction), yet there is no obvious Gaelic-based interpretation. The moor is rich in ancient remains of human activity (Taylor 1886 and a forthcoming survey by Lancaster University Archaeological Unit/ Oxford Archaeology North).

MOOREND NY0715 (hab.) Ennerdale.
From 1560 (*CW* 1931, 185).
MOOR END SD0899 (hab.) Irton.
Moore End 1662 (*PNCumb* 404).
MOOREND NY4725 (hab.) Barton.
From 1686 (*PNWestm* II, 213).
MOOREND FARM NY4023 Matterdale.
More End 1589 (*PNCumb* 257).
♦ Simply 'the edge of the upland waste'; **moor, end**. In Ennerdale, Low and Far M~ are distinguished on larger scale maps and are close to the present BROADMOOR. The Irton example is close to the parish boundary; it may possibly be named in relation to nearby WOOD END. The Barton farm is at the edge of Mill Moor (*Millnemore* 1588, *PNWestm* II, 213). **Farm** has been added to a pre-existing name in the Matterdale example.

MOOR FARM SD2893 Torver.
More 1599, *Moure* 1614 (PR).
♦ This is the same name as MOOR, with **farm** added later. The place is situated where the flat valley of the TORVER BECK starts to rise towards the peat moss of T~ LOW COMMON.

MOORGATE SD0998 (hab.) Irton.
Moor Yeatt 1675 (*PNCumb* 404).
♦ 'The gate onto the upland waste'; **moor, geat/gate**. The 1675 spelling points to 'gate' (from OE *geat*) rather than 'road' (ON *gata*).

MOOR HEAD NY4203 Windermere.
From 1859 (*PNWestm* I, 196).
♦ 'The top of the upland waste'; **moor, head**.

MOOR HOUSE SD2095 Ulpha.
From 1867 (OS).
♦ 'The house by the upland waste'; **moor, house**. The situation is similar to that of nearby NOOK; the fellside is presumably thought of as *moor*.

MOOR HOW SD3990 (hab.) Cartmel Fell.
Moor howe 1615/6 (Cartmel PR).
♦ This probably amounts to 'hill where the grazing is poor'; **moor, haugr/how(e)**. It is quite a rocky hillock.

MOORSIDE: HIGH M~ NY0701 (hab.) Gosforth, LOW M~ NY0701 (hab.).
Moorside 1608 (PR), 1867 (OS).
♦ 'The place beside the upland waste' (**moor, side**), later distinguished as **High** and **Low**.

MOOTA: M~ MOTEL NY1636 Blindcrake.
Mewthow, Mewtey (*Hill*) 1537, *Mewto* (*beacon*) 1576 (*PNCumb* 267), *Moothay* 1687–8 (Denton 143), 1777 (*PNCumb* 267).
♦ M~ is 'the assembly hill', from OE (*ge*)*mōt* 'meeting, assembly' and OE *hōh* 'hill-spur', possibly influenced by **how(e)** from ON **haugr** 'hill, mound'. OE (*ge*)*mōt* frequently compounds with 'hill' terms (Smith, *Elements*, OED: *mote-hill, moot-hill*), and Moota Hill (to which the early spellings refer) is mentioned as an assembly place in papers of Henry VIII (*PNCumb*). It was also known as a beacon site in the 17th cent. (Denton 1687–8, 58, Cowper 1897b, 140). The Motel is NE of the hill. *OED* quotes from the American *Hotel Monthly* in 1925 announcing the word *motel* for hotels designed for motorists.

MORTAR CRAG NY4017 Martindale.
From 1859 (*PNWestm* II, 219).
♦ If this is the word *mortar*, referring to a reddish clay (*EDD*) or to a mix of lime, sand and water, it derives from OFr *mortier*, appearing first in English in the late 13th cent.; see also **crag**.

MOSEDALE Examples at:
NY1418 Loweswater, M~ BECK NY1318. *Mosdarle(bek)* 1545 (CRO D/Law/1/239).
NY1710 Nether Wasdale, M~ BECK NY1809. *Mosedal(bek)* 1278, *Mosedale(bek)* 1338 (*PNCumb* 22).
NY2303 Ulpha. *Mosedale* c.1210 (*Furness Coucher* II, 565).
NY3532 (hab.) Mungrisdale. *Mosedale* (*in Allerdale*) 1285, *Mosdale* 1305, *Moss(e)dale* 1362 (*PNCumb* 304).
NY4909 Shap Rural, M~ BECK NY5010, LITTLE M~ BECK NY5008, M~ COTTAGE NY4909. *Mosdall* 1249, *Mosedale* 1612 (*PNWestm* II, 169); *Mosedale* (= M~ Beck); *Little Mosedale* (= Little M~ Beck); *Mosedale Cottage* all 1863 (OS).
MOSEDALE BECK NY3523 Threlkeld/Matterdale.
From 1867 (OS).
♦ 'The valley with a bog'; **mosi**, **dalr**, also **bekkr/beck** 'stream', **little**, **cottage**. All the Mosedales seem to share the same origin and meaning, and the name would have signalled a hollow among the fells that is bleak, damp and rushy, offering meagre grazing and supporting few or no farms. The Mungrisdale valley was formerly a glacial lake (Danby 1994, 29), but curiously this is the only example where the valley name is also that of a habitation; it was also an independent parish/township until absorbed into Mungrisdale in 1934. The M~ in Shap Rural is in fact the upper reach of SWINDALE, into which it descends over a great barrier of hanging valley. The two valley-names thus seem to signal the different opportunities for land-use: waste in the upper part and animal husbandry in the lower. (Pl. 3)

MOSERGH FARM/MOSER SD5299 Whitwell.
Moserga 1196 (*PNWestm* I, 149).
♦ M~ is 'the shieling by the bog', from ON **mosi** 'bog' and Gaelic-Norse **ærgi/erg** 'shieling, outlying farm'. The **farm** is situated just over 656ft/200m.

MOSS COTTAGE SD2885 Lowick.
Moss 1850 (OS).
♦ The place is among streams and springs; **moss**, **cottage**.

MOSS DYKE NY3631 (hab.) Mungrisdale.
Moss Dick in Murrer 1620, *the Mosse Dyke* 1623 (*PNCumb* 181).
♦ 'The embankment by the bog'; **moss**, **dyke**. The site is just above the boggy ground of Bowscale Moss; *dyke* probably referred to a boundary between the higher and lower ground.

MOSS ECCLES TARN SD3796 Claife.
Moss Eccles Moss 1851 (OS), *Moss Eccles Tarn (Fish Pond)* 1919 (OS).
♦ The **moss** or bog is evidently named from the Eccles family, who appear in Hawkshead PRs from 1756 to the early 19th cent.; the word order of M~ E~, with generic first, is striking. The **tarn** (from **tjǫrn**) is, as usual, a (relatively) high-level pool; in this case it has been created out of the moss (cf. THREE DUBS TARN).

MOSSER NY1124 Blindbothel, M~ FELL NY1223, MOSSERGATE FARM NY1124, MOSSER MAINS NY1125.
Moserg(e) 1203, 1220, *Mosergh(e)* 1279 to 1541 (*PNCumb* 422–3); *Mosser Fell* 1867

(OS); *Mosser Gate* 1857; *Mosser Mains* 1858 (Mosser PR).
♦ 'The shieling by the bog', from ON **mosi** 'peat bog' and Gaelic-Norse **ærgi/erg** 'shieling' (cf. MOSERGH); also **fell**, **gata/gate** 'road', **farm**, *mains* 'home farm, demesne farm' as in MAINS FARM. Mosser Fell is a stretch of high grazing land whose summit is FELLBARROW. M~ Mains is just N of Mosser on M~ Beck. Mossergate may contain the reflex of ON *gata* 'road', though that of OE **geat** 'gate' is also possible.

MOSS FORCE NY1917 Buttermere.
From 1867 (OS).
♦ 'The waterfall at Moss, the bog'; **moss**, **force**. This is a waterfall in Moss Beck, named either from the beck or from the boggy BUTTERMERE MOSS in which it rises.

MOSS RIGG WOOD NY3102 Coniston.
From 1850 (OS).
♦ 'The wood at Moss Rigg, the ridge by the bog'; **moss**, **rigg** (see **hryggr**), **wood**.

MOSS SIDE SD4490 (hab.) Crosthwaite.
Mossid(e) 1535 (in 1669 copy, *PNWestm* I, 85).
MOSS SIDE SD4895 (hab.) Strickland Ketel. From 1836 (*PNWestm* I, 155).
MOSS SIDE FARM SD2289 Broughton West.
Moss Side 1696 (Broughton-in-Furness PR).
♦ 'The place beside the bog'; **moss**, **side**, with **farm** added later in one instance. The farms each stand beside and slightly above a tract of wet ground. The Broughton example appears as a companion to MIRESIDE and the Crosthwaite one belongs in a cluster of names in *side* (see List of Common Elements).

MOSSY BECK NY4821 Askham.
From 1865 (*PNWestm* I, 11).
♦ 'The mossy stream.' The adj. *mossy* meaning 'moss-lined' is first recorded from the 16th cent. (*OED*). It only occurs here in this volume, and may be the only instance where **moss**, usually a bog in Lakeland p.ns, refers to plants of the Musci class. The **beck** flows to the S of TARN MOOR.

MOUNTAIN VIEW SD0999 (hab.) Irton.
From 1900 (OS).
♦ Both elements in this name are relatively recent additions to the onomastic vocabulary of Lakeland, only found in minor names and occurring only twice in the names in this volume: see CUMBRIAN MOUNTAINS and HALTCLIFF VIEW. The name is well-motivated, since MUNCASTER FELL and beyond it the Cumbrian Dome rise abruptly to the SE.

MOUNTJOY SD4693 (hab.) Underbarrow.
Mountiowe 1297, *Montjowe* 1383 (*PNWestm* I, 102).
♦ Apparently 'Hill of joy', the settlement being at the foot of a steep hill. This is a rarity in the Lake District, being linguistically of French origin (*mont* and *joie*). The early forms may show influence from OFr *jou(g)* 'summit' (*PNWestm* I, 102). Dickins' suggestion that the name is a humorous borrowing from a literary source — Laʒamon's *Brut* — is politely cited by Smith (*PNWestm* I, xv), but seems somewhat unnecessary in the case of a p.n. common in France. It is used there especially of hills surmounted by a cross (*PNWestm* I, 102). This M~ was part of the estate of the barony of Kendal, as documents of 1297 and 1409 show (*Kendale* I, 13, 37; I am indebted to Mary Atkin for drawing my attention to this).

MOUSTHWAITE: M~ COMB NY3427 Threlkeld/Mungrisdale.
Muscheweyt 1279 (p), *Mousthwait* 1332 (p) (*PNCumb* 252); *Mousethwaite Comb* 1867 (OS).
♦ M~ is 'mouse clearing' or 'Músi's clearing'. (a) If the original M~ was within the compact valley of M~ Comb, the 1st el. may be 'mouse' (ON *mús* and/or OE *mūs*) in the sense of 'tiny, mouse-sized' (rather than 'mouse-infested'). (b) An alternative possibility is an ON forename or nickname *Músi* (see Fellows-Jensen 1968, whose examples include a Yorks Mousethwaite; she notes the possibility that some names may contain the noun). The 2nd el. is clearly ON **þveit** 'clearing'. 'Comb' (**comb(e)¹**) well describes the situation: a short, enclosed, bowl-shaped side-valley.

MUNCASTER (parish): M~ CASTLE SD1096, M~ FELL SD1198 (732ft/223m.), M~ HEAD SD1498.
Mulcastre early 12th cent. (in 1412 copy) to 1509, *Molecastre* 1185 to 1278, *Mulcastr' et non* (and not) *Moncastel* 1292 (p), *Mulcastre vel* (or) *Moncastre* 1389 (*PNCumb* 423), *Mulcaster . . . vulgarly named Monkcastre* 1687–8 (Denton 78); *Moulcastre fell* 1578 (*PNCumb* 423), *Muncaster, Mulcaster Fell* 1802 (CN 1205, 1215); *Moncaster Head* 1774 (DHM).
♦ Probably 'the stronghold on the headland'. The 1st el. is uncertain, possibly (a) ON *múli* 'headland, promontory, jutting crag' (though according to *PNCumb* 423–4 this would be unparalleled in Cumberland hill-names); or conceivably (b) *Múli* as a nickname, though this is unlikely in conjunction with *cæster*. The 13th–14th-cent. documents allow a glimpse of the replacement of *-l-* by *-n-* (a fairly common phenomenon, especially under influence of French speakers), resisted then accepted by scribes. The 2nd el. is OE *cæster*, which was adopted from Lat *castra* 'camp', and frequently designates former Roman strongholds; here it may refer to the Roman fort of GLANNOVENTA, Ravenglass (Millward & Robinson 1970, 138). The **Castle** is 19th-cent., incorporating remains of a 13th-cent. fortified tower (Pevsner 1967, 165; LDHER 3981). The **fell** is the long 'island' of high ground which occupies much of the parish, and M~ **Head** is at the E. extremity of the parish.

MUNGRISDALE/MUNGRISEDALE NY3630 (hab. & parish), M~ COMMON NY3129.
G[r]isdale 1254 (*DEPN*), *Grisedal(e)* 1285 (p) to 1487, *Grysdale* 1292, *Mounge Grieesdell* 1600 (*PNCumb* 226), (*Chappell Grisedale and*) *Monk Grisdale* 1687–8 (Denton 304), *Grisedale or Mungrisedale* 1777 (*PNCumb* 226); *Mungrisdale Common* 1867 (OS).
♦ Gris(e)dale is 'the valley where young pigs graze', cf. GRISEDALE. Clarke reported finding references in old manuscripts to 'porklings' running wild on the hill-slopes above M~ (1789, vi). The 1st syllable was added later. (a) It may refer to St Mungo, the early British missionary also known as Kentigern, and dedicatee of M~ church, though the age of the dedication is uncertain. (b) Less likely is a connection with a local p.n. *Mungrane*, recorded from 1589 but of uncertain location and meaning (*PNCumb* 226). (c) Denton took the prefix as *monk*, explaining that there had been a pre-reformation chantry on Grisedale How (1687–8, 304). M~ **Common** is a stretch of high, rough grazing some way from M~ the settlement, but within the parish.

MURRAH NY3731 (hab.) Mungrisdale.
Morwra 1292, *(le) Murwra* 1323, *Murrah nr Greystock* 1811 (*PNCumb* 181).
♦ 'The nook by the upland waste'; **mór/mōr, vrá** 'corner, nook'.

MURT/MORT NY1304 (hab.) Nether Wasdale.
Moorte, (highe) morte 1578 (*PNCumb* 442), *Mourt* 1729, *Murt* 1749 (PR).
♦ The el. *mort* is unusual and problematic. The English *mort* and ON *murta* refer, among other things, to small fish (see *EDD*: *mort*, sb. 5, though *mort* sb. 1 'an abundance' is also attested in Cumberland), and dial. *murt*, sb. 1 and the ON nickname *murtr, murti* are used to refer to small persons and things. Smith, *Elements* suggests a connection with dial. *murt* 'short, stumpy person' and MHG *murz* 'a stump'. The common factor seems to be smallness, and since the name refers to a farm beside a low hillock by the R. IRT, one might suggest the sense 'pimple'.

MURTHWAITE NY5100 (hab.) Longsleddale.
Morethwayt 1395 (p) (*PNWestm* I, 161).
♦ 'The clearing by the waste'; **mór,** þ**veit**. The waste in this case is close to the R. SPRINT, rather than the high ground often implied by ON *mór*.

MURTON: M~ FELL NY0918 Lamplugh, KNOCK M~ NY0919 (1462ft/446m.).
Morton' 1203 to 1585, *Moreton(e)* 1294 to 1338 (*PNCumb* 406); *Murton Fell* 1867 (OS); *Knock morton* 1774 (DHM), *Knock Murton* 1867 (OS).
♦ M~ is probably 'the settlement by the upland waste', from OE **mōr** and OE **tūn**, as in other English Mor(e)tons. M~ is referred to as *villa* in 14th cent. and may date back to Anglian times. M~ **Fell** is a tract of high grazing, and Knock M~ is the distinct peak which crowns it. 'Knock' is *cnocc* 'hillock', which may have both Gaelic and OE origins, but is probably Gaelic here since Knock Murton has the word order of an inversion compound (Coates 2000, 349, *VEPN: cnocc*[1], *cnocc*[2]).

MYREGROUND/MIREGROUND SD1191 (hab.) Waberthwaite.
Mireground 1702 (*PNCumb* 365).
♦ 'The farm by the marsh'; **mýrr/ mire; ground**.

N

NAB: NAB COTTAGE NY3506 Rydal, NAB SCAR NY3506.
le/la Nab 13th cent., *the Nab(b)* 1655; *Nab Scar* 1801 (*PNWestm* I, 210); *Nab Cottage* 1920 (OS).
♦ 'The cottage and scarp at The Nab or hill-spur'; **nab, cottage, scar**. N~ S~ is the blunt, stony precipice N of RYDAL WATER. The cottage, built in 1702 (Danby 1994, 157), nestles beneath it.

NAB, THE NY4315 (1887ft/576m.) Martindale.
From 1865 (*PNWestm* II, 219).
♦ 'The hill-spur'; **nab**. A steep peak with elongated oval contours, this is one of the more substantial hills to be named Nab.

NAB CRAGS NY3112 St John's.
From 1867 (OS).
♦ These are rock outcrops clothing a long, steep-sided ridge, which differs from most other *nabs* in that it does not come to such a distinct point; **nab, crag**.

NABEND NY4125 (hab.) Matterdale.
From 1598 (*PNCumb* 257).
♦ 'The end of the hill-spur'; **nab, end**. The place occupies a shelf below the NW spur of LITTLE MELL FELL.

NABS MOOR NY5010 (1613ft/492m.) Shap Rural.
From 1860 (OSNB).
♦ This is the high ground above the rocky projection of Nabs Crag; **nab** 'hill-spur', **moor**.

NADDLE: N~ BECK NY2922 St John's.
Naddal(e) 1292 (p) (*PNCumb* 313), *Naddall* 1563/4, *Naddell* 1567 (Crosthwaite PR), *Nathdale* (*Fell*), c.1805 (publ. 1819, Wordsworth, *Poems*143); *Naddell becke* (*foote*) 1629 (*PNCumb* 22), *Naddle Beck* 1867 (OS).
NADDLE: N~ FARM NY5015 Shap Rural, N~ FOREST NY4914.
Naddale 1309 (p) to 1777 (*PNWestm* II, 169); *Nuddale Forest* 1823, *Naddle Forest* 1865 (*PNWestm* II, 176).
NADDLES BECK NY3829 Mungrisdale, N~ CRAGS NY3829.
Naddles Beck 1867 (OS), previously *riuulus de Bergher* 1292 (stream of Berrier), *Berryerbeke* 1589 (*PNCumb* 22); *Naddles Crags* 1867 (OS).
♦ Naddle is possibly 'the wedge-shaped valley'. Its 2nd el. is clearly ON **dalr** 'valley', while there are multiple possibilities for the 1st, though (a) is in my view the most convincing. (a) It may be from ON *naddr* 'point, wedge' (so Ekwall, *ERN* 112). This seems to occur, though not frequently, in Norw p.ns (*NSL*: Nadderud, cf. Naddvik). At Naddle, St John's (shown at NY2921 on larger scale maps), the valley itself or the higher ground flanking it, especially to the W, could be described as pointed, while N~ Forest occupies an elliptical rise between HAWESWATER and another Naddle Beck. Naddles Crags are also elliptical. (b) Smith suggested as an alternative that *naddr* might correspond to a fish name recorded in Norw dial. (*PNWestm* II, 176). However, that three valleys should include a rare fish name is highly unlikely. (c) ON *naðr* 'snake', cognate with English 'adder', is possible, but again not a common place-name element. (d) *PNCumb* 313 cites a remark from Professor Kenneth Jackson that 'it is possible that *n-* is the

Cumbric definite article...', but there is no suggestion as to what the remainder of the name would be in this case. (e) The suggestion of a Brit. root *nass(a) or ness(a) referring to 'wetness' or 'flowing water' (Roberts 1999, 86) for the St. John's example is not supported by the spellings. The pl. -s in the Mungrisdale names is puzzling. See also **beck** 'stream', **farm**, **forest**, **crag** 'rocky height'.

NAN BIELD PASS NY4509 Shap Rural/ Kentmere.
Pass of Nan Bield 1839 (Ford 129), *Nan Bield Pass* 1865 (Prior 23), *Nan Bield* 1770 (JefferysM, with what looks like a symbol for a boundary marker).
♦ 'The mountain pass by Nan's shelter', *Nan* being a pet form of the name *Anne*, plus **bield**, **pass**. The necessity for shelters on this important routeway is physically borne out by three substantial stone shelters which stand nearby (illustrated in Wainwright 1957, Mardale Ill Bell 7).

NAPES NEEDLE NY2009 Nether Wasdale.
Napes Needle 1894 (Smith 94), *The Gable Needle (or Napes Needle)* 1900 (Jones 146–7); cf. *Great Napes, White Napes* 1867 (OS).
♦ This is a rock spike on Great Gable, rising vertically from the scored wall of crags known as The Great Napes. The identity of the word 'Nape' is uncertain, but perhaps *naip*, for 'the highest part or ridge of a roof' (recorded from Scotland in *EDD*) is involved here. *OED*: *nape* n. 4 gives a single citation, from 1837, of *nape* in the sense of 'an ascending ridge of ground', but it is from a Devonshire dial. work. The first ascent of N~ N~ by W. P. Haskett-Smith in 1886 was among the critical moments in the early history of British rock-climbing,

and *needle* (from OE *nǣdl*) surely mimics Fr. *aiguille* 'needle', used of pinnacles in the Alps.

NARROW MOOR NY2317 Above Derwent.
From 1867 (OS).
♦ *Narrow* is from OE adj. *nearu*, *nearw-*, the only other occurrence in this volume being in LITTLE NARROWCOVE. The **moor** 'upland waste' lies S of MAIDEN MOOR, at a point where the ridge-top narrows between steep gradients on both sides.

NEAR as affix: see main name, e.g. for NEAR SAWREY see SAWREY

NEST NY2922 St John's.
From 1792 (*PNCumb* 317).
♦ 'High' and 'Low Nest' on 1:25,000 map. This is presumably to be taken at face value, as *nest* (ultimately from OE *nest*) without the qualifying bird-name seen in DOVE NEST, nearby Pietnest and numerous examples outside Cumbria (Smith, *Elements*). There is another Cumbrian place called simply Nest (*PNCumb* 178), and three named The Nest in *OSGazetteer*.

NETHER BECK NY1508 Nether Wasdale, OVER BECK NY1608.
Nether Beck 1839 (Ford 62), 1867 (OS), previously *le Blakecombek* (*PNCumb* 22); (*ad pedem de* [to the foot of]) *Overboutherdalebek* 1338, *Ouerbek* 1540 (*PNCumb* 23).
♦ 'The lower and upper streams'; **nether**, *over* probably from OE *uferra* 'upper, higher', **bekkr/beck**. The present name O~ B~ is a curtailed version of the one recorded in 1338 and meaning 'the upper BOWDERDALE stream'. The two becks flow into WAST WATER on opposite sides of BOWDERDALE, and with Nether B~ half a mile down-lake from Over B~.

NETHERCLOSE NY1421 Loweswater.
Nether Close 1774 (DHM).
♦ 'The lower farm or enclosure'; **nether, close**. It is in a low-lying situation by the R. COCKER.

NETHER HALL SD4384 Witherslack. From 1713 (PR).
♦ 'The lower hall'; **nether, hall**. This was the manor farm, built in the 16th cent., of WITHERSLACK HALL (Palmer 1944, 90), and is presumably named in relation to it.

NETHER HOUSE FARM NY5100 Longsleddale.
Netherhouse c.1692 (Machell II, 104, ed. Ewbank, 95), *Nether House* 1759 (PR).
♦ 'The lower house'; **nether, house**, with **farm** added fairly recently. This is somewhat farther down the valley than LOW HOUSE NY5100.

NETHERMOST PIKE NY3414 (2920ft/891m.) St John's/Patterdale, N~ COVE NY3414 Patterdale.
Nethermost Pike 1860 (OSNB); *Nethermost Cove* 1860 (*PNWestm* II, 226).
♦ Among the peaks of HELVELLYN, N~ Pike is lower than H~ itself and its LOWER MAN, but higher than HIGH CRAG and DOLLYWAGGON PIKE, hence not strictly the 'nethermost' or lowest in the range (see **nether**). Perhaps the comment of the Ordnance Surveyors, 'it is the nearest part of Helvellyn to Patterdale' (OSNB, Patterdale), hints at the explanation that N~ P~ was perceived as being nearest to civilisation. **Pike** 'peak' well describes this impressive mountain, though not its flat summit. N~ **Cove** is a large corrie NE of the Pike and is probably named from it since it is not the lowest of the Helvellyn coves.

NETHER ROW NY3237 Caldbeck.
Netherraw 1658 (*PNCumb* 279).
♦ 'The lower hamlet'; **nether, row**.

NETHERSCALE NY1730 Embleton.
Neitherscale 1692 (PR), *Nether Scales* 1774 (DHM).
♦ 'The lower shieling'; **nether, scale**.

NETTLESLACK NY4218 (hab.) Martindale.
Nettle Slack 1702 (PR).
♦ 'The hollow where nettle grows', with reflexes of OE *netele* and ON **slakki**.

NEWBIGGIN SD0994 Waberthwaite.
Newbigginge 1678 (*PNCumb* 440).
♦ 'The new building', with **new**, plus reflex of ON *bygging* 'building, farmstead'. This is a common post-Conquest name in the N. counties. The settlement may have been facilitated by drainage of the marshes around the ESK estuary.

NEW BUILDINGS SD1183 Whicham. From 1867 (OS).
♦ Self-explanatory; **new** plus *building*, unique in the names in this volume. The noun emerged in ME, formed from part of the OE verb *byldan* 'to encourage'. The name 'is given to a rather ruined and deserted farmstead' where traces of prehistoric artifacts and structures have been found (Cherry & Cherry 1987, 5–6 and 8).

NEWBY BRIDGE SD3786 (hab.) Staveley-in-Cartmel.
New bridge 1577, *Newbybridge* 1659 (*PNLancs* 217).
♦ 'The new bridge'; **new, bridge** from **brycg**. That 'Newby' rather than 'New' developed under influence of a family name (*PNLancs*) seems likely since p.ns in **-bý** are not characteristic

of this area. That the place is named from a bridge is fitting, since a substantial one is needed at this crossing over the broad LEVEN. The present bridge, five-arched and of slate, is 17th-cent. (Pevsner 1969, 182); the previous one was slightly upriver (as shown on 1:25,000 map).

NEWFIELD: N~ WOOD SD2295 Dunnerdale.
Newfield 1659 (Broughton-in-Furness PR), *New Field* 1774 (DHM); *Newfield Wood* 1850 (OS).
♦ Self-explanatory; **new, field, wood**. This is an outgrowth, to the SW, of SEATHWAITE hamlet.

NEW HALL SD4697 Nether Staveley.
From 1650 (*Kendale* I, 115).
♦ Self-explanatory; **new, hall**. The old hall was 'a capital messuage called Aswayt Hall, in Nether Staveley' in 1595 (or 'Astwhayt Haull', 1596, *Kendale* I, 334-5) belonging to Robert Bindlose, whose grandson Richard Braithwaite built a new hall a little to the N c.1640 (Scott 1995, 10); 'Richard Braithwaite's estate of Burneshead, Newhall &c.' is included in a list of offers made for 'estates of delinquents and Papists' in 1650 (*Kendale* I, 115).

NEWHOUSE FARM NY1523 Lorton.
New House 1774 (DHM).
♦ Simply **new, house**, and **farm** added later.

NEWLANDS NY2320 Above Derwent, N~ BECK NY2319, N~ HAUSE NY1917.
Neulandes 1318 (*PNCumb* 371); *Newlands Beck* 1867 (OS); *Newland Hose* 1783, *Newland hawse* 1784 (*PNCumb* 356).
♦ N~ is 'the land newly brought into cultivation'; **new, land**; also **beck** 'stream', **hause** 'neck of land'. The usable part of the valley was extended through drainage, probably in the 13th cent. (Winchester 1987, 40, who believes that *Rogersete* was the alternative medieval name for this tract; see also UZZICAR for *Usakredale* 1369). The name N~ Beck must replace an earlier one: see ORMATHWAITE. N~ Hause links BUTTERMERE with DERWENTWATER, via the N~ valley.

NEW MILL NY0504 Gosforth.
From 1867 (OS).
♦ The place is on a stream, where a corn mill and extensive mill races are indicated on the 1867 map; **new, mylen/mill**.

NEWSHAM NY3324 Threlkeld.
Noysom 1602, *Newsome* 1622/3 (Greystoke PR), *Newsham* 1774 (DHM).
♦ 'The new houses.' Although pre-modern spellings are lacking, this is probably of the same origin as other Newshams in N. England, which are from a phrase incorporating OE dat. pl. *nīw(um) hūsum* 'new houses' (see *DEPN*); **new, house**.

NEWTON: HIGH N~ SD4082 Upper Allithwaite, LOW N~ SD4082, N~ FELL SD4083.
Neutun 1086; *Over, Nether Newton* 1491 (*PNLancs* 199), *Higher Newton; Nether Newton; Newton Fell* all 1851 (OS).
♦ From OE **nīwe** 'new' and OE **tūn** 'farmstead, village', an appropriate name for a place situated in 'what is certainly the most infertile part of Cartmel' (Dickinson 1980, 74). Nevertheless, this 'new settlement' is, unusually for Lakeland, recorded as early as Domesday Book, 1086, as a manor of Earl Tostig at 1066. Ekwall believes (by implication) the reference there to be to **High** N~, the larger village (*PNLancs* 199). **Low** N~ was

still named Nether N~ on the One Inch map of 1966. The **Fell** is an extended tract of high ground.

NEWTOWN SD0995 Muncaster.
New(e)ton 1509–10 (*PNCumb* 424).
NEWTOWN/LOWTHER NEWTOWN NY5224 Lowther.
The New towne 1709, *Lowther New Town* 1741, *Lowther New Village* 1769, *Newtown* 1823 (*PNWestm* II, 185).
♦ Self-explanatory; **new**, **town**. The Muncaster hamlet is an outlier of RAVENGLASS, and the 16th-cent. *-ton* spellings suggest that a weakened pronunciation was current, and perhaps that the name, and hence the settlement, was already well established by then. The Lowther N~ is the earlier of two new villages established after the demolition of the previous L~ village by Sir John Lowther at the end of the 17th cent., the other being known as LOWTHER or L~ Village.

NIBTHWAITE: HIGH N~ SD2989 (hab.) Colton, N~ GRANGE SD2988.
Neubethayt 1246, *Neburthwait* 1336, probably = *Thornebuthwait* 1202, *Furnebuthetwayt* 1522 (*PNLancs* 217–8); *High Nibthwaite* 1851 (OS); *Nybthwaytgrange* 1539 (*PNLancs* 217).
♦ 'Thwaite' is from ON **þveit** 'clearing', and the specific may be a compound of ON *nýr* 'new' (see **nīwe**) and ON **búð** 'hut' or 'shieling', which may have formed a pre-existing p.n. The *Thor-* prefix in the 13th-cent. spellings could result from one of the many ON pers.ns beginning in *Þór-*. (See Lindkvist 1912, 125, *PNLancs* 217–8 and *SSNNW* 150 and 351 for discuss-ions of N~.) **High** N~ is distinguished from Low N~ (*Low Nibthwte* 1739, PR). N~ **Grange**, together with Colton and Bouth Granges, belonged to FURNESS Abbey (Baines 1870, II, 667; I am grateful to Mary Atkin for this reference).

NICHOL/NICKOL END NY2522 (hab.) Above Derwent.
Nickol End 1867 (OS).
♦ **End** is clear but the 1st el. uncertain. (a) The most obvious explanation is that it is the abbreviation of *Nicholas*, widespread in the Middle Ages, and popular locally, as suggested by the frequency of the surname *Nicholson*, e.g. in the Crosthwaite PRs. There would be a parallel in ROWLING END, two miles away. (b) However, places named **end** are normally named from a topographical feature, as is nearby HAWSE END, and the 1st el. in this case could be the word meaning 'knobbly, rounded hill' recorded since the 13th cent. in NICHOLS WOOD. (c) Parker took this name as meaning 'St. Nicholas' Landing', the departure point for medieval pilgrims going to ST HERBERT'S ISLAND (1903, 224).

NICHOLS: N~ WOOD SD4282 Witherslack.
le Knykeles 1280, 1354, *Knickles* 1631; *Nichols Wood* 1857 (*PNWestm* I, 78).
♦ 'The **wood** by the knobbly hills', referring to two neat hillocks beside the WINSTER. Scand comparative evidence suggests an ON **knykill* meaning 'hill-top', or the 1st el. may be OE *cnucel* 'knuckle', applied topographically (*PNWestm* I, 78, cf. Nichols at I, 203). The spelling *Nichols* seems to be influenced by the common pers.n. (see NICHOL END).

NITTING HAWS/HOWS NY2416 Above Derwent.
Nitting Haws 1862 (OSNB).
♦ A puzzling name. The 1st el. is

unidentified and even the 2nd, generic, one is unstraightforward, since neither **hause** 'col or ridge' (from **hals**) or **how(e)s** 'hills' (from **haugr**) is an obvious description of this craggy piece of fellside, though the latter is more appropriate.

NOOK SD2195 (hab.) Ulpha.
Nook in Ulpha 1707 (PR).
NOOK, THE SD0990 (hab.) Bootle.
the Nooke 1646 (*PNCumb* 348).
NOOK, THE NY4504 (hab.) Kentmere.
Nook 1836 (*PNWestm* I, 167).
♦ 'The secluded place or corner of land'; **nook**. The Ulpha one, for instance, lies between a slight bend in the R. DUDDON and the curved slopes of the fell above.

NOOK END FARM NY3705 Ambleside.
Nooke end 1668 (*PNWestm* I, 184).
♦ 'The end of the secluded place'; **nook**, **end**, with **farm**, as so often, added relatively recently.

NORAN/NORMAN BANK FARM NY3915 Patterdale.
Norhambank 1713, *Noranbank* 1800 (PR), *Norran Banks* 1828 (HodgsonM), *Noranbank* 1839, *Norenbank* 1860 (*PNWestm* II, 226). Seemingly = *Near Bank* 1770 (JefferysM).
♦ It is clear that *Noran* rather than *Norman* is the more authentic form, but its meaning remains obscure. The **bank** is the steep W side of PATTERDALE; **farm** has been added to the original name.

NORFOLK ISLAND NY3918 Patterdale.
Previously *Househoilme, Houson, House howme* c.1692 (Machell I, 728, II, 438), *House Holm* 1783 (CrosthwaiteM), 1920 (OS).
♦ The **island** is in ULLSWATER, opposite the manor of GLENCOYNE, which was in the possession of the Dukes of Norfolk (Hutchinson 1794, I, 428, n., Ford 1839, 136, and see LYULPH'S TOWER).

NORMAN NY2736 (hab.) Caldbeck.
Tenem[en]t called Norman 1591; cf. *Norman Close* 1560 (*PNCumb* 279).
♦ The place evidently derives from the surname *Norman*, well known in Cumbria (see next entry) though I have not found it in the Caldbeck PRs. It derives ultimately from *norðmaðr* 'North-man, Norwegian or Scandinavian', but was also applied to the Normans and used as a pers.n.

NORMAN CRAG NY3634 (hab.) Mungrisdale.
Normand Cragge 1568; *Ormant Cragg* 1581, *Normond Cragge* 1595, *Norman Cragg* 1602, *Ormond Cragg* 1617 (*PNCumb* 212), *Ormond Cragge* 1765, *Norman Crag* 1802 (Greystoke PR).
♦ The variable spellings of the 1st el. are too late for certainty, but are comparable to those for ORMATHWAITE, and could point to an ON pers.n. such as *Norðmaðr*, the equivalent appellative meaning 'Norwegian', or a later surname (cf. NORMAN), which occurs in the PRs for Greystoke though not, to judge from a rapid scan, for Mungrisdale. The 2nd el. is **crag(ge)** 'rocky height', though this is not a spectacular example. There is another Norman Crag at NY3922 in Matterdale.

NORMOSS SD1092 (hab.) Waberthwaite.
Northmosse 1190–1200 (*PNCumb* 365).
♦ 'The northern marsh', from ON *norðr*/OE *norð* and ON **mosi**/OE **mos**. It is perhaps so named since N of CORNEY, which was a separate parish before merger with Waberthwaite in 1934.

NORTH LODGE SD4287 Witherslack. From 1899 (OS).
♦ Self-explanatory; *north* from OE *norð*, plus **lodge**. This is situated at the N entrance to the WITHERSLACK HALL estate. Like LAWNS HOUSE, it does not appear on the 1862 OS map.

NORTH ROW NY2232 (hab.) Bassenthwaite.
Norrow 1663 (PR), *North Row* 1672 (PR), 1794 (Hutchinson II, map facing 153).
♦ *North* as in NORTH LODGE, plus **row**. The small *row* or hamlet is a NW outlier of BASSENTHWAITE village.

O

OAKBANK NY1422 (hab.) Loweswater. From 1867 (OS).
OAKBANK SD4795 (hab.) Crook.
the Oake banke 1535 (*PNWestm* I, 180).
♦ Simply 'the hill-slope where oak grows'; **oak**, **bank(e)** and compare AIKBANK.

OAK HEAD SD3883 (hab.) Staveley-in-Cartmel.
Ackehead 1603 (*PNLancs* 199), *Oak head, Staveley* 1779 (Finsthwaite PR), possibly = *Aykesheued* 1279 (p).
♦ 'The high place where oak grows'; **oak** plus **head**, here referring to a slight elevation. The early spellings have alternatively, though less credibly, been identified with nearby AYSIDE.

OAK HILL NY1525 (hab.) Lorton. From 1867 (OS).
♦ Presumably self-explanatory; **oak**, **hill**. This is on the outermost, SW, skirts of KIRK FELL.

OAK HOWE NY3005 (hab.) Langdales, O~ H~ CRAG NY3005.
Aikehowe 1564 (*PNWestm* I, 207), *oakhow(e)* 1571 (*CW* 1908, 200), *aykhowe* 1577 (*CW* 1908, 202); *Oakhowe Crag* 1862 (OS).
♦ 'The hill where oaks grow' (**oak** and **how(e)** from **haugr**), which has given its name to the farm below, and to the **Crag**, a scree and scarp above. The intermittent *aik-, ayk-* spellings reflect ON *eik(i)*. The 1571 document partly concerns rights to the use of oak woods 'within Baysbrown, Elterwater and Oakhow' for both timber and firewood.

OAKLAND SD4099 (hab.) Windermere. From 1865 (*PNWestm* I, 196).
♦ The name is probably to be taken at face value (**oak**, **land**), though confusion with the reflex of ON *lundr* 'grove' is a possibility (cf. Plumbland, *Plum(be)lund* 12th cent. (*PNCumb* 309–10), or Morland, *Morlund(ia)* 1133–47 (*PNWestm* II, 143)).

OAKS NY3404 (hab.) Rydal.
Oakes in Loughrigg 1689, *the Oakes* 1691 (Grasmere PR).
♦ Simply the pl. of **oak**. The name matches that of Ellers on the other (SE) side of LOUGHRIGG TARN, recorded from much the same time — 1717 (*PNWestm* I, 210); and see ASHES for other pl. tree names.

OLD BACKBARROW: see BACKBARROW

OLDCHURCH NY4421 (hab.) Matterdale.
a place called the Old Church 1606; cf. *New Church* 1674 (*PNCumb* 257–8).
♦ This is the lakeside site of the predecessor to WATERMILLOCK Church, the 'New Kirk' consecrated in 1558 (*PNCumb* 258); **old**, and *church* from OE *cirice*. James Clarke wrote that he owned a document mentioning a tenement at Old Church, which 'farther sets forth, that the parochial chapel and the burying-ground was then at Gowbarrow-Hall; and as it is dated in 1474, we must naturally conclude that the destruction of the church at this hamlet must have long preceded the reign of Edward the III' (1789, 26).

OLD COACH ROAD NY3423 Threlkeld/St John's.

Marked 'C.R.' but not named 1867, 1900 (OS).
♦ This is an ancient track across THRELKELD COMMON, though, as Hindle points out, 'its most likely origins are as a peat-road and for cart traffic' rather than horse-drawn coaches, which would have taken a lower, more northerly route (1998, 134–5); **old**, *coach*, an adoption from Fr first appearing in the 16th cent. (*OED*), plus *road*, from OE *rād* but only applied to routeways in modern times. The collocation *coach road* is recorded from 1712 (*OED*: *gravelled*) and 1715 (*OED*: *coach*).

OLD CORPSE ROAD NY4912 Shap Rural.
♦ The track leads W across the fell between MARDALE, now submerged under HAWESWATER, and SWINDALE HEAD, from whence the dead were taken via TAILBERT for burial at the parish church in SHAP. After 1731, burials were allowed at Mardale Chapel (*CW* 1902, 143). This is hence a reminder of the long and arduous journeys that the folk of central Lakeland often had to make to bury their dead at the larger centres of population in the lowland fringe (see Hindle 1984, 56–9 on corpse roads). Elements are **old**, *corpse*, a ME adoption from OFr *cors* in a spelling influenced by Lat *corpus*, plus *road* as in OLD COACH ROAD.

OLD HALL FARM SD3285 Colton.
Old Hall 1736/7 (PR).
♦ 'Anciently this was Colton Hall' (Cowper 1899, 58); **old, hall**, with **farm** added recently to the name as recorded on OS maps.

OLD MAN OF CONISTON, C~ OLD MAN SD2797 (2633ft/801m.) Coniston.
Old Man 1800 (*PNLancs* 193–4n.), *Coniston Fells & Old Man* 1800 (*CN* 801), *The Old Man of Coniston* 1851 (OS); cf. *Old Man Quarry* 1786 (YatesM; Ekwall suggests that *Old Man* is may also 'do service as the name of the hill-top', *PNLancs* 194n.).
♦ **Old**, though it cannot be certainly explained, suits this venerable peak, the highest in S. Lake District. **Man** (a) probably has the sense 'cairn, summit'. The mountain is a medieval beacon site (Cowper 1897b, 143). Ford reported, 'On the summit are three beacons of stone, the old man, his wife, and son' (1839, 9, revising this to 'the Old Man, his Wife and Son' in his 1843 edition, 11). Ruskin wrote to Ellwood, 'I have more correspondence upon my table than the bulk of the Old Man. I mean the cairn on the top, not the mountain' (Ellwood 1895, 65). (b) As exemplified here, whatever the original force of **man**, this fell probably competes with SKIDDAW as the most often personified in literary works. Hence there is a possibility that 'Man' is nothing but the word for a male person (OE *mann, monn*), which here could refer to the peak in general, or specifically to a cairn. Compare the use of **lad** and of ON *karl, kerling* '(old) man, old woman' in Icel and Norw mountain-names (cf. CARL SIDE, CARLING KNOTT). (c) Although 'Old man' occurs as a mining term applied to old workings there are none near the summit of this fell (so *PNLancs* 194n., citing Collingwood). (d) Derivation from Brit. *allt-maen* ('high rock or cairn') was suggested by Ferguson, quoting Dr Whittaker (1873, 201), but although REDMAIN supplies evidence of *-man* spellings for a presumed *maen*, there is no real case for this. See also CONISTON.

OLD PARK NY4123 (hab.) Matterdale.
Oldpark 1867 (OS).
♦ 'Old Park possibly started as Park House, the northern boundary of the Howard estate of GOWBARROW PARK' (Taylor 1995, 15); **old, park**.

OLD SCALES NY1928 Wythop.
Aldskell 1597, *Oldskell* 1619 (Lorton PR), *Oldscale* 1774 (DHM).
♦ 'The old shieling(s)' (**old, skáli/ scale**), later developed into a permanent farm. Winchester sees this as part of a transformation of the secluded valley of WYTHOP from outlying pasture to permanent settlement once it had been granted to John de Lucy c.1260 (1987, 39–40).

ORE GAP NY2307 Eskdale/Borrowdale.
Orscarth 1242 (*Furness Coucher* II, 564, Collingwood 1918, 96), *Ure Gate* 1805 (*CW* 1981, 67), *Ore Gap* 1867 (OS).
♦ *Ore* is from OE ōra 'ore', while original ON **skarð** 'gap, pass' has been replaced by **gap**, which occurs in two names in this corpus recorded from the 16th cent. (CLAY GAP and THE GAP). The most literally down-to-earth explanation, encouraged by patches of red ground showing haematite in the area, is that the name refers to the presence of iron ore. There has been some doubt as to whether this was a routeway for miners (and if so from where to where) or a mining site as such, but certainly it was the latter from c.1700 onwards (Johnson 1988, 247), with smelting in LANGSTRATH (Collingwood, citing Postlethwaite). On the other hand, the opinion of local historian C. M. Fair was that this was the gap through which iron ore was transported from Eskdale into Langdale (reported *PNCumb* 344).

ORMATHWAITE NY2625 (hab.) Underskiddaw.
Nordmanthait c.1160 (*Fountains Chartulary* I, 49), *Ormatwhat* 1564 (*PNCumb* 322).
♦ 'The clearing (**þveit**) of the Norwegians/Norwegian, or of a man named *Norðmaðr*.' It is a pity that the possibility of an ON pers.n. *Norðmaðr* or *Norðmann(r)* makes for uncertainty about the specific, since a direct reference to Norwegians (ON *norðmenn*) would be unique in the major names of NW England (*SSNNW* 308), and rather intriguing in an area where Norw origins would not be distinctive, unless perhaps the cultural identity lingered longer than usual. The traditional identification of the medieval spellings with this place is also somewhat problematic, since the charter bounds in the *Fountains Chartulary* descend from the *water of Huseker*, presumably NEWLANDS BECK, to *Nordmanthait*, which Collingwood reasonably concluded must be Bog, now BOG HOUSE (1921, 168; he dates the document 1210–16).

ORREST: FAR O~ NY4100 (hab.) Windermere, NEAR O~ NY4100 (hab.), O~ HEAD SD4199.
Orrest 1626 (*PNWestm* I, 196); (Far O~:) *Orrest* 1862 (OS); *Near Orrest* 1859 (OSNB); *Oresthede* 1530 (*PNWestm* I, 196), cf. *the nether Orrest* (Machell c.1692, II, 319).
♦ '(The site of a/the) battle', from ON *orrusta* 'battle', adopted into ME as *orrest*. This seems to be a rare case of a p.n. commemorating a particular event. Collingwood points out that proximity to a former Roman road makes this a plausible place for the meeting of armies (1933, 22). **Far** and

Near may indicate relative distance from the centre of Windermere parish. O~ **Head** is a famous viewpoint.

ORTHWAITE NY2534 (hab.) Ireby and Uldale.
Ouerthwait 1305 (*PNCumb* 328), *Or-thwaite* or *Ormesthwaite* 1687–8 (Denton 172), *Orthwait* 1703 (*PNCumb* 328).
♦ The 2nd el. is clearly ON **þveit** 'clearing'. The 1st is less certain, and *PNCumb* gives 'over **þveit**', adding 'Why "over" is not obvious' and noting the proximity to OVER WATER. The el. could be from (a) OE *ofer*, which could be a noun 'hill-slope'; (b) the OE adverb *ofer* 'above'; or possibly (c) ON *ofar*, the comparative adverb 'higher up'(cf. Overby, *SSNNW* 37), or OE *uferra* with the same meaning. If 'above' another point, that point could be the moated site which is of uncertain date but probably medieval (LDHER 885, SAM No. 23796).

OUSE BRIDGE NY1932 Setmurthy.
Hewsbridge 1560, *Oosbridge* 1578, *Usebrige* 1579, *Eusebridge* 1587, *Ouse Bridge* 1671 (*PNCumb* 434–5), *Euse* 1687–8 (Denton 56, in list of bridges), *Ewes-bridge* 1687–8 (Denton 138–9), *Ouse-Bridge (pronounced Ews-bridge)* 1770 (Gray 1095); cf. *a poole, or lough, cawlled Use* c.1539–45 (Leland V, 52, but reference is unclear), cf. *Usegarth* 1603–25 (a weir and pool in N of Bassenthwaite Lake, CRO D/Law/1/238).
♦ The bridge (see **brycg**) spans the DERWENT as it flows out of the N end of BASSENTHWAITE LAKE, and since an O~ B~ is also recorded in COCKERMOUTH (Hutchinson II, 118), O~ could have referred to this stretch of river and/or to a pool in B~ (see above; the Leland context is confusing). Various explanations of 'Ouse' are possible. (a) If this example is comparable with the Great Ouse or the Yorkshire Ouse, it could be a Brit. river-name based on an Indo-European root *udso meaning 'water' (*ERN* 317, *DEPN*). (b) If of the same origin as the Ouse Burn at Newcastle-upon-Tyne, it could be from a Germanic root meaning 'gushing' or a Brit. one 'boiling, seething' (*ERN* 318, *DEPN*). Breeze, developing the latter suggestion, favours a cognate of W *iesen* 'fair, sparkling' as the etymon for the Newcastle name (2000, 72–3). (c) A further possibility is that it could be an ON word meaning 'estuary, outflow' as in EUSEMERE; the two locations are very similar.

OUT DUBS TARN SD3694 Claife.
Out Dubs 1770 (JefferysM), *Out Dubs Tarn* 1851 (OS).
♦ A pool in CUNSEY BECK, which flows S out of ESTHWAITE WATER; *out*, ultimately from OE *ūt(e)*, **dub** 'pool (in stream)', **tjǫrn/tarn** '(mountain) pool'.

OUTERSIDE NY2121 (1864ft/568m.) Above Derwent.
From 1800 (*CN* 804).
♦ Doubtful origin and date, but if to be taken at its face value this is 'the hill-side which is further out' (from the NEWLANDS valley); *outer* from OE *ūtera* and variants, plus **side**.

OUTGATE SD3599 (hab.) Hawkshead.
Outyeate 1607/8, *ye Outyeate* 1645 (PR), *Out Yate* 1745 (West 1774, map).
♦ Seemingly 'the gate out', referring generally to the way out of HAWKSHEAD or more specifically to 'the point where commons began and enclosures ended' (Danby 1994, 203). The *yeate* spellings point to 'gate' from OE **geat** rather than 'road' from ON

gata. The word *out* derives ultimately from OE *ūt(e)*.

OUTLAW CRAG NY5112 Shap Rural.
From 1859 (*PNWestm* II, 176).
♦ The word *outlaw* derives from ON *útlági* (*PNWestm* II, 176), but the date of this **crag** name, and the reason for it, are impossible to determine.

OUTRUN NOOK SD4396 (hab.) Nether Staveley.
Thoue tarne 1170–84, *Tawterne* 1390–4 (p), *Tovetarne* 1558–69 (*PNWestm* I, 173), *Toutren* 1646 (Windermere PR), *Tootern Nook* 1828 (HodgsonM), *Towtron nook* 1836 (*PNWestm* I, 173).
♦ Originally this appears to have been 'Tofi's tarn', from *Tófa*, gen. sing. of the ON pers.n. *Tófi*, and ON **tjǫrn** 'tarn, small lake' (*PNWestm*), but both landscape and language have undergone transformation. The tarn has dwindled to a boggy area NW of O~ N~, while *tarn* has become unrecognisable through metathesis of *r*, **nook** has been added and the 19th-cent. *Towtron* has shed its initial consonant, presumably through misdivision of the phrase 'at Towtron' as 'at Owtron, Outrun'.

OUT SCAR: see SCAR

OVER BECK: see NETHER BECK

OVER CAVE/COVE NY4308 Kentmere.
Over Cove 1857 (*PNWestm* I, 167), *Over Cave* 1994 (One Inch map), *Over Cove* 1998 (1:25,000 map).
♦ This is a **cove** or stony recess high in the fells, with 'Over' originating in the OE adverb *ofer* 'above', reinforced in this locality by the ON comparative adverb *ofar* 'higher up', or possibly from OE *uferra* 'upper, higher'.

OVEREND NY4605 (hab.) Kentmere.
Dever End [sic] 1770 (JefferysM), *Overend* 1836 (*PNWestm* I, 167).
♦ Perhaps 'the upper end' or 'the end of the slope'. The situation, near the foot of a slope flanking the R. KENT, leaves the reference of **end** unclear, and 'Over' could have any of the origins suggested for ORTHWAITE.

OVER WATER NY2535 Ireby and Uldale, OVERWATER HALL NY2434.
Orre Water, Orr-water 1687–8 (Denton 170, 172), *Orr water* 1777 (*PNCumb* 35), *Over Water* 1774 (DHM), 1774 (WestM); (Overwater Hall:) previously *Whitefield House* 1867, 1900 (OS).
♦ Possibly 'the lake where blackcock or grouse are found' or 'Orri's lake' (both *PNCumb*); also **hall**. ON *orri* is a bird of the grouse family, but hence also a nickname and pers.n. If the bird is meant, it may be matched in nearby COCKUP. The modern form in 'Over' may have been influenced by the older form of nearby ORTHWAITE. This is one of the smallest lakes named **water**.

OWSEN FELL NY1020 Lamplugh.
From 1867 (OS).
♦ 'The hill where oxen graze.' Probably the equivalent of OXEN FELL, with a dial. form of *oxen* (see *EDD*: *ouse*, sb., where NCy *awsen, ousen* are among the spellings), and cf. AUSIN FELL and Ousen Stand (*PNWestm* II, xiv). The **fell** rises to a distinct, if modest, summit at 1342ft/409m.

OXENDALE NY2605 Langdales.
From 1862 (OS).
♦ 'The valley where oxen are reared', from the reflexes of OE *oxa* (pl. *oxan*, gen. pl. *ox(e)na*) and ON **dalr**. Despite the lack of early forms, this obvious explanation is credible: the valley floor

is suitable for pasture and, in the past, arable farming, presumably using an ox-drawn plough. Of the references to oxen in this volume (see names in AUSIN, OWSEN and OXEN), this is the only one from central lakeland; cf. Winchester's comment that 'Oxen were much more common in the lowlands than the uplands . . . reflecting the greater importance of arable farming' (1987, 61).

OXEN FELL NY3202 (hab.) Skelwith.
Oxenfell 1598/9 (Hawkshead PR), *Oxen-Fell(-Park)* 1694 (Grasmere PR), *Oxenfell* 1745 (West 1774, map), *(High) Oxen Fell*, *(Low) Oxen Fell* both 1850 (OS).
♦ 'The high grazing land for oxen', and hence the settlement(s) nearby, distinguished as High and Low O~ F~ on larger scale maps. *Oxen* as in OXENDALE, plus **fell**.

OXEN HOUSE: O~ H~ BAY SD2892 Blawith.
Oxenhowse 1633 (Torver PR), *Oxness* 1809 (CrosthwaiteM); *Oxen House Bay* 1851 (OS).
♦ The bay, in CONISTON WATER, is close to Oxen House, which used to be marked on One Inch maps. *Ox(en)* as in OXENDALE, **house, bay**.

OXEN PARK SD3187 (hab.) Colton.
Oxen park c.1535 (West 1774, 104), *Oxenpke* 1623 (PR).
♦ *Oxen* as in OXENDALE, plus **park**. This name is one of a cluster of 'Park' names referring to granges of the Abbot of FURNESS and listed at the Dissolution; it is now a small village.

P

PADDIGILL NY2940 (hab.) Caldbeck.
Paddok keld c.1516, *Padokkold* 1543, *Paddygill* 1656 (*PNCumb* 279 and n. 1).
♦ If the 16th-cent. spellings are trustworthy, this is 'the spring where toads are found', from dial. *paddock* 'toad' and the reflex of ON **kelda** 'spring', which has been re-formed as **gill** 'ravine with stream'.

PADDOCK WRAY NY1801 (hab.) Eskdale.
Paddockwraye 1570 (*PNCumb* 392).
♦ 'Nook where toads are seen' or 'nook with a small field'. *Paddock* in Early ModE means either (a) 'toad, frog', a meaning which survives in Cumbrian dial., or (b) 'a small field' (a modified form of *parroc* from OE *pearroc*). *PNCumb* assumes 'toad' here. Like other examples of **wray** from **vrá**, this is quite secluded, sheltered by curving contours to the N.

PALACE HOW NY1027 (hab.) Blindbothel.
Palacehow 1774 (DHM).
♦ 'Palace' is (to me) obscure, unless it was an ironic naming of rather humble buildings; the word *palace* is an adoption from OFr into ME. **How(e)** is from ON **haugr** 'hill, mound', though the place is on a very modest rise.

PARK, THE SD3193 Colton.
From 1851 (OS).
♦ This is the high ground E of the middle reach of CONISTON WATER. For possible senses of **park**, see List of Common Elements.

PARKAMOOR: LOW P~ SD3092 (hab.) Colton.
Parkamore c.1535 (*PNLancs* 218), *Pkamoore* 1651, *Park o'moor* 1801 (PR); (Low P~:) *Parkamoor* 1851 (OS).
♦ Ekwall explains this as 'apparently . . . "the enclosure on the moor"' (*PNLancs*); **park, moor, low**. This is a former estate of FURNESS Abbey. Low P~ lies close to High P~, just S of THE PARK.

PARK BECK NY4402 Hugill/Kentmere.
From 1857 (*PNWestm* I, 167).
♦ This is the stream bounding KENTMERE Park; **park, beck**.

PARK COPPICE SD2995 Coniston.
From 1851 (OS).
♦ 'The coppice wood attached to the park'; **park, coppice**. This appears to refer to a deer park; it is named Hoghouse Park on a mid-19th-cent. map (Parker 1929, 275).

PARKEND NY3038 (hab.) Caldbeck.
Parkeend 1665 (PR); possibly cf. *parcus de Caldebek*' 1272 (*PNCumb* 276).
♦ 'The edge of the park' (**park, end**), 'where, formerly, there was a park of red-deer' (Hutchinson 1794, II, 379).

PARKERGATE NY2230 Bassenthwaite.
From 1761/2 (PR), 1774 (DHM).
♦ The specific is doubtless the widespread surname *Parker*, originally referring to a gamekeeper. Persons of that name are recorded in Bassenthwaite PRs in 1669–71, in locations close to P~. The sense of **gate** — whether 'gate' (**geat**), 'road' or 'pasture' (**gata**) — is unclear here. The situation, on the intersection of a green lane and what is now the A591, would be compatible with any of these.

PARK FELL NY3302 Skelwith, P~ HOUSE NY3303, LOW P~ NY3303 (hab.).
Park Fell 1850 (OS); *Parkhouse Skelwith* 1789; *Lowpark Skelwith* 1791 (Hawkshead PR); cf. *Park Skelwith* 1802 (Hawkshead PR).
♦ All self-explanatory; **park, fell, house, low**. These names presumably relate to ELTERWATER Park, shown nearby on the 1850 map. Park Fell is a distinct summit on BLACK FELL.

PARK FELL HEAD NY4209 Troutbeck.
From 1859 (*PNWestm* I, 191).
♦ This is the top of the tract of high grazing land N of TROUTBECK PARK; **park, fell, head**.

PARK FOOT NY4623 (hab.) Barton.
Parkfoot 1685 (*PNWestm* II, 213).
♦ The place is at the lower end of BARTON PARK; **park, foot**.

PARKGATE NY1100 (hab.) Irton, P~ TARN NY1100.
Parkyeate 1659 (*PNCumb* 404); *Parkgate Tarn* 1900 (OS).
PARKGATE FARM NY3921 Matterdale.
Parke Yeate 1663 (PR), *Park-Gate* 1704 (*PNCumb* 223); cf. *the ?Withy Sike at Gowberay now parke yeate* c.1692 (Machell II, 155).
♦ P~ is simply 'the gate to the park'; **park, geat/gate**, with the *yeate* spellings standardised to *gate* in the modern forms. The P~ in Irton is one of a cluster of names generated by IRTON Park. P~ **Tarn** (see **tjǫrn**) is the result of the damming of P~ Moss. The Matterdale example appears as *Park gate* on Donald's map of 1774, at the entrance to GOWBARROW PARKS, but it is on the other (W.) side of AIRA BECK from Gowbarrow, and was the entry to GLENCOYNE PARK (Keith Clark, pers.comm.). **Farm** has been added later.

PARK GROUND SD2893 Torver.
Parke ground (quarter) 1671 (PR, accounts).
♦ 'The farm of the Park family'; surname plus **ground**. The surname *Park(e)* is very common in the PRs, and people of that name are associated with P~ G~ in 1702/3 and with P~ G~ Quarter in 1671. Among the local places with names containing *ground*, this is relatively low-lying.

PARK HOUSE NY4700 Over Staveley.
Parkhouse 1596 (*PNWestm* I, 176).
PARK HOUSE NY4727 Dacre.
From 1691 (PR); cf. *Parke* c.1545 (*PNCumb* 189).
♦ The Over Staveley name refers to STAVELEY PARK, while in Dacre the reference is presumably to the DALEMAIN estate, where East and West Park are shown on the 1998 1:25,000 map; **park, house**.

PARKNOOK NY0803 (hab.) Gosforth.
From 1575 (*PNCumb* 396).
PARK NOOK SD1093 (hab.) Waberthwaite.
From 1702 (*PNCumb* 365).
♦ 'The secluded place or corner of land by the park'; **park, nook**. The Gosforth P~ N~ is almost enclosed by the R. BLENG and Capple Beck, and the nearest park on current maps is Bolton Head Park, at NY0804. The Waberthwaite name seems to relate to LANGLEY PARK.

PARK PLANTATION SD3295 Satterthwaite.
From 1851 (OS).
♦ One of the constituent **plantation**s of GRIZEDALE FOREST, this lies E of THE PARK and (slightly farther away) LAWSON PARK.

PARKSPRING WOOD SD4892 Underbarrow.
From 1836; cf. *the Highe Parke, the Lowe Parke* 1589 (*PNWestm* I, 105).

♦ P~ is 'park copse or plantation' (**park, spring**), referring to the park which lay to the NW. The el. **wood** seems to be a later addition, virtually duplicating the sense of *spring*.

PARKS WOOD SD3789 Colton.
From 1851 (OS).
♦ This lies just N of HIGH and LOW STOTT PARK; **park, wood**.

PARK WOOD Examples at:
   NY1634 Blindcrake.
From 1867 (OS).
   NY2333 Bassenthwaite.
From 1867 (OS); cf. *the park of Bassenthwaite* 1777 (*PNCumb* 264).
   SD3087 Colton.
From 1851 (OS).
   SD4387 Witherslack.
High Park Wood; Low Park Wood both 1862 (OS); cf. *Witherslack Park* 1654 (*PNWestm* I, 79).
♦ Self-explanatory; **park, wood**. The Blindcrake wood is presumably named from *Isell Park*, just to the N on the 1867 OS map, the Bassenthwaite one from B~ Park, and the Colton one from HILL PARK. In Witherslack, permission to empark 600 acres was granted to John de Haveringham in 1341 (*PNWestm* I, 79). High and Low P~ W~ are distinguished there on the 1:25,000 map of 1998.

PASTURE BECK NY4112 Patterdale.
From 1860 (*PNWestm* II, 226).
♦ Simply 'the stream flowing through the grazing land'; **pasture, beck**.

PATTERDALE NY3915 (valley, hab. & parish), P~ COMMON NY3615.
Patrichesdale pre-1184, *Paterickdale* c.1189, *Patredale* 1363 (*PNWestm* II, 221).
♦ 'Patrick's valley'; pers.n. plus **dalr**.

The Gaelic pers.n. *Patraic* was adopted into ME as *Patric*, and is well documented in medieval Cumbria, though not in the immediate locality. Smyth states that *Patr(a)ic* was rare in Ireland, but known in the Hebrides, in the Middle Ages (1975, 82). The village and parish are named from the valley. The church (present building 1853) is dedicated to St Patrick, but since no ancient link with the saint is proven and the church was merely called *capella de Patrikedale* in 1348, the dedication may have been inspired by the valley name rather than the reverse (so *PNWestm*, and cf. ST PATRICK'S WELL and MARTINDALE). P~ **Common** is the high grazing land at the head of GRISEDALE. (Pl. 21)

PAVEY ARK NY2808 (2288ft/697m.) Langdales.
From 1780 (West 103).
♦ Of uncertain origin, but possibly 'Pavia's shieling' and hence the peak above it, with a female pers.n. plus **ærgi/erg** 'shieling' (Collingwood, cited in *CW* 1903, 89n., and Collingwood 1933, 336). Two *Pavia*s and at least one *Pavy* appear in the *Pipe Rolls*, and female names appear quite frequently in shieling-names, including possibly BETHECAR and LANGLEY, both formed with *ærgi/erg*. Anglezark, Lancs illustrates the development of *ærgi/erg* to *-ark*, while GLARAMARA provides a parallel for a mountain name derived from a name in *ærgi/erg*, so overall this is plausible.

PEAGILL NY1004 (hab.) Gosforth.
From 1778 (original spelling not specified, Terrier).
♦ The place is beside a modest stream or **gill**. The 1st el. is obscure — conceivably *(mag)pie* as in PYE HOWE, but this is only a shot in the dark.

PEEL ISLAND SD2991 Colton.
From 1745 (West 1774, map).
♦ 'The fortified island'; *peel*, which refers to various defensive structures in ME and Early ModE (*OED*: *peel*, n.1), plus **island**. This was formerly Montagu Island, or the Gridiron, from its natural defences of rock ridges, between which 'ancient buildings and walls' were constructed (Cowper 1899, 53–4 and 141 (plan)).

PEGGY'S BRIDGE NY1814 Buttermere.
♦ This footbridge over WARNSCALE Beck, vital for walkers and farm traffic, commemorates Mrs Peggy Webb-Jones (1922–90). Built in 1991 to replace a previous bridge, it was commissioned by her husband Denis in collaboration with the Friends of the Lake District. (I am most grateful to Mr W. Richardson of GATESGARTH Farm and Mr Scott Henderson of the Lake District National Park for this information.)

PENN/THE PEN SD1890 (785ft/244m.) Ulpha.
Penn 1867 (OS).
PEN, THE NY4718 Bampton.
From 1859 (*PNWestm* II, 196).
♦ Despite the lack of early forms, these seem to reflect a Brittonic word for 'hill', Cumbric *pen(n)* 'head, top, end' (cf. PENRITH, PENRUDDOCK), which often appears elsewhere with an explanatory *hill*, hence Pen(n) Hill (see Smith, *Elements*). The Ulpha example is a small rocky peak, while the Bampton one is not an independent summit but the E spur of LOADPOT HILL (cf. comment on HOWE NY4919). For doubts about Brittonic origin at least in the Ulpha case, see Jones 1973. The OE and later English word *pen(n)* 'animal fold' does not occur in the names in this volume, and is virtually ruled out as the origin of the two hill-names.

PENNY HILL FARM NY1900 Eskdale.
Pennyhill 1769 (*JEPNS* 2, 58), *Penny Hill* 1773 (PR); cf. *Low Penny Hill* 1731 (PR), previously *Low Piet Nest* (Fair 1922, 77).
♦ 'Penny' may be a surname, as it is in Penny Bridge in Furness. A small **hill** rises S of the farm. The word **farm** has been added, as so often, to a pre-existing name.

PENRITH NY5130 (hab. & parish).
Penred 1167 to 1290, *Penereth* c.1185 (p), *Penreth* 1197 to 1597, *Penred'* 1214 to 1242, *Penrith'* 1242 (*PNCumb* 229–30).
♦ Probably 'the head of the ford' or 'headland by the ford'. This is a Brittonic name in which the 1st el. seems to be Cumbric *pen(n)* 'head, top, end', while the 2nd el. is Cumbric *rid* 'ford' (W *rhyd*). The reference may be to the major crossing at EAMONT Bridge, though this is over a mile from the present centre of P~ (*PNCumb* 230). Lying just outside the Lake District National Park, P~ is included in this volume because of its size and historical importance.

PENRUDDOCK NY4227 (hab.) Hutton.
Pendredoch 1276, *Penred(d)ok* 1278 to 1348, *Penruddok'* 1339 (p) to 1540 (*PNCumb* 213).
♦ Partly uncertain, perhaps 'the headland by the ford'. The 1st el. is as in PENRITH, but the 2nd el. is less obvious. (a) It may be Cumbric *rid* 'ford' (cf. W *rhyd*), with suffix (so Ekwall, *Scand & Celt* 112 but not *DEPN*); P~ is situated near a crossing of the R. Petteril. (b) It may alternatively be related to the W root *rhed-* 'run', which could have formed

a stream-name referring to the upper reaches of the Petteril (Dr Oliver Padel, pers.comm. gratefully acknowledged); (c) Coates also mentions W *rhedeg* 'to run' 'e.g. elliptically for something like *maes rhedeg* "racetrack", but this is very uncertain' (2000, 284).

PETER HOUSE FARM NY2432 Bassenthwaite.
Peter-House 1830 (PR), *Peters House* 1867 (OS); seemingly *Tenter How* 1823 (GreenwoodM).
♦ Simply 'Peter's **house**', with **farm** added subsequently.

PEWITS, THE NY2817 Borrowdale.
From 1867 (OS).
♦ Despite its position on a ridge at about 1600ft/nearly 500m. this place provides the wet moorland favoured as breeding habitat by the lapwing or pe(e)wit (on which, see TEWET TARN). The first occurrence of *pe(e)wit* in *OED* is from the 16th cent.

PHEASANT INN NY1930 Wythop.
From 1867 (OS).
♦ The inn-name may allude to the pheasantries which abounded nearby until recent times (Wainwright 1964, Sale Fell 2). This is the only occurrence of *pheasant*, an adoption from Fr, in this volume, and the word *inn* is also rare: see CASTLE INN.

PICKETT/PICKET HOWE (hab.) NY1521 Buttermere.
Pikehowe 1622 (Lorton PR), *Pichthow* 1653 (*PNCumb* 354), *Pik'd How* 1774 (DHM).
♦ 'The pointed hill', of the same origin as PIKED HOWE. For the spelling, cf. *Pikethowe* 1338 in Egremont (*PNCumb* 381).

PICKTHALL GROUND SD2090 Dunnerdale.
Picthall Gd. 1671 (original spelling not specified, Broughton-in-Furness PR).
♦ 'The farm of the Pic(k)thall family'; surname plus **ground**. The place is associated with persons named *Picthall* in the 1671 and 1711 PRs. The surname derives from a p.n. which is quite common in Lakeland (see PICTHALL).

PICTHALL SD2888 (hab.) Blawith.
Pickthowe 1609, *Pickthawe* 1644 (*PNLancs* 215), *Pickthall* 1729 (PR).
♦ 'The peaked or pointed hill', cf. PIKED HOWE(S).

PIERCE HOW BECK NY3101 Coniston/Skelwith.
From 1850 (OS).
♦ P~ H~ is probably 'Peter's hill', with pers.n. and **how(e)** from **haugr**, plus **beck** for the stream beside it. *Piers* is the standard shortening of *Peter* (cf. PIERS GILL) and this is perhaps more likely than the surname *Pierce, Pearce* which derives from it, since this does not appear in the index to Coniston PRs, and *Pearson* would be the more usual Lakeland form.

PIERS GILL NY2108 Nether Wasdale/Eskdale.
Peeregill(heade), Peares gill(heade) 1664 (Winchester 2000, 169), *Peers Gill* 1867 (OS).
♦ Although the **gill** is a spectacular right-angled gorge containing a lively stream, the 1st el. seems to be not descriptive but a pers.n., since *Piers* is a common pet form of *Peter*.

PIKE NY2821 (1177ft/359m.) St John's.
From 1867 (OS).
PIKE NY3031 Caldbeck.
From 1867 (OS).

♦ 'The peak'; **pike**. The Caldbeck P~ is a particularly good example, as the neat summit of a ridge called SNAB. The St John's one is a steep-sided outlying spur of CASTLERIGG FELL.

PIKEAWASSA NY4318 Martindale.
From 1920 (OS).
♦ Possibly 'the summit of Wassa, Wat's hill'. This rocky crown tops STEEL KNOTTS, and as 'the sharpest summit in Lakeland' (Wainwright 1957, Steel Knotts 3) is a true **pike** 'peak'. If the medial 'a' is equivalent to *of*, 'Wassa' would be an earlier name of Steel Knotts. Brearley 1974 compares PIKE O' BLISCO(E) and suggests a previous name 'Wat's **how**(e)' (from ON **haugr**); this is the solution adopted above.

PIKE DE BIELD: P~ DE B~ MOSS NY2306 Eskdale.
Both from 1867 (OS).
♦ P~ de B~ is 'the peak with a shelter or animal lair'; **pike**, Fr *de*, literally 'of', **bield**. At 2657ft/810m., it is a subsidiary summit of ESK PIKE; a 'bield' is also shown on maps, though there is no certainty that it is the one commemorated in the p.n. The **moss** is the boggy tract below.

PIKED HOWE SD4798 (hab.) Over Staveley.
pik'd how c.1692 (Machell II, 113).
PIKED HOWES NY4404 Kentmere.
From 1857 (*PNWestm* I, 167).
♦ 'The peaked or pointed hill(s)', with elements going back to ME past participle *pīced* 'peaked' and ON **haugr** 'hill'. For different outcomes from the same phrase, see PICKETT HOWE, PIKE HOWE, and PICTHALL.

PIKE HOW NY4008 (2072ft/632m.) Patterdale/Troutbeck.
the piked how 1614, *Pike How* 1859 (*PNWestm* I, 191).
♦ 'The peaked or pointed hill', of the same origin as nearby PIKED HOWES, though the noun **pike** 'peak' has taken over in this instance.

PIKE OF BLISCO/PIKE O' BLISCO NY2704 (2304ft/702m.) Langdales.
Pike of Blease 1828 (HodgsonM), *Pike of Blisco* 1865 (*PNWestm* I, 207).
♦ 'Blisco' is presumably the mountain of which the **pike** is the summit, but it is unexplained, though final -*o*, if authentic, may well derive from **haugr/how**(e) 'hill'.

PIKE OF CARRS NY3010 Grasmere.
From 1863 (OS).
♦ 'The summit above the crags'; **pike** 'peak', *carr* as in CARRS, GREAT and LITTLE. This is the E top of a long series of crags called Carrs.

PIKE OF STICKLE/PIKE O' STICKLE NY2707 (2323ft/708m.) Langdales.
'*Langdale-pike*, called *Pike-a-Stickle*, and *Steel Pike*, is an inaccessible pyramidal rock' 1780 (West 102), *Pike of Stickle* 1828 (HodgsonM).
♦ 'The peak crowning the steep mountain'; **pike** 'peak', *of*, and *stickle* as in STICKLE (PIKE). This mountain may itself have been known as Stickle (and there is a Stickle Breast below the Pike), rather than being named from HARRISON STICKLE to the E, as Smith assumes (*PNWestm* I, 207). (Pl. 26)

PIKE SIDE SD1893 (hab.) Ulpha.
From 1774 (DHM); cf. *Pike* 1737 (*PNCumb* 438).
♦ 'The place beside the Pike (peak)'; **pike**, **side**. The place is near the foot of The Pike, which rises to 1214ft/370 m., so that 'place beside' is the more likely sense of *side*, though 'slope' is also possible.

PILLAR NY1712 (2927ft/892m.) Ennerdale, P~ ROCK NY1712.
Pillar 1774 (DHM); *Pillar Stone* 1867, 1900 (OS).
♦ The mountain is named from the vertical climbing wall on its N face known now as Pillar Rock (cf. BLEA ROCK). The word *pillar* was ME *piler*, adopted from OFr *pilier*. P~ must have had an earlier name or names, and Collingwood believed that *(ad altum) Delhertgrene* in the 1322 Close Rolls corresponded with it (cited *PNCumb* 387).

PINNACLE BIELD NY2409 Borrowdale.
From 1867 (OS).
♦ 'The shelter or lair by the peak' — a rocky arc immediately below the summit of GLARAMARA. This is the sole occurrence of *pinnacle* (originally an adoption from OFr into ME) in this volume; plus **bield**.

PIOT CRAG NY4510 Shap Rural.
From 1859 (*PNWestm* II, 176).
♦ This looks like 'magpie **crag**', *piot*, *pyet* etc. being common in dial., though the remote and rocky situation seems an unlikely habitat for the bird, which at present mainly frequents the lowland, occasionally penetrating the remoter dales (Ratcliffe 2002, 105, 189); compare PYE HOWE.

PLACE FELL NY4016 (2154ft/657m.) Patterdale/Martindale.
Plesterfeld, Plessefeld 1256, *Plescefel* 1266 (*PNWestm* II, 224), *Plesfell* 1658 (PR), *pless fell alias place fell* c.1692 (Machell I, 704), *Place-Fell* 1769 (Gray 1077), *Martindale-fell, or Place-fell* 1786 (Gilpin II, 55).
♦ Possibly 'the mountain with open, level areas'. The 1st el. may be OE *plætse, plæce*, ME *place, plasce* 'open space' (*PNWestm*). This may have some technical sense, or could describe the relatively level areas N and especially S of the summit; the sense of space is also promoted by the vista over ULLSWATER. The 2nd el. is probably **fell**, despite the earliest spellings pointing to **field**.

PLANTATION BRIDGE SD4896 (hab.) Nether Staveley.
♦ Self-explanatory; **plantation, bridge** from **brycg**. The bridge is shown spanning the Lancaster and Carlisle railway on the 1862 OS map, but is not named there. RATHER HEATH Plantation lies to the S, and there are others to the N.

PLOUGH FELL SD1691 (1464 ft/446m.) Ulpha.
From 1867 (OS).
♦ If 'Plough' is to be taken at its face value (which is by no means guaranteed), it may refer to the shape of the **fell**, with a gradual slope on the W side, and a steep one on the E (cf. YOKE for a similar figurative usage). The word *plough* goes back to OE *plōh*.

PLUMGARTH NY1200 (hab.) Irton.
From 1615 (*PNCumb* 404).
PLUMGARTHS SD4994 (hab.) Strickland Ketel.
Plumgardes 1557, *Plumgarthes* 1613 (*PNWestm* I, 155).
♦ 'The plum orchard(s).' The co-occurrence of *plum* from OE *plūme* 'plum(-tree)' with **garth** presumably points to cultivation of plums, and the sense 'orchard' for *garth*. Ten bushels each of apples, pears and plums were among the tithe payments required of a large estate in GRASMERE in the 17th cent. (*CW* 1908, 176). The wild plum, *Prunus domestica*, rarely fruits in Cumbria (Halliday 1997, 280–1).

PONSONBY NY0505 (hab. & parish), P~ FELL NY0807 (1033ft/315m.), P~ OLD HALL NY0505.
Puncunesbi c.1160, c.1205 (p), *Punzunby* c.1170, *Punchunebi, -by* c.1180 to 1565, *Puncuneby* c.1185 (*PNCumb* 426–7); *Blaing & Ponsonby Fells* 1802 (*CN* 1205), *Ponsonby Fell*; *Ponsonby Old Hall* both 1867 (OS).
♦ 'Puncun's settlement', from an OFr pers.n. *Puncun*, recorded in England from the 12th cent., plus ON **bœr/bý**; see also **fell, old, hall**. A *Johannes filius Puncun, Punzun* (J~ son of P~) is recorded in the Pipe Rolls for 1178–85 (*PNCumb* 426–7). P~ Fell is a tract of high grazing, rising to a summit some two miles from P~ itself. P~ Hall was described as 'lately built' in 1794 (Hutchinson I, 593).

POOL, RIVER/UNDERBARROW POOL SD4690 Underbarrow.
Underbarrow Pool 1738 (*PNWestm* I, 106).
♦ **Pool** applies here, as elsewhere in S Lakeland, to a slow-moving river. The river is variously called UNDERBARROW Pool, Helsington Pool and River Pool (all recorded on the 1863 OS map) as it flows S to join the GILPIN.

POOL BANK SD4387 (hab.) Crosthwaite.
Powbancke & variants 1634 to 1726 (*PNWestm* I, 85).
♦ 'Pool' could refer to (a) a river or stream (as in R. POOL, though the place is not on the Pool but a smaller stream); or (b) a surname, which is recorded locally as *Poole, Powe* in 16th–18th-cent. documents (*PNWestm* I, 85). **Bank** does occur with a surname as 1st el. (e.g. GARNER BANK), though much more frequently with a topographical or descriptive word.

POOLEY: P~ BRIDGE NY4724 (hab.) Barton.
Pulhou(e) 1252 (p) to 1308, *Poulhou* 1284, *Pulhawe* 1292; *Powley-Bridge* 1671 (*PNWestm* II, 211).
♦ P~ is 'the hill by the pool', from OE *pōl*, **pull** (see **pool**), referring to a pool in the R. EAMONT, plus ON **haugr** 'hill, mound', with **bridge** (see **brycg**) added later. The village of P~ B~ is surrounded by modest slopes, and it is uncertain whether the name referred to them or to the prominent cone of Dunmallard Hill on the other side of the river.

POOL FOOT (hab.) SD3284 Haverthwaite. From 1851 (OS).
♦ 'The lower end of the stream' (**pool, foot**), where RUSLAND POOL joins the R. LEVEN.

POOL SCAR SD2691 Kirkby Ireleth. From 1851 (OS).
♦ 'The row of crags above the river'; **pool, scar**. The river in question is STEER'S POOL.

PORTINSCALE NY2423 (hab.) Above Derwent.
Porqeneschal c.1160, *Portewinscales* c.1265 (*PNCumb* 371).
♦ 'The hut of the harlot(s) or townswoman/women.' A rare and fascinating name. The 1st el. is OE *portcwēne* 'townswoman', possibly encouraged by, or replacing, the cognate ON *portkona*; but there is ambiguity as to whether the el. is sing. or pl., and dispute as to whether it implies harlots or merely, as suggested by Hough (1997), ladies of the town; the 'town' would presumably be CROSTHWAITE/ KESWICK. The 2nd el. clearly originates in ON **skáli** 'shieling or hut'.

POTTER: P~ FELL SD5099 Strickland Roger, P~ TARN SD4998.
Pot(t)ergh(a) 13th cent. (p), *Potherwe, Pottehergh* 1292 (p); *Potterfell* 1570 to 1777; *Potter Tarn* 1723 (*PNWestm* I, 157).
♦ (a) P~ is probably 'the shieling by the pool', from ME **potte** 'pool, pot, deep hole' and **ærgi/erg** 'shieling'. (b) Alternatively, P~ in these cases could be the surname derived from the p.n. (so Smith, who frequently prefers surname-based explanations, in *PNWestm*). The location of P~ is unknown, but the p.n. may be preserved in the names of P~ **Fell**, a tract of high, unenclosed grazing, and P~ **Tarn** from **tjǫrn** 'small mountain pool' (so *SSNNW* 67).

POTTS GILL/GHYLL NY3137 (hab.) Caldbeck.
Potskailles 1592, *Pott gills* 1680, *Pots Gill* 1697; cf. *Pottas* 1225 (*PNCumb* 279).
♦ 'The shielings by the pool'; **potte** and **skáli/scale**, which has been replaced by **gill**. Both *potte* and the secondary *gill* are appropriate. I am most grateful to the present owner, Dr D. W. K. Bird, for this description: 'A stream runs through the property and there is a small waterfall — a 5 or 6 ft drop with a pool below. On the other side of the house is a confluence of three small streams entering a pool or boggy area.'

POUND FARM SD4795 Crook.
Pound 1857 (*PNWestm* I, 180).
♦ The word *p(o)und* 'enclosure for (stray) animals, pound' appears in p.ns in Domesday Book, 1086, and in early ME compounds, but is otherwise first recorded in late ME; **farm** has been added recently.

POW BECK NY2424 Above Derwent.
From 1867 (OS).
♦ 'The **beck** or stream called Pol or Pow', which in turn probably means 'stream'; cf. Pow Beck near St Bees (*Pol* c.1175, *PNCumb* 24), and see **pool/pow**. The beck flows slowly down the very gentle gradient between DERWENTWATER and BASSENTHWAITE.

POWLEY'S HILL NY5013 (1555ft/474m.) Shap Rural.
From 1859 (*PNWestm* II, 176).
♦ Apparently from the surname *Powley, Pulley* or *Pooley*, common in the PRs 16th–18th cent., plus **hill**.

PRIEST POT SD3597 Claife.
Priest-pot 1802 (*CN* 1228), *Priest Pot* 1851 (OS).
♦ 'Priest's pool', with *priest* from OE *prēost*, an adoption from Lat, plus **pot(te)**, a deep hollow, in this case a deep pool in ground liable to flooding. Conjectures about the motivation of the name include Wordsworth's 'perhaps from some Ecclesiastic having been drowned in it' (*Prose* II, 333), Spencer's 'supposed to hold enough liquid to satisfy a priest's thirst' (1983, 51), and more soberly and plausibly, 'a private fishery pertaining to Hawkshead Hall' (Cowper 1899, 43, 90), which latter was a manor of the monks of FURNESS.

PRIEST'S CRAG NY4223 Matterdale.
Priests Cragg 1789 (Clarke 26).
♦ The **crag** may be named from its proximity to WATERMILLOCK Church and rectory, the church having stood here since the 16th cent. (see OLDCHURCH). Clarke, however, recounts an anecdote, set 120 years before his time, of a parson, though a keen hunter, rebuking his parishioners

for hunting and roistering there on Sundays (1789, 26). There was also a 'piece of Ground ... commonly called *Priest-Bound'* in GOWBARROW PARK (Visitation of Bishop Nicholson, quoted *CW* 1884, 28), i.e. part of the glebe lands. See PRIEST POT for the word *priest*.

PRIORLING NY0506 (hab.) St Bridget Beckermet.
Pear Ling 1774 (DHM), *Prior Ling* 1789 (PR).
♦ Like PRIOR SCALES, this lies NE of CALDER ABBEY, and appears to be an outlier of the abbey (see map, Fair *et al*. 1954, 83; Winchester 1987, 156–7). *Prior*, an OFr term adopted into late OE and ME, is either the head of a modest-sized religious community, or the deputy abbot in a larger one. If the 2nd el. is **ling** 'heather', its use as a generic, presumably meaning 'heathery place', is unusual, but there are parallels elsewhere, e.g. *Ling* 1836 under LINGMELL END, or Stainton Ling, SD1394. The place is not currently heathery, but a field is locally known as 'the Ling' (I am most grateful to Mr Howard Wilson, former owner of P~, for this information).

PRIOR PARK SD1490 Waberthwaite.
Prior Park (Corner) undated, ?early 19th cent. (*CW* 1966, 376), *Prior Park* 1867 (OS).
♦ *Prior* (also found in PRIORLING above) may here be a reference to the former priory at SEATON HALL. This **park** is a steep and stony tract, 'a great walled enclosure S.W. of BUCKBARROW' (Johnson 1966, 376), which is or was subdivided into Seaton Buckbarrow and KINMONT BUCKBARROW (Sykes 1926, 127).

PRIOR SCALES NY0607 (hab.) St Bridget Beckermet.
Prior Scale 1754 (*PNCumb* 340).
♦ This may have begun life as an outlier or shieling (**skáli/scale**) to CALDER ABBEY; cf. PRIORLING, PRIOR PARK (and see Winchester 1987, 156–7 on the Calder lands). High and Low Prior Scales are distinguished on larger scale maps.

PRISON BAND NY2700 Coniston.
From 1850 (OS).
PRISON CRAG NY4213 Patterdale.
From 1860 (*PNWestm* II, 226).
♦ These names presumably reflect the forbidding nature of the landscape; the word *prison* results from the adoption of OFr *prisun* into ME, plus **band** 'projecting ridge or stratum of rock', **crag** 'rocky height'. P~ Band, Coniston, is a rocky ridge close to (another) P~ Crag and 'The Prison', one of the four sheer, rock-strewn precipices enclosing LEVERS WATER. Gloomy associations are also present in DUNGEON GHYLL.

PROSPECT HOUSE NY1533 Setmurthy.
From 1867 (OS); cf. *Prospect* 1790 (Isel PR).
♦ If *prospect* (an adoption from Lat into late ME) refers to the view, it is apt, since there is a fine vista N over the lower DERWENT valley; see also **house**.

PULL: P~ GARTH WOOD NY3602 Skelwith, P~ WOODS NY3601, P~ WYKE NY3602 (bay & hab.).
poole (stong) 1631/2, *Pull* 1788 (Hawkshead PR), *Pool (House)* 1770 (JefferysM); *Pull Garth Wood* 1850 (OS); *Pull wike* 1783 (CrosthwaiteM).
♦ *Pull*, i.e. **pool** '(slow-flowing) river', must originally have referred to Pull

Beck, which joins WINDERMERE at the deep inlet of P~ Wyke. For the other elements, see **garth** 'enclosure', **wood**, and (THE) WYKE 'inlet, bay'. See also HUYTON HILL for the history of Pull Woods House.

PYE HOWE NY3006 (hab.) Langdales. *Pyehow* 1681 (Grasmere PR).

♦ 'The hill frequented by magpies', a low hill and the nearby farm. *Pie* or *pye* was the usual word for this striking if unpopular farmland bird, until displaced from c.1600 onwards by *magpie*, in which the 1st syllable is a shortened form of *Margaret* or *Margery*. Cf. PIOT CRAG. The 2nd el. is **how(e)** from ON **haugr** 'hill'.

Q

QUAGRIGG MOSS NY2004 Eskdale. From 1867 (OS).

♦ 'The bog at Q~, the marshy ridge'; *quag*, **hryggr/rigg**, **moss**. *Quag* is described in *OED* as 'a marshy or boggy spot, *esp.* one covered with a layer of turf which shakes or yields when walked on'. Citations are given from the 16th cent., but the word is recorded from Cumbria from the 13th cent. (*PNWestm* I, 94). Q~ presumably refers to one of the slopes framing the moss or bog.

RAINORS NY0903 (hab.) Gosforth.
Raynrosse 1597 (*PNCumb* 396), *Renros* 1756, *Renerhouse* 1763 (PR), *Rainors* 1772 (*PNCumb* 396), *Reynold Close* 1774 (DHM).
♦ An obscure name, and the early spellings discourage the impression that this could be the surname *Raynor*. (a) The 1st el. could conceivably be ON *rein* 'strip of land' (cf. BROAD-RAYNE), but there is no obvious ON or OE explanation for the 2nd el. (b) Conversely, the 2nd el. could be the Cumbric equivalent of W *rhos* 'moor' (perhaps cf. Fletchers, *PNCumb* 174), but there are no other examples in the p.ns in this volume, and the 1st el. would be unexplained. (c) Parker noted 'pronounced Renneray' and suggested 'the nook of rowan tree' (1926, 47), presumably thinking of ON *reynir*, but this would not account for the second syllable in the early spellings.

RAINSBARROW: R~ WOOD SD1893 Ulpha.
Ravenisberg 1273 (*PNCumb* 437); *Rainsbarrow Wood* 1867 (OS).
♦ R~ is probably 'the hill frequented by ravens, from ON *hrafn*/OE *hræfn* and ON/OE **berg**; cf. next entry and RAVEN'S BARROW. *Hrafn/Hræfn* could alternatively be the common pers.n. The **wood** has been described as 'a coppice wood famed for its nuts' (Johnson 1966, 381).

RAINSBORROW CRAG NY4406 Kentmere.
Rainsbarrow 1780 (*West* 74), *Rainsbarrow Cragg* 1823 (*Otley* 61), *Rangebarrow Crag* 1836, *Rainsborrow Crag* 1865 (*PNWestm* I, 167).
♦ R~ may well be 'the hill frequented by ravens', as above; Smith (*PNWestm*)

reported that ravens and buzzards nest there. A pers.n. such as ON *Hreinn*, or *Hrafn/Hræfn* as above, is less likely as the 1st el. **Crag** describes the rocky precipices on the E side.

RAISE NY3417 (2889ft/883m.) St John's/ Patterdale.
(*Highe*) *rase* 1589 (*PNCumb* 317), (*Top 'oth* [*sic*]) *Raise* c.1692 (Machell II, 155), *Raise* 1860 (*PNWestm* II, 226).
♦ Simply 'the cairn', with **raise** from ON **hreysi**. The fell itself could be thought of as cairn-shaped, and has a 'crown of rough rocks', so unusual on the HELVELLYN range as to make it 'deserve a special cheer' (Wainwright 1955, Raise 2). However, it stands on a parish and former county boundary, and is well endowed with cairns today, so the name may refer to an actual cairn as a territorial marker.

RAISE BECK NY3211 St John's/Grasmere.
Raisebeck c.1692 (Machell II, 127), *Raisbeck* 1777 (*PNWestm* I, 12).
♦ 'The stream by the Raise, the cairn' (**hreysi/raise, bekkr/beck**); it flows past DUNMAIL RAISE, which was also known as 'the Raise'.

RAISE GILL NY2816 Borrowdale.
raise gill (*Head*) 1805 (*PNCumb* 24).
♦ **Raise**, from ON **hreysi**, could be a cairn or cairn-like hill, and which one is meant here is unclear; **gil(l)** is a 'stream with a ravine'.

RAISTHWAITE SD2589 (hab.) Kirkby Ireleth.
Reisthuat(*bec*) 1319 (*PNLancs* 221).
♦ 'The clearing with a cairn' (**hreysi, þveit**), cf. ROSTHWAITE in Borrowdale.

RAKE NY2301 Ulpha.
From 1867 (OS).
♦ This is a steep but rather featureless tract of fellside above HARDKNOTT PASS, so that if **rake** here referred, as usual, to a track, it was a forbidding one.

RAKE CRAGS NY3111 St John's.
From 1867 (OS).
♦ 'The rock outcrops by the steep track' (**rake, crag**), though which track is meant is not clear.

RAKEFOOT NY2822 (hab.) St John's.
Rakefoote 1597 (*PNCumb* 317).
♦ 'The bottom of the steep track' (**rake, foot**) — probably the one that descends from WALLA CRAG. The lane of which this forms a part may be an ancient routeway. Clare suggests that it is 'reminiscent of a ridgeway' and 'should perhaps be considered part of the setting of [Castlerigg] stone circle' (1999, 74).

RAKEHEAD CRAG NY1906 Eskdale.
From 1867 (OS).
♦ 'The rocky height at Rakehead, the top of the steep track'; **rake, head, crag**. This is the top of a precipitous slope on the skirts of SCA FELL, so that a *rake* in the sense of 'track' or 'sheep-walk' may seem rather unlikely, yet it is close to routes authorised for the movement of sheep in 1664 (Winchester 2000, 169).

RALFLAND: R~ FOREST NY5313 Shap Rural.
Rafland [sic] c.1200; *forest of Rauphland* 1612 (*PNWestm* II, 169).
♦ *Ralf* is probably the popular ME pers.n. *Ralph, Ralf*, originally a shortening (according to *PNWestm*) of early Germanic *Radulf*; plus **land**, and **forest**, in its older sense of a protected hunting area.

RALLISS SD1584 (hab.) Whicham.
Rallies 1769, *Ralleys* 1774 (DHM), *the Rallies* 1824 (Kirkby Ireleth PR); cf. *Rallygreen* 1716, *Ralley-ground*, *Rallystreet* 1729 (*PNCumb* 445).
♦ This may have been the farm of a family named *Rally*, which occurs as a version of the surname *Raleigh*. It would be paralleled by other names consisting of surname plus *-s*, e.g. RAWSONS. However, the fact that the surname derives from a Devon or possibly Essex p.n. (Reaney 1976, Hanks & Hodges 1988: *Raleigh*) makes this explanation somewhat doubtful.

RAMP HOLME SD3995 Windermere.
Ramps-holme or Berkshire-island 1780 (West 57), *Ramp Holme* 1851 (OS), previously *Rogerholum* 1297, *Rogerholm(e)* 1323 to 1570 (*PNWestm* I, 193).
♦ 'The island where rams graze', with the reflexes of OE *ramm* and of ON **holmr** or alternatively, as Collingwood suggested, 'perhaps from *Ramp*, wild garlic' (OE *hramse, hramsa*; 1904, 276). *Berkshire-island*, recorded in 1780, commemorates ownership by an 18th-cent. earl of Suffolk and Berkshire.

RAMPSBECK HOTEL NY4523 Matterdale.
Ramps-beck 1839 (Ford 133), *Rampsbeck House* 1864 (Hodgson Plans 24), 1867 (OS); cf. *Ramps beck* c.1692 (Machell I, 725F), 1793 (= stream) and possibly *Rauenesgil(fot)* 1285 (*PNCumb* 24).
♦ The place is named from the stream Ramps **Beck** just to the N. (a) The 1st el. could be ON pers.n. *Hrafn* or the corresponding noun *hrafn* 'raven'. If *Ravenesgil* 1285 is this place, it would support the assumption of the pers.n. (b) Alternatively, it could be *ramm* 'male sheep' (with *-p-* as the commonly-occurring glide) or (c) dial.

ramp 'wild garlic'. The word *hotel* was adopted into English from Fr in the 17th cent., though not with the sense of 'an inn of a superior kind' until the 18th cent. (*OED*, sense 3).

RAMPS GILL/RAMS GILL NY4315 Martindale, RAMPSGILL BECK NY4315, R~ HEAD NY4412.
Ramskill 1588, *Ramsgill* 1860 (*PNWestm* II, 219), *Ramps Gill*; *Rampsgill Beck* both 1860 (OSNB).
♦ The 1st el. is probably from OE *ramm* 'ram', the 2nd either ON **gil** 'ravine with stream' or ON **skáli** 'shieling', the two being frequently confused; also **bekkr/beck** 'stream' and **head**.

RANDALE BECK NY4512 Bampton.
From 1859 (*PNWestm* II, 196).
♦ Without early spellings the 1st el. is obscure. The reflexes of ON *rann* 'building', ON *hrafn* 'raven' and OE *ramm* 'ram' are among the possibilities. The other elements are evidently from ON **dalr** 'valley' and ON **bekkr** 'stream'.

RANDEL CRAG NY2529 Bassenthwaite.
From 1867 (OS).
♦ (a) 'Randel' may be the shortened form of *Randolph* or the surname derived from it. (b) Alternatively, since R~ C~ stands between BARKBETHDALE and SOUTHERNDALE, it is not impossible that Randel was a valley name similar to RANDALE above. **Crag** is here a rugged height: 'Gibraltar Crag and Randel Crag are merely steep loose roughnesses. Neither would earn the name of crag in the Scafell area' (Wainwright 1962, Skiddaw 15).

RANDERSIDE NY3420 Matterdale.
From 1867 (OS).

♦ Uncertain. This is a rock-topped spur projecting NE from GREAT DODD. **Side** is of various origins, and without early spellings it is impossible to tell which applies here, and whether or not the meaning is the same as that of another Cumbrian Randerside, which was *Randolfsete* 1285 'Randolf's shieling' (*PNCumb* 203).

RANDY PIKE NY3601 (hab.) Skelwith.
From 1851 (Census, as hab.).
♦ (a) Connection with someone named *Randolph* or *Randal* seems likely: cf. a *Randlehow* 1632 in Eskdale PR, *Slack Randy* 1859 (*PNWestm* II, 161) and cf. *CW* 1918, 122 for a *Randal* known as *Randie*. (b) However, another possibility is suggested by Simpson's explanation of Randy Brow, Grasmere, as containing 'randy, dialect for Rough' (1928, 288). At a little over 230ft/70m., R~ P~ is not among the spectacular **pike**s.

RANKTHORNS PLANTATION SD4187 Cartmel Fell.
From 1851 (OS).
♦ Likely meanings of *rank* are '(over-)luxuriant' (recorded from 13th cent., *OED* adj. sense 5) and 'foul-smelling' (recorded from 16th cent., *OED* adj., sense 12). See also **þorn/thorn**, **plantation**.

RANNERDALE NY1618 Buttermere, R~ BECK NY1719, R~ KNOTTS NY1618.
Ranerdall 1508 (*PNCumb* 356); cf. *Ravenerhals* c.1170 (*PNCumb* 355); *Rannerdale Beck* 1867 (OS); *a rocky promontory*, Randon-knot, *or* Buttermere-hawse 1780 (West 133), *Rannerdale Knot* 1809 (CrosthwaiteM).
♦ 'Ranner' is 'the shieling frequented by the raven', from ON *hrafn* 'raven' and Gaelic-Norse **ærgi/erg** 'shieling'; this is clear from the form *Ravenerhals*,

which refers to what is now BUTTERMERE Hause. Although *Hrafn* could alternatively be a pers.n., the raven is still seen hereabouts. ON **dalr** 'valley' or its reflex has been added, and then **bekkr/beck** 'stream' and **knǫttr/knott**, referring to a small but steep and crag-topped hill.

RASP HOWE NY4603 Kentmere.
From 1857 (*PNWestm* I, 167–8).
♦ 'Rasp' is obscure, but co-occurs with 'hill' terms in Rasp Hill (*PNWestm* II, 111, NY7727) and Rasp Bank (NYorks, NZ1502). Of the possible senses, 'raspberry' (*EDD*: *rasp* sb. 2) seems the least unlikely (cf. Raspberry Hill, Kent, TQ8968). **How(e)** is from ON **haugr** 'hill'.

RATHER HEATH SD4796 Strickland Ketel.
Ratherheved 1348, *Rather head(e)* 1570, *Roderhead* 1584; pronounced [rædəriθ] (*PNWestm* I, 154).
♦ Either 'the head of the row or ridge' or 'high place associated with a man called *Rather*'. The 2nd el. is clearly OE **hēafod** 'head, high place', but the 1st is elusive. (a) *Raðar*, gen. sing. of ON *rǫð* 'row, ridge, bank' is possible, or (b) a pers.n. of early Germanic origin (see *PNWestm*). This is a tract of high ground above the R. GOWAN, hence re-interpretation as *heath* is fitting.

RATTEN ROW NY3140 Caldbeck.
Rattonrawe 1581, *Rattenrowe* 1664 (*PNCumb* 279).
♦ 'Rat Row', with N. dial. *ratton* 'rat' from ME *ratoun* and ultimately OFr *raton*, plus **row**. This semi-jocular name for a dismal row of dwellings is not unusual.

RAVEN CRAG Examples at:
SD1396 Muncaster (649ft/198m.).
From 1867 (OS).
NY1904 Eskdale.
From 1867 (OS).
NY2411 Borrowdale.
From 1867 (OS).
NY2806 Langdale.
R~ Cragg 1823 (*PNWestm* I, 207).
NY3018 St John's.
From 1803 (*CN* 1607).
NY3908 Ambleside (2493ft/760m).
From 1859 (*PNWestm* I, 184).
NY4104 Troutbeck.
From 1865 (*PNWestm* I, 191).
NY4211 Patterdale.
From 1860 (*PNWestm* II, 226).
NY4707 Longsleddale.
From 1857 (*PNWestm* I, 163).
RAVENCRAGG NY4521 (hab.) Barton.
Ravencrag 1859 (*PNWestm* II, 213).
RAVEN CRAGS NY3630 Mungrisdale.
Ravencrag 1794 (*PNCumb* 227).
♦ 'The rocky height(s) frequented by ravens', from OE *hræfn*, ON *hrafn* or their reflex, plus **crag**. Otley's list of Craggs (1823, 61) ends, 'and a Raven Cragg in almost every vale'. As seen above, most of these names are unrecorded until the mid-19th-cent. OS maps, and their age is difficult to ascertain. The raven, notorious as a predator on young lambs, was among the vermin whose extermination was ordered by 16th-cent. statutes. Hutchinson commented in 1794 that ravens breed among rocks and that 'where ever there is at present a raven's nest, there has always been one in the same place, or in the neighbourhood, for time immemorial' (I, [7–8]), and nest sites in the high crags remain an important refuge for this threatened but currently recovering species (Hervey & Barnes 1970, 149, Ratcliffe 2002, 233). Compare EAGLE CRAG.

RAVENGLASS SD0896 (hab.) Muncaster, R~ & ESKDALE RAILWAY SD1198.
Rengles c.1180, Renglas 1208 to 1292, Ranglas(s) c.1240 (p) to 1323, Reynglas c.1240, Ravenglas 1297 to 1540 (PNCumb 425); Ravenglass and Eskdale Railway 1900 (OS).
♦ R~ is a Gaelic name, perhaps meaning 'Glas's share', cf. OIr *rann* 'share, part' and an Irish pers.n. *Glas* (so *PNCumb* and *DEPN*; Coates suggests either a Middle Irish pers.n. or *glas* 'blue-green', 2000, 287). The *Raven-* form seems to be due to folk etymology. R~ is the site of GLANNOVENTA. *OED's* first citation for *railway* is from 1776. The 7-mile R~ and E~ narrow gauge railway, founded in 1875, epitomises the transition of the Lakeland economy from small industry (transportation of iron and then granite) to tourism; see Davies 1981.

RAVEN HOWE NY4414 Martindale/Bampton.
From 1859 (*PNWestm* II, 196).
♦ 'The hill frequented by ravens'; *raven* as in RAVEN CRAG, and **how(e)** from ON **haugr**.

RAVENOAKS NY4422 (hab.) Matterdale.
Previously *Joyful Tree* 1793 (*PNCumb* 257), 1867, 1920 (OS).
♦ The name seems to be self-explanatory and recent, from *raven* as in RAVEN CRAG and **oak**.

RAVEN'S BARROW SD4087 Cartmel Fell.
From 1851 (OS).
♦ Presumably 'the hill frequented by ravens', cf. RAINSBARROW. Reference to the bird is more likely than a pers.n. in the name of this small, rock-spattered height.

RAVEN'S LODGE SD4685 Crosthwaite.
Ravens Lodge [sic] 1862 (OS).

♦ Probably self-explanatory, from *raven* as in RAVEN CRAG, plus **lodge** (cf. KESTREL LODGE). However, ravens are naturally more often associated with crags and other desolate haunts than with habitations, and, especially given the possessive *'s*, the possibility that this is from the surname *Raven* cannot be ruled out.

RAVENSTONE HOTEL NY2329 Bassenthwaite.
From 1900 (OS); previously *Slack House* 1867 (OS).
♦ Although the name R~ appears to be recent, it follows an older pattern: cf. Ravenstonedale, recorded from the later 12th cent. (*PNWestm* II, 30–1). The word *hotel* was adopted into English from Fr in the 17th cent.

RAVEN TOR SD2798 Coniston.
From 1850 (OS).
♦ 'The crag frequented by ravens'; *raven* as in RAVEN CRAG, plus *tor* 'rocky outcrop or peak', unique in this volume. The word *tor* occurs in OE, and in p.ns especially of SW England, which has led to the assumption that it is of Celtic, even specifically Cornish, origin. R~ T~ is a mighty rock buttress on BRIM FELL — a likely perch for ravens.

RAW: R~ HEAD NY3006 (hab.) Langdales, R~ PIKE NY3007.
Raw 1716, *Row* 1823 (*PNWestm* I, 207); *Rawhead* 1800 (*CN* 800); *Raw Pike* 1862 (OS).
♦ 'The height and peak above the row (of houses)'; **row**, **head**, **pike**.

RAWFOLD SD2089 (hab.) Dunnerdale.
Rawfold, Dun. 1711/2 (Broughton-in-Furness PR), *Rafold* 1774 (DHM).
♦ **Fold** is quite common in farm names, and it may here be qualified

either by the adj. *raw* 'cold, bleak and damp' or referring to poor, clayey soil (*EDD*: *raw*, sense 4), or by the surname *Raw*. The reflex of OE *rāw* 'row (of houses or trees)' (see **row**) is less likely in first position, as specific el.

RAWFOOT NY5217 (hab.) Shap Rural, RAWHEAD NY5216 (hab.).
Rowfoot 1739; *Rawhead* 1727, *Rowhead* 1736 (*PNWestm* II, 176).
♦ 'The bottom and top of the row'; **row**, **foot**, **head**. Since no line of buildings links the two farms, *row* may refer to a hedgerow, or it may be used rather loosely of the hamlet. Jefferys' map of 1770 shows *Row* at this point. That *Raw* is a surname, as suggested as an alternative in *PNWestm*, is unlikely given the compounding with *foot* and *head*.

RAW GHYLL/RAWGILL SD4599 (hab.) Hugill.
Rawgill 1738 (*PNWestm* I, 171).
♦ The 1st el. may be **row** (of houses or trees), but certainty is impossible. The 2nd el. is evidently from ON **gil** 'ravine with a stream'.

RAWLINSON NAB SD3893 Satterthwaite.
Rallinsons Nabb, Rallisons Nab c.1692 (Machell II, 150, 313), *Rawlinson's-nab* 1780 (West 59), *Rawlinson Nab* 1851 (OS). RAWLINSON'S INTAKE SD3790 Satterthwaite.
From 1851 (OS).
♦ 'The intake and headland associated with Rawlinson'; surname plus **intake**, **nab**. *Rawlinson* is a common Furness surname, recorded at least from 16th cent.; it occurs abundantly, e.g., in the 17th-cent. PRs for Colton, and is associated with at least eight different places there. The Rawlinsons were owners of Low GRAYTHWAITE, beside which R~ Intake lies; R~ Nab is a mile or two to the N. The surname is either based on *Raulin*, a diminutive of *Ralph*, *Ralf*, *Rauf*, or is based on or influenced by *Rowland*. R~ Intake is now wooded. R~ Nab is 'a high crowned promontory' projecting into WINDERMERE (West 1780, 59, also 65).

RAWSONS SD4686 (hab.) Crosthwaite.
Rawsons 1734 (PR), *Rowsons* 1770 (JeffreysM).
♦ An example of a name given in the modern period and based on a surname, in this case of a family recorded locally between 1394 and 1638 (*PNWestm* I, 85). The surname is a derivative of *Ralf*, *Rauf* and has a 'chiefly S Yorks.' distribution (Hanks & Hodges 1988: *Raw*).

RAYRIGG: R~ HALL SD4098 Windermere.
Rayrigge 1622 (PR); *Rayrigg Hall* 1823 (*PNWestm* I, 197).
♦ Seemingly 'the ridge frequented by deer', with reflex of OE *rā* or *rǣge* 'roe deer' (buck or doe), or possibly the cognate ON *rá*, plus reflex of **hryggr/hrycg**; also **hall**.

RAYSIDE NY5315 (hab.) Shap Rural.
Rasate c.1200, *Raset* 1343, *Raside* 1594 (*PNWestm* II, 169).
♦ Probably 'the shieling frequented by roe', from ON *rá* 'roe (deer)' and ON **sætr** 'shieling'. The 'deer' theme is continued in Buck Stone, just to the N.

RED BANK NY3305 Rydal.
Redbank (*Wood*) 1863 (OS).
♦ Probably to be taken at face value: 'the slope with reddish soil or vegetation'; **red**, **bank**.

RED CRAG NY4515 (2328ft/710m.) Martindale.
From 1860 (*PNWestm* II, 219).

♦ Presumably 'the reddish rocky height'; **red, crag**.

RED HOUSE NY2526 Underskiddaw.
From 1900 (OS), previously and presently *Oakfield House* 1867 (OS).
♦ Self-explanatory; **red, house**. Oakfield House, built in 1850, was renamed after it was extended, and its exterior rendered and coloured red to match the Penrith sandstone quoins, sills and lintels, in 1883–4. The name was changed back to Oakfield House after conversions in 1996–7, by the then owner, P. Nicholas Moor; I am greatly indebted to him for information about the house.

RED HOW NY2502 Ulpha.
From 1867 (OS).
♦ No early forms, but presumably 'red hill'; **red, how(e)** from **haugr**. This is a high shoulder of CRINKLE CRAGS.

REDHOW NY1422 (hab.) Loweswater.
Red How 1729 (PR).
♦ 'The reddish or reedy hill.' The 1st el. could be either **red** or else *reed*, as seems to be the case with another low-lying Redhow, the one in Lamplugh recorded as *Readhow* 1589, *the Readhowe* 1598 (*PNCumb* 407). The 2nd el. **how(e)** (from ON **haugr** 'hill, mound') here presumably refers to the modest but distinct height on which the farm is set.

REDMAIN NY1333 (hab.) Blindcrake.
Redeman 1188 to 1322 (p), *Redman(e)* 1235 (p) to 1555, *Redemayn* 1385 (p) (*PNCumb* 267).
♦ 'Stone ford', from Cumbric **rid* 'ford' (W *rhyd*) and **main* (W *maen*) 'rock, stone' (*PNCumb* 267, cf. Coates 2000, 284). R~ stands on a tributary of the DERWENT.

REDMIRE NY3729 (hab.) Mungrisdale.
Redmire 1323, *the Readmire beside Grisdaill* 1570, *Re(e)de Myre* 1589 (*PNCumb* 227).
♦ 'The reedy or rushy marsh', from OE *hrēod* 'reed' (suggested by the 16th-cent. spellings) and ON **mýrr**, or their ME or modern reflexes; cf. RUSHMIRE and RED SYKE. This is on the low-lying ground E of the R. GLENDERAMACKIN, and the name is still very apt. Rushes were in former times useful for thatching, floor coverings, and as wicks in rushlight candles (Winchester 2000, 137–8). The colour term **red**, though not the original meaning, well describes the surrounding common in autumn. (I am most grateful to Rev. Prof. Stephen Wright of Redmire for confirming the topography.)

RED NAB SD3899 Claife.
From 1851 (OS).
♦ 'The reddish headland'; **red, nab**. This is a minor projection into WINDERMERE, backed by a long and fairly steep slope. The soil hereabouts has a pinkish tinge, suggesting that the 1st el. is the colour term, especially since reeds — another possible explanation of 'red' in a modern form — are not abundant.

REDNESS POINT NY2226 Underskiddaw.
Redness; *Redness Point* both 1867 (OS).
♦ 'The point at, or called, Redness, the reedy or rushy promontory', with *reed* from *hrēod* as in REDMIRE, and **næss/nes**; also **point**. The nature of R~, which is just E of the Point, suggests a reference to reeds. The meaning of *ness* is duplicated by the addition of *point*, cf. MEARNESS POINT.

RED PIKE NY1610 Nether Wasdale.
Reede Pyke 1664 (Winchester 2000, 169), *Red Pike* 1867 (OS).

RED PIKE NY1615 (2479ft/755m.) Buttermere/Ennerdale.
le Rede Pike 1322 (*PNCumb* 387), *from its ferruginous colour, Red-pike* (West 1780, 129).
♦ 'The reddish peak'; **red, pike**. The two are often distinguished as the Wasdale and Buttermere Red Pikes. The latter, like Near and Far Ruddy Beck which rise on it, is coloured by syenite, in streaks made ever more visible by boot-induced erosion.

REDSCAR SD4592 (hab.) Underbarrow.
Redscar(r) 1671, 1716 (*PNWestm* I, 105).
♦ 'The reddish escarpment' (**red, scar**), though the dramatic 'scars' in this area are chiefly limestone.

RED SCREES NY3909 Patterdale.
From 1780 (West 74).
♦ The hill is named from its massive wall of reddish rock and loose stones, which frames the W side of KIRKSTONE PASS. See **red** and SCREES.

RED SYKE/REDSIKE FARM NY3626 Matterdale.
ye Red Sike 1705, *Reedsike* 1720 (*PNCumb* 223).
♦ '(The farm at) Red Syke, the reedy stream.' Despite the lack of early spellings, the 1st el. seems to have been *reed* (from OE *hrēod*) rather than *red*, for 'the area is characterised by an abundance of rushes (*Juncus* species), often erroneously called reeds', or locally 'seaves', whereas there is nothing unusually red about it (information from Mr Ivan Walsh of Red Syke Farm, to whom I am most grateful). The 2nd el. is **syke**, a small stream, often slow-running, and, as in many farm names, **farm** has been added in recent times. The site is close to Redsike Gill, while Red Syke/Sike itself is slightly to the E.

RED TARN NY2603 Langdale.
From 1865 (*PNWestm* I, 17).
RED TARN NY3415 Patterdale, R~ T~ BECK NY3515.
Red Tarn 1787 (*PNWestm* I, 17), *the Redtarn* 1805 (Scott, 'Hellvellyn', *Poetical Works* 703); *Redtarn Beck* 1860 (OSNB).
♦ 'The reddish mountain pool'; **red, tjǫrn/tarn**; also **beck** 'stream'. Although modern 'Red' in names of lakes and streams occasionally signifies 'reed', it is probably the colour term in R~ Tarn, Patterdale, which though it usually appears steely grey-blue has shallow reddish-brown waters around the edge, and Red Screes are not far away. The shallow Langdale **tarn** is reddish from haematite, which was mined at least in the 19th cent. (Johnson 1988, 247). R~ T~ Beck flows out of the Patterdale tarn. (Pl. 20)

REECASTLE CRAG NY2717 Borrowdale.
Ree-Castle 1794 (Hutchinson II, 154).
♦ (a) R~ may be 'roe-deer castle', with OE *rā*/ON *rá*, or OE *rǣge* for the female, as assumed in Rheabower (*PNWestm* II, 111). (b) A Scots dial. *ree* referring to an animal pen (*EDD*: *ree* sb. 2) might suggest an alternative interpretation. Meanwhile, the term **castle** is apt since this is 'a place of defence to guard the pass' (Hutchinson II, 154), where a fort, probably Iron Age, has been added to the natural defence of a projecting **crag** (LDHER 3002, SAM No. 23681, also Collingwood 1924, 83).

RENNY PARK COPPICE NY3501 Skelwith.
From 1850 (OS).
♦ R~ may be the surname partially originating in a diminutive of *Randolf* or *Reynold* (Hanks & Hodges 1988: *Rainey*, and cf. nearby RANDY PIKE and Renny Crags at SD3698). See also **park, coppice**.

RESERVOIR COTTAGE NY4407 Kentmere.
Not on OS maps up to 1920.
♦ This is a **cottage** at the S end of KENTMERE RESERVOIR.

REST DODD/DOD NY4313 (2278ft/ 696m.) Patterdale/Martindale.
Restdode late 12th cent., *Rostdode* 13th cent. (*PNWestm* II, 217).
♦ Possibly 'the summit used as a resting-place'. 'Rest' may imply 'where a rest is needed or taken', either from ON *rǫst* 'stage in a journey, resting-place' or ME *rest*. The hill is on a boundary and was one of the points on the bounds of the medieval manor of HARTSOP (see *CW* 1918, 151); it could therefore have been a resting-point for those walking the bounds. Elsewhere 'Resting Stones' were used by coffin-bearers on long routes. The **dod(d)** is a compact, slightly rounded summit.

RESTON SD4598 Hugill.
Rispeton, Ryspeton 1272 to 1570, *Respeton* 1307 to 1617, *Reston* 1738 (*PNWestm* I, 170).
♦ 'The settlement where brushwood grows', from OE ***hrispe** 'brushwood' and OE **tūn** 'farmstead, village'.

RIDDINGLEYS TOP/RIGGINGLEYS TOP NY4922 Askham.
Riddingleys Top 1859 (*PNWestm* II, 202).
♦ The 1st two elements appear to match **ridding** 'clearing', and *-ley* from OE **lēah** 'woodland clearing', though the combination is somewhat suspect, and without early spellings the name remains obscure. **Top** here refers to a point on ASKHAM FELL.

RIDDINGS, THE (hab.) NY3125 Threlkeld.
Redinges, Rydins' 1279 (p), *Riddings* 1574 (*PNCumb* 252).
♦ 'The clearings'; **ridding**.

RIDDING SIDE FARM SD3185 Colton.
Riddingside 1628 (PR).
♦ 'The side of the clearing' or 'slope by the clearing'; **ridding, side,** with **farm** added later. This is at the foot of Colton Heights.

RIGG, THE NY4711 Shap Rural.
From 1858 (OSNB); *del Rig* 1332, *Rygehead* 1610 (*PNWestm* II, 181) may be at different locations.
♦ 'The ridge'; **hryggr/rigg** — a high tongue of land projecting into the S of HAWESWATER.

RIGG BECK NY2120 Above Derwent, RIGG SCREE NY2120 (1821ft/555m.).
Both from 1867 (OS).
♦ 'The stream and screes by the ridge'; **hryggr/rigg, beck,** and *scree* as in SCREES. The ridge is formed by AIKIN KNOTT and ARD CRAGS.

RIGGHEAD NY3425 (hab.) Threlkeld.
Righead 1774 (DHM).
♦ 'The head of the ridge'; **hryggr/ rigg, head**. The site is on high ground which forms a slight spur, though not a classically elongated *rigg*.

RIGGINDALE NY4511 Bampton/Shap Rural, R~ BECK NY4511, R~ CRAG NY4411 Shap Rural, STRAIGHTS OF R~ NY4312 Patterdale/Shap Rural.
Regendale 1522, *Riggendale* & variants 1686 to 1754 (*PNWestm* II, 176); *Riggindale Beck* 1860 (OSNB); *Riggindale Crag* early 19th (*CW* 1902, 148); *Straits of Riggindale* 1860 (OSNB Shap).
♦ R~ is probably 'the valley below the ridge'. It is presumed to derive from *rigging*, ME and dial. 'ridge', which would describe the formidable rocky crest S of the valley, plus **dalr** 'valley'; also **beck** 'stream', **crag** 'rocky height' and *strai(gh)t*. The Straights are

a high defile between plunging slopes, 'a pass about 60 links wide . . . over which passes the old Roman Road' (Shap OSNB, 88). Hence this seems to be the word referring to a narrow place or way, more usually spelt *strait* (an adoption into ME from OFr).

RIGGS NY1830 (hab.) Wythop.
From 1608 (Lorton PR), 1774 (DHM).
RIGGS, THE SD3188 Colton.
From 1851 (OS).
RIGGS, THE NY4921 Askham.
From 1859 (*PNWestm* II, 202).
♦ 'The ridges'; **rigg** from ON **hryggr**. The Wythop example stands on long contours which climb towards SALE FELL; the Colton one is an elongated oval height above the upper COLTON BECK, and the Askham one is a small rise.

RIGGWOOD NY2134 (hab.) Bewaldeth.
From 1774 (DHM).
RIGG WOOD SD3092 Colton.
From 1851 (OS).
♦ 'The wood on the ridge'; **rigg** from ON **hryggr** plus **wood**. The Bewaldeth site is high, though not a classic spine-like *rigg* and not notably wooded at present — presumably it was more so in the past. The Colton wood is on the long contours above the SE shore of CONISTON WATER.

ROB RASH NY3304 Rydal.
From 1863 (OS).
♦ Presumably the *rash* (strip of rocky, uncultivated ground, often on a slope) belonging to an unidentified Rob (cf. Baxter Rash, *PNWestm* II, 285). Although *rash* is also a N. dial. form of *rush* (OE *risc*), which can mean 'rushy ground', this would not fit the location, a wooded slope E of ELTERWATER.

ROBIN HOOD NY2232 (hab.) Bassenthwaite.
From 1773 (PR), 1823 (GreenwoodM).
ROBIN HOOD NY5206 (hill) Shap Rural.
From 1858 (OSNB).
♦ Both places are presumably named after the legendary outlaw. The Bassenthwaite one, a habitation, is shown as R~ H~ House on larger scale maps. The Shap name designates a hill of 1617ft/493m., or more probably the beacon or cairn on the top, since other Cumbrian sites containing R~ H~'s name are tumuli (Godwin 1996), as are numerous examples of R~ H~'s Butts in NYorks, Shropshire and Somerset. The *OSGazetteer* lists over 40 British p.ns referring to R~ H~, including five more examples of R~ H~ as such, and several examples each of Robin Hood's Cave, Well, or Butts.

ROBINSON NY2016 (2417ft/737m.) Above Derwent, R~ CRAGS NY2017.
Robinson 1774 (DHM); *Robinson Crags* 1867 (OS).
♦ This unusual mountain name may well be associated with the family of Richard Robinson, who owned lands in the Buttermere area in the 16th cent. (Collingwood, 1918, 101 and sources there). R~ **Crags** fringe the NW side of the fell.

ROBINSON'S CAIRN NY1712 Ennerdale.
♦ This cairn or heap of stones on PILLAR mountain is a memorial to 'John Wilson Robinson (1853–1907), a pioneer cragsman of Lorton, who reputedly made a hundred ascents of Pillar' (Danby 1994, 90). Although thousands of cairns mark tracks and boundaries in the Lake District, this is the only instance of the word *cairn* (an adoption into Early ModE from Gaelic *carn*) in a p.n. on the One Inch map;

the local word in p.ns is **raise** from ON **hreysi**.

ROEHEAD NY4723 (hab.) Barton.
Rawhead 1669, 1702, *Row Head* 1747 (*PNWestm* II, 213).
♦ 'The top of the row'; **row**, **head**. The present layout of the hamlet, along the contour rather than stretching up the hill, makes the use of *head* rather incongruous. Perhaps a row of trees was meant.

ROGER GROUND SD3597 Hawkshead.
Roger Grounde 1582 (*CW* 1904, 148), *roger ground* 1598 (PR).
♦ 'The farm associated with Roger.' *Roger*, an early Germanic forename introduced by the Normans, also became an English surname, and since surnames normally qualify **ground**, this seems likely here, although I have not found evidence of a family of that name in the PRs.

ROGER/RODGER RIDDING FARM SD3489 Colton.
Roger ridding 1629/30 (PR).
♦ 'Roger's clearing'; pers.n. plus **ridding**, with **farm** specified later. The forename *Roger* is reasonably well represented in Colton PRs for the 17th cent., and see below.

ROGERSCALE FARM, LOW NY1426 Blindbothel.
Rogerscales 1260 (p), *Rogerscale* 1293 (*PNCumb* 447).
♦ R~ is 'Roger's shieling', from the pers.n. *Ro(d)ger* (Norman-French, though of Germanic origin) which was already common in Domesday Book, 1086, plus ON **skáli** 'shieling'. **Low** R~ is distinguished from High R~, a little higher and to the E; **farm** has been added relatively recently.

ROOKING NY4016 (hab.) Patterdale.
Roukin late 12th cent., *Rookings* 1787 (*PNWestm* II, 224).
♦ Probably 'the rough slope', from OE **rūh** 'rough' and ON *kinn* 'slope' (see KINN). Smith explains this as referring to the narrow gorge of GOLDRILL BECK (*PNWestm*), but the name may rather describe the stony slope above.

ROOKIN HOUSE FARM NY3825 Matterdale.
Buildings shown but unnamed 1867, 1900 (OS).
♦ R~ may be the surname *Rukin/Ruken(e)* found in the PRs from 1635 onwards, associated with farms including nearby TROUTBECK and Gill Head; see also **house** and **farm**.

ROSE CASTLE SD3399 Hawkshead.
From 1789 (Hawkshead PR).
♦ Not the famous residence of the bishops of Carlisle (*la Rose* 1230, *castrum de Rosa* 1404, *PNCumb* 134–5), but conceivably influenced by that name. From *rose*, adopted from Lat into Germanic languages including OE, and again from OFr into ME, plus **castle**. The present building is a picturesque 19th-cent. cottage.

ROSE COTTAGE NY0802 Gosforth.
From 1867 (OS).
♦ A relatively recent, and self-explanatory name; *rose* as in ROSE CASTLE, plus **cottage**.

ROSGILL NY5316 (hab.) Shap Rural, R~ HALL WOOD NY5416, R~ MOOR NY5115.
Rossegil(e) & variants late 12th cent. to 1372, *Rosegil(e)* & variants c.1240 to 1803 (*PNWestm* II, 170); *Rosgill Hall Wood* 1860 (OSNB), cf. *Rosegill hall* 1606; *Rosgill Moor* 1865 (*PNWestm* II, 176).

♦ R~ is 'the ravine where horses graze', from *hrossa*, gen. pl. of ON *hross* 'horse, mare', plus ON **gil** 'ravine with a stream'; also **hall, wood, moor**.

ROSSETT NY2906 (hab.) Langdales, R~ CRAG NY2507, R~ GILL NY2507 Langdales, R~ PIKE NY2407 Borrowdale/Langdales (2106ft/642m.).
Rosset 1684 (Grasmere PR); *Rossett Crag* 1859 (OSNB); *Rossett Gill* 1828 (HodgsonM); *Rossett Pike* 1862 (OS).
♦ R~ may be 'the high pastures where horses are kept', from ON *hross* 'horse, mare' and ON **sætr** 'hill-pasture, shieling' (so *PNWestm* I, 207). If so, it seems that a shieling (later a permanent farm) has given its name to the **gill** 'ravine with stream' and thence to the **crag** 'rocky height' and **pike** 'peak'. Although Rossett could be a shortening of ROSTHWAITE, the 16th–18th-cent. spellings of the Rosthwaites do not encourage this assumption.

ROSS'S CAMP SD1298 Muncaster.
From 1900 (OS).
♦ This is a point on MUNCASTER FELL at about 656ft/200m. A refreshment stop for shooting parties, it was built in 1883 (Danby 1994, 183). The identity of 'Ross' is unknown to me. This is the only p.n. containing *camp* (an adoption into Early ModE from Fr) in this volume.

ROSTHWAITE SD2490 (hab.) Broughton West.
Rosthwait 13th cent. (*PNLancs* 222).
ROSTHWAITE FARM SD4093 Cartmel Fell. From 1508–9 (*VCHLancs* 283, spelling not specified); *Rossethwayt* 1537 (*PNLancs* 200).
♦ 'The clearing where horses are kept', from ON *hross* 'horse, mare', probably in its gen. pl. form *hrossa*, plus ON **þveit** 'clearing'. **Farm** is a relatively recent addition to the Cartmel Fell name.

ROSTHWAITE NY2514 (hab.) Borrowdale, R~ FELL NY2511.
Rasethuate 1503, *Rastwhat* 1563 (*PNCumb* 353), *Rosthwait* 1786 (Gilpin I, 199); *Rosthwaite Fell* 1867 (OS).
♦ 'The clearing with/by a cairn'; **hreysi, þveit**, cf. RAISTHWAITE. *Hreysi* and its reflex *raise* normally signal a man-made cairn or heap of stones, but here it could refer to The How, a rocky knoll which rises steeply from the valley floor. R~ **Fell** is a bleak height some two miles S of the hamlet of R~.

ROTHAY, RIVER NY3308 Grasmere/Rydal/Ambleside.
Routha 13th cent. to 1681–4, *Rowthey* 1452 to 1614, *Rothay* 1793 (*PNWestm* I, 12).
♦ Possibly 'trout river', with 2nd el. '-ay' from ON **á** 'river'. The 1st el. seems to be related to ON *rauðr* 'red', but since neither water nor banks are red, it has been assumed to be an unrecorded **rauði* 'red one', referring to the trout (cf. TROUT BECK NY4104), which abounds in this river, whereas the char frequents the BRATHAY which the Rothay joins as it flows into WINDERMERE (*ERN* 336, *PNWestm*).

ROUDSEA: R~ WOOD SD3282 Haverthwaite.
Roudsea 1815 (Egton-cum-Newland PR); *Roudsea Wood* 1851 (OS).
♦ Perhaps 'the red lake', from ON *rauðr* 'red' and OE *sǣ* 'lake', encouraged by ON *sær* 'sea'; or with OE *rēad* 'red' as the original 1st el., later Scandinavianised; but without earlier forms all of this is highly speculative. A diminutive lake, R~ Tarn, stands in R~ **Wood**, but it is not

certain that this is what the name R~ refers to.

ROUGH CLOSE NY2736 (hab.) Caldbeck.
From 1652 (*PNCumb* 279).
♦ Probably 'the rough piece of enclosed land'; **rūh/rough, close**.

ROUGH CRAG Examples at:
  SD1697 Eskdale.
From 1867 (OS).
  NY3010 Grasmere.
From 1823 (GreenwoodM), 1859 (OSNB).
  NY4511 Shap Rural.
From 1860 (OSNB).
ROUGH CRAGS NY2802 Coniston.
From 1850 (OS).
♦ 'The rugged rocky height(s)'; **rūh/rough, crag**.

ROUGH EDGE NY4010 Patterdale.
From 1860 (OSNB), 1863 (OS).
♦ 'A rocky and stony brow falling towards Kirkstone Beck' (OSNB); **rūh/rough, edge**.

ROUGH HILL NY4919 (hab.) Bampton.
Rukhole 1285–90, *Rughol(e)* c.1316, *Rowgholl* 1462, *Roughill* 1658 (*PNWestm* II, 192).
♦ 'The rough hollow'; **rūh, hol(h)**, later replaced by **hill**.

ROUGH HOLME SD3997 Windermere.
Rough-holme c.1692 (Machell II, 342, ed. Ewbank, 87).
♦ 'The rough islet' (**rūh/rough, holmr/holme**), referring to an island in WINDERMERE.

ROUGHOLME SD1095 (hab.) Muncaster.
Rouholm 1278, *Rucholme* 1279, *Rugholm* 1279 (p) (*PNCumb* 425).
♦ This is 'the land beside water' (ON **holmr**) which is either 'rough' (OE **rūh**) or (less likely from the spellings)

'planted with rye' (ON *rugr*). The place is beside the R. ESK.

ROUGH MIRE NY2326 Above Derwent/Underskiddaw.
From 1867 (OS).
♦ Self-explanatory; **rough, mire**.

ROUGHTON/ROUGHTEN GILL NY3027 Threlkeld.
From 1867 (OS).
ROUGHTON GILL NY3034 Caldbeck.
From 1867 (OS).
♦ Probably both 'the roaring **gil(l)** or stream in the ravine', equivalent to ROUTENBECK, and apt since there are waterfalls in both places. The *-gh-* is non-historical, introduced into the spelling after *-gh-* in other words was no longer pronounced (cf. the spelling alternation in ROUTEN (FARM), and *EDD*: *rout* v. 3 and sb. 4, which has *rought* as NCy variant). The Threlkeld name and that of its neighbour SINEN GILL may have been deliberately contrasted.

ROUNDCLOSE HILL NY1328 Embleton.
From 1867 (OS).
♦ 'The hill at Roundclose, the rounded enclosure'; *round* from ME *ro(u)nd*, an adoption from OFr, **close, hill**. This is on a very modest slope above the R. COCKER. As with ROUND HOW, there is a theoretical possibility that 'Round' refers not to circular form but to rowan trees.

ROUNDHILL FARM NY3805 Ambleside.
Round Hill (hab.) 1865 (*PNWestm* I, 184).
♦ Probably 'the rounded hill'; *round* as in ROUNDCLOSE plus **hill**; **farm** added later. Though not forming a complete circle, the rising contours here are more gently rounded than some of the sharp spurs in the district.

This might, however, be a case where corruption of ON *raun* 'rowan' is conceivable (cf. the occurrence of *Roun(d)tree* as a variant of the surname *Rowntree*, Hanks & Hodges 1988).

ROUNDHOUSE NY3332 Mungrisdale.
♦ Simply from *round* as in ROUNDCLOSE plus **house**. 'The house is physically round in shape — actually twelve sides', and was named in 1915 by its owners, the Streatfields. I am pleased to thank the present owner, Mr R. D. Bucknall, for this information. He adds that the place was originally known as *Wellbank*.

ROUND HOW Examples at:
   NY2108 Eskdale.
From 1867 (OS); not = *the Roundhow upon Longrigg* 1660 (as claimed *PNCumb* 392), since too far from (LOW) LONGRIGG.
   NY3920 (1270ft/387m.) Matterdale.
From 1867 (OS).
   NY4016 Patterdale/Martindale.
From 1860 (*PNWestm* II, 226).
♦ 'The hill with the circular top', with *round* from ME *ro(u)nd*, an adoption from OFr, and **how(e)** from ON **haugr**. Modern 'Round' can result from ON **raun* 'rowan, mountain ash' as in Roundthwaite (*PNWestm* II, 51), but since the various Round Hows and R~ KNOTT could all reasonably be described as circular, the names probably have this obvious sense. The Eskdale name aptly describes a shape which contrasts with nearby LONG PIKE, GREAT END and BROAD CRAG. The Patterdale example is the SE top of PLACE FELL while the Matterdale one is a neatly formed **how(e)** or hill; cf. ROUNDHILL FARM.

ROUND KNOTT NY3333 Mungrisdale.
From 1867 (OS).

♦ 'The round craggy height'; *round* as in ROUNDCLOSE plus **knott**. This is a compact outcrop on the ridge between MITON HILL and CARROCK FELL.

ROUTENBECK NY1930 (hab.) Wythop.
(Stream:) *Rutenbec* 1195 (*PNCumb* 25); (hab.:) *Rowtanbecke* 1599/60 (Lorton PR), 1774 (DHM).
♦ 'The roaring stream', from OE *hrūtan* 'to snore', probably influenced by ON *rauta* 'to roar', which gives dial. *rowt* 'to bellow (of cattle)' (Ellwood 1895, 51) or 'to make any loud noise' (*EDD*: *rout* v. 3). The present participle (*-ing* in ModE) would have ended in *-and(e)* in northern ME. ROUTEN (FARM), ROUTING GILL and ROUGHTON GILL probably have the same meaning, but lack medieval forms. The 2nd el. is ON **bekkr** or its ME reflex.

ROUTEN FARM NY1016 Ennerdale.
Routon 1608 (*PNCumb* 387).
♦ *Routen* and variants occur with the meaning 'roaring' in combination with generics meaning 'stream' (see ROUTENBECK). The lack of such a generic here is puzzling, but the farm is close to the noisy Gill Beck. R~ is spelt *Roughton* on a nearby signpost and on OS maps (including 1867). **Farm** has been added to the pre-existing name.

ROUTING GILL NY4025 Hutton.
From 1867 (OS).
♦ Probably 'the roaring stream'; 1st el. as in ROUTENBECK, plus **gill** 'ravine with stream'. Strictly, *gill* must refer to the stream that roars as it cascades down the steep contours of GREAT MELL FELL, rather than the ravine; though on modern maps R~ G~ is printed in black while the watercourse running down it, Rowting Gill Beck, is named in blue.

ROW SD1094 (hab.) Waberthwaite, ROW FARM SD1093.
Row (= Row or Row Farm) 1774 (DHM), *Row* 1867 (OS, twice).
♦ Row, GLEBE HOUSE and Row Farm form a short **row** from NE to SW. **Farm**, so often added to an existing farm name without any distinguishing function (as in ROW FARM, Gosforth), here distinguishes two separate habitations, both of which are *Row* on the 1867 map.

ROW SD4589 (hab.) Crosthwaite.
(the) Row 1535, 1716, *Raw(e)* 1699, 1703 (*PNWestm* I, 85).
♦ 'The **row** (of houses)', now a substantial hamlet.

ROWAN'S GROUND NY3409 Grasmere.
♦ This is an intake alongside TONGUE GILL. I am most grateful to the owners for the following explanation: 'Rowan's Ground was acquired in 1988 to remember and celebrate the life of Rowan McCormick and his passionate knowledge of birds of prey and the natural world. Rowan lived most of his life in Grasmere and died there on 3rd September 1988, two months before his eighth birthday.'

ROWANTREE CRAG NY5212 Shap Rural. From 1859 (*PNWestm* II, 176).
♦ 'The rocky height where rowan grows'; *rowan, tree* from OE *trēo(w)*, **crag**. The word *rowan* is of Scand origin (cf. Norw *raun*, Icel *reynir*), and occurs in Cumbrian p.ns recorded from the 13th cent. (Roundthwaite, *PNWestm* II, 51, *Rauntreslak*, *PNWestm* II, 280; first citations in *OED* are 16th-cent.). The rowan or mountain ash, *Sorbus aucuparia*, is very common in Cumbria, both lowland and highland, where it grows to 2854ft/870m. especially when protected from grazing in places such as gills (Halliday 1997, 283–4).

ROWANTREE FORCE SD1493 Waberthwaite.
From 1867 (OS); possibly cf. *Rauntrehefd* c.1190–1200 (*CW* 1928, 155).
♦ 'The waterfall where rowan grows'; *rowan, tree* as in ROWANTREE CRAG, plus **force**. This is on R~ Gill, the upper reach of SAMGARTH BECK.

ROWAN TREE HILL SD3992 Cartmel Fell. From 1851 (OS).
♦ 'The hill where rowan grows'; *rowan, tree* as in ROWANTREE CRAG, plus **hill**.

ROWANTREE HOW SD1595 Muncaster/Ulpha.
From 1867 (OS).
♦ 'The hill where rowan grows'; *rowan, tree* as in ROWANTREE CRAG plus **how(e)** from ON **haugr** 'hill, mound'.

ROW END NY3023 St John's.
Row End 1604, *Raw End* 1739, 1743 (*PNCumb* 317).
♦ 'The end of the row of houses'; **row**, referring to the hamlet of St John's, which must have been more populous in former times, plus **end**.

ROW FARM NY0703 Gosforth.
Rowe 1600 (PR), *Rawe in Gosforth* 1600 (*PNCumb* 396).
♦ 'The **row** of buildings.' That this was part of a linear group in the 16th cent. is suggested by *Rowe Ende* 1593 (PR). **Farm** has been added relatively recently.

ROWLING END NY2220 (1422ft/433m.) Above Derwent.
Rawlingend (a brawny mountain) 1780 (West 126), *Rowling End* 1784 (*PNCumb*

373), (as hab.) 1792 (Thornythwaite PR).
♦ 'Hill-end associated with the Rawlin(g) family'; surname plus **end**. This is a spur of CAUSEY PIKE, and the farm below, probably named from the surname *Rawlin(g)*, which is recorded in the Newlands PR from 1750, one year after the registers begin. The name originated as a diminutive of *Ralph*, or a form of *Rowland*. Although *end* is rarely qualified by a personal name, NICHOL END may provide a parallel, and GREAT END parallels the use of *end* to refer to a mountain feature.

ROW RIDDING SD2489 (hab.) Kirkby Ireleth.
From 1649 (*PNLancs* 221).
♦ Probably 'the rough clearing'. 'Row' may be *rough*, ultimately from OE **rūh**, though **row** 'row of dwellings or trees' cannot be ruled out entirely. **Ridding** is a 'clearing'.

RUSHMIRE NY3923 (hab.) Matterdale.
Rashe Myer 1589, *Rushmire* 1749 (*PNCumb* 223).
♦ 'The marshy land where rushes grow', with *rush* derived from OE *risc* plus **mýrr/mire**, cf. REDMIRE. The land was brought into cultivation out of the 'rushy moorland' of MATTERDALE COMMON (Pearsall & Pennington 1973, 181), as is detectable from the 1867 OS map.

RUSLAND SD3389 (hab.) Colton, R~ CROSS SD3488 (hab.), R~ HEIGHTS SD3588, R~ POOL SD3487 (river) Colton/ Haverthwaite, R~ VALLEY SD3387.
Rolesland 1336, *Rwselande* 1537 (*PNLancs* 217); (R~ Cross:) *Rusland* 1851 (OS); *Rusland Heights* 1851 (OS); *Rusland Pool* 1787 (Finsthwaite PR).
♦ Probably 'Hrolf's land'. The pers.n. which forms the 1st el. cannot be determined with certainty, but may be *Hrolfr* or *Hróaldr* (*SSNNW* 352; *PNLancs* 217). The **Cross**, **Heights** and **Pool** (slow-moving stream) are secondarily named from Rusland. R~ Cross is a small cluster of homesteads, and the name is thought to refer to a crossing of ways rather than to a standing cross (I am most grateful to Canon Stephen Pye, Vicar of Rusland, for this). On the word *valley*, see LORTON VALE.

RUS MICKLE SD4587 (hab.) Crosthwaite.
Rusmickle 1535 (in 1669 copy) to 1823 (*PNWestm* I, 82).
♦ Uncertain, but possibly 'Michael's (wooded) hill'. If this is an inversion compound, the generic el. will be first and the specific second. For the 1st el., Smith considers OE *risc* 'rush-bed' unlikely, and suggests ON **rust* '(wooded) hill', which occurs in Norw p.ns (*NSL*: Rust). The 2nd el. is quite likely to be a pers.n., here *Michael*, probably in its ON form *Mikjáll*. Alternatively, it could be the adj. ON *mikill* 'large', as possibly also in Beckmickle (*PNWestm* I, 158).

RUTHWAITE NY2336 (hab.) Ireby and Uldale.
Rutheweyt 1242 (*PNCumb* 300), *Rugthwayt* 1255–6 (*Pipe Rolls* 177), *Ruthwayt* 1256, *Rughthweyt* 1296, *Routtweyt* 1300 (p) (*PNCumb* 300).
RUTHWAITE LODGE NY3513 Patterdale.
From 1860 (OSNB), cf. *Rewthwaite Cove* 1828 (HodgsonM); pronounced 'Ruthet' (Wainwright 1955, Dollywaggon Pike 7).
♦ The 1st el. in R~, Ireby and Uldale is either (a) OE **rūh** 'rough' (*SSNNW* 248) or (b) ON *rugr* 'rye' (Lindkvist 1912, 120 and n. 4, citing Scand parallels). The 2nd el. is **þveit/thwaite**

'clearing'. The R~ of R~ Lodge may also be 'the rough clearing'. Without early spellings this is far from certain, but the high, desolate location on the way to R~ Cove would suggest that any nearby farm called R~ would indeed have been a rough clearing. Built as a shooting lodge in 1854, R~ **Lodge** is now a climbing hut.

RYDAL NY3606 (hab. & parish), R~ BECK NY3608 Rydal/Ambleside, R~ FELL NY3509 Rydal, R~ HEAD NY3611, R~ MOUNT NY3606 (hab.), R~ WATER NY3506.
Rydal(e), Ridal(e), Rydall 1240 to 1777 (*PNWestm* I, 208); *Rydale-Beck* 1787 (*PNWestm* I, 211); *Ridal fell* c.1692 (Machell II, 125, 137); *Rydall-head* 1671 (*PNWestm* I, 211), seemingly = *le Dalheved* 1335, *Dall head* 1390-4 & variants to 1655 (*PNWestm* I, 209); *Rydal Mount* 1829 (Parson & White 617); *Rydal(l)-water* 1576 (in copy of 1681-4) to 1865, *Rydal-Lake* 1823, previously *Routh(e)mer(e)* 13th cent. (*PNWestm* I, 17), *Ridale-water (stiled sometimes anciently Routhe-mere)* 1671 (Fleming 15).
♦ 'The valley where rye is grown', from OE *ryge* and ON **dalr** (possibly with encouragement from OE *dæl*), hence the village and parish situated there. Though the valley of the R~ Beck is an unpromising area for arable cultivation, the lower parts of the parish boasted corn-mills for grinding oats and rye at TONGUE GILL and ELTERWATER (*CW* 1908, 154 and 160), and Dorothy Wordsworth observed 'the people busy getting in their corn' in nearby GRASMERE (*Journals*, Oct. 6, 1802). R~ **Head** is the corrie where several of R~ **Beck**'s contributory streams rise, while R~ **Fell** comprises the steep slopes on either side of the beck. R~ Mount is a 16th-cent. building. 'Originally a farmhouse called Keens, it was enlarged in the eighteenth century and renamed in 1803' (Lindop 1993, 60); it is famous for its tenants in 1813-50, the Wordsworths. On the word *mount*, see MOUNTJOY; on the mount itself, Collingwood comments, 'one suspects it of being a bit of landscape gardening of comparatively modern date' (1925, 62). R~ **Water** was formerly *Routhmere*, the lake of the ROTHAY, which runs eastwards through it.

SADDLE, THE NY1615 Loweswater. From 1867 (OS).
♦ The word **saddle** is from OE *sadol*; here it refers to the dip between RED PIKE and DODD.

SADDLEBACK or BLENCATHRA NY3227 (2847ft/868m.) Threlkeld/Mungrisdale. (*the Rackes of*) *Blenkarthure* 1589 (*PNCumb* 253), *Saddle-Back or Blenk-Arthur* 1704 (*PNCumb* lxxix), *Threlkate-fell, a part of which is called Saddle-back* 1786 (Gilpin II, 39–40), *Blencarter, Blenkarthur* 1794 (*PNCumb* 253).
♦ The name Blencathra appears to be of Brittonic origin, probably with the meaning 'the summit of the seat-like mountain'. The 1st el. is likely to be Cumbric **blain* (cf. W *blaen*) 'top, point' as in BLINDCRAKE (so Jackson 1953, 362n.1). The 2nd el. seems from the 16th–18th-cent. spellings to have been identified with the legendary Arthur, but may originally have been Cumbric **cadeir* 'seat, chair', whose W counterpart seems to occur in mountain names such as Cader Idris (see Smith, *Elements*: **cadeir*). The development of *d* to *th* would be matched locally in PENRITH and in Catherton, Shropshire, which may derive from **cadeir* as a hill-name (Watts 2004). An Irish pers.n. *Carthach*, as tentatively suggested by Coates (2000, 281), seems less likely with a Brittonic generic. The more recent and locally favoured name S~ refers to the saddle-shaped declivity between the summit and FOULE CRAG, which is especially marked when seen from E or W. *Saddleback* (noun or adj.) and *saddle-backed* (adj.) are recorded from Early Modern times, most often applied to animals or hills with concave, curving backs; 'a hill sadlebacked' occurs in 1599 (Hakluyt, cited in *OED*: *saddle-backed*); see also **saddle** in List of Common Elements. (Pl. 16)

SADDLE CRAGS NY5208 (1901ft/579m.) Shap Rural.
Saddle Crag 1859 (*PNWestm* II, 177).
♦ 'The rocky heights forming a saddle shape'; **saddle, crag**. The saddle is formed by Great and Little Saddle Crag (distinguished on larger scale maps) and the depression between them.

SADDLER'S KNOTT NY1018 Lamplugh.
Probably = *Saelamore Knott* c.1599 (*CW* 1925, 194), *Sadlemoore Knotte* 1609–10 (*PNCumb* 407), *Saddler's Knott* 1867 (OS).
♦ The 17th-cent. spelling suggests 'the hill by the saddle-shaped waste' (**saddle, mōr, knott**), after which *Saddlemoor* has been corrupted to Saddler's. *Knott* here seems to be used rather loosely as a 'hill' term, since S~ K~ is neither as compact nor as rocky as a typical *knott*.

SADGILL NY4805 (hab.) Longsleddale, S~ WOOD NY4805.
Sadgill, Sadgyl(l) 1238–46 to 1823, *Satgill* 1283, pre-1307 (*PNWestm* I, 161); *Sadgill Wood* 1863 (OS).
♦ S~ is probably 'the ravine/stream by the shieling'; ON **sætr** 'shieling', ON **gil** 'ravine with a stream'; also **wood**. For the 1st el., Fellows-Jensen suggests instead **sæt* 'shieling' or ON *sát* 'hiding place, ambush (for hunters)' (*SSNNW* 155), on the grounds that *sætr* would produce *Satter-* as in SATTERTHWAITE, but the loss of *-r* is paralleled in SETMURTHY

and Setmabanning (*PNCumb* 313–4), though these are inversion compounds. Sadgill was still used as a shieling for collection of hay in 1246, and is believed to have become a permanent settlement in the 14th cent. (Whyte 1985, 102, 112).

SAIL NY1920 (2530ft/773m.) Buttermere/ Above Derwent, S~ BECK NY1818 Buttermere.
Sail, Sale 1800 (*CN* 805); *Sail Beck* 1867 (OS).
♦ One of a number of hill-names including **sail, sale** (q.v.) which are recorded late and obscure in meaning. Out of the possible etymons the most likely seems to be ON *seyla* 'puddle, mire' which could describe the flat, poorly-drained top, unusual in the neighbourhood. S~ **Beck** 'stream' rises well below the summit.

SAIL HILLS NY1213 Ennerdale.
From 1867 (OS).
♦ S~ **Hills** is the name of a steep slope on ENNERDALE FELL, with rough vegetation, but not particularly wet, and not an obvious match for any of the possible meanings of **sail, sale**.

ST CATHERINE'S SD4099 (hab.) Windermere.
From 1823 (*PNWestm* I, 197); cf. *St Katheren Brow* 1646 (PR), *a Hill call'd St. Kathern's Brow* 1675 (*PNWestm* I, 197).
♦ Machell remarks, 'And in this Road, at a Place call'd St Katherines-Brow, was an old chaple now turn'd into a dwelling-house . . . and the name of the place doth show the dedication' (c.1692, II, 330; cf. Nicholson and Burn 1777, I, 180, Brydson 1911, 107 on the chapel).

ST HERBERT'S ISLAND NY2521 Borrowdale.

Island of Herbertholm 1343, *S. Herebertes isle* c.1540 (*PNCumb* 371), *St Herbert's Island* 1776 (Hutchinson 136).
♦ Self-explanatory; saint's name plus **island**. Herbert, OE *Hereberht*, a friend of St Cuthbert, lived as a hermit 'on an island in the large lake from which the sources of the River Derwent spring' (Bede, *Ecclesiastical History* iv, 29). A shrine to the saint built c.1374 became a place of pilgrimage; see also NICHOL END.

ST JOHN'S IN THE VALE NY3122 (parish), ST JOHN'S BECK NY3122.
St. John's Vale 1754 (Smith 1894, 131, citing *Gentleman's Magazine*), *valley of St John's* 1769, *vale of St John's* 1784 (*PNCumb* 311), cf. *chappell of Seynte John* 1554 (*PNCumb* 311); *St John's Beck* 1867 (OS), previously *Bure* 1777 (*ERN* 57, citing Nicholson & Burn).
♦ The valley and parish, and hence the **beck** 'stream', are named from the dedication of their church, to St John the Baptist, though tradition associates the church with the Knights Hospitallers of the Order of St John of Jerusalem (Danby 1994, 64), and a grant of land in THRELKELD c.1220–30 mentions 'the house of St John' (*Fountains Chartulary* I, 47), which has been taken as a reference to a hospice (Wiseman 1987, 88, citing Clay). On the word *vale*, see LORTON VALE. Part of the valley at least seems formerly to have been called *Buresdale* (see DALE HEAD). Smith attributed the general adoption of 'Vale of St. John/ St. John's Vale' to Sir Walter Scott's poem of 1813, *The Bridal of Triermain, or St. John's Vale* (Smith 1894, 16, 131); the poem itself uses the form *Valley of Saint John* (1904 edn, 553–84).

ST OSWALD'S NY3307 (hab.) Grasmere.
From 1862 (OS).

♦ The place is at the W end of GRAS-MERE village, close to the site of St Oswald's Well (mentioned by Machell c.1692, II, 125, and shown, e.g., on 1920 OS). The Northumbrian royal saint Oswald is the dedicatee of the parish church, which has 14th-cent. fabric and 12th-cent. sculpture, but the date of the dedications is uncertain (Graham & Collingwood 1925, 15, Binns 1995, 102, 162, 244, 258–9).

ST PATRICK'S WELL NY3816 Patterdale.
St. Patricks well c.1692 (Machell II, 436, 706, ed. Ewbank 148).
♦ Smith suggests that, like the dedication of nearby PATTERDALE church, this name has arisen from the name of the valley (*PNWestm* II, 221). Legend has it, nevertheless, that St Patrick baptised people in the **well**.

ST RAVEN'S EDGE NY4008 Patterdale/Troutbeck.
From 1860 (*PNWestm* II, 226).
♦ The charming oddity of this name makes one suspect that 'St Raven' is a corruption of Sattereven, close by across the ridge, which seems to belong to the group of inversion compounds with ON **sætr** 'shieling' followed by a pers.n. (*PNWestm* I, 191); *Hrafn* would be among the candidates; see also **edge**.

ST SUNDAY CRAG NY3613 (2756ft/841m.) Patterdale.
St Sunday's Cragg 1787 (*PNWestm* II, 226).
♦ This fine and **crag**gy fell seems to be named from St Dominic, who was also known, from the 15th cent., as St Sunday — a slight misunderstanding of the Lat for Sunday, *dies dominica*, literally 'the Lord's day' (*OED: Sunday*, n., sense 2). The reason for the name is unknown.

SALE FELL NY1929 (1170ft/357m.) Wythop.
From 1823 (GreenwoodM), 1867 (OS); perhaps cf. *Saw Crag* nearby 1785 (CrosthwaiteM).
♦ Possibly 'the **fell** or hill where willow grows'. Though early forms are lacking, as for most other **sail, sale** names, a derivation from ON *selja* or OE *salh* 'willow' seems possible, given the reference to willows in nearby WYTHOP and W~ BECK, and the fact that adjacent fells are identified by their vegetation: LING FELL, very much a partner to Sale Fell, and BROOM FELL, the main height in the parish. Cf. SALLOWS.

SALE HOW NY2728 Underskiddaw, SALEHOW BECK NY2828.
Both from 1867 (OS).
♦ S~ H~ may be 'the hill where willow grows' or 'the hill above the bog'. The 2nd el. is evidently **how(e)** from ON **haugr** 'hill, mound', but the interpretation of 'Sale' is highly uncertain (see **sail, sale**), especially in the absence of early spellings. Of the various possibilities, the most appropriate to the topography are: (a) ON *selja*, OE *salh* 'willow', since there is a significant growth of willows in the tiny valley formed by the beck; or (b) ON *seyla/seila* 'puddle, mire, stream', since the ground there is boggy, although the hill itself is not — Wainwright describes the route up SKIDDAW via Sale How as 'on dry grass' (1962, Skiddaw 21). S~ **Beck** rises on Salehow.

SALES BANK WOOD: see SAYLES FARM

SALETARN KNOTTS NY4405 Kentmere.
From 1858 (OSNB).
♦ The **knott**s 'craggy heights' comprise 'a rocky eminence on Kentmere

Fell, whose name is said to have derived from a small pool which formerly stood near it' (OSNB Kendal 1, 1858, 33). If so, **tarn** 'small mountain pool' can be taken at face value, despite the lack of standing water now. The 1st el. is problematic, as also in the names containing SAIL/SALE above. (a) If genuine, 'Sale-' could derive ultimately from ON *seyla* 'puddle, mire', referring to the bog which would result as the tarn filled up. (b) See **sail**, **sale** for other possibilities. (c) Smith compares *Scaletarn Knotts* 1857 and suggests 1st el. from ON **skáli** (*PNWestm* I, 168), but the 1857 form could be erroneous, influenced by nearby SCALE KNOTTS and SCALES.

SALKELD CLOSE NY1636 Blindcrake.
From 1777, cf. *Ground called Mr Salkeld Lees* 1777 (*PNCumb* 268).
♦ Seemingly 'enclosed land or farm belonging to an owner named Salkeld'; surname plus **close**. The surname would derive from the p.n. in E Cumbria.

SALLOWS NY4303 (1690ft/515m.) Kentmere.
From 1857 (*PNWestm* I, 168).
♦ Probably 'the willows', with the tree-name going back to OE *salh* 'willow'. Although willows are not most obviously associated with hills, some of the many species which abound in Cumbria grow at such altitudes as this, including *Salix herbacea*, dwarf willow (Halliday 1997, 198).

SALLY HILL NY0504 (hab.) Gosforth.
From 1774 (DHM).
♦ Possibly 'the hill where willow grows'. (a) 1st el. 'Sally' could be a dial. word for 'willow' (from OE *salh*, see *EDD*: *sally* sb.2, though the nearest citations to Lakeland are from NYorks and Isle of Man), hence the hill-name may be comparable with SALE FELL. (b) An alternative possibility is the pet-form of *Sarah*. The **hill** is a modest rise.

SALTCOATS SD0796 (hab.) Drigg.
From 1648 (PR); cf. *Saltcotecroft* 1578 (*PNCumb* 378).
♦ 'The salt-houses or huts'; *salt* from OE *salt*, plus **cot**. These were presumably for making and storing salt: see *OED*: *salt-cote*, with examples from 15th to 17th cent. The place is by the ESK estuary.

SAMGARTH BECK SD1294 Muncaster/Waberthwaite.
From 1867 (OS).
♦ Presumably 'the stream by Samgarth, Sam's enclosure'; pers.n. plus **garth**, and **beck**. *Garth* 'enclosure' occurs frequently with pers.ns and/or surnames in Early Modern documents and minor p.ns. See also ROWANTREE FORCE.

SAMPSON'S BRATFULL NY0908 Ennerdale.
From 1867 (OS).
♦ 'Samson's apron-full', *brat* being a dial. word for 'apron' (believed to be of Celtic origin, *OED*: *brat* n. 1). This is a 'long cairn', a heap of stones 96ft by 45ft (29m. by 14m.) at widest, and orientated E–W (Pevsner 1967, 86–7). Like certain similar sites, it attracted the folk legend that the devil, or here the Old Testament strong-man Samson, dropped stones when his apron-strings broke during some building work. Cf. Sampson's Stones, a scatter of erratics beside the ESK (at NY2105), and in Yorkshire, Apronful or Skirtful of Stones and The Devil's Apronful (*PNWYorks* VII, 73).

SAND GROUND SD3499 Hawkshead.
Sandground 1669/70 (PR).
♦ Presumably 'the farm of the Sand family' (surname plus **ground**), since the 'Ground' names of High Furness are usually qualified by a surname, and *Sandys, Sandes, Sands* is a famous surname in the parish. However, no Sand appears in the index to the PRs, so the common noun *sand* (from OE *sand*) cannot be ruled out entirely.

SAND HILL NY1821 Buttermere.
From 1867 (OS).
SAND HILL NY2328 (hab.) Bassenthwaite.
Sand Hill Farm 1785 (CrosthwaiteM), *Sand Hill* 1867 (OS); the current farm sign reads *Sandhills*.
♦ Simply *sand*, ultimately from OE *sand*, plus **hill**. The shape of the Buttermere hill, conical with a rounded top, may have inspired comparison with a hill of sand. The Bassenthwaite one contrasts with neighbouring GREEN HILL, and *sand* can here be taken literally, for larger scale maps show 'Sand Beds' and a Sandbeds Gill just to the NE. Cf. SANDYHILL.

SANDWICK NY4219 (hab.) Martindale.
Sandwic & variants 1200 to 1823 (*PNWestm* II, 217).
♦ 'The sandy bay', with ON *sandr*/OE *sand* and ON **vík** 'bay', from its proximity to what is now S~ Bay, ULLSWATER.

SANDYHILL SD4797 (hab.) Nether Staveley.
Sandhill 1618 (*PNWestm* I, 174), *Sandyhill* c.1692 (Machell II, 114, ed. Ewbank, 109), *Sand Hill* 1770 (JefferysM).
♦ Simply 'the sandy **hill**', with *sandy* ultimately from OE *sandig*, though this seems originally to have been another

SAND HILL. It lies close to the R. GOWAN.

SANTON NY1001 Irton, HALL S~ NY1001, S~ BRIDGE NY1101.
Santon c.1235 to 1381; *Hall Santon* 1710; *Santon Bridge* 1673 (*PNCumb* 402–3).
♦ 'The settlement on sandy ground', from OE *sand* and OE **tūn** 'farmstead, settlement, village'. There are patches of sand and a sand-pit in the district (*PNCumb* 403, citing Fair). The pattern of *hall* plus p.n. seen in Hall S~ is favoured locally: see **hall**. **bridge** is from **brycg**.

SATTERTHWAITE SD3392 (hab. & parish).
Saterthwayt 1336 (*PNLancs* 219).
♦ Probably (a) 'the clearing at the shieling' (**sætr, þveit**), the presumed shieling becoming a permanent farmstead and nucleus of a village and parish, from which cultivation of places like SCALE GREEN could be established. Final *-r* belongs to the stem of the ON word. (b) Fellows-Jensen suggests as an alternative that the 1st el. may be *sáttar*, gen. sing. of ON *sátt* 'agreement, settlement' and that the clearing may have been made after resolution of a dispute (*SSNNW* 156). (Pl. 30b)

SATURA CRAG NY4213 Martindale/Patterdale.
From 1828 (HodgsonM).
♦ **Crag** is, as usual, a 'rocky height', but the 1st el. is obscure. S~ could conceivably reflect a corruption of elements such as ON **sætr** 'shieling' (cf. SATTERTHWAITE) and ON **haugr** 'hill', but without earlier spellings any explanation is mere guesswork.

SAWMILL COTTAGE NY4501 Kentmere.
♦ Simply *sawmill*, for which *OED*'s first citation is from 1553, plus **cottage**.

It stands beside the largest plantation in the immediate vicinity (H. P. Wood, named from Henry Pattinson, who planted it). The building dates from the late 18th cent. and was formerly *Brery Close*; it is now known as Kentmere Pottery. (I am grateful to Mr Gordon Fox, owner, for this information.)

SAWREY: FAR S~ SD3795 (hab.) Claife, NEAR S~ SD3795 (hab.).
Sourer 1336; *Soray Extra* 1539 (*PNLancs* 219), *far Sawrey* 1654, cf. *betweene Sawreys* 1655 (Hawkshead PR); *Soray Infra* [sic] 1539, *Narr Sawrey* 1656 (*PNLancs* 219).
♦ S~ means 'the muddy, damp lands' from ON *saurar*, pl. of *saurr* 'mud, dirt, wet ground'. The name is still appropriate to the terrain hereabouts, which is marshy except for elevated sites such as HILL TOP, and it is paralleled in Norw examples of Sauren, Saura (*NSL*). As Fellows-Jensen points out, ON *saurr* can be an indicator of fertility, and is not necessarily uncomplimentary (1985a, 78). The *-ey* termination could be influenced by ON *ey* 'island', though the 1336 spelling does not point that way. The affixes **Far** (formerly *Extra* 'outside') and **Near** (*Intra* 'inside', seemingly miswritten as *Infra* 'below' in 1539) may result from division of an estate, and may specifically reflect distance from HAWKSHEAD.

SAWREY GROUND SD3399 Hawkshead.
Sawrey-ground 1661 (Hawkshead PR).
♦ 'The farm of the Sawrey family'; surname plus **ground**. Sawreys are recorded here c.1582 (Cowper 1904, 153), and the surname is common in Hawkshead PRs.

SAYLES FARM SD3086 Colton, SALES BANK WOOD SD3086.
Sayles c.1535 (West 1774, 104), 1537 (*PNLancs* 217), *Sayles* 1628/9, *Seiles* 1629 (PR), *Sales* 1851 (OS); *Sales Bank Wood* 1851 (OS).
♦ Possibly 'the pool(s), muddy place(s)', from an ON *seyla* 'mire, pool, puddle'. This does not obviously fit the topography, but poor quality ground may be suggested by the rating of Sayles lower than most of the other granges in Furness Fells assessed in the Survey of the Abbey of FURNESS of c.1535 (West 1774). See **sail, sale** for further possibilities. **Farm** has been added to the original farm name; see also **bank** '(hill-)slope' and **wood**.

SCA FELL/SCAFELL/SCAWFELL NY2006 (3162ft/964m.) Eskdale, SCAFELL PIKE(S) NY2107 (3210ft/977m.).
frithes or fences called Skallfeild 1578, *Scoffield-Rana* 1750, *the mountain Scofell or Scowfell* 1794 (*PNCumb* 390), *Sca fell* 1802 (*CN* 1205); *Scafell Pikes* (= Scafell Pike) 1867 (OS).
♦ Probably 'the bald (i.e. stony) mountain'. (a) The 1st el. is probably ON *skalli* 'bald head' (so *PNCumb* 390). Though not common, *skalli* occurs, e.g., in Scallow (*PNCumb* 407), and in Norw p.ns (*NG Indledning* 75; *NSL*: Skallelva explains *skalle* as *berg-rygg* 'mountain-ridge'). The W adj. *moel* 'bald' is similarly used of bare hills (see MELL FELL). *Skalli* is also favoured by the *ll* spelling, and by the topography: a bare, rocky wilderness. *Skalli* also occurs as a pers.n. (see Fellows-Jensen 1968, 244), but this is improbable here. (b) ON **skáli** 'shieling' is common in p.ns and is also linguistically possible here (so Ekwall, *DEPN*), though only the lowest slopes of the mountain would make a credible shieling site. (c) A further possibility, though again less well evidenced than (a) above, is ON

skál 'bowl', which is recorded as an early Icel p.n. (*Landnámabók* II, 326–7) and has been suggested in Norw p.ns, either referring to a hill like an upturned bowl or to hollows in the terrain (*NGIndledning* 74–5, *NG* VII, 453, XI, 309, 496, also *Scand & Celt* 27). The 2nd el. must be **fell**, despite the early spellings indicating **field**, and confusion of *fell* and *field* is fairly common. Scafell **Pike**, summit of all England, is the highest point on the grand mountain ridge including S~ and S~ Pike, which was probably all originally called Scafell. The pl. Pikes comprise S~ Pike, BROAD CRAG and ILL CRAG.

SCA FELL, GREAT NY2933 (2131ft/ 651m.) Caldbeck.
From 1867 (OS).
♦ This Sca Fell is probably 'the hill with the shieling'; **skáli, fell**. Derivation from *skalli* 'bald', likely for its grander counterparts SCAFELL and SCAFELL PIKE, is not supported by the landscape here. **Great** and Little S~ F~ are the dual peaks of what is locally known simply as Scafell (so Wainwright, who describes the grassy hill as offering 'excellent sheep pastures', 1962, Great Sca Fell 2, 1).

SCALDERSKEW NY0807 (hab.) St Bridget Beckermet, S~ WOOD NY0808.
Scalderscogh 1243, *Skelderischoth* 1287, *Skaldersko* 1303 (p) (*PNCumb* 339); *Scalderskew Wood* 1867 (OS).
♦ The 1st el. is based on ON *skjǫldr* 'shield', but it is unclear whether it is (a) *Skjaldar*, gen. sing. of the derived pers.n. *Skjǫldr* (so *PNCumb*, and probably cf. Skelderskew, which was *Schelderscoh* 1119 to 1190, *PNNYorks* 149); or (b) *skjaldara*, gen. pl. of the appellative *skjaldari* 'shield-maker' (so Fellows-Jensen, *SSNNW* 156–7), which would be suggested by the vowel spelt -*i*- in *Skelderischoth*. If (b) is correct, the name is a fascinatingly specific reference to the early arms industry. The 2nd el. is ON **skógr** 'wood', whose sense has been renewed by the relatively recent addition of **wood**.

SCALE/SCALES: see SCALES

SCALE BECK NY0905 Gosforth.
From 1867 (OS); cf. *Scale* 1365 (p) (*PNCumb* 396).
♦ 'The stream by Scale, the shieling'; **skáli, beck**.

SCALEBORROW/SCALEBARROW KNOTT NY5115 (1109ft/338m.) Shap Rural.
Scalbergh(*dik*) 12th cent., *Scalebarrow Knott* 1859 (*PNWestm* II, 177).
♦ S~ is 'the hill with a shieling'; **skáli, berg. Knott** seems to have been added later.

SCALECLOSE FORCE NY2414 Borrowdale.
From 1867 (OS).
♦ 'The waterfall by Scaleclose, the enclosure at the shieling'; **skáli/scale, close** and **force**.

SCALEGATE NY4820 (hab.) Askham.
From 1859 (*PNWestm* II, 202).
♦ 'The road or gate leading to Scales, the shieling site', from **skáli/scale** as in nearby SCALES FARM plus the reflex of ON **gata** 'way, road' or OE **geat** 'gate'. Brunskill notes that the farmsteads here and at Scales date from the 1760s and that their situation makes them likely former shieling sites for ASKHAM (1985, 157).

SCALEGILL NY1935 (hab.) Bewaldeth.
(Hab.:) *Scalegill* 1867 (OS), (stream:) *Scale Gill* 1796 (*PNCumb* 26).
♦ Probably 'the stream with the

shieling'; **skáli/scale, gil(l)**. The stream from which the settlement is named is now called Scalegill Beck. Remains of what are believed to be early medieval shieling huts have been found at another Scale Gill, in Eskdale (Winchester 1984, 267).

SCALE GREEN SD3292 Satterthwaite.
From 1851 (OS).
♦ Probably 'the grassy place by Scale, the shieling'; **skáli/scale; green**. This forms a link between SATTERTHWAITE and SCALE GREEN INTAKE in a succession of names showing use of land increasingly high up onto the fell, though it is currently forested.

SCALE HILL NY1521 (hab.) Buttermere.
Scalehill 1598/9 (Lorton PR), *Scale Hill* 1725 (Loweswater PR).
♦ 'The hill with the shieling'; **skáli/scale, hill**. This is on a modest incline.

SCALEHOW: S~ BECK NY4118 Martindale.
Scale howe 1838 (*PNWestm* II, 219); *Scalehow Beck* (1863 OS).
♦ S~ is 'the hill with the shieling'; **skáli/scale, haugr/how(e)**, also **beck** 'stream'.

SCALES NY1517 Loweswater, SCALE BECK NY1417, SCALE FORCE NY1517, SCALE KNOTT NY1517.
Scale or Grain Lousack c.1618/9 (CRO D/Law/1/242); *Scale Beck* 1867 (OS); *Scaleforce or Highforce* 1794 (Hutchinson II, 134), *Scale force* 1799 (CN 540); *Scale Knott* 1867 (OS).
SCALES NY1625 (hab.) Lorton.
Lortonscales 1517 (Winchester 1987, 148), *Skales* 1601/2 (PR), *Scale* 1774 (DHM).
SCALES NY3326 (hab.) Threlkeld, S~ FELL NY3327, S~ TARN NY3228.
Skales 1323 (*PNCumb* 253); *Scales-fell*; *Scales Tarn* both 1794 (Hutchinson I, 423).

SCALE/SCALES NY3733 (hab.) Mungrisdale.
the Skaill 1562 (*PNCumb* 212).
SCALES NY4505 (hab.) Kentmere, SCALE KNOTTS NY4505.
Scale 1836; *Scale Knotts* 1857 (*PNWestm* I, 168), possibly = *Scal-Knot* c.1692 (Machell II, 340).
SCALES FARM NY4820 Askham.
Scale 1577, 1795 (*PNWestm* II, 202), conceivably = *Bowskale* 1451 (*CW* 1921, 184, also in Heltondale).
♦ Scale(s) means 'the shieling(s) or shieling hut(s)', with **scale**, reflex of ON **skáli**, with or without ModE pl. -s. Alternation of sing. and pl. is common with this el., and the forms in -s reflect the absorption of the word into local dialect. Most of these sites are relatively high and remote (e.g. the Mungrisdale one at 984ft/300m.), and in all cases what was originally a summer pasture with hut(s) later developed into a permanent settlement, so ensuring survival of the name. In Loweswater, traces of structures close to the mouth of S~ B~ are interpreted as 'a rare survival in Cumbria of a [medieval] shieling settlement' (LDHER 1220, SAM No. 27674; also Valentine 1935, 40–1, Winchester 1987, 48). See **beck** 'stream', **force** 'waterfall' and **knott** 'rugged height' for the secondary names. The older alternative *High Force* for Scale Force is apt for a waterfall which, at 125ft/38m., is the longest single drop in Lakeland. The place of the Lorton Scales in the evolving land use in Lorton is discussed by Winchester (1987, 144–5, 148). In Threlkeld, the **fell** will have been named from the shielings, or from the name Scales: see BLEASE for its place in a larger pattern of naming on the S side of BLENCATHRA. The **tarn** (see **tjǫrn**), given its location, is probably named from the

fell; it was also known as *Threlkeld-Tarn*, e.g. 1705 (*CW* 1903, 14), and Coleridge seems to call it *Sattleback Tarn* (*CN* 793). The Kentmere site is very much an outlier to the main farms of the dale; the **Knott**s rise, a rugged hill, above it. The Askham name seems also to have generated nearby SCALEGATE, q.v.; **farm** is a relatively recent addition.

SCALY MOSS NY0513 Ennerdale.
From 1867 (OS).
♦ 'Scaly' is obscure. (a) It may be from **skáli/scale** 'shieling'; cf. Scaliber (*Scalberc* 1276, *Skailber* 1577, *PNWYorks* V, 31). (b) The adj. *scaly*, in the sense 'mean, stingy' (*EDD* sense 3) could possibly form a derogatory p.n., but I do not have evidence of the word in Cumbria. **Moss**, as usual, refers to low-lying, boggy ground.

SCANDALE: S~ BECK NY3707 Ambleside, S~ FELL NY3708, S~ HEAD NY3709, S~ PASS NY3809 Patterdale/Ambleside.
Scandal(e) 13th cent., *Scamdal* 1277; *Scamdalbec* 1274 (*PNWestm* I, 183); *Scandale Fell* 1859 (OSNB), *Scandale Head* 1825 (HodgsonM).
♦ S~ is 'the short valley', from ON *skammr* 'short' plus **dalr**. It is somewhat shorter than most of the neighbouring valleys, but the name may have referred to the upper part of the valley before it narrows above a bend in the beck at HIGH SWEDEN BRIDGE (cf. *SSNNW* 158). See also **bekkr/beck** 'stream', **fell** 'hill', **head** 'upper end' and **pass**.

SCAR: IN S~ NY5319 Lowther, OUT S~ NY5418 Thrimby.
Skarr 1582 (*PNWestm* II, 153); *In Scar* 1859 (*PNWestm* II, 184); *Out Scar* 1859 (*PNWestm* II, 153).
♦ The **scar** is a scarp or escarpment of carboniferous limestone edged in part by the spectacular KNIPE Scar, which rises above the LOWTHER valley. It is possible that *In* and *Out* (both of OE origin) distinguish the parts within and outside Lowther parish, though there are other possibilities. According to Noble, a wall divided the enclosed land of In Scar from Out Scar (1907, 211).

SCAR CRAG NY1920 Buttermere.
From 1867 (OS).
♦ Probably 'the rock outcrops on the escarpment'; **scar**, **crag**, cf. SCAR CRAGS. These crags, one of the three main sets on CRAG HILL, are quite linear, so that *scar* probably has its usual sense of 'escarpment'. A narrow passage below, however, is called The Scar — possibly a corruption of ON **skarð** 'gap, pass' — and this could have given its name to the Crag (cf. HIGH SCARTH CRAG).

SCAR CRAGS NY2120 (2205ft/672m.) Above Derwent.
From 1800 (*CN* 779).
♦ 'The rock outcrops on the escarpment'; **scar**, **crag**. This is a rugged ridge extending W from CAUSEY PIKE.

SCAR FOOT SD4892 (hab.) Underbarrow.
From 1662 (*PNWestm* I, 105).
♦ 'The foot of the escarpment'; **scar**, **foot**. Smith takes this as a reference to Underbarrow Scar but it appears closer to Scout Scar.

SCARGREEN NY0605 (hab.) Ponsonby.
Skargreene 1656; cf. *Scart* c.1212–20 (in late medieval copy), *Scarth* (p) 14th cent. (*St Bees* 303 and n.).
♦ 'The grassy place or hamlet at the cleft'; **skarð**, **green**. That 'Scar-' is the *Scart(h)* 'gap' of the medieval documents is suggested by the position of

S~, near to a cleft between two hills and close to the parish boundary.

SCAR/SCARR HEAD SD2894 (hab.) Torver.
Scarhead 1615/6, *Scarrehead* 1656/7 (PR).
♦ Not the **head** of a **scar** in its usual sense of an escarpment or scarp, but the point where the serious gradient begins above TORVER BECK.

SCAR LATHING NY2204 Eskdale.
From 1867 (OS).
♦ The 1st el., **scar**, is suitable to this rock outcrop on a bend in the R. ESK, but the 2nd is puzzling. Some connection with dial. *lathe* (ON *hlaða*) 'barn' is possible, though the farming here is definitely not arable. The remote situation seems completely to rule out a link with dial. *laithing*, referring to a neighbourhood from which people are invited to weddings and funerals.

SCARNESS NY2230 (hab.) Bassenthwaite.
Scarnas 1591 (PR), *Scarnes* 1663 (PR), *Scarnhouse* 1718 (PR), 1787 (*PNCumb* 264), *Scareness* 1780 (West 119), *Scarness* 1794 (Hutchinson I, 448).
♦ Possibly 'the rock-strewn point'. S~ is a hooked, beak-like promontory on the E of BASSENTHWAITE LAKE, and a farm close to it. The 2nd el. is most likely to be ON **nes**/OE **næss** 'headland', and the 1718 spelling *Scarnhouse* a misconstruction. The 1st el. is elusive. (a) Dial. **scar** from ON *sker* is possible, especially given the presence in the parish of a place called Scar (PR for 1671, 1683), and of rock outcrops at the place (I am very grateful to Mr Barritt and Mr Wilson for this information). (b) The shape makes derivation from OE *scearp* 'sharp' a tempting possibility, especially since 'sharp' co-occurs with 'ness' in two other p.ns (Smith, *Elements*: *scearp*).

Without spellings in *-p-*, however, this cannot be proven. (c) If the 2nd el. were not *-nes* but *-house* the 1st el. could be dial. *scarn* 'dung' (*PNCumb* 264). Clarke reports the local tradition that it refers to the build-up of cow-dung around a hut where cows were milked twice daily (1789, 98). However, this goes against the evidence of the earlier PRs.

SCARSIDE NY5318 (hab.) Bampton.
Scar(r), *Skarr* 1667, *Scarside* 1711 (*PNWestm* II, 196).
♦ 'The place beside the scarp' or 'the slope below the scarp'; **scar**, **side**. It is below IN SCAR.

SCARTH GAP PASS NY1813 Buttermere/Ennerdale.
Scarf Gapp 1821 (*PNCumb* 387), *Scarth Gap* 1867 (OS).
♦ S~ G~ is 'the gap called Scarth', from **skarð** 'a notch or col', here the one between HIGH CRAG and HAYSTACKS, duplicated in **gap**. **Pass**, although it originally also referred to gaps, came to refer to tracks running through them. Here an old pony track runs N and S through the gap, linking BUTTERMERE with ENNERDALE and WASDALE. (Pl. 4)

SCAW NY1215 Ennerdale, S~ WELL NY1216.
Both from 1867 (OS).
♦ In the absence of early forms the meaning of the name, which refers to the high, bare flanks of GREAT BORNE, is obscure. Possible etymons include (a) ON **skógr**/OE **sceaga** 'wood, copse' (as assumed in other Scaws, *PNCumb* 232, 400), or (b) ON **skáli** 'shieling', either of which would designate a feature below. (c) Wainwright refers to S~ as 'Herdus Scaw, or Scar' (1966, Great Borne 2), and if **Scar** were the

original form it would aptly describe the rocky site. S~ **Well** is at 1800ft/ 550m.

SCAWDEL/SCAWDALE, HIGH NY2315 Borrowdale, LOW S~ NY2416.
Hovedscaldale c.1210–12 (*PNCumb* 351).
♦ These are two rocky points on the slopes of HIGH SPY, which is also known as Scawdel Fell (Wainwright 1964, High Spy 1). (a) S~ is probably 'the valley with a shieling'; **skáli, dalr**, cf. the earliest spellings of SCALE-BORROW and KESKADALE; also **high, low**. The presence of a shieling implied by nearby SCALECLOSE FORCE would also favour the 'shieling' explanation. (b) The 1st el. could alternatively be explained as ON *skalli* 'bald head' (which may be present in SCA(W)FELL), as a description of the hill or possibly as the (nick)name of a landowner. This interpretation is favoured in *PNCumb* 351, as it is for SCAWTHWAITE. (c) A further possibility is ON *skál* 'bowl, hollow', again as mentioned under SCA(W)FELL. *Hoved* in the 13th-cent. spelling is from ON **hǫfuð** 'head', hence 'top, summit'. It seems to correspond with 'High' in the modern name.

SCAWGILL: S~ BRIDGE NY1725 Lorton.
Scawgill 1812 (PR); *Scawgill Bridge* 1900 (OS).
♦ S~, shown as a habitation on the 1867 OS map, is probably 'the ravine/ stream with the shieling' (**skáli/scale, gil(l)**), though without earlier forms certainty is impossible. The impressive, three-arched stone **bridge** (see **brycg**) is shown simply as 'New Br[idge]' on the 1867 map. It spans AIKEN BECK, and it may be that this was originally known as Scaw Gill.

SCAWTHWAITE: S~ CLOSE NY2435 Ireby and Uldale.
Scalethweit 1171–5 (in 1333 copy), *Scallethwayt* 1256 (*PNCumb* 300); *Skawthwat-Close* 1679/80 (Ireby PR).
♦ S~ is probably 'the clearing with a shieling', but 'the clearing in a hollow', or conceivably 'the bare clearing' or 'Skalli's clearing' are also possible. The 2nd el. is ON **þveit** 'clearing', as with other p.ns in the parish, but the 1st is doubtful, especially because of the variation between single and double *l* in the medieval spellings. (a) The *-l-* spelling suggests ON **skáli** 'shieling'. (b) ON *skál* 'bowl, hollow' is rare in p.ns, but not impossible; cf. SCA FELL. (c) The *-ll-* spelling would suggest the ON *skalli* 'bald head'. This could apply to a bare slope (cf. SCA FELL, main interpretation) but is unlikely here (so Fellows-Jensen, noting the present nature of the place, grassy and tree-covered, *SSNNW* 159). *Skalli* also functions as a nickname or pers.n. (see Fellows-Jensen 1968, 244 for the pers.n. and this interpretation is favoured in *SSNNW* and *PNCumb*). S~ **Close** is 'the enclosure or farm at S~'.

SCEUGH NY5019 Bampton.
Skeweghe early 14th cent. (in 18th-cent. printing, *CW* 1921, 180), *Sceugh* 1859; cf. *Medil Scogh* 1471 (*PNWestm* II, 196).
♦ 'The wood', from ON **skógr** 'wood', as is also nearby SKEWS. The area is not wooded now.

SCHOOL KNOTT SD4297 (761ft/232m.) Windermere.
From 1828 (HodgsonM), 1865 (*PNWestm* I, 197).
♦ Presumably 'the rocky height near the school'; *school* originally adopted from Lat into late OE, plus **knott**.

Machell describes a two-storey stone-built school 100 yds SE of Windermere church (c.1692, ed. Ewbank, 117), which would be just over half a mile from here, and a 'Schoole of Windermere' evidently existed by 1670 (*CW* 1964, 169, location unspecified). Otherwise, one might have imagined corruption of a pers.n. from ON *Skúli*, as in School Aycliffe, Durham (Watts 2002).

SCOAT FELL NY1511 Nether Wasdale, S~ TARN NY1510.
le Scote in Bouthdale 1338, *Leescote of Bowlderdale* 1578 (*PNCumb* 441).
♦ S~ is possibly from ON **skot** or *skotta*, related to *skjóta* 'to shoot', as suggested in *PNCumb*, citing Ekwall. The word is used in Norw p.ns of land which 'shoots' or projects in some direction, or of hills down which timber is sent (cf. *NSL*: Skotselv, and cf. Skotsberg, *NG* I, 184). **Fell** 'hill' was evidently added later, and the **tarn** (see **tjǫrn**) 'mountain pool' named from the hill.

SCOGARTH NY3426 (hab.) Threlkeld.
Scogarth 1773 (PR), *Scawgarth* 1774 (DHM).
♦ This may derive from ON **skógr** 'wood' and ON **garðr** 'enclosure', but in the absence of earlier forms the etymology is far from certain.

SCOPE: S~ END NY2218 Above Derwent, S~ BECK NY2118.
Skop 1774 (DHM); *Scope End*; *Scope Beck* both 1867 (OS).
♦ 'Scope' is interesting but enigmatic. S~ End is a point on the N ridge of HINDSCARTH known as Low Snab, and S~ **Beck** flows beside it. The situation, and the normal usage of **end**, would suggest that Scope refers to the ridge or its upper part, as an alternative to LOW SNAB. If so, *scope* could be the dial. form of *scalp*, in the sense 'bare, stony ground', sometimes used of hillsides (cf. The Scope, recorded from 1843, *PNWYorks* II, 285; *EDD: scalp*, sense 4). On the 1774 map, *Skop* is shown as a habitation below the ridge (cf. *Skep* 1823, GreenwoodM) and close to *Gold Skop Copper Mines*. This is Goldscope mine, originally named *Gottesgab* 'God's gift' (recorded 1569) by the German company mining there in the 16th cent. (*PNCumb* 370), but anglicised to *Gowd-Scalp* then *Gold Scalp* by 1709 (Collingwood 1912, 10) and to the present Goldscope. Hence either the anglicisation of *Gottesgab* was influenced by a pre-existing p.n. *Scope or *Scalp or (as Brearley 1974 suggests), Scope is 'named from the adjoining Goldscope mine'.

SCOT/SCOTT HALL NY0802 Gosforth.
Scott hooll 1597 (*PNCumb* 396), *Scothole* 1607 (PR), *Scot Hall* 1774 (DHM).
♦ Partly uncertain. 'Scot(t)' may be a surname, and a Rear-Admiral Francis Scott was influential in Gosforth parish in the mid-19th century (Parker 1926, 45, 59, 175), but I do not know of any earlier evidence for this family. 'Hall' seems to have developed from **hole** 'a hollow', though the buildings themselves are on a slight rise.

SCREES, THE NY1504 Eskdale.
Scrithes (*Edge*) 1537 (*CW* 1926, 96), *Screes* 1774 (DHM), *Eskdale Screes*, *Screes* 1794 (Hutchinson I, 580–1), *Screes* 1867 (OS).
♦ *Scree*, meaning a precipitous slope covered in loose stones, is from ON *skriða* 'land-slide, scree' (a derivative of *skríða* 'to slide'); it occurs four times in this volume. These are the screes on the SE side of WASTWATER, the most spectacular in England. (Pl. 24)

SCRITHWAITE FARM SD2191 Dunnerdale.
Skraithwaite 1615, *Scrythwaite* 1786 (*PNLancs* 223, which gives pronunciation [skraiþət], i.e. 1st vowel as in *sky*).
♦ S~ is 'the clearing by the stony slope', from ON *skriða* 'scree, slope with loose stones, landslide' and ON **þveit** 'clearing'. **Farm** appears on the One Inch map of 1994 but not, for instance, 1966.

SCROGGS FARM SD4699 Over Staveley.
Scroggs 1621 (*PNWestm* I, 176).
♦ S~ is from OE *scrogge* 'brushwood', with the pronunciation [sk] presumably due to Scand influence; **farm** has been added relatively recently.

SEAMEW CRAG NY3702 Windermere.
Seamow crag 1783 (CrosthwaiteM).
♦ 'Rock frequented by seagulls.' According to *OED*, *seamew* denotes specifically the common gull, *Larus canus*. The Lake District is towards the S. limit of the common gull's normal territory, but it winters inland there (Miles 1993, 44), and is listed by Hutchinson in 1794 as one of four breeding gulls found in Cumberland (I, [21]). Machell reported *seamaws* among the breeding birds in ANGLE TARN, Martindale, c.1692 (I, 704). This is a rocky islet in WINDERMERE — an unusual application of **crag**.

SEAT NY1813 Buttermere.
From 1867 (OS).
♦ 'The seat or high place'; **sæti/seat**. This is a lofty place, though barely a distinct summit, on the ridge between HIGH CRAG and SCARTH GAP. At 1841ft/561m., it is too exposed to be a shieling site (ON **sætr**).

SEAT, THE SD1697 Eskdale.
Seat 1867 (OS); cf. *Seat How* 1802 (*CN* 1225), possibly = *Satgodard* c.1205 (*Scand & Celt* 33).
♦ This is the ridge of which Seat How (not the one at NY2125) is the summit, and if the identification with *Satgodard* c.1205 is correct, 'Seat' probably refers to a shieling (ON **sætr**). The other main possibility is ON **sæti** 'seat, high place'.

SEATALLAN/SEAT ALLAN NY1308 (2270ft/692m.) Nether Wasdale.
Seatallan 1774 (DHM), *Settallian* 1783 (*PNCumb* 442), *Seat-allian* 1802 (*CN* 1205).
♦ 'Alan's shieling' or 'Alan's seat', seemingly an inversion compound with 1st el. either **sætr** 'shieling, hill-pasture' or **sæti** 'seat, high place', plus the pers.n. *Alein*, of Breton origin but popular in ME and also found in STARLING (DODD): see that entry for the possibility that *Alein* replaces an earlier Irish name.

SEAT FARM NY4622 Barton.
Seat 1754 (*PNWestm* II, 213).
♦ The situation is a quite high headland, so that the name could contain the reflex of either ON **sæti** 'seat, high place' or ON **sætr** 'shieling'. **Farm** has been added to the original name.

SEAT HOW NY2125 (1622ft/496m.) Above Derwent.
From 1867 (OS).
♦ Probably 'the seat-like hill' (**sæti/seat**, **haugr/how**(e)), unless the 1st el. is **sætr** 'shieling'.

SEATHWAITE SD2296 (hab., parish of Dunnerdale with S~), S~ FELLS SD2597, S~ TARN SD2598.
Seathwhot 1592, *Seathwhat* 1598 (*PNLancs* 223); *Seathwaite Fells* 1802 (*CN* 1228), 1851 (OS); *Seathwaite Tarn* 1745 (West 1774, map), 1850 (OS).

♦ (a) Probably 'the clearing by the lake'. The 2nd el. is from ON **þveit** 'clearing'. The 1st is probably ON *sær*, *sjár* 'sea, lake', referring to S~ **Tarn**, now a sizable reservoir. The village of S~, formerly known as Seathwaite Chapel (e.g. Yates' map of 1786, *CW* 1908, 352), is nearly two miles from the tarn, but perhaps there was a settlement named S~ nearer to the tarn, which gave its name to the parish. Meanwhile, other origins for 'Sea-' in this name are possible, but cannot be proved or discounted in the absence of medieval spellings: (b) ON *sef* 'sedge' as in the Borrowdale SEATHWAITE; (c) ON **sætr** 'shieling' (so Hoyte 1999, 268). The **fell**s (a tract of high grazing land) and **tarn** (from **tjǫrn**, small mountain pool) are probably named from S~ as a parish rather than as a nuclear settlement.

SEATHWAITE NY2312 (hab.) Borrowdale, S~ FELL NY2210.
Seuethwayt 1292, *Seuythwayt* 1340, *Sethwayt* 1542, *Seatwhait* 1566 (*PNCumb* 351); *Seathwaite Fell* 1867 (OS).
♦ 'The clearing where sedge grows', from ON *sef* 'sedge', preserved in local dial. as *seave*, plus ON **þveit** 'clearing'. The fortunate survival of the 1292 and 1340 spellings (containing ME *sevy* 'sedgy', *SSNNW* 159) shows that this cannot be 'clearing by a lake' (Lee 1998), despite being reputedly the wettest inhabited place in England, and anciently the site of a lake (Ward 1876, 86, Millward & Robinson 1970, 48). S~ **Fell**, some distance from the hamlet, indicates its designated grazing ground.

SEATHWAITE NY3804 (hab.) Ambleside.
Seathat, Seatwhat 1597, *Sealthwaite (green)* 1838 (*PNWestm* I, 184).
♦ The 2nd el. is 'clearing' (ON **þveit**), but in the absence of medieval spellings the 1st el. is elusive, as it is in SEATHWAITE SD2296. ON *sef* 'sedge' or to a lesser extent ON **sætr** 'shieling' are among the possibilities. There is no lake here, and scarcely room for there to have been one in the past, so ON *sær* seems unlikely.

SEATLE/SEATTLE SD3783 (hab.) Staveley-in-Cartmel, S~ PLANTATION SD3683.
Settyll 1491, *Seitill* 1508–9, *Seatle* 1593 (*PNLancs* 199–200); *Wakefield's or Seatle Plantation* 1851 (OS).
♦ Possibly 'seat-hill', with ME *seat* from ON **sæti** or late OE *sǣte*, plus **hill**; this would describe the flat-topped rise above S~. However, Ekwall was right that 'earlier material is needed' (*PNLancs* 200). Derivation from OE *setl* 'abode, dwelling' seems ruled out by the spellings pointing to a long vowel. See also **plantation**.

SEATOLLER NY2413 (hab.) Borrowdale, S~ FELL NY2213.
(*Comon of*) *Setower* 1555 (*CW* 1976, 110), *Settaller* 1563, *Seitaller, Seataller* 1566 (*PNCumb* 351), '*Settawre now called Seatallor Common*' 1759 (*CW* 1976, 106), *Seatholler in Borrowdale* 1792 (Thornthwaite PR); *Seatallor fell* 1777 (*PNCumb* 351).
♦ An interesting and problematic name, which may mean 'the shieling site with alders'. The forms are late but suggestive of an inversion compound with ON **sætr** 'shieling' as 1st, generic, el. (a) The 2nd el. may be OE *alor*, possibly replacing ON *elri(r)*, both 'alder(s)', which are present in the valley (so *PNCumb*). (b) Alternatively, a Gaelic or ON pers.n. is likely since they are extremely common in inversion compounds, and Collingwood claimed that the 2nd el. was formerly '-*haller*, possibly for

Halldor, a Norse name' (1933, 141). The -h- in the PR entry for 1792 gives some support to that, but the 1555 and 1759 spellings discourage it. The **fell** is not a specific peak but the high ground W of the hamlet.

SEATON: S~ HALL SD1089 Bootle.
Seton (*in Couplont*) 1184–90 to 1535, *Leakley now called Seaton* 1610 (*PNCumb* 347); *Seaton Hall* 1774 (DHM).
♦ A puzzling name, since the obvious derivation from OE *sǣ* 'sea' and OE **tūn** 'farmstead, settlement' hardly suits the situation, almost two miles inland. OE *sǣ* can refer to an inland lake or possibly a marsh in other English p.ns (Smith, *Elements*), and it may have had one of these senses here. The place was the site of Lekely Priory, a Benedictine nunnery founded between 1185 and 1210. The present S~ **Hall** is of 16th-cent. origin (Fair 1945, 134–5, Pevsner 1967, 187).

SEAT ROBERT NY5211 (1688ft/515m.) Shap Rural.
From 1859 (*PNWestm* II, 177).
♦ 'Robert's seat.' Although recorded so late, this seems to belong with a group of inversion compounds consisting of *Seat* and a pers.n. (see SEATALLAN, SEAT SANDAL, SETMURTHY). The 1st, generic, el. is probably from ON **sæti** 'seat', but ON **sætr** 'shieling' as probably in Seat Sandal cannot be ruled out. The pers.n. *Robert* is of both Continental Germanic and OE origin, and has been popular in England since the Norman Conquest (Withycombe 1977).

SEAT SANDAL NY3411 (2415ft/736m.) Patterdale/Grasmere.
Satsondolf 1274, *the Sate Sandall* 16th cent. (*PNWestm* I, 199), *Seat-Sandle* 1780 (West 80), *Seat Sandal* 1825 (HodgsonM).

♦ 'Sandulf's shieling', with ON **sætr** and an ON pers.n. *Sǫndulfr* forming an inversion compound. Remains of three old shelters on a 'natural shelf' at 1700 ft have been taken as evidence of the possibility of an old shieling site (Hay 1939, 16–17).

SEED HILL NY1001 (hab.) Irton.
From 1787 (*CW* 1974, 212), 1816 (PR).
♦ Presumably 'the hill where seed was grown'; OE *sǣd* 'seed, corn' plus **hill**. It is listed in 1787 as 'let to John Bragg farmer', who also had Irton Demesne and Park Yeat (*CW* 1974, 212), and a corn mill is indicated at the site on 1867 OS. Cf. Seedhowe, recorded from 1628 (*PNWestm* I, 171), several Yorkshire Seed Hills (*PNWYorks* VIII, 161), some of them also recorded from the 17th cent., and WINTERSEEDS in this volume.

SELDOM SEEN NY3718 (hab.) Patterdale.
Previously *Glencoin Cottages* (corrected from *Glencoyne C~*) 1860 (OSNB), 1920 (OS).
♦ This row of 19th-cent. cottages, with schoolhouse, hidden away in GLENCOYNE, was built for the employees of GREEN SIDE lead mines; they are still also known as Glencoyne Cottages. For another Cumbrian Seldom Seen, see *PNCumb* 179, and two further examples are listed in *OSGazetteer*. The elements of the name both go back to OE, while its morphology, a phrase consisting of adverb plus past participle, is unusual but matched in other relatively modern and minor names, e.g. Once and Twice Brewed, Northumberland.

SELE BOTTOM SD1591 Ulpha.
From 1867 (OS).
♦ 'The hollow with a shieling' or 'the hollow where willow grows'. Without

early spellings 'Sele' remains elusive: possible origins include ON *sel* 'shieling hut' (especially if this name is connected with LITTLECELL BOTTOM, which shows a short *e*), and ON *selja* or dial. *seel* 'willow'. **Bottom** here denotes a remote, symmetrical basin or corrie of the sort that in the central fells might be called a **comb(e)**[1] or **cove**.

SELLA SD1992 (hab.) Dunnerdale.
Sellaye 1584, *Sellowe* 1624 (*PNLancs* 223).
♦ Probably 'the hill with a shieling'. (a) The 1st el. is likely to be *sel* 'shieling', rather rare in Lakeland compared with other shieling terms (see **ærgi/erg**). The site, on a moderate slope above the DUDDON, is a likely place for a shieling. (b) Given the lack of medieval evidence, however, other origins such as ON *selja* 'sallow, willow' cannot be ruled out. The spellings of the 2nd el. are characteristic of reflexes of ON **haugr** 'hill'.

SELSIDE: S~ BROW NY4809 Shap Rural/Longsleddale, S~ END NY4911 (1806ft/550m.) Shap Rural, S~ PIKE NY4911 (2142ft/653m.).
Selside 1443, *Selsheyd* 1530 (*PNWestm* II, 170), *Selside Anciently Selshead* c.1692 (Machell II, 101); (S~ Brow:) *the highte of Sailesyde, Salsyde* 1578, *Sailesyde browe* 1633 (*PNWestm* I, 163); *Selside End* 1858 (OSNB); *Selside Pike* 1828 (HodgsonM).
♦ Interpretation is uncertain: there is an embarrassment of possibilities, compounded by doubt as to whether the name S~ Brow, applied to a point somewhat remote from the others, and characterised by rather different early spellings, is of the same origin as the others. The 1st el. of S~ could be ON

sel 'shieling hut', surprisingly rare in Lakeland but possibly found in SELLA; ON *seila* 'hollow'; or ON *seyla* 'wet ground' (see **sail, sale**). The 2nd el. could be ON **síða**/OE **sīde** 'hill-side', OE **hēafod** 'head, high ground', or ON **sætr** 'hill-pasture'. See further *PNWestm* II, 170.

SELSIDE: TOP O' SELSIDE SD3091 Colton.
Sellesec 1227 (possibly in error for *Selleset*, Wyld 1911, 229); *Top o' Selside* 1851 (OS).
♦ Uncertain. If the 1227 spelling belongs here, and *-sec* is a miswriting of *-set*, the 1st el. may be ON *selja* 'willow' and the 2nd ON **sætr** 'shieling' (cf. early spellings of AMBLESIDE, ANNASIDE, HAWKSHEAD etc.). If not, the 1st el. could be guessed to be the rather rare ON *sel* 'shieling', though see SELSIDE above for further possibilities for both 1st and 2nd elements.

SERGEANT MAN NY2808 Langdales.
Sergeant Crag c.1692 (Machell II, 154), *Sergeant Man* 1860 (OSNB), which also quotes *Serjent Man* from *Black's Guide to the Lakes*.
♦ Probably 'the land-sergeant's cairn'. This is a natural peak, but **man** probably refers primarily to the cairn on its summit; the present one is high and well-built, and close to a parish and former county boundary. *Sergeant* in ME and ModE is a noun designating a range of middle-ranking military or non-military positions, and hence also an occupational surname. Collingwood commented on S~ M~: 'No doubt so called as a cairn placed by the Land-serjeant of Egremont' (1933, 343). Parker similarly commented that p.ns containing *sergeant* in the fells of the Gosforth area 'seem to refer to the mediæval

officer whose duty it was to see to the boundaries' (1926, 85).

SERGEANT'S CRAG NY2711 (1874ft/ 571m.) Borrowdale.
Serjeant-crag 1780 (West 96).
♦ The specific 1st el. could be a noun, as probably in SERGEANT MAN, or a pers.n. *PNCumb* 352 suggests that this may be associated with the family of William Sargyante, whose name is in the Crosthwaite PR for 1602. This is a true **crag** — a rugged height faced with a rock wall.

SETMURTHY (parish): S~ COMMON NY1631.
Satmurdac 1195 (*DEPN*), *Setmerdac* 1195, *Satmyrthac* c.1220 (p), *Satmerdoc* c.1240 (p) (*PNCumb* 433–4, q.v. for further forms); *Setmurthy Common* 1867 (OS).
♦ S~ is 'Murdoch's shieling or summer pasture', from ON **sætr** 'shieling, outlying pasture' and a Gaelic pers.n. *Muiredach* which Smyth describes as 'characteristically Scottish' (1975, 82 and 91, n. 30, citing Ekwall). This is a classic inversion compound, with the generic el. first and specific second. On the history of S~, which contained 'a scatter of farms without any nucleated focus', see Winchester 1987, 158–61. See also **common** 'unenclosed pasture for communal use'.

SETTERAH: S~ PARK NY5121 Askham.
Satrou 1339 (Winchester 1987, 105); *parcus de Saterhou* 1289, (*murus* [wall]) *parci de Saterhowe* c.1291 (CRO D/Lons/ L/Deeds, AS.5, AS.2, pers.comm. Dr Angus Winchester), *Satrowe park(e)* 1420, *Sattrepark* 1488 (*PNWestm* II, 201).
♦ S~ is apparently 'the hill with the shieling'; ON **sætr**, ON **haugr**; also

park, here a medieval deer-park (Winchester 1987, 105–6).

SHAP: SHAP BLUE QUARRY NY5610 Shap Rural, SHAP PINK QUARRY NY5508, SHAP FELLS NY5208.
Hep 12th cent. to 1293, *Yhep* 1241, *Hepp(e)* 1246 to 1468, *S(c)hap* 1279 to 1823 (*PNWestm* II, 164); *Shap Blue Rock* 1920 (OS); *Shap Granite Quarries* 1920 (OS); *Shapp Fells* 1770 (JefferysM).
♦ Shap means 'the heap (of stones)', probably referring to the many ancient stone monuments in the area. The initial sounds of OE *hēap* 'pile' have produced modern *Sh*- by processes of sound change and substitution (cf. SHOULTHWAITE). The word *quarry*, from Medieval Lat *quare(r)ia* via OFr *quarriere*, is first recorded in an English p.n. from 1166 (*MED*: *Quarere*), but is rare in p.ns except for recent and minor names. The quarries are distinguished as Pink, yielding granite containing pink felspar crystals, and Blue, yielding a bluish rhyolite (Postlethwaite 1913, 157). The large upland tract of Shap **Fell**s links the Lake District fells with the N Pennines.

SHARP EDGE NY3228 Mungrisdale.
From 1823 (Otley 30), 1867 (OS).
♦ 'The narrow, steep ridge'; *sharp* from OE *scearp*, **edge**. A famously exposed rock arête leading to the summit of BLENCATHRA, this used to be called Razor Edge (so Wainwright 1962, Blencathra 25). Coleridge, describing a walk over Blencathra, writes, 'I had quite forgotten the fearfully sublime Precipice & striding Edge on its farther or Northern Side' (CN 1519, 1803). By chance, the phrase *to scea[r]pan æcge* 'to Sharp Edge/to the sharp edge' is recorded in an Anglo-Saxon charter (Smith, *Elements* I, 145).

SHARP KNOTT NY1020 Lamplugh.
From 1867 (OS).
♦ A compact peak on the NW flank of BLAKE FELL; *sharp* as in SHARP EDGE, plus **knott**.

SHARROW: SH~ BAY NY4521 Barton, SH~ COTTAGES NY4521.
Sharow 1678, *Sharrow* 1731 (*PNWestm* II, 213); *Sharrowbay* (hab.) 1863, *Sharrow Bay* (bay) 1900; *Sharrow Cottages* 1900.
♦ Sh~ is probably 'the boundary hill', with the reflexes of OE *scearu* 'boundary, share' and ON **haugr** 'hill'; the place is near the boundary of Barton with Martindale; the **Bay** is close by, an inlet in ULLSWATER. Sh~ **Cottages** are on the site named Sharrow on the 1863 map.

SHATTON: SH~ HALL NY1428 Embleton, SH~ LODGE NY1428.
S(c)haton 1322 to 1455 (*PNCumb* 384); *Shatton Lodge* 1867 (OS).
♦ Sh~ is possibly 'the farmstead in the corner of land', from OE *sceat* 'corner of land' and OE **tūn** 'farmstead, settlement'; **hall** and **lodge** added later. Sh~ Hall is simply *Shatton* on OS maps of 1867, 1900 and 1926.

SHAW BANK NY3021 (hab.) St John's.
Shabanke 1702, *Shaw Bank* 1739 (*PNCumb* 317).
♦ 'The slope with/by the wood'; **shaw** (see **skógr**), **bank**. The site is not significantly wooded now.

SHEEPBONE BUTTRESS NY1814 Buttermere.
♦ The first ascent of this renowned crag was in 1912 (Peascod & Rushworth 1949, 84), and presumably it was named then. Unique among the p.n. els in this volume, *buttress* (an adoption from Fr into ME) is a favoured term for mountain features and the climbing routes up them.

SHEFFIELD PIKE NY3618 Patterdale.
From 1859 (*PNWestm* II, 226), possibly = *Stone-cross-pike* 1780 (West 151).
♦ **Pike** is a pointed hill. The name may, like NORFOLK ISLAND and LYULPH'S TOWER, reflect the influence of the Dukes of Norfolk, who also had estates in Sheffield, though I do not know of direct evidence for this explanation.

SHELTER CRAGS NY2505 Langdales.
From 1867 (OS).
♦ The reference is perhaps to the shelter afforded by the small corrie which lies beneath these **crag**s or rocky heights. The word *shelter* is of uncertain origin, first appearing in the 16th cent.

SHERRY GILL NY5309 Shap Rural.
From 1860 (OSNB).
♦ *OED*'s first citation for the fortified wine *sherry* is from 1608, but it is not at all clear whether this is the 1st el. in the p.n. The 2nd el. is **gil(l)**.

SHIPMAN KNOTTS NY4706 (1926ft/587m.) Longsleddale.
From 1865; cf. *Sheipmantop(p)e* 1578 (*PNWestm* I, 163).
♦ 'Shepherd's rocky hills', with ME *shepman* 'shepherd' (unique in the names in this volume), or the surname derived from it, plus **knott**.

SHIVERY KNOTT NY2815 Borrowdale/St John's.
From 1805 (*PNCumb* 353).
♦ 'The stony compact hill'; *shivery*, **knott**. *Shivery* refers to loose flakes or shards of stone, as illustrated by Hutchinson's description of the mountains flanking WHINLATTER PASS:

most are grassy, while 'others are barren, bleak and shivery, sending down continued streams of sand, slates and stones, with every shower of rain' (1794, II, 121, cf. *EDD*: *shiver*, sense 4). Although Sh~ Kn~ does not match this description, it is a small, distinct island of rock amidst rough grass and bog.

SHORT STILE: see LONG STILE

SHOULTHWAITE FARM NY2920 St John's, SH~ GILL NY2919.
Heolthwaitis c.1280, *Shewltath* 1564, *Shoultwhait* 1567 (*PNCumb* 314).
♦ Sh~ is probably 'the clearing with a wheel' from ON *hjól*, perhaps referring to a mill-wheel driven by the NADDLE BECK (so *PNCumb* 314, also *SSNNW* 160), plus ON **þveit** 'clearing'; **farm** has been added later. The substitution of *sh-* for *hj-* (*j* pronounced like the 1st sound in modern *yes*) is paralleled in *Shetland*, from ON *Hjaltland*, and is comparable with the phenomenon seen in the name SHAP. The **gill** 'stream in a ravine' runs N past the farm.

SHUNDRAW NY3023 (hab.) St John's.
Shonderhowe, Shunderhowe 1571 (*PNCumb* 314).
♦ Possibly 'the look-out hill'. The 1st el. may be from *sjónar*, gen. sing. of ON *sjón* 'sight, view'(so *PNCumb*), and the 2nd is clearly **how(e)** from ON **haugr**, presumably referring to LOW RIGG, a viewpoint just above Shundraw. The 1571 spellings do not absolutely favour derivation from *sjónar*, however, inasmuch as they suggest that the *-d-* glide entered the pronunciation while *-er* from *-ar* was still pronounced with the vowel, whereas it would be more likely once the vowel had been lost, bringing *-nr-* together, since *nr* to *ndr* is a common development.

SIDE NY0609 (hab.) St Bridget Beckermet, S~ END NY0608.
Syde 1538 (*PNCumb* 340); *Side End* 1867 (OS).
♦ 'The hill-slope'(**síða, síde**) and its **end**. The place is on the NE slopes of COLD FELL. This was previously *Symonkeld Grange* (Winchester 1987, 155); cf. SIMON KELL.

SIDE, THE NY1113 Ennerdale.
From 1787 (PR), previously *Ennerdale Parke al[ia]s the ffence, fenced partlie with an old wall, and partlie with ye water called ye Broadwater* 1612 (*CW* 1931, 164).
♦ 'The hill-slope', with the reflex of OE **síde** and/or ON **síða**, referring to the long, steep contours rising from the S shore of ENNERDALE WATER. Since PRs normally record places of abode, the appearance of the name of a hill-side in the PR for 1787 is unusual. It occurs within a record of the burial of a destitute woman who died there in two feet of snow.

SIDE HOUSE NY2906 Langdales, S~ PIKE NY2905.
Sydhowse 1574, *Sidehouse* 1671; *Side Pike* 1823 (*PNWestm* I, 207).
SIDE HOUSE SD3187 Colton.
Sidehouse 1772 (PR), *Side House* 1851 (OS).
SIDE HOUSE SD4898 Strickland Roger.
From 1706 (*PNWestm* I, 159).
♦ (a) Almost certainly 'the house by the hill-side'; **side, house**. All three places are on or below significant slopes and S~ **Pike** towers above one. (b) A dial. adj. *side* meaning 'wide' or 'long' cannot be ruled out, but is much less likely.

SILECROFT SD1281 Whicham.
Selecrotf [sic] c.1205, *Selecroft* 1211 (p) to 1358 (*PNCumb* 444–5).

♦ Perhaps 'the enclosure with willows'. The 1st may be either 'willow' (ON *selja*, so *DEPN* and *SSNNW* 249) or, less likely 'hall' (OE *sele* 'hall, house', *PNCumb* 444); the 2nd el. is **croft** 'small enclosure, farm'.

SILLATHWAITE NY0512 (hab.) Ennerdale.
Sillithwate or Sirrithwate 1578 (*PNCumb* 387).
♦ Possibly 'Sigríð's clearing', from the ON female pers.n. *Sigríðr* and ON *þveit* 'clearing'(so *PNCumb*).

SILVER BIRCH SD4086 (hab.) Upper Allithwaite.
♦ Neither p.n. nor building is shown on the OS maps of 1851, 1892 and 1919, and the name appears to be a relatively recent application of the tree-name, using English *birch* rather than dial. **birk** (q.v.); see also **silver**.

SILVER COVE NY1311 Ennerdale, SILVERCOVE BECK NY1312.
Silfoucon, Silfhucone 1338, *Silver Cowe (or Coue)* 1578 (*PNCumb* 386); *Silvercove Beck* 1867 (OS).
♦ **Silver** was among the metals mined, on a modest scale, in Ennerdale (Littledale 1931, 189), yet the puzzling 14th-cent. spellings suggest that the later 'Silver' may be a reinterpretation of a name that is now unrecoverable. One possibility might be suggested by similar spellings for Silpho, NYorks: *Silfhou, Silfhow* 1155–65 to early 14th cent., *Silfho* 1231, *Silfow(e)* 1301 to 1577. These clearly point to **how** from ON **haugr** 'hill, mound' as 2nd el., and the 1st el. is believed to be 'the ODan pers.name *Sylve*' (*PNNYorks* 115; I am grateful to Prof. Richard Coates for drawing my attention to these spellings). Whether the name originally contained **cove** is similarly in doubt, though there are other Cove names nearby, and it is a true description of this high, rocky bowl at the head of Silvercove **Beck** 'stream'.

SILVER CRAG NY3918 Patterdale, S~ POINT NY3918.
Silvery Crag; Silvery Point both 1920 (OS).
♦ The **Crag** 'rocky height' and **Point** are close to Silver Bay in ULLSWATER. The significance of **silver(y)** here is obscure, but could refer to the mineral riches of the area, which included silver (Hay & Hay 1978, 12, 29, 67). Coleridge's 'how shall I express the Banks waters all fused Silver' describes this part of Ullswater in November sun (*CN* 549, 1799), but this is presumably coincidental.

SILVER HILL NY2522 (hab.) Above Derwent.
From 1797 (*PNCumb* 373).
♦ This is a classic example of a p.n. whose elements (**silver, hill**) seems clear enough, while the motivation for the name is uncertain. The situation, a small rise next to DERWENTWATER and close to Copperheap Hill and C~ Bay, might suggest that *silver* here has literal reference, perhaps connected with transportation of metal ore from the NEWLANDS mines to KESWICK.

SILVER HOLME SD3790 Windermere.
Silver holm 1783 (CrosthwaiteM).
♦ 'Silver island'; **silver, holm(e)**. Brydson reports 'an indistinct tradition that a hoard of silver is buried on the island' (1911, 103). Such traditions are not always wrong: Viking-Age silver brooches have been found in Silver Field near Penrith (Richardson 1996).

SILVER HOW SD1998 Eskdale.
From 1867 (OS).

SILVER HOW NY3206 (1292ft/394m.) Langdales/Grasmere.
From 1770 (JefferysM), 1800 (*CN* 769).
♦ 'Silver hill'; **silver**; **how(e)** from **haugr**. The motivation for the name is not obvious in either case. The hill at NY3206 is distinguished by attractive waterfalls and a band of junipers, the underside of whose leaves is grey-green — silvery?

SILVERYBIELD CRAG NY2204 Eskdale.
From 1867 (OS).
♦ 'The rocky height by S~, the silver-coloured shelter'; **silvery**, **bield**, **crag**. Why *silvery* is unclear.

SIMON KELL NY0610 (hab.) Ennerdale.
Simonkeld(e) 1279 (p), *Simundekelde* 1292 (*PNCumb* 386).
♦ 'Sigmund's spring', from an ON pers.n. *Sigmundr* plus ON **kelda** 'spring'; a spring is shown just to the S on larger scale maps. There is a cognate OE pers.n. *Sigemund*, but the ON name is more likely with an ON generic. (There is also an OE *celde* 'spring', but it is rare.) See also SIDE NY0609.

SIMPSON GROUND SD3986 Staveley-in-Cartmel, S~ G~ RESERVOIR SD3986.
Sympson Ground 1656 (Cartmel PR); *Simpson Ground Moss* 1851, 1892, 1919 (OS).
♦ 'The farm of the S~ family'; surname plus **ground**. *Simpson* is a patronymic surname derived from *Sim(m)*, a pet form of *Simon* common especially in Scotland and N. England. The **reservoir** was created out of a **moss** or bog.

SINEN GILL NY3028 Threlkeld.
Synin Gill (Otley 1823, 98), *Sinen Gill* 1867 (OS).
♦ This could mean 'the sining or draining stream' — one which seeps sluggishly across wet ground and is apt to soak through and run dry. For the 1st el., see *PNCumb* 35 and *EDD*: *sine* v., senses 2 and 4, and cf. SINEY TARN, also **gil(l)**. The stream rises on boggy ground and lacks the numerous feeder streams of noisy ROUGHTON GILL which runs parallel with it, though given the steepish gradient it is scarcely sluggish.

SINEY TARN NY1601 Eskdale.
the Sining Tarne 1587 (*PNCumb* 35), *Siney Tarn* 1867 (OS).
♦ 'The sining or draining pool' — one liable to dry up and become part of the generally wet, flat area of Sineytarn Moss; see SINEN GILL for the 1st el., plus **tjǫrn/tarn**.

SKEGGLES WATER NY4703 Kentmere/Longsleddale.
(*pool of*) *Skakeleswatre* 1375, *Skek(el)leswater* 1570, *Skeggleswater* 1777 (*PNWestm* I, 17).
♦ 'Skǫkul's lake'; pers.n. plus **water**. Like other lake-names, such as WINDERMERE and ULLSWATER, this seems to incorporate an early pers.n., here ON *Skǫkull*, originally a nickname referring to shaking or brandishing, and occurring in ME as *Scakel* (*PNWestm* I, xiv).

SKELGHYLL: HIGH S~ NY3902 (hab.) Ambleside, S~ WOOD NY3803.
Scalgill 1560; *High Skelgill* 1717 (*PNWestm* I, 184); *Skelgill Wood* 1863 (OS).
♦ S~ is either (a) 'the ravine/stream by the shieling' (**skáli/scale** 'shieling', **gil(l)** 'ravine with stream'), or possibly (b) 'the roaring stream', if the 1st el. is a stream-name from ON *skjallr* 'resounding, roaring', cf. SKELWITH NY3403; also **high**, **wood**. The stream

on which High and Low S~ stand is now called Hol Beck, but may previously have been Scale Gill or Skelgill. S~ Wood lies a little to the W.

SKELGILL NY2420 (hab.) Above Derwent.
Scalegayl 1260, *Scalgill in Derewentfell* 1334 (*PNCumb* 371).
♦ 'The ravine by the shieling', from ON **skáli** 'shieling' and ON *geil* 'ravine or narrow track', later influenced by the more common **gil(l)** 'ravine with stream'. This, with nearby KESKADALE and GUTHERSCALE, forms a row of shieling sites which became permanent settlements. (Pl. 12)

SKELLY NEB NY4320 Matterdale.
Skelling-nab 1780 (West 150), *Skelly Neb* 1789 (*PNCumb* 258).
♦ Probably 'the promontory where skelly-fish abound'. (a) The intriguing 'Skelly' may be a reference to *skelly/ schelly* (*Coregonus laveratus*), a fish found in the lake. 'Every few decades a large number of these fish are cast up onto the beach, though the number in the lake at present seems to be decreasing' (Keith Clark, pers.comm. gratefully acknowledged). (b) Alternatively, the 1780 spelling could point to the same origin as Skelling (*Scaling*' 1292, *PNCumb* 243), in ON **skáli** 'shieling, hut' and ON *eng* 'meadow', but the lakeside situation makes this uncertain. The 2nd el. 'Neb' here refers to one of the promontories in ULLSWATER, of which Hutchinson says, 'the local name of these spits is *nebs*, a word denoting the bill of a bird' (1794, I, 434 and n.). The word may be a topographical application of OE *neb* 'beak, nose', hence in ModE anything pointed (see *OED*: *neb*, sense 5a), but is probably also blended with **nab**.

SKELWITH: S~ BRIDGE NY3403 (hab.) Skelwith/Rydal, S~ FOLD NY3502 (hab.) Skelwith, S~ FORCE NY3403 Skelwith/ Rydal.
Schelwath 1246, *Skelwath* 1332 (p), *Skelwyth* 1537 (*PNLancs* 219); *Skelleth Bridge* 1651 (*PNWestm* I, 211), *Skelwaith bridge* 1688 (Grasmere PR), *Skelwith bridge* 1693 (*PNWestm* I, 211); *Skellathfold in the Parish of Hawkshead* 1778 (Underbarrow PR); *Skelwith Force* 1829 (Parson & White 616).
♦ 'The noisy ford.' (a) Ekwall convincingly proposed ON *Skjallr* 'loud, resounding' as the 1st el., referring to the roaring cascade of S~ **Force** which would have guided travellers to the ford (*PNLancs* 219; Smith in *PNWestm* and Fellows-Jensen, *SSNNW* 93, 160 agree). (b) Ekwall's alternative suggestion of ON *skjól* 'hut, refuge' as 1st el. seems less likely. The 2nd el. is ON **vað** 'ford', denoting a fording place on the R. BRATHAY later furnished with a **bridge** (see **brycg**). It seems to have been influenced by the reflex of ON **viðr** 'wood' as probably in nearby COLWITH. See also **fold** 'pen or farm'.

SKELWITH POOL SD3481 Haverthwaite/ Lower Allithwaite.
From 1851 (OS).
♦ S~ is possibly 'the wood by the shieling'. The absence of early spellings makes guesses hazardous, and the very different topography discourages comparison with SKELWITH NY3403. In this case the 1st el. could be from ON **skáli** 'shieling', and the 2nd from ON **viðr** 'wood' or else **vað** 'ford', referring to a crossing over S~ **Pool**, a slow-moving river.

SKEWS NY5018 (hab.) Bampton.
(*Banton*) *Skewes* 1564, *Skewes* 1626 (*PNWestm* II, 196).

♦ 'The woods', with the reflex of ON **skógr** 'wood, forest' and the ME pl. -(*e*)*s*; cf. nearby SCEUGH. The place is only lightly wooded now.

SKIDDAW NY2629 (3054ft/931m.) Bassenthwaite/Underskiddaw, S~ FOREST NY2729 Underskiddaw, S~ HOUSE NY2829 Bassenthwaite/ Underskiddaw.
Skydehow, Schydehow 1247, *Skythou* c.1260; *forest of Skithoc* 1230 (*PNCumb* 320, q.v. for further spellings); *Skiddaw House* 1867 (OS).
♦ Perhaps 'the mountain with the jutting crag'. S~ is probably of Scand origin, with 2nd el. ON **haugr** 'hill', and 1st el. one of: (a) A postulated word related to ON *skúti* 'projecting crag' or *skúta* 'to jut' (Fritzner, and *NG* I, 60, V, 227), though crags are relatively few on this shapely grass-covered slate mountain; (b) *skyti* 'archer', though the reason for such a reference would now be totally lost; (c) a pers.n., though ON *haugr* is normally accompanied by descriptive words rather than pers.ns; (d) ON *skítr* 'dung, filth, shit'. Whichever of the existing explanations is favoured, the variation in the early spellings between medial *d* and *t* remains problematic, and it is possible that a Cumbric solution is to be sought (Prof. Richard Coates pers.comm.). As for the secondary names, S~ **Forest** on the modern map is a useful reminder of the institution of the medieval forest: an enclosed hunting ground, which was not necessarily wooded. S~ **House** is by far the most remote instance of *house* in this volume. Situated on a major route across the S~ massif, it was a row of cottages for the game-keepers and shepherds of the Lonsdale estate (Ramshaw & Adams 1993, 36), and subsequently a Youth Hostel (illustrated in Wainwright 1962, Skiddaw 6). It is tempting to wonder whether the name is based on an original Skiddaw **Hause**. (Pl. 15)

SKINNER PASTURES SD3487 Colton.
From 1851 (OS).
♦ S~ is presumably a surname derived from the occupational term (which latter occurs, for instance, in Hawkshead PR for 1815); and see **pasture**.

SKIRSGILL: S~ HILL NY4923 Askham.
Scirsgil 1773 (Lowther PR); *Skirsgill Hill* 1859 (*PNWestm* II, 202).
♦ (a) This may be 'the haunted stream/ravine', from ON *skyrsi* 'phantom, monster', plus **gil(l)**, as suggested in *PNWestm*. This is a Romano-British settlement site (LDHER 2958, SAM No. 22517), and the name may reflect the widespread belief that such sites are haunted by supernatural beings. (b) ON *skyr*, referring to a type of curds like thick yoghurt, seems to occur in some Norw place- and river-names (*NG* III, 227; V, 171; XV, 132), and together with **gil(l)** would parallel SOUR MILK GILL, but the medial -*s*- would be unusual.

SLAPE SCAR SD3897 Claife.
From 1851 (OS).
♦ Probably 'slippery scarp', with dial. *slape* 'slippery, smooth' from ON *sleipr*, plus **scar**. This is a headland on WINDERMERE.

SLAPESTONE EDGE NY3008 Grasmere.
From 1847 (*PNWestm* I, 201).
♦ 'The scarp with slippery stones' or 'scarp at Slapestone', with *slape* as in SLAPE SCAR, **stone** and **edge**; cf. *le Slapestones* 1292 (*PNWestm* II, 286). This is the band of crags W of Easedale Tarn.

SLATE HILL SD4283 (hab.) Witherslack. From 1857 (*PNWestm* I, 79).
♦ This is the only p.n. in this volume containing the word *slate*, which first appears in the ME period. The **hill** is located on a small patch of slate. It is next to Fern Hill and, like it, does not have a particularly distinct summit.

SLEATHWAITE NY1200 (hab.) Irton.
Slapthwaite 1690, *Slatwhaite* 1774 (*PNCumb* 404), *Sleathwaite* 1774 (DHM).
♦ Probably 'the clearing where blackthorn grows', with reflexes of OE *slā(h)* 'sloe' and ON **þveit** 'clearing'. The sloe is the fruit of the blackthorn, *Prunus spinosa*, common nowadays in the hedgerows of lowland Cumbria (Halliday 1997, 280). The *-p-* spelling in 1690 is, however, puzzling.

SLEDBANK SD1282 (hab.) Whicham.
Sledbanke 1654 (*PNCumb* 445–6).
♦ Probably 'the slope beside the valley' (*slæd* as in SLEDDALE (q.v.), **bank(e)**); it is in a side-valley of WHICHAM VALLEY.

SLEDDALE: LONGSLEDDALE NY4903 (parish), S~ FELL NY4806, S~ FOREST NY4802.
Sleddal(e), *Sleddall* 1229 to 1622, *Sledale*, *Sledall* 1377 to 1652; *Langsleiddall* 1466, *Long Sleddal(e)* 1518–29 to 1777 (*PNWestm* I, 160); *Sleddale Fell* 1857 (*PNWestm* I, 163); *Sleddale Forest* 1770 (JefferysM).
♦ S~ is either 'the valley with a side-valley' or 'the valley called Slæd'. Both elements of Sleddal(e) mean 'valley': OE *slæd/slǣd*, rare in Lakeland, as the 1st el., and the ubiquitous ON **dalr** (possibly here encouraged by OE *dæl*) as the 2nd. *Slæd* in OE charters seems to designate 'flat-bottomed, especially wet-bottomed valleys' (Gelling & Cole 2000, 141, acknowledging Kitson forthcoming). They suggest that a common factor in its varied p.n. usage is that it 'probably never [applies to] main valleys', and propose that the *slæd* in Longsleddale refers to a short side-valley (141). The other explanation for the duplication of 'valley' terms would be that *Slæd* was the original name, so that Sleddale effectively means 'valley called Slæd' (Fellows-Jensen, *SSNNW* 250). Before being prefixed, aptly, with **long** (see **langr**) the valley was distinguished from WET S~ by the suffixes *Brunolf* (13th–14th cent.) or *Conyers* (14th cent.), referring to feudal owners. In recent times the valley has received fictional names: Long Whindale in Mrs Humphrey Ward's *Robert Elsmere* (1888), and Greendale in John Cunliffe's *Postman Pat* books (Lindop 1993, 30). S~ **Fell** is a tract of high ground at the head of the valley. S~ **Forest** is presumably an ancient hunting area rather than woodland; it is now a treeless expanse rising to over 1312ft/400m.

SLEDDALE: S~ BECK NY5109 Shap Rural, S~ GRANGE NY5411, S~ HALL NY5311, S~ PIKE NY5309 (1659ft/506m.), WET S~ RESERVOIR NY5411.
Sleddal(e) & variants 12th cent. to 1692 (*PNWestm* II, 170); *aquam de* (river of) *Sleddal* 1249, *Sledalebeck* 1279; *Sleddale Grange* 1612 (*PNWestm* II, 177); *Sleddale Hall* 1770 (JeffreysM), *Sleddale Pike* 1860 (OSNB); *Weat Sleddale* 1562 (*PNWestm* II, 170), *Wetsleddale* c.1692 (Machell II, 490).
♦ This S~ is of the same origin as (LONG)SLEDDALE, but distinguished from it by the prefix Wet- (from OE *wēt*/ON *vátr* 'wet'). 'It has not its distinctive appellation without reason, for it is said, "if any rain is stirring, the air scoops it surprisingly into the

hollow of the dale".' (Parson & White 1829, 603). The secondary names are formed from **bekkr/beck** 'stream', **grange** (an outlier of SHAP Abbey), **hall**, **pike** 'peak' and **reservoir**.

SLEET FELL NY4218 (1179ft/378m.) Martindale.
From 1859 (*PNWestm* II, 219).
♦ The meaning of 'Sleet' may be beyond recovery, given the lack of early spellings and the fact that the most obvious etymon of 'Sleet', ON *slétta* 'flat meadow or moor' (cf. SLIGHT SIDE) or the related adj. *sléttr*, 'level, smooth', does not match the landscape here or in SLEET HOW. *Sleet* in the sense of 'thawing snow' would be unusual at best. The 2nd el. is straightforwardly **fell** 'mountain'.

SLEET HOW NY2022 Above Derwent.
From 1867 (OS).
♦ As in SLEET FELL, the 1st el. is obscure. The 2nd el. is **how(e)** 'hill' from ON **haugr**, though as a spur of GRISEDALE PIKE, this does not have the distinct, rounded outline of a classic *haugr/how(e)*.

SLIGHT SIDE NY2005 (2499ft/748m.) Eskdale.
From 1578 (*PNCumb* 392).
♦ (a) Possibly 'the hill-slope above level ground'. (a) The 1st el. may be from ON *slétta* 'flat meadow or moor', referring to QUAGRIGG MOSS, a pocket of relatively level ground in these craggy parts; cf. names such as Sleightholme, also believed to be from *slétta* (*Slightholme* 1640, *PNCumb* 292). (b) Alternatively, it could be an adj. *slight* meaning 'smooth, slippery' (*EDD*, locating this usage in Scotland and Yorkshire). The 2nd el. is most likely from OE **sīde**/ON **síða** 'slope, hill-side', though this el. can be confused with ON **sætr** 'shieling' or OE **hēafod** 'high place'. This is a compact peak, and an excellent vantage point, at the end of the SCAFELL range.

SMAITHWAITE NY3119 (hab.) St John's.
Smathwaitis c.1280 (*PNCumb* 314).
♦ Probably 'the small clearing'; ON *smár* 'small', **þveit**.

SMALLSTONE BECK SD1898 Eskdale.
From 1867 (OS).
♦ Presumably 'the gravelly stream'; *small* from OE *smæl*, perhaps reinforced by ON *smár*, plus **stone** and **beck**.

SMALL WATER NY4509 Shap Rural.
From 1828 (HodgsonM).
♦ 'The small lake'; *small* as above, plus **water**. This lake is indeed modest compared with most others named *water*, including BLEA WATER just to the NW.

SMITHY BECK NY1214 Ennerdale.
From 1774 (DHM).
♦ This is a scheduled bloomery site, where iron ore from the haematite levels in CLEWS GILL was processed. A nearby *Sinderhill*, recorded in 1560, shows that it was already active by then (Fletcher & Fell 1987, 28). The word *smithy* (cf. OE *smiðða*) could refer to a place where metal is refined, as well as forged into implements; see also **beck** 'stream'.

SMITHY FELL NY1323 Loweswater.
From 1867 (OS).
♦ The **fell** is a minor summit on the FELLBARROW range. A *smithy* (cf. OE *smiðða*) site nearby is not unlikely, given the proximity of a drove-road (Wainwright 1966, Fellbarrow 6). Nearby Galloway Farm could also allude to cattle droving.

SNAB NY3031 Caldbeck.
Snabb 1719 (*PNCumb* 279).
SNAB: HIGH S~ NY2219 (hab.) Above Derwent, HIGH S~ BANK NY2118, LOW S~ NY2218 (hab.).
(*le*) *Snabb* 1503; *Hyesnabb* 1589 (*PNCumb* 373); *High Snab Bank* 1867 (OS); *Lawsnabb* 1573 (*PNCumb* 373).
♦ 'The projecting hill.' *Snab* in N. and Scots dial. means a steep, snout-like hill or projection from a hill. It is from ME *snabb*(*e*) (as in a 13th-cent. example, *PNWestm* II, 287) and is of obscure origin (see *PNCumb* 491, *OED*: *snab*, and cf. Norw *snav* 'spit of land'). The Caldbeck S~ is a snout-like ridge with PIKE at its summit. In Above Derwent, the settlements of **High** and **Low** S~ stand below the points of twin ridges projecting into the NEWLANDS valley, either side of SCOPE BECK, and H~ S~ **Bank** is the name of the higher ridge or *snab* itself.

SNARKER PIKE NY3807 (2096ft/639m.) Ambleside.
Snake Pike 1764 (*CW* 1906, 15), *Snarker Pike* 1863 (OS); cf. *Snaka Mosse* 1721, *Snake how moss* 1764 (*PNWestm* I, 184).
♦ S~ may have meant 'snake-hill'. It has somewhat sinuous contours, and may originally have been Snake How (OE *snaca* plus **haugr/how**(**e**)), later contracted to Snarker. The -*ar*- spelling is odd, but the identification of *Snake Pike* and *Snarker Pike* is secure. The **pike** is a steep subsidiary summit on RED SCREES.

SNECKYEAT NY1227 (hab.) Blindbothel.
Sneck gate 1774 (DHM), 1867 (OS).
♦ Seemingly 'the gate with a latch', most likely to control livestock movement. *Sneck* is a common word for 'latch' in N. dial., and *yeat* a regular outcome of OE **geat** 'gate'. For other Cumbrian examples of Sneckyeat or

Sneckgate, see *PNCumb* 379.

SNITTLEGARTH NY2137 (hab.) Bewaldeth.
Smyttlegarth [sic] 1580, *Snittlegarth* 1608 (*PNCumb* 265).
♦ The 1st el. may be dial. *snittle* 'noose, snare' (as suggested in *PNCumb*). The 2nd el. is **garth** 'enclosure'.

SORROWSTONES NY0902 (hab.) Irton.
Sorrowstone 1647 (*PNCumb* 404).
♦ 'Sorrow', if not a corruption, is from OE *sorh*, and see **stone**. Local tradition knows Sorrowstones as the place where those taken for execution were given a last drink (Parker 1926, 49). The place lies below Hanging How, which is not steep enough to mean 'overhanging', so there may be some connection between the two.

SOSGILL NY1023 (hab.) Loweswater.
Solrescales c.1203, *Saurescoles* 1230, *Soskell* 1609 (*PNCumb* 411).
♦ 'The muddy shieling(s).' The 1st el. is ON *saurr* 'mud, dirt, wet land' as in SAWREY (if it is correct to discount the -*l*- in the spelling of c.1203). The 2nd el. is ON **skáli** 'shieling, shieling hut', with or without ME pl. -(*e*)*s*. As often, the reflex of *skáli* has been reinterpreted as **gill** 'ravine with stream', with -*s*- now belonging to the previous syllable.

SOULAND GATE/SOURLANDGATE NY4625 (hab.) Dacre.
Sourlandgate 1867 (OS); cf. *Le Sower Lands* 1568 (*PNCumb* 189).
♦ 'The gate or road to Sourland, the infertile lands.' S~ is 'an early example of the phrase "sour lands"' (*PNCumb*), referring to poor, ungracious soil. The elements derive ultimately from OE *sūr* and/or ON *súrr*, both 'sour', plus

land. The recent form 'Souland' may be influenced by nearby SOULBY. 'Gate' could mean 'gate' from OE **geat**, or 'road' or 'pasture' from ON **gata**.

SOULBY NY4625 Dacre, S~ FELL NY4524 (904ft/276m.), S~ FELL FARM NY4525.
Suleby 1235 (p), *Souleby* 1279 to 1485, *Soulby* 1323; *Soulby Fell* 1487 (*PNCumb* 188).
◆ Probably 'the settlement at the fork or post', from ON *súla* 'post, fork' and ON **bœr, bý**. *PNCumb* proposed an ON pers.n. *Sūle* as 1st el. here (as did Ekwall for SUBBERTHWAITE, *PNLancs* 214). However, there is no direct evidence for this (it is not listed, e.g., in Fellows-Jensen 1968 or Insley 1994), and Fellows-Jensen is doubtless right that the three Soulbys in NW England are better explained as containing the appellative *súla* 'post, fork' (*SSNNW* 40–1), though it is not obvious what this would mean here. S~ **Fell** is not a distinct summit, but the high ground W of the hamlet; see also **farm**.

SOURFOOT FELL NY1323 Loweswater.
From 1867 (OS).
◆ The 1st el. could refer to muddy ground (ON *saurr*, as in nearby SOSGILL, or the word *sour* as in SOU(R)LAND), and to describe the place as the **foot** of this would be apt, since 'this is where the gradient levels out and a spring line creates wet boggy land'. I am very grateful to Mr Danny Leck for this information; he adds that '"sou" locally refers to such wet boggy land'. The **Fell** is a distinct, if minor, peak.

SOUR HOWES NY4203 (1568ft/478m.) Windermere.
From 1859 (*PNWestm* I, 197).
◆ 'The poor, boggy hills.' *Sour*, from OE *sūr* and/or ON *súrr*, often refers to 'poor, wet land' (cf. SOULAND), though derivation from ON *saurr* 'mud, dirt, wet ground' as in SAWREY is not impossible. **Howes** (from **haugr**) here denotes a bumpy hill-top.

SOUR MILK GILL Examples at:
NY1615 Buttermere.
From 1867 (OS); cf. *Sour-milk-force* 1786 (Gilpin I, 225).
NY2212 Borrowdale.
South Milk Gill 1759 (*CW* 1976, 106), *Sourmilk Gill* 1867 (OS); cf. *Sour Milk Force* 1799 (*CN* 541).
NY3108 Grasmere.
Sour Milk Gill 1828 (HodgsonM), *Sourmilk Gill* 1863 (OS); cf. *Churnmilk Force* 1801 (*CN* 516, Dorothy Wordsworth, *Journals*, 15 Nov.), *Sour Milk Force* 1803 (*CN* 1782).
◆ 'The foaming, white stream.' *Sour* is from OE *sūr* and/or ON *súrr*, and *milk* from OE *milc, me(o)luc*; **gill**, 'ravine with stream' but here especially referring to the streams, is of ON origin. All three gills are on a steep gradient, with cascades, and Gilpin's comment on the reason for the Buttermere name applies to them all: 'The people of the country, alluding to the whiteness of it's foam, call it *Sour-milk-force*' (1786, I, 225).

SOUTERSTEAD SD2793 Torver.
Sowtersted 1591 (*CW* 1958, 66), *Sowtestead* 1614, *Souterstead* 1684 (PR).
◆ (a) Probably 'the Souters' place or farm'; surname plus **stead**. *Stead* most commonly co-occurs with surnames, and *Souter* is a well-known surname (see Hanks & Hodges 1988: *Sauter*), though I have not noted it in the PRs. (b) The surname originates in the term for a shoemaker or cobbler, as probably in SOUTHER FELL, and this interpretation cannot be ruled out here.

SOUTH CRAG NY2510 Borrowdale. From 1867 (OS).
♦ Probably so named since this is the southernmost of a row of **crags** which clad the W side of LANGSTRATH. The Standard English *south* derives from OE *sūð*, though ON *sunnr/suðr* would also have been current in the Lake District for a time.

SOUTHER FELL NY3528 (hill) Mungrisdale, SOUTHERFELL NY3527 (hab.).
Souterfel 1323, *Souterfell in Grysedale* 1340, *Souther Fell* 1783 (*PNCumb* 227).
♦ 'Cobbler's hill', seemingly from OE *sūtere* 'cobbler, shoemaker' and **fell**, the early forms being continued in the modern pronunciation with [t] (e.g. Wainwright 1962, Souther Fell 1).

SOUTHERNDALE NY2429 Bassenthwaite.
Southern Dale 1867 (OS).
♦ 'The southern valley.' The 1st el. would be the reflex of ON *súðrænn* or OE *sūðerne*, perhaps referring to the situation in the S of Bassenthwaite parish. The **dale** is a secluded recess in the hills.

SOUTHFIELD SD1280 (hab.) Whicham.
Southfeyle 1570, *the Sowth-feild* 1590 (*PNCumb* 446).
♦ The place is at the S edge of Whicham parish, which may explain the name; *south* as in SOUTH CRAG plus **field**.

SOUTH LAKELAND DISTRICT SD4088 several parishes.
♦ The term Lakeland was first used in the 19th cent. (according to *OED*, with first citation from Southey, 1829); *south* as in SOUTH CRAG plus *district* as also in ALLERDALE DISTRICT.

SOUTHWAITE NY1228 (hab.) Embleton.
Southart (*Bridge*) 1774 (DHM).
SOUTHWAITE FARM NY4425 Dacre.
Southwaite c.1545 (*PNCumb* 189).
♦ Either 'sheep clearing' or 'south clearing'. The 1st el. may be either (a) ON *sauðr* 'sheep, ewe', as possibly in another Southwaite (*PNCumb* 379); or (b) ON *suðr*/OE, ME *sūð* 'south'. The Embleton place lies one mile S of COCKERMOUTH, and there is a parallel for a directional term coupled with *thwaite* in EASTHWAITE and possibly ESTHWAITE. The 2nd el. is from **þveit/ thwaite** 'clearing' (assuming that the spelling of the Embleton example is one of the many errors on Donald and Hodskinson's map of 1774). Higham sees the Dacre example as among the later names in *thwaite* which reflect population expansion and colonisation in the Inglewood Forest area (1986b, 94). **Farm** has been added to a pre-existing name.

SOWERMYRR NY0804 (hab.) Gosforth.
Sourrmyre 1596/7 (Gosforth PR), *Sour Mire* 1774 (DHM).
♦ 'The infertile, marshy place.' The 1st el. could descend from some such word as ON *súrr*/OE *sūr* 'sour, damp, poor' (applied to quality of ground) or possibly ON *saurr* 'mud, wet ground'. The current form of the 2nd el. bears a striking resemblance to ON *mýrr* 'marshy ground', and the earlier spellings confirm **mýrr/mire**. The farm stands beside Caple Beck.

SOW HOW SD3987 (hab.) Staveley-in-Cartmel.
Sowhowe 1598 (*PNLancs* 200).
♦ 'Sow hill', with ME or ModE *sow* from OE *sugu* (perhaps encouraged by *sú*, accusative of ON *sýr*), plus **haugr/ how(e)**.

SPARK BRIDGE SD3084 (hab.) Colton.
Sparkbridg 1642/3 (PR).
♦ The two most likely explanations of 'Spark' are: (a) It is the reflex of OE **sp(e)arca* which occurs in Spare Bridge, Essex (*Sperkebrige, Sparkebrige* 1216–72, *PNEssex* 267), SPARKET etc., and seems to mean 'brushwood, shrubs' (Smith, *Elements*). Watts (2004) favours this explanation, citing Risebridge, Essex (from OE *hrīs* 'brushwood') as supporting evidence that a brushwood causeway could be meant. (b) It is a surname, originally derived from the reflex of ON *sparkr* 'lively, sprightly'. Bridges (and hence settlements named from them) often take their names from those responsible for building them. For **bridge** see **brycg**.

SPARKET NY4325 (hab.) Matterdale, S~ MILL NY4326.
Sparcheued c.1250, *Sperkeheved* 1253, *Sparkhed(e)* 1419 to 1711, *Sparket* 1586 (*PNCumb* 256); *Sparket Miln(e)* 1631 (*CW* 1884, 34–5), *Sparket Mill* 1867 (OS).
♦ S~ is probably 'the high ground where brushwood grows', from OE **spearca* 'brushwood' (cf. SPARK BRIDGE) and OE **hēafod** 'headland, high ground' (as also in HESKET, just to the NE). **Mill** is from OE **mylen**. The 1631 record of the Mill requires the 'Milner' to make and maintain his dust hoops properly.

SPIGOT HOUSE SD4394 Crook.
Spriggot House 1836, *Spigot House* 1857 (*PNWestm* I, 181).
♦ Of uncertain origin. (a) A *spigot* is a peg or faucet for drawing off liquor from a container, but if this is the origin of the **house** name, its motivation is obscure — unless the name is more or less equivalent to SPOUT HOUSE. (b) If 'Spigot' is a corruption of some other element, the surname *Speight* found in the Kendal PRs, or the appellative *speight* 'green woodpecker', are among the candidates.

SPOON HALL SD2996 Coniston.
Spownhow 1648/9, *Spoonehow* 1679 (PR), *Spoon Hall* 1851 (OS).
♦ The early spellings suggest that **hall** has replaced **how**(e) 'hill, mound' as 2nd el., but the 1st el. is obscure. (a) *Spoon* in English p.ns can result from OE *spōn* 'chip, shaving of wood', while (b) *Spa(u)n-* can result from ON *spánn*, which can additionally refer to shingles or wooden tiles, so conceivably these were obtained here.

SPOUT CRAG SD2888 (hab.) Blawith.
From 1850 (OS).
♦ 'The rocky height by the water-source or waterfall'; *spout* as in SPOUT FORCE plus **crag**.

SPOUT FORCE NY1826 Lorton.
From 1867 (OS); cf. *Spute in Over Lorton* 1611 (PR).
♦ 'The cascading **force** or waterfall.' This is a dramatic series of waterfalls on AIKEN BECK, though hidden by tree planting (Wainwright 1964, Graystones 8). The word *spout* appears in later ME as *spoute* 'spout, water-outlet' and is of doubtful origin, but probably ON. As well as its normal application to a pipe or projecting lip, it is applied in 16th–17th cent. to a waterspout or cascade (*OED*, sense 6), which is clearly the sense here and in CAM SPOUT CRAG.

SPOUT HOUSE FARM NY1500 Eskdale.
Spout(e)house 1570 (*PNCumb* 392).
SPOUT HOUSE SD4491 Crosthwaite.
Spout(e) 1535 (in 1669 copy) to 1790 (*PNWestm* I, 85).

♦ This seems to be a house with a direct water-supply; *spout* as in SPOUT FORCE, **house, farm**. The Crosthwaite S~ H~ is by a stream, with a spring a short way to the S. Similarly, the Eskdale farm was, before the advent of mains water supplies, fed by a stream issuing from BLEA TARN which turns into a torrent after rain. I am most grateful to Mrs D. Postlethwaite of Spout House for this information.

SPRING BANK NY2224 (hab.) Above Derwent.
From 1867 (OS).
SPRING BANK NY4125 (hab.) Hutton.
Springbank 1867 (OS).
♦ 'The hill-slope with the spring or plantation'; **spring, bank**. *Spring* could have the sense either of 'spring, water-source' or 'plantation, copse'. The OS map of 1867 shows springs less than half a mile N of the Above Derwent example, and a plantation but no springs near the Hutton one.

SPRING HAG SD4898 Over Staveley.
Spring hagg 1836 (*PNWestm* I, 176).
♦ 'The place where plantation is cut down', with **spring** 'plantation, copse' and *hagg* from ON *hǫgg* 'cutting, clearing'. The name applies to a hillside that is presently wooded.

SPRING HOUSE SD1190 Waberthwaite.
From 1819 (Corney PR).
♦ A spring is marked nearby on larger scale maps, suggesting that **spring** here refers to a water-source, though a plantation is also possible, as in S~ HAGG; see also **house**.

SPRING KELD NY0804 (hab.) Gosforth.
Spring Kell 1818 (PR).
♦ (a) The name appears to be tautological, from **spring** 'water-source' plus ON or dial. **keld**(a)

'spring', cf., probably, Springkell in Dumfries (*SSNNW* 251). A well is shown nearby on the 1:25,000 map. (b) **Spring** in the Lake District normally refers to a plantation, but this seems less likely here.

SPRINKLING TARN NY2209 Borrowdale.
Sparkling Tarn 1774, 1821, *Sprinkling Tarn* c.1784 (*PNCumb* 35), probably = *Prentibiountern* 1322 (Smith 1903, 7, *PNCumb* 35).
♦ 'The sparkling pool', with the now obsolete verb *sprinkle* (*OED*, v. 2) and **tjǫrn/tarn**. Baines took 'Sprinkling' as a mountain-name, but was mistaken (1829, 146).

SPRINT, RIVER NY4904, SD5299 Longsleddale.
Sprit(t), *Sprytt*, (*aqua de* [river of]) *Spret(t)* 1186–1200 to 1777 (*PNWestm* I, 13–14), *Sleddal Bek* (*alias Sprit*) c.1692 (Machell II, 104).
♦ Perhaps 'the leaping one', from some such word as ON verb *spretta* or noun *sprettr* 'leap, spirt' (*PNWestm* I, 13–14, and xiii for Norw parallels). The -*nt*- does not appear in spellings until the 16th cent. so that, although it belongs historically in early forms of the ON root, it is attributed to the influence of neighbouring river-names — the KENT or Mint (*ERN* 377).

SPUNHAM SD2588 (hab.) Kirkby Ireleth.
Spoonham 1711 (*CW* 1926, 97), *Spnnam* [sic edition] 1761, *Spoonham* 1797, *Spunham* 1817 (Woodland PR), 1851 (OS).
♦ Uncertain. The 1st el. may be of the same origin as the specific of the equally obscure SPOON HALL, i.e. 'wood-shaving' or similar. The 2nd el. 'ham' seems unlikely to go back to OE *hām* 'homestead', given the lack of medieval spellings and the nature of the location, unlike the kind of prime

site favoured by and given *hām* names by Anglian incomers, but no better explanation is to hand. Another Spunham at SD1683, just outside the National Park, is *Spooname* 1624, *Sponenham* 1649, *Spunham* 1767, and is also unexplained (*PNCumb* 419).

ST = SAINT: Alphabetised as if SAINT

STABLE HARVEY SD2891 (hab.) Blawith, S~ H~ MOSS SD2791.
Stabel Hervy 1291 (Brydson [1908], 139), *Stableheruy* 1332 (p) (*PNLancs* 215), *stable arvie* 1605 (*CW* 1988, 122), *Stable-Harvy* 1780 (PR); *Stable Harvey Moss* 1851 (OS).
♦ 'The stable of the Harvey family', which then developed into an 'ancient hamlet' (*VCHLancs* 361, n. 27). ME *stable* is an adoption from OFr *estable*. *Harvey* is presumably the surname of Breton origin which came to England in 1066 (Hanks & Hodges 1988). The word order in this name may be influenced by Fr as well as by inversion compounds arising in a Gaelic-Scand environment. Brydson wrote of the place that 'a local story (maybe without foundation) ascribes it to a stable used by Monks of Furness Abbey' ([1908], 139). The Moss is a peat bog W of S~ H~.

STABLE HILLS NY2621 (hab.) Borrowdale.
Stable Hill 1734, *Stable-hills* 1770 (*PNCumb* 317).
♦ Simply *stable* as in S~ HARVEY plus hill. Standing beside a small rise opposite LORD'S ISLAND, and a mile W of CASTLERIGG, both onetime possessions of the Radcliffe family, this was a natural place to stable horses. The 1734 spelling occurs in documents relating to the former estate of the Radcliffes.

STAINTON SD1294 Muncaster, S~ BECK SD1394, S~ FELL SD1494, S~ PIKE SD1594, NETHER S~ SD1195.
(Stainton:) *magna* (Great) *Staneton* 1569, *Stainton* 1769 (*PNCumb* 426); *Stainton Beck* 1747 (PR); *Stainton Fell* 1823 (GreenwoodM), 1867 (OS); *Stainton Pike* 1867 (OS); *Nether Stenton, Lower Stanton, Middle Stenton* all 1774 (DHM).
♦ 'Stone settlement', from OE **stān** and/or ON **steinn** 'stone', plus **tūn** 'village, settlement, farmstead', though whether the 'stone' refers to the nature of the ground, to quarrying or early stone buildings is not clear. Nether S~, true to its name, stands some 262ft/80m. below S~. S~ Fell is a tract of high grazing land bounded by S~ Beck 'stream' and rising to a peak in S~ Pike.

STAINTON GROUND SD2192 Dunnerdale.
From 1688 (original spelling not specified, Broughton-in-Furness PR).
♦ 'The farm of the Stainton family'; surname plus **ground**, cf. CARTER GROUND. The surname S~ is well known in the area (*VCHLancs* 360), and is associated with S~ Ground in the 1688 record. It may have originated at Stainton (Gap) at SD2983.

STAIR NY2321 (hab.) Above Derwent.
Stayre 1565, *Stare* 1566 (*PNCumb* 373).
♦ 'Presumably so called from the rise in the road here' (*PNCumb* 373); and though the rise is far from stair-like, and somewhat removed from the hamlet, this is possible. *Stair* goes back to OE *stæger*. The 1565 spelling tells against the assumption of ME *star* or ON *storr*, both 'sedge', which would be appropriate to this marshy spot beside NEWLANDS BECK. Gaelic *stair, stoir*, which refers to stepping stones or a rough bridge across a bog, would

also be apt here, though purely Gaelic names are extremely rare in the locality.

STAKE: S~ BECK NY2609 Borrowdale, S~ PASS NY2608 Borrowdale/Langdales.
Over the Stake to Langdale 1775 (*CW* 1940, 13), *the foot of the* Stake *in Borrowdale, over the* Stake *of* Borrowdale *(a steep mountain so called)* 1780 (West 92, 99), *The Stakes* 1823 (*PNWestm* I, 207); *Stake Beck* 1867 (OS); *Stake Pass* 1859 (OSNB).
♦ *Stake* derives from OE *staca* 'stake', but whether in this instance it was of wood or stone, and whether a boundary or way-marker on this important routeway, or both, is uncertain. 'Stake' is still marked on larger scale maps at NY2609 near the parish and former county boundary, while Machell referred to *Langdale steake* which is probably to be identified with this Stake, adding in the margin 'a stone set up for a way make [*sic*]' (c.1692, II, 133). Stake apparently gave its name to the watershed between BORROWDALE and LANGDALE (as suggested by the West's reference to a mountain), to the **beck** 'stream' that rises there and to the **pass**; cf. STICKS PASS.

STAKES MOSS SD4684 (hab.) Crosthwaite.
From 1815 (*PNWestm* I, 85–6).
♦ Probably 'the bog marked out by stakes'; *stake* as in STAKE (PASS) above, plus **moss**; cf. (*the*) *sticks mosse* 1589 (*PNCumb* 318), close to STICKS PASS. The name refers both to the moss and to a settlement. Barnes, reporting the discovery here of two massive corduroy roads of unknown age, constructed of cross-timbers partly held in place by vertical stakes, conjectured that the name resulted from earlier discovery of these stakes (1904, 207).

STANAH NY3219 (hab.) St John's, S~ GILL NY3218.
Stanay 1606 (*PNCumb* 318); *Stanah Gill* 1867 (OS).
♦ Uncertain, possibly 'stony hill', with the reflex of **stān/steinn** 'stone' plus **haugr/how(e)** 'hill, mound', as in GREENAH and SETTARAH. This would be an apt description of the rising ground above the settlement; see also **gil(l)** 'ravine with stream'.

STAND CRAGS NY3810 Patterdale.
From 1860 (*PNWestm* II, 227).
♦ Presumably 'the jutting rocks with or near a platform-like feature'; *stand* as in BROAD STAND, plus **crag**. However, the position immediately next to a point called Stangs NY3810, could point to a corruption of that; see **stǫng/stang**.

STANDING CRAG NY2913 Borrowdale/St John's.
A standing Cragg 1805 (*CW* 1981, 67, cf. *Standing Crag* 1805, *PNCumb* 318, probably using same source).
♦ 'Standing' may refer to the commanding position of this **crag** or rocky height (captured in Wainwright's illustration, 1958, Ullscarf 9), or may possibly hint that it acts as a natural boundary marker akin to artificially raised standing stones. The verb *stand* is from OE *standan*.

STANDING STONES NY0513 (hab.) Ennerdale.
From 1677 (PR).
♦ *Standing* as above, plus **stān/stone**. Standing Stones on maps and in documents may be merely references to antiquities, without being proper names, but in this case S~ S~ has the

status of a name, as has Standing Stones, Kirksanton (mentioned as *lapides stantes* in a charter of 1260–80, *PNCumb* 417). There is no record of archaeological remains at NY0513 in LDHER.

STANEGARTH NY4917 (hab.) Bampton.
From 1538 (*PNWestm* II, 197).
♦ 'The stone or stony enclosure'; **stān/steinn/stone, garth.** 'Stanegarth is very aptly named, there is much stone lying about on the common near' (Noble 1901, 150).

STANG NY3517 Patterdale.
From 1859 (*PNWestm* II, 227).
♦ 'The long, narrow ridge.' ON **stǫng**, dial. **stang**, whose basic sense is 'pole', is here applied topographically to a high, snout-like projection E of RAISE which terminates in Stang End, NY3617 (see Whaley 2005).

STANG END NY3102 (hab.) Skelwith.
Stange End 1663–4 (*CW* 1920, 178), *Stang End* 1770 (JefferysM), *Stangend* 1792 (Hawkshead PR).
STANGENDS NY1103 (hab.) Irton.
Stangendes 1563 (*PNCumb* 404), *Stangend* 1733 (Nether Wasdale PR).
♦ Probably 'the (place at the) end of the plank bridge'; **stǫng/stang, end**. Both places are situated between a blunt projection of high ground and a river crossing: the Skelwith one is by an old ford on the upper BRATHAY (Ramshaw & Adams 1993, 121), and the Irton one by a crossing of the R. IRT. Although *stang* could mean 'projecting hill' (as in STANG, Patterdale), therefore, the dial. sense of 'bridge of poles or planks' seems more likely, cf. the four settlements named BRIDGE END.

STANGER NY1327 (hab.) Embleton.
Strangre 1298, 1332 (p), *Stanger* 1322 to 1777 (*PNCumb* 384).
♦ Probably 'the posts', from *stangir*, pl. of ON **stǫng** 'post, pole, stake', perhaps referring to waymarks, boundary markers, or, as suggested by Fellows-Jensen, a palisade. Her alternative suggestion of rock formations is less likely in this gentle terrain (*SSNNW* 59). Another Stanger formed a unit with STAIR in the Newlands valley (Denton 1687–8, 128).

STANGRAH SD1185 (hab.) Whicham.
Stangerhovet 1180–1210, *Stangarhawe, Stangarhay* 1526 (*PNCumb* 449), *Stangerey* 1666 (Whitbeck PR), *Stangrow* 1774 (DHM).
♦ The name seems to contain ON **stǫng** 'post' or 'projection' (gen. sing. *stangar*) and ON **hǫfuð** 'head', later replaced by a form of **how(e)** 'hill'. What feature the name originally referred to is not recoverable, but a landmark seems possible.

STANLEY FORCE SD1799 Eskdale.
From 1867 (OS); cf. *Stanley Gill* 1839 (Ford 165, in list of waterfalls).
♦ The **force** or waterfall is named from the Stanleys, lords of the manor of Austhwaite, later DALEGARTH, who took their name from Stanley, Staffordshire, in which they also held land (Nicolson & Burn 1777, II, 22; Parker 1926, 104).

STANTHWAITE NY2536 (hab.) Ireby and Uldale.
Stanthwaite 1643 (Uldale PR).
♦ Apparently 'the stony clearing'; **stān/steinn, þveit**.

STARLING DODD NY1415 (2085ft/ 633m.) Loweswater/Ennerdale.
From 1867 (OS).

♦ This compact hill or **dodd** seems to be named from the track *Styalein* which is mentioned in Loweswater bounds of 1230 (*PNCumb* 411). This in turn seems to be an inversion compound meaning 'Alan's path', from ON **stígr** or OE **stīg** 'steep path' and the pers.n. *Alein*, of Breton origin but popular in ME (cf. SEAT ALLAN). Ekwall suggested that *Alein* could replace an earlier name such as OIr *Ailène* (*Scand & Celt* 35).

STARNTHWAITE: S~ GHYLL SD4392 (hab.) Crosthwaite.
Starnthwaite (*Mill*) 1899 (OS), possibly = *Sternthwaite* (*meadowe*) 1716, *Starnthwaite* (*Hill*) 1815 (*PNWestm* I, 85, both cited in relation to Staithwaite Borrow).
♦ A local family named Starnthwaite is recorded from the 16th cent. (*PNWestm* I, 85), and may have given its name to the *ghyll* (**gill** 'ravine with a stream'), which is also the name of a settlement. As for the meaning of S~, the 2nd el. is from ON **þveit** 'clearing', but the 1st el. is obscure. 'Star' (ON *stjarna*/OE *stearn*) seems unlikely, as do dial. *starn* 'a grain, tiny particle' (*EDD*: *starn* sb.2) and OE *stearn* 'tern, sea-swallow' at this inland site. No ON or OE pers.n. would account for the spellings.

STAVELEY SD4698 (hab.) Hugill/Over Staveley/Nether Staveley, S~ HEAD FELL NY4701 Over Staveley, S~ PARK SD4798 Nether Staveley.
Staveley(*e*) & variants 1189–1200 to 1540 (*PNWestm* I, 172, under Nether S~), *Staveley*(*e*) & variants 1341 to 1633 (*PNWestm* I, 175, under Over S~); *Staveley Head Fell* 1859, cf. *Staveley Head* 1706; *Stavel' Parke* earlier 16th cent. (*PNWestm* I, 176).
STAVELEY-IN-CARTMEL SD3786 (hab. & parish).
Stavelay 1282 (p), *Staveley* 1491 (*PNLancs* 199).
♦ S~ is 'the clearing where staves or poles are obtained', from OE *stafa*, gen. pl. of *stæf* 'stave, staff' plus OE **lēah** 'clearing, woodland'. *Stæf* occurs 'usually in allusion to places where staves were obtained or where they were used as marks' (Smith, *Elements*); the former seems more likely in these areas, still heavily wooded. The two parts of S~ SD4698 were distinguished by the feudal affixes *Gamel* and *Godemund*, then from the 16th cent. by *Over* (etymologically from OE *uferra* 'upper') and **Nether** (OE *neoðera* 'lower'). See also **head**, **fell**, **park**, and CARTMEL. S~ Head Fell is the rough, unenclosed land at the N. edge of the parish.

STEEL END NY3212 (hab.) St John's, S~ FELL NY3111 (1811ft/552m.).
Steelle End 1570 (*PNCumb* 318); *Steel Fell* 1780 (West 80).
♦ 'The end of the steep ascent'; **steel/stile** from OE *stigel* 'stile, steep ascent', plus **end**. Here it could refer to the dramatic ridge which projects N from the summit of S~ **Fell**, or perhaps more likely to the old track which ran along the side of Steel Fell from GRASMERE to KESWICK over DUNMAIL RAISE (Simpson 1928, 274). S~ End nestles below the spur and is at the end of the path. Two similar 1570s forms identified with STILE END, Above Derwent, in *PNCumb* 373 are actually further spellings of STEEL END, St John's (I am indebted to Mrs Susan Dench of CRO, Carlisle, for this). The Above Derwent example is a hill-name, whereas this one refers to a hamlet (with a population large enough to include, for example, three Margaret Wilkinsons buried in the 1580s).

STEEL KNOTTS NY4418 Martindale.
Cf. *Steel End* 1859 (*PNWestm* II, 219–20).
♦ 'The rocky summits on the steep ridge'; **steel, knott**. The situation is closely comparable with STEEL FELL and END: a dramatic, beak-like ridge whose lower end is shown as S~ End on larger scale maps (at NY4419). Although *Steel* appears as a surname in the PRs for 1684 (as noted in *PNWestm*), it seems unlikely to be the source of S~ KNOTTS.

STEEL RIGG NY4708 Longsleddale.
From 1863 (OS); cf. *ye Steale Pikes* 1633 (*PNWestm* I, 164).
♦ Probably 'the ridge by the steep path or steep ascent'; **steel, hryggr/ rigg**. This hill and the neighbouring Steel Pike are close to GATESGARTH PASS as it climbs N out of LONGSLEDDALE. It seems less likely that S~ R~ is named from the Steel family, documented in the Kendal district from 1560 onwards (as suggested by Smith in *PNWestm*), though the surname could derive from this Steel or a similar p.n.

STEEPLE NY1511 (2687ft/841m.) Ennerdale.
From 1774 (DHM).
♦ This is a great rocky mass with an impressive cone (though not quite spire) as its summit. The word *steeple* goes back to OE *stēpel* 'tower', and *OED*'s first citation for ME *steple* applied to a spire is from 1473–4 (*steeple*, n.1), but the name may be more recent.

STEERS POOL SD2591 (river) Broughton West/Kirkby Ireleth.
Styrespol 1235 (*PNLancs* 191), *Sterespol* 1292 (*DEPN*).
♦ 'Styr's stream', from ON pers.n. *Styrr* and **pool**.

STENNERLEY: HIGH S~ SD2785 (hab.) Blawith, LOW S~ SD2785 (hab.).
Stainnerlid 1200–35 (p), *Stainerlith* 1251 (p), *Staynerlyth* 1285 (*PNLancs* 214); *Upper Stanerley* 1719, *High Stennerly* 1786 (Lowick PR), *High Stennerley* 1851 (OS); *Low Stanerlay* 1746/7 (Lowick PR).
♦ (a) Probably 'Steinar's hill-slope', from an ON pers.n. *Steinarr* plus ON *hlið* 'hill-slope'. The ending '-ley', elsewhere often derived from OE **lēah** 'woodland clearing', is therefore misleading. (b) Different OE origins are possible, however, if, as Fellows-Jensen suggests, the 13th-cent. spellings represent a remodelling of OE *stæner* 'stony, rocky ground' and OE *hlið* (*SSNNW* 165). The division into **High** and **Low** will be a secondary development.

STEP, THE NY3611 Patterdale.
From 1860 (OSNB).
♦ This is among the confusion of crags at the head of DEEPDALE. 'A narrow and rocky point having a steep precipice on either side, juts out from Fairfield and terminates in Greenhowe End' (OSNB). The word *step* goes back to OE *stæpe, stepe*.

STEPHENSON GROUND SD2393 Dunnerdale.
From 1670 (original spelling not specified, Broughton-in-Furness PR), *Stevenson ground* 1671 (Hawkshead PR), *Stephensonground Dunnerdale* 1747/8 (Seathwaite PR).
♦ 'The farm of the Stephenson family'; surname plus **ground**, cf. CARTER GROUND. The surname is well represented in the relevant PRs of

Broughton, though I have not found entries associating people of this name with S~ Ground.

STEPPING STONES NY3605 (hab.) Rydal.
From 1863 (as a feature), 1900 (as hab. & feature) (OS).
♦ Stepping stones cross the R. ROTHAY at this point, and give their name to a building on the W bank, which Lindop notes was previously Lanty Fleming's Cottage (1993, 55, cf. 56 for a charming 1907 sketch of the stones). Dorothy Wordsworth returned to GRASMERE from LOUGHRIGG FELL 'over the stepping stones' in 1800 (*Journals*, Oct. 19). *OED* has an instance of *stepping stones*, as the name of a small stream, from a Northumbrian document of 1550 (*OED*: siket).

STEPS HALL NY5513 Shap Rural.
From 1797 (*PNWestm* II, 177).
♦ 'Steps' (ultimately from OE *stæpe*, *stepe*) seems to refer to 'stepping stones across the Lowther, by which the hall could be reached from Shap' (*PNWestm*); see also **hall**.

STEWART HILL NY3634 (hab.) Mungrisdale.
Hill 1816, *Stewart Hill* 1822 (PR).
♦ Presumably 'the **hill** associated with the Stewart family', though I have not noted the surname in the PRs. The gradient is gentle, as is characteristic of the E of the CALDEW valley.

STICKLE: S~ PIKE SD2192 (1231ft/375m.) Dunnerdale, GREAT S~ SD2191 (1001ft/305m.).
Stickle 1669, *Stickle D[unnerdale]* 1700 (Broughton-in-Furness PR); *Stickle Pyke* 1786 (*PNLancs* 193, n. 2); *Great Stickle* 1850 (OS).

♦ *Stickle*, applied to steep, pointed hills, may go back to an OE **sticel(e)* 'steep place' which is assumed from OE adj. *sticol*, ME and ModE *stickle* 'steep' (Smith, *Elements*). S~ **Pike** is the highest point on the hill. Though less grand than the 'Stickles' in Langdale (HARRISON S~, PIKE O' S~), **Great** Stickle is larger than Little Stickle just to the S, and is a 'miniature Matterhorn' in Wainwright's view (1960, Dow Crag 2).

STICKLE TARN NY2807 Langdales.
From 1780 (West 103), *Stickel Tarn* 1800 (*PNWestm* I, 17).
♦ 'The mountain pool beneath the Stickle, the peak'; *stickle* as in STICKLE (PIKE), plus **tjǫrn/tarn**. It lies below HARRISON STICKLE.

STICKS: S~ PASS NY3418 St John's/ Patterdale.
Stikes, (the) sticks (mosse) 1589 (*PNCumb* 318), *(Top oth') Sticks* c.1692 (Machell II, 155), *Styx (Top)* 1800 (*CN* 798); *Sticks Pass* 1920 (OS).
♦ *Stick* goes back to OE *sticca*, plus **pass**. 'Sticks Pass is said to be so named from a number of sticks which mark the route' (Barber & Atkinson 1928, 32, *PNWestm* II, 227, and cf. STAKE PASS). This seems likely, on this highest of Lakeland passes, even if the markers reported in 1928 were not the original ones. The streams flowing W and E from this watershed are both called Sticks Gill.

STILE END NY2221 Above Derwent.
From 1839 (Ford 55), 1867 (OS), = *Steil Point* 1800 (*CN* 804–5).
STILE END NY4604 Kentmere (hab.).
Steel end 1836, *Stile End* 1857 (*PNWestm* I, 168).
♦ 'The end of the steep place'; **steel/stile**, **end**. S~ E~ in Above Derwent is a

small summit below a steep descent from OUTERSIDE, though the ridge continues downwards from it. For early forms wrongly identified with this place, see STEEL END. The Kentmere S~ E~ lies near the foot of a long, steep ridge running N and S, but also at the foot of a track running E and W between Kentmere and Longsleddale, so that it is not certain which of them *stile* refers to here.

STIRRUP CRAG NY1709 Nether Wasdale. From 1867 (OS).
♦ 'The stirrup-shaped rocky height'; *stirrup*, **crag**. A rock spike on the crag juts vertically from the N spur of YEWBARROW, forming a deep U shape which presumably gave rise to the name (for illustration, see Wainwright 1966, Red Pike (W) 6 and 10). The word *stirrup*, from OE *stīg-rāp* 'step-rope', is unusual in p.ns (Smith, *Elements*, offers only one example), but *saddle* is more common: see SADDLE CRAGS etc.

STOCK SD3088 (hab.) Colton, S~ WOOD SD3188.
Stock 1640 (PR), *Stocks* 1851 (OS), *Stock* 1893 (OS); *Stocks Wood* 1851 (OS), *Stock Wood* 1893 (OS).
♦ Possibly 'the log bridge'. Neither early spellings nor topography (a situation beside Bletherbarrow Gill) are decisive, but an origin in ON *stokkr* or OE *stocc* 'tree-trunk, log, or structure made of logs' seems most likely. Smith suggests the sense 'wooden bridge' in Millbeck Stock, Bowness, noting the parallel usage of Swed *stock* (*PNWestm* I, xvi), and this would be supported by STOCKBRIDGE below and by a similar application of **stǫng/stang** to bridges. It could well be the sense in the Colton example; see also **wood**.

STOCKBRIDGE SD0993 (hab.) Bootle.
Stock Bridge 1867 (OS).
♦ Despite the lack of early forms, this is presumably 'the bridge made of logs', with elements derived from OE *stocc* (cf. STOCK) and **brycg**. This would refer to a crossing over nearby ESKMEALS Pool. Another Stockbridge is recorded from 1251 (*PNCumb* 398), and there are at least three other English examples (Smith, *Elements*: *stocc*).

STOCKDALE NY2534 (hab.) Ireby and Uldale.
Stakedale in Uluedale 1292, *Stokedale* 1539 (*PNCumb* 328), *Stockdale* 1643 (Uldale PR).
♦ Probably 'the valley where (hay)stacks are made'. The 1st el. may be *stakka*, gen. pl. of ON *stakkr* 'haystack, heap (e.g. of dung)', as suggested in *SSNNW* 165, since the alternative suggestion, of a pillar of rock (*PNCumb* 328) is not borne out by the local topography. The 2nd el. is clearly ON **dalr** 'valley'.

STOCKDALE NY4905 (hab.) Longsleddale.
Stokdale 1377 (p) (*PNWestm* I, 161).
STOCKDALE MOOR NY0908 Ennerdale.
Stockdale Moor 1826 (CRO D/Lec/94); cf. *Stockdalebek* 1540 (*PNCumb* 387).
♦ Probably 'the valley where tree-trunks or stakes are found', from ON *stokkr*/OE *stocc* as in STOCK, and ON **dalr** 'valley'; also **moor**.

STOCK GHYLL NY3906 Ambleside, STOCKGHYLL FORCE NY3804.
Stockgill 1639 (*PNWestm* I, 184), *Stockgill Force* 1839 (Ford 28), *Stockghyll Force* 1865 (*PNWestm* I, 184); cf. *Stock* 1630 (*PNWestm* I, 184).
♦ Stock appears to have been an independent name, referring to some wooden feature as in STOCK, Colton, and tenants 'above' and 'below

Stock(e)' are referred to c.1630 (*CW* 1906, 11). See also **gil(l)** 'ravine with stream'. Armitt suggested that *stock* might have a specific meaning associated with the fulling industry (cf. *EDD*: *stock*, sense 9), quoting a 16th-cent. regulation that cloth must not be carried 'to the walk-stock or fulling-mill' on Sundays. She also believed that S~ G~ may have superseded 'an earlier name "Sleddal-beck"' (1908, 154; cf. Simpson 1928, 276).

STOCKLEY: S~ BRIDGE NY2310 Borrowdale.
Stokeleye 1285 (*PNCumb* 351); *Stockley Bridge* 1753 (*CW* 1899, 121).
♦ S~ is 'the clearing with the tree-trunks or logs', from OE *stocc* 'tree-trunk, log, stake' and OE **lēah** 'woodland clearing, meadow'. The name is of OE or early ME origin, and as such is a rarity in this inner part of BORROWDALE, but, situated on the important route over STY HEAD to WASDALE, it must have been less remote than now appears, and the packhorse **bridge** (see **brycg**) is presumably a successor to more ancient bridges. Although trees now survive only among the crags, recent excavations just N of S~ B~ discovered 157 worked stakes, a horizontal brushwood layer, and wood chips that suggested that 'scrub growth was being coppiced to fashion stakes, most probably for use elsewhere'. The structure at S~ appears to have been a very substantial livestock fence. The investigators date the material to the end of the medieval period, and tentatively associate it with expansion of sheep farming by the Cistercians of FURNESS Abbey, i.e. to a later period than the putative clearance in the Scandinavian-Gaelic phase indicated by the proliferation of *thwaite* names in the area, and than the Anglian phase possibly indicated by the name Stockley. The date when the name Stockley was given cannot be recovered, but at the very least the archaeological evidence points to the presence and exploitation of woodland at this site late in the medieval period (Wild *et al.*, 2001). (Pl. 14)

STODDAH: S~ BANK NY4126 (hab.) Hutton, S~ FARM NY4126.
Stothow 1371 (p), *Stodhowe* 1383 (p), *Stodday* 1559 (*PNCumb* 213); *Stoddo banke* 1575-6 (*CW* 1969, 119).
♦ 'The hill where stud horses or bullocks are kept', from OE *stōd* 'stud' and OE *hōh* 'hill-spur' or ON *stóð* 'stud' and ON **haugr** 'hill'. Another possibility, judging from the 1371 spelling, is ME *stot* 'bullock' (cf. STOTT PARK). **Bank** will have been added later, and designates both the hill and the building which is set slightly above S~ **Farm**.

STONE ARTHUR NY3409 Grasmere.
From 1843 (Wordsworth, *Poetical Works* ed. de Selincourt, II, 488), 'sometimes referred to as Arthur's Chair' (Wainwright 1955, Stone Arthur 1).
♦ 'Arthur's rock'; **stān/stone**, plus pers.n. *Arthur* as in ARTHUR'S PIKE, with word order seemingly influenced by the inversion compounds of the Scand-speaking period. This is a point at 1652ft/514m. among the rock outcrops on a spur of GREAT RIGG. The context of the first record of this name is a note dictated by Wordsworth in 1843, explaining that 'Stone-Arthur' was the peak referred to in his poem 'There is an Eminence'. The poem concludes with a reference to the Wordsworths' private naming: 'And She who dwells with me, whom I have

loved ... I Hath to this lonely Summit given my Name' ('Poems on the Naming of Places III', *Poems* 118).

STONE COVE NY3510 Rydal.
From 1859 (OSNB).
♦ This is a feature high on GREAT RIGG described in OSNB as 'a flat piece of ground covered with loose stones'; **stān/stone, cove**.

STONE ENDS NY3533 (hab.) Mungrisdale.
Stane Ends 1568 (*PNCumb* 305).
♦ The name presumably refers to the situation below CARROCK FELL, where the rocky slopes around Apronful of Stones (NY3533 on larger scale maps) level off. In the modern form, Standard English **stone** has replaced local *stane*; see also **end**.

STONE PIKE NY0707 (1056ft/322m.) St Bridget Beckermet.
Staynpik' 1243 (*PNCumb* 340).
♦ 'The stony peak' (**stān/steinn, pik(e)**) — a small summit on SWAINSON KNOTT. ModE *Stone* has replaced the distinctly Norse *Stayn*.

STONERAISE NY3426 (hab.) Threlkeld.
Stoneraise 1774 (DHM), *Stone raise* 1803 (*CN* 1518).
♦ 'The stone cairn' (**stān/stone, hreysi/raise**), and hence the farm there. Although there are no known records of this S~ until the 18th cent., other examples appear in the 13th cent. (*PNWestm* II, 264). No archaeological remains are recorded here (LDHER).

STONESIDE HILL SD1489 (1383ft/422m.) Bootle/Millom Without.
Stones Head Fell 1774 (DHM), 1802 (*CN* 1224), *Stoneside Fell* 1823 (GreenwoodM), *Stoneside Hill* 1867 (OS); cf. *Stoneside Crag* c.1850–60 (*CW* 1966, 378).
♦ If the 1774 and 1802 forms are correct, the name may originally have been 'Stones Head, the stony height' (**stān/stone, head**), appropriate to this steep and rocky summit. **Fell** was added later, only to be replaced by **hill**, and *head* has been replaced by **side** — a common process, though it seems to have happened unusually late in this case.

STONESTAR SD2091 (hab.) Dunnerdale.
Stonescarre 1584 (*PNLancs* 223), *Stone Star in Dunnerdale* 1712 (Broughton-in-Furness PR), *Stonester* 1786 (*PNLancs* 223).
♦ Possibly 'the rocky scarp'; **stone, scar**. The 16th-cent. spellings suggest that the 2nd el. is *scar*, and the topography is compatible with that. Corruption of *-scar* to *-star* under the influence of preceding *stone-* would be plausible.

STONESTY PIKE NY2403 Ulpha.
From 1867 (OS).
♦ 'The peak by Stonesty, the stony track' (**stān/stone, stígr/stīg, pike**), though without early forms certainty is impossible.

STONETHWAITE NY2513 (hab.) Borrowdale, S~ BECK NY2613, S~ FELL NY2610 and NY2713.
Staynethwayt 1190s, *Stontwhat* 1563 (*PNCumb* 351); *Stonethwaite Beck*; *Stonethwaite Fell* both 1867 (OS).
♦ 'The stony clearing', with **stone** replacing its ON cognate **stein(n)**, plus ON **þveit** 'clearing' — one of several instances of the el. in Borrowdale; see also **beck** 'stream'. The two tracts of S~ Fell illustrate the application of **fell** to upland pasture, rather than to a specific peak. 'Stonethwaite Fell is the

same as Langstrath Fell' (Johnson 1981, 66, presumably referring to the S~ Fell at NY2610).

STONEYCROFT NY2321 (hab.) Above Derwent.
Stanycroft 1505, *Stonicrofte* 1563 (*PNCumb* 373).
♦ 'The stony enclosure or farm', with *ston(e)y* from OE *stānig*, plus **croft**.

STONY COVE PIKE NY4110 Patterdale/Troutbeck.
From 1614 (*PNWestm* I, 191).
♦ 'The peak by Stony Cove, the stony recess.' This is the earliest-recorded of five occurrences of *ston(e)y* (ultimately from OE *stānig*) in this volume; also **cove, pike**.

STONY DALE SD3981 (hab.) Broughton East.
From 1851 (OS).
♦ 'The stony valley'; *ston(e)y* as in STONEYCROFT plus **dalr**.

STONY TARN NY1902 Eskdale.
From 1867 (OS).
♦ 'The stony mountain pool'; *ston(e)y* as in STONEYCROFT plus **tjǫrn/tarn**. Heaton Cooper praises its 'heather-clad rocky promontories and bays of clear water that show the stones on the floor' (1983, 43).

STONYTHWAITE SD2196 (hab.) Ulpha.
From 1707 (PR).
♦ 'The stony clearing'; *ston(e)y* as in STONEYCROFT plus **þveit/thwaite**. The whole area is extremely stony (see 1:25,000 OS map, on which High and Low S~ are also distinguished).

STOOL END NY2705 (hab.) Langdales.
The Stoolend, Stool-end 1695 (Grasmere PR), *Stealend* 1706 (*PNWestm* I, 207), *Style End* 1770 (JefferysM), *Stile End* 1823, 1839, *Stool End* 1865 (*PNWestm* I, 207).
♦ 'The **end** of the stool-shaped hill.' *Stool*, from OE *stōl*, is not common in p.ns. This is the point where THE BAND projects into GREAT LANGDALE. As Smith explains, the *stile* spellings are probably due to the fact that *steal* was the dial. form of both *stōl* and **stile** (*PNWestm*).

STORDS NY0708 St Bridget Beckermet.
From 1867 (OS).
STORRS SD3994 (hab.) Bowness.
(*ye*) *Storthes* 1292 (p), 1553, *Storr(e)s* 1606 (*PNWestm* I, 186).
STORTHES SD1693 Ulpha.
From 1867 (OS).
♦ All three may be 'the plantation(s) or area of brushwood', ultimately from ON *storð* 'area of brushwood, young wood, plantation'. The St Bridget example is the W brow of SWAINSON KNOTT; the Ulpha one is situated around the feeder streams of HOLEHOUSE GILL, where brushwood is likely.

STOTT/STOT PARK: HIGH S~ P~ SD3788 (hab.) Colton, LOW S~ P~ SD3788 (hab.).
Stot parke 1535, *Stotparke* 1537 (*PNLancs* 217); *High Stotpark* 1822, *Low Stottpark* 1838 (Finsthwaite PR).
♦ 'The enclosure for bullocks', from ME *stot*, which refers either to a young bull or ox or to a horse, but is more common in the former sense in N. dial. (so Ekwall, *PNLancs* 217). This was among the outlying farms of FURNESS Abbey (see **park**; also **high, low**).

STOTT GHYLL/STOTGILL NY3437 (hab.) Caldbeck.
the Stotte gill 1587 (*PNCumb* 279).
♦ Apparently 'the ravine/stream where bullocks graze', with *stot* as in STOTT PARK, plus **gil(l)**.

STOUPDALE CRAGS SD1587 (1548ft/ 472m.) Whicham.
From 1867 (OS).
♦ 'The rocky heights above S~', which may mean 'stake valley'. Without early spellings the etymology of the 1st el. is elusive, but it could go back to ON *stolpi* 'stake, post', which was absorbed into Cumbrian dialect. The Elizabethan Paines at HUTTON JOHN, for instance, decree that 'it belongeth to the Lord of the manor to uphold the pale and stowpps of the Common pasture' (*CW* 1969, 121). The 2nd el. is **dale**, referring to the narrow valley formed by Stoupdale Beck, while S~ **Crags** form the rocky head of the valley.

STRAWBERRY BANK SD4189 Cartmel Fell.
From 1851 (OS).
♦ 'The slope where strawberries grow'; reflex of OE *strēaw-berige* plus **bank**. The wild strawberry, *Fragaria vesca*, is common throughout Cumbria except in the high fells (Halliday 1997, 269), and the name S~ B~ occurs elsewhere in Cumbria (from 1717 and 1836, *PNWestm* I, 133, 147 respectively; also *Strawberrybank, Ambleside, Westmorland* 1837 (Finsthwaite PR)).

STREET HEAD NY3338 (hab.) Caldbeck.
From 1695; cf. *the Street* 1656 (*PNCumb* 279).
♦ The hamlet is situated on The Street (now the B5299) at the point before it descends E to HESKET NEWMARKET; *street* as in HIGH STREET, plus **head**.

STRIBERS SD3581 (hab.) Haverthwaite.
'*A close called Stribus*' c.1576 (*VCHLancs* c.277), *Stribus* 1612 (Cartmel PR).
♦ The name is obscure.

STRICKLAND HILL SD4285 (hab.) Witherslack.

Strickland Hill 1720 (PR), *Strickland How* 1770 (JefferysM), 1823 (*PNWestm* I, 79), *Strickland Hill* 1862 (OS).
♦ The name is from the S~ family, recorded as living here in 1720 (PR) and documented locally between 1662 and 1741 (*PNWestm* I, 79). They presumably came from one of the places of that name in the former Westmorland (see *PNWestm* index). As the spellings above show, **hill** and **how(e)** have alternated in the p.n. S~ H~.

STRIDING EDGE NY3415 Patterdale.
Striden-edge 1805 (Scott 'Hellvellyn', *Poetical Works* 703), *Striding Edge* 1823 (Otley 45), *Strathon Edge called locally Striding Edge* 1828 (HodgsonM).
♦ *Stride* is ultimately from OE *strīdan*, plus **edge**. Together with SWIRRAL EDGE, this forms part of the mighty horseshoe of HELVELLYN. The precipitous sides and occasional difficult steps make this an *edge* for striding with great care. Coleridge uses the same phrase in talking about SHARP EDGE (q.v.). The Scott and Hodgson forms suggest that *striding* may have replaced some earlier el.

STRUDDA BANK NY0508 (hab.) St Bridget Beckermet.
Strowd bank 1594, *the Strothie bank* 1595, *Struddebanke* 1658 (*PNCumb* 340).
♦ Perhaps 'the slope by the brushwood or marsh', with 1st el. derived from an OE word such as **strōðer* 'place overgrown with brushwood' or *strōd* 'marshy land overgrown with brushwood' (Smith, *Elements*; cf. LANGSTRATH), plus **bank(e)**.

STUBB/STUB PLACE SD0890 Bootle.
Stub Place 1646 (*PNCumb* 348).
♦ OE *stubb* meant 'tree-stump', a meaning that continues in local usage

(Dilley 1970, 202). This name could contain that or, more likely, a surname derived from it, given the frequency with which **place** 'plot of ground, farm' is qualified by a surname, although a rapid survey has not revealed the surname *Stub(b)* in Bootle PRs, and Stub Place is associated with other family names in the late 18th cent.

STURDY'S SD3781 (hab.) Broughton East. From 1851 (OS).
♦ This is presumably from the surname *Sturdy*, which originated in the ME *st(o)urdi* 'reckless, rash', given as a nickname (so Hanks & Hodges 1988). I have not found this as a surname in the relevant PRs, those of Cartmel.

STYBARROW CRAG NY3817 Patterdale. *Sty(e)braycragge* 1573 (*PNWestm* II, 227), *Stibarrow* c.1692 (Machell I, 708), *Stiburrow Crag* (corrected from *Stiveray Crag*) c.1692 (Machell II, 438), *Stayborcrag* 1780 (West 157), *Styboar-crag* (West 157), *Stibrah* 1787 (*PNWestm* II, 227), *Stibrow cragg* 1794 (Hutchinson, I, 336 and 436, quoting Ritson and Gray).
♦ S~ is 'the hill with a steep path', from **stígr/sty(e)** 'steep path' and **barrow** 'hill', possibly replacing an earlier el.; also **crag(ge)** 'rocky height'. A road used to pass over the crag, which Machell described as 'a monstrous stony Rock ... on the west Side of Ulswater, & In the way to Barton and penrith - where there is a Strait passage about 2 yards broad, haueing an High hanging-Rock on the one Side, wh[ich] threatens passengers to fall vpon them' (c.1692, I, 708).

STYBARROW: S~ DODD NY3418 (2756ft/ 843m.) St John's.

Stibarro (gill head) 1589, *Stybrow* 1794 (*PNCumb* 318), *Styborough* 1800 (*CN* 777); *Stiveray Dod* c.1692 (Machell II, 155), *Stybarrow Dodd* 1867 (OS).
♦ S~ is probably 'the hill by the steep path', from **stígr/sty(e)** and **berg/ barrow**. **Dodd** 'compact, rounded summit' may have been added after S~ no longer referred specifically to the hill, or it may designate the top in particular.

STYBECK FARM NY3118 St John's. *Styebecke* 1600, *Stibeck* 1618 (*PNCumb* 318).
♦ S~ is 'the stream by the steep path'; **stígr/sty(e)**, **bekkr/beck**. The **farm** is close to the track which ascends, as STICKS PASS, onto STYBARROW DODD.

STY HEAD NY2209 Borrowdale/Nether Wasdale, STYHEAD GILL NY2210 Borrowdale, STYHEAD TARN NY2209. *The Stey heade* 1540 (*PNCumb* 351); *Styhead Gill; Styhead Tarn* both 1867 (OS).
♦ 'The top of the pass'; **stígr/sty(e)**, **head**; also **gil(l)**, **tjǫrn/tarn**. This is an important col on an ancient routeway — 'the only Passage from the Vale of Borrodale into Warsdale and so to Ravenglass a very rocky bad one' to an 18th-cent. visitor (*CW* 1918, 50). An earlier name seems to have been *Edderlanghals* 1209–10, in which *Edder* may have meant 'swift', referring to the cascades of Styhead Gill (*PNCumb* 351, *ERN* 156).

SUBBERTHWAITE (hab., parish of Blawith and S~): S~ COMMON SD2686. *Sulbythwayt* 1284 (p), *Sulbithwayt* 1346, *Soelbythwayt* 1489 (*PNLancs* 214); *Subberthwaite Common* 1851 (OS).
♦ S~ seems to have been a secondary settlement or clearing (ON **þveit**) from a place called Sulby or similar: cf.

SOULBY. The **common** is the high grazing land belonging to S~, which for a time was a separate township.

SUMMER HILL SD1182 (hab.) Whicham.
From 1867 (OS).
SUMMER HILL SD4482 (hab.) Witherslack.
From 1862 (OS).
♦ Self-explanatory; *summer* (ultimately from OE *sumor*), **hill**. Both sites are very low rises, one (Whicham) near the coast and the other on level land just NW of ULPHA FELL SD4581, hence both are plausible sites of summer grazing. Cf. field-names in which *summer* is taken to refer to usage in summer (Field 1972, 222–3).

SUMMER SIDES WOOD SD3687 Colton.
From 1851 (OS).
♦ (a) Despite the lack of early forms, the site, on Furness Heights, the specific el. *summer* (ultimately from OE *sumor*/ON *sumar*) and the pl. -*s* would favour the assumption that 'Sides' is from ON **sætr** 'shieling'. Other references to seasonal grazing may be found in the local names Summer House Knott SD3786 and Wintering Park SD3686. (b) 'Side' here could alternatively be from OE **sīde**/ON **síða** 'hill-side'; see also **wood**.

SUNDERLAND NY1735 (hab.) Blindcrake, S~ HEADS (744ft/228m.) NY1735.
Sonderland in Blankrayk 1278 (*PNCumb* 324); *Sunderland Heads* 1867 (OS), perhaps cf. *Heads* 1804 (name of a close in Sunderland, CRO D/Law/1/212).
♦ S~ is 'the separate, remote, or private tract of land', from OE *sundorland*, referring to the situation, on a tongue of land between two streams, or to some feature of its early ownership or use. S~ **Head**s is a small height NW of the village.

SUNNY BANK SD2892 (hab.) Torver.
From 1695 (PR).
♦ This is an east-facing slope; *sunny*, which first appears in ME, derived from OE, ME *sunne* 'sun', plus **bank**.

SUNNY BROW NY3400 (hab.) Hawkshead.
From 1851 (OS).
SUNNYBROW SD4495 (hab.) Crook.
From 1857 (*PNWestm* I, 181).
♦ These are east- or south-facing slopes; *sunny* as in SUNNY BANK, plus **brow**.

SUNNY GREEN SD4182 (hab.) Upper Allithwaite.
From 1850 (OS).
♦ The site is in the valley of the WINSTER, in the lee of the E slopes of NEWTON FELL; *sunny* as in SUNNY BANK, plus **green** 'grassy place'.

SUNNYSIDE FARM SD4479 Meathop.
Seemingly = *Meathop Marsh Farm* 1899, 1920 (OS).
♦ 'The farm on the sunny slope'; *sunny* as in SUNNY BANK, plus **side** and **farm**.

SWAINSON KNOTT NY0708 St Bridget Beckermet.
From 1867 (OS).
♦ A typical **knott** — a rocky, compact hill, which appears to be named from *Swainson*, a surname originally meaning 'son of Sveinn' or 'son of the servant or lad (ON *sveinn*)'. It is recorded locally (PR 1718, a burial).

SWALLOWHURST SD1091 (hab.) Waberthwaite.
Sualewehurst 1228–43 (p), *Swaleweherst* 1255 (p), 1257 (p) (*PNCumb* 347–8).
♦ 'The wooded hill frequented by swallows'; OE *swalwe* 'swallow' plus **hyrst**. Hutchinson commented on the bird in Cumberland (1794, I, [13] and

457), and swallows and martins have nested at S~ in recent times. Although *swallow* can refer to an underground stream, as in the following name, there is no such landscape feature at S~ (local information from Mrs Rachel Curry of S~ Hall, gratefully acknowledged. I am also grateful to Mr Roy Vincent for pointing out that swallows abound throughout the area at present, so that an onomastic reference to them would not be particularly meaningful; perhaps the distribution was more concentrated in the past).

SWALLOW MIRE/SWALLOWMIRE SD4187 (hab.) Cartmel Fell.
Swallomire 1703 (CW 1966, 239), *Swallowmyre* 1728 (CW 1979, 71), *Swallow Mire* 1851 (OS).
♦ 'The bog with an underground stream' or 'the bog frequented by swallows'. Swallows are seen at S~ M~. However, the place also boasts a stream which disappears underground for some of its length. This is known in dial. as a *swallow* (cf. *OED*: swallow n. 1, sense 1b), which could be the source of the p.n. (I am grateful to Mrs M. Hodgson of Swallowmire for local information.) 2nd el. is **mýrr/ mire**.

SWAN BECK, MIDDLE NY3819 Matterdale.
From 1900 (OS).
♦ Self-explanatory, assuming that *swan* can be taken at face value, as from OE *swan*; **middle, swan, beck**. Flanked by Near and Far Swan Beck, the stream flows into ULLSWATER, where swans are occasionally seen (as Mr Keith Clark kindly informs me, adding that the name Middle Swan Beck seems not to be used locally). This is one of only three p.ns in this volume referring to swans (cf. ELTERWATER and SWAN HOTEL).

SWAN HOTEL NY2226 Above Derwent.
The Swan 1860 (OSNB).
♦ Situated on what was formerly the main road to COCKERMOUTH and the coast, and considered 'a middling large Inn' in 1860 (OSNB), this is associated with THE BISHOP. 'Swan' (OE *swan*) is among the numerous common inn names which originated in heraldic signs, and examples from other parts of the country are known from the 15th cent. (Cox 1994, 81). On the word *hotel*, see BEECH HILL HOTEL.

SWARTHBECK NY4520 (hab.) Martindale, SWARTHBANK NY4419 (hab.), SWARTH FELL NY4520, SWARTHFIELD NY4420.
Swart(e)bec, -beck(e) 1278 to 1629 (*PNWestm* II, 217); *Bank* 1787 (*PNWestm* II, 220), 1863 (OS), *Bank House* 1899, 1920 (OS); *Swarthfell* 1220–47, *Swartefell* 1247 (II, 217); *Swarthfield* 1863 (OS), possibly = *Swartfield* 1256 (CW 1913, 67–8).
♦ 'The black stream', from ON *svartr* 'black' and ON **bekkr** 'stream'. The farm is named from the stream, and the el. *svartr* 'black' seems also to have generated Swarth **Fell** (on which, see BONSCALE PIKE) and Swarth**field**, then secondarily Swarth**bank**, but what was first described as 'black' is unclear. Ekwall's guess that the stream was originally *Svarta* 'black one', and that it gave rise to the other names (*ERN* 385) may be correct.

SWEDEN: HIGH S~ BRIDGE NY3706 Ambleside.
le Swythene 13th cent. (*PNWestm* I, 183); *High Sweden Bridge* 1863 (OS).
♦ S~ is 'the burnt area'. The 13th-cent. spelling seems to refer to an area that has been cleared by burning (ON

sviðinn 'burnt, singed', dial. *swidden, swithen*, used of moorland cleared by burning, *EDD*); see also **high, brycg/bridge**. This packhorse bridge stands a good half mile higher up SCANDALE BECK than Low S~ B~.

SWINBURN'S PARK NY4221 Matterdale. From 1900 (OS).
♦ *Swinburn* is presumably the surname derived from a Northumberland p.n. (Reaney 1976, Hanks & Hodges 1988). With GLENCOYNE and GOWBARROW, this is one of the **parks** along the NW shore of ULLSWATER.

SWINDALE: S~ BECK NY5113 Shap Rural, S~ COMMON NY4913, S~ FOOT NY5213 (hab.), S~ HEAD NY5012 (hab.).
Swindal(e) & variants c.1200 to 1777 (*PNWestm* II, 171); *Swyndellbeck* 1249 (*PNWestm* II, 177); *Swindale Common* 1858 (OSNB); *Swindale Foot* 1865; *Swindall head* 1693 (*PNWestm* II, 177).
♦ 'The valley where pigs graze'; **svín/swīn, dalr**. Mixed woodland still clothes much of the W side of the valley, and oaks providing pig-fodder may well have flourished here in the past. See also **bekkr/beck** 'stream', **common** 'unenclosed pasture in communal use', **foot, hēafod/head**. (Pl. 2)

SWINE CRAGS NY3608 Rydal.
Far Swine Crag 1828 (HodgsonM), *Near Swine Crag, Far Swine Crag* both 1859 (OSNB).
♦ 'Rocky heights associated with pigs' (**swine, crag**). Since these are at or above 1150ft/350m., actual pannage (pig grazing) seems rather unlikely. *Near* and *Far* in the names of the constituent crags refer to relative distance from RYDAL village.

SWINESCALES NY4127 (hab.) Hutton.
Swineskills 1692/3, *Swinescales in Hutton soyle* 1759 (Greystoke PR).
♦ Probably 'the shielings or high pasture where pigs graze'; **svín/swīn/swine, skáli/scale.** Cf. most examples of SWIN(E)SIDE.

SWINESIDE NY3432 (hab.) Mungrisdale.
Swynesyde 1568, *Swansted* 1794 (*PNCumb* 305), *Swineshead* 1704 (*CW* 1902, 199).
SWINESIDE: S~ KNOTT NY3719 Matterdale.
Swynesyd 1589, *Swinesett* 1604 (*PNCumb* 223); *Swineside Knott* 1900 (OS).
SWINSIDE Examples at:
   NY0614 (hab.) Ennerdale, S~ END (hab.) NY0514.
Swinsyde, Swainside, Swanside 1578, *Swain-side* 1770 (*PNCumb* 387); *Swinside-end* 1648 (PR).
   NY1624 (hab.) Lorton.
Swynesheued 1260, 1292, *Swyneside* (*high close*) 1578 (*PNCumb* 409).
   NY1723 (hill) Buttermere.
Swynsed, Swynsyde 1570 (*PNCumb* 356).
   SD1688 (hab.) Millom Without, S~ FELL SD1588.
Swynesat 1242, *Swinside* 1608 (*PNCumb* 417); *Swinside Fell* 1867 (OS).
   NY2422 (hill, 803ft/245m.) Above Derwent, NY2421 (hab.).
le Swynesyde 1507 (*PNCumb* 373).
♦ Most of these are probably 'the shieling or hill for pig grazing (pannage)'. The 1st el. is in most or all cases 'swine, pig'. Whether it originates in ON **svín** or OE **swīn**, or both, is difficult to tell, although it is most likely to be of the same linguistic origin as the 2nd el. The early spellings for the Ennerdale S~ could point to some other origins, such as the ON pers.n. *Sveinn* or the ON *sveinn*/OE *swān* 'lad'. For the 2nd el. there are at least three possible origins: (a) ON **sætr** 'shieling, high pasture'; (b) OE **sīde**/ON **síða** 'slope, hill-side'; (c) OE

hēafod 'head, high place'; and discriminating between them even tentatively is only possible in the presence of early, preferably medieval, spellings, as when *-sat* 1242 in the Millom example or *-set* 1604 in the Matterdale one point to *sætr* 'shieling'. In the Lorton S~ the 2nd el. is clearly 'head' (OE **hēafod**), which has given rise to the suggestion that this is an instance of a figurative 'swine's head', named for its resemblance to the shape of the animal head in relation to its body (Gelling 1984, 159–60, Gelling & Cole 2000, 175–6). For the elements in the derived names, see also **knott** 'rugged height', **end** and **fell**. Swinside Fell is not a distinct summit, but the hill-sides above S~.

SWINKLEBANK NY4904 (hab.) Longsleddale, S~ CRAG NY4904.
Swin-, Swynkel(l)ban(c)k 1579 to 1692 (*PNWestm* I, 162); *Swinklebank Crag* 1920 (OS).
♦ Possibly 'the slope by *Swinkeld, the spring frequented by pigs', with the reflexes of ON **svín**/OE **swīn** 'pig', ON **kelda** 'spring' and ODan/ME **banke** 'bank, slope'. High and Low S~ are distinguished on larger scale maps. S~ **Crag** is a line of rocks high up on the slope.

SWINSIDE: see SWINESIDE

SWIRL HOW NY2700 (2630ft/802m.) Coniston.
The summit recently called Swirl How 1929 (*CW* 1929, 41); cf. *Swirl Band* 1850 (OS).
♦ Uncertain, but conceivably 'the precipitous mountain', which would describe its E face, or else 'wind-swept'; 1st el. as in SWIRRAL EDGE plus **haugr/how(e)**.

SWIRRAL EDGE NY3415 Patterdale.
Swirrel Edge 1823 (Otley 45), 1828 (HodgsonM).
♦ Conceivably 'the precipitous, giddy-making ridge'. *EDD: swirl*, sense 6 includes the variant form *swir(r)el* in Scotland and Lancs and cites Scots *swirling* 'giddiness, vertigo'. This would well describe this exposed **edge** or arête on HELVELLYN. Another possibility is *EDD*'s sense 8, 'a place among mountains where the wind or snow eddies; a whirling gust of wind', also Scots and NCy. SWIRL HOW also has a dramatic drop and could be of the same origin, but the theory is not supported by the gentler bluff of The Swirls at NY3116 near THIRLMERE and not far from Swirral Edge. *Swirl* and *swirrel* are also Lakeland dial. forms of 'squirrel', but they can hardly apply in Swirral Edge and Swirl How even in the light of Hutchinson's claim that squirrels abounded on the Wasdale mountain-tops before deforestation (1794, I, 581). (Pl. 20)

SYKE BECK/SIKEBECK FARM SD0987 Bootle.
Sykebeck 1702 (*PNCumb* 348).
♦ A **syke** is a small, usually sluggish stream, **beck** normally a more lively one; the **farm** is close to a minor watercourse.

SYKES NY2922 (hab.) St John's/ Castlerigg.
Sikes 1777 (PR), 1867 (OS).
♦ Probably 'the slow streams'; **syke**. The place is on low, flat land meshed by small streams. It is not, however, possible to rule out reference to a family of this name, such as the one documented 1775 to 1811 on a tombstone in St John's churchyard; for p.ns consisting of the pl. of a surname, cf. INMANS.

T

TAILBERT NY5314 (hab.), Shap Rural, T~ HEAD NY5214 (hab.).
Thambord c.1200, *Thannellbord* 1339, *Taylleborth* 1357, *Taylebord* 1384, *Taylbart* 1589–97 (*PNWestm* II, 171–2, q.v. for further forms); *Tailbert head* 1616 (*PNWestm* II, 177).
♦ T~ is a major puzzle, but the spellings in the 2nd el. point to ON *borð* and/or OE *bord* 'board, table', and the whole has been interpreted as ON *tafl-borð* 'playing board', perhaps describing square pieces of land, as with the cognate name in Denmark (*PNWestm, SSNNW* 167).

TAIL O' LING NY4716 Bampton.
Tail o'Lang 1859 (*PNWestm* II, 197).
♦ A quaint and puzzling name for the scarp N of MEASAND BECK. Partial explanations might involve OE *tægl*, referring to a projection of land, **ling** 'heather', or corruption of the surname *Taylor* found in the PRs, but without earlier forms, guesswork is probably in vain.

TARBARREL MOSS NY2025 Lorton.
From 1862 (OSNB).
♦ An unusually fanciful name for what the Ordnance Surveyors more soberly described as a 'small peat moss' (OSNB 1862). The noun *tarbarrel* is recorded from c.1450 (*OED*), and see **moss**.

TARN, THE SD0789 (hab.) Bootle, T~ BAY SD0789.
Tarne 1646 (*PNCumb* 348), *the Tarn* 1663, *Tarn* 1667 (PR), *Tarn House* 1867 (OS); *Tarn Bay* 1867 (OS).
♦ A **tarn** (see **tjǫrn**) is most often a mountain pool, in contrast to the low-lying coastal situation here, but Craigbank Tarn, Pinnermoor T~ and Inman's T~ all lie about a mile to the SE. No pool is marked at this site on the 1867 map or currently, but a deep trench leading from here to the sea may indicate drainage in the past. (I am most grateful to the present owner, T. M. Bowes, for this information.) See also **bay**.

TARN AT LEAVES NY2512 Borrowdale.
Tarn of Leaves 1823 (Otley 30), *Tarn at Leaves* 1867 (OS).
♦ This small mountain pool (**tarn** from **tjǫrn**) 'has a lovely name but no other appeal' (Wainwright 1960, Rosthwaite Fell 1), and the name is obscure. The tarn is characterised in modern times by rushes but not leaves (Heaton Cooper 1983, 97).

TARN BECK SD2397 Dunnerdale.
From 1851 (OS).
♦ 'The stream from the mountain pool'; **tjǫrn/tarn, beck**. It flows out of SEATHWAITE TARN down to the DUDDON. Another Tarn Beck is recorded as *Tarnebeck* 1256 (*PNCumb* 28).

TARN CLOSE SD4695 (hab.) Crook.
From 1836 (*PNWestm* I, 181).
♦ 'The enclosure or farm near the small lake'; **tjǫrn/tarn, close**. The farmhouse was built by Quakers c.1820, on a site next to a wet field, with spring, which was made only in the 1980s into a pond, and no pool appears on the 1st edition OS map. *Tarn* therefore seems to have referred, at most, to a very small pool. (I am most grateful to Mr Nick Barnes-Batty, owner, for this local information.)

TARN CRAG Examples at:
SD1999 Eskdale (referring to LOW BIRKER Tarn).
From 1867 (OS).
NY2907 Langdales (STICKLE TARN).
From 1865 (*PNWestm* I, 208).
NY3009 (1801ft/549m.) Grasmere (EASEDALE TARN).
From 1863; cf. *Tarnhouse* 1716, *Tarnpotts* 1838 (*PNWestm* I, 202).
NY3512 Patterdale (GRISEDALE TARN).
Tarn Crag 1799 (*CN* 515).
NY4807 Longsleddale (GREYCRAG TARN).
Tarn Crag 1857; cf. *Tarn(e)hil(l)crag(ge)* 1580 (*PNWestm* I, 164).
TARN CRAGS Examples at:
NY3013 St John's (HARROP TARN).
From 1867 (OS).
NY3228 Threlkeld (SCALES TARN).
From 1867 (OS).
NY3331 Mungrisdale (BOWSCALE TARN).
Tarn-Cragg 1704 (*CW* 1902, 199).
♦ 'The rocky height(s) by the small mountain pool'; **tjǫrn/tarn, crag(ge)**. The tarns in question are all still in evidence, except for GREYCRAG TARN, which is now merely a marshy watershed. The Threlkeld and Mungrisdale sets of crags form similar arcs to the W of their respective tarns.

TARN FOOT NY3404 (hab.) Rydal.
th' Tarnefoot 1656 (*CW* 1908, 150).
♦ The place lies below LOUGHRIGG TARN; **tarn/tjǫrn, foot**.

TARN GREEN: HIGH T~ G~ SD4185 (hab.) Upper Allithwaite, LOW T~ G~ SD4185 (hab.).
Turnegreene 1627, *Ternegreene* 1627/8 (Cartmel PR); *High Tarn Green*; *Low Tarn Green* both 1851 (OS).
♦ 'The grassy place by the small lake'; **tjǫrn/tarn, green**, also **high, low**. These are situated quite close to HELTON TARN, which, judging from the modern embankments and drainage ditches, may have been larger in the past.

TARN HOWS SD3299 Coniston/ Hawkshead, T~ H~ HOTEL SD3399 Hawkshead, T~ H~ WOOD SD3199 Coniston.
Ternehowys 1538 (*PNLancs* 194); *Tarn Hows Wood* 1850 (OS).
♦ 'The hills by the pool(s)'; **tjǫrn/tarn, haugr/how(e)**. There are five or six low hills, most of which form an **intake**. The pools, now amalgamated through damming, used to be separated by Tarn Moss and called *The Tarns* (*High, Middle* and *Low Tarn*); the nearby farm was *Tarn Hows* (e.g. on the 1850 OS map, and cf. *Tarnhouse* 1598, Hawkshead PR). For the word *hotel* see BEECH HILL HOTEL; see also **wood**.

TARN MOOR NY4821 Askham.
From 1859 (*PNWestm* II, 202).
♦ 'The marsh around the pool'; **tjǫrn/ tarn, moor**. There is no longer a named or defined tarn here, but a general marshy area.

TARN RIGGS SD2790 (536ft/163m.) Blawith/Kirkby Ireleth.
From 1851 (OS).
♦ 'The ridge by the pool'; **tjǫrn/tarn, hryggr/rigg**. This is the short ridge above BEACON TARN. The pl. -s does not seem to have particular significance.

TARNSIDE SD4390 (hab.) Crosthwaite.
Tarn(e)side 1535 (in 1669 copy) to 1815; cf. *del Terne* 1374 (p) (*PNWestm* I, 86).
♦ 'The place beside the pool' (**tjǫrn/tarn, side**), an apt description. See **side** for other local examples.

TAW HOUSE NY2101 Eskdale.
Taythes 1631, *Taythes* 1642, *Toes* 1666, *Thoes* 1667, *Teaths* 1677, *Tyths* 1756, *Toughs* 1773 (PR), *Toes* 1774 (DHM), 1802 (CN 1220), *Tawhouse* 1786 (PR); see also *JEPNS* 2, 58 and Brierley 1974.
♦ Possibly 'manured lands', from dial. *tath(e)* 'dung of sheep or cattle, esp. when pastured on a field in order to manure it' (*EDD*), or some other word related to ON *tað* 'dung' or *taða* '(manured) homefield or the hay from it' (cf. *le Tathes, PNWestm* I, 114), though the development to *Taw* is problematic. The plural -(*e*)s has then given rise to a form of the name containing **house**, cf. nearby WHA HOUSE. Teathes is 'a common field name' (Warriner 1926, 99).

TAYLOR GILL: TAYLORGILL FORCE NY2211 Borrowdale.
Taylor's Gill 1839 (Ford 60); *Taylorgill Force* 1867 (OS).
♦ *PNCumb* 352 suggests an association with the family of William Taylor, mentioned in the Crosthwaite PR for 1718. The **Force** is a cascade at the point where Taylor **Gill** flows into STYHEAD GILL.

TENT LODGE SD3197 Coniston.
From 1851 (OS).
♦ The name indirectly commemorates the brilliant young polyglot scholar Elizabeth Smith, who died of tuberculosis in 1806. She is said to have slept in a tent on the lawn in the hopes that fresh air would benefit her (Lindop 1993, 365–6). *Tent* was adopted from OFr into ME as also, by chance, was **lodge**.

TEWET TARN NY3023 St John's.
From 1867 (OS).
♦ 'The pool frequented by lapwings', with dial. *tewit, tewet* plus **tjǫrn/tarn**. The description still holds true. Perhaps the birds were commemorated here and in THE PEWITS and TEWIT HOW not only for their striking appearance and flight but also because 'their eggs are esteemed a dainty' (Hutchinson 1794, I, 456). Writing on the present distribution of birds in the Lakeland fells, Ratcliffe counts the lapwing among the species that are 'scarce among the higher fells and [found] mainly on the lower moors or in the dale bottoms' (2002, 243). The first citation for *lapwing* in *OED* is c.1050, for *tewit* 15th cent. and for *pe(e)wit* 16th cent., but an earlier occurrence of *tewit* is in *Tequitmos* 1279 (*Furness Coucher* II, 537).

TEWIT HOW NY1412 Ennerdale.
From 1867 (OS).
♦ 'The hill frequented by lapwings', with *tewit, tewet* as in the above name, plus **haugr/how(e)**. This steeply contoured and rather dry hill is a less obvious habitat for the bird than TEWET TARN (q.v.) or THE PEWITS.

THACKTHWAITE NY1423 (hab.) Loweswater.
Thacthwait 1220 (*PNCumb* 412).
THACKTHWAITE NY4225 (hab.) Matterdale.
Thactwyt 1285, *Thakthweyt* 1339 to 1561 (*PNCumb* 257).
♦ 'The clearing where thatching material is found', from ON *þak* 'thatch, thatching material, roof' (cf. OE *þæc*) plus ON **þveit** 'clearing'. (The symbol *þ* in OE and ON represents the initial sound of *thorn*.)

THE: For names in THE, see under the first letter of the next, significant, word

THIRLMERE NY3116 St John's.
water of Thyrlmere 1573; also = *Witheburn Water* 1671, *Leathes-Water called also Thirlmere or Wythburn-water* 1769, *Brackmeer* 1777 (*PNCumb* 35-6).
♦ Probably 'the lake with/at the narrowing', from OE *þyrel* 'aperture, pierced hole' plus OE **mere**[1] 'lake'. The lake had an especially narrow 'waist', spanned by a causeway and bridges, until dammed to form a reservoir in the late 19th cent. (Pl. 19)

THIRLSPOT NY3117 (hab.) St John's.
Thirspott 1616, *Thrispott* 1622, *Thirlspott* 1628 (*PNCumb* 315).
♦ Usually explained as 'giant's pool', from OE *þyrs* 'giant' and ME **potte** 'deep hole, pool'. Although unusual, a Northumbrian *Therspettes* recorded in 1256 may be of the same origin (so *PNCumb*). The modern form in *Thirl-* is influenced by the name of nearby THIRLMERE, and that is presumably the pool referred to.

THISTLETON NY0904 Gosforth.
Thystilton 1318 (*PNCumb* 395).
♦ 'The farmstead where thistles grow', from OE *þistel* 'thistle' and OE **tūn** 'farmstead, village'. The most common species of thistle in present-day Cumbria are the marsh thistle (*Cirsium palustre*), spear thistle (*C. vulgare*) and creeping thistle (*C. arvense*). They flourish in almost all parts of Cumbria, including Gosforth (Halliday 1997, 413-5). High and Low Th~ are distinguished on larger scale maps.

THOMPSON'S HOLME SD3997 Windermere.
Thompson's Holm 1783 (CrosthwaiteM), *Thompson's Holme* 1851 (OS).
♦ 'The island (**holme**) associated with the T(h)om(p)son family', who are recorded at UNDERMILLBECK in documents of the 14th, 16th and 17th cent. (*PNWestm* I, 193).

THORNBANK NY0702 (hab.) Gosforth.
Thornebanc, Thornbank c.1230 (*PNCumb* 395).
THORNEY BANK NY5512 (hab.) Shap Rural.
Thornybancke 1712 (*PNWestm* II, 178).
♦ 'The slope where thorns grow'; **þorn/thorn(ey)**, **bank(e)**. As with other names in *thorn*, this is doubtless a reference to hawthorns.

THORN COTTAGE NY5401 Fawcett Forest.
Previously *High Jock Scar* 1863, 1899 (OS).
♦ Presumably 'the cottage by the (haw)thorns'; **thorn, cottage**.

THORNFLATT/THORNFLAT SD0897 (hab.) Drigg.
Thornflatt 1656 (*PNCumb* 378).
♦ Presumably 'the level ground where (haw)thorns grow', from the reflexes of ON **þorn** (possibly supported by OE **þorn**) and ON *flǫt*.

THORNHOLME NY0608 (hab.) St Bridget Beckermet.
Thorneholme earlier 16th cent. (*PNCumb* 340).
♦ 'The land beside water, where (haw)thorns grow'; **þorn/thorn, holmr**. The site is in the angle formed by WORM GILL as it flows into the R. CALDER.

THORNSGILL BECK NY3723 Matterdale. From 1900 (OS).
♦ Thornsgill was presumably the 'ravine where (haw)thorn bushes grow' (**thorn, gil(l)**), and through which the **beck** flows, but no early evidence is known to me.

THORNSHIP: TH~ GILL NY5513 Shap Rural.
Fornhep 1226, 1232, *Fornischap(p)* 1292, *Thorn(e)shapp(e), -shap* 1395 (p) to 1842 (*PNWestm* II, 166); *Thornship Gill* 1860 (OSNB).
♦ Th~ is a mile SW of SHAP (both just outside the National Park), and the 2nd el. of the name is a version of Shap. The 1st el. has been taken as ON *forn* 'old', but Smith points out that Thornship was not the original settlement at Shap, and proposes instead the pers.n. *Forni* (*PNWestm*); see also **gil(l)** 'ravine with stream'.

THORNTHWAITE NY2225 (hab.) Above Derwent.
Thorn(e)thwayt, -thweyt, -thwait 1230 (*PNCumb* 371).
THORNTHWAITE CRAG NY4309 (2569ft/783m.) Patterdale/Troutbeck/Kentmere.
Thornthatcrag 1614 (*PNWestm* I, 191).
THORNTHWAITE: TH~ HALL NY5116 Bampton.
Thornethwaite & variants 1274 to 1777 (*PNWestm* I, 192); *Thornthwaite Hall* 1863 (OS).
♦ 'The clearing by the (haw)thorn trees', from ON/OE þorn and ON þveit (þ representing the initial sound in *thorn*); see also **crag** and **hall**. The Above Derwent name no longer describes the landscape now that Th~ is the centre of a major Forestry Commission plantation, mainly conifer. Th~ **Hall** is 'L-shaped, with one wing a former pele tower' (Pevsner 1967, 223, and see Curwen 1907, 137–42).

THORNYFIELDS SD4393 (hab.) Crook.
From 1857 (*PNWestm* I, 181).
♦ Presumably to be taken at face value; **thorn/thorn(e)y; field**.

THORNYTHWAITE: TH~ FARM NY2413, TH~ FELL NY2412 Borrowdale.
Thornythwaite (*Liberties*) 1759 (*CW* 1976, 106), *Thornythwaite* 1782 (PR); *Thornythwaite Fell* 1867 (OS).
THORNYTHWAITE NY3922 (hab.) Matterdale.
Thornythwait 1570 (*PNCumb* 223).
♦ 'The thorny clearing', with reflexes of OE þornig and ON þveit (þ representing the initial sound in *th*), though the date of the names is unclear. **Farm** has been added later to the Borrowdale example, and **Fell** has the sense 'upland grazing' there.

THORPHINSTY: TH~ HALL SD4186 Cartmel Fell.
Thorfinsty 1275 (*PNLancs* 200), *Thorphinsty Hall* 1851 (OS).
♦ Th~ is 'Thorfinn's steep path', from the ON pers.n. Þorfinnr and ON **stígr** 'path' (cf. MANESTY). A Thorfinsty is also recorded in Hawkshead PR for 1687. Thorphinsty is among a number of p.ns containing compound ON pers.ns which (on evidence from E. England) are considered to be relatively old, compared with short forms such as *Þorfi in TORVER (*SSNNW* 331). The -*ph*- spelling may be due to classical influence; cf. ULPHA. The **Hall** has features dating from the 16th cent.

THRANG CRAG: TH~ C~ WOOD SD2891 Blawith.
Thrang Cragge 1664 (Torver PR); *Thrang Crag Wood* 1851 (OS).
THRANG CRAG NY4317 Martindale.
Thronge cragge 1588, *Thrang Crage* 1636 (*PNWestm* II, 218).
♦ Th~ C~ is 'the rocky height at the narrow place', with reflex of ON þrǫng 'throng, defile', plus **crag**; also **wood**. The Torver name refers to the defile between the crag and CONISTON

WATER (through which the present A5084 runs). The wood occupies the slope above.

THRANG FARM NY3105 Langdales.
Thrange 1574, *Throng* 1706, *Thrang* 1839 (*PNWestm* I, 204).
♦ 'The narrow place', from ON *þrǫng* 'throng, defile' — apt where two hillocks produce a narrowing in GREAT LANGDALE; **farm** added recently.

THREE DUBS TARN SD3797 Claife, TH~ D~ CRAGS SD3897.
Three Dubs 1851, *Three Dubs Tarn* (*Fish Pond*) 1919 (OS); *Three Dubs Crags* 1851 (OS).
♦ From *three* (ultimately from OE *þrēo*), **dub(b)** 'pool' and **tarn** (from **tjǫrn**). Th~ D~ Tarn is the partly artificial product of three former pools; cf. the neighbouring MOSS ECCLES TARN and WISE EEN TARN, created out of mosses. The **crag**s or rock outcrops rise just to the N.

THREE SHIRE STONE NY2702 Ulpha/Langdale/Dunnerdale/Coniston.
Shirestones upon Wrenose 1576 (*PNCumb* 437), *Three Shire Stones* 1774 (DHM), *the Shire stones* 1777 (*PNWestm* I, 208), *Three Shire Stones or Threefoot Brandreth* 1850 (OS).
♦ All three constituent words are of OE origin: *þrēo*, *scīr*, **stān**, but the name is late medieval or early modern. Before the creation of Cumbria in 1974, this point at the WRYNOSE watershed was the meeting of three counties and of four parishes within them. It was marked by 'three stones called shire stones which are but a foot from each other yet stand in three counties, viz.: Lancashire, Cumberland and Westmoreland' (Fleming 1671, *Description of Cumberland*, cited Rawnsley 1957, 155, though not found in Fleming ed. Hughes). Machell described them as 'long stones pitched up at about a yard distance [with] a spring in the midst of them, which are called the Shire Stones being in the place where Westmorland, Cumberland and Lancashire meet' (c.1692, ed. Ewbank, p. 139; cf. also Hutchinson 1794, I, 571). These were replaced in the 19th cent. by a single limestone monolith marking Lancashire, to which were added, in 1998, three flat stones incised with the letters C, W and L and set into the turf, together with an explanatory plaque (see further Haszeldine 1998).

THREE TARNS NY2405 Eskdale.
Three Tarns or Tarns of Buscoe 1862 (OS).
♦ These are the tiny mountain pools lying in the high col between BOWFELL and CRINKLE CRAGS; *three* from OE *þrēo*, plus **tjǫrn/tarn**. see further BUSCOE.

THRELKELD NY3125 (hab. & parish), TH~ COMMON NY3424, TH~ KNOTTS NY3223 (St John's).
Trellekell' 1197 (p), *Threlkeld(e)* c.1240, *Trellekeld* 1278 (*PNCumb* 252); *Threlkeld Common* 1867 (OS), *Threlkeld Knotts* 1900 (OS).
♦ 'The thralls' spring', from ON *þræll* 'thrall, slave', probably in its gen. pl. form *þræla*, plus ON **kelda** 'spring'. There are mineral springs in the area (Ward 1876, 59). Th~ **Common** is a tract of unenclosed pasture, and Th~ **Knotts** is a rugged height.

THRELKELD LEYS NY1127 (hab.) Blindbothel.
Threlkeld Place 1802 (Mosser PR), 1867 (OS).
♦ Since this is far from THRELKELD

parish, the specific is probably a surname. *Leys* derives from OE *lǣs* 'meadow' (or occasionally from OE **lēah** 'woodland clearing'), but here seems to be a modern replacement of **Place** 'plot of ground, farm', perhaps inspired by p.ns such as (High, Middle, Low) Leys in nearby Lamplugh (*PNCumb* 407).

THRESHTHWAITE COVE NY4210 Patterdale, TH~ MOUTH NY4210 Patterdale/Troutbeck.
Threshthwaite Cove 1860 (OSNB); *a strait place between two dales called the Threshward* [sic], *Threshold mouth* c.1692 (Machell II, 319), *Threshold or Threshthwaite Mouth* 1828 (HodgsonM), *Threshthwaite Mouth* 1860 (OSNB); pronounced [þreʃit] (*PNWestm* I, 191).
♦ (a) The modern form Threshthwaite suggests a settlement-name whose 2nd el. is ON **þveit** 'clearing' or its reflex *thwaite* and whose 1st el. could perhaps be a pers.n. such as ON *Þrǫstr*, which is believed to occur in Norw p.ns including some derived from *þveit* (Rygh 1901, 267, Lind 1905–15, 1225–6). However, if Th~ did refer to a settlement it can hardly have been close to this high wilderness. (b) Machell's *Threshward, Threshold* match standard and dial. forms of *threshold*, a word which as early as 888 AD could be used in reference to a geographical border or limit (*OED*). We must assume either that Machell's form is authentic, and the *-thwaite* form a later rationalisation, or the reverse: Machell's form is a mistaken attempt to make sense of the name in relation to the location. *Mouth* derives from OE *mūð* 'mouth' or from *mūða*, normally 'estuary' and common in p.ns including COCKERMOUTH; here the reference is to 'a narrow pass at the head of Pasture Beck' (OSNB). The **Cove** is a stony arc also at the head of PASTURE BECK.

THROSTLE GARTH NY2204 Eskdale.
From 1867 (OS).
♦ 'Thrush enclosure.' Both words are common in modern dial. *Throstle* is from OE *þrostle* 'thrush', and could refer either to the fell throstle (*Turdus torquatus*) or the song thrush (*T. musicus*), cf. Hutchinson 1794, I, 457 and [10]), and **garth** is from ON *garðr*. This is a spur of Throstlehow Crag by the River ESK, and a rugged and remote spot for a *garth*, which usually refers to habitations, though WHIN GARTH is also a wild spot.

THROSTLE HALL NY3339 Caldbeck.
Throssel-hall 1807 (PR).
♦ Dial. *throstle* is 'thrush' (see above); bird-names quite frequently co-occur with **hall** (see MOORCOCK HALL).

THUNACAR KNOTT NY2708 (2351ft/ 723m.) Langdales.
From 1859 (OSNB).
♦ This **knott** is 'a rocky prominence in Langdale Fell', near the county boundary (OSNB). The origin of the 1st el. is obscure, but Brearley suggested an ON nickname *Thunkarr*, which is backed up by its occurrence in the Icel *Landnámabók*, though the sense would probably be 'with thin, wavy hair', rather than 'thin as a card' as Brearley suggests (1974).

THURSTON SD3196 Coniston.
Thurston 1932 (map relating to sale of the BRANTWOOD estate). Previously *Coniston Bank* 1851, 1892 (OS).
♦ Built in the mid 19th cent., the house stands beside CONISTON WATER (q.v.). It seems to have been renamed, deliberately recalling the older name

of the lake, 'Thorstein's water', and there are similar names nearby, such as Thurston Bank and the house-name Thurstonville (not the one below). Coniston Bank was added to Ruskin's BRANTWOOD estate in 1897 by his relatives the Severns, and may have been renamed about that time — perhaps under the antiquarian influence of John Ruskin's amanuensis W.G. Collingwood, author of *Thorstein of the Mere*. I am grateful to John Batchelor, Vicky Slowe, Andy Stubbs and Stephen Wildman for help in pursuing this line of enquiry.

THURSTON VILLE SD3084 Lowick.
Thurstonvill 1807 (Egton-cum-Newland PR), *Thirston Villa*, p. *Ulverston* 1819 (Kirkby Ireleth PR).
♦ There are several places named Thurston (*OSGazetteer*), e.g. one in Suffolk assumed to be the estate of Þori or Thur (*DEPN*, Watts 2004), and a surname *Thurston* arose either from the p.n. or from the ON pers.n. *Þorsteinn*. In the absence of earlier spellings, it seems most likely that the Lowick name is a relatively modern one containing the surname or a transferred p.n. Alternatively, it is possible that the name could allude to Thurston Water, the old name for CONISTON WATER, though the lake is nearly four miles to the N. The el. *ville* is unique in this volume, and is either a modern borrowing from Fr *ville* 'town' or an adaptation of OE, ME **feld**, ModE **field**, found in Thurstonfield 'Þorsteinn's field', recorded from 13th cent. (*PNCumb* 128).

THWAITE: TH~ HALL NY3735 Mungrisdale.
Thwayte 1509; *Thwait Hall* 1562 (*PNCumb* 212).

♦ 'The hall at Thwaite, the clearing'; **þveit/thwaite, hall**. Thwaite was an alternative designation for the township of HUTTON ROOF in the 16th–17th centuries (*PNCumb* 210).

THWAITE HEAD (hab.) SD3590 Colton, TH~ H~ FELL SD3591, TH~ MOSS SD3389 (hab.).
Whaithead 1628/9, *Tweathead* 1632/3 (PR); *Thwaite Head Fell* 1851 (OS); *Twaitmoss* 1629, *Whaitmosse* 1632 (PR).
♦ 'The upper end of Thwaite, the clearing'; **þveit/thwaite, hēafod/head**, also **fell, moss**. The 'Thwaite' must have lain to the S, since only there is the ground lower than TH~ H~ (this precludes GRAYTHWAITE, suggested by Sephton, 1913, 77). *Thwaite* 'clearing' is common hereabouts, and together with elements such as **ridding** and **intake** signals a centuries-long process of reclamation of woodland and waste in FURNESS FELLS. *Fell* here refers to the unenclosed grazing land above TH~ H~, rather than a distinct peak.

THWAITEHILL/THWAITE HILL NY4521 (hab.) Barton.
Whaithill 1693, *Thwaite Hill* 1721 (*PNWestm* II, 213).
♦ 'The hill at Thwaite, the clearing'; **þveit/thwaite, hill**.

THWAITE(S): THWAITE YEAT SD1888 (hab.) Millom Without, THWAITES FELL SD1690.
Thueites c.1170 (*PNCumb* 417), *Thwayts* 1607/8 (PR); *Thwet-yeat* 1610 (*PNCumb* 417), *Thwaitegate in Parish of Millom Without* 1813 (Nether Wasdale PR); *Thwaites Fell* 1825 (spelling not specified, *CW* 1966, 369), 1867 (OS).
♦ 'Thwaites, the clearings, and the gate and common upland grazing belonging to it'; **þveit, gate** from OE **geat, fell**. Thwaite(s) was the name of

a manor and is now the N part of Millom Without parish. Th~ Y~ is in a valley leading onto the **Fell**.

TILBERTHWAITE: HIGH T~ NY3001 (hab.) Coniston, T~ FELLS NY2801, T~ HIGH FELLS NY2801.
Tildesburgthwait 1196, c.1200, *Tilburthwait* pre-1412, cf. *tillesburc* 1157–63 *(PNLancs* 216); *High Tilberthwaite; T~ Fells; T~ High Fells* all 1850 (OS).
♦ 'The clearing at *Tillesburh*, Tilli/ Tilhere's fort.' A ME form of ON þveit 'clearing' has apparently been added to a p.n. consisting of a pers.n. such as OE *Tilhere* or the shortened *Tilli*, plus OE *burh* 'fortified place' (so Lindkvist 1912, 124, n. 6, *PNLancs* 216, *SSNNW* 254). Collingwood & Collingwood 1923 suggest a site for the fortification, and LDHER 1871 records signs of human exploitation of 'a naturally defensible site'. T~ **Fells** and T~ **High Fells** are all part of the same range, sloping down to the N of WETHERLAM.

TILL'S HOLE NY4805 (hab.) Longsleddale.
Tillzhole 1578, *Till's Hole* 1865 *(PNWestm* I, 164).
♦ Seemingly 'the hollow associated with Till', from a pet form of *Matilda*, and **hol**. The farm is a short distance from TOM'S HOWE, and the two names seem to be counterparts.

TOADPOOL SD4994 (hab.) Strickland Ketel.
Toad Pool 1836, *Tadpool* 1857 *(PNWestm* I, 154–5).
♦ Without earlier forms, this can only tentatively be taken at its face value, as is Toadpool in Derbyshire (recorded from 1714, *PNDerbs* 251). OE *tādige,* *tāde 'toad' and its reflexes make occasional appearances in p.ns from before the Conquest (Smith, *Elements*), and see PADDOCK WRAY for another possible reference to toads; see also **pool**.

TOATHMAIN NY5216 (hab.) Shap Rural.
Todman, Toteman 1279 (p), *Totheman(feld)* 1429, 1625 *(PNWestm* II, 172).
♦ A difficult name, not least because the medieval spellings would not normally develop into the modern form. *Todman* is recorded as a surname, probably meaning 'fox-man' i.e. 'fox-hunter' (cf. p.ns in 'TOD(D)' below), though not in the Shap PRs. Smith suggests that this was originally coupled with the generic el. **f(i)eld,** which was then dropped *(PNWestm).* A County Durham field-name which consistently shows *totheman* in 13th– 14th-cent. documents is also unexplained. It is prefixed *Over-* and *Nethir-* (PNDu I, 74).

TOD CRAGS NY5210 Shap Rural.
Todd Crags 1859 *(PNWestm* II, 179).
TODD CRAG NY3603 (696ft/212m.) Rydal.
From 1865 *(PNWestm* I, 211).
♦ 'The **crag(s)** or rocky height(s) frequented by foxes', with N. dial. *tod(d)*, which seems to have developed from OE *todd 'bushy mass' to a term for 'fox' in or around the 12th cent. (Smith, *Elements*; on the fox in Lakeland see FOX CRAGS). FOX GHYLL and Fox How lie just to the N of the Rydal crag.

TOD FELL NY5101 Longsleddale.
Tod Fell 1770 (JefferysM); cf. *Todcragge* 1629 *(PNWestm* I, 164).
♦ Either (a) 'the hill frequented by foxes', from the dial. word *tod(d)* 'fox' as in TOD(D) CRAG(S), or (b) 'hill associated with the Todd family', who are documented locally from the

16th–17th cent. (*PNWestm*). The **Fell** rises to a distinct summit.

TODGILL NY4022 (hab.) Matterdale.
Todgil(house) 1676 (*CW* 1884, 40), *Toddgill* 1754 (*PNCumb* 258); Gambles cites Tod Gill from 1279, but it is not clear whether this Todgill is meant (1989, 116).
♦ Probably 'the ravine frequented by foxes', with dial. *tod(d)* 'fox' as in TOD(D) CRAG(S) plus **gil(l)** (so also Taylor 1995, 15). The farm is named from the neat ravine through which Todgill Sike flows.

TODRIGG: LOW T~ (hab.) NY4126 Hutton.
Todrigge 1573 (*PNCumb* 213–4); *Low Todrigg* 1867 (OS).
♦ T~ is 'the ridge frequented by foxes', from dial. *tod(d)* 'fox' as in TOD(D) CRAG(S) and **rigg** from ON **hryggr**. The hill Tod Rigg is just N of the farm at Low Todrigg, which is distinguished from Todrigg on the 1867 OS map.

TOM HEIGHTS NY3200 Coniston.
From 1850 (OS).
♦ **Heights** in reference to low ranges of hills is quite common in the former Lancs part of Lakeland, but it is rare for the el. to be qualified, as apparently here, by a pers.n.

TOM RUDD BECK NY1728 Embleton.
From 1867 (OS).
♦ A **beck** or stream presumably named after a locally renowned figure. The surname *Rudd* appears frequently in the Lorton PRs and infrequently in Embleton PRs, which record the baptism of a Thomas Rudd in 1698/9.

TOM'S HOWE NY4804 (hab.) Longsleddale.
Tomshow 1713 (PR), *Thoms how* 1836,

Tom's Howe 1865 (*PNWestm* I, 164).
♦ 'Tom's hill', with **how(e)** (from **haugr**) referring, as often in this part of former Westmorland, to a linear slope rather than a distinct hill. Other local names comprising a pers.n. and a 'hill' term are TILL'S HOLE and WAD'S HOWE.

TONGUE NY2207 Eskdale.
From 1802 (*CN* 1219); perhaps cf. *Tonge bridge* 1540 (*PNCumb* 392).
TONGUE NY4324 (hab.) Matterdale.
Tong 1487, *Tounge* 1580 (*PNCumb* 258).
TONGUE, THE Examples at:
    NY3430 Mungrisdale.
Tongue(-Topp) 1704 (*CW2* 50, 1951, 126), *The Tongue* 1794 (*PNCumb* 227).
    NY3513 Patterdale.
From 1863 (OS).
    NY4206 Troutbeck.
an Hill call'd the Tongue in the midst of the [Troutbeck] parke c.1692 (Machell II, 324).
♦ 'The tongue of land'; **tongue** from ON **tunga**. The Eskdale example is a classic *tongue*: a long, narrow triangle of high ground between Calfcove Gill and the highest reaches of the R. ESK. The Matterdale place lies just W of a modest but perceptible tongue of land between two becks. The Mungrisdale one projects between two branches of the GLENDERAMACKIN, and the Troutbeck one between Hagg Gill and TROUT BECK. The Patterdale one projects NE from DOLLYWAGGON PIKE and, unusually, is not directly framed by streams.

TONGUE GILL NY3410 Grasmere.
From 1859 (OSNB).
TONGUE GILL, NEAR NY2216 Above Derwent.
From 1867 (OS).
♦ 'The stream/ravine by the tongue of land'; **tongue**, **gil(l)**. The Grasmere

stream, together with Little Tongue Gill, frames and forms GREAT TONGUE. **Near** Tongue Gill in Above Derwent similarly forms a tongue of land with Far T~ G~.

TONGUE HEAD NY2408 Borrowdale.
caput de Tunghe (head of Tongue) 1242 (*PNCumb* 352), *Tongue Head* 1867 (OS).
♦ 'The top of the tongue of land', from ON **tunga** and OE **hēafod** or its reflex, perhaps replacing ON *hǫfuð*. The *tongue* lies between ANGLE TARN and ALLEN CRAGS Gill. See also ESK PIKE.

TONGUE HOUSE SD2397 Dunnerdale, T~ H~ HIGH CLOSE SD2397.
Tongue house, tounge house 1681 (CW 1908, 353), *Tongue House* 1774 (West, map); *Tongue House High Close* 1851 (OS).
TONGUE HOUSE NY4506 Kentmere.
Tongue house (field) 1836 (*PNWestm* I, 168).
♦ 'The house by the tongue of land'; **tongue, house**, also **high, close**. T~ H~, Dunnerdale, stands beside HIGH TONGUE, and T~ H~ Close and T~ H~ High Close are named in relation to it. The Kentmere tongue is formed by the R. KENT and its tributary, Ullstone Gill; the house is now a ruin.

TONGUE HOW NY0709 Ennerdale.
From 1867 (OS).
♦ This is the high ground above the tongue of land between WORM GILL and the R. CALDER; **tongue, how**(e) from **haugr**. ON *haugr* frequently referred to burial mounds, and a cairn and ancient enclosures are sited on T~ H~ (Spence 1938, 64), but it is more likely that *how* here simply has its normal Lakeland sense of 'hill'.

TONGUE MOOR NY1603 Eskdale.
From 1867 (OS); cf. *ye Tounge feelde* 1578 (*PNCumb* 392).
♦ 'The upland waste on the tongue of land'; **tongue, moor**. This is a tract of rough upland between two streams feeding the R. MITE.

TONGUE RIGG NY5210 Shap Rural.
From 1859; cf. *Tung(e)fel(l)* 12th cent. (*PNWestm* II, 178).
♦ 'The ridge on the tongue of land'; **tunga, hryggr/rigg**. It lies between SLEDDALE BECK and Tonguerigg Gill.

TORVER SD2894 (hab. & parish), T~ BACK COMMON SD2993, T~ HIGH COMMON SD2695, T~ LOW COMMON SD2792, T~ COMMON WOOD SD2994.
Thoruergh 1190–9, *Thorfergh*, *Torver(e)gh* 1246, *Torweg* 1252, *Toruerg* 1272–80 (*PNLancs* 215); *Torver Back Common*; T~ *High* C~; T~ *Low* C~; T~ C~ *Wood* all 1851 (OS).
♦ 'Torfi's shieling' or 'turf shieling'. The 2nd el. is clearly Gaelic-Norse **ærgi/erg** 'shieling', but the 1st is doubtful, especially because of the alternation of *Th-* and *T-* in the early spellings. (a) It may be an ON pers.n. such as *Torfi* or an unrecorded **Þorfi*, a shortening of some such name as *Þorfinnr* or *Þórólfr* (the preferred solution of Fellows-Jensen, *SSNNW* 72). (b) Alternatively, the 1st el. may be ON *torf* 'turf', referring to peat-cutting or to sods used for the shieling hut itself (so *DEPN*). The shieling grew into a permanent, thriving settlement, and hence belongs to a number of quite favourable sites which appear not to have been exploited in the pre-Norse period (cf. *SSNNW* 351–2, 374). T~ Back is a range of low fells on the edge of the

parish, parallel with CONISTON WATER; the word *back* appears to be used as an adj. in p.ns only from the late medieval period (*VEPN: bæc*). See also **common** 'unenclosed pasture for communal use', **high**, **low**, **wood**.

TOTTLE BANK SD2688 (hab.) Blawith.
Totlbank 1612 (*PNLancs* 214).
TOTTLEBANK SD3184 (hab.) Colton.
Totlebanke c.1535 (West 1774, 104), *Totilbanke* 1537 (*PNLancs* 217).
♦ T~ is 'the look-out hill', with the reflex of OE *tōt-hyll* (cf. Smith, *Elements*). The Blawith site is at the foot of Tottlebank Height and close to BLAWITH KNOTT on the parish boundary. The Colton example is a modest but well-shaped height, and cf. nearby Spy Hill SD3188 (656ft/ 200m.). The addition of **bank**, here meaning 'hill-slope', would become necessary as the original el. **hill** was effaced.

TOWER PLANTATION, GREAT SD3991 Cartmel Fell.
From 1851 (OS); cf. *Towerwood in Cartmelffells* 1605 (Cartmel PR), *Tower wood* 1783 (CrosthwaiteM).
♦ *Tower* is an adoption into ME from OFr, only otherwise found in the p.ns in this volume in LYULPH'S TOWER. Whatever structure the name referred to evidently predated the 17th cent. The alternative possibility that 'Tower' is a surname cannot be ruled out, though *Towers* is usual in local documents. **Great** qualifies the **Plantation**, not the tower.

TOWN BANK NY0710 Ennerdale.
From 1867 (OS).
♦ **Bank** is appropriate to this hill-slope on KINNISIDE COMMON, while **town** may designate a unit of land rather than a nucleated settlement since the place is a mile from FARTHWAITE and other settlements in the CALDER valley.

TOWN END Examples at:
 NY3406 Grasmere.
Towne end 1647 (PR).
 SD3598 Claife.
From 1770 (JefferysM).
 NY3635 Caldbeck.
the towne end of Haltley 1652 (*PNCumb* 280).
 SD3687 Colton.
From 1739 (Finsthwaite PR).
 SD3795 Claife.
From 1851 (OS).
 NY4001 Troutbeck.
Troutbeck Towne End c.1692 (Machell II, 313), *Town End* 1865 (*PNWestm* I, 191).
 SD4483 Witherslack.
ye Town end 1701 (*PNWestm* I, 79).
♦ 'The edge of the village'; **town**, **end**. Several villages or hamlets have their 'Town End', which also occurs as a field-name. The Grasmere example is the hamlet containing the famous DOVE COTTAGE, a southern outlier to G~ village; it contrasts with TOWN HEAD NY3309. T~ E~ in Caldbeck was the S. limit of HALTCLIFF, while the Colton example is at FINSTHWAITE, over half a mile from the church, and the Claife ones are at the outer limit of old HAWKSHEAD (SD3598), and at FAR SAWREY, by the church. The Troutbeck one is the lower end of TROUTBECK village, and is the name of a fine 17th-cent. yeoman's house; the Witherslack one is the end of W~ village.

TOWN HEAD Examples at:
 NY2538 Ireby and Uldale.
Aughertree Townhead 1867 (OS).
 NY3239 Caldbeck.
Townehead 1720 (*PNCumb* 280).
 NY3309 Grasmere.
Towne head 1653 (PR).

SD3887 Staveley-in-Cartmel.
Town head 1783 (CrosthwaiteM).
NY4103 Troutbeck.
From 1865 (*PNWestm* I, 191).
♦ 'The upper end of the village'; **town, head**. The Ireby and Uldale example is a small outlier of AUGHERTREE, to its S; the Caldbeck one is an outlier of CALDBECK village, situated above UPTON. The Grasmere and Troutbeck places are at the upper (and northernmost) extremities of those villages, and both are matched by a TOWN END. T~ H~ in Staveley seems too far from the village to refer to that, and here **town** may refer to a unit of land rather than a village.

TOWN YEAT SD4591 (hab.) Underbarrow.
Town(e)yeat(e) 1535 to 1769 (*PNWestm* I, 86).
♦ 'The gate of the hamlet'; **town, geat**. The place is midway between the hamlets of UNDERBARROW and CROSTHWAITE, and very close to the boundary of the two corresponding townships.

TOWTOP KIRK NY4917 Bampton.
From 1903 (Noble 265); cf. *Tautop* 1770 (which, however, may be Towtop in Cartmel Fell, *PNWestm* II, 192–3, and II, xiv).
♦ 'Towtop' is obscure. (a) Derivation from *t'howe top* ('the hill-top', **how(e), top**) is tentatively suggested in *PNWestm* II, 192–3, and would be paralleled by TOW TOP PLANTATION (below) and a Tow Top in NYorks (SD9646; not in *PNNYorks*). But these two are on steeply sloping sites whereas there is no prominent hill-top at Towtop Kirk. (b) If the 1st el. were from ON *taufr* 'witchcraft' (*PNWestm* II, xiv), the 2nd el. would be unexplained. (c) The area just to the S is *Towthwaite* (e.g. OS 1863); if this is connected with Towtop, it argues against interpretation (a). As to the final el., *kirk* from ON *kirkja* 'church' (and possibly OE *cirice*) is frequently used for ancient stone monuments in Cumbria and Yorkshire, and The Kirk here (described in Noble 1903) is a circular earthwork inset with stones.

TOW TOP PLANTATION SD4183 Upper Allithwaite.
From 1851 (OS).
♦ T~ T~ may be 'the top of the hill' as possibly in TOWTOP KIRK; plus **plantation**.

TRANEARTH SD2895 (hab.) Torver.
Tranneth (House) 1824 (PR), *Tranearth* 1851 (OS), *Trannorth* 1900 (Collingwood 100).
♦ Possibly 'ground frequented by cranes or herons', but without earlier spellings this is highly conjectural. The 1st el. resembles, and may be connected with, ON *trani, trana* 'crane', found in several Lakeland p.ns (Gambles 1989, 148–9). Though now extinct in Britain, the crane was well documented in the past, and the heron was also vulgarly called *crane* (Hutchinson 1794, I, 455). If the 2nd el. is 'earth' from OE *eorðe* 'ground, soil' or (less likely here on the fell above TORVER BECK) 'ploughed land', the name may be comparable with Hawkearth (recorded from 1397, *PNWestm* I, 180). Corruption of some other el., e.g. **wath** 'ford' as in the undated *Tranewath* in Lancaster (*PNLancs* 253), is also possible. Fleming T~, Frank T~ and Matthew T~ are distinguished on the 1:25,000 map (and were in 1851).

TRANTHWAITE: T~ HALL SD4793 Underbarrow.
Tranthwett 1170–84, *Tran(e)thwait(e)* &

variants 1186–1200 to c.1722 (*PNWestm* I, 103); *Tranthwaite Hall* 1858 (OSNB).
♦ 'The clearing frequented by crane(s) or heron(s)', or 'Trani's clearing'. The 1st el. is ON *trani* or *trana* 'crane', or the pers.n. *Trani*. Since herons are among the local wildlife, the reference may be to them (cf. TRANEARTH above). The 2nd el. is ON þveit 'clearing'. The **hall** is a substantial and ancient farmhouse.

TROUGHTON: T~ HALL SD2591 Broughton West.
Troughtona 1422 (p), *Troughton Hall* 1599 (*PNLancs* 221).
♦ 'The homestead in a valley.' Unless appearances deceive, this is a rare instance in FURNESS FELLS of an OE name, from OE *trog* 'trough', hence 'valley' and OE **tūn** 'farmstead, settlement, village'; see also **hall**.

TROUTAL SD2398 (hab.) Dunnerdale, T~ FELL SD2599, T~ TONGUE SD2398.
Trutehil 1157–63 (*PNLancs* 223); *Troutal Fell* 1850; *Troutal Tongue* 1851 (OS).
♦ T~ is 'the trout pool', from OE *truht* 'trout' (with spelling, like that of *trout*, influenced by OFr *troute, troite*) and ON *hylr* 'pool'. In the 1157–63 document, *Trutehil* refers to a pool in the DUDDON, and salmon and sea-trout frequent the deep pools in the river at this point (information from Troutal Farm, gratefully acknowledged). *Lonnin(g) Side* is reported as an old name for the farm at Troutal (*JEPNS* 2, 60, 'SGJ' 1961). In T~ Fell senses (a) and (b) of **fell** (q.v.) may well be combined. For T~ **Tongue** see HIGH TONGUE.

TROUT BECK NY3825 (stream) Matterdale/Mungrisdale/Hutton, TROUTBECK NY3827 (hab.) Matterdale.
(Stream:) *Trutebek, Trytebek* 1292 (*PNCumb* 29); (hab.:) *Troutbek* 1332 (p) (*PNCumb* 222).
TROUT BECK NY4104 (stream) Ambleside/Troutbeck, TROUTBECK NY4003 (hab. & parish), TROUTBECK BRIDGE NY4000 (hab.), T~ PARK NY4205 (hab.), T~ PARK NY4207 (area of fell).
(Stream:) *Trutebyk* 1292, *Trowtbeke* 1535 (*PNWestm* I, 14); (hab.:) *Trutebek* 1272, *Trout(e)bek* & variants 1324 to 1656, *Trought(e)bek* 1437 to 1451 (*PNWestm* I, 188); *Trowtbeke brigge* 1453 (as hab., CW 1906, 18), *Troutbeck bridge* 1686 (as hab., Grasmere PR); *Trout(e)-, Trowt(e)be(c)kpark(e)* 1376 to 1823 (*PNWestm* I, 189).
♦ 'The trout stream', with *trout* as in TROUTAL, plus **bekkr**, and hence the settlement beside it; also **brycg/bridge, park**. The Ambleside/Troutbeck river, as Machell observed, 'has plenty of Troutes and some very large one[s] about Michaelmas Time' (c.1692, II, 422, ed. Ewbank, 124). T~ Park, with Old Park, was one of two deer parks enclosed, probably, in the later 13th cent., being turned into an enclosed farm in the 16th cent. (Parsons 1993, 115–7). The then T~ Park is drawn on Jefferys' map of 1770.

TROUTDALE COTTAGES NY2517 Borrowdale.
From 1901 (OS).
♦ 'The cottages in Troutdale, the trout valley'; *trout* as in TROUTAL, **dale, cottage**. This secluded place was the site of a salmon and brown trout hatchery in the 19th cent. (Danby 1994, 105), and the name appears to be relatively recent, since neither buildings nor name appear on the 1867 OS map.

TRUSMADOOR NY2733 Ireby and Uldale.
From 1867 (OS).
♦ 'The pass called Trusma, the door-like place.' Despite its late appearance

in the records, this name contains Brittonic elements rare or unparalleled in northern p.ns. *Trus* is a Cumbric counterpart to W *drws* 'doorway, pass, gap in mountains', also found in TRUSS GAP; *-ma* is a Brittonic suffix meaning 'place', and English **door** must have been added subsequently (see Whaley 2002). This is a spectacular gap between MEAL FELL and GREAT COCKUP. (Pl. 1)

TRUSS GAP NY5113 (hab.) Shap Rural.
From 1703, 1728 (*Later Records* 363, 365).
♦ Both T~ and G~ denote gaps in the landscape, here referring to a V-shaped narrowing in SWINDALE. T~ appears to be a Cumbric variant of W *drws* 'doorway, pass', also found in TRUSMADOOR, and **gap** has probably been added in modern times (Whaley 2002). (Pl. 2)

TULLYTHWAITE: TULLITHWAITE [*sic*] HALL SD4790 Underbarrow, TULLYTHWAITE [*sic*] HOUSE SD4791.
Tillouthwayt & variants 1326 to 1401, *Tillithwayt* 1372, *Tullyth(a)wayte* & variants 1373 to 1670; *Tullywhait Hall* 1558 (*PNWestm* I, 103); *Tulithwaite House* 1829 (Parson & White 656), *Tullythwaite House* 1858 (OSNB).
♦ T~ is believed to be 'Tillaug's clearing', from an ON pers.n. such as *Týlaugr* plus ON **þveit** 'clearing' (*PNWestm*); also **hall** and **house**.

TURNER HALL FARM SD2396 Dunnerdale.
Turner How 1709 (original spelling not specified, Broughton-in-Furness PR), *Turnerhow* 1772, *Turnerhall* 1814, *Turnerhall Turnerhow* [*sic*] 1816 (Seathwaite PR), *Turner Hall* 1851 (OS).
♦ T~ is presumably the surname. **Hall** seems, as often in this locality, to replace **how(e)** 'hill' (see **haugr**), which could refer to the hillock by which the **farm** stands.

U

ULCAT ROW (hab.) NY4022 Matterdale. *Ulcotewra* c.1250, *Ulcotwra* 1487, *Ulcatrow* 1610 (*PNCumb* 257).

♦ U~ is 'the cottage(s) frequented by owls', from OE *ūle* 'owl' and **cot(e)** 'cottage, hut' (also found in Ullcoats, *PNCumb* 341). To this ON **vrá** (later **wray**) 'nook, secluded place' was added but subsequently replaced by **row**, which well describes the hamlet which grew up along the line of the road.

ULDALE NY2436 (hab., parish of Ireby and U~), U~ FELLS NY2634, U~ MILL FARM NY2337.
Ulvesdal' 1216, *Uluedal(e)* 1228 to 1399, *Ulledale* 1332 (*PNCumb* 327); *Uldale Mill* 1755 (Uldale PR).

♦ 'Ulf's valley' or 'wolf's/wolves' valley'. The 1216 form suggests the ON pers.n. *Ulfr* as the 1st el. but *ulfs* or *ulfa*, gen. sing. or gen. pl. of the corresponding appellative *ulfr* meaning 'wolf', are also possible. The 2nd el. is ON **dalr** 'valley'; see also **fell**, **mill** from **mylen**, **farm**. U~ Fells comprise a range of sizable hills N of (or 'Back o'') SKIDDAW.

ULGRAVES SD5199 (1090ft/332m.) Strickland Roger.
From 1857 (*PNWestm* I, 159).

♦ Possibly, as Smith suggests, 'wolf-pits' from ON *ulfr* 'wolf' and ON *grof* 'pit', referring to pits for trapping wolves (*PNWestm* I, 159 and xvi); but the high situation perhaps makes this unlikely, and without earlier forms certainty is impossible.

ULLOCK NY2423 (hab.) Above Derwent. Probably = *Hulueleyc* (p) early 13th cent. (*Fountains Chartulary* I, 47), *Uloke* 1564 (PR), certainly = *Ullock* 1774 (DHM); cf. *Ullaikmire* 1304 (*PNCumb* 373).

♦ 'Wolves' play, the place where wolves play', from *ulfa*, gen. pl. of ON *ulfr* 'wolf' and ON *leikr* m. 'play, sport'. On the wolf in Lakeland, see WOLF CRAGS.

ULLOCK PIKE NY2428 Bassenthwaite. *Gloomy Ullock; a descendant hill of parent Skiddaw* 1780 (West 124), *Ullock Pike* 1867 (OS).

♦ 'The peak by Ullock', with p.n. plus **pike**. This hill can hardly be named from ULLOCK NY2423, above, which is over three miles to the S. It is, however, immediately E of MIRE HOUSE, so that it is possible that the 'mire' in question was *Ullaikmire* 1304 (again see ULLOCK above).

ULLSCARF NY2912 (2370ft/726m.) Borrowdale/St John's, U~ GILL NY3012 St John's.
Both from 1867 (OS).

♦ A distinctive and puzzling name, possibly meaning 'wolf gap'. No early forms are known, though Brearley (1974) cites *Scarthulfe* 1190 and *Ulvescarth* 1203 as Cumbrian comparanda. The 1st el. appears to be either the ON noun *ulfr* 'wolf' or its counterpart, the pers.n. *Ulfr*. The 2nd el. looks like a reflex of ON **skarð** 'gap, col', for *scarth* can appear as *scarf*: see SCARTH GAP; Collingwood in fact gives this hill-name as Ullscarth (1933, 36, 349). It is not clear, however, what gap would be meant, since U~ is notable, if anything, for 'the few breaches in its seven-mile circumference' (Wainwright 1958, Ullscarf 2). See also **gill**.

ULL STONE NY4508 Kentmere.
From 1857 (*PNWestm* I, 168).
♦ A huge boulder (illustrated in Wainwright 1957, Harter Fell 3, 4). Why 'Ull' is unknown — conceivably from ON *ulfr* 'wolf' or the pers.n. *Ulfr* (cf. ULLSCARF), plus **stone**.

ULLSWATER NY4220 Dacre/Barton/Martindale/Patterdale/Matterdale.
Ulueswater 1220–47, *Ulleswat(e)r* 1357 to 1671 (*PNWestm* I, 17–18; see also *PNCumb* 36).
♦ 'Ulf's lake', from ON pers.n. *Ulfr* plus ME *water* influenced in usage by ON *vatn* 'water, lake' (see **wæter**). *Ulfr* is also the ON noun meaning 'wolf', and Hutchinson thought that the name might refer to the lake as a resort of wolves, or 'still more probably' (to him though not to modern scholarship) to its elbow-shaped bend (citing a Celtic *ulle*, 1794, I, 434 and n.).

ULPHA SD1993 (hab. & parish), U~ FELL SD1795, U~ FELL NY2502, ULPHA PARK SD1890.
Wolfhou 1279 (p), *Ulfhou* 1337, *Ulpho* 1449, *Ulpha* 1625 (*PNCumb* 437); *Ulpha Fell* 1825 (= SD1795, CW 1966, 370), *Ulpha Fell* 1867 (= SD1795, NY2502); *Uffay Park* 1576, *Uffay or Woolfhay Park* 1610 (*PNCumb* 437).
♦ 'The hill frequented by wolves', from ON *ulfr* and ON **haugr** 'hill, mound'. The 1279 spelling appears influenced by early English *wulf*, while the modern *-ph-* rather than *-f-* or *-v-* may have been introduced by scribes educated in the classics (cf. THORPHINSTY). Two tracts of upland grazing are shown as U~ Fell on the 1994 One Inch map, one ten miles N of Ulpha village, the other towards the S of this elongated parish, and both exemplify **fell** sense (b); see also **park**. The hamlet of Ulpha was formerly referred to as Ulpha Kirk.

ULPHA SD4581 (hab.) Meathop, U~ FELL SD4581.
Ulvey(pol) 1280, 1354, *Ullvay*, *Ulluay* 1420, 1612, *Ulva(pool)* 1566 (*PNWestm* I, 76); *Ulpha Fell* 1862 (OS).
♦ Possibly 'wolf-enclosure' or 'wolf-trap', from OE *wulf-hege* (so *PNWestm*, following Ekwall *DEPN*). The initial el. has been influenced by ON *ulfr*, and the spelling may be due to influence of the ULPHA above. U~ **Fell** is a modest but prominent hill.

ULTHWAITE/ULLTHWAITE RIGG NY5109 (1648ft/502m.) Shap Rural.
Ulthwaite Rigg 1858 (OSNB), *Ullthwaite Rigg* 1865 (*PNWestm* II, 178).
♦ 'The ridge above Ulthwaite.' U~ may be 'wolf clearing' (ON *ulfr* and ON **þveit**), as is another Ullthwaite (*Ulvethwayt* 1301, *PNWestm* I, 170), but without early spellings certainty is not possible. **Rigg** is ultimately from ON **hryggr**.

UNDERBARROW SD4691 (hab., parish of U~ and Bradleyfield).
Underbarro(e) & variants 1517 to 1736 (*PNWestm* I, 100).
♦ '(The place) under the hill', with the preposition *under* (from OE *under*) as 1st el., plus **barrow** from OE **berg** 'hill', perhaps referring specifically to Helsington Barrows (so *DEPN*), recorded as *Helsington barrey* 1170–84, (*le*) *Berghes* 1301, *Le Bergh de Helsington* 1332 (*PNWestm* I, 109). As Smith points out, this kind of elliptical p.n. in *under* is common in the former Westmorland (*PNWestm* I, 100), though there are also Cumberland and Lancashire examples among the following entries.

UNDER CRAG Examples at:
   SD2396 (hab.) Dunnerdale.
Under Cragg 1724 (*CW* 1961, 238), *Undercragge* 1733/4 (Seathwaite PR), *Under Crag* 1774 (DHM), 1851 (OS).
   SD2794 (hab.) Torver.
Undercrag 1591 (*CW* 1958, 66), 1599 (PR).
   NY3630 (hab.) Mungrisdale.
Undercragge in Grysdell 1588 (*PNCumb* 227).

♦ '(The place) below the rocky height', with preposition *under* as in UNDERBARROW forming an elliptical p.n. with **crag**. The crags in question are Hollin House Haw (Dunnerdale), Banks (Torver), and RAVEN CRAGS (Mungrisdale).

UNDER HOWE NY3728 (hab.) Mungrisdale.
Underhow 1752 (*PNCumb* 227).
♦ '(The place) below the hill'; *under* as in UNDERBARROW, plus **how(e)** from **haugr**, referring to the same neat elevation as HOWE, NEAR & FAR.

UNDERMILLBECK: U~ COMMON SD4295 Crook.
Undermilnebek & variants 1376 to 1636; *Undermillbeck Common* 1865 (*PNWestm* I, 178); cf. *Mulnebec* 1220–46 (*PNWestm* I, 11).
♦ U~ is '(the place) S of the millstream' — as Machell said, it 'takes [its] name from the river above [it] . . . which drives a mill here' (Machell c.1692, ed. Ewbank, 117); *under* from OE *under*, **mylen** 'mill', **bekkr** 'stream'. Ekwall is correct in assuming that *under* here means 'south of' (*DEPN*); the beck rises at about SD4298. See also **common** 'unenclosed pasture for communal use'.

UNDERSCAR NY2725 (hab.) Underskiddaw.
From 1867 (OS).
♦ '(The place) beneath the scarp or crag'; *under* as in UNDERBARROW, plus **scar**. It nestles below the steep SE contours of the SKIDDAW massif, and the name may be influenced by Underskiddaw (*subtus Skedow* 1508, *Undr Skedow* 1567, *PNCumb* 321).

UNDERWOOD NY1024 (hab.) Blindbothel.
From 1774 (DHM).
UNDERWOOD NY4223 (hab.) Matterdale.
From 1900 (OS), previously *Woodhouse* 1867 (OS).
♦ U~ is probably a construction of preposition *under* (cf. UNDERBARROW) and noun **wood**, hence '(the place) below the wood', an apt description of the Matterdale example. Alternatively, it could be a reference to *underwood* in the sense of coppice-wood or brushwood growing beneath taller trees (recorded from before 1325, *OED*).

UPTON NY3239 Caldbeck.
Caldebeck Upton 1519, *Uptoune* 1540 (*PNCumb* 276).
♦ 'The upper village' or 'higher settlement', with elements derived from OE *uppe* 'above' and OE **tūn** 'settlement, village'. This is an outlying part of CALDBECK, a short way below TOWNHEAD. Hutchinson remarked that 'that part of [C~] which lay near the church, being higher than the rest, was called *C~ Upperton, Uppeton* or *Upton* whilst the part nearer the mountains naturally got the name of *C~ Under-Fell*'; a more recent subdivision was the *East-End* (1794, II, 377).

UZZICAR NY2321 (hab.) Above Derwent.
Huseker 1160, *Husaker* 1160 to 1578; cf. *Usakredale* 1369 (*PNCumb* 372), *Husacre* 1774 (DHM).
♦ 'The cultivated land by the building' — which must have been

notable by local standards. This is a classic case of the difficulty of distinguishing Norse from English (or mixed) origins, since the elements are ON **hús** or OE **hūs** 'building, house' and ON *akr* or OE *æcer* 'cultivated land, field'. U~ stands slightly above the flat floor of the NEWLANDS valley which before drainage contained a tarn.

V

VAUGH STEEL NY4918 (hab.) Bampton. *Vah Stile* 1681, *Vah Steel(e)* 1687 to 1732 (*PNWestm* II, 197), *Steel* 1770 (JefferysM), 1860 (OSNB), *Vaugh or Vah Stile* 1901 (Noble 149), *Vaugh Steel* 1920 (OS).
♦ The 1st el. has not been identified. Possibilities might include the reflex of OE *fāh* 'variegated' or of OE/ON *fall* 'clearing, felled woodland' as in FAWE PARK, but in either case the voicing of *f* to *v* would be puzzling. The 2nd el. appears to be from OE *stigel* 'steep place, stile' (see **steel, stile**).

VICTORIA BAY NY2520 Above Derwent. From 1862 (OSNB); previously *Mutton Pie Bay* 1787 (*PNCumb* 373).
♦ Presumably one of the numerous 19th-cent. tributes to the redoubtable queen who reigned 1837–1901. The **bay** is on the W side of DERWENTWATER.

WABERTHWAITE SD1093 (hab. & parish), W~ FELL SD1393, HALL W~ SD1095. *Waybyrthwayt'* c.1210 (p), *Wayburthwayt* c.1215, *Warthebuthewaite, Weibuththwait* c.1230 (p), *Waybuthethwait, Waybouthueit* c.1250 (p), *Waithbuthwait* c.1255 (p), *Waythebuthwayt* c.1260 (p) (*PNCumb* 439); *Waberthwaite Fell* 1867 (OS); *Hallwaberthwaite* 1702 (PR).
♦ Probably 'the clearing by the hunting or fishing hut'. The early forms, of which this is only a selection, do not guarantee a single etymology, but they point unequivocally to ON **þveit** 'clearing' as the generic, and probably to ON **veiði-búð* 'hunting or fishing hut' as the specific, or possibly a synonymous *veiði-búr*, as suggested by Gillian Fellows-Jensen (*SSNNW* 173). W~ **Fell** is a tract of upland grazing. The pattern of *hall* plus p.n. seen in Hall W~ is favoured locally: see **hall**.

WADCRAG NY1829 (hab.) Embleton. *Wadcragg* 1709, *Wadda Crag* 1726 (PR), *Wadd Crag* 1774 (DHM).
♦ Obscure. *Wad* can refer in dial. to black lead (plumbago, graphite), or to a mark, measuring or dividing line, or direction (*EDD, OED: wad* n. 2, 3). It is not clear which, if either, of these applies here, but there is a local tradition that the reference is to graphite (I am grateful to Dorothy Graves for this information), and the place is close to a current parish boundary. As with THE CRAG in nearby Setmurthy, the **crag** here is a rocky eminence that is modest but rather rare in the locality.

WAD'S/WADS HOWE NY4903 (hab.) Longsleddale.

Waddshow 1647 (*PNWestm* I, 164).
♦ 'Wad(de)'s hill', from a pers.n. or derived surname which may have originated as a shortening of some name such as *Waltheof* (Hanks & Hodges 1988: *Wadds*), plus **how(e)** from ON **haugr**, here, as in nearby TOM'S HOWE, applied to a linear rather than rounded slope.

WAINGAP SD4596 (hab.) Crook. From 1857 (*PNWestm* I, 181).
♦ 'The pass wide enough for a wagon.' Despite the lack of early forms, Smith's suggestion seems plausible, that this is a compound of *wain* (OE *wægen* 'cart, wagon'), and **gap** (*PNWestm*).

WAKEBARROW SD4487 Crosthwaite. From 1857 (*PNWestm* I, 86).
♦ Probably 'look-out hill', with elements from OE *wacu* 'watch, look-out' and OE **berg** 'hill'; cf. also Wakeburgh Scar (*PNWestm* I, 61). Though not the highest point on WHITBARROW, this has extensive views to the E.

WALLA CRAG/WALLOW CRAG NY2721 (1234ft/376m.) Borrowdale. *Walla-crag* 1769 (Gray 1079), *Wallow Crag* 1776 (Hutchinson 158), *Wallow Crag* 1867 (OS); cf. *Willow's craggy brow* 1794 (Cumberland 1777, quoted by Hutchinson 1794, II, 168, though his own form is *Wallow Crag*, II, 176).
WALLOW CRAG NY4915 Shap Rural. *Walla Crag* (1799 CN 510).
♦ Both **Crag**s are rock-topped hills. The 1st el. is problematic, but (a) could be the dial. adj. *wallow, walla* in some such sense as 'indistinct in colour' (*EDD: wallow*, sense 4) or 'of the

weather: blowing a cold, strong and hollow wind' (sense 5). Other remote possibilities are (b) OE *walu* 'ridge, embankment'; (c) *wallows*, baths of mud used by red deer along their lines of travel (Hervey & Barnes 1970, 187–8); (d) OE *wealh* 'Briton'.

WALL END Examples at:
SD2287 (hab.) Broughton West. From 1669/70 (Broughton-in-Furness PR).
NY2805 (hab.) Langdales.
ye wallend in langdaill 1571 (*CW* 1908, 200), *the Walend* 1612 (*CW* 1986, 143), *Wall-End* 1693 (Grasmere PR).
NY3913 (hab.) Patterdale.
Wallend 1860 (*PNWestm* II, 227).
♦ Self-explanatory; *wall* from OE *wall*, plus **end**. Concerning the Langdale example, Smith, citing RCHM, notes that it is near the end of an ancient enclosure, hence accounting for *wall* as the 1st el. (*PNWestm* I, 208). The Patterdale name may reflect proximity to DEEPDALE Park.

WALLENRIGG SD2289 (hab.) Broughton West.
Wallenrigg(s), *Whallenriggs*, *Walingrigg* 1668–9 (Broughton-in-Furness PR), *Whallenrigg* 1701, 1703 (original spelling not specified, *CW* 1932, 65–6).
♦ The 1st el. is uncertain; 2nd is presumably **hryggr/rigg** 'ridge'.

WALLOW CRAG: see WALLA CRAG

WALLOWAY NY4125 (hab.) Hutton.
Wallow Way (*Grene*) 1649 (*PNCumb* 214), *Wallaway*(*green*) 1867 (OS).
♦ Uncertain. That the 2nd el. is *way* from OE *weg* is suggested by the situation, on an old routeway between GREAT and LITTLE MELL FELL (e.g. Hutchinson 1794, I, 412). The 1st el. is highly problematic, with possibilities including: (a) OE *walu* 'ridge, bank' (so Sedgefield 1915); (b) pale or indistinct in colour; (c) a muddy place; (d) of the Welsh (for all four see WALLA CRAG). Two solutions which would involve a different interpretation of '-way' are: (e) a corruption of *Galloway*, referring to pack-horses or cattle; (f) a variant of *walaway* 'an exclamation of sorrow' (*EDD*, sense 1). This is the site of what may be a Roman fort (Clark 1995, 16), though evidence is lacking (LDHER 11845).

WALLOWBARROW SD2196 (hab.) Ulpha, W~ HEALD SD2197 (1336ft/407m.).
Wallobarrow 1646, *the Walley barrey* 1660 (*PNCumb* 438), *Wallabarrow* (*in*) (*Ulpha*) 1737, *Wallbarrow Ulpha* 1757, *Wallawbarrow in Ulpha* 1768 (Seathwaite PR), *Wallabarrow* 1774 (DHM); *Wallowbarrow Heald* 1867 (OS).
♦ A difficult name: possibly 'pale hill' or 'hill with an embankment or deer wallow' (see WALLA CRAG and **berg**). W~ Heald is the long fell above W~; *heald*, from OE *helde* 'hill-slope' is used of contours rising in long lines; cf. H~ BROW, PASTURE and WOOD.

WALLOW CRAG: see WALLA CRAG

WALLS SD0895 (antiquity) Muncaster.
Waw (*castell*), *Wawes* (*closse*) 1578 (*PNCumb* 426), *Walls* 1774 (DHM).
♦ Walls Castle is the name given to the unusually well-preserved Roman bath-house at GLANNOVENTA (see survey report in Brann 1985). The word *wall* is simply from OE *wall*. (Pl. 28)

WALLTHWAITE NY3526 (hab.) Matterdale.
(*le*) *Wakethwayt*& variants 1359 to 1576, *Walltwhate* 1487, *Walcktwaith* 1571 (*PNCumb* 222).
♦ Probably 'the clearing where fulling took place', from OE or ME

walc-, walk- 'fulling, preparation of cloth', plus **þveit** 'clearing'. This is suggested by the early spellings and by the situation beside a ford in the MOSEDALE BECK. The processes of soaking and beating the cloth were often undertaken in water-powered mills in Lakeland valleys, evidence for which exists from the 13th cent. In Matterdale parish, there is 16th-cent. evidence for a fulling mill (*Walke milne, PNCumb* 222).

WALMGATE NY5117 (hab.) Bampton, W~ HEAD NY5116 (hab.), WALM HOW NY5117 (hab.).
Walm(e)gat(e) 1366 to 1865 (*PNWestm* II, 193); *Walmgatehead* 1860 (OSNB), cf. *Walmgate foote* 1688 (*PNWestm* II, 197); *Worm(e) how* 1690 to 1865 (*PNWestm* II, 193).
♦ In Walmgate, the 1st el. may be the reflex of OE *wælm, welm* 'spring', as suggested by the spellings and the presence of small springs (so *PNWestm* II, 193); the 2nd el. appears to be **gata** 'road', referring to the track linking BAMPTON with the foot of HAWESWATER (Parson & White 1829, 577). See also **head**. Walm How may contain the same 1st el. as in Walmgate or else *worm*, the reflex of OE *wyrm* 'snake, reptile', influenced by *Walm-*; **how(e)** is from ON **haugr** 'hill'.

WALNA SCAR SD2596 Dunnerdale/Torver, W~ S~ ROAD SD2696.
Walneyscar 1774 (Seathwaite PR), *Scar* 1786, *Walna Scar* 1822, *Walney Scar* 1830 (*PNLancs* 193, n. 2); *Walney Scar Road* 1851 (OS).
♦ If W~ S~ contains the word **scar** (escarpment), it is typical in its linearity, though unusual in being a summit ridge. The meaning of W~ is unknown, and there is no apparent reason for a connection with Walney Island, Furness. W~ S~ Road is a green lane, once used for transporting slate and other goods. The word *road* is from OE *rād* but only applied to routeways in modern times; it otherwise occurs in IRTON R~ STATION, OLD COACH R~ and OLD CORPSE R~.

WALTHWAITE NY3205 (hab.) Langdales.
Walthait 1591 (*CW* 1908, 148), *Walethwaite* 1722 (*PNWestm* I, 208).
♦ The 2nd el. is from **þveit** 'clearing', but the 1st is uncertain. (a) It could refer to fulling (*walk*, as in WALLTHWAITE), especially since the 1591 form occurs in a document referring to a fulling mill here. Alternatively, it could be (b) *wall* from OE *wall*; or (c) ON *vǫllr* 'meadow', as assumed 'probably' by Ekwall for another Walthwaite (*Walthwayt* 1260–80 (p); *PNLancs* 211).

WANDOPE NY1819 (2533ft/772m.) Buttermere.
Wanlope [sic] 1867 (OS), *Wandope, Wandup a century ago* 1964 (Wainwright, Wandope 2).
♦ Obscure. The second syllable could well result from OE *hop* 'valley', but this hill is hemmed in, the narrow groove of Sail Beck below hardly constituting a valley (so that ON *vand-hop* 'osier valley', suggested in Brearley 1974, is doubtful). One is left, appropriately, with ME *wanhope* 'lack of hope', but an intrusive *-d-* would be problematic. (Pl. 5)

WANLASS HOWE NY3703 (hab.) Ambleside.
From 1863; cf. *Wandlase yeat* 1721 (*PNWestm* I, 184).
♦ Possibly 'the hill with winding gear'. **How(e)** is from ON **haugr** 'hill', while the intriguing *wanlass/wandlase*

(attached to *yeat* 'gate' in the 1721 form) may refer to a windlass or winding gear. The word *windlass* is of obscure origin, appearing first in the early 15th cent. (*OED*). It can hardly be unconnected with the postulated OE **windels* 'winch' (see WINDSOR), though its form, here and elsewhere, may be influenced by *wanlace, wanless,* recorded in late ME and Early ModE as a hunting term referring to a circling movement to intercept game, hence any kind of interception or trick. W~ H~ rises above WATERHEAD at the N end of WINDERMERE, a traditional jetty site, and not far from 'Clappersgate port, from the wharves of which nearby quarried slate was transported' (Danby 1994, 162). If connected with a landing, W~ H~ would match LANDING HOW at the S end of the lake, and cf. Winlass/Wynlas Beck (*Winlass Beck* corrected from *Windlass B~*, OSNB Windermere, 1859, 132), which joins WINDERMERE at MILLERGROUND LANDING. I am most grateful to Mary Atkin for pointing out that (*lez*) *Wyndeles* appears frequently in the 14th-cent. records in *Kendale* (see vol. I, index), and for her alternative suggestion that a winch could have had agricultural purposes, for instance a hoist for hay or grass. (b) Brearley (1974) suggests ON *vandlauss* 'not difficult (to ascend)', which is an attractive solution (though the standard ON form would be *vandalauss* rather than *vandlauss,* Fritzner; *ONP*). (c) Were it not for the *Wand-* spelling, the specific might have been suspected to be the surname *Wanless*.

WANSFELL NY3904 Ambleside/Troutbeck, W~ PIKE NY3904 (1581ft/484m.), W~ HOLME NY3802 (hab.) Ambleside.

Wansfel (*Wall*) c.1692 (Machell II, 151, ed. Ewbank, 131); *Wansfell, the Pike of Wansfell* 1839 (Ford 38, 22), *Wansfell Pike* 1859 (OSNB); *Wansfell Holme* 1859 (OSNB).

♦ The origin and meaning of the 1st el. 'Wans' is obscure. ON *hvǫnn* 'angelica' (cf. WANTHWAITE) is among the possibilities, though the medial *-s-* might rather suggest a pers.n. The **fell** rises as a distinct height E of AMBLESIDE, capped by W~ **Pike** 'peak'. W~ Holme, 'a neat mansion' (OSNB), is some distance from W~ but **holme** 'island, land beside water' well describes its situation beside WINDERMERE.

WANTHWAITE NY3023 (hab.) St John's, W~ CRAGS NY3222.
Wannethwayth 1301 (p), *Wanthwait* 1303 (*PNCumb* 315); *Wanthwaite Crag* 1800 (*CN* 782).

♦ 'The clearing where angelica grows', with 2nd el. certainly from ON **þveit** 'clearing' and 1st el. probably from ON *hvǫnn* 'angelica', a word rare in p.ns. *Angelica sylvestris,* wild angelica, is found in wet places throughout Lakeland (Halliday 1997, 342), and the situation here is close to ST JOHN'S BECK. (*A. archangelica,* mentioned in *PNCumb,* quoting Ekwall, is the cultivated angelica). The **crags** form part of a rocky curtain E of ST JOHN'S IN THE VALE.

WARDLESS NY4303 Kentmere.
From 1857 (*PNWestm* I, 168).
♦ A problematic name. For the 1st el. (a) a connection with ON *varði* or *varða* 'a cairn, used as memorial or marker' is favoured by the position near the boundary between Troutbeck, Kentmere and Hugill, and *varði, varða* occurs in Ward Hall (with ON *hóll* 'hill', *PNCumb* 310) and Warth Hill

(*PNWestm* I, 64). However, this is hardly a peak, rather a basin-shaped grouse-moor which drains into the KENT. (b) 'Ward' could have the sense 'defence' or 'enclosure', if the reflex of OE *w(e)ard* 'protection'. A set of charter bounds for Docker and Grayrigg, former Westmorland, includes a stretch *usque subter Wardas, et a Wardis* . . . 'up to below Wards, and from Wards . . .' (Ragg 1924, 339). The 2nd el. is also problematic, as it is in WHITELESS (PIKE).

WARDWARROW NY0801 (hab.) Irton.
Wardwarey 1676 (*PNCumb* 404), *Wardwarrow in Irton* 1806 (Eskdale PR).
♦ Obscure. The 1st el. resembles the reflex of OE *w(e)ard* 'watch, protection', or its ON cognate *varða, varði*, which co-occur in p.ns especially with words for 'hill' (Smith, *Elements*), but this is a low plateau above the R. IRT, rather than a notable viewpoint. The 2nd el. is puzzling, as are the *-warrow(e)* spellings for the 2nd el. of WHINNERAY. (a) ON **vrá** 'nook, corner' could be the origin, though *warrow* would be an unusual outcome — perhaps influenced by *barrow* from OE **beorg** 'hill' — and the situation is not a typical *vrá*. (b) An application of OE *waru* 'defence', assumed to occur in Warden, Northumberland, would be another possibility, but it is open to the same objection as above, and does not have OE forms in **w(e)arwe* which would give rise to *warrow*.

WARNSCALE: W~ BOTTOM NY1913 Buttermere.
From 1867 (OS); cf. *Warnscale Hollow* 1799 (*CN* 544).
♦ Warnscale is part of BUTTERMERE FELL, and a possible location for a **scale** (ON **skáli**), a shieling. The meaning of the 1st el. is elusive. A connection with ON *vǫrn* 'defence', vb. *varna* 'warn, protect' is possible. Brearley (1974) suggests a place safe against cattle-raiders. W~ **Bottom** is the valley below W~.

WARTCHES NY4620 Askham.
Wattshares (*Gate*) 1588, *Wartches* 1859 (*PNWestm* II, 203).
♦ Judging from the 1588 spelling, this may be 'Watt's share of land', with forename or surname *Wat(t)*, an abbreviation of *Walter*, and the reflex of OE *scearu* 'share, division of land, boundary', with pl. *-(e)s* (so *PNWestm* and cf. SHARROW). This is a tract of fell on WHITESTONE MOOR.

WASDALE: NETHER W~ (hab. & parish) NY1204, NETHER W~ COMMON NY1307, W~ COMMON NY1207, W~ FELL NY1909, W~ HALL NY1404, W~ HEAD (hab.) NY1808, W~ HEAD HALL FARM NY1806 Eskdale.
Wastedal(e), *Wassedale* 1279; *Netherwacedal*, *Netherwasdale* 1338 (*PNCumb* 390, 440); *Nether Wasdale Common*; *Wasdale Common* both 1867 (OS); *Wasdall fells* 1687–8 (Denton 51), *Wasdale Fell*; *Wasdale Hall* both 1867 (OS); *Wascedaleheved* 1334, *Wasdale Hede* 1448 (*PNCumb* 390); (W~ Head Hall Farm:) previously *Wasdale Hall* 1867 (OS).
♦ Wasdale is 'the valley of the lake' or possibly 'the valley of Vatnsá, the river of the lake', from ON *vatn* 'lake' (i.e. WASTWATER, and see **wæter**) and ON **dalr** 'valley', cf. Vatnsdalr and its descendants in Norway and Iceland. The medial *-e-* which appears in several of the early spellings for this Wasdale and the one in Shap may reflect ON **á** 'river' (cf. *PNWestm* II, 172–3), which appears in other valley names including both instances of BORROWDALE. **Nether** 'lower'

distinguished the township of N~ Wasdale from that of Wasdale before they were united in 1934, but the **Common**s and **Fell** refer, presumably, to common grazing areas in each of the two former townships. In a rather remarkable substitution, Nether Wasdale, as a village name, replaces Strands, which still appeared on the 1966 One Inch map. This was *Strand* (shore, or possibly stream) *of Irt* in 1578 (*PNCumb* 442), and *Stran(d)s* is frequent in the 18th-cent. PRs. W~ **Hall** was built mainly in 1829. Two W~ Halls are shown on 1867 map, which may explain the change of name of one of them to W~ **Head** Hall **Farm**.

WASDALE: W~ BECK NY5508 Shap Rural, W~ HEAD NY5408 (hab.), W~ PIKE NY5308 (1852ft/564m.).
Wassadala 12th cent., *Wacedal(e)*, *Wascedal(e)* 1235 to 1292, *Wastedal(e)* 1282 (*PNWestm* II, 172); *Wa(s)cedalebec* 12th cent.; *Was(t)dayl(l)hed* 16th cent.; *Wasdale Pike* 1823 (*PNWestm* II, 178).
♦ 'The valley of the lake' or possibly 'the valley of Vatnsá, the river of the lake', cf. (Nether) WASDALE. No lake is now in evidence, but *Wascedalterne* (from ON **tjǫrn** 'mountain pool') is recorded from the 12th cent. (*PNWestm* II, 172) and Hodgson's map of 1828 shows a ribbon-like tarn in the valley. The secondary names are from **bekkr** 'stream', **head**, **pike** 'peak'. W~ Head is not at the innermost point of the valley, but it lies well above W~ Foot.

WAST WATER NY1505 Nether Wasdale.
Wassewater 1294, *Waswater*, *Wastewater* 1322, *Wastwater* 1338, *Wasdale Water* 1671 (*PNCumb* 36).
♦ 'Wasdale lake' or 'the Lake of Vatnsá, lake river'. The present name rather curiously contains the reflexes of both ON *vatn* 'water, lake', and OE **wæter** 'water', with the meaning 'lake' probably influenced by the ON *vatn*. (a) It is assumed in *Cumb* 36 that the lake (*vatn*) gave its name to the valley of *Vatnsdal(r)* ((Nether) WASDALE, q.v.), which may have been shortened to give the 1294 and subsequent forms of the lake name. (b) Alternatively, the first *-e-* in the 1294 spelling could reflect ON **á** 'river' (cf. again (Nether) WASDALE). (Pl. 23)

WATCH HILL NY1431 Setmurthy.
From 1867 (OS).
♦ The noun *watch* is from OE *wæcce*, and see **hill**. This a prominent spur, and is one of a number of Watch Hills in N Cumbria which provide good look-out positions for warning of border raiders.

WATENDLATH NY2716 (hab.) Borrowdale, W~ BECK NY2617, W~ FELL NY2814, W~ TARN NY2716.
Wattendlane 1190s, *Watendelair* 1209–10, *Wattendelan* early 13th cent., *Wathendeland* c.1250, *Wat(t)endleth* 1564 (*PNCumb* 352, q.v., for full range of forms); *Watendlath Beck*; *Watendlath Fell*; *Watendlath Tarn* all 1867 (OS).
♦ The name comprises three elements and has a complex history. The first two mean 'lake-end', with ON **vatn** 'water, lake' (see **wæter**), here referring to Watendlath Tarn, plus ON **endi** 'end'; cf. WATEREND and the Norw *Vassenden* (several examples). The 3rd el. appears to have been understood differently at different periods, possibly as OE *lanu* 'lane', which could refer to the sunken, slow-moving course of a stream (*DEPN*, *SSNNW* 257), then **land** 'land', then dial. *lath(e)* from ON *hlaða* 'barn'. The **beck** flows out of the

Tarn, and the **fell** is a stretch of high grazing.

WATER CRAG SD1597 Eskdale.
From 1867 (OS).
♦ 'The rocky height above the lake'; **water**, **crag**. It is just above DEVOKE WATER.

WATEREND NY1222 (hab.) Loweswater.
dil (of the) *Waterende* 1292 (p), *atte* (at the) *Waterende* 1302 (p) (*PNCumb* 412).
♦ 'The end of the lake' (**wæter**, **endi**) — NW of LOWESWATER. Wainwright notes that **end** is used curiously in this case, of the head rather than outflow of the lake (1966, Fellbarrow 1), and Loweswater is unique among the major lakes in draining towards the centre of Lakeland, not away from it.

WATERFOOT NY4524 (hab.) Dacre.
From 1693 (PR).
♦ 'The lower end of the lake' (**water**, **foot**), since close to the N end of ULLSWATER.

WATERGATE FARM NY1221 Loweswater.
Watergate 1725 (PR), *Wateryeat* 1729 (PR), *Water Gate* 1774 (DHM), *Watergate* 1794 (Hutchinson II, map facing p. 153).
♦ (a) W~ may be the compound *watergate*, in its northern sense of 'water-course' or 'channel' (*OED*: *watergate* 2, sense 1); cf. WATER YEAT; or, since there is no obvious trace of an engineered water-course here, a more modest sheep-barrier across a stream could be meant. (b) Alternatively, this could be simply 'the way or gate to the lake' (see **water**, **gata** or **geat**). The place is near the shore of LOWESWATER, and either way the name probably points to management of resources on the Loweswater estate. **Farm** has been added to a pre-existing name.

WATERHEAD SD3197 (hab.) Coniston, HIGH WATER HEAD SD3198 (hab.).
Waterhed 1537 (*PNLancs* 218); *High Water Head* 1851 (OS) is not shown in the same place as the present H~ W~ H~.
WATERHEAD NY3703 (hab.) Ambleside.
Watterhead 1597, *ye Waterhead* 1604 (*PNWestm* I, 184).
♦ W~ is 'the upper end of the lake'; **water**, **head**. The Coniston example is almost at the head of CONISTON WATER, with **High** W~ H~ just to the N. The Ambleside one is on WINDERMERE. *The water head* is common as a phrase in documents of the Early Modern period.

WATERMILLOCK NY4422 (hab.) Matterdale, W~ COMMON NY3720, W~ HOUSE NY4422.
Weþermeloc early 13th cent., *Wethermelok* 1253 to 1682, *Wedermelok'* 1285 (p), *Wattermannock* 1541, *Waltermelocke* 1568 (*PNCumb* 254); *Watermillock Common*; W~ *House* both 1867 (OS).
♦ W~ is close to LITTLE MELL FELL, and 'Millock' (*meloc*, *melok* in the early documents) seems to refer to it or perhaps to it and GREAT MELL FELL. *Meloc* probably contains, like them, Cumbric *$m\bar{e}l$* 'bald, bare hill' (see **mell**), plus a suffix which Ekwall took as a diminutive (*DEPN*, also *PNCumb*, Mills 1998, Watts 2004). Ekwall assumed that 'originally Great Mell Fell was *Mell* (Welsh *Moel*), Little Mell Fell being *Meloc* (Welsh **Moelog*)'. Coates (2000, 284, 353) instead regards the suffix (which he gives as Brit. -$\bar{o}g$) as adjectival, as does Dr Oliver Padel who adds that the adj. could function as a noun meaning 'place of', hence here 'place with hills, hilly place' referring to both Great and Little Mell Fells (pers.comm. and Padel 1985, 90

on the suffix). Meanwhile, Mellock Hill in the Ochil Hills (*mons de* [hill of] *Melloch* 1615) has been explained as Gaelic *meallach*, an adj. from *meall* 'lump, hill' (Watson 1995), which might raise the possibility of Gaelic origins for (Water)millock. The el. now appearing as 'Water' cannot be explained as Cumbric and must have been added, for whatever reason, to the pre-existing name. It was evidently OE/ME *weðer* 'young castrated ram' as in WETHER HILL, but this was replaced, in or before the 16th cent., by **water**, encouraged by proximity to ULLSWATER. There was also a variant *Walter-* (to which Watts 2004 draws attention). W~ **Common** must have been the grazing land of the former parish of W~, which was incorporated into Matterdale in 1934 (*PNCumb* 221); see also **house**.

WATERNOOK NY4319 (hab.) Martindale.
ye Waternooke 1633, *Waternooke* 1640/1 (PR).
♦ The farm stands beside an inlet on the E shore of ULLSWATER; **water, nook**.

WATER PARK SD2990 (hab.) Colton.
Waterside park c.1535 (West 1774, 104), *Water Park* 1632–3 (West 1774, 180), 1851 (OS).
♦ 'The estate by the lake' (**water, park,** and originally **side**), formerly a possession of FURNESS Abbey. Brydson vouches for the equivalence of *Waterside Park* and W~ P~ ([1908], 51). The place lies close to the S tip of CONISTON WATER.

WATERSIDE: HIGH W~ (hab.) NY0415 Ennerdale, LOW W~ (hab.) NY0415.
Watersyde 1322; *Highwaterside* 1675 (*PNCumb* 387); *Lowwater side in Kiniside* 1674/5 (PR).

WATER SIDE: W~ S~ HOUSE SD3686 Colton.
Waterside 1627/8 (PR); *Water Side House* 1851 (OS).
WATERSIDE HOUSE NY4623 Barton.
the Water Sid 1578, *Waterside* 1676 (*PNWestm* II, 213).
WATER SIDE WOODS SD3697 Claife.
Water Side Woods 1851 (OS); cf. *Easthwaite Water-syde* 1624/5 (Hawkshead PR).
♦ Waterside is in each case 'the place beside the river or water'; **water, side**, also **high, low, house, wood**. The Ennerdale settlements are beside the R. EHEN, respectively higher and lower on its course. The Colton W~ House is beside the R. LEVEN as it flows out of WINDERMERE, while the Barton one is on the E shore of ULLSWATER. W~ Woods, Claife, fringe the E side of ESTHWAITE WATER.

WATER YEAT SD2889 (hab.) Blawith.
Wateryate (*myll*) 1539, *Wotteryait* 1597 (*PNLancs* 215), *Water Yeat* 1729 (PR).
♦ 'The water gate' (**water, geat**), referring to a sluice or watercourse, still in evidence at this old (corn-)mill site.

WATSON'S DODD NY3319 (2584ft/ 789m.) St John's/Matterdale.
Watson Dod 1800 (*CN* 777).
♦ This is one of a chain of three grand **dodd**s (q.v.). The Watson commemorated is unidentified, but if the same as in Watson's Park, also in St John's, and recorded from 1734 (*PNCumb* 318), he must have lived in the early 18th cent. at latest.

WELCOME NOOK SD1193 (hab.) Waberthwaite.
From 1702 (*PNCumb* 365).
♦ *Welcome* is of OE origin but with a somewhat complicated history (see

OED). Whether it is an adj. here, implying a welcome refuge, or a noun, implying that the place offers a welcome, is unclear. A **nook** is a secluded place or corner of land.

WELL WOOD SD2486 (hab.) Kirkby Ireleth.
Wellwood 1701 (PR), *Well Wood* 1851 (OS).
♦ 'The wood by the well or spring'; **well**, **wood**. There are springs in the neighbourhood, and a well appears on the 1851 OS map. The place stands beside Kirkby Park Wood (1851 OS and current larger scale maps).

WELLFOOT NY4903 (hab.) Longsleddale.
Well-foot 1794 (PR), *Wellfoot* 1865 (*PNWestm* I, 164).
♦ 'The low-lying place by the well or spring'; **well**, **foot**. A well is marked at this spot on larger scale maps.

WELLINGTON NY0704 (hab.) Gosforth.
From 1829 (Parson & White 210), 1867 (OS).
♦ The hamlet forms the NE outlier of GOSFORTH. Wellington Close and Waterloo Park are among 'several curiously named fields' in Gosforth parish listed in *CW* 1904, 195. The combination of Wellington and Waterloo may suggest commemoration of Arthur Wellesley, created Duke of Wellington in 1814, and his famous defeat of Napoleon at Waterloo in 1815. Wellington was a Gosforth inn-name: 'Agnes Jackson, vict.' is listed there in 1829 and the Wellington Inn appears in the 1901 Census. The hamlet may have taken its name from the inn, as happened elsewhere, e.g. Nelson, Lancs.

WESCOE NY3025 (hab.) Threlkeld.
Weastkoe 1571 (*PNCumb* 253), *Wescow* 1867 (OS).
♦ Probably 'the west wood', from ON *vestr* (or the ME word resulting from this and OE *west*) and ON **skógr** 'wood' or its reflex. The name could refer to the location W of THRELKELD.

WEST FELL NY3335 Caldbeck.
Westfell 1800 (*CN* 798).
♦ Simply *west* from OE *west*/ON *vestr* plus **fell**. This is the high ground lying W of HALTCLIFF, and this may be the reason for the name.

WESTING NY3729 (hab.) Mungrisdale.
From 1774 (DHM).
♦ Uncertain, in the absence of earlier forms, but conceivably 'the west meadow', with *west* from OE *west*/ON *vestr* and dial. *ing* from ON *eng* 'meadow, water-meadow'. The flat, stream-side site would be compatible with this.

WESTRAY NY1530 (hab.) Embleton.
From 1703 (PR), 1774 (DHM), possibly = *Westwra* 1292 (p) (*PNCumb* 384).
♦ If the identification with the 1292 spelling is correct, the meaning is 'western nook', from ON *vestr* and ON **vrá** 'nook, secluded place'. It may be significant that the place is on the W side of Embleton parish (cf. EAST HOUSE).

WESTSIDE/WETSIDE NY3816 (hab.) Patterdale.
Wetside 1860 (*PNWestm* II, 227).
♦ Originally 'the wet slope', cf. WET SIDE (EDGE). The present name Westside is also apt since the site is W of a small beck and of ULLSWATER.

WETHER HILL NY4516 (2174ft/663m.) Martindale/Bampton.
Weather Hill 1722 (*PNWestm* II, 197).
♦ (a) Apparently 'the **hill** of wethers', with the reflex of OE *weðer* 'wether,

castrated ram' or its ON cognate *veðr*, though the situation, high and quite remote, makes it rather surprising that young animals should have grazed here. (b) ON *veðr* 'weather, gale' is not impossible (cf. Marwick 1952, 38 on a similar ambiguity in the Orkney Weatherness).

WETHER HOWE, HIGH NY5110 (1703ft/ 519m.) Shap Rural.
High Weather Howe 1863 (OS), *Weather Howe* 1865 (*PNWestm* II, 178).
♦ Presumably 'the high hill frequented by wethers or castrated rams'; **high**, *wether* as probably in WETHER HILL, plus **haugr/how(e)**.

WETHERLAM NY2801 (2502ft/763m.) Coniston.
Weatherlom 1786, *Wetherlam* 1822 (*PNLancs* 193, n. 2).
♦ Obscure, and Ekwall wisely declined to offer an explanation in the absence of early forms (*PNLancs* 193–4). (a) The 1st el., however, may well be from OE *weðer* 'wether, castrated ram', since **ewe** is frequent in hill and crag names, and High and Low Wether Crag lie just to the south. (There are comparable spellings for WETHER HILL, but that name is slightly uncertain.) Brearley (1974) hazarded '"the wether lamb", from some fancied resemblance', and this would be supported by the occurrence of ON *veðrar-lamb* 'wether lamb', though Fritzner and *ONP* offer only one occurrence. (b) That the 1st el. instead referred to 'weather' was proposed by Wordsworth: 'The last syllable is doubtless a corruption, &, like the Wetterhorn of the Alps, the name is probably taken from the storms to which this highest point is exposed' (*Prose* II, 319). (c) Collingwood tentatively suggested an ON *veðr-*

hjálmr 'weather-helmet', referring to a hood of cloud; he added that the name 'in the Coniston boundary-roll is "Little Walls"' (1918, 93).

WET SIDE EDGE NY2702 Coniston.
From 1850 (OS).
♦ Presumably 'the escarpment on the wet hill-side', with *wet* derived from OE *wēt*, possibly encouraged by ON *vátr*, plus **side** and **edge**. The Edge itself is a steep, narrow spur above WRYNOSE PASS, but easy walking and 'in spite of its name, quite dry underfoot' (Wainwright 1960, Grey Friar 7); however *wet* refers here to the *side*, not the *edge*. The word *dry* is more common than *wet* in Lakeland p.ns, probably because names tend to register features that are locally distinctive, and because there is a plentiful vocabulary to describe boggy ground.

WHA HOUSE FARM NY2000 Eskdale.
Whawes 1570, *Whaes* 1587, *Whose* 1700; cf. *the Whawbottom* 1587 (*PNCumb* 392).
♦ An obscure name. 'Wha' has been explained from ON *vað* 'ford' (Warriner 1926, 101), but the 'Wh-' suggests derivation from an ON word in *hv-* (pronounced *hw-*), e.g. *hváll* 'small rounded hill' (as in WHALE), or **hwæl*, the OE cognate of *hváll* (see WHALE and WHASDIKE). **House** may be a reinterpretation of a plural, given that it does not appear in the 16th-cent. spellings, or if the *-es* was a contraction of *house*, the el. has been restored; **farm** is a recent addition. This is a low, level site between Hare Crag and the R. ESK.

WHALE NY5221 (hab.) Lowther, WHALEMOOR NY5320 (hab.).
Vwal 1178, *Wal'* 1179 to 1279 (p), *Vhala* c.1230 (p), *Whale* 1234–6 (p) to 1823,

Qual 1249, *Quale, Qwale* 1252 (p) to 1478 (*PNWestm* II, 183–4); *Whailemoor(e)* 1671, *Whalemoor* 1704 (*PNWestm* II, 185).
♦ W~ is 'the hill', from ON *hváll* 'small rounded hill'. It is just SW of the hill occupied by the LOWTHER estate; see also **moor**.

WHASDIKE SD4397 (hab.) Hugill.
Whausdike(?) [sic] 1625 (Windermere PR), *Whansedike* 1621, *Wawsdyke* 1770 (JefferysM), *Whase dyke* 1836 (*PNWestm* I, 171).
♦ The 1st el. is obscure, possibly derived from ON *hváll* 'small, rounded hill' found in WHALE (or a cognate OE *hwæl* which may occur in English p.ns such as Whal(l)ey and Whalton, Smith, *Elements*), though the -s- would be puzzling. The terrain is hilly. **Dike** probably refers to an embankment of some kind.

WHATSHAW COMMON NY5405 Shap Rural.
Watshaw Common 1865 (*PNWestm* II, 178).
♦ W~ may be 'Wat's wood'; pers.n. plus **sceaga**. The 19th-cent. spelling could point to the forename or surname *Wat-*, the pet form of *Walter*, as 1st el., and the 2nd is probably the reflex of OE *sceaga* 'wood' (see **skógr**), but without earlier spellings certainty is impossible. See also **common** 'unenclosed pasture for communal use'.

WHEEL FELL NY0707 St Bridget Beckermet.
From 1867 (OS).
♦ Uncertain, perhaps 'rounded hill-spur'. Especially in the absence of early spellings it is unclear whether or not 'wheel' is to be taken at face value, as the reflex of OE *hwēol* 'wheel'. This is applied in p.ns to circular hills, and the **fell** is rounded, though since it is a lower spur of SWAINSON KNOTT its contours do not describe a complete circle. The only other probable 'wheel' in this volume is in SHOULTHWAITE, from ON *hjól*.

WHELPO NY3039 (hab.) Caldbeck.
Cuelpou 1278 (p), *Quelphou* 1279 (p), *Whelphou* 1285 to 1344 (*PNCumb* 278).
♦ 'Hill associated with young dogs or wolves', from ON *hvelpr* 'young dog, wolf' and ON **haugr** 'hill'. Sedgefield, as frequently, assumed a pers.n. here and in WHELP SIDE (1915, 124), but *haugr* is rarely compounded with a pers.n.

WHELP SIDE NY3314 St John's.
Whelpside 1800 (*CN* 802).
♦ Possibly 'hill-side associated with young dogs or wolves'; *whelp*, plus **sīde**. *Whelp*, from OE *hwelp* and/or ON *hvelpr*, could mean a young dog (cf. WHELPO and HUNDITH HILL for names associating hills with dogs), or if the name is pre-16th-cent., a young wolf could be meant (cf. WOLF CRAGS). This is the SW buttress of HELVELLYN, hence a steep hill-slope (*side*).

WHELTER: W~ CRAGS NY4613 Bampton.
Quilter 1366, *Whelter* 1549; *Whelter Crags* 1859 (*PNWestm* II, 193).
♦ W~ is from the rare ON *hwilftar*, pl. of *hwilft* 'hollow, combe' (see Smith 1966–9). The place is a spectacular natural basin framed by W~ **Crags**. If the pl. form is significant, it may reflect the fact that there is one main hollow and a subsidiary one.

WHICHAM SD1382 (hab. & parish), W~ BECK SD1583, W~ HALL SD1483, W~ MILL SD1585, W~ VALLEY SD1583.
Witingham 1086 to 1291, *Wintinga ham*

c.1130, *Whittingham* 1279, *Wycheham* 1550 (*PNCumb* 443–4); *Whicham Beck* 1867 (OS), seemingly = *Layrwatpul* 1260–80 (*CW* 1926, 135); *Whittingham Hall* 1504–15 (*PNCumb* 443–4); *Whicham Mill* 1867 (OS); *Whicham Valley* 1900 (OS).
♦ 'The village or homestead associated with Hwīta.' The 1st el. is probably the OE pers.n. *Hwīta*, and the last is OE *hām* 'village, homestead'. (a) *-ing-* is used to form p.ns, hence probably **Hwīting*, the place associated with Hwīta, and *hām* may then have been added to this (cf. Watts 2004). (b) Without stronger evidence for gen. pl. *-inga-* a reference to an OE group or tribal name *Hwītingas* 'people of Hwīta' as assumed in *PNCumb* is less likely. The name is nevertheless of interest. English p.ns in *-ing(a)hām* are often associated with the earlier phases of Anglo-Saxon settlement, and with favourable sites. They are therefore rare in the Lake District except, as here, in the lowland fringes. It is also characteristic that such a site should become a parish centre, and be mentioned in Domesday Book, 1086 (cf. BOOTLE, KENDAL and NEWTON). See also **beck** 'stream', **hall**, **mill** from **mylen** and, on the word *valley*, LORTON VALE. The Mill is shown as a corn mill on 1867 OS.

WHILLAN BECK NY1802 Eskdale.
Whillon 1587, *Whillan Beck* 1794 (*PNCumb* 30).
♦ The identity of W~ is unclear. It has the appearance of a surname (cf., perhaps, *Whelan*, an Anglicised Irish name), and surnames do occur with **beck** 'stream' (cf. TOM RUDD BECK); however, instances are extremely rare, for *beck* is usually qualified by a p.n. or a descriptive term.

WHINCOP SD1799 (hab.) Eskdale.
From 1627 (*PNCumb* 344).
♦ 'The summit where gorse grows', from **whin** and the reflex of OE *cop(p)* 'top, summit', which otherwise only occurs in this volume in COP STONE and HIGH/LOW KOP, though it also appears, e.g., in the hill-name Kinniside Cop, NY0414.

WHINERAY GROUND SD2090 Dunnerdale.
Whinerah Ground 1670 (PR), *Winneray Ground in Dunerdale* 1692, *Whinrey Ground in Dunerdale* 1696 (Broughton-in-Furness PR).
♦ 'The farm of the Whineray or Whinerah family'; surname plus **ground**. The place is associated with persons named *Whinerah* in the Dunnerdale PR for 1670 and 1676, and the surname is quite frequent throughout the PRs. It must derive from WHINNERAY in Gosforth or another such p.n., unless it existed as a p.n. here in Dunnerdale, where the reference to gorse would certainly be appropriate. High and Low W~ G~ are distinguished on larger scale maps.

WHIN FELL NY1325 Blindbothel, WHINFELL HALL NY1425.
(Hab.:) *Wynfell* c.1170, *Whinnefelde* 1230, *Quinfel* 1268; (hill:) *the Fell* 1658 (*PNCumb* 446–7), *Whinfield Fell* 1774 (DHM); *Whinfell Halle* 1602 (*PNCumb* 446–7).
♦ 'The hill where gorse grows'; **whin**, **fell**, also **hall**. W~ Fell is the ground rising to HATTERINGILL HEAD and FELLBARROW. The c.1170 spelling in *W-* as opposed to *Wh-* is paralled in another Whinfell (*PNWestm* II, 132).

WHIN GARTH NY0905 Gosforth.
Wind Garth 1774 (DHM), *Wind Guards* 1784 (PR).

♦ Originally 'the windswept enclosure'. The 18th-cent. spellings suggest *wind* from OE *wind* and/or ON *vindr*, plus **garth**, while **whin** 'gorse' seems to be secondary. This is currently a tract of upland, not a farm.

WHINLATTER (hill) NY1925 Lorton, W~ PASS NY1924 Lorton/Above Derwent.
Whynlater 1505, *Whinlatter* 16th cent. (*PNCumb* 409); *the Whinlater road, an alpine pass* (Hutchinson 1776, 200), *Whinlatter Pass* 1900 (OS).
♦ W~ is (a) probably 'the slope where gorse grows', from ON **hvin* 'gorse' (see **whin**) and Gaelic *lettir* or *leitir* 'hill, slope' (so *PNCumb*). Halliday notes that the hill-slopes around LORTON are golden with flowers of *Ulex galii*, Western gorse, in late August and September (1997, 306). (b) ON *látr* 'lair, shelter' (see **latter**) is less likely as 2nd el. when the 1st refers to vegetation. (c) Coates tentatively favours a wholly Gaelic solution, citing MIr *fi(o)nn* 'white, fair' and MIr *lettir* 'slope' and comparing a lost *Findlater* in Banffshire (2000, 287). The **pass** follows the approximate line of a Roman road (Hindle 1998, 27, 29–30).

WHINNERAH/WHINNERAY NY0805 (hab.) Gosforth.
Wynwarrowe 1599 (*PNCumb* 396), *Whinbarrow(e)* 1606/7, 1726, *Whinwray* 1691/2, *Whinwarrow* 1708/9, 1732 (PR), *Whinnery* 1778 (Terrier).
♦ Obscure. The 1st el. may be **whin** 'gorse'. The 2nd might be guessed to be ON **vrá**, later **wray** 'nook, secluded corner of land', but the spellings in *-warrow(e), -barrow* and the situation on a steep slope W of the R. BLENG tell against that (cf. WARDWARROW). The p.n. may have given rise to the surname incorporated in WHINERAY GROUND.

WHINNY CRAG NY4520 Barton.
From 1865; cf. *a stone in the Whinye hole called Grayestone* 1588 (*PNWestm* II, 213).
♦ 'The rugged height where gorse grows'; *whinny* from **whin**, **crag**. This is the only occurrence in this volume of the adj. *whinny*, recorded from 1482–3 (*OED*).

WHIN RIGG NY1503 Eskdale.
From 1867 (OS).
♦ Seemingly 'the ridge where gorse grows' (**whin**, **hryggr/rigg**), though given its exposed position — a bluff at 1755ft/535m. —, one might suspect corruption of *wind* as in the WHIN GARTH.

WHINS/WHINNS NY0916 (hab.) Ennerdale.
Wyines 1647, *Whines in Casehill* 1674 (PR), 1774 (DHM).
♦ The presence of **whin** or gorse bushes locally is attested by Coleridge, who notes 'a nice ground, now whins . . . just at the foot of the Lake [Ennerdale Water]' (*CN* 1209, 1802). This name belongs formally to a type consisting of the pl. of a vegetation name; cf. BIRKS, HOLLINS.

WHINSCALES NY1903 Eskdale.
From 1867 (OS).
♦ Presumably 'the shielings where gorse grows' (**whin**, **skáli/scale**). This is a remote, rugged spot without buildings.

WHIRL HOWE NY4805 Longsleddale.
Wharlehowe(myer) 1577 (*PNWestm* I, 164), *whorlehow* c.1692 (Machell II, 104), *Whirl Howe* 1836 (*PNWestm* I, 164).
♦ Probably 'the circular mound', with the reflex of OE *hwerfel* 'circle', and **howe**. This is 'a small hill of conical form with clumps of fir trees on the

top. It is pasture to the top and very remarkable, but I cannot learn that it is an antiquity of any kind' (OSNB Kendal, 1858, 1, 49). It may be a tumulus, though it has not been excavated (LDHER 17109).

WHIRLPIPPIN/WHOLE PIPPIN SD1685 (hab.) Whicham.
Whirleppin, *Whirnepepin* 1646, *Whinnepepin* 1651, *Whirlepippen* 1659 (*PNCumb* 445), *Hurle Pippin* 1774 (DHM).
♦ Apparently 'the pot with the whirlpool'. *Whirl* meaning an eddy or vortex of water is recorded from 1547 (*OED*: whirl, n., II, sense 8b). *Pippen* or *pippin* is a dial. word for a deep earthenware pot or a milk pail (*EDD*, though Lakeland counties are not specified). The place is close to STOUPDALE Beck.

WHISTLING GREEN SD1992 (hab.) Dunnerdale.
Wissleton Green, *Wisselton Green* 1814 (Kirkby Ireleth PR), *Whistleton Green* 1850 (OS), *Whistling Green* 1891 (OS).
♦ **Green** is 'grassy place, green', but the 1st el. is puzzling. The 19th-cent. spellings of W~ have the look of an Anglian p.n., but such a name here and recorded so late is highly unlikely; perhaps it is a surname derived from a p.n. elsewhere.

WHITBARROW SD4486 Crosthwaite, W~ SCAR SD4387.
Witeber, *Wyteber* 1186–1200, *Whitberg(h)* 1187–1200 (*PNWestm* I, 82); *Whitbarrow Scar* 1770 (JefferysM).
♦ W~ is 'the white hill'; OE **hwīt**, OE **berg** and/or their ON cognates. Despite the 12th-cent. spellings in W- as well as Wh-, this etymology can be fairly certain, given the pale colour of the rock on this large limestone oval.

The **Scar** is the long escarpment which gashes the W side of Whitbarrow (cf. WHITE SCAR); Machell may have had it in mind when writing of 'the horrid long Rock wh[ich] is called Whitbarrow' (Machell c.1692, II, 394). (Pl. 32)

WHIT BECK NY1725 Lorton.
Wythebek' 1247 (*PNCumb* 30).
WHIT BECK NY2826 Underskiddaw.
Withebech c.1220 (*PNCumb* 30).
♦ (a) Probably 'the stream where willows grow', with OE *wīðig*, suggested by the -*th*- spellings (so *PNCumb* 30 and cf. WYTHOP), plus **bekkr**. (b) Ekwall suggested 'the white, foaming stream', an apt description for both becks (*ERN* 456–7), but did not have early spellings for the Lorton example.

WHITBECK SD1184 (hab.) Whicham.
Witebec c.1160 (in early copy), *Whitebec* c.1200, *Witbec* 1228–43 (*PNCumb* 447–8).
♦ 'The white, foaming stream'; **hvítr**, **bekkr**. The place is close to waterfalls on MILLERGILL BECK.

WHITBY STEAD/WHITBYSTEADS NY5022 (hab.) Askham.
Whitbysteed 1723, *Whitbystead* 1823 (*PNWestm* II, 203).
♦ (a) Probably 'The Whitbys' farm or place'. Smith takes 'Whitby' as a surname recorded locally in the 13th cent., and derived from the NYorks p.n.; he adds that Whitby Abbey held land some eight miles away (*PNWestm*). **Stead(s)** frequently occurs with a surname as specific. (b) Brearley 1974, citing *Quitteby* c.1250 (though without specifying the source), seems instead to envisage a local p.n. 'Whitby', in a location now unknown.

WHITEBECK SD4589 (hab.) Crosthwaite. *Whitbeck* 1535 to 1823 (*PNWestm* I, 86).
♦ 'The clear stream', and hence the settlement by it; **hvítr/white, beck**. Smith is doubtless correct in assuming this rather unusual sense of 'white' (*PNWestm*), since the slope of the LYTH VALLEY down which White Beck flows is too gentle to produce foaming 'white water'. The 1535 and modern spellings illustrate the replacement of a shortened form by the fuller one, at least in writing.

WHITE BOG NY4617 Bampton.
From 1860 (OSNB).
♦ 'The pale-coloured bog'; **white** (see **hwīt**), presumably referring to vegetation, plus *bog* as in BOG HOUSE. This was in 1860 'a large tract of moss ... from which the people ... get their peats' (OSNB).

WHITE BORRAN SD2689 Kirkby Ireleth/Blawith.
From 1851 (OS).
♦ 'The pale cairn'; **white** (see **hwīt**), **borran(s)**. Two large cairns stand here, both presumed Bronze Age and clearly funerary in one case, together with further remains (LDHER 2491, 2157); there are more sites nearby.

WHITE COMBE SD1586 (1361ft/417m.) Whicham.
From 1867 (OS).
♦ 'The pale crest' (**white** (see **hwīt**); **comb(e)**^{2}), contrasting with its immediate neighbour BLACK COMBE.

WHITE CRAG NY2612 Borrowdale.
From 1867 (OS).
♦ 'The pale rock outcrop'; **white** (see **hwīt**), **crag**. Seen as part of the skyline from LANGSTRATH, this linear outcrop is not noticeably paler than others in the valley, though the largish vertical slabs on the SE side catch the light.

WHITE CROSS BAY NY3800 Troutbeck.
Seemingly = *Craams Bay* 1915 (*CW* 1973, 114, 118).
♦ This is on the E shore of WINDERMERE, and is named from a white cross commemorating Ralph Thickness, drowned in the bay in 1853 (Kipling 1973, 118); **white** (see **hwīt**), **kross, bay**.

WHITE GILL: WHITEGILL CRAG NY2907 Langdales.
Whitegill 1574 (*PNWestm* I, 208); *Whitegill-crag* 1780 (West 103).
♦ 'The ravine with a white, foaming stream'; **white** (see **hwīt**), **gil(l)**. The **crag**, a set of outcrops part way up the contours, stands next to it.

WHITE HAUSE NY2732 Ireby and Uldale.
From 1867 (OS).
♦ 'The pale-coloured pass or ridge' (**white** from **hwīt, hause**), though this is not a classic hause.

WHITE HORSE BENT NY3428 Mungrisdale.
From 1867 (OS); cf. *the White horse* 1800 (*CN* 797).
♦ (a) Coleridge, recording a descent from BANNERDALE CRAGS in 1800 (*CN* 797), is referring to the White Horse Inn, and the p.n. may derive from the inn-name. (b) Alternatively, *horse* may be a form of **hals/hause**, as it seems to be in BURNT HORSE, and **white** (from **hwīt**) may refer to the pale vegetation. In either case, *bent*, from OE *beonet*, is wild, coarse, grass, hence a tract of open grassland, here on the steep slope above the R. GLENDERAMACKIN.

WHITE HOWE NY3807 Ambleside.
From 1865 (*PNWestm* I, 184).
WHITE HOWE NY5204 (1739ft/530m.) Fawcett Forest.

White how hill 1836 (*PNWestm* I, 139).
♦ 'The pale hill'; **white** (see **hwīt**), **howe**. With its compact, roundish summit, the Fawcett Forest example is a characteristic *how(e)*; the Ambleside one is less so.

WHITE KNOTT NY4621 (1378ft/420m.) Barton.
White Knotts 1859 (*PNWestm* II, 213).
♦ 'The pale rocky height(s)' (**white** from **hwīt**, **knott**) — a moderately rocky spur on BARTON FELL.

WHITELESS BREAST NY1818 Buttermere, W~ PIKE NY1818 (2158ft/ 568m.).
Whiteless Breast 1867 (OS); *Whiteless Pike* 1803 (*CN* 1518).
♦ 'Whiteless' is clearly the primary hill-name, but is obscure. The landscape offers little help. 'White' would describe the striking flashes of light-coloured rock which outcrop on W~ Breast (and to a lesser extent the Pike), but this would leave '-less' unexplained. Otherwise, it could conceivably be of the same origin as the puzzling WARDLESS. *Breast*, ultimately from OE *breost*, can denote the fore part of a hill (see *EDD*: *breast*, sense 1). The only other certain occurrence in this volume is WRYNOSE BREAST but BREASTY HAW may be comparable. The 17th-cent. clergyman Machell described another hill as 'ris[ing] with an easy ascent, like a woman's breast' (ed. Ewbank, 101). The **pike** 'peak' towers above the more rounded top of W~ Breast. (Pl. 5)

WHITE MOSS COMMON NY3406 Grasmere.
Grasmere Common calld the white-moss, white-mose c.1692 (Machell II, 137, 139, ed. Ewbank, 134), *Whitemoss (foot)* 1731 (*PNWestm* I, 202), *White Moss* 1863, 1920 (OS).

♦ **White** (see **hwīt**) applied to the **moss** 'bog' may refer to vegetation (for instance bog cotton, or the silver birch growing there today), less likely to 'a strange Mountain lightness... almost like a peculiar *sort* of light' which Dorothy Wordsworth remarked there in the evenings (*Journals*, Feb. 8 1802). The **common** was the shared land of Grasmere township (*CW* 1908, 142).

WHITEOAK BECK NY1219 Loweswater, W~ MOSS NY1217.
Whiteoak Beck; Whiteoak Moss both 1867 (OS).
♦ Unless there has been undetectable corruption, these are from **white** (see **hwīt**), **oak**, **beck**, **moss**. The Beck flows from the boggy W~ Moss, close to the small summit called White Oak. Presumably a distinctive tree grew on it (*OED* has *white oak* among its collocations of *white* with vegetation terms (11b)). The white oak is a tree species, but is not native to N. Europe.

WHITE PIKE Examples at:
    SD1595 Muncaster/Ulpha.
From 1867 (OS).
    SD2495 (1962ft/598m.) Dunnerdale/Broughton/Torver.
From 1851 (OS).
    NY3323 St John's.
From 1774 (DHM).
♦ 'The pale summit'; **white** (see **hwīt**), **pike**. The W~ P~ in Muncaster/Ulpha stands on BIRKBY FELL, and is perhaps so called because of the cairns, outcrops and piles of stones strewn over its surface. The one at SD2495 stands just W of White Maiden, clustering with hills named 'Green', 'Black', and 'Brown'. The one at NY3323 is a subsidiary peak, marked with cairns, to the NE of RED SCREES, and perhaps named in contrast to them.

WHITE SCAR SD4585 (613ft/187m.) Crosthwaite.
From 1857 (*PNWestm* I, 86).
♦ This whitish limestone escarpment forms the SE side of WHITBARROW; **white** (see **hwīt**), **scar**, and cf. WHITBARROW SCAR.

WHITESIDE NY1621 (2317ft/716m.) Buttermere.
White-side 1784 (West 138, corrected from *Silverside* West 1780).
WHITE SIDE NY3317 (2832ft/863m.) St John's.
White Side 1774 (DHM), *Whiteside* 1800 (CN 777).
WHITESIDE PIKE NY5201 (1302ft/397m.) Whitwell.
Whiteside 1635 (*PNWestm* I, 151).
♦ 'The pale hill-slope'; **white** (see **hwīt**), **side**. The St John's name strictly applies to the lower, western slopes of the mountain W~ S~, and may refer to 'splashes of quartz on many of the stones' (Wainwright 1955, White Side 1) or 'its top so rugged with white cliffs' (*CN* 798, 1800). As with WHITE PIKE, a small scatter of p.ns including colour terms surrounds this name, including a GREEN SIDE, though this is nearly two miles to the NE. W~ **Pike**, Whitwell, is the summit of another Whiteside.

WHITESTOCK HALL SD3289 Colton.
Whitstockhowe 1597 (*PNLancs* 217), *Whitestockhow* 1630, *Whitestockhall* 1632/3 (PR), *White Stock Hall* 1850 (OS).
♦ Originally 'the hill at Whitestock, the white ?footbridge'. 'Whitestock' has elements going back to OE **hwīt**/ON **hvítr** 'white' and to OE *stocc*/ON *stokkr* which could mean 'tree-trunk, log' or perhaps 'wooden bridge' (see STOCK; a footbridge is marked just to the N on the 1850 map). **Hall** has evidently replaced **how(e)** from ON **haugr** 'hill', as often in S Lakeland. Both words are appropriate to the present W~ H~, a fine five-bayed house of 1806 set beneath a hill-side.

WHITESTONE ENCLOSURE SD3884 Staveley-in-Cartmel.
White Stones Enclosure 1851 (OS).
♦ This tract of ground lies just below the small peak of White Stone; **white** (see **hwīt**), **stone**. This is the sole occurrence in this volume of *enclosure*, an early 16th-cent. adoption from Fr.

WHITESTONE MOOR NY4720 Askham.
Whitestean Moor c.1692 (Machell I, 655A).
♦ As well as being descriptive (**white** from **hwīt**, **stone**, **moor**), it is possible that this name, like some examples of Greystones, has associations with boundaries. The site is close to the parish boundary with Barton, as well as to the course of the Roman Road HIGH STREET. Machell's record of the name is in the context of a boundary, and boundary posts and stones, and piles of stones, are marked on larger scale maps (cf. the fact that White Stones on STYBARROW DODD marks the boundary of Matterdale and Patterdale).

WHITESTONES SD1193 (hab.) Waberthwaite.
From 1660 (*PNCumb* 365).
♦ The name perhaps refers to the slight outcrops nearby; **white** (see **hwīt**), **stone**.

WHITEWATER DASH NY2731 Bassenthwaite/Ireby and Uldale/Underskiddaw.
le white waterdashe 1594 (*PNCumb* 264), *White-water Dash* 1656 (hab., Uldale PR), *the head of White water dash . . . otherwise called Cassbeck* 1687–8 (Denton 172).

♦ 'The waterfall with (spectacularly) white water.' *OED* records *white water* n. referring to waterfalls from the 16th cent., and *dash* 'a violent throwing and breaking of water' also first appears in the Early Modern period (*OED: dash* n. 1, sense 4a). This is a fine chain of cascades, also known as Dash Falls, which gives its name to DASH BECK, the stream of which it is a part. (Pl. 11a)

WHITFELL SD1592 (1876ft/572m.) Waberthwaite/Ulpha.
From 1867 (OS).
♦ Presumably 'the pale-coloured hill'; **white** (see **hwīt**), **fell**.

WHITROW BECK SD1293 Waberthwaite.
From 1867 (OS).
♦ The elements are apparently **white** (see **hwīt**), **row** (of buildings or trees?), and the unproblematic **beck** 'stream', but the motivation for the name is unclear.

WHITTAS PARK NY2136 Bewaldeth.
Whitt House 1777 (*PNCumb* 326).
♦ 'The **park** or enclosure associated with W~', which seems to mean 'the white house' (**white** from **hwīt**, **house**). This is a tract of high ground.

WHOAP NY0913 Ennerdale.
Hope 1826 (CRO D/Lec 94), *Whoap* 1867 (OS).
♦ Obscure. This is a neat triangular hill-top rising to 1677ft/511m. on the N of KINNISIDE COMMON.

WHORNEY SIDE NY2505 Langdales.
Whorney Side (Force) 1865 (*PNWestm* I, 208).
♦ **Side** is a 'slope', but 'Whorney' is uncertain. Brearley (1974) suggested origins in OE *cweorn-ēa* 'quern-river' i.e. a stream with stones suitable for mill-stones. Wh- spellings for 'quern' do occur (*Whirnestone* 1578, 'quern-stone, mill-stone', *PNCumb* 383, and cf. *OED: quern* 1). If *quern* is possible as the 1st el., however, OE *ēa* 'river' is less likely as the origin of 2nd el. '-ey', given its rarity in p.ns hereabouts, and another explanation, such as ON **haugr** 'hill', may be preferable (cf. POOLEY).

WIDEPOT NY5209 Shap Rural.
(le) Whytepot, Wy(h)tpot 1279 (*PNWestm* II, 178).
♦ 'The pale or white hollow'; **hwīt**, **potte**. This is a steep-sided bowl rimmed on two sides by crags.

WIDEWATH NY5021 (hab.) Askham.
Wythewat 1289, *Wythwayt* 1328, *Withwath* & variants 1420 to 1741 (*PNWestm* II, 201).
♦ 'The ford where willow grows', from ON *víðir* 'willow, withy' and ON **vað** 'ford'. The situation is beside a ford across HELTONDALE BECK. The current form of the 1st el. seems to be influenced by *wide*, while in the 2nd el. the 1328 spelling seems to have been influenced by **þveit/thwaite** 'clearing'.

WIDOW HAUSE NY1826 Lorton.
From 1867 (OS).
♦ **Hause** is a neck of land, but why *widow* (from OE *widewe*) is obscure. The place lies on the watershed roughly midway between WYTHOP and LORTON, and the name may commemorate a corpse road between these places (Ramshaw & Adams 1993, 52). Certainly, a deposition of 1601 notes that the people of Wythop 'come to the parishe Church of Lorton when they have occasion to burie, christen or to come to receive the holy Communion haveinge their health' (George 1995, 4), since their chapel at

KELSWICK lacked these rights. However, as Ramshaw & Adams point out, the boggy Widow Hause route is far from ideal, and a Corpse Road shown at NY1729 on larger scale maps would have provided an easier way between Wythop and Lorton or the distant mother church of Brigham, near Cockermouth. Perhaps there has been corruption, for instance of a word for 'willow' (cf. WIDEWATH), in this name.

WILEY GILL NY2931 Ireby and Uldale/Caldbeck/Underskiddaw.
From 1867 (OS), possibly = *Grane beck* 1589 (*PNCumb* 30).
♦ W~ may well be the surname which appears, for instance, in Crosthwaite and Ennerdale PRs; **gil(l)** is a ravine with a stream.

WILKINSYKE FARM NY1716 Buttermere.
Wilkinsyke 1966 (OS One Inch), previously *White Hall* 1867, 1900 (OS).
♦ 'Wilkin' is most likely the pet form of *William*, and could originate in a forename or surname. A small **syke** or trickling stream flows past the **farm** into BUTTERMERE. This seems to be a renaming of White Hall (I am grateful to Mr A. D. Beard of Wilkinsyke Farm for confirming this).

WILLDALE NY4817 Bampton.
Wildale c.1160 (*PNWestm* II, 197).
♦ Probably 'the wild valley', from OE *wilde*/ON *vildr* 'wild', plus **dalr**. Notwithstanding the name, it is rich in evidence of ancient settlement including TOWTOP KIRK.

WILLIAMSON'S MONUMENT NY4500 Hugill.
From 1858 (OSNB).
♦ This is 'a circular pillar about 15ft high' on High Knott, built in 1803 'In memory of Thomas Williamson of Heights in Hugill. Gent. who died February 19 1797. Aged 66' (OSNB Kendal, 1858, 2, 5). It was 'erected ... by the Rev. Thomas Williamson, in memory of his father, who for many years walked every day to this eminence' (Parson & White 1829, 651).

WILLIE WIFE MOOR NY3312 St John's.
Willy-wife-moore c.1692 (Machell II, 154).
♦ This name, in which **moor** applies to the SW slopes of DOLLYWAGGON PIKE, is one of several p.ns referring to wives (using the reflex of OE *wīf* 'woman, wife'). Most refer to minor features and are confined to larger scale maps (e.g. Fisher's Wife's Rake NY3222, Hodge Wife Gill SD2690). The identity of the people concerned is unknown, but from the Machell evidence above, they lived in the 17th cent. or earlier. He mentions 'Bounder Stones w[hi]ch stand near together on Willy-wife-moore'.

WILLY KNOTT NY2836 Caldbeck.
From 1867 (OS).
♦ 'Willy' may be the common pet-form of *William*. This is not a classic **knott**: no rocks are shown on the map, and the place is halfway up BRAE FELL, rather than an independent summit.

WILSON HOUSE SD4281 Upper Allithwaite.
From 1850 (OS).
♦ Presumably 'the **house** of the W~ family'. Wilson Hill lies to the N.

WILSON PLACE/WILSON'S PLACE NY3103 Langdales.
Wilson place in Litle Langdale 1698/9 (Grasmere PR).
♦ 'The Wilson farm.' The pattern of surname plus **place** (q. v.) is common

in the locality. On the identity of the W~ family, see Forsyth 1998, esp. 207, 229.

WILTON HILL NY3628 (hab.) Mungrisdale.
Wiltonhill 1727 (*PNCumb* 227), *Welton Hill* 1774 (DHM), 1823 (GreenwoodM).
♦ W~, like the specifics in other **hill**-names in Mungrisdale, EYCOTT H~ and MITON H~, has the appearance of a p.n. but is not found as such locally. As with them (and STEWART H~), the most likely, if unproven, explanation is that this is a surname derived from a p.n.

WINDER: HIGH W~ NY4923 (hab.) Barton, W~ GREEN NY4923 (hab.), W~ HALL FARM NY4924.
Winderge 1170–80, *Winderhe* c.1240 (p); *Over Winder* 1333, *High(e) Winder* 1572 to 1823; *Winder Green* 1865; *Winder Hall* 1676, previously *Low Winder* 1572 (*PNWestm* II, 211–3).
♦ W~ is 'the windy shieling', probably from ON *vindr* and Gaelic-Norse **ærgi/erg** 'shieling'. In the late 17th cent., Machell explained **High** and Low W~ as 'both so-calld because of their high and windy location . . . High Winder is in the more lofty place' (I, 688). **Green** refers to a grassy spot; see also **hall**, **farm**.

WINDERMERE SD3894 (lake), SD4198 (hab. & parish).
(Lake:) *Winendermer* 1154–89, *Winandremer(e)* & variants 1157–63 to 1770, *Wenandewatir* 1278 (p) (*PNWestm* I, 18; cf. similar forms for parish name *PNWestm* I, 192, and cf. *PNLancs* 193).
♦ 'Winand or Vinand's lake', with 2nd el. **mere**[1] 'lake', 'the great lake being so remarkable that it gives its name both to church and parish' (Machell c.1692, II, 312), and then to the village which burgeoned with the coming of the railway in 1844 (formerly Birthwaite, Lindop 1993, 39). The specific has usually been identified with an Old Swed pers.n. *Vinandr*, gen. sing. *Vinandar*, the *-ar* being preserved as *-er-* in the modern name. Winderwath in the Penrith area is believed to commemorate another man with the same name (*PNWestm* II, 132–3), and there was also a *Wynanderthwayte* 1337 (Lindkvist 1912, 127); but this is rather disconcerting since the pers.n. is of very restricted distribution even in Sweden. Dr John Insley argues instead for a Continental Germanic pers.n. *Wīnand* (2005; I am most grateful to him for informing me of this). Since this name could not have been current until the 12th cent., the fact that the ON gen. sing. *-ar* has been added to it would suggest that ON still survived as a living language at that time. (Pl. 31)

WIND GAP NY1611 Ennerdale/Wasdale. From 1867 (OS); previously *le Windʒate* 1322 (*PNCumb* 388); cf. *W(h)ynyat Cove* 1802 (*CN* 1208).
♦ This exposed pass carries the track between STEEPLE and PILLAR. **Gate** from OE **geat** seems to have been replaced by **gap** fairly recently. Cf. WINDY GAP.

WIND HALL NY0704 Gosforth.
Wyndhowe 1597, *Windhal* 1605 (PR), 1639 (*PNCumb* 396).
♦ 'The windy hill.' In this quite exposed site, the meaning of the 1st el. is probably the obvious one: *wind* from OE **wind**. **Hall** evidently replaces **how(e)** from ON **haugr** 'hill, mound', and Parker called the name 'a refinement of Wyndehowe' (1904, 195).

WINDSOR FARM NY1205 Nether Wasdale.
Wyndsore 1570, *Windsor* 1578 (*PNCumb* 442).
♦ This gives the impression of being a curiously displaced name. Other Windsors, including the famous Berkshire one, are presumed to derive from OE *windels-ōra*, in which the 1st el., otherwise unrecorded in OE, refers to a winch or windlass for pulling boats or carts up banks or through difficult ground (see Smith, *Elements* and cf. WANLASS HOWE). The 2nd is a word for 'hill-slope' which is normally confined to S. England. Gelling (1984, 182) counts the Wasdale name among the 'possible' examples, adding that this would involve assuming a wider currency for *windels-ōra* than for *ōra* itself. Alternatively, the Berkshire name may have been transferred to Cumbria, as it seems to have been to Pembrokeshire (Gelling 1984, 181). **Farm** has, as often, been added to a pre-existing name.

WINDY GAP NY2110 Ennerdale/Borrowdale.
Windy Gap 1867 (OS), possibly = *summitatem montis qui vocatur Windeg*, *Windheg* (summit of the mountain called *W~*) 1209–10 (*Furness Coucher* II, 570), *le Egge* 1338 (Collingwood 1920, 244).
♦ This is the exposed and dramatic **gap** or col between GREAT GABLE and GREEN GABLE. *Windeg* ('wind-?edge') in the *Furness Coucher* was tentatively identified with Great Gable by the editor but Haskett-Smith argued for identification with W~ Gap (accepted in Collingwood 1920, 244). Even if modern 'Windy' is based on partial misunderstanding, it is extremely apt.

WINDY SLACK SD1788 Millom Without.
Windislacke 1607 (*PNCumb* 419).
♦ 'The windy hollow'; *windy* from OE *windig* plus **slakki**.

WINSTER, RIVER SD4285 Crook/Cartmel Fell etc., WINSTER SD4193 (hab.) Crook, W~ HOUSE SD4192 Cartmel Fell.
(River:) (*rivulum de*) *Winster, Wynster* 1170–84, *Winstar* 1577; cf. *Winstirt(h)wayt(e)s* 1240–9 (*PNLancs* 190, *PNWestm* I, 14–15 and *ERN* 463); (hab.:) *Winster* & variants 13th cent. to 1777 (*PNWestm* I, 178).
♦ Two main explanations of W~ exist: (a) It could be '(the river on) the left' (ON *vinstra* 'left'), cf. two Norw rivers called Vinstra. The Cumbrian river is left of the LEVEN when facing their estuaries, or of the GILPIN facing the source (so Ekwall, *PNLancs* 190 and *ERN* 463). (b) The river has been identified with *Gwensteri* 'white stream', the name of a battle site in the Welsh *Book of Taliesin*. See further *PNWestm* I, 14, where Smith does not arbitrate between the two theories. Fellows-Jensen favours (b) (*SSNNW* 425), while for objections to it, see Jones 1973, 356–7. The river has given its name to the village. W~ House is not shown or named on the OS maps of 1851, 1892 or 1919.

WINTER CRAG NY4318 (hab.) Martindale.
Wintercragge 1588 (*PNWestm* II, 220).
♦ The farm is named from the **crag** 'rocky height' above it. Whether 'Winter' is the noun (from OE *winter*) or a surname is unclear; if the former it could refer to snow remaining relatively late into spring.

WINTERSEEDS NY3308 Grasmere.
From 1590 (*PNWestm* I, 202).
♦ Probably 'the area where winter grass is sown', with *winter* ultimately from OE *winter*, plus dial. *seed* 'area of sown grass' (*PNWestm*).

WISE EEN TARN SD3797 Claife.
Wise Een Moss 1851 (OS), *Wise Een Tarn (Fish Pond)* 1919 (OS).
♦ The **tarn** (from **tjǫrn** 'small high-level pool') has been created out of a moss (cf. nearby THREE DUBS TARN). *Een* is a dial. form of 'eyes' which goes back to the OE pl. *ēagan*, but whether this is 'wise eyes', and if so, why, is obscure.

WISENHOLME BECK NY0920 Lamplugh.
Wizzenholme Beck 1867 (OS).
♦ The **holm(e)** 'land beside water' is probably the tract between W~ B~ and MEREGILL BECK. 'Wisen-' is puzzling. In form it could be a variant of *wizzen* 'wither(ed), parch(ed) or wrinkle(d)', recorded in *EDD* from Westmorland, Lancashire and Scotland, but this seems an unlikely description of a **holm(e)**, though perhaps cf. WITHERED HOWE. See also **beck**.

WITHE BOTTOM SD1493 Waberthwaite.
Wisebottom 1867 (OS).
♦ Possibly 'the valley where willow grows'. The 1st el. may derive from OE *wīðig* 'willow', which also appears in WYTHOP (q.v.), in Withe Sike (*Wythesik*' 1292, *PNCumb* 30), and in two WHIT BECKS, to judge from their early spellings. The 1867 form may be erroneous. The 2nd el. is **bottom** here refers to a corrie on the NW side of WHITFELL.

WITHERED HOWE NY4606 Kentmere.
From 1857 (*PNWestm* I, 168).
♦ Certainty is impossible without earlier forms, but **How(e)** is presumably from ON **haugr** 'hill'. (a) 'Withered' might refer to the pitted appearance of this rock-strewn hill; the verb *wither* appears from the 14th cent. onwards. (b) Smith, quoting Dickins, suggests Early ModE *whitred*,

whitret 'weasel, stoat' (*PNWestm* I, xvi; cf. *OED: whitret*).

WITHERSLACK SD4384 (hab. & parish), W~ HALL SD4386.
Witherslak & variants 1186–1200 to 1672 (*PNWestm* I, 77); *Wither-slack-hall* c.1692 (Machell I, 106), *Witherslack Hall* 1770 (JefferysM).
♦ Either 'the hollow in the wood' or 'the hollow where willow grows'. The 1st el. is either from (a) *viðar*, gen. sing. of ON **viðr** 'wood', or (b) *viðjar*, gen. sing. of ON *við* 'willow, withy' (both mentioned in *PNWestm* I, 77). Either way, the modern form preserves a reduced form of the Scand. gen. sing. -(j)ar. The 2nd el. is ON **slakki** 'hollow'. Machell wrote 'there [are] two slacks or gills, one on each side of Witherslack Park' (Machell c.1692, ed. Ewbank, 74), and see his comment under **slakki**. The **hall** was referred to as 'a mansion house commonly called *Wither Slacke*' in 1654 (*PNWestm* I, 79); the present building is from 1874 (Pevsner 1967, 299). (Pl. 32)

WOLF CRAGS NY3522 Matterdale.
Wolf Cragg 1787 (*PNCumb* 223).
♦ 'The **crags** or rocky heights frequented by wolves'; ModE *wolf* is from OE *wulf*, whose ON cognate *úlfr* also occurs in Lakeland p.ns (see ULLOCK, ULPHA). If to be taken at face value, the name would predate the extinction of the wolf in Lakeland, which has been dated to the 16th cent. (see Gambles 1989, 141).

WOOD END/WOODEND Examples at:
    SD1099 (hab.) Irton.
le Wodend 1363 (*PNCumb* 404).
    SD1696 (hab.) Ulpha, W~ HEIGHT SD1595 (1597ft/494m.).
Woodend 1606, possibly = *the Woodhend* c.1441 (p) (*PNCumb* 438); *Woodend*

(*Height*) 1867 (OS).
    NY2127 (hab.) Above Derwent.
Woodend 1564 (*PNCumb* 373).
♦ 'The **end** of the **wood**' or possibly, as Watts 2004 suggests as one explanation of the Ulpha example, 'the wood end of the township', i.e. the end of the township (now civil parish) where the wood is. The Above Derwent example presently stands at the end of a wood, while the Irton and Ulpha ones, judging from the map, do not. In both places very small woods are shown on the 1867 OS map; they are, however, both close to township/parish boundaries. The **height** is a prominent hill a little to the SW.

WOOD FARM NY1126 Blindbothel.
Wood 1867 (OS).
WOOD FARM SD4091 Cartmel Fell.
the Wood 1604 (Cartmel PR), 1851 (OS).
♦ 'The (place by the) **wood**', with **farm** added relatively recently in both cases. The Blindbothel place is next to what is nowadays a sizable mixed wood along Sandy Beck; the Cartmel Fell one is just S of Grubbins Wood (shown on 1851 map).

WOODFOOT NY5018 (hab.) Bampton.
(*th'*) *Woodfoot* 1642 (*PNWestm* II, 197).
♦ 'The lower end of the wood'; **wood**, **foot**. It lies below SKEWS, which means 'woodland', but is now only very lightly wooded, according to current maps, as it is on the 1863 OS map.

WOODGATE SD1194 (hab.) Waberthwaite.
Wood Gate 1774 (DHM), *Woodyeat* 1785 (Corney PR).
♦ 'The gate to the **wood**', since the local spelling of 1785 points to the reflex of OE **geat** 'gate' rather than of ON **gata** 'road'. It is close to WOODSIDE SD1194.

WOOD HALL NY3437 Caldbeck.
Wodhall 1507, *Woodhaulle* 1595 (*PNCumb* 280).
♦ 'The **hall** by the **wood**', unless the 16th-cent. spellings are totally deceptive.

WOOD HOUSE Examples at:
    SD1282 Whicham.
From 1774 (DHM).
    NY1617 Buttermere.
Woode 1547 (CRO D/Lec/314/38, pers.comm. Dr Angus Winchester), *Wood House* 1774 (DHM).
    NY3537 Caldbeck.
From 1766 (PR). Buttermere.
♦ Simply 'the **house** by the **wood**'. The Buttermere name is still appropriate, for the house is sheltered by woodland; in the other two cases there are no significant woods today.

WOODHOW NY1304 (hab.) Nether Wasdale.
Woodhawe 1570 (*PNCumb* 442).
♦ 'The hill by the wood' or 'wooded hill'; **wood**, **haugr/how(e)**.

WOOD KNOTTS SD1795 Ulpha.
From 1867 (OS).
♦ 'The craggy heights near the wood'; **wood**, **knott**. The immediate vicinity is not wooded, either on current maps or on the 1867 one, though it could have been in the past. Similarly, this flank of Hesk Fell is stony, but not a distinct hill such as is usually denoted by *knott*.

WOODLAND SD2489 (hab.) Kirkby Ireleth, W~ FELL SD2689, W~ GROVE SD2490 (hab.), W~ HALL SD2488.
(*Kirkeby*) *wodelands* 1544, *Wodland* (*chap.*) 1577 (*PNLancs* 221); *Woodland Fell* 1851 (OS); *Woodland grove* 1835 (Woodland PR), 1851 (OS); *Woodland Hall* 1893 (OS).

♦ W~ (**wood, land**) was a division of Kirkby Ireleth parish. The **Fell** is the upland grazing belonging to it, rather than a distinct hill. See also **grove** 'copse' and **hall**.

WOODSIDE Examples at:
SD1194 (hab.) Waberthwaite.
Wood Side 1774 (DHM), 1780 (PR).
SD2888 (hab.) Blawith.
From 1851 (OS).
SD4289 (hab.) Crosthwaite.
From 1535 (*PNWestm* I, 86).
♦ Simply '(the place) beside the wood' (**wood, side**) — still an apt description for the Blawith example, less so for the Waberthwaite and Crosthwaite ones. W~ in Waberthwaite is close to WOODGATE. Crosthwaite has a cluster of names in *-side*: see List of Common Elements. Further examples of Woodside are given in *PNCumb* 558–9.

WOOF CRAG NY4912 Shap Rural.
From 1865 (*PNWestm* II, 178).
♦ The meaning of W~ here is unknown, but (a) given that other **crag**s or rocky heights in the neighbourhood are named Ritchie C~ and Geordie Greathead C~, the surname *Woof* (a variant of *Wolf*) is not impossible. (b) Corruption of the animal name *wolf* is also possible.

WOOL KNOTT SD2789 Blawith.
Woolknot in Blawith 1722 (Lowick PR); *Wool Knott* 1851 (OS).
♦ If the forms are not deceptive (concealing, for instance, the word *wolf*), this may be 'the rugged height on which sheep's wool catches', with *wool* ultimately from OE *wull*, plus **knott**. Ellwood includes *Woo craggs or oo craggs* in the Supplement to his Glossary, explaining, 'the names of rocks or craggs in Lakeland, over which sheep having passed, have left some of their wool cleaving to the craggs' (1895, 84). A *Woolrig* is recorded from 1626 (Cartmel PR).

WORM CRAG, GREAT SD1996 (1400ft/427m.) Ulpha.
From 1867 (OS).
♦ Without early spellings, the name of this stony height or **crag** remains enigmatic. Various explanations of 'Worm' here and in nearby WORMSHELL HOW are possible, including the following. (a) *Worm*, from OE *wyrm* 'snake, reptile', perhaps referring to one of the types listed by Hutchinson in nearby Eskdale: 'slow-worm, asp, and hag-worm or snake, of which latter, some are of a large size' (1794, I, 581; cf. Hervey & Barnes 1970, 148). (b) Corruption of some other el. such as *walm* 'spring' (cf. WALMGATE) is possible, though no springs appear on the current map. **Great** W~ C~ is distinguished from Little W~ C~ immediately to the NE.

WORM GILL NY0809 Ennerdale.
Worm Gill 1867 (OS), previously *Frithebec* 1243, *Le Wrongyl(hed)* 14th cent. (in Elizabethan copy, *PNCumb* 31), *Wormdale River* 1826, *Wormdale Beck* 1827 (plans in CRO D/Lec/94).
♦ Formerly 'the winding stream'. For the 1st el., the 14th-cent. spelling suggests *wrong* from ON (*v*)*rangr* 'crooked', which would describe the river's course quite well; cf. WREN GILL. The modern spellings seem to have substituted *worm* 'worm, snake', as possibly found in WORM CRAG. **Gill** is, as usual, the Norse-derived 'ravine with a stream'.

WORMSHELL HOW SD2097 Ulpha.
From 1867 (OS).

♦ In the specific el., 'Worm' has a similar range of possibilities to nearby GREAT WORM CRAG, and 'shell' is still more baffling, unless from **scale** or its equivalent ME *schele* 'shieling'. **How(e)** from ON **haugr** is a hill.

WOUNDALE NY4107 Troutbeck, W~ BECK NY4107.
Woundall, Woundale 1560; *Woundale Beck* 1865 (*PNWestm* I, 191).
♦ The 1st el. is obscure — perhaps OE *wunden* 'twisted, curving', which would well describe the course of the **dale** 'valley' and **beck** 'stream'.

WRAY: HIGH WR~ SD3799 (hab.) Claife, LOW WR~ NY3701 (hab.), HIGH WR~ BAY NY3700, LOW WR~ BAY NY3701, WR~ CASTLE NY3700.
Wraye c.1535; *the Heywray* 1619; *Lowrey* 1656 (*PNLancs* 219); *High Wray Bay*; *Low Wray Bay*; *Wray Castle* all 1850 (OS).
♦ **Wray** from ON **vrá** 'secluded or remote place' aptly describes the triangular nook which is bounded by WINDERMERE, BLELHAM Beck with its marshy surrounds, and CLAIFE HEIGHTS, and remote from the traditional route to the WINDERMERE ferry; see also **high**, **low**, **bay**. Wray Castle dates from the 1840s, a 'great, foolish toy of gray stone' to Hawthorne (quoted in Lindop 1993, 393), but a 'dignified feature' to Wordsworth.

WREAKS END SD2286 (hab.) Broughton West.
From 1669 (original spelling not specified, Broughton-in-Furness PR).
♦ Wr~ may well be '(the area subject to) flood debris'. *Wreak* is a variant of *wreck*, with the sense 'a drifted or tossed-up mass' (*OED*: *wreck* n.1, sense 5, cf. *EDD*: *wrack* sb.1). Hence it could apply to a debris-strewn area in the flood-plain, here the very low, wet land, now artificially drained, beside KIRKBY POOL. Wr~ **End** stands on the rising ground at the edge of this area, with Wr~ Causeway and Moss just to the E on the valley floor. This interpretation is supported by (and in turn confirms) the proposed explanation of a Yorks Wreaks, recorded as (*les*) *Wrekes* in the late 12th cent., for which Smith tentatively suggested ON **vrek(i)* 'wreck, wreckage', an assumed earlier version of ON *rek(i)* 'jetsam' (*PNWYorks* V, 131). He wondered about an 'untidy place' (*PNWYorks* VII, 72), but since this too is a riverside site — an unusually flat stretch beside the Nidd — it could in particular be strewn with debris left behind by floods. *Wrack, wreck* means 'seaweed' or the right to harvest from the sea-shore in Cumberland documents (Dilley 1970, 203, Denton 1687–8, 80, 83).

WREAY NY4423 (hab.) Matterdale.
Wra 1487, *The Wraye* 1581 (*PNCumb* 258).
♦ 'The nook or secluded place'; **vrá/wray**, cf. WRAY. Watts (2004) remarks on the situation at the edge of Watermillock township.

WREN GILL NY4608 Longsleddale.
Wrangdale gill 1728 (*CW* 1987, 231), *Rangle-Gill* 1829 (Parson & White 652); cf. *Wrangdayll* 1577, *Wrangdale, -dall* 1580 to 1633 (*PNWestm* I, 162).
♦ Wr~ was originally 'the crooked valley', from ON (*v*)*rangr* 'crooked, awry' or its OE counterpart *wrang*, plus ON **dalr**/OE **dæl** 'valley', or their descendants, but **gil(l)** 'ravine with stream' has been added and the original name contracted and reinterpreted as *wren* — not unreasonably, since 'wrens range high

into the fells' (Ratcliffe 2002, 245). A side-valley to LONGSLEDDALE, this is small and narrow, with two pronounced bends.

WRYNOSE: WR~ BREAST NY2602, WR~ BOTTOM NY2501 Ulpha/Dunnerdale, WR~ FELL NY2704 Langdales, WR~ PASS NY2702 Ulpha/Dunnerdale. (On the boundaries here, see THREE SHIRE STONE.)
Wreineshals, Wrenhalse 1157–63, *Wrainshals* 1170–84, *Wrenose* 1576, *the mountain Wrynose* 1610, *Wrey Nose a great fell* 1671 (*PNCumb* 437–8); *Wrynose Breast* 1867 (OS); *Wrynose Bottom* 1850 (OS); *Wrynose pass* 1865 (Prior 60).
♦ Wr~ probably means 'stallion's pass'. It is at the watershed of the DUDDON and BRATHAY, and the 2nd el., from ON **hals** 'neck of land', is entirely appropriate. For the 1st el. ON (*v*)*reini* 'stallion' seems the most likely origin, perhaps implying that a strong horse is needed on the pull over the **pass**. The -s- in the spellings, however, is problematic, since the gen. case would be *vreina* not *vreins*. Perhaps a strong form **vreinn* existed alongside *vreini*. Collingwood suggested the tautological *vrein-hests-hals* 'stallion-horse-pass' (1918, 94); *PNCumb* 438 suggests a medial el. *nes* 'headland'; and *PNWestm* I, 205 suggests a different 1st el. altogether: OE *wrīgan-nes*, giving ME *wrie-nes* 'twisted headland', but the 12th-cent. *ei, ai* spellings would be unexpected, and the suitability of **nes** here is somewhat doubtful. Breast (as in WHITELESS BREAST) here refers to an unlovely tract of scree and rubble, while **bottom** 'valley' denotes to a broad, flat, damp expanse. Wr~ **Fell** is a stretch of high unenclosed ground nearly a mile N of the Pass.

WYKE, THE NY3306 (hab.) Grasmere.
Wicke 1580, *Wyke, Wike* 1683, *The Wyke* 1802 (*PNWestm* I, 202).
♦ 'The bay', ultimately from ON **vík**. The farm stands on Wyke Gill, which flows the short distance to GRASMERE lake. The bay on the shoreline is not pronounced.

WYTHBURN (hab.) NY3213 St John's, WYTH BURN (stream) NY3011, W~ FELLS NY3012.
Withebotine c.1280, *Wytheboten* 1345, *Wythbottom* 1552, *Wyeborne* 1564, *Wythbourne* 1568 (*PNCumb* 315); *Wytheburn fells* 1687–8 (Denton 51).
♦ 'The valley where withies or willows grow', from ON *viðir* 'willow' and ON **botn** 'inner valley', possibly replacing the cognate OE *wīðig* and *botm* (Cole 1987–8, 40); also **fell**. The 2nd el. in any case was replaced by **burn** from OE *burna* 'stream', and W~ was adopted as the stream name. The stream may previously have been called *Kaltre* (c.1280, see *ERN* 60). The original place W~ is partially submerged by THIRLMERE since its enlargement into a reservoir. (Pl. 19)

WYTHOP (hab. & parish), BECK W~ NY2128 (stream & hab.), W~ BECK NY1829, W~ HALL NY2028, W~ MILL NY1729 (hab.) Embleton, W~ MOSS NY1827 Wythop, W~ WOODS NY2029.
Wizope 1195, *Wythorp*(*e*) 1260, 1383, *Wythope* 1279 (*PNCumb* 457); *Bechwythop'* 1247 (*PNCumb* 31); *Wythop Beck* 1867 (OS); *Wythopall* 1619; *Wythoppe mill* 1578; (W~ Moss:) probably = *mussa* 1195; *boscus de* (wood of) *Wydehop'* 1292 (*PNCumb* 457).
♦ W~ is 'the willow valley', from OE *wīðig* 'withy, osier, willow' and OE **hop** 'blind valley'. At the present time, 'with 17 species and 20 hybrids Cumbria has an exceptionally rich

willow flora' (Halliday 1997, 189(–198)), though by no means all of these occur in the Lake District. W~ **Beck**, which runs through the hamlet of W~ **Mill** and is larger than Beck W~, is presumably meant by *Bechwythop'* in the 1247 document. Watermills for fulling, corn, timber and paper have flourished at W~ Mill, one being operative until very recently. See also **hall**, **moss** 'bog' and **woods**. Hutchinson refers to 'a capital mansion or hall-house, called *Wythorp-Hall'* (1794, II, 125); this was formerly the seat of the lord of the manor.

Y

YARD STEEL NY2934 Caldbeck.
From 1867 (OS).
♦ **Yard** may well go back to OE *geard* 'enclosure' (cf. **garðr**), though it is not clear what it would have referred to in this uninhabited tract of high fell. **Steel** 'a steep place' is appropriate to this N. spur of GREAT SCA FELL.

YARLSIDE: Y~ CRAG NY5207 Shap Rural, GREAT Y~ NY5207, LITTLE Y~ NY5307.
Ierlesete 12th cent., *Yarlside* 1859; *Yarlside Crag* 1859 (*PNWestm* II, 178); *Great Yarlside*; *Little Yarlside* both 1860 (OSNB).
♦ Y~ is probably 'the lord's seat', from ON *jarl* 'lord, earl' and ON **sæti** 'seat'. Y~ **Crag** is a sloping natural balcony which, like nearby LORD'S SEAT, forms a rocky vantage point; see also **great**, **little**.

YEASTYRIGG CRAGS NY2306 Eskdale.
Yeastyrigg Crags 1867 (OS).
♦ The row of **crags** 'rock outcrops' forms a high-level **rigg** or ridge (see **hryggr**), but without early spellings 'Yeasty' is obscure. If the adj. *yeasty*, it could perhaps describe the serrated appearance of the crags, comparing them to the bubbling of yeast (the word can refer to foaming water, *OED*, sense 3, while *EDD* records the sense 'gusty, stormy', though only in Sussex).

YEW BANK NY2303 (1637ft/499m.) Eskdale.
From 1867 (OS).
YEW BANK SD2690 (678ft/207m.) Kirkby Ireleth.
From 1851 (OS).
♦ Probably 'the hill-slope grazed by ewes' (reflex of OE *eowu* plus **bank**), but reference to **yew** (trees) is also possible, at least for the lower-lying Y~ B~ in Kirkby Ireleth where I have noted one young specimen of yew.

YEWBARROW NY1708 (2058ft/627m.) Nether Wasdale.
(*le Mikeldor de*) *Yowberg* 1332, (*le Durre de*) *Youbergh* 1338, *Yeweberrowe* 1540 (*PNCumb* 441), *Ewe berray* 1664 (Winchester 2000, 169), *Yeabarrow* 1802 (*CN* 1213).
YEW BARROW SD3587 (795ft/242m.) Colton.
From 1851 (OS).
♦ Probably 'the hill grazed by ewes'; OE *eowu* and **berg**. This is suggested by the medieval spellings for the Wasdale example (in which *Mikeldor* and *Durre* refer to the present DORE HEAD.) The Colton example lacks early spellings but may be of the same origin, though **yew** cannot be ruled out. (Pl. 23)

YEWBARROW HALL NY5002 Longsleddale.
Ubarrehall 1518–29 (*PNWestm* I, 162, but same forms also attached to Yewbarrow, Witherslack at I, 79), *Yewbarrow hall* 1843 (Ford 140), *Ubarrow Hall, Yewbarrow Hall* 1858 (OSNB).
♦ (a) Y~ may well be 'the hill grazed by ewes', with reflex of OE *eowu* and **berg**, cf. YEWBARROW Nether Wasdale. (b) **Yew** is also possible. Machell, e.g., mixes *Ew-* with *Yew-* spellings, but uses both for the tree name and is clear that the p.n. reflects the wooded nature of Longsleddale. The **hall** is described by Ford as 'the most interesting mansion in the dale . . . having an ancient tower, whose walls

are several feet thick' (1843, 140); it is shown on OS maps as an Antiquity.

YEW CRAG Examples at:
NY2214 Buttermere.
From 1774 (DHM).
NY2615 Borrowdale.
From 1867 (OS).
NY3020 St John's.
From 1867 (OS).
NY3107 Grasmere.
Not 1863, 1899 (OS).
NY4120 Matterdale (1434ft/437m).
Yew Crag 1784, *Ewe-Cragg* 1789 (*PNCumb* 258).
YEW CRAGS NY2202 Eskdale.
From 1867 (OS).
♦ Some of these at least may be 'rocky height(s) frequented by ewes', with *ewe* from OE *eowu* (cf. EWE CRAGS), plus **crag**. Female sheep are more likely than **yew**s, since these trees are not usually indigenous in N. Lakeland. In some cases, however, hardy yews are in evidence: on the exposed top of Yew Crag, Borrowdale (with others clustering in the mixed woodland just below), and at Yew Crags, Eskdale. Below the latter is a Yew Bank, which Coleridge wrote of as 'Ewe bank (or Yewbank, for there are still Yew-Trees, & a tradition that it was covered with them)' (*CN* 1220, 1802).

YEWDALE: HIGH Y~ SD3199 (hab.) Coniston, LOW Y~ SD3199 (hab.), Y~ BECK SD3098, Y~ FELLS SD3099.
Ywedale(bec) 1196 (*PNLancs* 192); *Upermorewdaile* 1645, *Upr Youdall* 1690 (PR), *High Yewdale* 1850 (OS); *Neither Vdall* 1645, *Low Youdall* 1684 (PR), *Low Yewdale* 1850 (OS); *Ywedalebec* 1196 (*PNLancs* 192); *Yewdale Fells* 1802 (*CN* 1228), 1850 (OS).

♦ 'The valley with yew trees'; **yew**, **dalr**. The 12th-cent. spelling in this case virtually removes the possibility of OE *eowu* 'ewe' which exists for other names in 'Yew-'. Rollinson records a legend that an 18th-cent. farmer at High Y~ planted one yew for each of his fifteen children (1997, 192–3), while Crosthwaite's map of 1788 has a drawing of 'The large YEW TREE, near Coniston Water head, 9 feet diameter', and 'The Old Yew' is marked just NE of High Y~ on the OS map of 1850. Other features are then named from the valley: the farms sited relatively **high** and **low** along it, the **beck** (from **bekkr**) running through it, and the **fell(s)** to the NW.

YEW PIKE SD2092 Dunnerdale.
From 1850 (OS).
YEW PIKE SD2998 Coniston.
From 1850 (OS).
♦ The Coniston mountain lies above YEWDALE and has yews on its lower slopes; otherwise one might have suspected **yew** to be a corruption of *ewe* (see EWE CRAGS). In the Dunnerdale example, either derivation is possible. Both are steep heights worthy of the name **pike**.

YEWS: LOW Y~ SD4291 (hab.) Crosthwaite.
Yew(e)s 1535 (in 1669 copy) to 1815 (*PNWestm* I, 86); *Low Yews* 1863 (OS).
♦ An original 'place where yews grow' has evidently been divided into two, with High Yews just to the N, together with another arboreal pair, Low and HIGH BIRKS; **low**, **yew**.

YEWS SD4397 (hab.) Nether Staveley.
Yews 1836 (*PNWestm* I, 174).
♦ Simply 'the **yew** trees'. For similar names, cf. (THE) ASHES.

YEW TREE/YEWTREE (hab.) NY1105 Nether Wasdale.
Ewtree 1578 (*PNCumb* 442), *Yewtree* 1774 (DHM).
YEW TREE FARM NY3023 St John's.
Yewtree 1592 (*PNCumb* 318).
YEW TREE FARM SD4595 Crook.
Yewtree 1836 (*PNWestm* I, 181).
♦ Simply 'the yew tree, place where a distinctive yew tree grows'; **yew** plus *tree* from OE *trēo(w)*. In St John's and Crook the addition of **farm** was registered on the 1994 One Inch map but not the 1966.

YEW TREE TARN NY3200 Coniston.
From 1850 (OS).
♦ 'The high pool with the yew tree'; **yew**, *tree* ultimately from OE *trēo(w)*, plus **tjǫrn/tarn**.

YOADCASTLE SD1595 Ulpha.
From 1867 (OS).
♦ 'Horse fort.' 'Yoad' may be the dial. *yoad, yad, yaad* (*EDD*: *yad*, sb. 1), '(old) mare or work-horse, nag'. The **Castle** is a natural rocky fortification.

YOKE NY4306 (2309ft/706m.) Kentmere.
(*Pike of*) *Yoak* c.1692 (Machell II, 306), *the Yoak* 1780 (West 74); *Yoke* 1828 (HodgsonM), *Yolk* 1858 (OSNB), 1920 (OS).
♦ This name can plausibly be taken at face value, since the summit of Yoke is linked to the slightly higher one of ILL BELL by a long ridge; or the name could describe the main summit, a subsidiary one to the SW and the short, bowed ridge between. The 'yoke' metaphor is implicit in hill-names elsewhere, e.g. the Icelandic *Ok*. The word *yoke* derives from OE *geoc*, and the spelling *yolk* is found in the 19th cent.

LIST OF COMMON ELEMENTS
IN LAKE DISTRICT PLACE-NAMES

Most of the place-names of England are compounds consisting of two elements, with a 'specific' or defining element first and a 'generic' last, e.g. BUTTERMERE, STONETHWAITE or EAGLE CRAG. In the Lake District and elsewhere in NW England 'inversion compounds' with the reverse order are also found (see p. xxii). Simplex or single-element names such as DODD or (THE) WYKE form a numerous minority, while there are also many names, simplex or compound, which are extended by the addition of an affix, e.g. HIGH and LOW LORTON or OLD BRANDLEHOW, or which give rise to secondary names, e.g. LORTON FELLS or BRANDLEHOW PARK.

The notes below cover the more common name-forming elements, explaining their origin and usage, and giving examples to illustrate typical spellings and patterns of combination with other elements. An indication of approximate frequency, period at which productive, potential for confusion with other elements, and other points of interest are also noted as appropriate. These notes refer, unless otherwise specified, to the elements as used in the Lake District place-names in this volume, and are primarily based on my own analysis of names and sites; all statements about the earliest occurrence of elements should be read as being 'to my knowledge', since further evidence could always come to light, and there are, of course, hundreds of Lakeland place-names not covered here. In compiling the notes, however, reference has also been made to other sources and a broader picture can be obtained from these: the great surveys of the elements of English place-names in Smith, *Elements*, and *VEPN* (in progress), the lists of elements in county surveys, especially *PNCumb*, *PNLancs* and *PNWestm*, dictionaries including those listed below, and works on specific aspects of place-names such as Coates and Breeze 2000, Fellows-Jensen's *SSNNW* (1985), Field 1972, Gelling 1984, Gelling & Cole 2000, and Padel 1985.

In referring to particular linguistic periods, the following conventions are used:

a. The terms Old English (OE) or Old Norse (ON) refer to the varieties of language spoken approximately up to the 11th cent. On the difficulty of separating OE from ON see p. xxiv.

b. The term Middle English (ME) is applied to the language current c.1100–c.1500. ME words are of diverse origin, mainly OE, ON, French (often via Anglo-Norman) and Latin.

c. The term Modern English (ModE) refers to the period from the 16th cent. to the present, and is used below both for modern forms of older words, and for elements which do not appear until this period. These forms are Standard English, while 'dial.' (dialect) signals forms (in ModE and often also ME) that are more or less restricted in distribution.

Forms of headwords normally follow Smith, *Elements*, supplemented by Cameron 1968–9 and Jackson 1968–9, using the form appropriate for Cumbria where there is variation (e.g. the Anglian forms of OE elements). As is traditional, the headforms of adjectives are given in their masculine nominative singular forms; hence LONTHWAITE is explained from **langr, þveit**, even though the ON adj. *langr* 'long' is masculine and the ON noun *þveit* 'clearing' is feminine. ON elements are given, as is also customary, in normalised Old Icelandic forms, this being the best-recorded branch of the early Scandinavian languages. It should be noted that the consonant which appears as *w* in the Lakeland names is spelt *v* in the ON etymons (e.g. WATENDLATH, with 1st element from *vatn* 'water'). The following have also been consulted, especially where no entry is available in Smith, *Elements*: Padel 1985 for Brittonic elements; Fritzner 1886–96 and *ONP* (in progress) for ON; Bosworth & Toller 1898, the *Microfiche Concordance* and *DOE* (in progress) for OE; *MED* for ME; *OED* for ModE and *EDD* for dialect forms.

A few considerations relating to Lakeland place-names and their constituent elements (as to other toponymic material) are extremely important to bear in mind.

First, since most written documents containing Lakeland place-names are no earlier than the 12th or 13th cent., and often much more recent than that, many of the names are extremely difficult to date. Some names appearing in, say, 16th-cent. documents or 19th-cent. maps will be genuinely recent; others may have been centuries old when first written down (and some documents are lost).

Second, many OE and ON name-elements were absorbed into everyday dialect speech and continued to be productive in generating new place-names, so that names containing the reflexes or descendants of ON words such as **bekkr/beck** 'stream', **dalr/dale** 'valley', or **fjall/fell**, 'hill, mountain, unenclosed high ground' could have been formed in the Norse-speaking era, or in the ME and ModE periods. The date of the first known written record, the remainder of the name, or other evidence may help to limit the possible date-range, but there may still be doubt.

This uncertainty about the dating of names can entail uncertainty about the meaning of their constituent elements, since this could change through time. It also makes it difficult to select the appropriate form of the elements, and a common convention in the place-name literature is to refer to the oldest forms, e.g. OE or ON, even in the case of names that may well be post-Conquest or even post-medieval. The citation of an OE or ON etymon can therefore have the sense 'OE/ON *xxxx* or its later reflex'. This system is modified somewhat in this volume, as follows.

In the List of Common Elements:

   a. Headwords for elements that were productive over a long period are given in both older and newer forms, always with the oldest first, e.g. **bekkr ON, beck dial. 'stream'** or **hara OE, hare ModE**. ME forms are also included where helpful, e.g. **tjǫrn ON, terne ME, tarn ModE 'small mountain pool'**, since the ME *terne* is common and rather unlike the normalised ON *tjǫrn*.

b. Elements that do not belong to the OE or ON periods are given headwords as appropriate. Some of these are words not recorded in English documents until the ME or ModE periods, such as **grange**, an adoption from French into ME. Others may be linguistically of OE or ON origin, but do not appear in Lakeland place-names until later. For instance, the entry **ground ModE 'farm'** indicates that there is no evidence in this volume of the element occurring in pre-modern Lakeland names, though the word *ground* derives from ON *grund* and/or OE *grund*; further examples are **great**, **hall**, and **syke**.

c. Although most headwords in the List of Common Elements begin with the oldest relevant forms, some take particularly difficult modern forms as their starting point, e.g. **red in modern forms of names**, since this can represent either 'red' (from OE *rēad*) or 'reed' (from OE *hrēod*).

In the Dictionary entries for individual place-names:

a. Elements are referred to in the form that seems most appropriate. Thus a stream name ending in 'Beck' and recorded in medieval documents is designated as containing **bekkr**, whereas one first known from 19th cent. maps is designated as containing **beck** — though note the possibility mentioned above that it is older. In especially doubtful cases phrasing such as 'ON *xxxx* or its reflex' is used, or both older and newer forms of the element may be given, e.g. **bekkr/beck**.

b. Alternative forms, as well as being used where the age of the name is particularly uncertain, are also given where the older and newer forms differ sufficiently to make the newer one difficult to find in the alphabetical listing. Thus **hryggr/rigg** ('ridge') points the reader to **hryggr**, and phrasing such as '**how(e)** from **haugr**' ('hill') points the reader to **haugr** in the List of Common Elements.

c. Translations of elements in the Dictionary entries for individual place-names match those in the List of Common Elements, with some accommodation to individual topography. It should also be mentioned that many habitations (farms, hamlets, villages or towns) are named from adjacent natural features.

ELEMENTS

á ON 'river'
This occurs in some dozen names of important rivers, usually coupled with a word describing the water or its surroundings: AIRA (with 1st el. meaning 'gravelly spit of land'), BRATHAY ('broad'), GRETA ('rock, stone'), LIZA ('bright') and ROTHAY (?a fish species). *Á* also appears in AMBLESIDE and AYSIDE, and in its gen. sing. form *ár* it may have formed part of valley names such as the two BORROWDALES, DUNNERDALE and MITERDALE, as it still does in several Icel names (e.g. Laxárdalur 'Salmon-river-valley'). Medieval spellings of AIRA and LIZA with *-h-*, *-ch*, *-gh-* or *-th* interestingly preserve an early form of the word ending with a spirant consonant.

ærgi, erg Gaelic-Norse 'shieling, summer pasture'
A distinctive el., originating in Gaelic *áirge* and evidently introduced into NW England and SW Scotland by Gaelic-Scandinavian incomers. (Its only occurrence in ON sources is in an Orkney place-name.) There are eight certain examples of *ærgi* in this volume, recorded from the late 12th or 13th centuries. The specifics include **moss** in MOSERGH, MOSSER, 'turf' (possibly) in TORVER, 'wind' in WINDER, and probably a female pers.n. in LANGLEY; (LITTLE) ARROW deploys the el. as a simplex. Other possible examples include ARKLID and BETHECAR.

Along with ON **skáli** 'shieling (hut), summer pasture', ON **sætr** 'shieling, summer pasture', ON *sel* 'shieling hut' (see SELLA) and probably **búð** 'hut', *ærgi* commemorates the medieval and Early Modern practice of transhumance whereby, as in Scandinavia, the Alps and elsewhere, sheep, goats and cows were grazed on high, unenclosed pasture or other marginal land during the summer months, leaving the homefields for hay and crops. The shieling sites, furnished with groups of huts for accommodation and dairying, in many cases developed into permanent farms and hamlets, especially at times of population growth such as the 13th cent. (see, e.g., OLD SCALES). The system continued well into the 16th cent., but was already declining in Lakeland from c.1350 (Winchester 1987, 92–6). It appears that the *ærgi* names tend to denote lower, better, and probably older-established sites than the **sætr** and, still more, the **skáli** names, which are characteristic of high, remote sites such as small tributary valleys or hillsides (Fellows-Jensen 1980 and *SSNNW* 49–50, Whyte 1985). It has been further suggested that the *ærgi* sites may have been specific types of vaccaries (cattle farms, Higham 1977–8; for a recent study of the el. in NW England and SW Scotland, see Grant 2004, ch. 3).

aik: see oak

askr, eski ON, æsc OE, ash ModE
The tree-name in Lake District p.ns results from ON *askr* 'ash-tree', ON *eski* 'ash-trees, ash-copse' or 'place with ash-trees', and/or from OE *æsc* (pronounced roughly as ModE *ash*). Of some 14 occurrences in this volume, five have 13th-cent. spellings: ASHLACK, ASHNESS, ASKHAM and HESKET (two instances). The forms of all five except ASKHAM point to ON *esk-* as the origin. The ash, *Fraxinus excelsior*, is currently 'probably the commonest tree in [Cumbria]' (Halliday 1997, 376). It is found up to 1608ft/490m. above sea level, and the specimens that flourish along the sides of gills are relics of much more widespread upland forests. The trees traditionally provided shelter and boundary-markers, their wood was used for fencing, tools and fuel, and their leaves for winter fodder. They were often pollarded for these purposes. Ash trees grew beside many a Viking-Age farm in Norway, and in Norse mythology the sacred world-tree is the ash Yggdrasill, but if these cultural traditions were current in Lakeland, they have no known impact on the p.ns.

band ModE 'projecting ridge, stratum of rock'
This el. occurs in the names of five hill-features, none recorded before the 19th cent. THE BAND illustrates its application to a ridge projecting from a larger mountain, while LONG BAND is a horizontal feature. The origin of the word is uncertain, but it may owe something to OFr *bande* 'strip' as well as to OE *bend*, ON *band* 'tie, binding device'.

banke ME, bank ModE '(hill)-slope, hill, bank'
Originally from ODan *banke* (rather than OWN *bakki*, for which there is no evidence in the present volume). There are over 80 occurrences, most often as a generic, and referring to a habitation, usually part-way up a slope, e.g. Wordsworth's onetime home

ALLAN BANK. The specific often describes the bank itself or refers to vegetation, animals, or a structure; or when the name denotes the bank of a river or lake, the specific may refer to that (e.g. BURNBANK, GILPIN BANK). The earliest occurrences are *Ellerbank* c.1215, now ELLERBECK, and THORNBANK, c.1230. **Bank(e)** can also function as specific in names such as BANK END and BANK HOUSE.

barrow: see berg 'hill'

bay ModE
The word derives, via ME *bay(e)* etc., from OFr *baie*. TARN BAY is a coastal bay, but the remainder in this volume are all wide inlets in lakes: five in DERWENTWATER, three in WINDERMERE and one each in CONISTON and ULLSWATER. Most are only documented in the last two centuries. *Bay* may in some places have been in competition with *wyke* from ON **vík**.

bekkr ON, beck dial. 'stream'
This is the usual Lakeland word for 'stream', occurring some 200 times, and taking its place as the intermediate stage in an informal hierarchy of terms for watercourses (see **gill**). It is most often used as a generic, with a specific referring to the hamlet or valley passed by the stream (e.g. SEATHWAITE BECK, or various MOSEDALE BECKS), but some names also highlight attributes of the *beck* itself, e.g. MERE BECK, TROUT BECK, and the picturesque GOLDRILL BECK. The el. also occurs as a specific in BECKFOOT, BECKSIDE etc. In what happen to be the earliest recorded occurrences (CALDBECK and WHITBECK, from the 12th cent.), a settlement takes its name from a nearby stream. Many of the *beck* names are difficult to date, and could have emerged either during the Norse-speaking period or much later.

belle OE, ME, bell ModE 'rounded, bell-shaped hill'
Seven of the nine occurrences are in hill-names (two BELL HILLS, one recorded from the 13th cent., two ILL BELLS, BELL RIB, CAT BELLS and THE BELL), and BELL CRAGS and BALL HALL probably also refer to a hill. ON *bjalli* is used in this way in Norway and Iceland, OE *belle* may have been used of hills (so Smith, *Elements*, *VEPN*), and hill-names in *bell* are common in the Borders. BELL GROVE is of uncertain meaning.

berg OE, ON, barrow, barf dial. 'hill'
Normally used of relatively low hills, rounded rather than sharply peaked. It appears not to be applied, as in S. England, to man-made tumuli (with a possible exception in LICKBARROW). *Berg* is the Angl counterpart to WS *beorg*. The nom. (subject) case of OE *berg* commonly gives modern BARF (in minor p.ns and in dial.), while dat. (prepositional) *berge* gives BARROW, by far the most common form. ON *berg* 'mountain, rock' and ON *bjarg* 'rock, precipice' can also give rise to names in -*barrow*, and these are difficult to separate from OE *berg*, though in names such as KELBARROW or SCALEBARROW/-BORROW where the specific is definitely ON, the generic is also likely to be of ON origin. Some of the four dozen examples are recorded from medieval times, including SCALEBARROW and WHITBARROW from the 12th cent.

bield dial. 'shelter, lair'
This refers either to an animal's lair, usually among rocks (e.g. CAT BIELDS, DEER BIELDS), or to a man-made shelter, as presumably in NAN BIELD (PASS). The seven instances are typically in high, remote situations, and only THE BIELD (1697) has been found before the 19th cent. Common in N. dial., and not recorded before ME *belde*, the word is of uncertain origin, but it may be connected with OE *beldo*, ME *belde* 'boldness, security' or conceivably with OE *bold*, *boðl* 'dwelling' (*VEPN*).

birki ON, birk dial. 'birch'
The tree-name occurs in some 30 p.ns, including BIRKER, BIRK FELL and BIRKRIGG which are recorded from 13th cent. Dial. *birk* is from ON *birki* 'birches, birch-copse' (*NGIndledning* 44), probably reinforced by ON *bjǫrk* and OE *birce*, both 'birch-tree'. It can appear as a simplex (BIRKS), or in the compound names of natural features (BIRK CRAG, BIRK MOSS) or settlements ((HIGH) BIRCHCLOSE). The last name is one of three in which Standard English *birch* appears to have replaced original *birk*, which also appears disguised in BURTNESS. BIRKBY (FELL) is of different origin. The downy birch (*Betula pubescens*) is currently more common than silver birch (*Betula pendula*), flourishing especially in damp habitats to an altitude of 2247ft/685m. (Halliday 1997, 140). One of the ancient native trees of Cumbria, its wood and bark were prized for a range of purposes.

blæc OE, black ModE
From OE *blæc* 'black' or possibly (in the case of BLAKEBECK and BLAKE HILLS) its alternative form *blaca*. Some 30 names in this volume contain the adj., including nine **Crag**s and six or seven **Beck**s. Streams described as black are often peaty, while hill features may have been heather-covered. ON *blakkr* 'dusky black' exists but is of very restricted use, and 'black' is normally expressed in ON by *svartr* as in SWARTHBECK, or by **blár**. There is a certain danger of confusion between *blæc* and the following el.

blake in modern forms of names
Six or seven occurrences, none with medieval spellings, so that the various possible origins and meanings are not easy to disentangle.
1. Probably 'pale' or 'yellow' in most cases, from ON *bleikr* or its OE cognate *blāc* 'pale, shining': 'Blake in these dialects is that pale yellow colour which corn assumes when it first begins to ripen' (Clarke 1789, xxvi).
2. The spelling *blake* can also represent a dial. pronunciation of *bleak* 'bare, exposed, wind-swept', a word which first appears in the 16th cent. and which may have developed from ME *blaik* and variants, which is in turn from ON *bleikr*.
3. 'Black, dark' is possible in BLAKEBECK and BLAKE HILLS: see **blæc**.

blár ON, blea dial. 'dark, blue, black'
Applied to a range of landscape features, especially crags and pools, in a dozen names. The names spelt *blea* are not recorded until the modern period, but BLOWICK appears in the 12th cent. and BLAWITH in the 13th.

bœr, bý ON 'farmstead, village, settlement'
ONorw *bœr*, usually referring to a farmstead, or ODan *bý*, a farmstead, hamlet or village. There are ten instances when SUBBERTHWAITE is included, all with medieval spellings except LACONBY and all around the perimeter of the Lake District. Norwegian origins are more likely where some Irish presence is indicated (IREBY, LACONBY), but the *-by* names in E. Lakeland and the Cumbrian coastal plain may bear witness to Danish influence, whether or not in the form of settlement by partly Danish groups. BIRKBY suggests contact between British and Norse speakers, while PONSONBY suggests continued use of *-by* after the Norman Conquest. For full analysis of the el. in the NW, see Fellows-Jensen, *SSNNW* 10–43, 287–90; and see also Grant 2004, ch. 5. In KIRKBY HOUSE, NEWBY BRIDGE, WHITBY STEADS and probably CROSBY GILL, the '-by' name is not primary.

borrans, borran ModE 'cairn, heap of stones'
This seems to derive from an OE **burgæsn* 'burial, cairn', ME *burghan(es)*, *borghan(es)*. The final *-s* is due to reversal of *-sn* to *-ns* and is not a pl., but it has been understood as

one, resulting in a sing. ME *burghan*, ModE *borran* as in WHITE BORRAN. It is especially common in the former Westmorland (*PNWestm* II, 320), though in this volume the only certain examples are BORRANS and WHITE BORRAN. It may, however, lie behind certain modern forms containing *burn* (see **burn**), especially when these refer to known prehistoric sites or places close to sites, such as BURNBANK FELL, BURNBANKS, BURNEY and BURNMOOR (both Eskdale and Waberthwaite ones).

botn ON 'head or innermost part of a valley'
Only certainly found in GREENBURN (Coniston) and WYTHBURN, both recorded from the 13th cent., but it may have affected the Lakeland usage of **bottom**.

bottom ModE 'valley'
The word probably originates in OE *botm* 'valley bottom', which was applied especially to damp valley floors, notably those with abruptly rising sides and prone to flooding (Cole 1987–8, esp. 46), but since most of the six Lakeland examples are heads of valleys, including the high corries LITTLECELL, SELE and WITHE BOTTOM, this may suggest that *bottom* has inherited the usage of ON **botn**, or in fact replaced the word in some cases. DALE BOTTOM first appears in 1605, the other five examples in the 19th cent.

brād OE, broad ModE
The adj., cognate with ON **breiðr**, appears in only one clearly medieval name, BRADLEYFIELD, and in a further 16 which are more recent or difficult to date, including BROAD OAK(S) (four instances), BROAD END (three) and BROADGATE (two).

braken ME, bracken ModE
The fern *Pteridium aquilinum*, with its bright spring green and autumnal russet, is extremely widespread (Halliday 1997, 99–100), but was formerly less so, and therefore more distinctive. Though a bane to sheep-farmers, the plant was useful in the past, for fuel, bedding, thatching, as a source of potash in soap-making, and for medicinal purposes (Gambles 1989, 13), and was a fiercely protected resource (e.g. Dilley 1967, 136, Winchester 2000, 133–6). The word first appears in the Lake District in the ME period (BRACKENTHWAITE, 1230, being the earliest of the four occurrences), and is of uncertain linguistic origin (see *VEPN: braken*). A Brit. word for bracken may be preserved in GLENRIDDING.

breiðr ON 'broad'
BRAE FELL, BRAITHWAITE, BRATHAY and BROADMOOR all have medieval spellings indicating this adj., and BRACELET probably contains it. BROADMOOR illustrates the replacement of *breiðr* by English **broad** (see **brād**).

bridge: see brycg, bryggja

brock dial. 'badger'
Brock is the older and dial. word, ultimately of Celtic origin, for this most distinctive mammal, while *badger* first appears as an animal-name in the 16th cent., and is not represented in this volume. *Brock* appears in some ten names, mostly referring to the haunts of the badger as **crag** or **hole**. The earliest documented is BROCKSTONE, from 1605, though there are earlier Cumbrian examples (*PNWestm* II, 101 and 52, *PNCumb* 331). Classed as vermin in the Statutes of Henry VIII, hunted, baited and sometimes eaten until the early 19th cent., the badger is still the victim of illegal digging and baiting (Gambles 1989, 105).

brōm OE, broom ModE
The plant-name appears in BROOM FELL, THE BROOM, two BROOMHILLS and possibly in BRANTHWAITE (the only one with medieval spellings), BRAMLEY and BRAM CRAG. Broom,

Low). These last preserve a usage with pers.ns similar to that in Early Modern documents, e.g. *John Robinson his close att the Howe* (1633, *CW* 1903, 155).

cocc OE, cock ModE
The OE word referred principally to the male domestic fowl, but in p.ns it and its reflex often refer to the woodcock, grouse or other wild species, and this is probably the meaning in most of the nine instances in this volume, including MOORCOCK HALL. *Cock* usually appears as specific el. with generics referring to natural features, as in COCKUP and LOW COCK HOW. COCKLAW (recorded from the 13th cent.) and possibly COCKLEY refer to places where black grouse 'play' or display, while COCKSHOTT refers to game birds (though the 2nd el. does not refer directly to shooting). COCKPIT alludes to the sport of cock-fighting.

comb(e)¹, coomb dial. 'corrie, rounded valley'
This derives from Brit. *cumbā* (cf. W *cwm*) or OE *cumb* 'valley', which was adopted from it, possibly encouraged by OE *cumb* 'cup, vessel' (Gelling & Cole 2000, 107). It refers to a valley, usually short, enclosed and basin-shaped as in MOUSETHWAITE COMB; some are high, glaciated corries or cirques backed by rocky walls, such as LING COMB or BURTNESS COMBE. There are some nine probable examples, but in some names such as COMB BECK and COMB PLANTATION this el. is difficult or impossible to distinguish from the next, not least since medieval spellings are not available.

comb(e)², co(o)mb, cam dial. 'ridge, crest'
From ON *kambr* and/or OE *camb*, this is present in BLACK COMBE and WHITE COMB(E), COOMB and probably HEN COMB; also CATSTYE CAM and probably CAM CRAG. There is a possibility of confusion with the preceding el. and perhaps also with *cam* 'crooked' from Brit. **cambo*- (W *cam*) 'crooked', found in CAM SPOUT and conceivably in CAM CRAG.

common ModE 'unenclosed pasture for communal use'
This el. was adopted into ME from OFr *commune*, and first recorded in the sense of 'common, undivided land' in the 15th cent. (*OED*: *common* n. 3.). The two dozen examples almost all have a township or parish name as their specific, e.g. BAMPTON COMMON, GRASMERE COMMON, and refer, as **fell** often does, to unenclosed, relatively poor grazing land. ('The inclosed lands lie in the depth of the vales, and bear a very small proportion to the commons', as Housman reported, quoted in Hutchinson 1794, II, 181.) *Common* is especially frequent in the former Westmorland and in lower-lying parts of Cumberland, **fell** being preferred elsewhere. The two terms alternate in ENNERDALE, while WASDALE is the only parish within which both *Fell* and *Common* appear on the One Inch map. Names in *Common* are relatively poorly documented before the 19th cent., though THE COMMON appears in 1611 and APPLETHWAITE COMMON in 1702.

coppice ModE
Adopted from OFr *copeiz*, the word was applied, especially from the 16th cent. onwards, to small woods intended for periodic cutting, allowing for regrowth from the trunk. The five examples in this volume are all in High Furness and all recorded no earlier than the mid 19th cent.

cot(e) OE, ModE 'cottage'
ULCAT (ROW), recorded c.1250, provides the oldest example of OE *cot*; the other seven examples, from the 16th cent. onwards, include AUGHERTREE, BECKCOTE (FARM), COTE HOW and possibly EYCOTT. An ON *kot* existed, but cannot be certainly established in the names in this volume.

ELEMENTS

cottage ModE
Adopted into ME from OFr, this appears in 14 names, usually qualified by a p.n. (e.g. TROUTDALE COTTAGES), landscape feature (FELL COTTAGE) or industrial reference (SAWMILL COTTAGE). Most of the names are not listed in EPNS volumes, and none has a form earlier than 20th cent.

cove ModE 'corrie, recess in a hillside, small rounded valley'
Ultimately from OE *cofa* 'chamber, cove', this el. is similar in usage to **comb(e)**[1]. It occurs, usually as a generic, in some 18 names. Most are not recorded until the 19th cent. (e.g. KEPPEL COVE, NETHERMOST COVE), but there are 13th-cent. spellings for LINGCOVE. *Cove* typically refers to corries gouged out by glaciation in high, remote situations (see, e.g., Ward 1876, 85), though COVE, Matterdale does not fit the description, and since the Scand cognates (from ON *kofi*) refer to 'hut, shed', a man-made shelter might have been meant in this and one or two other names. COFA PIKE is an interesting case.

cragge ME, crag dial. 'rocky height, major outcrop or wall of rock'
Probably from Gaelic *creag*, brought by Scandinavians in the 10th cent., the word became a classic Lakeland el., occurring some 280 times in this volume, including ten LONG CRAGS and eight TARN CRAGS. The majority of the names are first recorded in the mid 19th cent., on the earliest OS maps, but may well be centuries older (see the 13th-cent. forms of BLACK CRAG Eskdale, BUCK CRAG Martindale, DOW CRAG and ERNE CRAG). It usually appears as a 2nd el., sing. or pl., often qualified by a descriptive word, as in the examples above. It is applied in some cases to a whole mountain. *Crag* also appears as a simplex (three instances of CRAG, and three of CRAGG FARM where **farm** has only been added recently) or as specific in a handful of names including CRAG HOUSE(S) and CRAG WOOD.

croft OE, ME, ModE 'enclosure, farm'
Originally a small enclosed field, arable or pasture, often near a house, but also applied to the inhabited site as a whole. That the primary meaning is an area of land rather than a building is clear from its frequent co-occurrence with words such as **end**, **head** and **brow**. Most of the 13 occurrences are on the lowland fringes of the Lake District, the earliest recorded being SILECROFT, c.1205.

crook: see krókr 'bend, corner of land'

cros(s): see kross

cumb: see comb(e)[1] 'corrie, rounded valley'

dalr ON, dale dial. 'valley'
An important el., occurring some 70 times in names of valleys, many of which appear in 12th- or 13th-cent. documents; several are parish names. First elements may describe the valley itself (LANGDALE, SCANDALE) or refer to a major river (DUNNERDALE, ENNERDALE, ESKDALE), or an attribute of the valley such as animals (GRISEDALE, SWINDALE but not BOREDALE), vegetation (MOSEDALE) or a named person (EASEDALE, ULDALE). Spellings such as NADDLE, RYDAL and SCAWDEL show a regular development to a reduced pronunciation of *-dale*, which is also current locally for names spelt *-dale*, though fuller, hypercorrect, pronunciations of these are gaining currency. About a dozen names have *dale* as 1st el., e.g. DALE HEAD, DALE END, Longsleddale, and DALE FOOT; the DALE END in Grasmere, however, is of different origin.

dike: see dyke 'embankment, ditch'

dod(d) dial. 'a compact, rounded summit'
Of uncertain origin, but common in ME and later dial. Over 20 instances in this volume. REST DODD, DODD in Loweswater and BROWN DODD appear in medieval documents, but many are unrecorded until the 19th cent. **Dodd** often appears either unqualified, or with another p.n. as 1st el. (e.g. STARLING DODD, HARTSOP DODD). Three **dodd**s, GREAT DODD, STYBARROW DODD and WATSON'S DODD, are distinct summits on a single mighty ridge which forms a parish boundary, but other examples are smaller summits, either free-standing (e.g. DODD in Underskiddaw) or subsidiary shoulders to a higher neighbour (e.g. MALLEN DODD).

dove ME, ModE, dow dial.
The bird name, referring to various species of *Columbidæ*, occurs in nine names. Five of these are **crag** names, which might suggest that the rock dove was more widespread in the past than recently (though Ratcliffe only refers to it on Cumbrian sea-cliffs, and then as having been rare in the past and now vanished as such through interbreeding with feral domestic pigeons, 2002, 95). The famous DOVE COTTAGE originates in the inn name 'Dove and Olive Bough'. OE *dūfe* and ON *dúfa* are recorded, but it is unclear whether **dove** (first recorded c.1200, *OED*) originates from the OE word.

dub(b) dial. 'pool'
A dial. word of uncertain origin, referring to a pool, sometimes in a river, and often muddy or stagnant. Some five occurrences, three of these in the pl., e.g. GURNAL DUBBS. Recorded from 13th cent. onwards, as probably in DUBWATH.

duru, dor OE, door, dore ModE 'door, col, pass between hills'
Occurring five times in this volume, this is one of several architectural terms put to topographical use. Clear examples, recorded in medieval documents, are LODORE and DORE HEAD.

dyke ModE 'embankment, ditch'
The word is from OE *dīc* 'ditch, dike'; the form *dyke/dike*, as opposed to *ditch*, may well have been reinforced by ON *dík, díki*. OE *dīc* had the sense 'drainage ditch' or 'defensive trench', and hence in later English 'embankment', as in the 'head-dykes', which separated enclosed land from open fell or moor (Winchester 2000, 52). A *dyke* could take various forms, including a hedge, a steep embankment and ditch or a four- or five-foot stone wall such as those that farmers are instructed to build or maintain in the Elizabethan 'paines' of Shap (*CW* 1903, 155–6 etc.). In DEER DYKE MOSS, *dyke* may be a small watercourse. The el. occurs in eight names, FELL DYKE being the earliest recorded (1581).

edge ModE 'escarpment, ridge, edge'
A steep, linear mountain feature: either an escarpment where the land falls away more steeply to one side than the other (e.g. GREENUP EDGE) or an arête, a narrow ascending ridge with a steep drop on both sides (e.g. SHARP EDGE, STRIDING EDGE). The word, which occurs most often as a 2nd el., derives from OE *ecg*, possibly encouraged by ON *egg*, both of them being used topographically (see SHARP EDGE). FELL EDGE (recorded from 1535) is the earliest example in this volume, and most of the other twelve examples are first recorded in the 19th cent., but a document of 1338 refers to *le egge of Kirkefelle* (*PNCumb* 392).

eiki: see oak

eller ME, dial. 'alder'
Dial. *eller* is from ON *elri* 'alder, alder copse' or the rarer *elrir*. In most of the eleven names concerned, *eller* is the specific, while the generic refers to the place where it

grows, e.g. ELLERAY, ELLER BECK, ELLERMIRE; in SEATOLLER it may be the specific in an inversion compound. The earliest record is of ELLERBANK, c.1215. There is some possibility of confusion with *elder*, as in ELDER BECK. The alder, *Alnus glutinosa*, is extremely common in the lake shores and damp valley bottoms of Cumbria, and was probably still more so before widespread drainage (Halliday 1997, 140–1). The wood was valuable for a range of purposes including the making of gunpowder and footwear, and the bark yielded a coarse black dye (Gambles 1989, 59).

endi ON, ende OE, end ModE
This most often refers to a settlement at the end of a natural feature (DALE, FELL, FIELD, KNOTT, MOOR END etc.) or a man-made one (LANE, TOWN, WALL END); for exceptions, see LIND END. There are some 77 occcurrences, always as a 2nd el. Most are not recorded until the Early Modern period at earliest, and the use of the definite article for TOWN END Caldbeck in 1652 and TOWN END Witherslack 1701 might suggest that these were descriptive phrases in transition to names at that time. On the other hand, WATENDLATH appears from 1209–10 and WATEREND in 1292.

erg: see ærgi 'shieling'

eski: see askr 'ash'

far ModE
The word, from OE *feor*, appears in ten names, none recorded before the 16th cent., e.g. FARTHWAITE, FAR END. The reference may often be to distance from the township or parish centre, as in the contrastive pair FAR and NEAR SAWREY. *Near* is from OE *nēar(r)* 'nearer', comparative of *nēah* 'near'; it occurs five times in this volume.

farm ModE
Adopted from OFr *ferme* in the ME period, the word was initially applied to a fixed rent or a lease, then in the 16th cent. to land held on lease, and hence to a tract of land used for agriculture or pasturage. *Farm* appears in some 170 names in this volume, usually as an addition to an existing p.n. (e.g. BROWNRIGG FARM cf. *le Brounrigg* 1323). This is a relatively recent process, at least in the written form as it appears on maps. There are examples as early as the 1850s (LONGLANDS FARM), but many farm-names appear without *farm* on the earliest OS maps (e.g. GROVEFOOT, RAW and PETER HOUSE, all 1867) and have since had it added, sometimes very recently. Several more farms whose name does not include *farm* on the 1994 map show it on the farm sign (e.g. LANE HEAD Mungrisdale, DODDICK or FORNSIDE), which is doubtless a useful device for distinguishing farmhouses from cottages or converted barns. In some cases the addition of *farm* is due to map–makers or administrators rather than farmers (see HILL PARK FARM).

fell, fjall ON, fell dial. 'hill, mountain, tract of high unenclosed land, high ground'
Among the most characteristic Lakeland elements, this occurs in well over 180 names in this volume. Many *fell* names are recorded relatively recently (and BIRKETT FELL is a late 20th-cent. coinage) but there are several instances from the 12th–13th centuries, including WHIN FELL (as a settlement named from a hill), BOW FELL, BRAE FELL, DERWENTFELLS, FURNESS FELLS, and MELL FELL. In many instances the age of the name is extremely difficult to judge. ON *fell* and *fjall* referred, respectively, to single hills and 'mountainous country' (Fellows-Jensen 1985a, 79), though the two usages were not entirely separate or consistent. Both resulted in English *fell*, whose usage is also somewhat complex: (a) an individual mountain or hill (e.g. BOW FELL, SCA FELL); (b) (in sing. or pl.) a range of hills or tract of high ground, as in ULDALE FELLS or in the record of 'a person unknown, found dead upon the Fell' (Borrowdale PR, 1819); or more specifically, (c) high unenclosed land, often straddling valleys and named from

townships or manors, and hence not always directly adjacent to the settlements of the same name, e.g. ROSTHWAITE, SEATHWAITE, STONETHWAITE (two locations) and THORNYTHWAITE FELL, all in BORROWDALE; cf. also BRACKENTHWAITE FELL. STAINTON FELL has this sense, its highest point being STAINTON PIKE. In these cases *fell* is apparently used for designated grazing areas, much as **common** in lower-lying areas, or in eastern Lakeland (see Winchester 1987, 88 on the division of waste between townships in the decades around 1500). There is some confusion between *fell* and **field**, as in HAMPSFIELD, PLACE FELL, SCA FELL.

field ModE
From OE *feld*, which referred to large stretches of open, as opposed to wooded, land, and hence to land used for agriculture or pasturage. With the progressive enclosure of land, from the 14th cent. onwards but especially in the 18th–19th centuries, the word took on its modern sense of 'enclosed plot of land' (Smith, *Elements*), and this probably applies in most of the 30 or so Lakeland instances, which are all recorded from the 16th cent. (e.g. BRADLEYFIELD, FIELDSIDE) or later. SWARTHFIELD is a possible exception. In one third of instances, *field* is the specific, e.g. four cases of FIELD HEAD. *Field* as a generic is qualified by a variety of terms, including descriptive adjectives (FAIRFIELD, LOWFIELD). *Field* can be confused with **fell** (q.v.).

fold ModE 'pen, farm'
This derives from OE *fal(o)d*, but is rare in northern names except in post-medieval times. Originally referring to an animal pen or small compound, it occurs in 14 names, many of them farm-names where *fold* is qualified by a descriptive adj. as in BROADFOLD, or by a surname as in CARTMELL FOLD and DAWSON FOLD, both in Crosthwaite and recorded from 1535, and hence the earliest examples in this volume. The spelling is *fold* except in FAULDS, which is also one of three instances of the pl. In both examples of FOLDGATE (q.v.) the *Fold-* seems to be of different origin.

foot ModE 'foot, lower end'
This word, the reflex of OE *fōt* and/or ON *fótr*, occurs in some 30 names, all first appearing in the 16th cent. or later. It usually designates the lower end of a stream, track, pool, hill, valley, wood or settlement which is specified in the 1st el., e.g. BECKFOOT (four examples), RAKEFOOT or SCAR FOOT. LAMBFOOT is of different origin, and SOURFOOT is puzzling.

force dial. 'waterfall'
From ON *fors*, whose later form *foss* may occur in MONK FOSS. The 16 certain examples make their first documentary appearances from the 16th to (more usually) the 19th cent. Several contain a p.n. as 1st el., e.g. AIRA FORCE, SKELWITH FORCE, though *force* can itself appear as 1st el., in FORCE CRAG, FORCE FORGE and possibly FORESLACK.

ford OE, ford, forth ModE
Only six names, including GOSFORTH and EASTWARD (both mentioned in 12th-cent. documents), ESP FORD, and FORD HOUSE. Fords are also designated by ON **vað**.

forest ME, ModE
There are 14 instances in this volume, *forest of Skithoc* (SKIDDAW) 1230 being the earliest record. Adopted from OFr into ME, this originally denoted an area under forest law in which hunting rights and beasts of the chase were protected. Huge tracts of the Cumbrian uplands were designated as baronial *forests* in medieval times, usually linked with lowlands under the same ownership (Winchester 1987, 20–1), and the names of COPELAND FOREST, GRISEDALE FOREST, RALFLAND FOREST, SKIDDAW FOREST and SLEDDALE FOREST survive on the modern map, while DERWENTFELLS was also a medieval

forest (see map, Millward & Robinson 1970, 162). Although not necessarily wooded, parts at least often would be, providing cover for deer and other large animals. The pressure of population, however, led to a gradual displacement of forest by settlement and sheep-farming from the 12th to 16th centuries — the 'chief theme of landscape history in the Lake District' during this period (Millward & Robinson 1970, 164–8, especially 168; Winchester 1987, 22 and 39–44). This, meanwhile, encouraged the creation of physically enclosed deer **park**s which enjoyed long use. In recent names, *forest* has its modern sense of a large area of woodland, GRIZEDALE FOREST being a spectacular example.

gap ModE
ON *gap* was adopted into ME, and is recorded referring to gaps or cols in mountains from 1555 (*OED*: *gap* n. 1). The earliest-recorded of the nine examples in this volume are also 16th-cent.: THE GAP and CLAY GAP. In some names, *gap* replaces or supplements an earlier word for such a gap (e.g. ORE GAP, TRUSS GAP, WIND GAP). The word was also used in connection with access rights, as in the 17th-cent. stipulation 'that Willm Alexander doe Suffer Willm Measond to have his gapp or Yaite between the Meare Stone & the Intake Nooke to the Water & Common . . .' (*CW* 1903, 153), and this is conceivably the meaning where the landscape contains no notable gap, e.g. in JACK GAP PLANTATION. For Coleridge's rough sketch of a 'gap', see *CN* 1211 (1802).

garðr ON, garth, g(u)ard dial. 'enclosure, farm'
An enclosure, sometimes of a specialised sort, e.g. an orchard or animal enclosure (PLUMGARTH, CALGARTH), and hence, as also in Scandinavia, a farm. The el. was especially productive in the Early Modern period, though there are late medieval spellings for CALGARTH, DALEGARTH and GUARDHOUSE, while ELLERGARTH seems very recent. Of some 27 examples, most are outside the high fells, and refer to habitations, though THROSTLE GARTH and WHIN GARTH do not, at least now. *G(u)ard* is the minority spelling, seemingly favoured in plurals, e.g. GRASSGUARDS, formerly *Grasgarth*, and three others, compared with only one *garths*: PLUMGARTHS. The two instances of GATESGARTH are from **skarð** 'gap', not *garðr*.

gata ON, gate dial. 'road, way, track'
This occurs in BROADGATE and ISELGATE and probably or possibly in up to eleven more names, but it is very difficult to distinguish from the following el., especially since most of the spellings are 16th-cent. at earliest (WALMGATE being an exception). A dial. word *gate* meaning 'pasturage rights, tract of permitted grazing land' is normally assumed to derive from the ON *gata*, but whether this occurs in any of the p.ns in this volume is uncertain.

[geat OE], yeat ME, dial., gate ModE 'gate, opening'
This el. is difficult to separate from **gata** 'road, track', **geit** 'she-goat' and their reflexes, but is likely to occur in over 20 names, many of them first recorded in the Early Modern period. The el. refers either to a man-made gate, often designed to regulate access to grazing land or otherwise control livestock movement, as probably in FELL GATE and MOORGATE, or to a gap or gorge in the landscape. Sing. forms of OE *geat* contained the initial consonant preserved in N. dial. *yeat*, while the pl. had the hard [g] of the standard ModE *gate*, but either sound could spread by analogy, and the ON *gata* would encourage the hard [g]. Alternation between *g*- and *y*- spellings can often be seen in the early forms (e.g. PARKGATE or BROADGATE, Millom), but *yeat* only survives in five names as they appear on recent maps, including SNECKYEAT and WATER YEAT.

geil ON 'ravine, narrow track': see under gil

geit ON 'she-goat', gāt OE, goat ModE
The age and origin (whether ON, OE, ME or ModE) of these 'goat' terms can be difficult to determine, in addition to the danger of confusion with ON **gata** or OE **geat**. However, there are probably a dozen references to goats in high places, including five GOAT/ GATE/GAIT CRAG(S), GATESCARTH, two GATESGARTHS and GASKETH, a few with medieval spellings. These are presumably wild goats in all or most cases.

gil ON, ME, gill, ghyll dial. 'ravine with stream, stream, ravine'
A steep, narrow valley or ravine, usually with a stream, hence often the stream itself (and *gills* are usually marked in blue on OS maps). There are over 120 instances, concentrated especially in the high fells; some are recorded as early as the decades around 1200 (ROSGILL, SADGILL). *Gills* often feed into, and change their name to, **beck** before turning into true rivers known by an individual name, often Brit. or more ancient, e.g. ESK or DERWENT. *Gill* is usually qualified by a 1st el. referring to location (BLINDTARN GILL, ROSSETT GILL) or a descriptive term (DEEP GILL, WREN GILL — not the bird!). The el. is 'sometimes wrote Ghyll to ensure the hard sound of the G' (Otley 1823, 32). This spelling is often credited to the Wordsworths (e.g. *OED*: *gill* n.2), and is still in occasional use, though the Ordnance Surveyors deemed it 'evidently incorrect & contrasts to all northern Glossaries' (OSNB Grasmere, 1859, 29). Some apparent *gills* are of different origin: from ON *geil* 'ravine or narrow track' in HUGILL and probably SKELGILL or from ON **skáli** 'shieling (hut)' in POTTS GILL, SOSGILL.

glen in modern forms of names
The element **glen** is almost certainly the Brittonic or Gaelic term for 'valley': Cumbric **glinn*, Gaelic *gle(a)nn*. It occurs in GLENCOYNE and GLENRIDDING, both near ULLSWATER and recorded from the 13th cent., and probably in nearby GLENAMARA. The exact genesis of these names is difficult to trace, however. They have usually been assumed to be Cumbric rather than Gaelic, and the 2nd elements (modern '-coyne' and '-ridding') have been taken as the Cumbric equivalents of, respectively, W *cawn* 'reeds' and W *rhedyn* 'bracken', while Gaelic solutions are harder to find. However, *glinn* is not common in Brittonic p.ns and normally appears as a simplex (Padel 1985, 104–5 and pers. comm., gratefully acknowledged), whereas its Gaelic equivalent *gleann* is extremely prolific in Scottish and Irish p.n.s, often qualified by following adjs or nouns (see Basden 1997: *gleann* for references to Watson 1926, also Flanagan & Flanagan 1994, 92 and McKay 199, 153). This introduces the possibility that *glen* in the Lakeland names was added by Norse-Gaelic speakers to pre-existing Brittonic stream-names, or in the case of Glencoyne that the name may be wholly Gaelic. Although the *Glen-* names in this volume are northerly in distribution, a lost *Glanscalan* or *Glensalan* is recorded from 1170–84 in Furness, and this Ekwall regards as Brittonic or possibly Gaelic (*PNLancs* 226). Meanwhile, GLENDERAMACKIN and GLENDERATERRA, referring to two tributaries of the GRETA, have never been satisfactorily explained, and have complications additional to those already noted. The *Glen-* names are of great historical importance and would repay further study in a wider geographical and onomastic context. THE GLEN, Nether Staveley, is probably modern.

goat: see geit

grain dial. 'fork, confluence'
This is normally the fork of a river or valley, a confluence of two streams, from ON *grein* in that sense. There are five examples, the earliest from 1569 (GRAINSGILL BECK). In LONG GRAIN and Short Grain, Ennerdale, the word applies to individual branches of a stream, while LONG GRAIN Bampton is puzzling, being a hill name.

grass: see gres

gra(u)nge OFr, grange ME, ModE
An outlying farm or store, often a granary, and usually part of a large monastic estate, as was GRANGE-IN-BORROWDALE, recorded from 1396, and the *granges* at BAMPTON, NIBTHWAITE and SLEDDALE. In later usage it refers to a substantial country house, with or without attached farm, as in BELLE GRANGE. There are seven instances in this volume.

great ME, ModE
The word goes back to OE *grēat* 'massive', but it only gradually ousted *mickle* (from ON *mikill*, OE *micel*) as the main local word for 'large', and *mickle* survives in dial. and in some Lakeland p.ns — three in this volume. The replacement process can be seen in GREAT GABLE (*Mykelgavel* 1338) and GREAT LANGDALE (*Micklelangdaile* 1564), though *great* appears as early as c.1220 in GREAT CROSTHWAITE. *Great* occurs some 48 times in this volume, often qualifying a pre-existing p.n, and often with a **little** counterpart.

greave: see gro(o)ve

grēne OE, ME, grœnn ON, green ModE
1. Adj. 'green' occurs as a specific el. in some 40 names, including two instances of GREENRIGG(S), recorded from the 12th and 13th centuries.
2. Noun 'grassy place, (village) green' already appears (outside Cumbria) in Domesday Book 1086, though of the 20 examples in this volume BARBER GREEN (1594) is the earliest recorded. *Green* in this sense is normally a generic el. referring to a habitation and co-occurring with a variety of specifics: p.ns as in WINDER GREEN, descriptive epithets as in FERNEY GREEN, and surnames as in BARBER GREEN.
GREEN QUARTER, Kentmere, could be either 1. or 2.

gres ON, gres, græs OE, grass ModE
Grass is the specific el. in ten names — good pastureland was distinctive. Medieval forms for GRASMERE, GRASSOMS and GRASSTHWAITE retain the ON/OE spelling *gres-*, later standardised to *gras(s)*.

grey, gray ME, ModE
The word is from OE *grǣg*, though in some cases this could have replaced the cognate ON *grár*. It appears in a dozen names, spelt either *gray* or *grey*, mainly referring to rocks and crags and mainly recorded from the 19th cent., though GRAYTHWAITE, Satterthwaite appears in 1336 and GREY CRAG, Longsleddale in 1597. The latter example illustrates its occasional application to limestone scenery. In most cases, as here, it can be assumed to refer simply to colour, but 'gray stone', like Icel *Grásteinn*, can denote a boundary stone (as in *the cloven gray stone*, i.e. Cloven Stone in Eskdale, *PNCumb* 476), and at least GRAYSTONES in Lorton may have this significance (cf. *PNWestm* I, xvi and 163). Similar names in the Faeroes are touched by folk-beliefs associating grey rocks with supernatural beings (*PNWestm* I, 163, citing Matras), but whether this might be the case in Cumbria is impossible to prove.

gríss ON, grise, grice dial. '(young) pig'
As Machell remarked, in connection with the Patterdale Grisedale, 'the word grice or grise is yet use[d] amongst northern People to signify swine' (c.1692, I, 718). Reference to the animal (rather than the ON pers.n. derived from the animal term) can be assumed in the p.ns: GRIZEBECK, GRISEDALE (two examples), GRIZEDALE, MUNGRISDALE and possibly GRICE CROFT. The el. is suggestive of pannage — pig grazing — and hence of oak forest, a major source of fodder. Pannage, and p.ns referring to it, flourished

especially 10th–12th cent. (Winchester 1987, 100–2), and all the instances above have medieval spellings except GRICE CROFT.

gro(o)ve, greave in modern forms of names

These elements are sometimes difficult to disentangle, especially since most of the few instances are first recorded in the 19th cent.

1. **Grove** is ultimately from OE *grāf, grāfa, grāfe* 'grove, copse'; e.g. LOW GROVE, MIDDLE GROVE recorded from 16th cent., and probably other names in *grove* except for DRYGROVE (below).

2. **Greave** is ultimately from the related OE *grǣfe* 'copse'; it can be assumed in GREAVES (two), HIGH MERE GREAVE.

3. **Groove** 'pit, hollow, stream-bed or stream' is from ON *gróf*; e.g. DRYGROVE and possibly GROOVE BECK.

ground ModE 'farm'

ON *grund* refers to fertile, low-lying land and is common in Icel farm-names, but not Norw ones; OE *grund* refers to a valley bottom or an outlying farm, but is not common in early English p.ns., and the use of *ground* in the Lakeland p.ns may be a later, separate development. Some 30 names in *ground* are shown on the One Inch map which provides the corpus for this volume. They are concentrated in S. Lakeland, especially Furness — a distribution which almost complements that of **place**. The *ground* names constitute a fascinating group of farm-names which seem to mark a stage in the continuous process of reclaiming land from waste (as suggested by Waterson Ground, Waterson Intake and Waterson Moss N of Hawkshead), yet their exact interpretation is elusive. The earliest record of *ground* in the present corpus is ROGER GROUND (1582), followed by several in the 17th cent., e.g. KEEN, SIMPSON, and HOLME GROUND, though for possible late 14th-cent. examples of *ground*, see *PNCumb* 364. In the great majority of these names, the 1st el. is a surname, often one documented locally in the early parish registers, even at the farm itself (e.g. Hartleys at HARTLEY GROUND and Kitchens at KITCHEN GROUND). In light of the fact that so many of the farms concerned are on former lands of FURNESS Abbey it has often been assumed that the 'Ground' farmsteads were created when Furness Abbey permitted its tenant farmers to enclose 'grounds' or pockets of common pasture early in the 16th cent. (Collingwood 1900, 65–6, Millward and Robinson 1970, 177–8), the process perhaps accelerating in the disarray of the Dissolution period (Cowper 1899, 92–3). However, Winchester points out that 'the agreements themselves placed a strict limit on further intaking and they contain no evidence of the establishment of new farms', and that there is little evidence of such expansion after 1450 in Cumbria except for Inglewood and Westward (1987, 54). He suggests instead that the new farms 'were founded in the colonisation of the thirteenth century, even if their modern names were coined only in the sixteenth'. Such widescale renaming would be striking but not impossible, and the whole question needs further investigation.

hall ME, ModE

Although of OE origin (Angl *hall*), the el. is rare in English p.ns until after the Norman Conquest. Among the earliest instances in this volume are BLACK HALL (1398–9) and BOLTON HALL (1497). The word is often applied to a manor-house or other substantial residence such as ISEL HALL or DALEGARTH HALL, hence the majority of over 90 occurrences are in the lowland fringe; but it is also applied to farmhouses. *Hall* is usually appended to a pre-existing p.n., while in an interesting group of names in SW Lakeland, it precedes the p.n., e.g. HALL BOLTON (in contrast to nearby Bolton Hall), and see under CARLETON, DUNNERDALE, SANTON and WABERTHWAITE. If the reference is primarily to the building itself, so that *hall* is the generic, the word order is reminiscent of older inversion compounds and possibly indirectly influenced by them (*PNCumb* III, xxiv). Alternatively (and I thank Dr Oliver Padel for this suggestion), *hall* could be the specific, distinguishing a subdivision of the place referred to — 'the part of X with the hall'. HALL BOLTON, as the name of a manor-house and a township, illustrates the ambiguity. Some apparent *hall* names, including WIND HALL, result from interchange with *how(e)*, *haw(e)* from ON **haugr** 'hill, mound'. Most of these are in the former Lancashire, e.g. FIDDLER HALL and PICTHALL. BAWD HALL, meanwhile, has developed from -*hole*.

hals ON, hause, hawse dial. 'neck of land, col, ridge between valleys'
ON *hals* (perhaps encouraged by OE *hals*), whose basic sense was the anatomical 'neck', gave rise to *hause* through standard vocalisation of -*l*- between vowel and consonant. There are over 20 occurrences, mainly spelt *hause*, though see also BURNT HORSE, HAWSE END, and possibly HASSNESS, and there is some confusion between this el. and *how(e)* from **haugr** 'hill'. The word can designate a dip in high ground, as in WRYNOSE (with 12th-cent. *hals* spellings), an important watershed and pass between Eskdale and Langdale, or in BOREDALE HAUSE, GRISEDALE HAUSE and NEWLANDS HAUSE, which carry high-level routes from the heads of the valleys whose names they bear. HAWSE END on the other hand illustrates the use of the term to refer to a neck of land rising from lower ground.

hara OE, hare ModE
This volume contains two HARE CRAG(S), as well as HARECROFT, HARE GILL, HARE HALL, HARE HILL and HARE SHAW, and two HARROPS, and in at least some cases the mammal (OE *hara*) may be meant: hares used to be 'everywhere plentiful', according to Hutchinson (1794, I, 3). However, since even where there are medieval spellings (as for both HARROPS) they are not unambiguous, and since the native brown hare is a mainly lowland species (Ratcliffe 2002, 103), other possibilities cannot be ruled out, such as derivation from OE *hār* 'grey, hoary (perhaps with lichen)' or conceivably 'boundary', or OE **hær* 'cairn, heap of stones' (see HARE CRAG(S) and HARROP).

haugr ON, how(e) dial. 'mountain, hill, mound'
A *haugr* or *how(e)* is characteristically compact and free-standing, with relatively steep, roughly round or oval, contours. Size varies: compare HOW in Above Derwent with its massive neighbours GREAT CALVA and probably SKIDDAW. ON *haugr* frequently referred to burial mounds, and TONGUE HOW and BOAT HOW NY0810 have prehistoric cairns and ELVA a stone circle. In general, however, the Lakeland material does not match the statement in Smith, *Elements* that *haugr* mostly applies in the N to artificial structures. The el. is tremendously prolific, present in at least 130 names, occasionally as a specific (as in HOWTOWN), but more often as a generic compounded with a 1st el. describing the hill itself (several GREAT HOWS, PIKED HOW and variants), its situation (WALMGATE HOW) or other attributes (CROW HOW). As seen here, *how(e)* is the most usual reflex of ON *haugr*, but *haw* also appears, especially in S. Lakeland (former Lancashire), e.g. BLEABERRY HAWS, BULL HAW MOSS, or as **hall** (q.v.). The present and early forms of ULPHA SD1993, STODDAH, POOLEY and CRACOE illustrate the variability of spelling, and the first three are among the *haugr*/*howe* names recorded in medieval documents. The pl. of *how(e)* from *haugr* can be confused with *howse* or *house* (e.g. ARMENT HOUSE).

hause, hawse: see hals 'neck of land, col, ridge between valleys'

hazel ModE
Ultimately from OE *hæsel* (possibly reinforced by ON *hesli*), this occurs in seven names, the earliest-recorded being HAZELRIGG (1508). The hazel (*Corylus avellana*) is widespread in the Lake District, growing to 1640ft/500m., though it is less common in the N and on the high fells (Halliday 1997, 141). A favourite coppice tree in the past, it had many uses including fences and baskets, and the nuts were sufficiently plentiful to be exported from Barrow-in-Furness in the 19th cent. (Gambles 1989, 83). BLENNERHAZEL seems to be a modern application of the tree-name in a name modelled on an older one.

hēafod OE, hǫfuð ON, heved ME, head ModE
This applies to the top or upper end of a ridge, stream, valley, hamlet or other feature, or to a hill or high place (and see SWIN(E)SIDE for another possibility). There are nearly 100 instances in this volume, almost all as generic elements and in the form *head*, though other outcomes are seen, e.g., in ARMASIDE and SWINSIDE (both in Lorton), HESKET and RATHER HEATH. The earliest instances are 13th-cent. spellings of KIRKHEAD and LATTERHEAD, both in Loweswater, though *caput Cawdell* in the late 12th cent. may translate an early version of CAUDLE HEAD. A few apparent 'head' names are not, e.g. HAWKSHEAD. The el. occurs widely, but especially in NW England.

hegh ME, high ModE
The adj., from OE *hēah*, but possibly in some cases replacing ON *hár*, occurs in over 120 names. In many it is a recent affix, distinguishing 'high' and 'low' natural features, or parts of an estate or settlement that has grown and divided. The adj. occurs in names mainly recorded from the 16th cent. onwards, but HIGH IREBY appears in 1279. *High* sometimes replaces other affixes: *over* in HIGH KNIPE, HIGH LORTON and HIGH NEWTON, or *upper* in HIGH YEWDALE. Occasionally, *high* and *low* relate to location upstream or downstream in a valley, rather than a marked difference of altitude (e.g. HIGH HOUSE). *High* also occurs as a noun referring to a height or (modest) hill in some Cumbrian names including THE HIGH (FARM), Crosthwaite, cf. the phrase 'h(e)igh and howe', common in Scots (*EDD*: *high*, sense 8).

height(s) ModE
The word is applied in the sing. to modest hills, in the pl. to a tract of relatively high ground, especially in Furness, where the low hills between ESTHWAITE WATER and WINDERMERE are distinguished as COLTHOUSE HEIGHTS and CLAIFE HEIGHTS; cf. also FINSTHWAITE, RUSLAND and TOM HEIGHTS, the latter name exceptional in not deriving from a p.n. For all these examples the first appearance known to me is on the OS maps of the 1850s. The word derives from OE *hēhðu*, but the earliest of the dozen records in this volume is from 1652 (THE HEIGHT).

hen ModE
From OE *hen(n)*, *hæn(n)*, the word refers to the female of the domestic fowl and hence (from ME onwards) to the female of other birds, especially moorland birds (as presumably in HEN COMB, HEN CRAG) and waterhens of either sex (as in HEN HOLME). HENHOW (1588) is the earliest of the five examples.

high: see hegh

hill ModE
Although the standard English 'hill' word (from OE *hyll*), and extremely common in English-language p.ns of all periods (Gelling 1984, 169–71), this is only about half as common in the Lakeland names as the originally Scand **fell**. There are some 80 occurrences (sometimes as 1st el., e.g. HILL TOP), mainly around the fringes of the high fells. Many names in *hill* appear first on the 19th-cent. OS maps, but where the name is

shared by a settlement it may appear in the 16th cent., e.g. HILL, HIGH HILL. Qualifying 1st elements are normally quite transparent, e.g. GREEN HILL, SAND HILL, WATCH HILL.

hlāford: see laverd, lord

hollin, hollen dial. 'holly'
The word goes back to OE *holegn*, but the twelve p.ns containing it are recorded no earlier than 1573 (HOLLIN ROOT, St John's). The p.n. HOLLINS is particularly common in NYorks and in Cumbria. The widespread evergreen *Ilex aquifolium* (Halliday 1997, 317), visually so distinctive in the landscape, provided young leaves valuable as sheep- and cattle-fodder in the winter, and the distribution of *hollin* p.ns has been found to correlate well with known drovers' routes (Atkin 1988–9).

holmr ON, holm(e) dial. 'islet, land beside water, water-meadow, patch of dry ground'
This is used of islets (several in Derwentwater and Windermere), higher ground almost surrounded by streams or lakes, and water-meadows. There are over 30 examples, usually as the 2nd el. in compounds, often coupled with descriptive adjs, e.g. LOW HOLME, or nouns referring to vegetation, e.g. LINGHOLM(E). The earliest records, from the 13th cent., are of BRIGHAM, GRASSOMS, and BUCKHOLME in Muncaster.

hol(r) ON, hol(h) OE, hole, hollow ModE
(a) The el. can be a noun going back to ON *hol* or OE *hol* or *holh* 'hole, hollow', often referring to a rounded depression or small valley (e.g. FROST HOLE, HOLE HOUSE). The noun form *hollow* is from *holwe*, dat. sing. of OE *holh*, and only occurs in the pl., in three instances of (HIGH) HOLLOWS. (b) The related adj. goes back to ON *holr* or OE *hol* 'hollowed out' (e.g. HOLE GILL). The adj. form *hollow* which arose in ME also occurs in five names. The el. (or group of elements) appears in over a dozen names, and is occasionally reduced to *-le*, as in BROCKLE as opposed to BROCK HOLE (OE *brocc-hol* 'badger sett'), or confused with **hall** (e.g. BAWD HALL), **hill** (e.g. ROUGH HILL), or **how(e)** (HOWBURN).

hop OE 'blind valley'
This usually refers to a secluded blind valley, often rounded and branching off or overhanging a larger valley, e.g. HARROP above Thirlmere or HARTSOP above Brothers Water. MEATHOP, however, has the sense common elsewhere in England of a patch of better land next to a marsh. Medieval spellings are available for both HARROPS, HARTSOP, MEATHOP and WYTHOP, and HOPE BECK, COCKUP and GREENUP are probable examples, though only documented from modern times.

house: see hūs

how(e): see haugr 'mountain, hill, mound'

hraca OE: see rake 'track'

hreysi ON, raise dial. 'cairn, heap of stones'
A *raise* could serve to mark a path, a boundary (e.g. *PNWestm* II, 168), or possibly a burial, as at HIGH RAISE, Martindale/Bampton, which is topped by a Bronze Age cairn, cf. 'a heap of stones called the Raise at Spying How in Troutbeck forest . . . in the midst of [which] they found a chest of four stones with large bones in it' (Machell c.1692, II, 323; ed. Ewbank, 127–8). Some stony heaps may have no other purpose than to rid the pasture of stones. In some cases (e.g. RAISE) the el. refers to a mountain which is both cairn-topped and cairn-like, making it uncertain whether the name refers primarily to the cairn or the whole mountain. The el. is usually spelt *raise*, except in ROSTHWAITE,

Borrowdale, which is of the same origin as RAISTHWAITE, the earliest recorded of the eleven names (from 1319).

hryggr ON, hrycg OE, rigg dial. 'ridge'

The ON and OE words both mean 'spine, ridge', and the dial. *rigg* may be a product of either or both, though the plosive *gg* suggests that the Norse influence is dominant. The el. occurs in some 60 names, usually applied to lowish ridges or similar landforms; it can also refer to cultivated strips of ground (*PNWestm* II, 264), but this application is unlikely in the high fells. Some examples are recorded from the 13th cent. and may well be pre-Conquest (e.g. BROWNRIGG and GREENRIGG, both Caldbeck, LOUGHRIGG), while CASTLERIGG is recorded from 1256 but must be post-Conquest.

hūs OE, hús ON, house ModE

This el. usually refers to a single dwelling, though COLTHOUSE is, at least now, a hamlet, and the pl. occurs, e.g. in BIRKET(T) HOUSES and in AYNSOME, which descends from an old dat. pl. With the exception of UZZICAR, the el. appears 2nd in compounds, often coupled with some reference to situation (e.g. BANK, CHAPEL, HIGH, HINNING, and HOLE HOUSE, all occurring more than once). Most of the 140 or so examples are recorded from the 16th cent. or later, but UZZICAR appears in 1160 and GUARDHOUSE in 1332. *House* has often been added fairly recently to a pre-existing name, e.g. TAW HOUSE, WATERSIDE HOUSE. WHITTAS and the hill-name HERDUS seem to preserve a weakened pronunciation of *house* which in other names been 'corrected', at least in the written form, to *-house*, as in *Miras* 1717, now MIREHOUSE in Bassenthwaite. On the other hand, *house* in some modern forms of names results from re-interpretation of a different word altogether: see ARMONT HOUSE or the 1763 spelling *Renerhouse* for RAINORS.

hwīt OE, hvítr ON, white ModE

The el. *white* in Cumbria results from ON *hvítr* and OE *hwīt* or some confluence of the two. In names given in the medieval period, the vowel shortens, so that names in *Whit-* may be older than those in *White*. There is also a danger of confusion here, and two of the three WHITBECKs probably refer to willows rather than whiteness. This apart, the el. is fairly easy to identify, and occurs as the specific to some three dozen names, especially of hills and becks, but its meaning is often much less obvious. In some cases the reference is to pale or light-reflecting rock, as with WHITBARROW and WHITE SCAR on the limestone belt in the SE of the National Park, or WHITE SIDE, where there are streaks of quartz. Old woodland or rock can also take on a white or 'hoary' appearance when covered with lichens, e.g. *Lecanora*, especially on limestone; large stretches of cotton grass can make flat banket bogs white in appearance; and elsewhere pale grasses can produce whiteness. When applied to water, *white* normally denotes foaming cascades (in contrast to Scandinavia, where 'white river' often signals the milky appearance of glacier water), though in WHITEBECK, Crosthwaite, the sense must be 'clear'.

hyrst OE, hurst, hirst ModE 'wooded hillock, copse or hillock'

There are some seven examples, out of which SWALLOWHURST first appears in the 13th cent. and BANNEST in the 16th; the others are later and all have the modern standard spelling *hurst* rather than the Northern *hirst*. As with other woodland terms, the description may or may not still apply.

ill dial. 'treacherous'

ON *illr* is an adj. used mainly to designate humans as morally bad or evil, but its dial. descendant *ill* characterises, among other things, landscape features which are steep, rough or otherwise treacherous. The el. occurs in up to six names; it can seemingly be confused with **hill** (see ILL BELL and MARDALE ILL BELL), and probably appears in the two EEL CRAG(S). ILLGILL HEAD is recorded from the 14th cent., but other examples only from modern times.

intake dial. 'piece of land enclosed from moor or waste'
Coleridge writes of 'the connection by Intakes of the smooth bowling Green Vale with the steep Mountain' (*CN*, note 1782). A notable cluster of *Intake* names occurs in the low hills of Furness, including five out of the seven instances in this volume, all first recorded, to my knowledge, on the mid-19th-cent. OS maps, although the heyday of intaking in the Lakeland valleys was earlier, c.1450 to c.1600 (Winchester 2000, 68–9). Derived from ON *inntak*, the word is often spelt *intakk* or *intack* in Early Modern documents, but is standardised to *intake* in current spellings. HIGH INTACK, Caldbeck (*the Newe Intacke* 1652) is an exception in its location, spelling and early documentary appearance.

island, isle ME, ModE
Island is from OE *ēgland* 'island', with some confusion with OE *ēaland* 'land by river, island'. The modern spelling is influenced by *isle*, which was adopted into ME from OFr *isle, ile*. There are five instances of *island*, from an early name for LORD'S ISLAND in 1230 to the very recent NORFOLK ISLAND, and two of *isle* (BELLE and DERWENT).

kambr: see comb(e)² 'ridge, crest'

kelda ON, keld dial. 'spring'
There are 13 certain or possible occurrences, including THRELKELD (recorded 1197), SIMON KELL (1279) and KELBARROW (1375). A cognate OE *celde* is evidenced in charters from SE England, but would be indistinguishable from *kelda* in Angl areas, so it is uncertain whether it may have reinforced use of the ON *kelda*.

knǫttr ON, knott dial. 'compact hill, craggy or rugged height'
This el. is applied to (a) a compact hill, often an independent peak unattached to another hill, so that when mapped, the contours form an oval or circle, looking (by chance) like a knot in wood; or (b) a rocky outcrop or outcrops (and hence sometimes pl.). Several of the nearly 60 instances combine the two senses, being compact and rocky hills. The el. derives from ON *knǫttr* 'ball, hard rounded mass' (cf. late OE *cnotta* 'knot, hillock'), which also gives rise to Norw hill-names (*NSL*: Knatten), but also partly from ON *knút(r)* 'knot, hard lump', seen in the early spellings of HARD KNOTT. It occurs alone as (THE) KNOTT(S), but is often qualified by a descriptive term (GREY KNOTTS, HANGING KNOTT), or else a p.n. (RANNERDALE KNOTTS), or surname (BARKER KNOTT). The particular examples here are recorded no earlier than 15th cent. (THE KNOTTS, Matterdale) but the el. is known from the 12th–13th centuries (see KNOTT, Patterdale). Most examples are in the high fells but there are several outlying ones, e.g. GREAT KNOTT, SWAINSON KNOTT, and proportionately more in the former Westmorland and Lancashire than in Cumberland.

krókr ON, crok(e) ME, crook dial. 'bend, corner of land'
The ON word or its dial. descendant (*EDD*: *crook*, sense 7, though the examples are from Scotland and Yorks) occurs in some nine names, normally of places in bends of streams. The parish name CROOK is recorded from the 12th cent., and CROOKWATH from the 14th, while the other, apparently later, examples include CROOK-A-DYKE, CROOK-A-BECK, CROOKAFLEET.

kross ON, cros ME, cross ModE
The word *cros* was adopted from Gaelic into ON as *kross*, and thence into English, displacing the native words for cross, from OE *rōd* 'rood' and late OE *crūc* 'crouch'. This

el. is an excellent illustration of the truth that identifying an el. is often easier than discovering what it refers to. The possibilities include: (a) a standing cross marking a place of worship (e.g. CROSS), a boundary, a resting place on a corpse route or possibly a market or meeting place; or a cross as a memorial (WHITE CROSS BAY); (b) 'a crossroad or the cross-shaped form of a place or parcel of land' (Lindkvist 1918, 115), as when HIGH CROSS is probably a crossroads (two examples); (c) land lying crosswise (Smith, *Elements* II, 220). Parker concluded a survey of cross sites and *cross* names in W Cumberland by suggesting that they correlate more strongly with raised stone crosses than with crossroads (1909, 118). The el. occurs, as *Cross*, in ten names from the modern period, also in two examples of CROSTHWAITE recorded from 12th cent., and in CROASDALE from the 13th. CROSBYTHWAITE is puzzling, and CROSS DORMONT and CROSS GATE only have *Cross* by late re-formation.

lad in modern forms of names

Lad occurs in LAD CRAGS, LAD HOWS and LADSTONES (two instances); the earliest record is 1803 (*Ladhow*). Although the el. is used as a specific in these cases, it also appears as a generic, e.g., in Lambert Lad, a boundary stone at NY5418 (*PNWestm* II, 213; Machell referred c.1692 to 'a Raise of Stones called Lambert Lad', I, 655A), and cf. Two Lads (*PNWYorks* III, 168). Of the various possible explanations, two are most likely.
1. 'A pile or stack', probably from ON *hlaði* and/or OE *hlæð* (*EDD*: *lad* sb. 2), as found especially in *Ladstones*. Ellwood noted that the Coniston Lad Stones are '*stones piled up*. There is the same idea in the p.n. Lad Cragg . . .' (1895, 36).
2. 'Boy, youth', from ME *ladde*, this sense being recorded from c.1440 (*OED*). This might be comparable with other terms which seem implicitly to compare mountain tops with human figures, e.g. CARL SIDE, CARLING KNOTT, and various instances of **man**.
It is possible that sense 1. 'pile' was original, but the term came to be understood as having human reference. Writing of a Yorks Lad Stone (which is not far from a Maiden Stones) and similar names, Smith considers such names to be 'of English origin and usually not older than the later middle ages' (*PNWYorks* III, 54–5 cf. 64).

lágr: see low

land ON, OE, ME, ModE

A tract of land, district, estate, or (esp. when pl., in field-names and in lowland areas) 'strip of arable land in the common field'. *Land* occurs some two dozen times, eight of them in the pl. The earliest records (of COPELAND, c.1125, RALFLAND, c.1200 and RUSLAND, 1336) probably refer to very large units, as do the former county names Cumberland and Westmorland. Some names originate in compounds or fixed phrases (KIRKLAND(S), SOULAND, SUNDERLAND). NEWLANDS was a valley floor reclaimed by drainage, and in other names the usage of *land* may be rather specific and worth further investigation. *Land* is the generic in some 2,000 Norw farm-names, concentrated especially in central districts and seemingly applied in pre-Viking times to sites not of the highest value or prestige (see *NSL: land*), but the Lakeland usage may be not directly connected with this.

lane ModE

The word derives from OE *lane, lone, lanu*, but Lakeland p.ns containing it first appear in documents from the later 17th cent. (e.g. LANE END, Broughton), with the possible exception of an early version of WATENDLATH. Except for HEGGLE LANE, all the examples are of LANE END (two instances), LANE FOOT or HEAD (three each); it can be difficult to determine which routeway is referred to in these names. In some cases *lane* can be seen to have replaced dial. *lon(n)ing, lon(n)en*.

langr ON, lang OE, dial., long ModE
The adj. 'long' is applied in over 40 names referring to linear features (e.g. LONG GRAIN, LONG RIGG) or expanses of bog, moor, etc. (e.g. two LONGMIRES). The form with -*o*- prevails in Standard English and dominates in the Lakeland names, at least in their written forms. This has clearly replaced the ON and dial. form with -*a*- in some names including LONG CLOSE, Underskiddaw, but -*a*- survives in some eight names including the famous LANGDALE, the earliest-recorded of the names in 'long' (1179), and in LAMBFIT and LANCRIGG.

latter, lat- in modern forms of names
These spellings are usually compounded with an el. meaning 'hill': in three instances of LATTERBARROW (*Lat'ber(brigg)* 1280 in one of them), two of LATRIGG (*Laterhayheved* 13th cent. in one), and in LATTERHAW CRAG and LATTERHEAD (*Laterheued* 1260). WHINLATTER (*Whynlat(t)er* 16th cent.) differs in having **latter** as 2nd el., with a plant-name or adj. as 1st el. The two possible etymologies of **latter, lat-** are:
1. ON *lát(t)r* 'lair', which could, on the evidence of its descendant in Norw dial., also mean a shed or shelter for animals. It is found in Icel and Norw p.ns. The -*r* belongs to the root, which would account for *Lat(t)er* spellings.
2. A Gaelic *lettir* 'hill, slope' (OIr *lettir*; cf. various forms in Watson 1926, 264, 487 and 510, and examples in Scottish hill-names in Drummond 1991, 34).
Commentators have agreed that both are probably represented in Cumbrian names, especially since WHINLATTER seems likely to contain the Gaelic el., but they differ over the detail. Ekwall in *PNLancs* 194, modifying his earlier view in *Scand & Celt* 91–2, favoured the ON word more strongly than the editors of *PNCumb*, and since in most cases the el. co-occurs with a Germanic, probably ON, el., the ON word seems more likely. LATRIGG, Underskiddaw, defies plausible explanation.

laverd ME, lord ModE
Derived from OE *hlāford* 'lord', the word occurs in ten names, mainly in conjunction with terms referring to mountain features, e.g. LORD CRAG and LORD'S SEAT NY2026, both recorded from the 13th cent. and both seemingly giving glimpses of territorial history. The 'lord' in question is not usually identifiable (and in a few cases there is a possibility that *Lord* is a surname) but LORD'S ISLAND refers specifically to the Earls of Derwentwater.

lēah OE 'woodland clearing'
An extremely common el. in English p.ns generally, referring to woodland, hence a woodland glade or clearing, a meadow or pasture, or a settlement in a partially wooded area. Of the few Lake District examples in this volume, only BRADLEY(FIELD), STOCKLEY (BRIDGE) and STAVELEY have so far been found in medieval sources, while with names only recorded in modern times it is unclear whether the name has OE roots or is secondary, for instance based on a surname such as ASHLEY, BLAKELEY, or MEADLEY. Almost certainly modern are HIGH LEYS, LEAGATE and LONG LEA, while COCKLEY probably does not contain this el.

ling: see lyng 'heather'

lítill ON, lȳtel OE, litel ME, little ModE
The adj. occurs almost 30 times in this volume, usually as an affix distinguishing the smaller of two features, e.g. LITTLE GRASSOMS and GREAT GRASSOMS, GOWDER CRAG and LITTLE GOWDER CRAG. In a few cases, however, it is an integral to the name, as in LITTLEDALE (recorded from 1332) or LITTLEWATER (1289), which together with LITTLE ARROW (pre 1220) are the earliest examples.

lodge ME, ModE
From OFr *loge* 'hut, arbour', the el. in ME and early ModE denoted a hut or small house, often made of wood, situated in a wood, and used by fishermen, foresters, travellers etc. The 13 examples in this volume refer to such a hut (e.g. BURNMOOR LODGE), to a groundsman's house at the entrance of a large estate, or to a private house, which may be quite grand (BROUGHTON LODGE, ESTHWAITE LODGE). The Lakeland examples all first appear in the 19th or 20th cent., but there are also records from 1277 and 1340 in *PNWestm* II, 272.

long: see langr

lord: see laverd

low ME, ModE
The adj. was adopted from ON *lágr*, as is clear from the 1209–10 spelling of LODORE. It occurs in over 90 p.ns. It is frequent as a specific in names such as LOW HOLME or LOW HOUSE (seven instances), and still more so as an affix, contrasted with **high** to distinguish nearby settlements or natural features from each other; cf. also **nether**. In LOW FARMS, *low* might instead reflect OE *hlāw* 'hill, tumulus', though this el. is rare in the Lake District, and LOWCRAY and some of the three LOWTHWAITES certainly or probably do not contain *low*.

lyng ON, ling ModE 'heather'; ling(e)y ModE 'heathery'
This occurs in some 14 names, a few with further derivatives, and most referring to upland features, e.g. LING CRAG(S) or the three examples of LINGMELL; LINGCOVE is the earliest recorded, from 1242. The adj. *ling(e)y* appears in three, of which LINGEYBANK is earliest recorded, from 1569. Despite Collingwood's remark on *ling* 'meaning any kind of whins' (1918, 96), many of the places bearing *ling* names are still clad in heather (*Calluna vulgaris*), which is extremely widespread, growing on the fells up to 2822ft/860m., though less abundant than formerly as a result of sheep grazing (Halliday 1997, 222; Hervey and Barnes speak of decline over the previous fifty years, 1970, 47). The wiry stems had a range of practical uses, including roofing, bedding and fuel, although it was also burned in order to improve pasture (see Winchester 2000, 136–7).

lȳtel: see lítill, little

man dial. 'cairn'
'The provincial term for one of those rude obelisks, or piles of stones, which are commonly built by the country people upon the summits of remarkable hills' (Otley 1823, 44). There are five examples in this volume, none recorded before 1800. Some of these are associated with boundaries, as is SERGEANT MAN. The word may also be applied to the mountain peaks themselves, especially where there is more than one: see LOWER MAN and (SKIDDAW) LITTLE MAN. There are two possible sources of the word: (a) A specialised use of 'man, human', since there are Scand examples of a similar use of the cognate word, as well as of *karl* 'old man' and *kerling* 'old woman' (see CARL SIDE, CARLING KNOTT, and Drummond 1991, 26 for a similar phenomenon in Gaelic). (b) Cumbric **main* (W *maen*) 'rock, stone'. Whatever the origin, there are signs of personification, as in CONISTON OLD MAN, or in Wordsworth's 'Rural Architecture' (1800), where three schoolboys climb GREAT HOW and build, 'without mortar or lime, | A Man on the peak of the Crag'; they name him Ralph Jones (*Poems* 68).

mell in modern forms of names
This element occurs in MELL FELL (GREAT & LITTLE) (*Melfel* 1279, *Mele Fell* 1487) and probably MEAL FELL (1867), three instances of LING MELL (*Lingmale* 1578, otherwise only

19th-cent. spellings), MELLBREAK (1780) and WATERMILLOCK (*Weþermeloc* early 13th cent.). A Celtic word meaning 'hill' is probable in all these cases, but its exact origin is uncertain. A Brittonic adj. **mēl* 'bald, bare' is found in several English p.ns, often collocating with **bre(ʒ)* 'hill', e.g. Malvern, Mallerstang and two cases of Mellor, but it also seems to be used as a noun with the sense 'bare hill' at least in Fontmell (see Jackson 1953, 326, Smith *Elements*: *mēl*²; also Watts 2004 for the individual names). This word may be the source, with *fell* added later, of the Mell/Meal Fells, and of the generic el. in WATERMILLOCK (q.v.). Meanwhile, the repeated compounding of *mell* with **lyng, ling**, 'heather', of ON origin, is striking, and the Lingmell names could instead be products of a Norse-Gaelic language situation. If so, their 2nd el. could be Gaelic *maol* 'bald, bare' (the cognate of W *moel*), which appears as a qualifier in Scottish hill names (Watson 1926, 182, Johnston 1934: Mulben, cf. McKay 1999, 65, 92 for Ulster examples) but can also appear as a generic in its own right meaning '(round) hill' (Watson 1926, 480, McKay 1999, 114). This gives rise to a diminutive *maoilinn* 'bare round hillock' (Watson 1926, 146) which might conceivably be present in MALLEN DODD. An alternative Gaelic possibility is the noun *meall* 'lump, hill', which is found in the p.ns of Scotland (Watson 1926, 399, 402, 505), Ulster (McKay 1999, 106) and Munster (Flanagan & Flanagan 1994, 120–1), and the numerous Munster examples include a possible parallel to MELLBREAK (q.v.). See also WATERMILLOCK for Mellock Hill, in the Ochil Hills, which has been explained as a derivative of Gaelic *meall*. I am most grateful to Dr Oliver Padel for encouraging the Gaelic line of investigation in relation to this el.

mere¹ OE, ModE 'lake, pool'

The el. refers to major lakes: BUTTERMERE, GRASMERE, KENTMERE, THIRLMERE and WINDERMERE, all except THIRLMERE recorded from the late 12th or 13th cent., though **water** is the more usual designation for the larger lakes of Cumbria. ELTERMERE and MEREGARTH appear to be recent applications of the word.

mere² dial. 'boundary'

This el., which derives from OE *(ge)mǣre*, co-occurs with **beck, gil(l)** and **crag(ge)** in the names of six places, some of them definitely located on ancient boundaries. (HIGH) MEREBECK is the earliest record known to me, from 1540. The use of *mere* 'boundary' in dial. is illustrated when in 1619 a woman paid 12d for 'slandering John Norman sayinge he did remove meere stakes or merestones betweene her husband and him' (Littledale 1931, 193).

middel ME, middle ModE

The word derives from OE *middel*, sometimes reinforced or influenced by ON *meðal* 'between, among' or (in compounds) 'middle'. It appears 13 times, often as an affix distinguishing a place from **high** and **low** counterparts (e.g. MIDDLE GROVE), or **far** and *near* ones (e.g. MIDDLE SWAN BECK). MIDDLE TONGUE (recorded c.1260) is the only instance with a medieval spelling.

mill: see mylen

mire: see mýrr

mōr OE (mór ON), moor ModE

A tract of upland waste, usually quite level and often boggy (GRASMOOR is unusual in being a mountain-name). The el. is probably of mainly OE origin, since *mýrr* is the usual word in ON. It features in at least 50 names, often as 2nd el. to a pre-existing p.n., e.g. STOCKDALE MOOR, and such names are often first recorded in the 19th cent. There are, however, earlier records, e.g. MURRAH, MURTON, MONK MOORS and MOOR DIVOCK from 13th cent., and some instances of MOOREND from the Early Modern period.

Commenting on Westmorland and its name, Sir Daniel Fleming wrote, 'For such barren places w[hi]ch cannot easily, by ye painful labour of ye husbandman be brought to fruitfulness, ye northern Englishmen call Moores; and Westmoreland is nothing else with us but a westerne moorish countrey' (1671, 3).

mosi ON, mos OE, moss dial. 'bog'
Normally, in northern p.ns, a tract of bog rather than vegetation of the class *Musci*. In the Lake District mosses are mainly peaty and acidic, except in limestone areas, and mostly of the 'raised' rather than 'blanket' sort (Hervey and Barnes 1970, 50). There are some 50 occurrences in this volume, including five MOSEDALES, most of them with 13th-cent. records, as have MOSSER and NORMOSS. Among the more recent names *moss* is common as a generic qualified by a p.n. or descriptive term, e.g. MEATHOP MOSS, LOW MOSS (two instances).

mylen OE, mill ModE; miller ModE
Mill occurs in some 26 names, most recorded no earlier than the 17th cent., though two MILLBECKs and UNDERMILLBECK have 13th- or 14th-cent. spellings which preserve the *-n* of OE *mylen*. All the mills were water-powered, but they served a variety of purposes, often changing through time (see WYTHOP MILL). Grinding grain, fulling cloth and providing power for forges were probably the most common uses. M. L. Armitt provides a useful map of mill sites in the ancient parish of Grasmere at the height of the fulling industry in the 16th cent. (*CW* 1908, 158). Hinde counts 144 mills, mainly water-mills, on Donald and Hodskinson's 1774 map of Cumberland, and considers this an under-representation (2001, 142, 144). *Miller* (from OE *mylnere*, though the more usual term was *mylenweard*) occurs six times in this volume, sometimes definitely as a surname (e.g. MILLERGROUND), but in other cases the occupational term may be intended.

mýrr ON, mire ModE 'bog, swampy ground'
This occurs some 22 times, both as 1st, specific el. (e.g. four MIRE SIDES) and as 2nd, generic el. (BROADMOOR, COWMIRE, EUSEMERE, REDMIRE). The last four have medieval spellings while most of the other examples are 16th- or 17th-cent. ON *mýrr* is common in names of wet, treeless sites in Iceland, which at a time of widespread tree cover could have been seen as advantageous to settlers.

nab dial. 'hill-spur, knob or promontory'
Ultimately from ON *nabbr/nabbi* 'knob', possibly encouraged by ME *knabbe* 'hill-top' or by *neb* (nose, beak), this appears eight times (nine including SKELLY NEB). Only NAB in Rydal appears in a medieval document (13th-cent.). The names concerned are of prominent hills and hill-spurs (e.g. NABEND) or of promontories in lakes (e.g. FERRY NAB).

næss OE, nes ON 'headland, promontory'
Usually applied to land jutting into a lake. Twelve instances, always as a generic, and either with a descriptive term as 1st el. or a reference to a nearby feature. There are medieval spellings for ASHNESS, BOWNESS (-ON-WINDERMERE) and FURNESS.

near: see far

nether ModE 'lower'
The word, from OE *neoðera* and/or ON *neðri*, competes with the Norse-derived **low** in referring to the lower part of something. It occurs in seven modern name-forms, the earliest recorded being NETHER ROW (1658), though there are earlier instances for WINDER (1203, *PNWestm* II, 211), and LOW NEWTON (1491, q.v.).

nīwe OE, new(e) ME, ModE

The adj. occurs in some 13 names, most often qualifying a term for a structure or settlement, but NEWLANDS (recorded from 1318) refers to land reclaimed by drainage. The 'new settlement' of NEWTON in Upper Allithwaite is paradoxically one of the earliest-recorded names in this volume, appearing in Domesday Book, 1086. There are few clear traces of the ON cognate *nýr*, but NIBTHWAITE appears to be one.

nook ModE

A secluded place or corner of land. From ME *nōk(e)*, though the earliest record in this volume is of PARKNOOK, Gosforth (1575). It occurs in some 16 names, always as a simplex or as the generic in a compound, except for NOOK END FARM.

oak ModE, eik(i) ON, aik dial.

Standard English *oak* from OE *āc* is found in the current form of 14 p.ns, while the ON cognates *eik* 'oak-tree' or *eiki* '(place with) oak-trees, oak-wood' or their dial. descendant *aik/ake* appear in four names including LYZZICK, recorded c.1220; some names containing 'Aiken, Aikin' may also belong here. The past and present spellings of names such as AIKBANK, OAK HOWE and possibly OAK HEAD show the competition between the Norse- and English-derived forms. Oak woodland is, as Halliday puts it, 'the natural woodland of most of the Lake District', and is still found on a reduced scale throughout all Lakeland except the high fells, although natural woods are difficult to distinguish from plantings. Also practically indistinguishable are the two native species, *Quercus robor* and *Quercus petraea* (Halliday 1997, 139). The trees were coppiced in S. Lakeland, being a rich source both for bark used in tanning and timber used in building, furniture-making and charcoal-burning.

old ModE

From OE, ME *ald*, the adj. occurs in 13 names, referring to three Halls, two Parks, two Roads, one Church and one Scales. It also qualifies three established p.ns, and one of this group, OLD HYTON, provides the only medieval spelling (from 1358). The OLD MAN OF CONISTON is the only natural feature to be designated *old*.

park ME, ModE 'enclosed hunting ground or pasture, estate, parkland'

ME *park(e)* is from OFr *parc* 'land enclosed for beasts of the chase'. Celia Fiennes, passing along the W shore of ULLSWATER in 1698, 'had a little Course' while riding 'through a fine forest or parke where was deer skipping about and haires' (p. 169); this must have been GLENCOYNE or GOWBARROW PARK. *Park* is also applied to large pieces of enclosed pasture (influenced by OE *pearroc* according to Smith, *Elements*), e.g. the tracts of fell at MECKLIN PARK and TROUTBECK PARK, and to parkland attached to large manor houses, e.g. IRTON PARK, LOWTHER PARK. Six *parks* in High Furness, listed in documents of Furness Abbey at the time of its dissolution c.1535, had been enclosed by royal licence to the Abbey in 1338: ABBOT PARK, HILL PARK, LAWSON PARK, PARKAMOOR, HIGH and LOW STOTT PARK and WATER PARK (Millward and Robinson 1970, 161). Of some 60 instances of *park* in this volume BARTON PARK is the earliest recorded (1279, in a documentary Latin form). *Parks* often give rise to further p.ns such as PARK FELL, PARK HOUSE and PARK WOOD.

pass ME, ModE

Adopted into ME and possibly again into early ModE from (O)Fr *pas* 'step, passage', this is recorded in English from the 14th cent. onwards, though *a pass called Kirkestone* (1671) is the earliest trace in this volume. The el. referred initially to an opening in difficult country, and hence a track passing through such an opening. It features in 13 names, all relating to high routes between valleys, many of them ancient and important, e.g. GARBURN PASS and STICKS PASS.

pasture ModE
The word was adopted from OFr into ME, with much the same meaning, of 'grassland for grazing', as its present one. It appears in Lakeland names from the 16th cent. at latest, though all six occurrences in this volume are first recorded in the 19th cent.

pīc OE, pík ON, pike ModE 'peak, summit, pointed hill'
Used in over 60 names for the summits of major fells, and hence often the fell itself, though SCAFELL PIKE, summit of all England, is not strictly the summit of SCA FELL. Some *pikes* are quite modest in height, e.g. PIKE in St John's. Though most names in *pike* first appear in documents of the 18th–19th centuries, often building on pre-existing p.ns (e.g. FLEETWITH PIKE, ROSSETT PIKE), the el. was clearly in use much earlier: see the medieval forms for STONE PIKE and the Buttermere RED PIKE.

place ME, ModE 'plot of ground, farm'
Adopted into ME from OFr *place* 'open space in a town, surrounded by buildings', this is used in Cumbrian rural names to refer to a plot of ground, hence a farm. Winchester more specifically links farms named *place* with the holdings carved out of the great medieval upland forests and referred to in Lat documents as *placeæ* or *loca* 'places' (1987, 62). Although the el. is recorded from the 14th cent. (*PNWestm* II, 278), all the eleven examples in this volume are first recorded in the 16th cent. or later, and most are formed from surnames: FISHER, HARRY, MIDDLEFELL, ROBINSON and WILSON('S) PLACE. The last four are all in Langdales, typifying the central and northerly distribution of *place*, which may to some extent be regarded as a northerly alternative to **ground**. PLACE FELL, a hill-name already recorded in the 13th cent., may be of similar origin. *Place* has been replaced in THRELKELD LEYS.

plantation ModE
Adopted from Lat *plantatio(nem)* and initially applied to the act of planting, the word refers to a 'wood of planted trees' from the 17th cent. (*OED* sense 2b., 1669). Most of the 14 examples in this volume make their first appearance on the mid-19th-cent. OS maps.

point ModE
A promontory, normally in the Lake District jutting into a lake; a few such as DRIGG POINT are at river estuaries. Adopted from OFr *point*, the word is first used in English in the sense 'tapering promontory' from the 16th cent. (*OED: point* n. 1, sense B2b), but all the 13 instances in this corpus occur no earlier than the First Edition OS maps. All are generics, usually qualified by a p.n. which may itself contain an older word for 'promontory' e.g. AIRA POINT, REDNESS POINT.

pool ModE, pow dial.
This occurs in a dozen names, mainly with the sense 'slow-flowing river or stream', especially in S. Lakeland, e.g. STEERS POOL (recorded from 1235), RUSLAND POOL and POOL BANK (*Powbancke* & variants 17th cent.); cf. also POW BECK and PULL WYKE. Its application to rivers might suggest that it is partially of Brit. origin (cf. W *pwll* 'pool, stream', *ERN* 329–31), but is difficult to separate from OE *pōl*, *pull*, ME *pol(l)* 'pool'.

potte ME, pot(t) dial. 'pit, hollow, pool in river bed'
Possibly of Scand origin (see *SSNNW* 67; also *EDD: pot* sb. 2). The earliest-recorded of the eight examples are POTTER, WIDEPOT (13th cent.) and BLACK POTS (1338).

raise: see hreysi 'cairn'

rake dial. 'track'

This refers especially to a narrow path on a hill. It is also used of designated grazing areas (e.g., Winchester 1987, 90 and *CW* 1906, 12), but this sense is not clearly present in any of the seven p.ns, some of which refer to spectacularly rugged routes (e.g. LADY'S RAKE, LORD'S RAKE). The earliest occurrence in this volume is of RAKEFOOT (1597), and the word is of uncertain origin, from ON *reik* f. 'parting of the hair' and/or OE *hraca*, *hrace* 'throat, pass', or from ON *rák* 'cattle-track, pasture-ground' (Ekwall, *PNLancs* 204).

red in modern forms of names

The 14 names containing 'Red' on the modern map are potentially of two different origins:
1. In most, the colour term *red*, from OE *hrēad*, seems likely, referring to rock or soil, as in RED SCREES and the two RED PIKES, of which the Buttermere one is recorded as early as 1322. The ON *rauðr* 'red' or a derivative of it seems only to occur in ROTHAY and possibly in ROUDSEA (where the lack of early spellings is a difficulty).
2. *Reed* from OE *hrēod*, especially in names of streams or marshes, is found in REDMIRE and perhaps REDNESS, REDSYKE and the Loweswater RED HOW.

reservoir ModE

Of French origin, the word is applied to artificially constructed lakes used for storage and supply of water from the late 17th cent. onwards. It appears in nine reservoir names in this volume, the earliest-recorded being KENTMERE RESERVOIR, from 1858, and in RESERVOIR COTTAGE.

ridding, rydding, rudding ME 'clearing'

The ME word means 'the action of clearing or removing', hence 'a cleared piece of ground, a clearing' (*OED*: *ridding* vbl. n., sense 1, first citation 1347–8, and sense 2, first citation 1586, but these are predated by the p.n. evidence). The word is traced by some back to an OE **rydding* (Smith, *Elements*). For five out of the six certain instances in this volume (which include ABBOTS READING and DEER RUDDING), the earliest known records are 16th–17th-cent., but THE RIDDINGS appears in the 13th cent., and the term is common in documents of that period.

rigg: see hryggr 'ridge'

river ModE

An adoption into ME from OFr *riv(i)ere*, the word *river* is attached for clarification to the names of 27 major rivers, including the BLENG, GRETA and KENT. It acts like a modern equivalent to Lat *aqua* or *rivulum* in medieval documents (see ELLEN, IRT, WINSTER) or *wat(t)er* in Early Modern ones (GILPIN).

rough: see rūh

row, raw ModE

Usually a row or line of houses, and hence a hamlet; some Lakeland examples are not obviously linear, e.g. GREENROW, or may be linear but not continuous. The word derives from OE *rāw* 'row, line', which often became confused with OE *rǣw* 'row, line, hedge', hence hedgerows may sometimes be meant. Certain or possible examples are post-medieval, number about 20 and include ROW in Crosthwaite (recorded from 1535) and HIGH ROW (FARM), Threlkeld (1574). Early Modern forms in *raw* are often found, and sometimes survive (e.g. RAWFOOT, RAWHEAD), and ROEHEAD also contains *row*. Modern 'Row' can, however, also result from confusion with **vrá, wray** 'nook' (e.g. BIRK ROW, ULCAT ROW), or possibly with the word *rough* (ROW RIDDING).

rūh OE, rough ModE
The adj. describes rugged hill features, especially **crags** (four examples) or lower-lying sites that are 'rough' in other senses, perhaps uneven or overgrown. The earliest-recorded of about twelve occurrences are ROUGH HILL and ROUGHOLME, both in 13th-cent. documents (though Rougholme may alternatively contain ON *rugr* 'rye'). One or both of the places named RUTHWAITE may also be a 'rough clearing'.

saddle ModE
The geographers' use of *saddle* (from OE *sadol*) for a depression in a hill or mountain ridge is matched in six names in this volume, of which SADDLER'S KNOTT is the earliest recorded (17th cent.). In all the examples the depression is bounded by two clear peaks.

sæti ON 'seat', seat ModE 'a natural seat or high place'
ON *sæti* (from which ModE *seat* derives, perhaps via late OE *sǣte*) applies to 'lofty places and hilltops, especially . . . natural seat-shaped outcrops of rock' (*PNWestm* II, 12), as probably in YARLSIDE and LORD'S SEAT, Above Derwent, both with medieval spellings. A group of names containing modern SEAT followed by a pers.n. (ALLAN, ROBERT, SANDAL) may contain this el. or the following one, **sætr** 'shieling', and there is potential for confusion with other elements including those mentioned below.

sætr ON 'shieling, summer pasture'
This is one of a set of words relating to the transhumance (summer pasturage) system discussed under **ærgi, erg**. *Sætr* appears in diverse guises in SATTERTHWAITE, SETMURTHY, SETTERAH, and in several quite early names in *-side*, including AMBLESIDE, FORNSIDE, RAYSIDE and at least some examples of SWIN(E)SIDE. If Whyte is correct that the *sætr* shielings were typically lower-lying than those with names in *scale(s)* (1985, 105), some doubtful examples in high locations may instead be from ON **sæti** (above), but the two sets of names are so difficult to distinguish that the argument can easily become circular. In the interpretations offered in the Dictionary section, it is assumed that spellings suggesting a development to short *a* (e.g. *Satsondulf* 1274, later SEAT SANDAL) are more likely to result from *sætr* than *sæti*. There is also potential for confusion with OE *(ge)set* 'dwelling, fold, pastures' or ON *setr* 'residence, settlement', both suggested by Fellows-Jensen as possible origins for HAWKSHEAD, with *sæt* 'shieling' or ON *sát* 'hiding place, ambush (for hunters)', suggested by her in SADGILL, or with **sīde, síða** 'slope, side'. The possibility of *(ge)set, setr* and *sīde, síða* can at least be eliminated where there are spellings with root vowel *-a-*, and the certain occurrence of *sætr* in some medieval names gives some encouragement (though not proof) for assuming it elsewhere.

sail, sale in modern forms of names
A handful of hill-names, unrecorded until relatively recent times, contain an el. with one of these spellings: BLACK SAILS, SAIL (two examples), SAIL HILLS, SALE FELL and SALE HOW; SALETARN KNOTTS and SALLY HILL may conceivably also belong here. There are several possible origins, though it is usually impossible to judge between them. The most likely are:
1. ON *seila* 'hollow', though poorly evidenced (Cleasby & Vigfusson 1957 but not Fritzner, *ONP*);
2. ON *seyla* 'puddle, mire, stream', assumed in p.ns such as BLACK SAIL (*le Blacksayl* 1322) and SAYLES (FARM). The *ONP* database has one example, as a p.n.;
3. ON *selja*, OE *salh* 'willow';
4. ModE *sail* 'sail' from OE or ON *segl*, since the latter occurs in Norw p.ns believed to refer to pale sail-like rocks or mountain shapes (see *NSL*: Segelstein).

5. Gaelic *sail* 'heel, hill shaped like an inverted heel', applied to mountains of the NW Highlands (see Drummond 1991, 92; Watson has only one example, 1926, 92).

scale: see skáli 'shieling'

scar dial. 'escarpment, scarp, crag'
From ON *sker*, which was often used of coastal skerries, the el. occurs nearly 20 times, as generic or specific. It is recorded from the 13th cent. onwards (e.g. *Claterandsker*, *PNWestm* II, 285), but since it often refers to remote, minor, features, the examples in this volume are all from 17th cent. (e.g. SCAR HEAD) or later. Several examples are of crags forming a long gash, especially in the limestone region of SE Lakeland, e.g. WHITE SCAR and CUNSWICK SCAR, and one wonders whether the modern usage has been influenced by the word *scar* 'scab, trace of a wound, often linear', which was adopted into late ME from OFr *escharre*. Some modern forms in *scar* are of quite different origins, e.g. SCARGREEN.

sceaga: see skógr '(small) wood'

seat: see sæti 'seat' or sætr 'shieling'

shaw: see skógr '(small) wood'

síða ON, sīde OE, side ME, ModE 'slope, hill-side, side'
(a) The sense 'slope, hill-side' seems operative in names such as GREEN SIDE, LOW SIDE. The contours are usually long, and Smith gives 'the land extending alongside a river or lake, the edge of a wood or village' as the sense of the ME word. (b) When the el. follows a noun referring to a natural feature, it can also mean 'side, the place beside', as in a cluster of names in Crosthwaite & Lyth: BORDER SIDE, FELLSIDE, MIRESIDE, MOSS SIDE, TARNSIDE and WOODSIDE. If the 1st el. refers to a mountain, e.g. FELLSIDE, the senses of 'slope' and 'side' cannot be disentangled, but such names probably belong with the BECKSIDES, MIRESIDES and WATERSIDES, with the sense 'side' dominant. Although the el. is of OE and/or ON origin, the certain examples of names containing it are mainly 16th-cent. or later. WATERSIDE, Ennerdale is the earliest, from 1332. The modern spelling *side* in the p.ns is of multiple origins, and further possibilities include: **sætr** 'shieling'; **hēafod** 'head, upper end, high place'; **sæti** ON 'seat'. It is therefore difficult to estimate the frequency of the el., but there are some 50–60 more or less certain examples.

sike: see syke 'small stream'

silver(y)
Silver derives from OE *seolfor*, ON *silfr*, but with the (doubtful) exception of SILVER COVE, the six examples are all found no earlier than the 18th cent. *Silvery*, whose first occurrence outside p.ns is dated c.1600 in *OED*, appears in SILVERYBIELD CRAG and formerly in SILVER CRAG and POINT. Most of the names concerned are of hill features. Short of undetectable corruption, *silver* is probably to be taken at face value, but its meaning is tantalisingly unclear. In SILVER HILL, SILVER BAY and SILVER POINT it may refer literally to the precious metal, but elsewhere the appearance of rock or vegetation may have inspired the name, SILVER HOW being a case in point.

skáli ON, dial. scale 'shieling, summer pasture with hut, hut'
This occurs over 40 times, as simplex, specific or generic. ON *skáli* can refer to a hall or a main room in a farmhouse, but in the Lake District it normally refers to a modest, seasonal structure. It is a common and important reminder of the shieling system of pasturage (see **ærgi, erg**), and must in most cases refer to a tract of summer pasture with

its hut, but it is not a completely reliable indicator, since *scale* can refer to other types of hut or shed, including milking sheds, peat stores, fisherman's huts or miner's cottages (see Winchester 2000, 90–3), and stone remains can be extremely difficult to interpret (Whyte 1985, 111). PORTINSCALE is both the earliest-recorded instance (c.1160) and arguably the most interesting. The el. normally appears as modern *scale*, sometimes with pl. *-s* which shows its naturalisation as an English dial. word, but SCAWTHWAITE and SKELGILL illustrate different outcomes, and it can be re-formed as **gill**, as in POTTS GILL and SOSGILL. Variation between sing. and pl. often occurs (e.g. COCKENSKELL, GUTHERSCALE); Fellows-Jensen regards the pl. as secondary (*SSNNW* 209).

skarð ON 'gap, mountain pass, col'
This occurs in some ten names, including GATESGARTH, Buttermere, recorded from 1211, and SCARTH GAP. It is almost certainly preserved in SCARGREEN and probably in BUSCOE; it has been replaced in ORE GAP.

skógr ON, sceaga OE, scaw, shaw dial. '(small) wood'
ON *skógr* and OE *sceaga* are related, and they or their reflexes occur in a handful of names, some referring to high ground which has not been wooded for centuries. The OE form regularly gives *sh-* forms (e.g. FOULSHAW, recorded 17th cent.) while the ON one gives *sk-* or *sc-* spellings (FLASKA, SCALDERSKEW and SCEUGH, all recorded in medieval documents), but in some cases there has been alternation between the forms, and in others it is doubtful whether either is present (e.g. LAUNCHY, LISCO).

slakki ON, slack dial. 'small valley'
The sense 'small, shallow valley, a hollow in the ground' (Smith, *Elements*) fits the topography of, for example, Furness (GREENSLACK), but the el. often refers to features which are steep and/or high (e.g. AARON SLACK, HARROW SLACK). Another possible interpretation (which again does not cover all instances) is suggested by Machell: 'In these Rocky Countryes where there is abundance of Craggy stones, such places as are exempted from them are called Slaks, or Gill (and wee use the word yet much in the like sense ... of intervals in showery or stormy weather)' (c.1692, II, 396). WITHERSLACK and ASHLACK both appear in medieval documents, but the el. seems to remain productive into the modern period.

spring ModE 'spring', dial. 'copse'
From OE *spring*, but all eight examples in this volume are recorded no earlier than the 19th cent. (a) It has its standard modern sense of 'water-spring' in SPRING KELD and possibly SPRING HOUSE and at least one SPRING BANK; but (b) it can also refer to new growth on plants and hence 'plantation, copse', an early example being *Heselspring* 'hazel copse' 1285 (*CW* 1909, 32). This usage is common in woodland names in the former S. Westmorland, examples being ASH SPRING, BROWNSPRING COPPICE, PARKSPRING WOOD and SPRING HAG.

stān OE, steinn ON, sta(y)ne dial., stone ModE
Standard English *stone*, from OE *stān*, has evidently ousted its ON cognate *stein(n)* in some names, such as STONETHWAITE (*Stayne-* in 1190s), while modern forms such as STANEGARTH could be a dial. form descending from either. Of over 40 names containing the el., several refer to massive and celebrated single rocks (BOWDER STONE, KIRK STONE or ULL STONE), some of them on boundaries (CLOVEN STONE and the sculpted THREE SHIRE STONE), and others to scatters of stones which are distinctive (SMALLSTONE BECK) or useful for particular purposes (BECKSTONES, KILNSTONES). In nine names, including the curious STONE ARTHUR, *stone* is the 1st el.

stang: see stǫng 'pole, pointed ridge, wooden bridge'

stede ME, stead(s) ModE 'place, farmstead'
From OE *stede* 'place, site of a house', this is used in ten farm names recorded from the 16th cent. onwards. Five of them have a surname as specific, e.g. BAKERSTEAD, BOWMANSTEAD and WHITBYSTEADS. HUBBERSTY HEAD, recorded from 1283, has a complex history. LOOKING STEAD is an exception, as the only hill-name in this volume containing **stead**.

steel, stile ModE 'steep ascent'
This is ultimately from OE *stigel* 'stile', though the earliest record here is of STEEL END, from 1570. There are nine modern names containing *Stile* and five containing *Steel*, but CHAPEL STILE, CLIMB STILE and both STILE ENDs have earlier spellings in *steel* or equivalent; only VAUGH STEEL reverses the trend with a *Stile* spelling in 1681. *Steel* in Scots refers to 'a steep bank, esp. a spur of a hill-ridge' (*SND: steel* n. 2; I am grateful to Dr Simon Taylor for drawing my attention to this), and several of the Lakeland examples are long, steep, beak-like ridges. Since these often carry tracks, there is some ambiguity as to whether *steel, stile* refers to the ridge along or beside which a steep track passes, or the track itself. Except in the special case of CHAPEL STILE, it is unlikely that the el. ever refers literally to stiles in the modern sense, and Smith's assumption that Steel is a surname in STEEL END, FELL and RIGG is also questionable.

steinn: see stān, stone

stígr ON, stīg OE, sty(e) dial. 'steep path'
Related ultimately to verbs of climbing, the el. denotes a steep mountain path (and in dial., a ladder). It occurs in some 13 names, STARLING (DODD) and THORPHINSTY recorded from the 13th cent. and the remainder from the 16th cent. (STYE HEAD, STYBARROW CRAG, etc.) or later. Four seem to contain the names of animals as specifics, including CATSTYCAM, while MANESTY and THORPHINSTY contain pers.ns probably introduced during the Viking period.

stone: see stān

stǫng ON, stang dial. 'pole, pointed ridge, wooden bridge'
The ON pl. *stangir* seems to be preserved in STANGER (recorded 1298), where it presumably refers to stakes marking a routeway or boundary or forming a palisade. STANG is a pointed ridge, and hence illustrates the topographical application of the word. The two STANG END(S) may contain this sense but since they stand near minor bridging points the sense 'bridge' is probable here. In STANGRAH (recorded 1180–1210) the meaning is unclear. See further Whaley 2005.

svín ON, swīn OE, swine ModE
While ON **gríss** '(young) pig' occurs in four GRISEDALES (or variants), *svín/swīn* appears in seven SWIN(E)SIDES, two of which appear in medieval documents, as well as SWINDALE and probably three other p.ns.

syke, sike dial. 'small stream, often slow-running'
The word derives from OE *sīc* 'small stream, especially in flat marshland', and/or ON *sík(i)* 'ditch, stream'. It occurs in nine names, none further E than Matterdale (RED SYKE), and most applied to a slow-running stream and a nearby settlement (BUSCOE SIKE being an exception on both counts). Most appear first in Early Modern times, FOULSYKE as early as 1545, though elsewhere in Cumbria an 'Eel Sike' is recorded c.1176 (*Elesic*, *PNCumb* 12). KELSWICK may also contain this el.

tarn: see tjǫrn 'small mountain pool'

þorn ON, OE, thorn ModE; thorn(e)y ModE

The noun *þorn, thorn* occurs in eleven names, and the adj. *thorn(e)y* (ultimately from OE *þornig*) in four. It is compounded with *þveit* 'clearing' or its reflex in five instances, including two THORNTHWAITES recorded from c.1230. Other names, including the eccentric BEAUTHORN, may be much more recent. The word *þorn* is usually taken to mean 'hawthorn' (*Crataegus monogyna*) and the plant is common throughout Cumbria, in hedgerows, woodland and often as the last survivor of fellside grazing (Halliday 1997, 289–90).

þveit ON, thwaite dial. 'clearing, meadow, settlement'

This is the most prolific single el. to refer to habitations in the Lake District. It seems at least originally to have designated a particular topographical situation, and hence is often regarded as 'quasi-habitative'. ON *þveit*, from a root referring to cutting, meant 'a cut-off piece of land' or one cleared by cutting down of woodland, often relatively low-lying or on a slope, and near water. The el. occurs in over 100 settlement names, including those of parishes such as the two CROSTHWAITES, recorded from the 12th cent. Although many or most *thwaite* names can be assumed to be from the main period of Scand settlement, some are clearly later, e.g. BASSENTHWAITE, which contains a post-Conquest pers.n., and there is good reason to believe that the el. continued to be used 'well into the thirteenth century' for 'clearings ... on the edges of an early nucleus' (Winchester 1987, 41). Qualifying 1st elements usually refer to vegetation (e.g. THORN(Y)THWAITE and the Borrowdale SEATHWAITE), a natural feature or a man-made feature (the Dunnerdale SEATHWAITE, TILBERTHWAITE) or a person (FINSTHWAITE). The modern spelling is usually *thwaite*, but pronunciation is often [θət], and this is often registered in earlier spellings, e.g. *Thornthatcrag* 1614. Most modern names in *-thwaite* derive from ON *þveit*, but not quite all: see LAYTHWAITE (CRAGS) for an example.

tjǫrn ON, terne ME, tarn dial. 'small mountain pool'

A classic Lakeland el. occurring in over 70 names. Though derived from ON *tjǫrn* 'small lake', *Ternis* 1185 (*PNCumb* 296) shows early naturalisation into ME by the addition of the English *-(i)s* pl. Other examples, such as ANGLE TARN, Martindale, DOCK TARN and HELTON TARN, are recorded as early as the 13th cent., but since most of the *Tarn* names refer to pools situated high among the fells (THE TARN, Bootle, being a notable exception), they are often not recorded until recent times. Still more than the larger lakes, tarns are somewhat evanescent features, formed mainly by glaciation and susceptible to infilling by the build-up of alluvial sand and gravel: see the comments on TARN CRAG (Greycrag Tarn) and TARN MOOR. In an almost opposite process and by human intervention, tarns have been created out of the moss on Claife Heights: see THREE DUBS TARN. *Tarn* is most common as a generic, in names whose 1st el. specifies the attributes of the pool itself (BLEA, BLIND, DOCK, EEL TARN) or its surroundings (BLEABERRY, TEWET TARN); a number have a p.n. as specific, e.g. GRIZEDALE, LOUGHRIGG TARN. It also occurs as specific in nine different names referring to a nearby feature, e.g. TARN BECK (two examples) or TARN CRAG (eight).

tongue: see tunga

top ModE

From OE *topp* 'summit, highest point', there are eleven probable instances of this, including five HILL TOPS, of which the one in St John's is the earliest recorded, from 1595. These are habitations, but all the other examples are simply hill features.

tūn OE, toune ME, town ModE 'farmstead, settlement, village'

OE *tūn* seems originally to have meant 'enclosure', hence 'farmstead, estate' and, especially as settlements expanded, 'village'. This el., the most common in OE p.ns, and used long after the Norman Conquest, occurs especially in the lower-lying, fertile areas around the fringe of Lakeland, where early Angl settlement took place. Out of over 30 names with modern forms end in *-ton*, the classic examples are parish centres such as BAMPTON, CONISTON or BROUGHTON (two instances), whose names are recorded in medieval documents and whose 1st elements are common nouns. A few names of minor places such as HUTTON in Dunnerdale are puzzling exceptions, while in MIDDLETON PLACE and STAINTON GROUND the *-ton* is a surname derived from a p.n., which acts as specific to another p.n. WELLINGTON by a different route also commemorates a person. The full form *town* occurs in the names of four hamlets: HOW TOWN, LITTLE TOWN and NEWTOWN (two), and in seven examples of TOWN END and four of TOWN HEAD; none of these is recorded before Early Modern times. In a few names, there is presently no nucleated settlement to which *town* would obviously refer and it may designate a unit of land: see TOWN BANK and TOWN HEAD, Staveley. The *town* forms contrast with names given in the OE period, which tend to be pronounced with a reduced vowel [ə] and spelt *-ton*. The 16th-cent. *-ton* spellings for LITTLETOWN and NEWTOWN might, however, suggest that the names were given in the ME period. Some village or hamlet names sporadically had 'town' added in earlier times, e.g. *Seathwaite Town* and *Seatallor* (= SEATOLLER) *Town* on Hetherington's map of 1759 (*CW* 1976, 106).

tunga ON, tongue ModE
A raised tongue of land, usually in the sharp angle between two streams which meet, though LONG TONGUE projects into a lake. As a p.n. el., modern *tongue* probably owes more to ON *tunga*, which is applied to topography in early Norw and Icel names, than to OE or ME *tunge*, which seem not to be so used. The earliest-recorded of some 20 instances in this volume are TONGUE HEAD and MIDDLE TONGUE, both mid 13th-cent.

vað ON 'ford', wath dial. 'ford'
The ON word usually appears in English p.ns as *wath*, the initial consonant probably reflecting the pronunciation at the time of the Scand settlement. CROOKWATH, SKELWITH NY3403 and WIDEWATH are certain examples with medieval spellings, and another four are probable or possible. In DUBWATH this el. may have replaced the reflex of ON **vrá** 'nook'. 'Ford' is also expressed by English **ford**.

vatn ON: see wæter, water

viðr ON 'wood, timber'
This is found in a handful of names with early spellings: BLAWITH, WITHERSLACK (which preserves the gen. sing. form *viðar*), the Martindale BANNERDALE and BENNETHEAD (both of which contain the compound *beinviðr* 'holly', literally 'bone-wood'). COLWITH may refer to charcoal-burning. In SKELWITH NY3403 *viðr/with* seems to have replaced **vað** 'ford', while for the other SKELWITH and FLEETWITH the evidence is too late for certainty.

vík ON, wyke ModE 'inlet, bay'
The el. is found in Viken, Norway, Reykjavík, Iceland, and probably the word *viking*. In the Lake District it is certain in SANDWICK, PULL WYKE and THE WYKE, and may occur in LOWICK, where the sense 'bend in a river' has been suggested, and in FROSWICK and LOTHWAITE, where the reference could be to some other curving feature in the landscape.

vrá ON, wray dial. 'nook of land'
The el. is used in Norw and Swed p.ns, for places in secluded or remote situations. It is documented from the 13th cent. in the Lakeland names DOCKRAY, MURRAH and ULCAT ROW, which illustrate some of the modern spellings, though *wr(e)ay* is usual in names recorded more recently. There are ten certain examples in this volume and a further four possible ones.

wæter OE, water ModE
The dominant term for 'lake', occurring in ELTERWATER, HAWESWATER and HAYESWATER (all documented from late 12th) and in CRUMMOCK WATER, DERWENTWATER, DEVOKE WATER, LITTLE WATER, LOWESWATER and ULLSWATER (early 13th cent.). It derives from OE *wæter*, with the meaning probably influenced by its ON relative *vatn* (identifiable in WAST WATER and WATENDLATH). The synonymous **mere** and **lake** (as in GRASMERE, BASSENTHWAITE LAKE) are much less well represented, though there was formerly more variability in lake names, as seen in 18th- and 19th-cent. guidebooks which have 'Derwent Lake' etc. (partly preserved in local usage). There is also some alternation between *water* and **tarn**, e.g. in the history of FLOUTERN TARN, GOATS WATER and LEVERS WATER. When *water* appears as a specific it is also mainly in reference to lakes, e.g. the WATERHEADS by Lake Coniston and Windermere. ME, ModE *wa(t)ter* is also used of rivers in local documents (as was OE *wæter*), though WATERSIDE, Ennerdale, is the only unambiguous instance. *Water* in WATER YEAT and WATERGATE FARM probably has something like its usual modern sense.

wath: see vað 'ford'

well ModE
The el. goes back to OE *w(i)ella* 'well, spring', and is common in English p.ns, but to my knowledge the earliest-recorded Lakeland example is GOOSEWELL (1567). It refers to natural springs both with man-made superstructures (e.g. ST PATRICK'S WELL) and without (e.g. BROWNSPRING WELL). There are nine occurrences, all as generics except for WELL WOOD and WELLFOOT.

whin, whinny ME, ModE 'whin, gorse, furze'
The el. is of uncertain origin; an ON **hvin* has been postulated, cf. Norw *kvein* 'thin grasses', but even that is extremely rare (*NG* XVI, 5). *Whin* appears in seven names and *whinny* in one, usually referring to upland features. WHIN FELL is recorded from c.1170. In WHIN GARTH the el. seems to have replaced *wind*. In Cumbria nowadays *Ulex europaeus*, the common gorse, flourishes on rough ground, especially below 984ft/300m., while the less common *Ulex gallii*, the Western gorse, grows at medium altitude on still poorer soils (Halliday 1997, 306). The brilliant yellow flowers of the plant blaze out of the landscape, and were useful as dyestuffs.

white: see hwīt, hvítr

wīc OE '(specialised, dependent) farm, settlement'
One of the classic elements in English p.ns, this ultimately derives from Lat *vicus*, and elsewhere in England is often attached to sites which had been Romano-British civilian settlements. It can often be shown to denote a dependent settlement with a specific purpose (see Coates 1999). There are only three certainly 'authentic' cases in this volume: BUTTERWICK, CUNSWICK and KESWICK, all with medieval spellings. Other *-wick* spellings including LOWICK and possibly LOTHWAITE may result from ON **vík**, while some late-recorded cases are difficult to account for: DUDDERWICK, FENWICK.

wode ME, wood ModE
Deriving from OE *wudu*, which meant both 'wood, forest' and 'timber', the word is very common in English p.ns from Anglo-Saxon times onwards. Most of the over 90 Lakeland examples appear to be recent, though *Brentwode* (possibly = BRANTWOOD) is recorded in 1356. Most of the names refer simply to a tract of woodland, with a p.n. as 1st el., e.g. BOLTON WOOD, RAINSBARROW WOOD, but some are settlement names, e.g. BOONWOOD, LOW WOOD. *Wood* also gives rise to names such as WOODEND, WOOD

HOUSE. Other woodland terms include ON **skógr,** ON **viðr** and OE **hyrst**, as well as the more recent **coppice** and **plantation.**

wray: see vrá 'nook of land'

wyke: see vík 'bay, inlet'

yeat: see geat 'gate'

yew in modern forms of names
Current forms containing 'Yew' refer either to the tree-name *yew* (OE *īw, ēow*) or to the *ewe* or female sheep (OE *ē(o)wu*). The two can be difficult to tell apart, since many of these names lack pre-modern spellings, and Early Modern spellings can be misleading, as when YEW TREE NY1105 was spelt *Ewtree* in 1578, while two writers in the 1780s spelt the same Crag as *Yew* and *Ewe* (YEW CRAG NY4120).

1. Nevertheless, some nine names, such as GREEN YEW or those containing 'Yew Tree', unambiguously refer to the tree, and for YEWDALE the 1196 spelling *Ywedale* favours this. The tree *Taxus baccata*, though hardy and viable up to 1500ft/457m. especially on limestone scars, is found mainly in valleys and, except in S. Lakeland, is normally planted rather than indigenous (Halliday 1997, 115).

2. It is reasonable to think that most names of crags and hills containing 'Yew' refer not to the tree but to the female sheep which frequent them (cf. also GIMMER CRAG), and in the case of YEWBARROW there are 14th-cent. spellings in *Yow-, You-* which favour 'ewe'. Local pronunciation as 'yow' can also be an indicator.